Praise for *Life Cycle Driven Structures*

The level of detail and overall quality of this book are world-class, making it an essential resource for professionals in the construction industry. The building case studies are exceptionally thorough, offering valuable insight into how projects can be designed and constructed, while clearly illustrating the importance of life cycle driven design approaches.

Joseph Bonica
Principal Structural Engineer at Robert Bird Group
Sydney, Australia

This publication makes a valuable contribution to the design and engineering community by capturing the evolving state of practice in life cycle driven structural design across an international range of projects. Of particular focus, it highlights how hybrid structural systems combining timber, steel, and concrete can advance architectural ambition while reducing embodied carbon, helping to chart pathways toward sustainable building practice.

Carol Phillips
Partner, Moriyama Teshima Architects,
Toronto, Canada

Whole life carbon is reshaping how building structures are designed and delivered. This book offers a practical, real-world approach alongside a thorough grounding in the underlying theory and reasoning, making it an essential tool in delivering low-carbon buildings.

Tom Carr
Associate Director, Structural Engineer at Buro Happold
London, UK

This world-class book offers a comprehensive and practice-oriented approach to sustainable structural design, seamlessly integrating low-carbon, resilient, and efficient solutions. Through exceptionally thorough case studies of buildings, bridges, and special structures, it vividly illustrates how thoughtful design can deliver meaningful value across the built environment whilst highlighting the critical importance of life cycle driven design approaches, making it an essential and inspiring resource for every professional in the construction industry.

José Oliveira Pedro
Associate Professor, Civil Engineering, Instituto Superior Técnico
Senior Engineer, GRID International
Lisbon, Portugal

Circular economy is so important to achieving our carbon goals to help address the climate emergency, so it is great to see a book that brings the theory together with the practice so well. This book encourages people to go and do what needs to be done.

Roy Fishwick
Managing Director at Cleveland Steel & Tubes,
North Yorkshire, UK

This book offers thoughtful analysis of diverse material approaches and stands as a testament to how collaboration, innovation, and creative design can drive more regenerative and responsible outcomes. It includes a great collection of projects from around the world, showcasing how whole life cycle thinking can be meaningfully integrated into holistic design.

Golnaz Ighany
Sustainability Director at RIOS,
London, UK

Defining the role of design and engineering in sustainability and positive climate change action is ever more valuable. It is the structural engineer's responsibility to design elegant and sustainable structures, and we need ways to measure and compare the parameters for these values (at least conceptually). Thanks to these efforts, we are capable of positively measuring embodied carbon in the construction of buildings. Wonderful to be a part of this state-of-the-art review and to be a case study in the framework.

Ahmet Topbaş
Director at ATTEC
Istanbul, Türkiye

By bringing sustainable design principles together with the realities of project delivery, this book provides valuable insight into the challenges of delivering low-carbon, resilient, and efficient structures. The detailed case studies demonstrate how thoughtful design can create meaningful change across the built environment, while offering designers practical examples of how to turn the competing demands of modern projects into added value.

Ricardo Pimentel
Divisional Director, Buildings at Pell Frischmann
London, UK

This book presents an invaluable collection of background data, case studies, and practical guidance for designers, demystifying embodied carbon analysis and reduction strategies. The level of detail is unsurpassed among current resources available to designers, making it an essential reference that deserves a place in every architect's, engineer's, and contractor's library.

Kirby Beegles
Associate, Structural Engineer at Martin/Martin Inc.
Denver, USA

Structural engineering plays an extremely important role in the carbon outcome of our buildings. This is not simply because the structural components of buildings use large quantities of materials. In fact, the engineers have the "computing capacity" to understand the material and environmental consequences of design proposals brought up by architects and inspire optimization, improvement, or, in some cases, just say "no!" This book will contribute to empower students and engineering practitioners in understanding their role and responsibilities in the battle against embodied carbon emissions.

Dario Trabucco
Associate Professor, Building technology, Venice IUAV University
Council on Vertical Urbanism (CVU) Research Manager
Venice, Italy

Life Cycle Driven Structures

Life Cycle Driven Structures

Alper Kanyılmaz

Department of Architecture, Built Environment and Construction Engineering,
Politecnico di Milano,
Italy

WILEY

Published by John Wiley & Sons, Inc., Hoboken, New Jersey.

Library of Congress Cataloging-in-Publication Data:

Names: Kanyılmaz, Alper author
Title: Life cycle driven structures / Alper Kanyılmaz.
Description: Hoboken, New Jersey : John Wiley & Sons, Inc, [2026] |
Includes bibliographical references and index.
Identifiers: LCCN 2026015320 | ISBN 9781394300525 hardback | ISBN 9781394300549 adobe pdf | ISBN 9781394300532 epub | ISBN 9781394300556 obook
Subjects: LCSH: Sustainable construction | Structural engineering
Classification: LCC TH880 .K335 2026
LC record available at https://lccn.loc.gov/2026015320

Cover Design: Wiley
Cover Image: © Paha_L/Getty Images

SKY10164030_072926

I am dedicating this book to my family:

To my wife, whose confidence in me sustained this project from its earliest idea to its final pages.

To my children, who made sure I never forgot what all this building is actually for.

To my father, whose faith in education made everything possible for me.

To my mother, whose presence behind every decision I took (no matter how far it took me from home) was the strongest foundation I have ever known.

Contents

Foreword

The built environment shapes the natural environment across timescales and geographies no single project ever fully accounts for. Structural engineers carry more of that climate and environmental consequence than the profession has historically been asked to own (or been given the data and tools to act on). That is changing fast.

The signals are coming from every direction. Procurement teams want Environmental Product Declarations (EPDs). Policymakers are setting embodied carbon limits. Corporate clients are asking biodiversity and whole-life carbon questions before a shovel hits the dirt. And as buildings become more energy-efficient in operation, the greenhouse gas emissions locked into their structure at the outset represent an ever-larger share of the total. The structural system, typically the single largest contributor, is also one of the greatest opportunities.

The standards, data, and tools have come a long way. The data infrastructure has been catching up to the policy moment faster than most anticipated. EC3, the open-access platform I oversee at Building Transparency, now holds more than 200,000 EPDs from manufacturers across 84 countries. That vantage point makes one thing clear: having data is only part of the solution. Getting from a verified number in an EPD to a structural design decision requires real expertise at every step. That expertise is exactly what this book builds.

Professor Kanyılmaz treats embodied carbon not as a compliance metric appended to a project but as a structural design parameter that shapes choices engineers are already making every day: material selection, structural system, and seismic design strategy. He extends that lens into resilience, adaptability, repair, reuse, and end-of-life planning: the full arc of a structure's life. Practitioners who contributed case studies from firms across 15 countries show what that integration looks like in the field.

This book arrives at a critical moment. Most of the construction still ahead of us is concentrated in regions where embodied carbon awareness, data, tools, and regulatory frameworks remain comparatively underdeveloped. Those markets have a real chance to leapfrog emissions-intensive materials and technologies rather than replicating the carbon-intensive path others took, but only if engineers there, and the profession globally, share a common language for measuring and communicating environmental performance. This book is part of building that language. The engineers practicing today will determine, structure by structure, whether the construction boom ahead of us deepens the climate crisis or begins to bend the curve.

Arlington, Virginia, United States, June 2026 *Alison Kinn Bennett*

Alison Kinn Bennett is Executive Director of Building Transparency, the nonprofit stewarding the Embodied Carbon in Construction Calculator (EC3), an open-access environmental product life cycle data platform for decarbonizing buildings and infrastructure. She joined Building Transparency in October 2025 after 28 years with the U.S. Environmental Protection Agency (EPA), where she led federal initiatives on environmentally preferable purchasing, embodied carbon, and sustainable buildings. She is a member of the American Center for Life Cycle Assessment and the ISO/Greenhouse Gas Protocol Product-Level Carbon Accounting Joint Working Group.

Preface

Structural engineering has always been a profession defined by responsibility. We design the buildings, bridges, and infrastructure that keep people safe, and for over a century, we have refined our methods around strength, stiffness, and stability. The question that now confronts the profession is whether designing for safety and serviceability alone is still enough, when every choice carries an environmental consequence that will outlast the engineer who made it.

The climate crisis has fundamentally changed what it means to design a structure. Construction consumes a large amount of the world's raw materials and accounts for a comparable share of global carbon emissions, and a significant portion of that total is locked in before a building ever opens its doors (embodied in the steel, cement, and timber). Operational carbon has received sustained attention from the energy efficiency community for decades, and rightly so. But as buildings become more energy-efficient, the relative weight of embodied carbon in the life-cycle equation grows. This is the structural engineer's carbon, and it is time we took ownership of it. Much of the construction anticipated in the coming decades will be concentrated in regions where awareness of construction-related carbon emissions remains less developed. This makes a globally shared knowledge base not merely useful but urgent.

An essential dimension that emerging sustainability practices often overlooks is resilience. When a building is damaged beyond repair by an earthquake or flood, or rendered obsolete long before the end of its life, the embodied carbon invested in its materials and construction is lost, and the cycle of extraction, manufacturing, transport, and erection begins again. Designing structures that endure, accommodate change, and withstand extreme events is therefore not only a matter of safety but also one of the most powerful carbon reduction strategies available to the profession. Our ability to verify structural condition after an extreme event is part of the same equation: buildings demolished due to uncertain damage status, rather than confirmed structural failure, represent wasted embodied carbon that could have been preserved through better information. The rest of this preface sets out why I wrote this book, what it offers, and how it is organized.

Why This Book

As sustainability becomes central to design and construction practices, professionals need Life-cycle Assessment (LCA) to measure and minimize embodied carbon. Yet a critical knowledge gap persists in Architecture, Engineering, and Construction (AEC) curricula and practice: while LCA methodology is already well established, its systematic integration into structural design decision-making remains limited in most resources. Valuable guides oriented to professionals already working in practice do exist, and they have done important work in raising awareness and providing clear frameworks for project teams. What I aimed for is a textbook designed with and for students: one that builds knowledge progressively from fundamentals, includes worked examples and exercises at every stage, includes case studies from practice, and prepares readers to apply life-cycle methods before they enter the profession rather than after. It also recognizes that life-cycle-driven design demands skills that extend beyond standard curricula, and that digital tools must evolve to link structural analysis with fabrication processes, maintenance planning, and carbon accounting.

This textbook offers a structured, progressive guide to assessing and reducing embodied carbon across all building life-cycle stages. It integrates life-cycle thinking directly into structural design decisions, equipping readers with the knowledge, analytical methods, and practical tools to perform robust LCA, interpret and compare results critically, and translate environmental data into design decisions about materials, structural systems, and construction methods.

What You Will Learn

By the time you reach the final chapter, you will be able to:

- Develop a complete picture of life-cycle-driven structural design, where safety, serviceability, resilience, and embodied carbon performance are weighed together across every life-cycle assessment (LCA) stage from material sourcing to end-of-life.
- Perform life-cycle assessment across all its stages for steel, reinforced concrete, and mass timber structures, understanding how each material carries its carbon through the supply chain.
- Calculate and compare embodied carbon using Environmental Product Declarations (EPDs), regional databases, and carbon factor ranges drawn from over 1000 published EPDs.
- Normalize and benchmark results across building types, and apply embodied carbon rating systems.
- Conduct systematic "what-if" parameter analyses at both the component level and the building level, identifying which design variables most significantly impact a project's environmental outcome.
- Apply multi-objective optimization to generate and evaluate hundreds of candidate designs simultaneously, navigating the trade-offs between cost, span, and embodied carbon through Pareto front analysis.
- Evaluate earthquake-resistant design through a life-cycle lens, understanding how different seismic design philosophies (conventional, damage-controlled, and no-damage) create fundamentally different carbon profiles across a building's service life.
- Learn from 25 real-world case studies from 15 countries and 6 continents to guide your own projects with proven strategies in adaptive reuse and circularity, material efficiency, modular and hybrid systems, regenerative design, and resilient infrastructure in extreme seismic and soil conditions.

Each chapter includes worked examples, review, and discussion questions to reinforce understanding and build applied skills.

Who This Book Is For

This book is written first and foremost as a university textbook. It is designed to serve as a primary text for upper-level undergraduate and graduate courses in civil and structural engineering, whether in dedicated modules on sustainable structural design and embodied carbon or as a companion to conventional courses in structural analysis, steel and concrete design, or earthquake engineering.

Students in architecture, building engineering, and construction management programs with a good technical focus will also find it valuable, particularly where structural systems and quantitative evaluation form part of the curriculum.

The depth of the book's technical content, the parametric studies based on real design scenarios, and the 25 industry case studies make it equally valuable for practicing structural engineers, architects, sustainability consultants, and policy-makers who need to integrate life-cycle thinking into their work. For AEC professionals already in practice, it offers a reference for applying LCA methodology to their projects, supported by the comparative data, sensitivity analyses, and optimization tools that make the transition from theory to application as direct as possible.

A Note on Learning in the Age of Generative Artificial Intelligence (AI)

There is a good chance you have already used generative AI to help you understand a concept, check your work, or draft a solution.

Generative AI is extending the reach of the teacher (making personalized explanation and feedback available at any hour, to any student, without the constraints of a single classroom). In many ways, this makes learning easier and more accessible than it has ever been.

A conversation with AI is driven by the user's questions; it returns what is asked for. A textbook, by contrast, guides the reader through a landscape of information, revealing relationships between ideas that the reader may not yet know to ask about. The progression in this book is designed so that each chapter equips the reader with the judgment needed to engage critically with the next.

There is also something subtler at work: when a reader opens a textbook to a specific section, their eyes inevitably glance at what comes before and after. They see a figure on the facing page, a table of assumptions from the previous study, and a heading that hints at what the next chapter will demand of them. This peripheral learning (unplanned and accidental) builds context and connections that no targeted query can replicate.

That said, the two are genuinely complementary: once a textbook has built the foundation, AI becomes a powerful amplifier, enabling readers to extend exercises to new scenarios, test alternative assumptions rapidly, and deepen their exploration of any topic with the confidence that comes from structured understanding.

Let this book teach you how to think about the problem, and let AI help you think further.

How This Book Is Organized

The book is structured as a learning journey from foundational concepts to advanced methods and real-world applications:

Chapter 1 establishes the climate context for the Architecture, Engineering, and Construction (AEC) industry, defining the major components of the crisis (greenhouse gas emissions, resource consumption, deforestation, biodiversity loss, and feedback loops), and the risk of carbon tunnel vision, where an exclusive focus on a single metric leads designers to overlook durability, cost, or broader environmental harm. It quantifies the impact of structural systems on a building's carbon footprint, showing how the attention shift from operational to embodied carbon has made the structural frame (often hidden behind finishes) the single greatest point of leverage for reducing whole-life emissions. The chapter examines the role of steel, concrete, and timber, presenting their structural strengths, environmental risks, and mitigation opportunities side by side, and demonstrating through a generic sensitivity analysis that no material is inherently sustainable without verifying its specific sourcing, production, and end-of-life route. It discusses the evolving regulatory landscape, and redefines the professional responsibilities of structural engineers and architects in the context of climate action. The chapter closes by introducing a four-pillar framework with 12 strategies for life-cycle-driven design.

Chapter 2 provides a comprehensive summary of life-cycle assessment (LCA) compatible with EN 15978:2026, walking through every stage from pre-construction (A0) through production (A1–A3), construction (A4–A5), in-use (B1–B5), end-of-life (C1–C4), and beyond-life (D). It establishes the methodological foundations of LCA, including system boundaries, functional units, reference study period, and key environmental indicators such as global warming potential, enabling consistent interpretation and comparison of results across products, buildings, and infrastructure. The chapter develops detailed upfront carbon profiles for steel (BF-BOF, EAF, and DRI-EAF routes), concrete (cement types and supplementary cementitious materials), and timber (biogenic carbon and adhesive contributions) and introduces embodied carbon intensity rating systems. For each material, the chapter breaks down how carbon emissions arise and can be mitigated at every stage of the life cycle, highlighting both common patterns and material-specific processes.

Chapter 3 introduces the fundamental embodied carbon equation, explains how to read and use Environmental Product Declarations (EPDs), clarifies the structure and content of EPDs, including product information, life-cycle scope, and reported impact indicators, and discusses the differences between project-specific and generic data, the role of global and regional databases, and the implications of data variability and uncertainty for design and assessment. It introduces normalization methods based on appropriate functional units, enabling consistent comparison of embodied carbon across buildings, bridges, stadia, and energy infrastructure. The chapter presents 10 strategies for embodied carbon reduction. It also features a comparative exercise across reinforced concrete, steel, and timber systems, demonstrating how embodied carbon responds to changes in material sourcing scenarios, structural configurations, and sensitivity to key structural design parameters.

Chapter 4 introduces life-cycle parameter analysis (LCPA) at the component level: systematic "what-if" investigations that move beyond a single carbon score to reveal which design decisions most influence a structure's environmental outcome. The chapter addresses both design uncertainty (which material, section type, or height drives the result) and data uncertainty (whether switching from a generic database value to a specific EPD changes the conclusion). Two detailed case studies form the core of the chapter: columns (comparing steel open sections, timber, reinforced concrete, and steel-concrete composite across varying heights, concrete strengths, and reinforcement ratios) and beams (comparing open section girders, steel trusses with open and tubular sections, timber, and reinforced concrete across increasing span lengths). For each component type, the analysis tracks both mass and embodied carbon as design parameters change, showing that sourcing and manufacturing practices can be as influential as material choice itself, and how study conclusions are highly sensitive to input carbon factor data.

Chapter 5 introduces life-cycle parameter analysis (LCPA) at the building level, evaluating 25 design configurations for a 10-story structure across reinforced concrete, structural steel, and hybrid timber–steel systems. The chapter introduces the representative portion method, showing that a single bay can approximate full-building embodied carbon with limited deviation, enabling rapid early-stage comparisons. It then systematically varies span length, composite versus noncomposite action, column section type, and material grade (including high-strength steel), before examining how changing assumptions about steel carbon factors, steel sourcing (virgin, scrap, reclaimed), transportation distances, and connection complexity alter not only the absolute results but also the ranking between alternatives. All options are classified using the SCORS embodied carbon rating system. The chapter closes by discussing how uncertainty propagates through LCA calculations and introduces Monte Carlo simulation as a probabilistic alternative to single-value estimates.

Chapter 6 introduces life-cycle optimization (LCO), beginning with the key decisions that shape a building's carbon profile at the conceptual stage: material selection, structural grid, floor system, and lateral load resisting system. It presents genetic algorithms as a method for exploring large design spaces where cost, open space, and embodied carbon must be balanced simultaneously. A multi-objective optimization tool automates the design of floor systems, beams, columns, foundations, and lateral stability systems, generating hundreds of candidate solutions and presenting them as Pareto fronts that map the trade-off surface between competing objectives. Worked examples illustrate how the method can operate in practice. Parametric studies examine how span length, building height, building shape, steel recycling rates, and GGBS content in concrete shift the trade-offs between economic and environmental performance. The chapter closes by discussing future trends in data-driven conceptual design, including neural networks for rapid structural assessment, interactive web-based platforms, and integration with industry-standard design software.

Chapter 7 addresses earthquake-resistant design through a life-cycle lens. It maps three seismic design philosophies (conventional ductility-based, damage-controlled with repairable fuses, and no-damage using seismic isolation) across all life-cycle stages, and shows how each philosophy shifts the balance between upfront embodied carbon and long-term repair-related emissions. It examines seismic design characteristics of steel, reinforced concrete, and timber structures from production through end-of-life and reuse (modules A1–A5 through D), explores resilience-versus-sustainability trade-offs across four strategic dimensions (embodied carbon vs. repair frequency, design for deconstruction, material intensity vs. innovative hybrid systems, and economic trade-offs), and covers nonstructural component resilience. The chapter treats seismic isolation and retrofitting as a circular economy strategy, compares how design codes across Europe, the US, Japan, China, New Zealand, Turkey, India, Latin America, and performance-based frameworks like ATC-58/FEMA P-58 integrate seismic performance with sustainability targets, and considers the broader impact of seismic design decisions on community resilience and long-term recovery. It also introduces the emerging concept of dual-purposing telecommunications fiber-optic infrastructure for scalable post-earthquake structural assessment.

Chapter 8 presents 25 case studies from 15 countries and 6 continents, covering a wide range of building and infrastructure types (including offices, hospitals, stadiums, cultural buildings, airports, bridges, viaducts, shopping malls, mixed-use towers, residential buildings, data centers, and temporary construction facilities), organized across six themes: adaptive reuse and circular construction, material efficiency and technology, low-carbon concrete and industrialized construction, hybrid and modular systems, resilience in challenging seismic and geotechnical conditions, and sustainable infrastructure. Each project is documented with quantified outcomes such as embodied carbon reductions, material savings, and reuse rates, and direct links to the methods developed in earlier chapters, including LCA benchmarking, life-cycle optimization, and design-for-deconstruction strategies. The selection covers diverse structural materials (steel, concrete, timber, and hybrid combinations), climatic zones, and regulatory frameworks, demonstrating that solutions are transferable rather than context-dependent. These case studies show that life-cycle-driven design is not a theoretical aspiration but an emerging professional practice, already delivering measurable results across a wide range of materials and regulatory contexts.

The Role of Practice in This Book

A defining feature of this book is its engagement with industry through detailed case studies: evidence that life-cycle-driven design is already taking shape on construction sites around the world. These contributions (drawn from the work of leading international firms including Martin/Martin, Inc., Bureau Greisch, Cleveland Steel & Tubes, ATI Project, Redesco Progetti, Buro Happold, Columbia Property Trust, Pell Frischmann, Ergodomus Timber Engineering, Gregull & Spang, Moriyama Teshima Architects, Robert Bird Group, RIOS, ATTEC, Turkish Association for Seismic Isolation, GRID International, Proges Engineering, Cimolai S.p.A., and Nikken Sekkei) demonstrate the range of contexts in which life-cycle thinking is already delivering measurable results. I am deeply grateful to the people and firms who opened their project files, shared their quantified outcomes, and allowed their work to serve as teaching material for the next generation of AEC professionals.

Looking Forward

The construction industry is at an inflection point. Regulations on embodied carbon are tightening. Clients are demanding whole-life carbon assessments, and rating systems are rewarding life-cycle performance. The engineers and architects who will lead this transition will be those who can translate carbon metrics into structural design choices, balancing safety, cost, durability, and environmental impact within everyday design decisions. This transition will also require that design platforms close the gap between structural modeling, fabrication planning, and carbon accounting, and that engineers are trained not only to select low-carbon materials but to detail, fabricate, erect, and maintain them well.

I believe that lasting change in our industry will come not from mandates alone but from education: from equipping the next generation of engineers with the knowledge, methods, and conviction to design structures that are safe, resilient, and environmentally responsible across their entire life cycle. My hope is that this book provides them with the foundation to do that.

This book was born in the classroom. It carries the fingerprints of every student who pushed me to explain something more clearly than I thought necessary. I am thankful to them.

Milan, 2026

Alper Kanyılmaz

Acknowledgments

This book would not exist without the generous support of many people.

I am grateful to the many collaborators across universities and firms worldwide who have shaped my thinking over the years. The ideas in this book grew out of conversations, joint projects, and shared curiosity with them.

A large number of individuals made direct contributions to this book, and I wish to acknowledge them here.

Several industry professionals read selected chapters, reviewed my assumptions, and offered valuable perspectives from practice. I am grateful to Tolga Tutar, Andrea Imbrenda, Bayezid Özden, Tom Carr, Tito Cudini, Bahadır Şadan, Emre Emanetoğlu, Marina D'Antimo, Tunç Deniz Uludağ, Hasan Ceylan, Carlos del Castillo, Yıldız Ağaya Çağan, Aleksei Kondratenko, and Matteo Romano for their careful reading and expert feedback on several sections. In particular, Matteo Romano, Marina D'Antimo, Emre Emanetoğlu, Monica Lavagna, Barış Çiftçi, Franco Piva, and Tito Cudini offered detailed feedback for Chapters 2 and 3. Bahadır Şadan, Massimo Latour, Andrea Imbrenda, and Tunç Deniz Uludağ provided photographs, technical suggestions, and an in-depth review of Chapter 7.

Among the student reviewers, I owe special thanks to Defne Dengizek, who read the draft and offered suggestions that helped with the book's clarity and accessibility across disciplinary boundaries, and to Oğuzhan Gököz, who generously provided help with sketches, analysis, and drawings I requested throughout the writing process.

I thank Murat Güven for helping me with some of the visual content.

Saber Zolfaghari worked patiently through the book's worked examples, testing calculations and refining the presentation material. Mahdi Chadede, in cooperation with Jon Carr, contributed directly to the parameter analysis of Chapter 3.

The case studies in Chapter 8 owe a great deal to the dedication of the many individuals with whom I was in direct contact for the preparation of the material, and I am equally grateful to the colleagues and team members who supported them throughout this process. I wish to thank Tunç Deniz Uludağ, Kirby Beegles, Amaury Leroy, Roy Fishwick, Alessio Bacci, Manuel Quagliaroli, Andrea Riva, Paolo de Angelis, Mauro Eugenio Giuliani, Rob May, Julie Janiski, Kristen Brozowski, Ricardo Pimentel, Stacy McMahon, Franco Piva, Marian Kempkes, Carol Phillips, Katie Weber, Joseph Bonica, Ben Radjenovic, Golnaz Ighany, Ahmet Topbaş, Bahadır Şadan, José Olivier Pedro, Andrea Imbrenda, Alessandro Catanzano, Eleonora Lucchese, Giorgio Campanile, Sofia Lucchese, David Varcoe, Giovanni Brambilla,

Antonio Condello, Ania Khodabakhshian, and Raffaella Bedini. Each of them gave generously of their time, sharing project data, photographs, technical narratives, and multiple rounds of review to ensure that the case studies met the standards of both engineering and pedagogical clarity.

I owe a singular debt to Carlo Andrea Castiglioni, who supervised my MSc and PhD theses and who is the reason I came to build my academic career in Italy. Over the years, he has been a supervisor, colleague, and friend, and his influence runs through every chapter of this book. His red pen has accompanied my entire academic life: it corrected my MSc and PhD dissertations and academic papers, and now it has passed through every chapter of this book.

Finally, I thank my family for their patience and support during the long evenings, weekends, and holidays this book required. This work is dedicated to them.

Milan, 2026 *Alper Kanyılmaz*

1

Introduction

LIST OF CONTENTS

In the face of the global climate crisis, the construction industry stands at a critical position. As the built environment accounts for a major portion of the greenhouse gas (GHG) emissions, and consumes vast quantities of natural resources, there is a pressing need for architects, engineers, and construction professionals to adopt sustainable practices. The traditional approach of designing structures solely for cost and structural performance is no longer sufficient. On top of this, the global population is expected to increase by 2.5 billion people by 2050, leading to an estimated 230 billion square meters of new construction (Figures 1.1 and 1.2) (IEA, 2018; LETI, 2020b; UNFCCC, 2025), leading to an estimated 120 gigatons of embodied GHG emissions from the materials and construction processes alone: this is roughly one-third of the entire remaining global carbon budget for limiting warming to 1.5 °C before the year 2050 (IPCC, 2018). This figure accounts only for new construction; when the maintenance, refurbishment, and retrofit of the existing building stock are considered, the true scale of material demand (and associated emissions) is likely to be larger. Moreover, the majority of this population growth is projected to occur in Asia and Africa (United Nations Department of Economic and Social Affairs, 2018), regions where awareness and regulatory frameworks for embodied carbon in construction remain comparatively underdeveloped, highlighting the urgent need for the global community to accelerate knowledge transfer and collective action. This surge in construction activities presents both a significant

Figure 1.1 A view from a construction site (zhu difeng/Adobe Stock Photos).

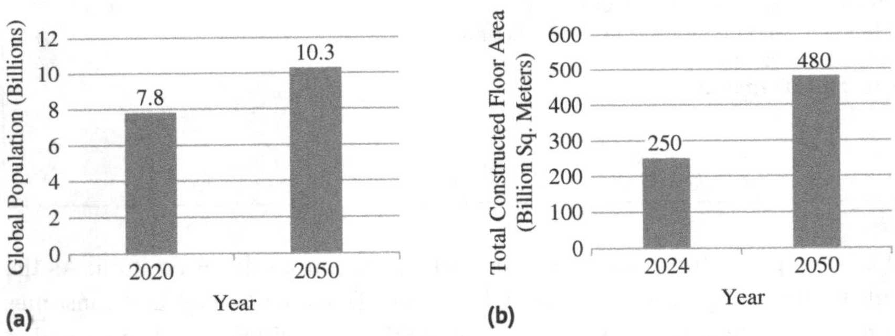

Figure 1.2 Projected growth in global population and floor area by 2050: (a) global population trend between 2020 and 2050 and (b) global floor area trend between 2024 and 2050.

challenge and an opportunity for the construction community to implement sustainable practices and reduce GHG emissions.

The growth in construction is not limited to buildings but extends to various construction products such as bridges, tunnels, highways, railways, and energy infrastructures (Figure 1.3). This widespread expansion means that efforts to reduce emissions must be comprehensive and integrated across all types of construction projects.

Meeting this challenge requires both a fundamental shift in how we design and construct our structures, and a fundamental update to the education of future architects and engineers. This book is designed to enable those by introducing a concept titled "life-cycle-driven structures." This new framework places environmental impact, particularly embodied carbon, at the very center of structural design decision-making, providing a guideline for both students and practicing professionals.

Figure 1.3 (a) Steel bridge construction (Randy Hergenrether/Shutterstock); (b) solar plant construction (rh2010/Adobe Stock Photo); (c) reinforced concrete bridge construction (ThamKC/Adobe Stock Photos); and (d) transmission towers (chuyuss/Shutterstock).

In this life-cycle-driven framework, one needs to incorporate quantitative environmental impact assessments into traditional structural design procedures, treating carbon emissions as a critical performance metric alongside safety and cost (treating member sizing, material grades, connection strategies, fabrication, logistics, maintenance, and end-of-life as central decisions assessed with consistent data and boundaries). The workflow employs parametric models, sensitivity studies, and multi-criteria optimization to quantify trade-offs across safety, serviceability, cost, embodied carbon, adaptability, and reuse potential over the whole life-cycle stages. In short, the aim is to meet structural performance targets while cutting GHG emissions, enabling future reuse, and making carbon visible in every design choice (as a shift from high-emission outcomes toward low-emission outcomes across the life cycle, Figure 1.4).

This chapter maps the problem space by defining GHGs, resource use, deforestation, biodiversity loss, feedback loops, and tunnel vision risk; locates structural systems and materials within the structures' carbon footprint; clarifies the role of designers and regulations; and traces the shift from conventional design to life-cycle thinking, preparing the reader for the life-cycle-driven structures framework used throughout the book.

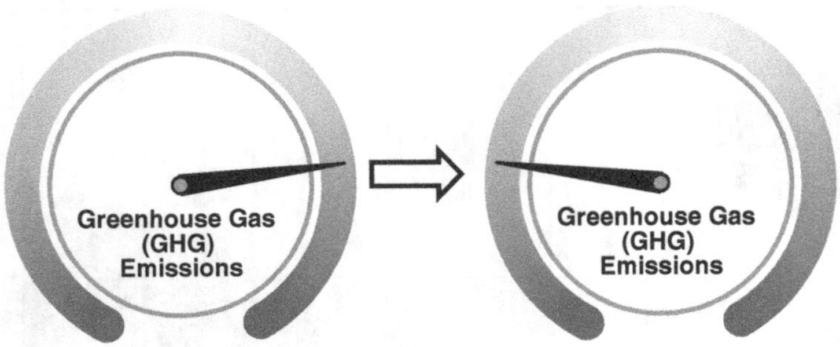

Figure 1.4 Conceptual illustration of the life-cycle-driven objective: shifting greenhouse gas (GHG) emissions from higher levels toward lower levels through design and delivery choices.

1.1 Definitions of Major Components of Climate Crisis

The climate crisis is shaped by six linked drivers: GHG emissions, resource consumption, deforestation, biodiversity loss, feedback loops, and tunnel vision risk. Together, these drivers intensify warming and ecosystem damage, so we need to map their interactions and mitigation levers in order to prioritize low-carbon materials and structural decisions, and circular practices within the life-cycle-driven structures framework.

1.1.1 Greenhouse Gas (GHG) Emissions and Global Warming Potential (GWP)

Greenhouse Gas Emissions (GHGs) contribute to the warming of the Earth by trapping heat in the atmosphere, a phenomenon known as the greenhouse effect. Identified in the 1800s, this natural process has been amplified by human activities (Figure 1.5), leading to unprecedented levels of global warming. In response, the Paris Agreement, adopted in 2015 by 196 parties at COP21 (Conference of the Parties), aimed to limit the global temperature increase to 1.5 °C by the year 2050. Countries are expected to submit updated GHG emission reduction targets every five years, aiming to reach these commitments by 2050 (UNFCCC, 2025). For this reason, the COP has been held every year since its beginning.

Major GHGs include:

Carbon Dioxide (CO_2): Primarily produced from the burning of fossil fuels, waste, and biological sources, as well as from chemical reactions in industries such as steel and cement production.

Methane (CH_4): Results from coal, natural gas, and oil production, transport, agricultural practices, land use, and the decay of organic waste.

Nitrous Oxide (N_2O): Results from agricultural activities, land use, industrial processes, combustion of fossil fuels and solid waste, and wastewater treatment.

Fluorinated Gases: Synthetic gases used in household, commercial, and industrial applications. Although emitted in smaller quantities, they have high global warming potentials (GWPs), trapping significantly more heat than CO_2.

Global Warming Potential (GWP) is a measure used to compare the relative impact of different GHGs over a specific period. It considers both the efficiency of the gas at trapping heat and its lifetime in the atmosphere (Figure 1.6). Table 1.1 summarizes the GWP values of the key GHGs, along with their respective atmospheric lifetimes.

Figure 1.5 Industrial emissions on ecosystems (TR Stok/Shutterstock.com).

Although CO_2 accounts for 79% of global GHG emissions by volume, the remaining gases (methane, nitrous oxide, and fluorinated gases) carry disproportionately high warming potentials per unit emitted, making them critical targets for mitigation alongside CO_2. Figure 1.7 illustrates the relative atmospheric shares of major GHGs.

To enable meaningful comparison, reporting, and regulation of diverse GHG emissions, the industrial and policy sectors adopt CO_2 equivalent (CO_2e or CO_2-eq) as the universal metric. This standardized unit converts all GHG emissions into the equivalent amount of CO_2 that would produce the same warming effect over the specified time period. Throughout this book, we will also reference CO_2 equivalent as our primary unit of measurement when discussing GHG emissions.

1.1.2 Resource Consumption

Resource consumption represents the second critical component of the climate crisis, creating both direct environmental damage and indirect GHG emissions throughout extraction, processing, transportation, and disposal phases. This pattern is particularly relevant to the construction industry, which consumes approximately 40% of global raw materials and 36% of global energy (UNEP, 2023). Resource consumption in the construction industry involves the extensive use of raw materials (e.g. steel, concrete, wood), water (Figure 1.8), and energy (burning of fossil fuels for electricity, heat, and transportation).

Such consumption leads to significant environmental impacts, including habitat damage, resource depletion, and increased CO_2 emissions from high energy use. Therefore, from a resource consumption standpoint, addressing the climate crisis requires a dual focus: transitioning to clean energy sources and simultaneously promoting a more sustainable, circular economy that reduces overall resource demand.

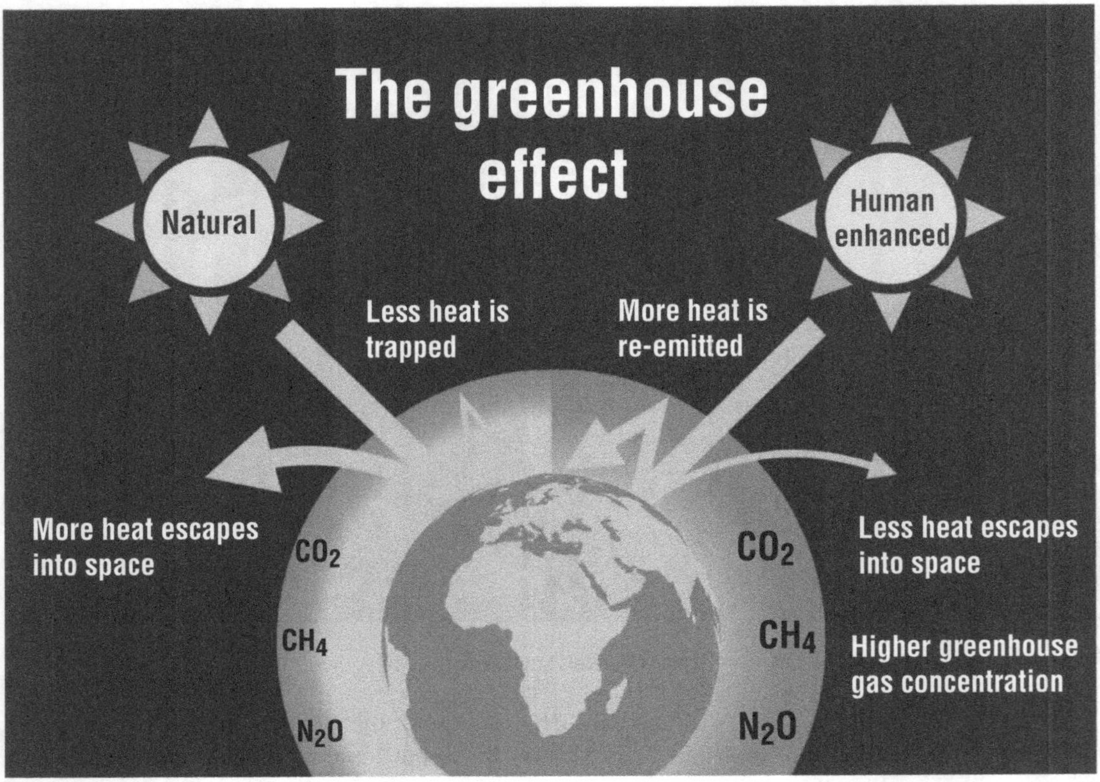

Figure 1.6 GHG effect on the earth, and the influence of the human factor to enhance it (Dimitrios/stock.adobe.com).

Table 1.1 GWP and atmospheric lifetimes of major GHGs. The GWP is a dimensionless factor that expresses how much heat a given mass of a greenhouse gas traps over a chosen period (typically 100 years) relative to the same mass of CO_2, which is assigned a value of 1 as reference; multiplying each gas's emitted mass by its GWP converts a mixture of gases into a single CO_2-equivalent (CO_2e) figure.

GHG	GWP	Atmospheric lifetime
Carbon dioxide (CO_2)	1	Thousands of years
Methane (CH_4)	28	Tens of years
Nitrous oxide (N_2O)	265	Hundreds of years
Fluorinated gases	Up to 24 000	Varies, up to 50 000 years

Source: U.S. Environmental Protection Agency (2023)/U.S. Environmental Protection Agency/Public Domain.

For both steel and concrete, mining and quarrying (Figure 1.9) involve land-use changes that can degrade or remove natural habitats, while also contributing to acid rain and water pollution, thereby affecting river ecosystems. Landfill sites (these are designated areas for waste disposal) can leak toxic chemicals into surrounding groundwater, potentially making nearby soils unfit for many species.

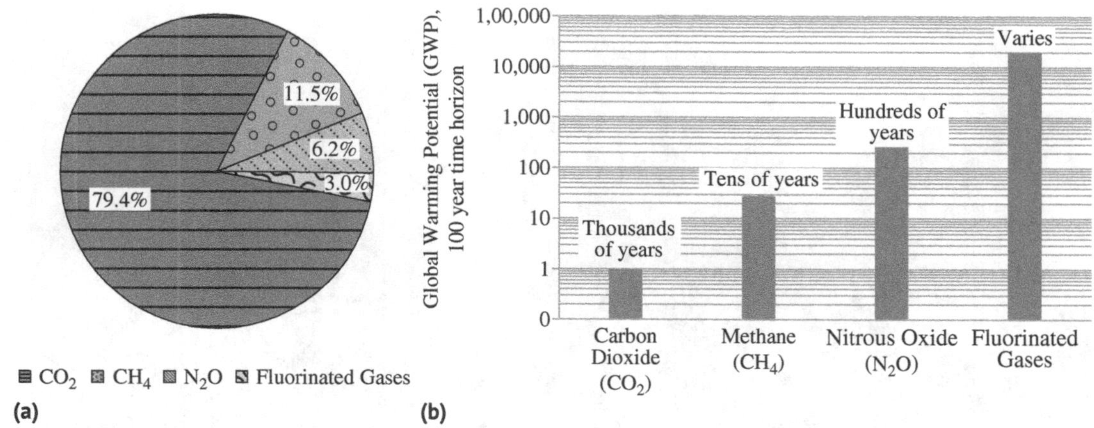

Figure 1.7 (a) Distribution of major GHGs in the atmosphere and (b) atmospheric lifetime of major GHGs (U.S. Environmental Protection Agency, 2023/U.S. Environmental Protection Agency/Public Domain).

Figure 1.8 Dried lakebed with cracked soil and receding water, illustrating growing pressure on freshwater resources. The construction industry is a major contributor to water stress, with concrete production, steel manufacturing, and site operations all requiring large volumes of water. As climate change intensifies drought cycles, competition for freshwater among construction, agriculture, and municipal needs is expected to increase (Piyaset/Shutterstock).

Figure 1.9 Aerial view of a large-scale open-pit mine, showing terraced extraction levels and haul roads. Mining operations for construction-related raw materials (such as iron ore, limestone, and aggregates) are among the most land-intensive human activities, generating significant GHG emissions through excavation, material transport, and processing (Dmitriy Kuzmichev/Shutterstock).

1.1.3 Deforestation

Deforestation is the third critical driver of the climate crisis. Forests act as natural carbon sinks, absorbing CO_2; when cleared or degraded, they release stored carbon back into the atmosphere while eliminating future absorption capacity. A common argument in favor of timber as a construction material is that the carbon captured during tree growth (biogenic carbon) can be considered climate-neutral, provided the harvested forest is replanted and sustainably managed over successive cycles. Under this logic, wood-based construction could function as a temporary carbon store, offsetting emissions from other materials. However, this reasoning hinges entirely on one condition: that forests are, in fact, sustainably managed.

Only around 9% of global forests are certified for sustainable management (PEFC, n.d.), and the vast majority of certified forests are located in Europe and North America (Figure 1.10). Tropical forests in regions like the Amazon Basin, the Congo Basin, and Southeast Asia (which are at the highest risk of deforestation and biodiversity loss) have a much lower percentage of certified area.

Unsustainable forestry practices often replace diverse forests with monocultures (forests where only a single species of tree is grown, often for commercial harvesting), reducing habitat variety and

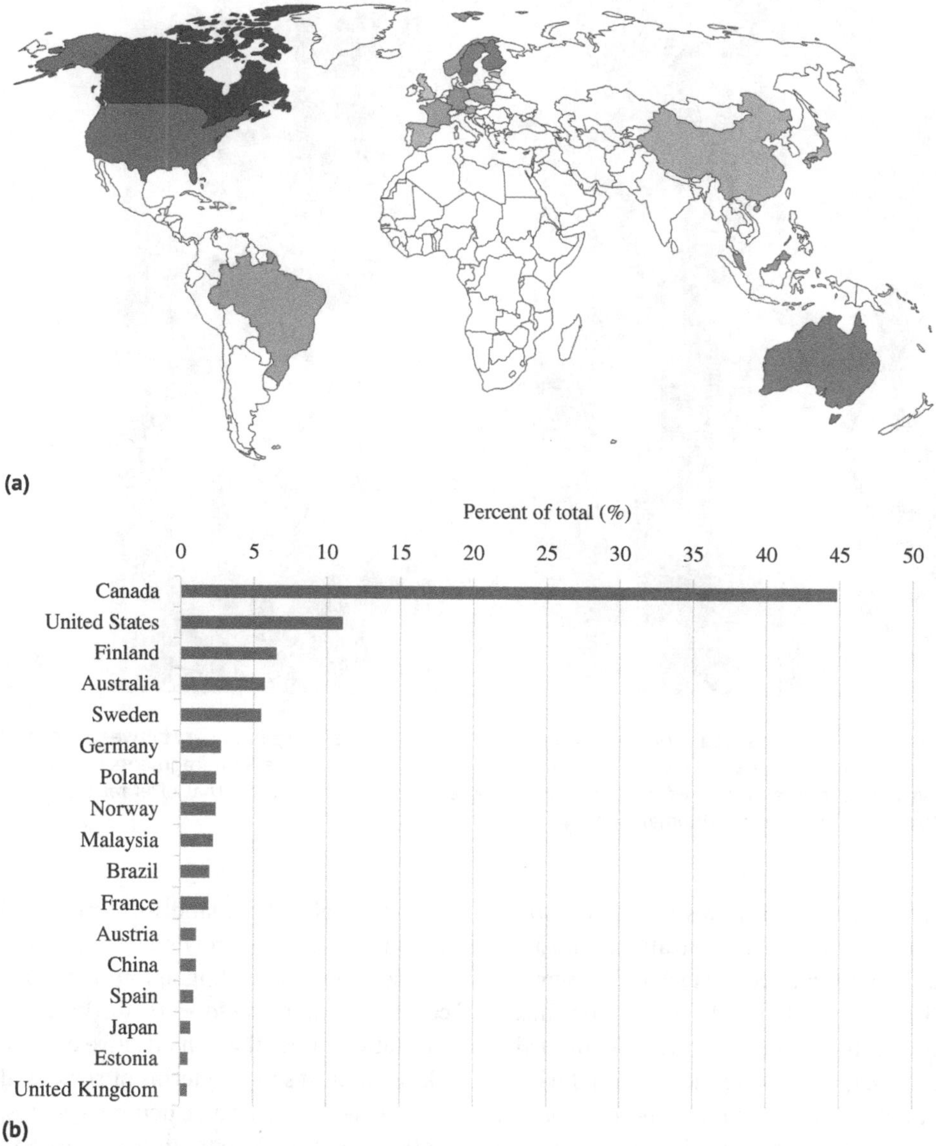

Figure 1.10 Global distribution of sustainably certified forest area and the share of total certified area held by selected countries. (a) Global distribution of sustainably certified forest area by selected countries (darker shading indicates larger certified area). (b) Share of global sustainably certified forest area by selected countries (% of total certified area).

introducing invasive species that spread beyond the site boundary. Together, these processes disrupt ecosystems, reduce biodiversity, and highlight the importance of responsible resource extraction, waste management, and sustainable forestry (Figure 1.11).

Figure 1.11 Aerial view of a cleared area carved into dense forest, showing the sharp boundary between intact canopy and stripped land. Such large-scale clearing for timber harvesting or land conversion fragments habitats, disrupts local ecosystems, and often precedes replanting with single-species monocultures that offer far less ecological value than the original forest (luoman/Getty Images).

These facts carry direct implications for the construction sector. As global demand for timber and land continues to rise (driven by urbanization, infrastructure expansion, and the growing promotion of wood as a low-carbon alternative to steel and concrete), the pressure on already fragile forest systems will intensify. Without a parallel and verifiable expansion of certified sustainable forestry, the biogenic carbon argument risks becoming a justification for further deforestation rather than an effective climate mitigation strategy. That said, using each material where it makes the most sense in terms of structural performance, durability, fire resistance, and environmental impact, within hybrid structural systems, while expanding the share of sustainably managed forests, can offer a credible path to unlocking wood's climate benefits without amplifying deforestation risk.

1.1.4 Biodiversity Loss

Construction materials can affect biodiversity throughout their life cycle. Concrete production, for instance, involves cement manufacturing that emits harmful pollutants and degrades surrounding habitats. Iron and steel making requires mining iron ore and coking coal, disrupting ecosystems through deforestation, soil erosion, and water contamination. These activities, combined with the

Figure 1.12 A deer standing at the boundary between a thriving forest and a degraded, lifeless landscape, symbolizing the fragile divide between healthy ecosystems and habitat destruction. Construction-related activities (mining, cement production, land clearing) are among the leading causes of biodiversity loss, threatening species worldwide (dottedyeti/Adobe Stock).

broader impacts of the built environment, contribute to around 30% of threatened species worldwide (Hutchins, 2024), highlighting the urgency for more careful material choices (Figure 1.12).

Timber sourcing can also alter biodiversity outcomes, especially when forests are managed as monocultures that limit habitat variety. By contrast, procuring wood from certified providers (such as those endorsed by FSC or PEFC) helps retain a healthier balance of plant and animal species. Natural forests support complex vertical ecosystems with distinct layers (from the forest floor through the understory and canopy to the emergent layer), each hosting unique communities of flora and fauna that monoculture plantations cannot replicate (Figure 1.13). Preserving this multilayered forest structure through responsible sourcing, minimizing material volumes, and prioritizing reuse help maintain these diverse habitats while still meeting construction needs.

1.1.5 Feedback Loops

In the context of the climate crisis, a feedback loop is a self-reinforcing cycle where an initial climate-related change triggers a series of subsequent effects that ultimately amplify the original change, leading to accelerated environmental degradation (Ripple et al., 2023; Thayer and MacCarthy, 2025). For example:

- Increased resource extraction drives habitat destruction, reducing natural carbon sinks while simultaneously increasing emissions from processing and transportation.
- Deforestation for material extraction eliminates CO_2-absorbing forests while generating emissions from land-use change.
- Climate change impacts drive increased demand for climate adaptation infrastructure (e.g. flood barriers, cooling systems), creating additional resource consumption and emissions in a self-reinforcing cycle.
- Climate-related disruptions to resource supply chains often lead to longer transportation distances and higher-emission alternative materials, compounding both consumption and emission impacts.

Because such loops (Figure 1.14) dangerously amplify environmental damage, breaking these cycles is a fundamental climate priority.

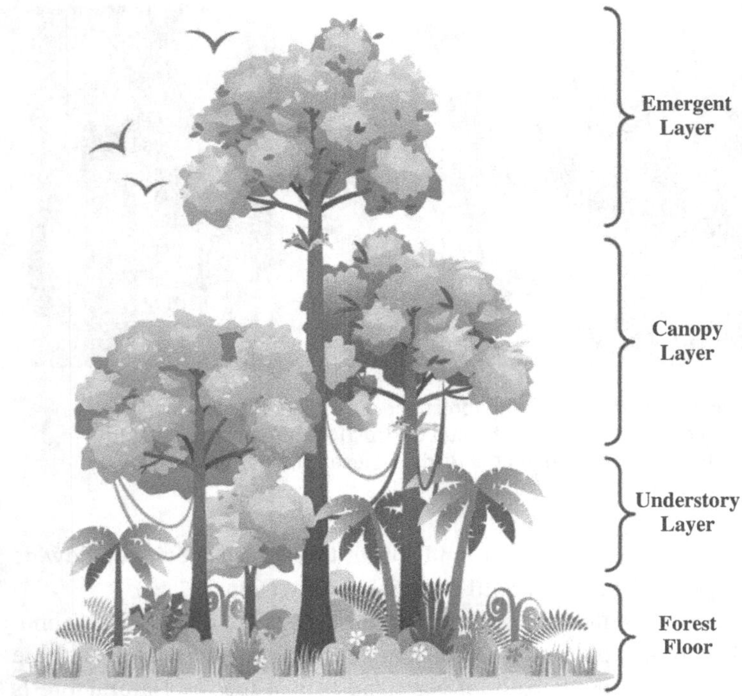

Figure 1.13 Schematic illustration of the vertical stratification of a natural forest, showing distinct layers (forest floor, understory, canopy, and emergent layer), each supporting unique communities of flora and fauna. Monoculture plantations grown for commercial timber harvesting cannot replicate this ecological complexity, underscoring the importance of certified sustainable sourcing (VectorMine/Shutterstock).

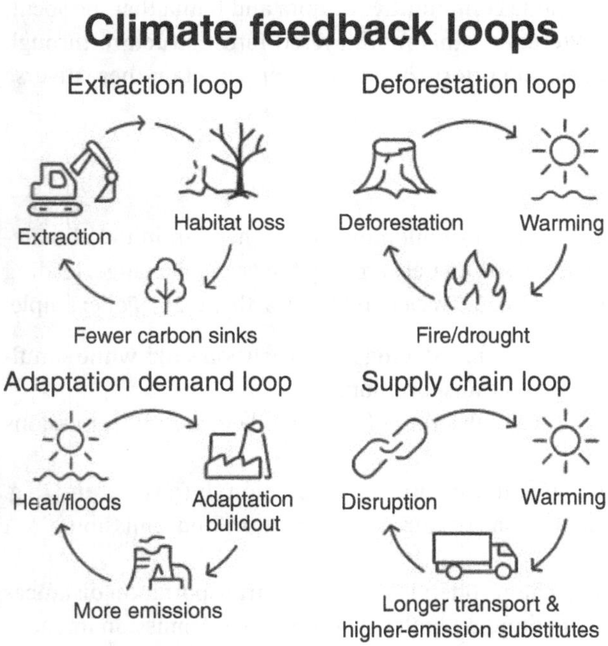

Figure 1.14 Feedback loops accelerating climate crisis.

1.1.6 Tunnel Vision Risk

While GHG emissions are a primary driver of the climate crisis, an exclusive focus on a single metric can lead to a critical oversight known as "tunnel vision" (Nurmaulina, 2024) (Figure 1.15). In the context of carbon, this occurs when a design team prioritizes GHG reduction so intensely that they inadvertently ignore (or even compromise) other vital project goals. These include not only wider environmental factors like resource depletion and biodiversity loss, but also crucial performance metrics such as durability, long-term service life, and overall life cycle cost.

The construction industry is particularly susceptible to this danger. For example, a project might specify a mass timber system based solely on its carbon benefits, but if that timber is sourced from an unsustainably managed forest, the net environmental outcome is negative. Similarly, a novel, ultra-low-carbon material might be chosen without adequate consideration for its higher cost, unproven durability, or shorter service life compared to conventional alternatives. In both cases, a narrow focus on a single metric leads to a solution that is either not genuinely sustainable or economically and functionally unviable in the long term.

Figure 1.15 "Carbon tunnel vision," a common danger of overemphasizing CO_2 while neglecting other vital project metrics like cost, durability, biodiversity, and service life (generated with AI using Google Gemini).

The interconnected nature of design demands a more holistic perspective. The life-cycle-driven structures framework presented in this book is designed as a direct countermeasure to this kind of tunnel vision. It forces a multi-criteria approach, ensuring that decisions are evaluated not just on their carbon impact but also on their performance across the building's entire service life, balancing environmental metrics with durability, adaptability, and life cycle cost to guide professionals toward truly resilient and sustainable outcomes.

1.2 Impact of Structural Systems on a Construction Product's Carbon Footprint

The structural system is often overlooked yet carries a decisive share of a construction product's carbon footprint. While this chapter focuses on buildings to illustrate this impact, the principles discussed are broadly applicable to all construction products.

The carbon footprint of a building is composed of two components: "operational carbon" and "embodied carbon." Operational carbon refers to emissions generated from energy use during the building's service phase, including heating, cooling, lighting, and maintenance. Embodied carbon refers to emissions associated with the extraction, manufacturing, transportation, construction, and disposal of building materials (LETI, 2020b). Traditionally, efforts to fight the climate crisis have focused on reducing operational carbon. However, as buildings become more energy-efficient, the relative importance of embodied carbon increases (Figure 1.16): it can account for 40–70% of the total life-cycle carbon emissions of new buildings (LETI, 2020b).

As shown in Figure 1.17, operational carbon during the in-use stage dominates, while embodied carbon impacts across product, transport, construction, maintenance, replacements, and end-of-life. This explains why early material choices, service life planning, and repair strategies shape embodied carbon outcomes.

While operational carbon emissions from building energy use have received significant attention over the past decades, the embodied carbon within structural materials has remained largely invisible to designers and clients. Figure 1.18 shows a building's structural skeleton (columns, beams, slabs, and other load-bearing elements) that remains normally hidden behind exterior finishes. Despite being out of sight, a structural frame can comprise a significant portion of a building's total materials, which often carry a higher embodied carbon footprint compared to more visible components like façades or interior finishes.

Understanding how different components of a building (Table 1.2) contribute to its embodied carbon is crucial for identifying effective areas for carbon reduction. Structures and foundations are the most long-lasting parts of a building and account for more than half of the building's embodied carbon.

The components in Table 1.2 can vary depending on the type of building, such as medium- to large-scale housing, small-scale housing, commercial offices, and school buildings. Figure 1.19 shows the estimated service life of building components, while Figure 1.20 shows the average distribution of embodied carbon across these components. Because the structural elements are designed to last for the entire life of the building, the high carbon emissions associated with them are effectively "locked in" from day one. This makes the initial design of the foundation and superstructure the single greatest point of leverage for reducing a building's total lifetime carbon footprint.

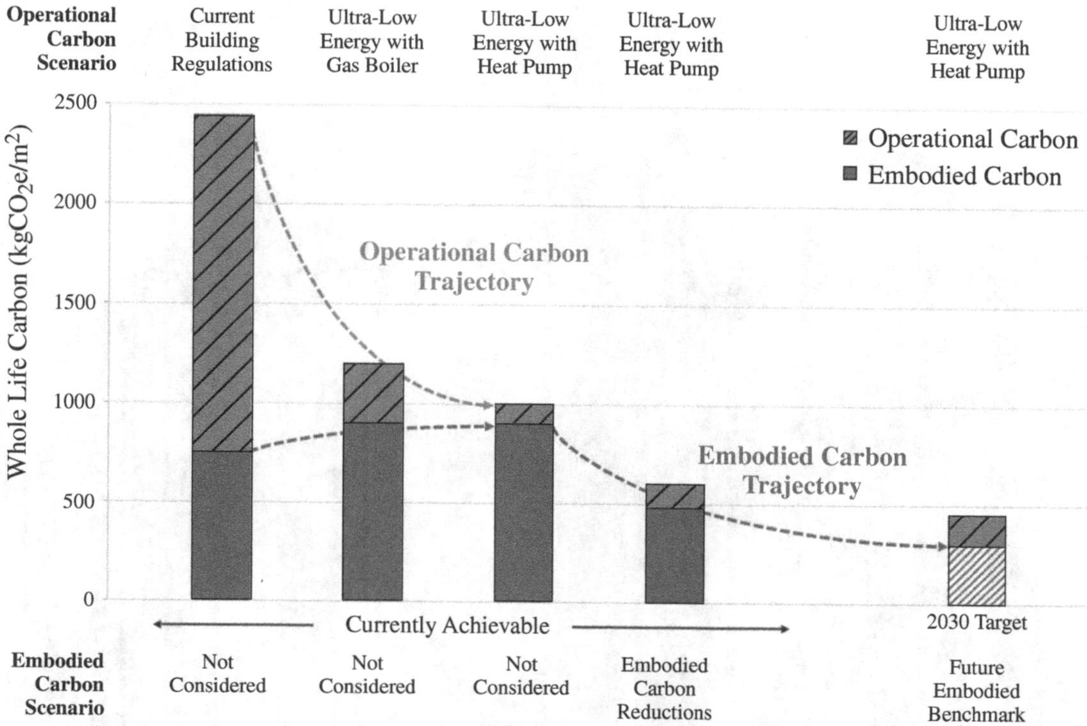

Figure 1.16 Focus shift from operational carbon to embodied carbon (graph reproduced according to LETI, 2020b/LETI).

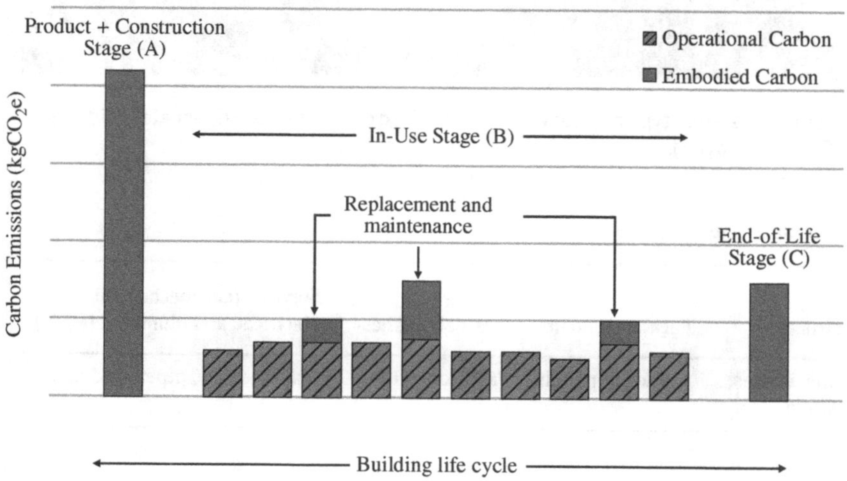

Figure 1.17 Breakdown of the building's carbon footprint over its life cycle (adapted from LETI, 2020b).

Figure 1.18 A building's structural skeleton, typically concealed behind exterior finishes, shown alongside its completed form (courtesy of Chapman Taylor).

Table 1.2 Example building components.

Substructures	Superstructure	Façade elements	Internal finishes	Services (e.g. mechanical, electrical, and plumbing (MEP))
Foundations and basements	Columns, beams, slabs, walls	Cladding panels	Partitions, floors	Plants, ducts, pipework, cables

Understanding the sources and scale of embodied carbon within these structural elements would help designers, engineers, and property owners to identify opportunities for reductions, whether through leaner design (i.e. optimizing the structure to use fewer materials and minimize waste without

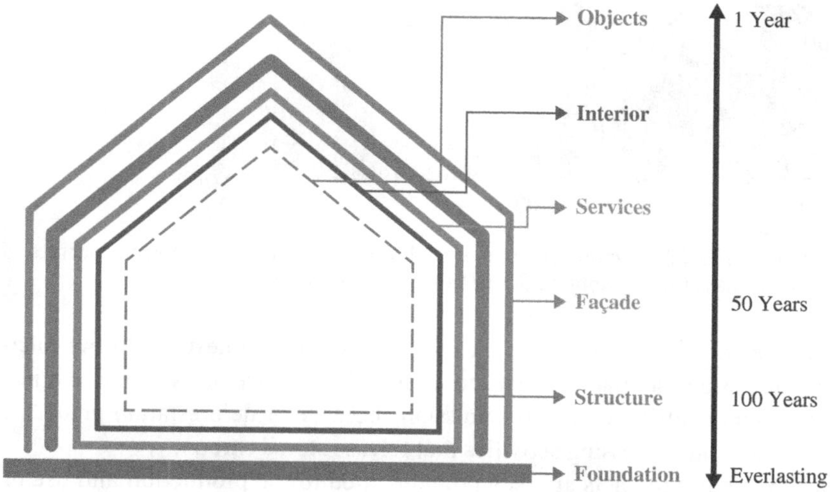

Figure 1.19 Building components and their estimated service life before requiring replacement (adapted from Adams et al., 2019).

Figure 1.20 Average distribution of embodied carbon across different building components. The data is aggregated from four building types: medium-scale housing, small-scale housing, commercial offices, and school buildings (adapted from LETI, 2020a).

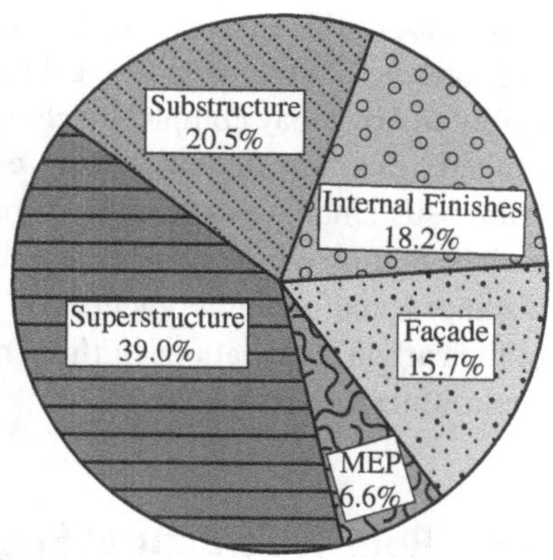

compromising performance), alternate materials, or more efficient construction methods. This book is about quantifying and reducing the environmental impact of buildings, focusing on the entire life cycle of structural systems.

1.3 Role of Construction Materials in the Climate Crisis

Construction materials (Figure 1.21) play a significant role in both causing and mitigating the climate crisis. Steel is a widely used material in construction due to its strength and durability; however, its production process is highly energy-intensive and contributes significantly to global GHG emissions. Concrete, the most widely used construction material globally, is renowned for its durability and versatility but has a substantial environmental impact because cement production releases CO_2 both from

(a) (b) (c)

Figure 1.21 Major construction materials: (a) structural steel (gooutsight/Shutterstock); (b) reinforced concrete (marius_popa/Adobe Stock Photo); (c) mass timber (Roberto Sorin/Adobe Stock Photos).

the energy-intensive heating process and from the chemical decomposition of limestone during calcination. Timber is often promoted as a sustainable construction material because of its carbon sequestration capabilities; however, it can contribute to GHG emissions because of deforestation practices, transportation over long distances, and processing activities that release stored carbon.

Table 1.3 outlines the environmental impacts and key points related to the production and use of the major construction materials. These observations and data are available in the literature in several resources (Arup, 2023a, b, c; *EPD International*, n.d.; ICE, 2022; Chehade, 2023; LETI, 2020b; PEFC, n.d.).

Because of the reasons stated in Table 1.3, there is no simple way to claim one material is sustainable over another. Based on how steel and concrete are produced, or how timber is harvested, the sustainability metrics may completely change. Figure 1.22 shows a generic example of how embodied carbon can vary dramatically for a steel, reinforced concrete, and timber frame when comparing sources that follow sustainable practices vs. those that do not (these values are illustrative and actual embodied carbon figures will vary significantly based on specific suppliers, production methods, transportation distances, and regional practices). The modern structural engineer and architect must develop fluency in the carbon implications of these materials while understanding how design decisions influence their environmental performance.

1.4 Role of Structural Engineers and Architects in Climate Action

In the times of the Roman Empire, Vitruvius defined the core principles of design as "firmitas, utilitas, and venustas" ("firmness, utility, and beauty") (*On Architecture, Volume I*, 1914). For centuries, these principles guided the design and construction of structures. Today, while these classical virtues remain relevant, they are being fundamentally redefined by the urgency of the climate crisis. The modern focus has shifted toward creating a sustainable built environment that balances performance and aesthetics with profound environmental and social responsibility (Figure 1.23).

Structural engineers and architects take a central position in the fight against the climate crisis. Their influence goes beyond answering the fundamental question "Why buildings stand up?" (Salvadori, 1982): it extends deep into the supply chain and forward into the building's entire life cycle. This pivotal role includes both expanded technical duties and crucial leadership responsibilities. Today, the eight key responsibilities of structural engineers and architects for a sustainable built environment can be described as shown in Figure 1.24.

Table 1.3 Structural performance benefits, environmental impact risks, and climate mitigation opportunities of primary construction materials.

Material	Strengths and benefits	Environmental impact risks	Mitigation opportunities
Steel	• Good strength-to-weight ratio enabling long spans and tall structures, while requiring moderately sized foundations • Excellent tensile strength and ductility (235–460 MPa, typical) • Predictable structural behavior and well-established design codes • Recyclable and reusable with minimal quality loss • Fast construction with prefabricated components • Highly suitable for seismic zones due to high strength and ductility • Minimal on-site waste generation	• Accounts for ~7% of global GHG emissions • Energy-intensive manufacturing requiring high temperatures (1500 °C+) • CO_2 release from chemical reactions with coal and iron ore • Resource-intensive fabrication (up to 30–40% of resources spent on joints) • Vulnerable to corrosion without protective coatings (except weathering steel) • Mining impacts for iron ore extraction • Thermal bridging issues affecting building energy performance • Low fire resistance (requires separate fire protection in most instances such as intumescent paint or fireboard).	• 100% recyclable without quality loss and directly reusable (i.e. reclaimed steel) • Properly designed steel structures can last 100+ years, reducing replacement frequency • High strength allows for minimal material use in optimized designs (also available up to 1000 MPa) • Steel connections enable disassembly and reuse of entire structural elements • Easily modifiable for changing building functions (versatility) • Off-site manufacturing reduces construction waste and site impact
Concrete	• Good compressive strength (20–60 MPa typical) • Moldable into complex geometric forms helping architecture ambitions • Inherent fire resistance and durability • Cost-effective with locally available materials • Thermal mass properties for energy efficiency • Low maintenance requirements • Good acoustic and vibration properties • Resistant to weathering and chemical attack	• Responsible for ~8% of global GHG emissions • High CO_2 emissions from clinker production • Massive global production scale (14 billion m^3 annually), though, some regions with high construction activity (such as the Middle East) have to import sand for concrete, as locally available desert sand lacks the particle shape and grading required for structural use. • Limited recyclability and reusability – typically downcycled to aggregate • Water consumption in production and curing • Aggregate mining environmental impacts • Urban heat island effect from thermal properties • While clinker replacement materials exist (e.g. GGBS), their global supply is limited and unevenly distributed, meaning they cannot always be relied upon as a scalable decarbonization strategy. • Relatively large weight of concrete structures may demand larger and deeper foundations, increasing material consumption (particularly concrete and steel reinforcement in the substructure)	• Well-designed reinforced concrete structures can last 50+ years with minimal maintenance • Reduces operational energy consumption for heating/cooling, thanks to its high thermal mass • Aggregates typically available within 50 km radius, reducing transportation (in many parts of the world) • Optimized designs minimize material quantities while maximizing performance • A number of promising low-carbon concrete technologies are currently in production or at advanced research and testing stages, which should significantly reduce the reliance on clinker in the near future.

(Continued)

Table 1.3 (Continued)

Material	Strengths and benefits	Environmental impact risks	Mitigation opportunities
Timber	• Good strength-to-weight ratio allows for smaller and shallower foundations, reducing substructure material consumption and the associated embodied carbon. • Fast construction with prefabricated CLT/GLT systems • Good insulation properties (low thermal conductivity) • Aesthetic and biophilic appeal • Carbon storage potential • Renewable resource when sustainably managed	• Potential for deforestation and biodiversity loss • Only ~9% of global forests are certified for sustainable management (FSC/PEFC) • Carbon neutrality claims often misleading without LCA • Stored carbon released if timber degrades, burns, or ends up in landfill (currently, approximately 90% of timber follows one of these end-of-life routes, reflecting very poor recycling rates across the industry) • Non-local sourcing is often required, as structural timber must meet specific grade and strength requirements that are not always available from local forests (for instance, the UK imports the majority of its structural timber from Europe) • Highly dependent on sustainable forest management certification • Moisture sensitivity and potential for decay • Pest susceptibility (termites, fungi) • Limited global availability of engineered timber products	• May reduce operational energy demands compared to other materials, since its low thermal conductivity improves insulation and limits heat loss • Regrows and sequesters more carbon when sustainably harvested • Prefabricated timber reduces construction time and associated emissions • Can be reused structurally, repurposed into secondary products, or used for bioenergy at end of life, while remaining carbon neutral when sourced and managed sustainably, although structural reuse of reclaimed timber is much less common than with steel.

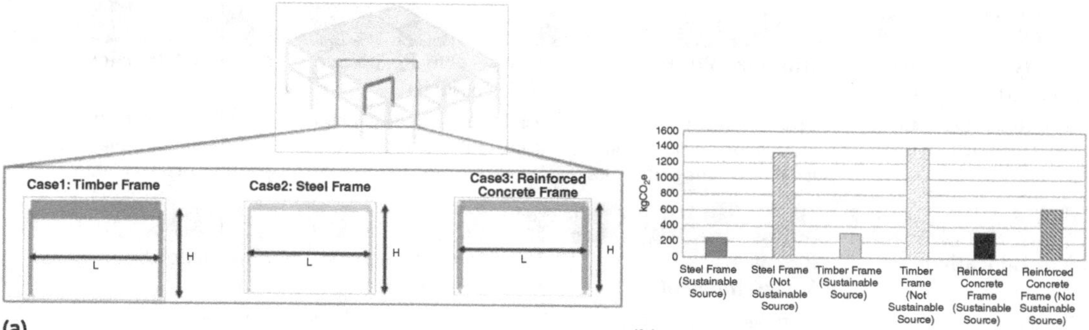

(a) (b)

Figure 1.22 A conceptual comparison of embodied carbon for timber, steel, and reinforced concrete frames, highlighting the influence of material sourcing and production practices. For each material, values associated with more sustainably managed supply chains are contrasted with those from less sustainable sources, emphasizing that outcomes can vary substantially within the same material category. The values shown are indicative only, as actual embodied carbon depends on factors such as supplier practices, manufacturing processes, transport distances, and regional conditions. (a) Three structural frames – steel, concrete, and timber – used for comparison and (b) embodied carbon levels under different production assumptions.

(a) (b)

Figure 1.23 (a) Vitruvian architecture emphasizing large and impressive structures and (b) modern sustainable design, balancing size, functionality, and environmental considerations (AI-generated).

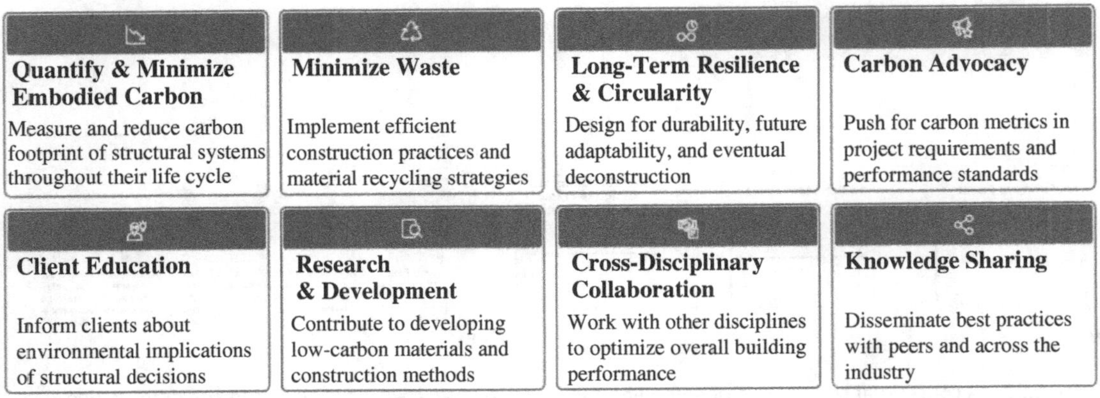

Figure 1.24 Eight key responsibilities of engineers and architects for sustainable design and construction of structures.

Climate action requires us to expand our traditional scope of practice, developing new competencies in environmental assessment, material science, and systems thinking. In this book, we introduce life-cycle parametric analysis (LCPA), a systematic approach to understanding how design variables influence environmental performance throughout the building life cycle, and life-cycle optimization (LCO), a data-driven, multi-objective process that uses the principles of LCPA to also score and rank hundreds of computer-generated design variations. We also show how to apply the life-cycle thinking to particularly demanding areas of structural engineering, such as earthquake-resistant design, where conventional safety measures often appear to conflict with sustainability objectives.

1.5 Regulatory Push for Decarbonization

Over the years, several global initiatives, such as the US Green New Deal (GND) (H.Res.109 – Recognizing the Duty of the Federal Government to Create a Green New Deal, 2019), the European Green Deal (European Commission, 2019), have been launched to address the climate crisis. These initiatives aim to significantly reduce GHG emissions and promote sustainable development practices. The foundation for many of these efforts was laid by the Paris Agreement (UNFCCC, 2015), the first legally binding international treaty in which nations committed to limiting global warming to well below 2 °C (and pursuing efforts toward 1.5 °C) above pre-industrial levels, establishing a shared framework for national climate action.

The regulatory landscape surrounding GHGs is also rapidly evolving worldwide. Cities and countries worldwide are implementing embodied carbon limits, carbon pricing mechanisms, and mandatory life-cycle assessments (LCA) for buildings. This trend is creating a new reality for the construction sector, as seen in the following examples:

European Union: The Energy Performance of Buildings Directive (EPBD, Directive (EU) 2024/1275) mandates whole-life-cycle GWP disclosure in Energy Performance Certificates from 2028 (buildings above 1,000 m^2) and from 2030 (all new buildings), with limit values to be set by Member States through national roadmaps by January 2027. Its Annex III prescribes that calculations follow EN 15978, with the scope of building elements defined by the Level(s) common EU framework (Indicator 1.2) (Directive (EU) 2024/1275). Moreover, please note that the 2011 version of EN 15978 has been recently updated: EN 15978:2026 (publication scheduled within 2026) will succeed EN 15978:2011 as the methodological basis underpinning the EPBD's life-cycle GWP requirement. The revision aligns the indicator set with EN 15804+A2:2019 and the EN 15643:2021 framework; restructures the modular life-cycle diagram (introducing a pre-construction Module A0, finer-grained sub-modules for stages A4 and A5, sub-divisions of operational energy and water use modules B6 and B7, an optional Module B8 for users' activities including commuting, and the splitting of Module D into D1 and D2); adds a new Clause linking data-quality requirements to four design-stage assessment types (from concept design to as-built); and expands the indicator scope to include biogenic carbon, carbon storage, local environmental aspects (land use, soil sealing, local emissions), and end-of-life ease-of-disassembly and reuse/recycling potential.

Denmark: As of 2023, all new buildings must include a 50-year LCA in their permit applications. A strict carbon limit of 12 $kgCO_2e/m^2/yr$ was introduced in 2023 for large buildings, with typology-specific limits introduced in 2025 and progressive tightening to a range of 3.2–6.4 $kgCO_2e/m^2/yr$ by 2029 depending on building type (*Reduction Roadmap*, n.d.).

The Netherlands: New office and residential buildings must comply with a mandatory environmental performance calculation (MPG). The MPG translates the results of a whole-life LCA into a single indicator expressed as a "shadow price" (the estimated cost of the building's environmental impact over its life cycle). The government has already tightened this cap and proposes further reductions (*Nationale Milieudatabase|Environmental Performance of Buildings*, n.d.).

United Kingdom: The UK has recently introduced a voluntary whole-life carbon framework (the Net Zero Carbon Buildings Standard, NZCBS) for major developments, reinforcing its position as one of the leading regulatory environments for embodied carbon. The NZCBS sets upfront embodied carbon limits (A1–A5) differentiated by building type and year of commencement (e.g. 500 $kgCO_2e/m^2$ for new homes and ~600 $kgCO_2e/m^2$ for new offices in 2025, tightening to ~350 $kgCO_2e/m^2$ for homes by 2031, with retrofit limits at ~50% of new-build values, and progressive tightening to 2050), alongside whole-life carbon reporting (modules A–D, without binding limits). The Greater London Authority already requires whole-life carbon assessments for referable planning applications, and national-level policy is progressing toward stricter limits aligned with the country's legally binding net-zero 2050 target (LETI, n.d.).

United States (federal): The Inflation Reduction Act (IRA) of 2022 funded the GSA Buy Clean program (setting EPD requirements and GWP limits for federally procured cement, concrete, steel, flat glass, and asphalt), but the One Big Beautiful Bill Act (signed 4 July 2025) rescinded unobligated IRA funds and effectively ended the program, shifting the centre of gravity to state and city level (California Buy Clean Act, New York City LL97, Colorado, Oregon).

Australia: Shifting focus to corporate accountability, Australia will require large businesses and financial institutions to publish annual sustainability reports with mandatory climate-related financial disclosures starting in 2025 (*24-205MR ASIC Urges Businesses to Prepare for Mandatory Climate Reporting*, n.d.). In parallel, the National Construction Code (NCC) 2025 establishes a voluntary pathway for commercial buildings to measure and report upfront embodied carbon (A1 to A5) using the NABERS Embodied Carbon rating tool as the primary methodology, with sector targets driven by state and territory net-zero 2050 commitments and varying interim 2030 reduction goals.

It is worth noting that regulatory momentum is not limited to governments alone. Large public-sector organizations are independently setting ambitious carbon targets for their built assets. Similarly, private-sector clients are increasingly dictating low-carbon requirements for their projects, motivated by a range of factors including corporate reputation, access to green financing and ESG-linked investment, and a growing conviction that decarbonization is both an environmental and a business imperative.

The categorization of GHG emissions into Scopes 1, 2, and 3 (Greenhouse Gas Protocol, 2011) has been a strategic effort to bring clarity and a standardized approach to corporate climate action. Scope 1 covers an organization's direct emissions, Scope 2 covers purchased energy, and Scope 3 includes all other indirect emissions from its value chain (including embodied carbon of structural systems). This approach provides a clear framework for businesses and other organizations to measure, manage, and ultimately reduce their climate impact in a way that aligns with global mitigation efforts. A good example of it can be read here (*Scope 3|Mission Zero*, n.d.).

Professional engineering organizations are updating standards and guidelines to incorporate sustainability considerations. These items are discussed in detail in Chapter 3 of this book. Annex 2 provides a detailed regulatory map worldwide.

1.6 Life-cycle-driven Structures Framework

Structural engineering has evolved through several paradigm shifts in history (Figure 1.25). Until the early 20th century, the primary focus was on strength and safety. The latter half of the century introduced limit state design and probabilistic approaches, emphasizing reliability and performance. The 21st century has seen the integration of seismic resilience, sustainability considerations, and increasingly sophisticated computational tools.

We stand at the threshold of the next evolution: life-cycle-driven design. This approach maintains the fundamental requirements of structural safety and performance while incorporating comprehensive environmental impact assessment, especially focusing on embodied carbon from material extraction, manufacturing, transportation, and construction, throughout the entire building life cycle (Figure 1.26).

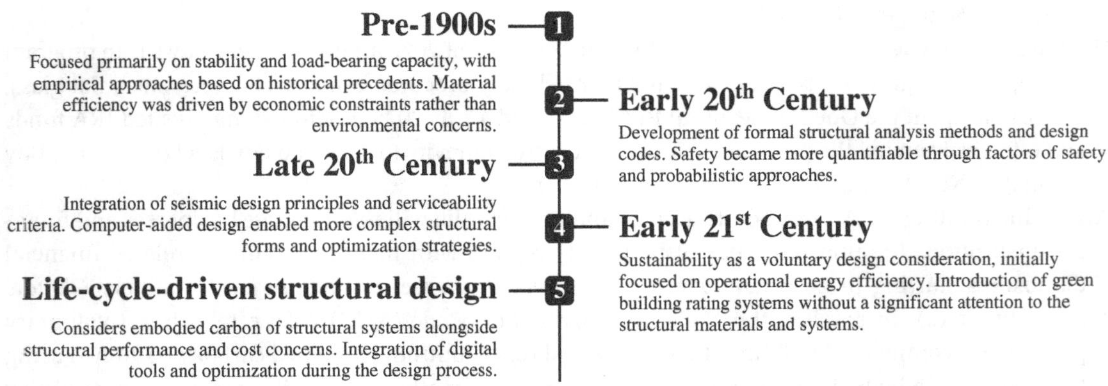

Figure 1.25 Structural design toward life-cycle thinking.

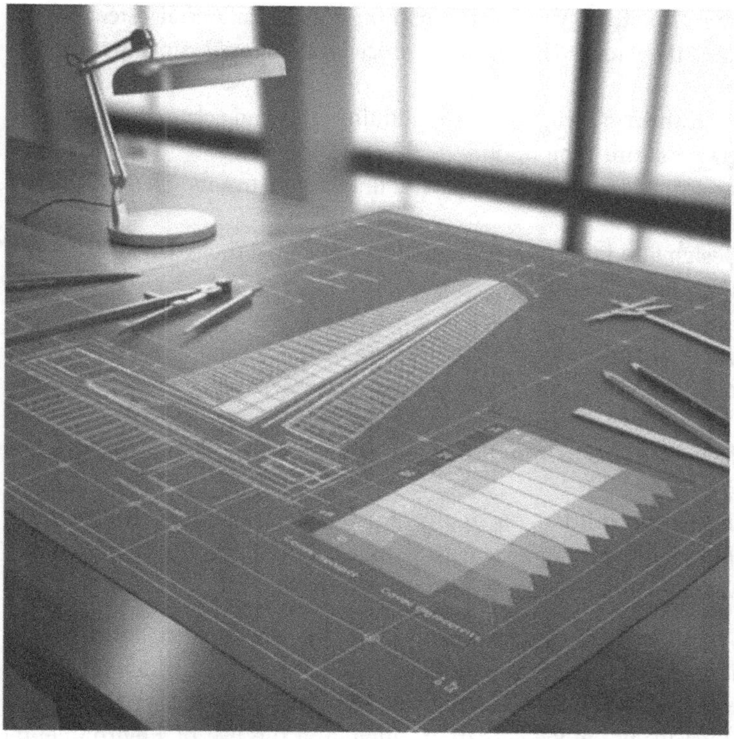

Figure 1.26 Future building design frameworks integrating embodied carbon ratings directly into structural design (generated by Gamma AI).

Life-cycle-driven structures stand on four pillars, including 12 specific strategies:

I. **BE OPEN-MINDED AND TRANSPARENT**

Challenge Default Solutions: Instead of assuming one material is always superior, conduct a holistic assessment for each project. Evaluate materials and structural systems, based on a wide range of criteria, including structural performance, cost, availability, and resilience.

Promote Transparency: Clearly document and communicate the rationale behind design and material decisions. This will help avoid "green-washing."

Engage Stakeholders Early: By involving contractors and fabricators early, designs can be aligned with standard material sizes and efficient fabrication methods, which directly reduces material waste and optimizes resource use.

II. **BUILD WITH THE RIGHT RESOURCES**

Prioritize Reclaimed Materials and structures: This is the "design *with* circularity" aspect, where the "right" resource is one that is reclaimed or reused from a previous structure, avoiding virgin material production, or even reusing an existing structure (adapting it to its new use).

Low-carbon Material and Systems: When and where possible, focus on choosing resources manufactured with lower emissions. Evaluate higher-strength materials that may allow you to use less material overall to achieve the same structural performance.

Hybrid and Composite Systems: This strategy is about combining different materials (steel, timber and concrete), in a way that uses the strengths of each to reduce total material consumption.

III. DESIGN TO LAST

Design for Durability and Adaptability: Design resilient buildings that can accommodate future changes in use, thus avoiding premature demolition.

Design for Robustness: Design structures that can withstand unexpected events, preserving their integrity and extending their service life.

Design for Deconstruction: Design buildings and their components so that they can "last" beyond the building's initial lifespan and be used again.

IV. USE TECHNOLOGY

Digital, AI-driven, and Automated Methods: This is the core of the category, using tools like BIM, AI optimization, and robotic fabrication to create precise, waste-free components.

Conceptual Optioneering: This strategy relies on computational technology to rapidly analyze and compare a large set of design options to find the most efficient solution before construction begins.

Prefabrication and Modular Construction: This approach uses the technology of controlled factory environments (including automation and precision machinery) to produce high-quality components with minimal waste.

This book presents the concepts, methods, and applications of life-cycle-driven structural design. It begins by establishing the foundational knowledge of life-cycle assessment focusing on the embodied carbon characteristics of steel, mass timber, and reinforced concrete. Building on this, it provides practical guidance for calculating and measuring embodied carbon, including the use of Environmental Product Declarations. The text then introduces LCPA at both component and building levels, demonstrating systematic approaches to carbon optimization, before advancing to multi-objective optimization methods that balance cost and carbon considerations (Life-cycle Optimization, LCO). Special attention is given to the challenges of earthquake-resistant design within this framework. To demonstrate real-world applications, comprehensive case studies from practice are presented across diverse project types (offices, hospitals, stadiums, cultural buildings, airports, bridges, viaducts, shopping malls, mixed-use towers, residential buildings, data centers, and temporary construction facilities). Each chapter includes practical exercises, discussion questions, and references to support both learning and professional application.

1.7 Conclusion

The journey toward life-cycle-driven structures begins with understanding that every design decision carries environmental consequences. This book provides the roadmap for students and professionals ready to embrace this responsibility and lead the transformation of structural engineering for the 21st century and beyond.

Structural engineers and architects have always been entrusted with public safety and welfare. Today, that trust extends to the environmental stewardship of our planet. The structures we design today will stand for decades, carrying within them the carbon footprint of our decisions. The time has come to ensure that the footprint is as light as possible while maintaining the structural integrity and performance that society demands.

A challenge is always matched by an opportunity. The life-cycle-driven approach offers a powerful, proven, and practical path forward, structured around four pillars that address the full spectrum of decisions a designer faces. By being open-minded and transparent (challenging default material assumptions, promoting data-backed rationale, and engaging fabricators early), by building with the right resources (prioritizing reclaimed materials, selecting low-carbon supply chains, and combining materials in hybrid systems that exploit the strengths of each), by designing to last (embedding durability, adaptability, and deconstructability so that structures and their components can serve well beyond a single use cycle), and by using technology (leveraging digital optimization, conceptual optioneering, and prefabrication to minimize waste at every stage), the construction community can break the feedback loops that currently tie new construction to rising emissions. As the regulatory landscape tightens, these strategies are no longer aspirational; they are becoming prerequisites for practice. The following chapters will develop each pillar into specific methods, from embodied-carbon quantification and life-cycle parameter analysis to life-cycle optimization and sustainable earthquake-resistant design, equipping both students and practitioners with the tools to make carbon visible in every design choice. The final chapter brings these principles to life through 25 case studies from around the world, demonstrating how diverse project teams have already translated life-cycle-driven thinking into built reality.

1.8 Questions

Answer the following questions to enhance your understanding of the concepts covered in this chapter.

Scenario Question 1: Operational vs. Embodied Carbon

Imagine you are part of a sustainability team working on a new commercial real estate project. Below is a list of activities, materials, and systems involved in the project. Your task is to classify each item as contributing primarily to operational carbon (related to day-to-day energy use and maintenance) or embodied carbon (arising from production, construction, and end-of-life processes).

A. Manufacturing concrete for the building foundation.
B. Energy consumption of HVAC systems during daily operation.
C. Transporting steel beams to the construction site.
D. Lighting energy usage over the building's lifespan.
E. Installation of triple-glazed windows.
F. Routine maintenance of building systems (e.g. elevators).
G. Carbon emitted during demolition of the structure.
H. Insulation material production and delivery.
I. Running electric appliances (e.g. computers) in the office.
J. Recycling bricks from a deconstructed wall.

Determine which entries contribute primarily to operational carbon and which contribute to embodied carbon.

Briefly explain the reasoning behind each classification. Consider where and when the emissions occur during manufacturing, transport, or construction (embodied), or throughout the building's functioning and upkeep (operational).

Scenario Question 2: Early-stage decision-making

You are comparing two office superstructures for a 60-year design life:

- Option S: Steel frame with composite slabs. Upfront GWP is higher than timber, but members are reusable and corrosion protection is specified for a 60-year interval with minor maintenance.
- Option T: Mass-timber frame with CLT floors sourced from a distant region. Lower upfront GWP, but moisture risk requires façade replacement at year 25 and selective strengthening at year 30.

Identify tunnel vision risks for each option: note potential trade-offs with durability, service life, biodiversity, and whole-life cost.

Scenario Question 3: The "Lock-in" vs. Long-term Use

An institutional client is commissioning a new university laboratory building meant to last 100 years.

Team A proposes a highly optimized, ultra-low carbon reinforced concrete structure. It uses very little material, achieving excellent upfront embodied carbon figures. However, to achieve this optimization, the floor plates have irregular shapes and small span column grids specific to the currently planned laboratory layout.

Team B proposes a robust, standardized steel frame grid with higher upfront embodied carbon. However, the open plan and high load capacity mean it could easily be converted into classrooms, offices, or different lab types in 30 years without structural alteration.

Using the "Life-cycle-driven structures framework" (specifically Pillar III: Design to Last) and the concept of emissions being "locked in" early (Section 1.2), analyze the long-term sustainability risks of Team A's proposal vs. Team B's proposal. Which approach might ultimately lead to a lower whole-life carbon footprint if the building's function changes in year 40?

Scenario Question 4: Interpreting Material Variability

You are the lead structural engineer defending your choice of materials to a sustainability consultant:

- You have chosen a reinforced concrete frame. The consultant argues a generic timber frame would be more sustainable.
- Construct a technical argument defending how your reinforced concrete design could achieve a lower embodied carbon footprint than a "business as usual" timber design. What specific supply chain and design factors must you control to make your argument valid? (e.g. cement replacement levels, energy source for rebar production, transport distances vs. the specific timber source's forestry practices).

Scenario Question 5: Recycling and Local Sources

Your design team specified locally sourced recycled steel (85% recycled content, 120 km transport) for a warehouse project. Two months before construction, your supplier announces they cannot fulfill the order. The alternatives are:

Alternative 1: Virgin steel from a domestic mill (1200 km transport) with BF-BOF route (see Chapter 2 for details).
Alternative 2: Recycled steel from an overseas supplier (8500 km sea freight + 200 km land transport).
Alternative 3: Redesign the structure using precast concrete with local aggregates.

Explain how you would evaluate these alternatives. What data would you need? How does this scenario illustrate the feedback loop between climate disruption and construction emissions?

Scenario Question 6: The Coastal Exhibition Hall (Material Trade-offs)

You are the lead structural engineer for a new maritime museum located on an exposed coastline. The architectural brief has three nonnegotiable requirements:

Long Spans: The main gallery requires massive, column-free spaces to display boats.
Exposed Structure: The client wants the structural elements visible as part of the aesthetic.
100-Year Life: The building is intended to be a permanent civic landmark.

Answer the following questions based on Table 1.3 of this chapter:

1. You initially select steel because Table 1.3 identifies its "high strength-to-weight ratio enabling long spans" as a key benefit. However, given the coastal location, what specific environmental impact risk listed in the table could compromise the 100-year life goal, and what "resource-intensive" trade-off does the table warn about regarding its protection?
2. The architect asks if timber would be a better sustainable choice. While Table 1.3 notes timber's "aesthetic appeal" and "carbon storage potential," what specific environmental impact risk regarding the supply chain (certification) must you verify to ensure the "carbon neutrality" claim is not misleading?
3. If you proceed with steel, the client wants to know why it might be arguably "greener" than concrete at the building's end-of-life. Use the mitigation opportunities column in Table 1.3 to explain the difference in recyclability/reusability between the two materials.

Scenario Question 7: The "Future-proof" Headquarters

A fast-growing technology company is commissioning a new headquarters in a seismically active region (e.g. California or Japan). They have an unusual set of requirements for the structural design:

Seismic Safety: The building must perform exceptionally well during earthquakes.
Flexibility: They anticipate their workflow will change rapidly, requiring the removal of walls and the repurposing of entire floors every 5–10 years.
Asset Liquidity: The client views the building materials as a "bank." They want the option to disassemble the structure and sell the components or move them to a new site in 50 years, rather than demolishing the building.

You are the structural engineer proposing steel as the primary material. Using Table 1.3, construct an argument specifically addressing the client's three requirements addressing:

1. **Safety:** How does the "strengths and benefits" column of Table 1.3 support the choice of steel for a seismic zone compared to a brittle material? Identify the specific property mentioned.

2. **Flexibility:** The client needs large, open spaces that can be reconfigured. Which "strength/benefit" allows steel to achieve long spans with minimal intrusion, and which "mitigation opportunity" mentions the ease of modifying the frame for changing functions?
3. **Liquidity:** Contrast the end-of-life potential of steel vs. concrete. According to the "mitigation opportunities" column, what specific characteristic of steel connections enables the client's goal of "disassembly and reuse" rather than demolition?

Short Answer Questions

1 Why has embodied carbon become more important as buildings become more energy-efficient?

2 Which building components typically have the longest service life and account for the largest portion of embodied carbon?

3 Which components usually drive upfront impacts and why are these "locked in" early?

4 Explain how resource extraction creates a self-reinforcing cycle of environmental damage.

5 Explain why there is "not a simple way to claim one material is sustainable over another."

6 Define "carbon tunnel vision" in structural design and list two project risks it can cause.

7 Give two examples of climate feedback loops that connect construction activity to rising emissions.

True or False

1 Timber is always the most sustainable construction material choice.

2 Operational carbon includes emissions generated during the construction phase of a building.

3 Embodied carbon refers to emissions from the extraction, manufacturing, transportation, construction, and disposal of building materials.

4 Structural systems remain hidden behind finishes but can comprise a significant portion of total building materials.

5 As buildings become more energy-efficient, operational carbon becomes more important than embodied carbon.

6 Structures and foundations can account for more than half of a building's embodied carbon.

7 Replacing a concrete structure with a mass timber one automatically guarantees a lower carbon footprint, regardless of where the timber is sourced or how it is transported.

8 In the context of climate feedback loops, the disruption of supply chains due to climate events can lead to higher emissions, as builders are forced to transport materials over longer distances or use higher-impact alternatives.

9 According to standard definitions, the emissions released from heating, cooling, and lighting a building during its daily operation are included in the embodied carbon calculation.

10 "Carbon tunnel vision" describes a risk where a design team focuses so intensely on reducing greenhouse gas emissions that they inadvertently compromise other critical factors like durability, biodiversity, or life cycle cost.

11 Although steel production is energy-intensive, a key mitigation opportunity is that steel structural members can potentially be disassembled and reused directly in new projects without being melted down.

12 Global warming potential (GWP) is a metric used to measure the impact of carbon dioxide (CO_2) only; it does not apply to other gases like methane or nitrous oxide.

Discussion Questions

1 Discuss how the construction industry can align itself with global initiatives like the European Green Deal to promote sustainable development.

2 Evaluate the trade-offs for steel, concrete, and timber. How should a structural engineer or architect approach material selection given these competing factors?

3 Beyond material choice, what other aspects of a construction project (e.g. site selection, construction methods, building footprint) could be managed to better protect or even enhance local biodiversity?

4 Argue for a situation where a reinforced concrete structure might have a better overall sustainability profile than a mass timber structure.

5 Discuss how mandatory reporting and life-cycle assessment regulations create both a challenge and an opportunity for design and construction firms.

6 Analyze the relationship between material choice in construction and biodiversity conservation. How can structural engineers balance performance requirements with ecological protection?

7 Analyze why structural systems have remained "invisible" to designers and clients despite their significant carbon impact. What strategies could make this impact more visible and actionable?

8 Examine the evolution from Vitruvian principles to modern sustainable design. How can engineers maintain "firmness, utility, and beauty" while addressing climate responsibilities?

9 Discuss the challenges structural engineers face when conventional safety measures appear to conflict with sustainability objectives, particularly in earthquake-resistant design. How might LCPA and LCO help resolve these conflicts?

10 Describe a realistic scenario where a design team, despite good intentions, falls into "tunnel vision." What specific material or system choice might they make, and what negative consequences could be overlooked?

11 How does the introduction of embodied carbon as a critical metric constrain the traditional pursuit of venustas? Conversely, how might "visible sustainability" (e.g. exposed low-carbon structures) create a new aesthetic paradigm for venustas?

12 Explain how a hyper-focus on Pillar IV ("use technology," specifically using complex, robotically fabricated, highly optimized unique components to minimize waste) could inadvertently create significant conflicts with Pillar III ("design to last," specifically the sub-strategy of adaptability/reuse). Why might highly optimized, bespoke components be harder to adapt later than standardized ones?

References

24-205MR ASIC urges businesses to prepare for mandatory climate reporting. (n.d.). [Media release]. Retrieved August 11, 2025, from https://www.asic.gov.au/

Adams, M., Burrows, V., & Richardson, S. (2019). *Bringing Embodied Carbon Upfront: Coordinated Action for the Building and Construction Sector to Tackle Embodied Carbon*. World Green Building Council.

Arup. (2023a). *Buildings & Infrastructure Priority Actions for Sustainability—Embodied Carbon: Concrete* (No. 07762000-RP-SUS-0003). https://www.istructe.org/IStructE/media/Public/Resources/ARUP-Embodied-carbon-concrete_1.pdf

Arup. (2023b). *Buildings & Infrastructure Priority Actions for Sustainability—Embodied Carbon: Steel* (No. 07762000-RP-SUS-0001). https://www.istructe.org/IStructE/media/Public/Resources/ARUP-Embodied-carbon-steel_1.pdf

Arup. (2023c). *Buildings & Infrastructure Priority Actions for Sustainability—Embodied Carbon: Timber* (No. 07762000-RP-SUS-0002). https://www.shareyourgreendesign.com/research/buildings-infrastructure-priority-actions-for-sustainability-embodied-carbon-timber/

Chapman Taylor. (2025). *Why BIM matters* (https://www.chapmantaylor.com/zh/) [Graphic]. Chapman Taylor. https://www.chapmantaylor.com/zh/insights/why-bim-matters

Chehade, M. (2023). *Strategies for Embodied Carbon Intensity Reduction [MSc Thesis, Supervised by A. Kanyilmaz; co-supervised by J. Carr and R. Harpin]*. Politecnico di Milano

EPD International. (n.d.). Retrieved August 11, 2025, from https://www.environdec.com/home

European Commission (2019). *The European Green Deal* (No. COM(2019) 640 final). https://eur-lex.europa.eu/resource.html?uri=cellar:b828d165-1c22-11ea-8c1f-01aa75ed71a1.0002.02/DOC_1&format=PDF

Greenhouse Gas Protocol (2011). *Corporate Value Chain (Scope 3) Accounting and Reporting Standard: Supplement to the GHG Protocol Corporate Accounting and Reporting Standard*. World Resources Institute and World Business Council for Sustainable Development. https://ghgprotocol.org/sites/default/files/standards/Corporate-Value-Chain-Accounting-Reporing-Standard_041613_2.pdf

H.Res.109 – Recognizing the Duty of the Federal Government to Create a Green New Deal, H.Res.109, United States House of Representatives 2019, 116th Congress (2019–2020) (2019). https://www.congress.gov/bill/116th-congress/house-resolution/109

Hutchins, L. (2024, November 12). Construction is Driving Global Biodiversity Loss – It's Time for Government to Act. UKGBC. https://ukgbc.org/news/construction-is-driving-global-biodiversity-loss-its-time-for-government-to-act/

ICE. (2022). *Low Carbon Concrete Routemap.* https://www.ice.org.uk/media/q12jkljj/low-carbon-concrete-routemap.pdf

IEA. (2018). *2018 Global Status Report*. International Energy Agency, IEA.

LETI. (2020a). Climate Emergency Design Guide. https://www.leti.uk/_files/ugd/252d09_3b0f2acf2bb24c019f5ed9173fc5d9f4.pdf

LETI. (2020b). Embodied Carbon Primer. https://www.leti.uk/_files/ugd/252d09_8ceffcbcafdb43cf8a19ab9af5073b92.pdf

London Energy Transformation Initiative. (n.d.). Home. Retrieved April 1, 2026, from https://www.leti.uk/

Masson-Delmotte, V., Zhai, P., Pörtner, H.-O., et al. (eds.). *IPCC, 2018: Global Warming of 1.5 °C. An IPCC Special Report on the impacts of global warming of 1.5 °C above pre-industrial levels and related global greenhouse gas emission pathways, in the context of strengthening the global response to the threat of climate change, sustainable development, and efforts to eradicate poverty.* Cambridge University Press, Cambridge, UK and New York, NY, USA, 616 pp. https://doi.org/10.1017/9781009157940

Nationale Milieudatabase. (n.d). Environmental Performance of Buildings. Retrieved August 11, 2025, from https://milieudatabase.nl/en/

Nurmaulina, D. (2024, August 30). Carbon Tunnel Vision: A Wider Lens on Carbon Emissions. Carbonethics. https://www.carbonethics.co/post/carbon-tunnel-vision-a-wider-lens-on-carbon-emissions

On Architecture, Volume I. (1914). Retrieved August 11, 2025, from https://www.loebclassics.com/view/LCL251/1931/volume.xml

PEFC. (n.d.). Facts and Figures (Forest Management) [PEFC]. Retrieved August 30, 2025, from https://www.pefc.org/discover-pefc/facts-and-figures

Reduction Roadmap. (n.d.). Reduction Roadmap. Retrieved August 11, 2025, from https://reductionroadmap.dk

Ripple, W. J., Wolf, C., Lenton, T. M., Gregg, J. W., Natali, S. M., Duffy, P. B., Rockström, J., & Schellnhuber, H. J. (2023). Many risky feedback loops amplify the need for climate action. *One Earth*, *6*(2), 86–91. https://doi.org/10.1016/j.oneear.2023.01.004

Salvadori, M. (1982). *Why Buildings Stand Up: The Strength of Architecture* (Reprint, 1. McGraw-Hill paperback ed). McGraw-Hill.

Scope 3 | Mission Zero (n.d.). Retrieved August 11, 2025, from https://www.lendlease.com/au/sustainability/climate-and-environment/scope-3-emissions-protocol/

Thayer, K., & MacCarthy, J. (2025). 5 Graphics Explain the Climate-Fire Feedback Loop. https://www.wri.org/insights/climate-fire-feedback-loop-explained

United Nations Environment Programme (UNEP). (2023). *Building materials and the climate: Constructing a new future.* United Nations Environment Programme.

UNFCCC. (2025). The Paris Agreement | UNFCCC. https://unfccc.int/process-and-meetings/the-paris-agreement

United Nations Framework Convention on Climate Change. (n.d.). The Paris Agreement. Retrieved April 1, 2026, from https://unfccc.int/process-and-meetings/the-paris-agreement

United Nations Department of Economic and Social Affairs. (2018, May 16). 68% of the world population projected to live in urban areas by 2050, says UN. https://www.un.org/development/desa/en/news/population/2018-revision-of-world-urbanization-prospects.html

U.S. Environmental Protection Agency. (2023). Inventory of U.S. Greenhouse Gas Emissions and Sinks: 1990–2021. https://www.epa.gov/ghgemissions/inventory-us-greenhouse-gas-emissions-and-sinks-1990-2021

2

A Summary of Life-cycle Assessment Focusing on Embodied Carbon of Steel, Timber, and Concrete

LIST OF CONTENTS

In this chapter, the key concepts and strategies of life-cycle assessment (LCA) for the building structures are explored. The chapter presents in detail the stages of LCA for steel, concrete, and timber, covering material production, emissions during construction, end-of-life phases, and emissions extending beyond the conventional building life cycle.

Life-cycle assessment (LCA) is a methodology used to evaluate the environmental performance of a building or product throughout its entire life cycle (EN 15978, 2026; ASTM International, 2022; International Organization for Standardization, 2022). By systematically assessing each stage from raw material extraction to end-of-life disposal, LCA helps identify the major sources of greenhouse gas emissions and other environmental impacts. In this way, the design and construction processes can be guided toward more sustainable practices. For example, an LCA conducted for a building project may reveal that over half of the greenhouse gas (GHG) emissions occur during cement and steel production. This awareness can guide the design team to optimize the structure's design for greater efficiency. The life cycle of a building is split into stages, usually grouped and sorted into modules, as presented in Figure 2.1, which shows civil engineering life-cycle stages and modules (adapted from EN 15643, 2021; EN 15978, 2026; EN 17472, 2022; RICS, 2023).

LCA methodologies are standardized under the ISO 14000 (ISO 14000, 2025) series, which provide guidelines for inventory analysis, impact assessment, and interpretation of impacts. Inventory analysis involves collecting data on energy and material inputs and outputs throughout the life cycle. Impact assessment focuses on evaluating the potential environmental impacts based on the inventory data. Interpretation of impacts involves analyzing the results to make decisions that can reduce environmental impacts. LCA approaches vary based on the scope of the assessment, each focusing on different stages of the product or building's life cycle.

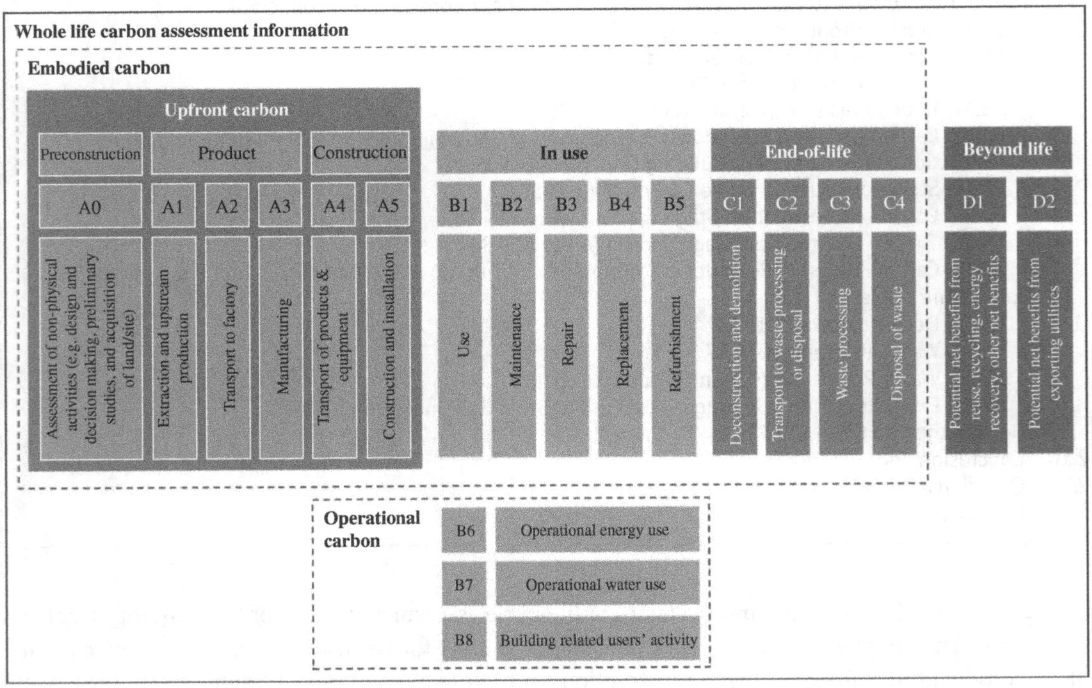

Figure 2.1 Building and infrastructure life-cycle stages (adapted from EN 15643, 2021; EN 15978, 2026; EN 17472, 2022; RICS, 2023).

2.1 Product-level vs. Building-level Life-cycle Assessment

LCA can be performed at two distinct scales: product level and building level.

Product-level LCA evaluates the environmental impacts of individual construction products, such as a steel beam, a cubic meter of concrete, or a timber slab. These assessments are typically documented in accordance to the EN 15804:2012+A2:2019, in Environmental Product Declarations (EPDs), which report impacts for modules A1–A3 (cradle to gate) and, in many cases, modules C and D. However, product-level data alone cannot capture the full environmental profile of a building.

Building-level LCA combines the environmental impacts of all materials, components, and processes across the entire structure. This approach is governed by EN 15978 (2026) for buildings and EN 17472 (2022) for civil engineering works such as bridges and infrastructure. It requires combining EPD data with project-specific information, including quantities from bills of quantities (BoQ) or building information models (BIM), actual transportation distances, on-site energy consumption, and anticipated maintenance and replacement schedules. Building-level assessments provide a holistic view of environmental performance and are essential for benchmarking against carbon intensity targets, such as those defined by the SCORS rating system (please see Chapter 3).

2.2 Declared and Functional Units, Functional Equivalent, and Reference Study Period

Four fundamental parameters must be established at the outset of any LCA: the declared and functional units, the functional equivalent, and the reference study period.

The functional unit defines the basis for comparison and quantification. According to ISO 14044 (2006), the functional unit is "the quantified performance of a product system for use as a reference unit." It specifies the unit to which all environmental impacts are normalized, enabling meaningful comparisons between different design options or materials. For construction products, EN 15804:2012+A2:2019 distinguishes between a **declared unit** (used when the precise function is not defined, e.g. per kg or per m^3 of material) and a functional unit (used when the function and performance are specified, e.g. per ton of reinforcement with defined yield strength over 60 years). For building-level assessments, EN 15643 (2021) and EN 15978 (2026) apply the concept of **functional equivalent**, which defines the building's required technical and functional performance as the basis for comparison.

Common bases of comparison in building and infrastructure LCA include:

- Per kilogram or cubic meter of a construction product (declared unit) – commonly used in EPDs for product-level reporting, as specified in EN 15804:2012+A2:2019
- Per square meter of gross internal area (GIA) – typically used for building-level benchmarking (e.g. $kgCO_2e/m^2$), as specified in EN 15978 (2026)
- Per structural element – useful for comparing alternative designs for specific components
- Per lane-kilometer or per square meter of deck area – used for infrastructure such as roads and bridges, as specified in EN 17472 (2022)

The choice of functional unit significantly affects the interpretation of results. For example, a high-strength concrete may have higher emissions per kilogram than standard concrete, but if it enables

Table 2.1 Interaction between the several standards mentioned in this chapter.

Standard	Scope	Purpose
ISO 14044	General LCA Methodology	The global baseline for Life-cycle Assessment (LCA) used by all the below.
EN 15643	Overarching Framework for Construction Works	The "umbrella" standard. It defines the general principles for assessing sustainability across buildings and civil engineering.
EN 15804	Construction Products	Rules for creating Environmental Product Declarations (EPDs) for individual materials.
EN 15978	Buildings	Specific LCA calculation rules for the environmental performance of a whole building.
EN 17472	Civil Engineering works	Specific LCA calculation rules for infrastructure (bridges, roads, etc.).

a reduction in total concrete volume, the emissions per square meter of floor area may be lower. Similarly, a steel bridge deck may show higher embodied carbon per square meter than a concrete alternative, but when assessed per lane-kilometer the comparison may yield different conclusions.

The reference study period (RSP) defines the timeframe over which the building's life cycle is assessed. In accordance with EN 15978 (2026) and EN 17472 (2022), the RSP is typically 50 to 60 years for buildings and can be up to 100 to 120 years for infrastructure assets such as bridges, tunnels, and highways, reflecting their longer design lives and higher durability requirements. The RSP directly influences the calculation of use-stage impacts (modules B1–B5), as it determines how many replacement cycles are required for components with service lives shorter than the building itself. For instance, if a building has an RSP of 60 years and a cladding system has an estimated service life of 30 years, one full replacement must be accounted for in module B4.

Table 2.1 shows how general ISO methodologies inform the overarching construction framework, which in turn dictates how we assess products, buildings, and infrastructure.

2.3 Defining the Scope and System Boundaries

The scope of a Life-cycle Assessment (LCA) defines which life-cycle stages, processes, and environmental impacts are included in the assessment. System boundaries define where the assessment begins and ends, determining which inputs, outputs, and processes fall within the analysis. Clear and consistent boundary definitions are essential for comparability between different assessments and for ensuring that all significant environmental impacts are captured.

According to EN 15978 (2026) and EN 15804:2012+A2:2019, system boundaries must be explicitly stated and should include all processes that contribute meaningfully to the environmental profile of the product or building. For structural engineering applications, the system boundary typically includes raw material extraction (A1), manufacturing (A3), transportation (A2 and A4), construction (A5), use-phase interventions (B), and end-of-life treatment (C), and in some cases beyond life (module D). The choice of system boundary directly influences the LCA results.

When defining the scope, practitioners must also specify which environmental impact categories will be assessed. While a comprehensive LCA may address multiple indicators (including global warming potential, acidification, resource depletion, and water use) this book focuses primarily on embodied carbon due to its direct relevance to climate change mitigation in the construction sector.

The LCA approaches include (RICS, 2023):

Cradle to Gate (A1 to A3): Evaluates the environmental impact from raw material extraction (cradle) to the factory gate before the product is transported to the construction site. This approach focuses on the upfront carbon emissions associated with material production.

Cradle to Site (A1 to A5): Extends the Cradle to Gate approach by including the impacts of transporting materials to the construction site and the construction process itself.

Cradle to Grave (A to C): Covers the entire life cycle of the product or building, from raw material extraction to its disposal at the end of its useful life.

Cradle to Cradle (A to C and D): Extends the Cradle to Grave by including recycling and reuse stages, aiming for a circular economy where materials are continually repurposed, minimizing waste and environmental impact.

Whole Life Carbon Assessment (WLCA): WLCA considers both embodied carbon (steps A + C) and operational carbon (steps B) to provide a comprehensive view of the environmental impact of buildings. In some cases, Module D is included in a WLCA (and in others, it is considered optional).

Dynamic Life Cycle Assessment (DLCA): It considers the timing of emissions to evaluate their environmental impact more accurately. Unlike traditional LCA methods, dynamic LCA accounts for the fact that the earlier an emission occurs, the more harm it causes due to the prolonged presence of greenhouse gases in the atmosphere. This method also incorporates the benefits of biogenic carbon, which is temporarily stored in buildings and can mitigate some of the immediate impacts. In a dynamic LCA, a heavier weight is assigned to emissions generated in the present (A1–A5 steps) compared to those projected for the future (stages C and D). This weighting reflects the higher immediate impact of current emissions versus those occurring later in the life cycle. For instance, the production and manufacturing phases, which happen early in the life cycle, are scrutinized more heavily for their immediate environmental consequences.

Note: There is also a preliminary stage known as Module A0, which consists of non-physical pre-construction steps such as design studies, surveys, and planning activities. While this stage can carry measurable impacts for large infrastructure projects, it is generally assumed to be negligible for standard building construction and is typically excluded from the LCA.

For example, consider a residential building project (Figure 2.2) where the design team applies different LCA approaches to measure its environmental effects. In the Cradle to Gate stage, the analysis covers emissions from raw material extraction and process up to when the product leaves the factory (e.g. the emissions from concrete production). In the Cradle to Site approach, emissions from transporting the products to the construction site (where the building will be constructed) and the building installation process (e.g. the diesel used by a crane) are added. In the Cradle to Grave method, the evaluation continues through the building's full lifespan (e.g. including the electricity consumed over building's life span, embodied carbon from maintenance of the structure), and its disposal (e.g. the fuel used by an excavator during the building's demolition when this building will have to be replaced in

Figure 2.2 A view of a construction site (Mny-Jhee/Adobe Stock Photos).

future, with a new one probably because of aging materials, increased maintenance demands, or evolving performance standards). The Cradle to Cradle approach goes a step further by accounting for the recycling and reuse of materials after the building's life ends (e.g. designing a steel frame to be unbolted and reused in a new structure, avoiding the need for new steel production).

2.4 LCA Reporting and Environmental Impact Indicators

LCA reports typically address the following components:

Primary Energy Demand (GJ): The total energy required from primary sources – such as fossil fuels and renewables – over the life cycle of a product or building.

Global Warming Potential (GWP): A measure of how much heat a greenhouse gas traps in the atmosphere over a specific time period, relative to carbon dioxide.

Acidification: The increase in acidity in soils and water bodies caused by emissions of pollutants such as sulfur dioxide and nitrogen oxides.

Resource Depletion: A metric that quantifies the use and eventual depletion of natural resources by combining both renewable and nonrenewable resource consumption into a single index.

Renewable/Nonrenewable Resource Use: The consumption of resources, distinguishing between those that can be replenished and those that cannot.

Water Use: The volume of water used throughout the life cycle of the product or building.

Ozone Depletion: The reduction of the ozone layer due to emissions of ozone-depleting substances.

Toxic Releases to Air, Water, and Land: The emission of toxic substances that can harm human health and the environment.

Solid Waste: The total amount of waste generated that is not recycled or repurposed, including byproducts and residues from production, construction, and end-of-life stages.

It is important to keep in mind the following definitions about GWP of construction products (EN 15804:2012+A2, 2019):

Fossil Global Warming Potential (GWP-fossil): This indicator measures the global warming potential (GWP) from greenhouse gas emissions and removals originating from the oxidation or reduction of fossil fuels or materials containing fossil carbon through transformation or degradation processes such as combustion, incineration, and landfilling. It also accounts for GWP from greenhouse gas emissions from sources like peat and calcination, as well as GHG removals from processes such as the carbonation of cement-based materials and lime.

Biogenic Global Warming Potential (GWP-biogenic): Biogenic carbon refers to the carbon that is sequestered or stored in biological materials, such as plants, trees, and soil. This carbon is originally captured from the atmosphere through the process of photosynthesis and becomes part of the biomass. GWP-biogenic measures the impact of CO_2 removals and emissions from biomass throughout the product life cycle (in other words, it assesses how the absorption and release of CO_2 by biological materials like wood or plant matter contribute to global warming):

- Includes CO_2 removals into biomass from all sources except native forests.
- Accounts for GWP from transfers of biogenic carbon from previous systems.
- Covers biogenic emissions to air from biomass due to oxidation or degradation.
- Considers transfers of biogenic carbon into subsequent product systems.
- Characterizes removals of biogenic CO_2 as -1 kg CO_2 eq. per kg CO_2 entering the product system, reflecting the temporary storage benefit; conversely, biogenic CO_2 emissions are characterized as $+1$ kg CO_2 eq. per kg CO_2 released.

Land Use and Land Use Change Global Warming Potential (GWP-luluc): This indicator accounts for greenhouse gas emissions and removals originating from changes in carbon stocks due to land use and land use changes:

- Includes biogenic carbon exchanges from deforestation and soil activities.
- All CO_2 emissions from native forests are modeled under this subcategory.
- CO_2 uptake from biomass in native forests is set to zero.
- Calculation rules exclude net increase in carbon stocks from biomass-based sources.
- Soil carbon storage may be included as additional information if proof is provided.
- GWP-luluc is included in GWP-total, and if it contributes less than 5% to GWP-total, it may be declared as not applicable (ND).

Total Global Warming Potential (GWP-total): This indicator is the sum of the above-mentioned three subcategories of climate change.

Table 2.2 provides an example of an LCA output for various building components, showing how different impacts are quantified. This example highlights the detailed breakdown of various environmental impacts for different building components, helping project teams identify significant areas for improvement and potential design optimizations to lower.

The process of reporting on LCA involves several key steps, as shown in Figure 2.3. Each step is crucial for a thorough assessment and for compiling a comprehensive report (RICS, 2023).

As an example to this process, imagine a new hospital building project that follows the sequence in Figure 2.3. The team first sets the boundaries for the assessment (they decide which stages of the building's life cycle are included – such as material extraction, manufacturing, transportation,

Table 2.2 Example of an LCA output (a similar example is also shown in Kibert (2022)).

Building component	Primary energy demand (GJ)	Global warming potential (CO_2e, metric tons)	Solid waste (metric tons)	Recycled content (% by mass)
Foundations	14 500	5700	420	12
Structural frame	24 000	9100	680	20
Envelope & cladding	12 300	4900	310	15
Roof assembly	8200	3300	200	18
Interior partitions	7600	2700	180	10
Finishes & fit-out	5900	2100	140	8
Mechanical & electrical	9400	3700	260	22
Total	81 900	31 500	2190	–

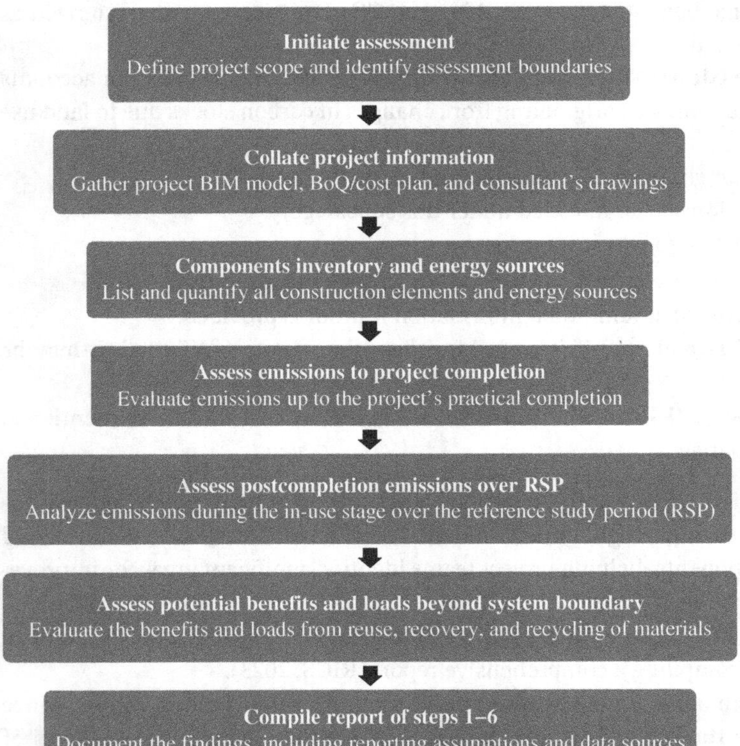

Figure 2.3 Life-cycle assessment (LCA) reporting methodology (author's/with permission of RICS).

construction, operation, and disposal – and select the impact categories to be evaluated, like global warming potential or water consumption), then gathers the building information model (BIM), cost data, and consultant reports (they compile the digital building model that includes all architectural and engineering details, the financial estimates for materials and labor, and technical advice from specialists). Next, they compile details on the materials and energy sources (they list and quantify everything from the concrete and steel to the energy used for on-site equipment, ensuring these inputs are captured accurately for the life-cycle calculations). After that, they measure emissions from the start of construction through completion. They continue by tracking emissions during the hospital's operation over its expected lifespan. Finally, they explore options for reusing or recycling materials beyond the service boundary. All these findings are gathered into a report that supports low-carbon choices in design and construction.

While LCA evaluates various environmental impacts, including resource depletion, pollution, and carbon emissions, in this book, we specifically focus on the Embodied Carbon aspect, as it reflects the climate impact associated with construction materials across their entire life cycle, from raw material extraction and processing to manufacturing, transportation, installation, maintenance, and end-of-life scenarios. Embodied carbon is quantified using the "Carbon Factor," representing the amount of carbon dioxide equivalent (CO_2e) emitted per unit mass of a material throughout its life cycle (e.g. how much CO_2e is generated to produce 1 kg of a material, as explained more in detail in the Chapter 3). Using CO_2e as a metric allows us to account not only for carbon dioxide but also other greenhouse gases, providing a more accurate measure of the material's total climate impact.

2.5 The Stages of Life-cycle Assessment for Building Structures

In this subsection, the stages of life-cycle assessment for building structures are described, covering pre-construction, use, end-of-life, and beyond-life phases (EN 15643, 2021; EN 15978, 2026; EN 17472, 2022). Figure 2.4 summarizes the LC steps involved in the construction industry.

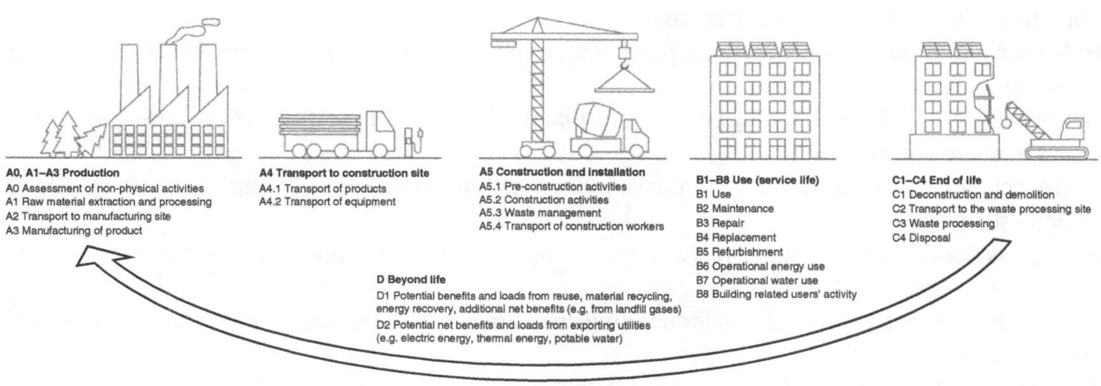

Figure 2.4 Life-cycle assessment steps relevant to the construction industry.

2.5.1 Life-cycle Stage A (A0 to A5): Preconstruction, Production, and Construction Stages

Life-cycle stage A covers all carbon emissions and removals associated with the activities necessary to complete the construction of an asset. This stage is also referred to as "Upfront Carbon" and is subdivided into three main phases: preconstruction, product, and construction. Each phase includes specific activities that contribute to the overall carbon footprint of the construction process.

A0: This stage consists of the nonphysical steps before construction. It involves studies, surveys, and activities associated with the design of the building and has much lower direct environmental impacts compared to the other stages. It is generally assumed negligible for buildings. For infrastructure projects, however, stage A0 includes extensive feasibility studies, site investigations, planning processes, and large-scale design activities, making its impacts nonnegligible but still less significant compared to subsequent construction stages.

A1, A2, A3: These are the steps needed to create a construction product. Module A1 covers the extraction and processing of all raw materials used to produce the construction materials. A2 represents the transport of the raw material to manufacturing unit whether by road, rail, or sea. Module A3 represents the manufacturing phase construction products.

A4: This stage considers the transport of the construction products and equipment from the manufacturing mill to the construction site.

A5: This stage considers any pre-construction, construction and installation-related activities.

2.5.2 In-use Stage (B1 to B8)

This stage highlights the environmental impacts related to the operation of the building, from the project completion. This includes emissions from installed products during use (B1), maintenance (B2), repair (B3), replacement (B4), planned refurbishment of built asset components and systems (B5), operational energy use (B6), operational water use (B7), and building related users' activities not covered in B1–B7 (B8).

B1: Environmental impacts from the installed product during operation, excluding energy and water use (e.g. refrigerant emissions, surface treatments releasing Volatile Organic Compounds, VOCs).

B2: Impacts from activities required to maintain the functional and technical performance of the building (e.g. cleaning, protective coatings).

B3: Impacts from restoring or replacing parts of a product to return it to an acceptable condition after damage or wear.

B4: Impacts from replacing building components that reach the end of their service life during the building's reference study period.

B5: Impacts from planned upgrading, renovation, or restoration activities to improve performance or extend service life.

B6: Impacts from energy consumed for heating, cooling, ventilation, lighting, and other building systems during operation.

B7: Impacts from water consumption for building operations, including potable water supply and wastewater treatment.

B8: Impacts of the users' activities associated with the building's intended use during its normal operation, that are not related to energy and water use addressed in modules B1–B7.

Modules B1–B7 are typically excluded from embodied carbon assessments of building structures, as primary structural elements are designed to last the entire building lifespan without significant emissions during use or major interventions. Exceptions include B2 for periodic recoating of exposed steel structures, B4 for replacement of components with shorter service lives such as seismic dampers or sacrificial elements, and B6–B7 which, while representing operational rather than embodied carbon, may be indirectly influenced by structural choices affecting thermal mass.

2.5.3 End-of-life Stage (C1 to C4): Demolition, Waste Management, Recycling

The end-of-life modules (C1 to C4) cover demolition, waste management, material recovery, and recycling processes.

C1: Represents the deconstruction phase and covers all site activities needed to dismantle, deconstruct, and/or demolish.

C2: Represents the transportation of deconstructed and demolished material to the final location. The waste processing or disposal facilities are likely to be local to the site; thus, the transportation distances are usually short.

C3, C4: These two stages are usually considered together as the scenarios taking place are mutually exclusive. During C3, different activities are implemented for the preparation of waste for their disposal. C4 represents waste disposal and involves any processing before the disposal and from the degradation of landfilled materials, or incineration without energy recovery.

2.5.4 Beyond-life Stage (D)

This phase involves the environmental benefits and/or loads occurring beyond the building system boundary resulting from the net flows of raw materials or products (D1), and environmental benefits and loads occurring beyond system boundary resulting from utilities (e.g. power and heat generated during building operation) (D2). Usually, stage D is reported separately (unless stated otherwise), as it is outside the system boundary of the project and comprises uncertainties about the future treatments of the components. Stage D is correlated to stages C1–C4 through the scenarios occurring in these modules.

2.6 Upfront Carbon (A1 to A3) for Steel, Concrete, and Timber Construction Products

This section covers the carbon emissions produced during the initial life-cycle stages (A1 to A3) for steel, concrete, and timber construction products (EN 15643, 2021; EN 15978, 2026; EN 17472, 2022).

2.6.1 Steel Upfront Carbon (A1 to A3)

This subsection focuses on the carbon emissions from the steel production process, including raw material extraction, transport, and manufacturing.

2.6.1.1 Extraction and Processing of Raw Materials for Steel (A1)

Module A1 covers the extraction and processing of all raw materials used to produce the steel, including iron ore, coal, and scrap metal. This process is energy-intensive and a major contributor to the overall carbon footprint of steel products. Iron ore and coal are the major components:

Figure 2.5 (a) Processing iron ore (A.PAES/Shutterstock), (b) iron ore (Kletr/Adobe Stock Photos).

Figure 2.6 (a) Explosion breaking apart rock layers during open-cast coal mining operations (agnormark/Adobe Stock Photos). (b) Raw extracted coal material (zhengchengbao/Shutterstock).

Iron Ore (Mining and Processing): Most virgin steel starts with iron ore mined from the earth (Figure 2.5). Iron ore extraction is typically done in large open-pit mines using drilling, blasting, and heavy mechanical equipment. The ore is then hauled (often by diesel-powered trucks or conveyors) to onsite facilities for crushing and washing, which remove waste rock and concentrate the iron content. The extraction of iron ore accounts for 34 million tons (Mt) of CO_2e emissions per year. This is due to the substantial energy requirements involved in mining and processing the ore. Of these emissions, 55% are attributed to the use of diesel in mining operations, while 35% result from the consumption of nonrenewable electricity (Arup, 2023b). Efforts to reduce emissions from iron ore extraction include decarbonizing the electricity grid and adopting hydrogen fuel cells and electric vehicles in mining operations, which can significantly lower the carbon footprint by reducing reliance on fossil fuels and promoting the use of renewable energy sources.

Coal (Mining and Processing): The extraction of coal (Figure 2.6) is another major source of emissions, accounting for 2000 Mt of CO_2e per year. A significant portion of these emissions, approximately 70%, are attributed to methane released into the atmosphere during underground extraction (Arup, 2023b). Methane is a potent greenhouse gas, and its release contributes substantially to the carbon footprint of coal mining. Carbon capture technology can help mitigate these emissions by preventing methane from reaching the atmosphere, thereby reducing the overall environmental impact of coal extraction.

All such upstream emissions and resource extractions are included in Module A1 assessment, contributing to indicators like abiotic resource depletion (for iron ore, coal, limestone) and climate change potential.

Scrap Steel Recycling: Steel is highly recyclable, and scrap can substitute a large portion of virgin ore in steel production (especially in electric arc furnaces). An opportunity to reduce A1 impact lies in the use of recycled steel. Recycling steel can significantly lower the carbon footprint associated with raw material extraction. This topic is explained in Section 2.6.1.4.

Please note that in some Environmental Product Declarations (EPDs), producers report the environmental impacts of raw material supply, transportation, and manufacturing collectively as A1–A3, without separately detailing the individual contributions from raw material processing (A1) and manufacturing (A3), the reason being explained in Section 2.6.4.

2.6.1.2 Transport (Road, Rail, Sea) of the Raw Material to Manufacturing Unit for Steel (A2)

The transport stage (A2) covers the emissions generated by transporting raw materials (such as iron ore, coal, and scrap) to the manufacturing unit. The primary modes of transport for raw materials include sea, rail, and road, each with varying carbon emissions (Figure 2.7).

Steel production often involves global supply chains. Major steel exporting countries include China, the Republic of Korea, Japan, and Germany, with China having the highest export value and longest transport distances. Major importers include the United States, Turkey, Italy, and Germany, with the United States having a significant import value and long transport distances (Trade Map, 2023a, 2023b). This highlights the global nature of steel trade and the importance of improving transportation efficiency to reduce its carbon footprint. Two significant developments have the potential to reduce carbon emissions from transport:

Electrification of Railways: Shifting from diesel to electric trains can significantly cut emissions.

Hydrogen-powered Heavy Goods Vehicles (HGVs): The development and adoption of hydrogen fuel cells for HGVs offer a promising reduction in carbon emissions. These advancements could lead to near-zero carbon emissions for the transportation of steel in the future.

2.6.1.3 Manufacturing (Iron Making, Steel Making, Casting, Rolling, Steel Profile Production) (A3)

Module A3 represents the manufacturing phase of the steel product – all processes at the steel plant (and associated facilities) that transform raw materials into the finished steel product ready to leave the factory gate. Steel manufacturing processes include iron making, steel making, casting, rolling, and steel profile production, which are contributors to the A3 carbon emissions in steel products. This stage includes both the processing and postprocessing of steel (in 2019, the steel sector emitted 2.6 Gt of CO_2, more than twice the emissions in 2000, accounting for about a quarter of industrial CO_2 emissions (IEA, 2020). In general, A3 operations can be divided into two categories: "processing" and "postprocessing."

(a) By sea: 17 $kgCO_2$/t/1000 km (Federico Rostagno/Shutterstock)

(b) By rails: 27 $kgCO_2$/t/1000 km (zennie/Getty Images)

(c) By road: 107 $kgCO_2$/t/1000 km (am/Adobe Stock Photos)

Figure 2.7 Primary modes of transport for raw materials: (a) sea, (b) rail, (c) road, and their respective emissions (Arup, 2023b).

2.6.1.3.1 Processing

The processing stage involves several energy-intensive activities using as input iron ore, energy (mainly coal, natural gas, and electricity), limestone, and steel scrap. These processes are characterized by high energy consumption and significant CO_2 emissions due to the combustion of fossil fuels and the chemical reduction of iron ore. The current pathways to produce the steel are the blast furnace-basic oxygen furnace (BF-BOF) route (Figure 2.8), the electric arc furnace (EAF) fed by scrap (Figure 2.9), and the direct reduced iron-electric arc furnace (DRI-EAF) route (Figure 2.10).

i. **Blast Furnace-Basic Oxygen Furnace (BF-BOF) Route:** In 2025, this is the most common way for producing steel products from virgin iron ore, accounting for around 70% of global steel production (IEA, 2020). This process includes four main stages:

   Coking: Using a coke oven to heat coal up to 1100 °C and convert it into coke (coal processed at high temperature to remove volatiles & improve carbon content and mechanical strength).

   Sintering/Pelletizing: This phase is required to agglomerate the iron ore fines coming from the mines, using heat and pressure to form nodules (sinter) or pebble-sized particles (pellets). The specific shape is needed to allow the reducing gas flowing through the particles when stacked in the blast furnace.

   Iron Making: Producing pig iron from coke, iron ore sinter or pellets, and limestone pellets in a blast furnace, heated up to a temperature of 1800 °C.

   Steel Making: Obtaining liquid steel by blowing oxygen into the Basic Oxygen Furnace (BOF) containing pig iron, lime, and recycled steel, thereby reducing the carbon content through oxidation.

In Europe, the BF-BOF route is typically used to flat steel products such as plates, coils, and sheets, which are subsequently processed into cladding, decking, and other building envelope products. To produce 1 ton of crude steel, the BF-BOF route typically consumes approximately 1.4 ton of iron ore, 0.8 ton of metallurgical coal, 0.30 ton of limestone, and 0.13 ton of recycled steel. In terms of CO_2 emissions, the process emits around on average 1.2 t CO_2 per ton of crude steel, plus additional 1.0 t CO_2 in emissions from electricity and heat generation (deriving from the combustion of carbon-rich steel plant off-gases), for a global emission of 2.2 t/CO_2 (World Steel, 2025d).

(a) (b)

Figure 2.8 Steel making with blast furnace. (a) Moving hot metal into a basic oxygen furnace to be made into steel. (b) BF control room.

Figure 2.9 A close-up view of electrodes within an Electric Arc Furnace (EAF) during the steelmaking process. In an EAF, electricity passes through graphite electrodes (the large cylindrical elements), creating intense heat through electric arcs. This heat melts steel scrap and other materials inside the furnace. *Source*: Scanrail1/ Shutterstock.

Figure 2.10 Iron ore pellets in a DRI plant (M.Khebra/Shutterstock).

In the integrated BF-BOF route, up to around 30% recycled steel can be incorporated into the molten charge in the basic oxygen furnace without compromising product quality (World Steel, 2025d). Although advances in electric arc furnace (EAF) and direct reduced iron (DRI) technologies are accelerating, the BF-BOF process is expected to remain part of the industry since: its deep-rooted infrastructure, ability to produce high-quality and specialty steels, and inherent flexibility in feedstock combinations ensure it continues to serve key market segments even as decarbonization efforts intensify (IEA, 2023).

ii. **Electric Arc Furnace (EAF):** EAF is the main route used for producing steel products from recycled steel elements ("scrap," Figure 2.11). In 2025, EAF accounts for around 25% of global steel production worldwide (IEA, 2020). This process typically follows four main stages:

Scrap Collection and Sorting: High-quality scrap is collected and sorted to remove impurities and isolate different scrap grades. In some cases, additional virgin material is blended – usually kept below 10–15% – to adjust the chemical composition when scrap quality is insufficient.

Charging and Additive Dosing: The sorted scrap, along with minor quantities of limestone and other additives, is charged into an electric arc furnace.

Melting: An intense electric arc generated by graphite electrodes rapidly melts the scrap. The furnace heats the charge up to temperatures of approximately 1800 °C.

Refining: Once melted, the liquid steel is refined to adjust its chemical composition and remove nonmetallic inclusions.

To produce 1 ton of crude steel, the EAF process typically uses on average 0.9–1.1 tons of steel scrap, small quantities of additives such as limestone and carbon for slag formation and chemical adjustment,

Figure 2.11 Scrap metal and iron ore ready to be sorted and turned into new steel in the electric arc furnace (World Steel, 2025e). *Source:* World Steel Association (worldsteel).

and approximately 2.3 GJ electricity (World Steel, 2025k). Most carbon emissions from EAF are due to electricity use, which can decrease with the decarbonization of the grid. For scrap-based EAF production, average emissions are approximately 0.3 t CO_2 per ton of crude steel, though this value varies considerably depending on the carbon intensity of the local electricity grid (IEA, 2020).

While the BF-BOF route relies on virgin iron ore and includes upstream processes such as coking, sintering/pelletizing, blast furnace reduction to produce pig iron, and oxygen blowing in the BOF to refine the iron, the EAF route primarily melts recycled steel scrap using electrical energy, with only minor additions of virgin material when needed. Consequently, the BF-BOF process is more energy-intensive and carbon-heavy, whereas the EAF route offers lower emissions by capitalizing on recycled inputs and grid decarbonization. However, some argue that despite its higher energy consumption, the BF-BOF route benefits from decades of optimization and can exploit (in future) emerging technologies – such as hydrogen-based reduction and carbon capture – to potentially narrow its environmental gap with the EAF process. Moreover, reliance on recycled scrap for EAF production may be limited by availability and quality constraints. In conclusion, the optimal path forward will not be a single global choice but a strategic balance between the different routes, guided by regional factors such as raw material availability, grid decarbonization, and the pace of technological adoption.

iii. **Direct Reduced Iron and Electric Arc Furnace (DRI-EAF) Route:** This process converts iron ore pellets into direct-reduced iron (DRI) by using coal or natural gas to generate the reducing agents (H_2 and CO) needed to remove oxygen from the iron ore. The DRI is then converted into steel in a downstream steelmaking process, typically an EAF. Currently, DRI-EAF route accounts for around 5% of global steel production (IEA, 2020). The main difference of this route from the conventional EAF route is that while a standard EAF typically melts mostly recycled steel scrap (with minimal virgin material), the DRI-EAF route obtains metallic iron directly from iron ore, bypassing coking and sintering stages seen in BF-BOF production. DRI-EAF route can use natural gas or coal.

The main steps of the DRI-EAF route consist of:

Pelletizing: the process uses mainly high-quality pellets produced in pelletizing plants where these ore fines are agglomerated in pebble-sized particles (pellets).
Iron Making: the iron ore pellets are reduced in a solid state in the DRI furnace at high temperature (around 1100 °C), reacting with the process gas that is a mixture of hydrogen, carbon monoxide, and methane.
Steel Making: DRI is melted in an electric arc furnace, obtaining liquid steel by mixing it with lime.

Table 2.3 compares the three major steel-making processes in terms of energy consumption, carbon emissions, and product suitability.

2.6.1.3.2 Rolling, Forming, and Finishing Processes (Postprocessing Steps)

Once crude steel is made (e.g. BF-BOF or EAF), it undergoes downstream manufacturing steps to turn it into final products. These steps are also part of A3, as they occur before the product leaves the plant. The steel is typically first cast into semifinished forms (slabs, blooms, or billets). These then go through rolling mills to produce finished shapes. Postprocessing involves shaping and finishing of steel into various products such as coils (Figure 2.12), beams (Figure 2.13), tubes (Figure 2.14), cold

Table 2.3 Comparison of three major steel-making processes, based on IEA (2020, 2023) and World Steel (2025d, 2025k).

Aspect	BF-BOF route	EAF process	DRI-EAF route
Energy consumption	Uses high-energy coal in blast furnaces (~21–23 GJ/t) due to high-temperature reduction.	Relies primarily on electricity to melt scrap (~2.5 GJ/t), making it more energy-efficient.	Combines gas-based reduction (900–1100 °C) with EAF melting (1600–1700 °C), resulting in intermediate energy use (~17–22 GJ/t).
Carbon emissions	Highest upfront emissions (~2.2 tCO_2e/t steel) because of coal combustion in both the blast furnace and BOF stages.	Low emissions (~0.3 tCO_2e/t steel), especially when powered by renewable electricity.	Moderate emissions (~1.4 tCO_2e/t steel with natural gas); potential for near-zero if using green hydrogen. Carbon Capture Utilization and Storage (CCUS) can also be applied to reduce significantly (and at lower cost) CO_2 emissions from DRI ironmaking.
Product suitability	Produces a wide range of high-quality, specialty steels with tight chemical control.	Vary by geography and producer depending on regional market demands, local scrap quality, and producer strategy	Offers high purity and can produce premium grades by blending DRI with scrap; suitable for many high-strength and specialty steels.

Figure 2.12 Steel coils (Oleksandr Fagaulin/Shutterstock).

(a) (b)

Figure 2.13 Hot rolling (*Source*: ABCDstock/Shutterstock, same with Figure 2.18) and hot-rolled steel profiles (World Steel, 2025f; *Source*: World Steel Association (worldsteel)).

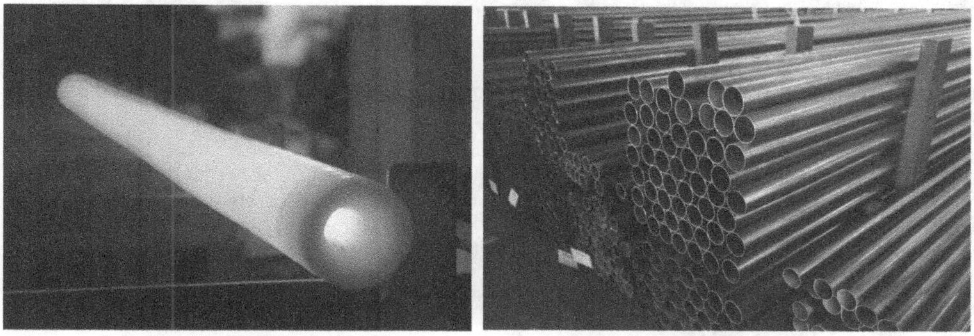

(a) Hot rolling for tube profiles (tdhster/Shutterstock.com)

(b) Finished tube profiles (นววิช ว่องไว/Adobe Stock Photos)

Figure 2.14 Manufacturing stages of steel tube profiles.

rolling (Figure 2.15), and bars (Figure 2.16). This includes rolling, casting, and other forming techniques that require energy and contribute to the carbon footprint of the product (World Steel, 2025g).

To produce steel sections, reinforcing bars, and metal deck panels, steel is either hot-rolled or cold-rolled. Hot rolling involves heating steel (often to 1100+ °C) to make it workable and then using rolling stands to reduce it to the desired thickness/shape. This process consumes significant energy, often through natural gas-fired furnaces to reheat the steel and electricity to run the rolls (Figure 2.13).

Cold rolling is a process where steel is passed between rollers at room temperature to reduce its thickness and improve its mechanical properties.

Fabrication Processes: Fabrication processes in steel manufacturing include joining, machining, bending, and sometimes surface cleaning (e.g. reclaimed steel). As an example, joining can be achieved through arc welding, which fuses steel pieces together to form a solid structure; machining involves cutting and drilling the steel to achieve the desired shape and dimensions; bending can be performed either cold or hot, depending on the specific requirements and properties of the steel being processed (Figures 2.17, 2.18, and 2.20) (World Steel, 2025g).

(a) (b)

Figure 2.15 (a) Cold-rolling steel sheet (*Source*: World Steel Association (worldsteel)) and (b) cold-rolled profiles (*Source*: SARIN KUNTHONG/Shutterstock.com).

Figure 2.16 Steel reinforcement bars (used in reinforced concrete applications), from Wiley Library.

(a) Joining: by arc welding (Zahovaev K/Shutterstock)

(b) Machining: Cutting and drilling (Bogdan VASILESCU/Shutterstock)

(c) Surface cleaning (dimik_777/Adobe Stock Photos)

Figure 2.17 Steel fabrication processes including joining, machining, and surface cleaning.

Figure 2.18 Steel fabrication techniques for shaping (e.g. hot bending). *Source*: Elchin/Adobe Systems Incorporated.

(a) Coating steel sheet with paint (World Steel, 2025a; *Source:* World Steel Association (worldsteel)).

(b) Galvanizing steel sheets (World Steel, 2025b; *Source:* World Steel Association (worldsteel)).

Figure 2.19 Photos of (a) painting and (b) galvanizing processes for steel.

Painting and Galvanizing: Painting involves the airless spraying of atomized particles onto the steel surface to provide protection against corrosion and enhance its longevity. Galvanizing consists of immersing the steel in molten zinc, creating a protective coating (Figures 2.19 and 2.20).

2.6.1.4 Opportunities to Reduce Upfront Carbon (A1–A3) Impact of Steel

Reducing the upfront carbon impact in steel production is critical for lowering the overall environmental footprint of steel constructions. This section examines several opportunities to cut emissions across the raw material supply, transportation, and manufacturing stages (A1–A3).

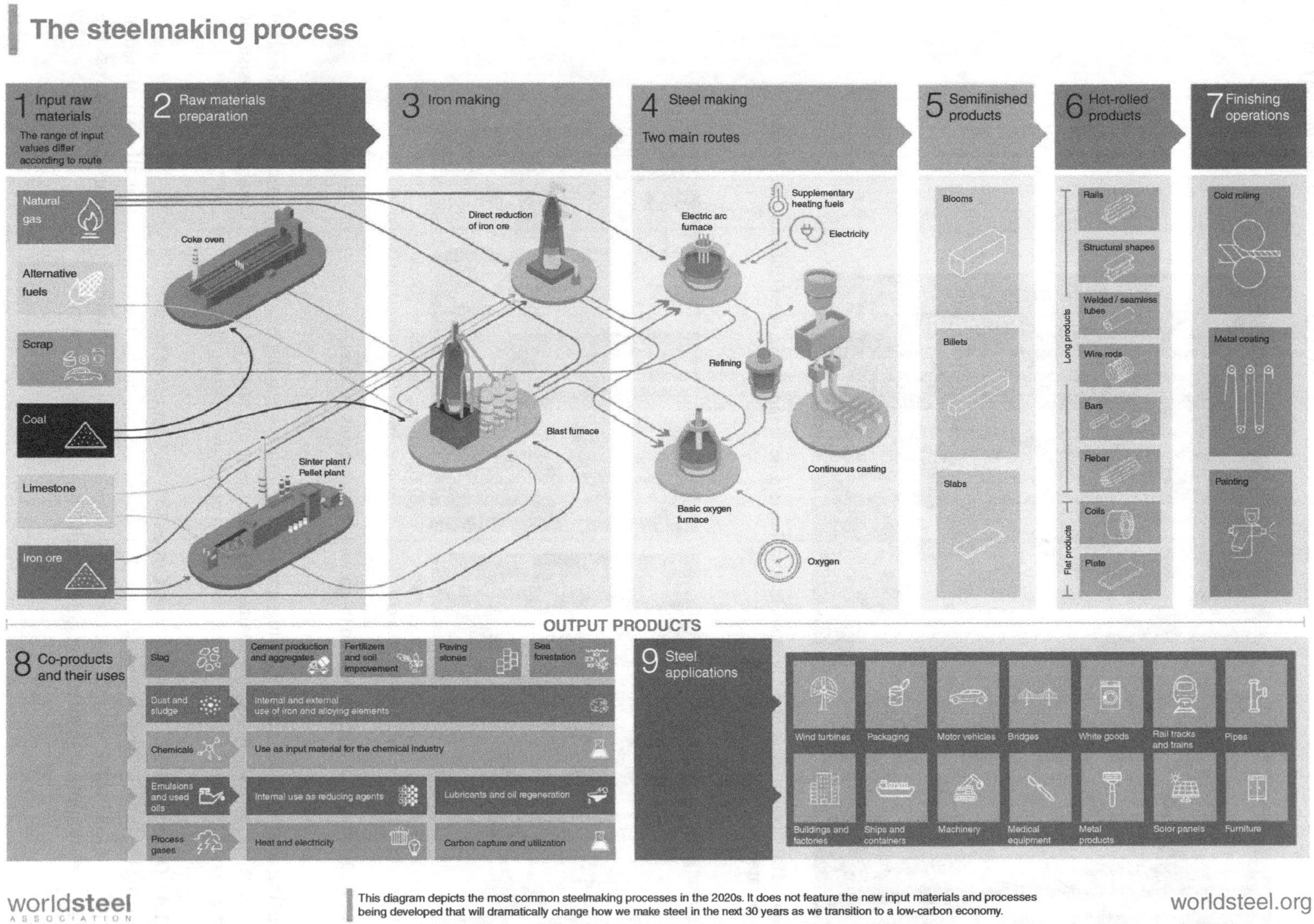

Figure 2.20 A comprehensive view of steel-making process (World Steel, 2025g). *Source*: World Steel Association (worldsteel).

Figure 2.21 Unloading post-consumer steel scrap to be recycled in an electric arc furnace (World Steel, 2025i). *Source*: World Steel Association (worldsteel).

2.6.1.4.1 Recycled Steel (Scrap) and EAF

Scrap refers to steel that is no longer in service – either as "postconsumer scrap" from end-of-life products (Figure 2.21) or as "pre-consumer scrap" generated during manufacturing (steel that never became a finished product because it was removed or unused during manufacturing). Thanks to its natural magnetism, steel can be easily separated and recycled, making it the most recycled material globally. Although the word "scrap" might suggest a discarded material, it is, in fact, a valuable raw input essential to every steel-making process. By remelting scrap steel, manufacturers can produce new steel products with adjusted chemical compositions and shapes to meet specific requirements (World Steel, 2025d).

The increase of recycled steel (scrap) can be a strong asset to reduce the upfront carbon emissions of steel products. Scrap can come from local or regional sources including manufacturing and construction waste, demolition waste, and the recycling of electrical goods, and is used to produce new steel using electric arc furnaces. Scrap is sourced from excess material generated in steel facilities (home scrap), from downstream production processes (industrial scrap), and from discarded products (obsolete scrap). The availability of home and industrial scrap depends on current domestic steel production levels, while the amount of obsolete scrap is influenced by past production, product lifespans, and the efficiency of recycling programs (Figure 2.22).

If the global steel sector could significantly expand its use of EAF technology – ideally powered by 100% renewable electricity – steel production could be achieved with a dramatically lower carbon footprint. However, a full transition to scrap-based EAF steelmaking is not feasible, as there will always be demand for primary steel production unless demolition rates exceed new construction, which is not realistic.

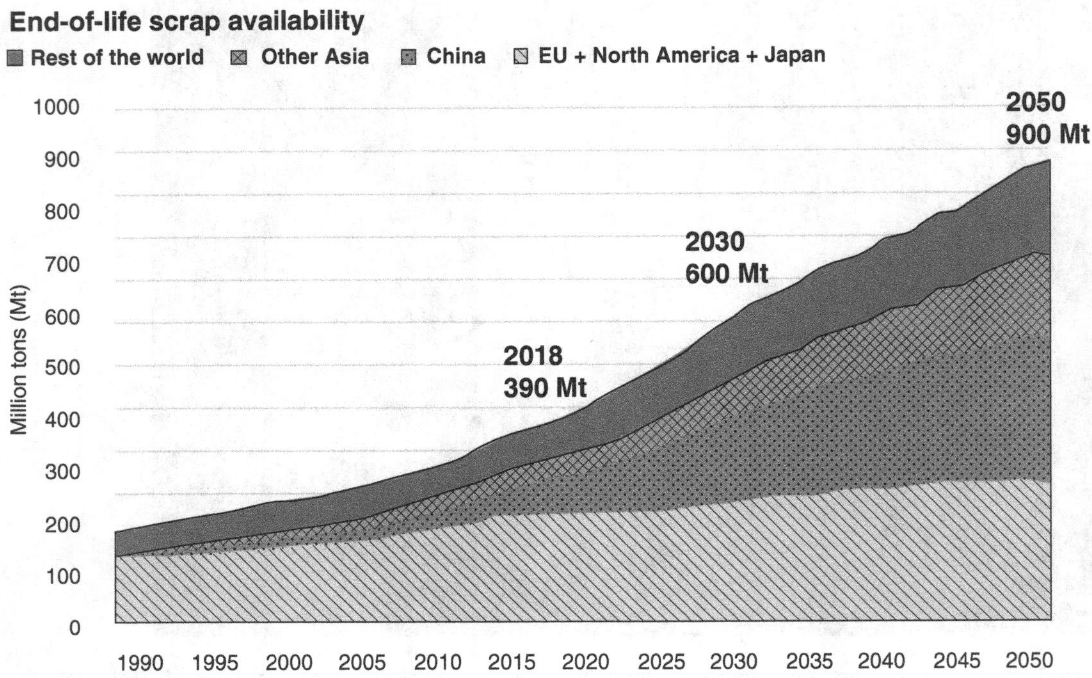

Figure 2.22 Scrap availability projection. *Source*: World Steel Association (2021).

Currently, only 30% of steel plants use EAF worldwide (World Steel Association, 2025), and these plants are partly powered by fossil fuels (International Energy Agency, 2020). Moreover, global demand for steel is approximately three times greater than the supply of scrap (UK Steel, 2025), meaning that primary steelmaking routes (BF-BOF or DRI-EAF) will remain essential to meet demand. Addressing this disparity through improved scrap collection, better sorting technologies, and supportive policy measures – such as extended producer responsibility and favorable carbon pricing – could further increase the adoption of recycled steel production.

2.6.1.4.2 Decarbonization of the Grid and Vehicles

Decarbonizing grids (i.e. reducing the carbon emissions associated with electricity generation by shifting from fossil fuels to renewable and low-carbon energy sources, such as solar, wind, or hydroelectric power) and vehicles (i.e. transitioning electricity grids to renewable and low-carbon energy sources, and replacing traditional fossil-fuel-powered vehicles with electric or other low-emission alternatives) (e.g. Figures 2.23 and 2.24) can cut a high quantity of carbon footprint from steel production (World Economic Forum, 2024). Electrifying mining trucks and freight can eliminate emissions in raw material extraction and delivery (Mission Possible Partnership, 2021; Rocky Mountain Institute, 2019).

Moreover, policies such as carbon pricing, clean power standards, and incentives for clean tech (renewables, hydrogen, EVs) drive the shift to low-carbon electricity and transport (World Economic Forum, 2024). These measures make fossil-based steel and logistics less competitive, improving the business case for green steel and zero-emission vehicles.

Figure 2.23 Types of vehicles used in mining operations (Mr. Tempter/Shutterstock).

Figure 2.24 Types of vehicles used in mining operations (John/Adobe Stock Photos).

2.6.1.4.3 Low Emission DRI with EAF

The natural gas-based DRI coupled with an EAF would allow significant cuts in direct emissions of CO_2 for a ton of crude steel. The DRI technology is developing quickly, and there are ongoing projects (ArcelorMittal, n.d.; HBI S.p.A., n.d.; HYBRIT Development, n.d.; HYDRA Project IPCEI n.d.) in China, Europe, Middle East, and Center/North America that would allow the natural gas to be substituted with hydrogen, once this is available at large scale and a competitive price from water electrolysis. The challenge on this path is the need for a significant volume of storage to provide a constant supply of hydrogen to the reduction furnace. In this case, the emissions would depend on the carbon intensity of electricity used and could be reduced almost to zero in case of full use of renewable energy (Hydrogen Europe, 2022).

2.6.1.4.4 Carbon Capture, Use, and Storage (CCUS) Integration in the Iron and Steel Making

Carbon capture, utilization, and storage (CCUS), which is also useful for cement production, may become an essential technology for reducing emissions from existing global blast furnaces, which are expected to remain a significant component of steel production in the coming decades (IEA, 2020; World Economic Forum, 2024).

2.6.1.4.5 Replacement of Fossil Fuel with Biomass

In areas with high availability of biomass resources, sustainable biomass and biochar can be used to substitute coal in existing steel-making process, injecting for instance biomass into both existing and newly constructed blast furnaces.

2.6.2 Concrete Upfront Carbon (A1 to A3)

This subsection focuses on the carbon emissions from the concrete production process, including raw material extraction, transport, and manufacturing.

2.6.2.1 Extraction and Manufacturing of Raw Materials (Cement, Aggregates, Water, Admixtures) (A1)

Module A1 covers the extraction and processing of all raw materials used to produce cement and concrete, including limestone, clay, and aggregates (Figure 2.25). The main processes involve quarrying limestone and aggregates, production of clinker through calcination in cement kilns, and preparation of supplementary materials such as fly ash or slag. These operations significantly affect resource depletion, energy consumption, and carbon emissions – especially the calcination of limestone, which releases substantial amounts of CO_2. For concrete, the A1 stage typically generates the highest environmental impact, particularly in terms of embodied carbon. The following list outlines various components of concrete and their contribution to its environmental impact (Arup, 2023a; Mason and Lea, 2026).

Cement: Cement constitutes up to 20% of the mass of concrete, but it is the predominant contributor to the A1–A3 carbon emissions. This high impact is largely due to the energy-intensive process of calcination, where limestone ($CaCO_3$) is heated to produce lime (CaO) and CO_2. The energy required to heat the kiln (explained in the next page) to 1500 °C, as well as the energy needed for grinding, further adds to the carbon footprint, making cement production a significant environmental concern.

Aggregates: Aggregates constitute about 75% of the concrete volume. Although less carbon-intensive than cement, their extraction and transportation still contribute to overall emissions.

Water: Although it has a smaller impact compared to other concrete ingredients, efficient management of water sourcing and usage is necessary to ensure sustainability and reduce environmental impacts.

(a) Cement (Okaycm'Stocker/Shutterstock, Konstantin Faraktinov/Shutterstock)

(b) Aggregates (Kriengsuk Prasroetsung/Shutterstock.com)

Figure 2.25 Raw materials for concrete including cement and aggregates.

Figure 2.26 Carbon emissions of concrete components (author's/with permission of Ove Arup & Partners Limited).

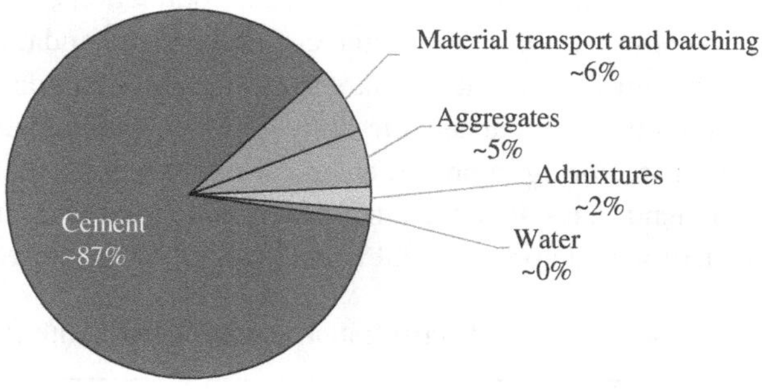

Admixtures: Admixtures are employed to calibrate the concrete mix by reducing water usage and increasing durability. While they play a crucial role in improving the workability and overall performance of concrete, they still contribute to A1 emissions.

Supplementary Cementitious Materials (SCMs): Industry by-products that can significantly replace clinker in cement, lowering concrete's carbon footprint.

Figure 2.26 presents a detailed breakdown of the carbon emissions associated with different components of concrete, highlighting the substantial impact of cement production compared to aggregates, water, and admixtures.

Understanding the emission sources of cement is important for identifying potential areas for improvement and innovation in reducing the environmental impact of cement and concrete production.

Calcination Reaction in the Kiln: The process begins with the calcination reaction, where limestone ($CaCO_3$) is heated to produce lime (CaO) and carbon dioxide (CO_2) (Equation 2.1). This chemical reaction is a major source of CO_2 emissions in cement production.

$$CaCO_3 + \text{Heat (up to 860°C)} \rightarrow CaO + CO_2 \tag{2.1}$$

Clinker Making: The calcined raw meal must be heated to very high temperatures, up to 1500 °C. The energy required to achieve and maintain these temperatures is substantial, contributing significantly to the overall emissions.

Energy for Grinders: After calcination, the produced clinker needs to be ground into a fine powder. This grinding process requires a significant amount of energy, further adding to the carbon footprint of cement production.

The most significant factor influencing greenhouse gas (GHG) emissions in cement production is the ratio of clinker used. This ratio varies considerably depending on the type of cement, and it significantly impacts the GHG emissions associated with cement production. Table 2.4 shows a summarized classification of cement types (CEM I to CEM V), along with their clinker content ranges and corresponding descriptions of their composition and use referring to (EN 197-1, 2011).

Please note that (EN 197-1, 2011; Hewlett and Liska, 2019; Neville, 2011) CEM I is often used in reinforced concrete structures, foundations, precast elements, because it provides high early strength and predictable performance, which is useful when fast setting, rapid formwork removal, or high load capacity is required. CEM II, with a lower clinker factor, is widely used for general construction since it offers good strength with lower environmental impact; in residential buildings it is commonly applied for cast-in-place floor slabs, beams, columns, and foundations, while in infrastructure it serves bridge decks, road pavements, retaining walls, and culverts under standard environmental conditions. CEM III, containing significant slag, is suited for large-scale (massive foundations) or marine structures due to reduced heat of hydration and enhanced durability. CEM IV, which includes pozzolanic materials, is preferred for applications requiring improved chemical resistance, like dams or water-retaining structures. Lastly, CEM V is a blend that offers moderate heat of hydration and durability, making it a practical choice for various structural and nonstructural works, though it has limited commercial availability today.

Table 2.4 Summary of classification of cement types with clinker content.

Cement type	Cement description	Clinker (%) – Min	Clinker (%) – Max
CEM I	Portland cement	95	100
CEM II	Portland-composite cement (e.g. fly ash, slag, pozzolana)	65	94
CEM III	Blast furnace cement	5	64
CEM IV	Pozzolanic cement	45	69
CEM V	Composite cement	20	49

2.6.2.2 Transport of the Raw Material to Batching Plant (A2)

The A2 carbon emission quantity is based on the distance between the quarries and factories to the concrete batching plants (i.e. transport of cement, aggregates, sand, SCMs, and admixtures to the concrete

Table 2.5 Carbon emissions based on distance and means of transport.

Transport scope	Emissions ($kgCO_2e/kg$)
Local (<50 km):	0.005
National (<300 km):	0.032
Regional (<1500 km):	0.161
Global (>1500 km):	0.183

Source: Author's/with permission of Ove Arup & Partners Limited.

batching plant), which can be categorized from local to global distances for the concrete's components (Arup, 2023a) (Table 2.5). Portland cement is transported from cement factories, often located near quarries where raw materials are extracted, to batching plants near towns or cities that require concrete. Supplementary cementitious materials (SCMs), such as GGBS, are frequently transported directly from the steel or coal plants to concrete batching plants. This direct supply route can support the production of lower-carbon concrete by simplifying logistics and reducing intermediate handling, especially when substitutes are locally available, though imports may be considered depending on their carbon impact. Aggregates are typically locally or regionally sourced and have shorter transportation distances compared to cement products. In regions such as Middle-East, aggregates are often imported from nearby countries due to limited local natural resources, which can increase transport distances. However, even in these cases, transport emissions for aggregates tend to remain relatively low compared to the higher embodied carbon associated with cement production.

2.6.2.3 Manufacturing (Batching) of Concrete (A3)

The manufacturing process of concrete, commonly referred to as batching, involves combining raw materials such as cement, supplementary cementitious materials (SCMs), and aggregates in specified proportions. These components are stored at the batching plant in silos (Figure 2.27), mixed either in a plant mixer or truck mixer, and then transported to the construction site. The batching stage is predominantly powered by electricity or diesel, depending on the specific batching facility. The choice of fuel significantly impacts the carbon emissions associated with this phase of concrete production. Therefore, the main contributor to CO_2 emissions during the batching process is the energy consumption required to operate the mixers and other equipment, such as conveyor belts, pumps, compressors, and control systems.

The emissions associated with concrete production can vary based on the method of batching. For the concrete mixed to use on-site (in situ), the emissions are relatively lower compared to precast concrete (Figure 2.28) (ICE states that the emissions for concrete in situ are approximately 0.0007 $kgCO_2e$ per kilogram of concrete, while precast concrete emits about 0.0142 $kgCO_2e$ per kilogram (ICE, 2022)). The higher emissions for precast concrete are primarily due to the additional energy required for the manufacturing and curing processes since precast concrete components are produced in a factory setting, where they undergo controlled curing conditions, often involving steam or heated environments to accelerate the curing process. While precast concrete may have higher emissions during manufacturing, its advantages include faster construction times, reduced on-site labor and waste (that will enter to the life-cycle stage A5), and consistent quality due to controlled production conditions.

Figure 2.27 Cement factory (ESB Professional/Shutterstock).

2.6.2.4 Opportunities to Reduce Upfront Carbon (A1–A3) Impact of Concrete

This section introduces various strategies aimed at lowering the upfront carbon emissions linked to concrete production.

Therefore, while it is essential to use global supplies of SCMs to their fullest extent, relying heavily on GGBS or other SCMs at a local level is unlikely to significantly reduce global emissions. The focus should be on maximizing the efficient use of available materials while continuing to explore and develop new SCMs and techniques for reducing the environmental impact of concrete production.

I. **Supplementary Cementitious Materials (SCMs):** SCMs (such as ground granulated blast furnace slag (GGBS), pulverized fuel ash (PFA or fly ash), and silica fume) are by-products of other industries that can replace a substantial portion of clinker in the cement mix. Replacing clinker can reduce the carbon footprint of concrete production. The global availability and use of Portland cement and supplementary cementitious materials (SCMs) play a crucial role in determining the environmental impact of concrete production. While structural concrete has the potential to incorporate up to 50% cement replacements, the global supply of SCMs is currently limited (Figure 2.29), comprising around 15% of cement usage today (Arup, 2023a). This limitation poses challenges for significantly reducing the carbon footprint of concrete on a global scale (increasing the ground granulated blast furnace slag (GGBS) content in concrete for specific projects may lead to the reduced availability of GGBS for other projects, thereby limiting its overall effectiveness in emission reduction). Higher GGBS content can result in longer setting times compared to conventional Portland cement (CEM I) that is another parameter to consider for GHG emissions of the project.

(a) (Dark Caramel/Adobe Stock Photos)

(b) (Wrangler/Adobe Stock Photos)

(c) (graemenicholson/Getty Images)

Figure 2.28 Precast concrete elements and a building under construction with precast elements. (a) Precast segmental lining units. (b) Precast hollow-core floor slabs. (c) Precast structural wall and frame panels during construction.

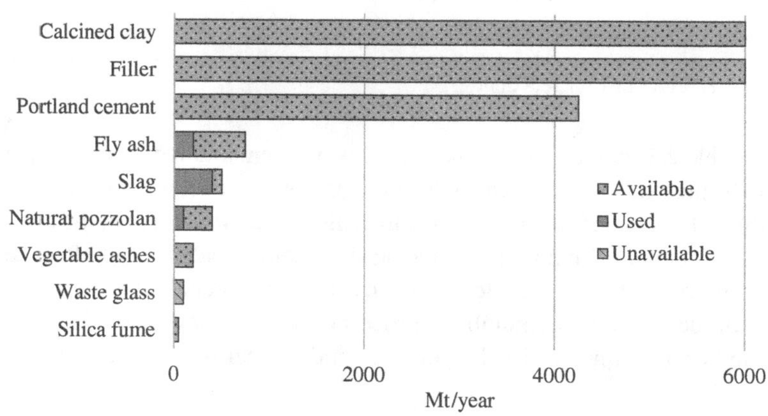

Figure 2.29 Chart showing amount of supplementary cementitious materials (Mt/year) (Scrivener, K., John, V. M., & Gartner, E./with permission of Elsevier).

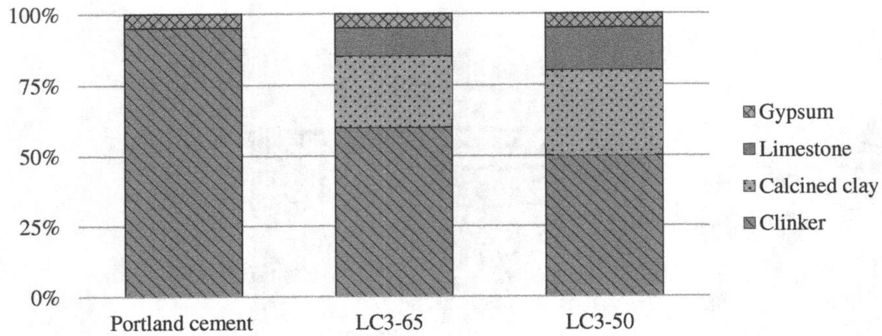

Figure 2.30 Comparison of the contents of different cement types (including LC3 cement) (LC3, 2025).

II. **LC³ Cement:** Limestone calcined clay cement (LC³) is demonstrated to reduce the clinker content by half, offering a promising alternative to traditional Portland cement (LC3, 2025). It is in the research stage, and it has seen some applications in India and Latin America (Figure 2.30).

III. **Recycled Aggregates:** Recycled aggregates (RCA) are obtained by crushing and processing existing concrete structures that have reached the end of their service life. This process involves breaking down the old concrete into smaller pieces and then cleaning and grading the material to ensure it meets specific standards for reuse.

Table 2.6 Carbon emissions for virgin and recycled aggregates at varying transport distances.

Material and delivery distance	+/− % CO_2 with respect to virgin aggregates
+58.5 km (delivery and return distance by road)	—
Used on-site, 0 km transport	−15%
+5 km (delivery and return distance by road)	−10%
+10 km (delivery and return distance by road)	−5%
+15 km (delivery and return distance by road)	0%
+20 km (delivery and return distance by road)	5%
+58.5 km (delivery and return distance by road)	14%

Source: MPA The Concrete Centre.

Table 2.6 compares the CO_2 emissions of virgin and recycled aggregates at different delivery distances. RCA present some challenges. The fine materials generated during the crushing process tend to increase the water demand of the concrete mix. This increased water demand can affect the workability and strength of the concrete if not properly managed. To counterbalance this effect, adjustments in the cement content of the concrete mix are often necessary. This may include increasing the cement content to maintain the required strength and durability properties. Furthermore, the quantity and quality of recycled aggregates can be constrained by local regulations and standards. For instance, in nonstructural applications such as pavements, the use of 100% recycled aggregates is often permissible. In structural applications, the allowable percentage of recycled aggregates typically ranges from 0% to 20%, depending on the concrete class and specific requirements. Table 2.7 presents allowable percentage of recycled aggregates for different concrete types (The Concrete Center, 2020).

Table 2.7 Allowable percentage in the structural applications of recycled aggregates for different concrete types.

Concrete type	Allowable percentage of recycled aggregates
GEN0 to GEN3	100%
RC20/25 to RC40/45	20%*
FND1	20%
RC40/50XF	0%
PAV1 and PAV2	0%
FND2 to FND4	0%

Source: MPA The Concrete Centre. *except where the specifications allow higher proportions

IV. **High Strength Concrete:** Higher strength concrete requires a higher cement ratio, which typically increases its carbon footprint during production. Table 2.8 outlines the different strength classes of concrete according to (EN 1992-1-1, 2023), detailing the type of cement and the composition of the binder, which includes both cement and secondary cementitious materials (SCM).

Higher concrete strength may enable the use of smaller volumes of concrete to achieve the same structural safety, potentially reducing the overall carbon footprint of the structure. Therefore, whether higher strength concrete increases or decreases the carbon footprint depends on the specific project requirements

Table 2.8 Nationally determined parameters (NDP) – strength development classes of concrete.

	Composition and properties of the binder		
Class	**Type of cement**	**Cement strength class**[a]	**Composition of the binder (cement and SCM**[b]**)**
CS	CEM III CEM II/B	32.5 N; 42.5 N	Portland cement clinker and more than 65 M.-% of ground granulated blast furnace slag (GGBS) or more than 35 M.-% of fly ash (FA)
CN	CEM II, CEM I	42.5 N; 32.5 R	Portland cement clinker and more than 35 M.-% but less than 65 M.-% of ground granulated blast furnace slag (GGBS) or more than 20 M.-% but less than 35 M.-% of fly ash (FA)
CR	CEM I	42.5 N; 52.5 N; 52.5 R	–

Source: British Standards Institution.
[a]According to EN 197-1.
[b]Secondary cementitious materials (SCM) considered as cement replacement for the water-binder ratio according to EN 206.

and design. An interesting study about the impact of high strength concrete on carbon footprint of different structural configurations is presented in this article (Fantilli et al., 2019).

V. **Electrification of Mining Operations:** Regarding transport (A2), the electrification of railways and development of hydrogen-powered heavy goods vehicles (HGVs) (instead of the traditional ones, e.g. Figure 2.31) are two changes that could see their carbon contributions fall over the coming years.

Figure 2.31 Traditional types of vehicles involved in mining operations (monstersparrow/Adobe Stock Photos).

VI. **Waste Heat Recovery:** In a typical cement plant, roughly 30–40% of the input heat is lost through hot exhaust gases from the preheater and clinker cooler (Gupta and Kaul, 2011). Waste heat recovery (WHR) in a cement plant captures excess heat from the kiln and clinker cooling processes and converts it into electricity via a Rankine cycle system (often using an Organic Rankine cycle). This improves overall energy efficiency by utilizing heat that would otherwise be lost to the environment. WHR systems are designed to tap into this wasted thermal energy and turn it into usable power, thereby reducing the need for purchased electricity (Turbines, 2024) and lowering the plant's electrical energy intensity. Many large cement plants have commissioned WHR systems in the range of 5–15 MW capacity, which directly offset an equivalent amount of grid or captive power. For instance, the cement industry in Turkey installed 12 WHR power plants totaling ~120 MW over 7 years (2011–2018) (Santarossa, 2018). Implementing waste heat recovery in cement manufacturing faces several challenges such as ensuring adequate waste heat availability, managing high capital costs, addressing technical integration issues like dust fouling and system maintenance, and navigating variable energy prices and grid reliability.

In conclusion, reducing the upfront carbon impact of concrete production requires balancing multiple strategies. Increasing the use of SCMs like GGBS, PFA, and silica fume can lower CO_2 emissions by reducing clinker content, yet their limited supply and associated issues – such as longer setting times – need careful consideration. Emerging alternatives like LC3 cement, which can halve the clinker content, and the incorporation of recycled aggregates offer further opportunities to cut emissions, provided that challenges such as water demand and material quality are addressed. Moreover, optimizing concrete strength to use less material, along with innovations like electrification of mining operations and waste heat recovery systems, contributes to overall emission reductions. Ultimately, combining these approaches while tailoring them to project-specific requirements is key to significantly reducing the carbon footprint of concrete production, and continued advances in material technology and processing methods remain vital for long-term sustainability.

2.6.3 Timber Upfront Carbon (A1 to A3)

This subsection focuses on the carbon emissions from the timber production process, including wood harvesting, transport from the forest to the sawmill, manufacturing and machining. Quantifying A1–A3 of timber involves understanding biogenic carbon, as explained in Section 2.4.

2.6.3.1 Harvesting Wood and Production of Adhesives (A1)

Timber is known for its carbon sequestration properties. However, its extraction and processing contribute to carbon emissions. The availability of timber varies across regions due to differences in forest area and distribution (Figures 2.32 and 2.33). Therefore, geographic factors should be considered when harvesting raw materials. Besides, the sourcing and production of adhesives also play a role in environmental impact.

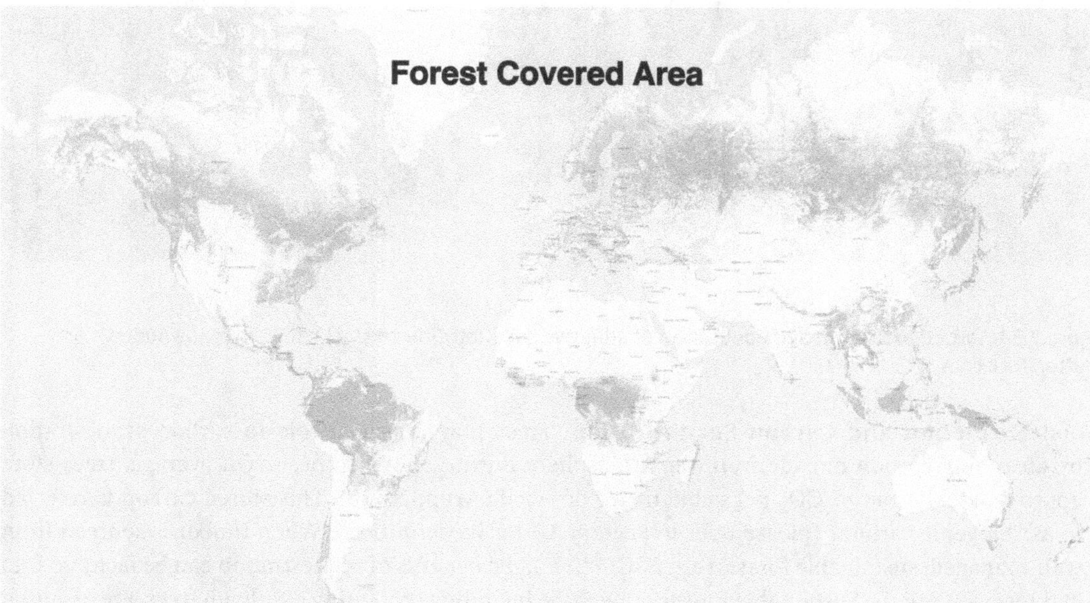

Figure 2.32 Global forest distribution showing boreal, temperate, subtropical, and tropical forests (author/with permission of Geneva Environment Network). *Source*: https://en.wikipedia.org/wiki/List_of_countries_by_forest_area#/media/File:Forest_Covered_Area_World_Map.webp.

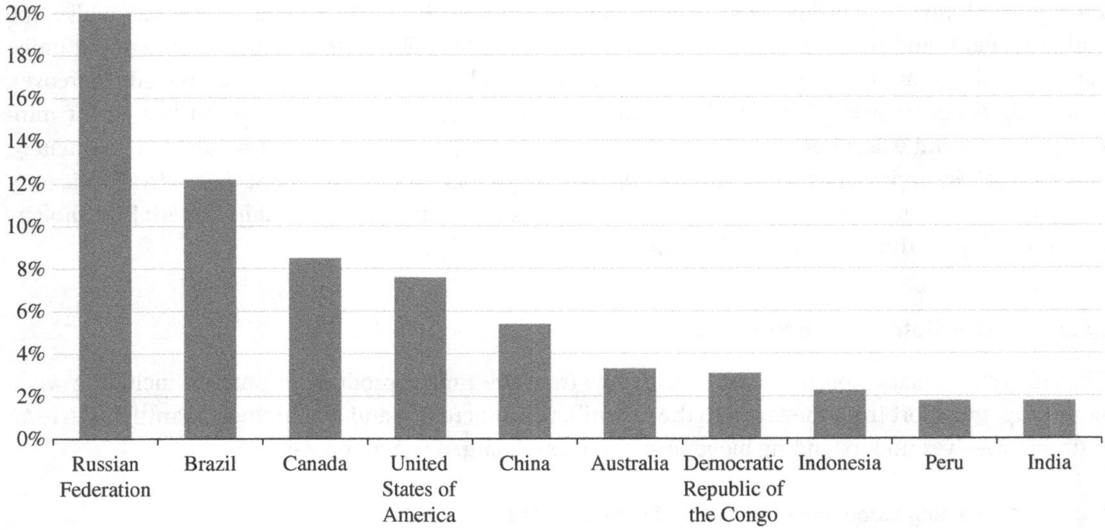

Figure 2.33 Top 10 countries by forest area in 2020 (% of world's forests) (author/with permission of Geneva Environment Network).

(a)

(b)

Figure 2.34 Wood extraction and application of adhesive. (a): Kletr/Shutterstock); (b): Guryanov Andrey/Shutterstock.com.

Wood Extraction and Carbon Sequestration: Trees play a crucial role in carbon sequestration by absorbing carbon dioxide from the atmosphere during photosynthesis. On average, trees store approximately 1 ton of CO_2 per cubic meter of wood (Arup, 2023c). This stored carbon is referred to as "biogenic carbon" (please refer to Section 2.4 for its definition). When timber is sourced from well-managed, sustainable forests (e.g. FSC, PEFC), the benefits of sequestration can be factored into carbon calculations. Sustainable forestry practices, including replanting cut-down trees, are essential for maintaining this carbon balance. However, today only around 9% of global forests are certified for sustainable management (PEFC, n.d.), and the vast majority of certified forests are located in

Europe and North America (tropical forests in regions like the Amazon Basin, the Congo Basin, and Southeast Asia have a much lower percentage of certified area). Moreover, the process of harvesting wood carries its own carbon emissions. The machinery used in logging, such as chainsaws, harvesters, and forwarders, typically operates on fossil fuels (Figure 2.34a), contributing to greenhouse gas emissions. Furthermore, the construction and maintenance of forest roads, essential for accessing logging sites, also involve significant carbon outputs (Kazama et al., 2021). In some cases, the use of more advanced equipment and techniques, such as selective logging and reduced-impact logging (RIL), can mitigate these emissions. Moreover, the handling of logging residues, such as branches and treetops, can influence carbon emissions. If left to decompose naturally, these residues can emit methane, a potent greenhouse gas. Alternatively, they can be collected and used for bioenergy, which, if managed sustainably, can offset some of the carbon emissions associated with extraction (Repo et al., 2020).

Adhesives in Timber Products: Timber products like cross-laminated timber (CLT) and glulam (GLT) typically require oil-based adhesives (Figure 2.34b) that consume significant energy during production. Although CLT and glulam may only contain 1–2% or 1.5–2.5% adhesive by volume, the GWP of these binders can be much higher than wood on a per-kilogram basis (Arup, 2023c). Some emerging bio-based adhesives aim to reduce fossil carbon content and lower overall emissions, though they are still developing in the market (European Commission's Joint Research Centre (JRC), 2024).

2.6.3.2 Transport (Road, Rail, Sea) of the Wood to Manufacturing Unit (A2)

This stage involves moving wood from the forest to the manufacturing units via sea, rail, or road. Each mode of transport has a different carbon footprint, influencing the overall environmental impact of timber products (Figure 2.35) (Arup, 2023c).

- Transporting timber by sea emits approximately 17 kg of CO_2 per ton per 1000 km.
- Rail transport of timber results in 27 kg of CO_2 per ton per 1000 km.
- Road transport has the highest carbon footprint, emitting 107 kg of CO_2 per ton per 1000 km.

The electrification of railways and the development of hydrogen-powered heavy goods vehicles (HGVs) are two innovations that can help reduce the carbon footprint of timber logistics.

(a) By sea: 17 $kgCO_2$/t/1000 km (Dawid K Photography/ Shutterstock.com)

(b) By rail: 27 $kgCO_2$/t/1000 km (Kletr/Shutterstock.com)

(c) By road: 107 $kgCO_2$/t/1000 km (am/Adobe Stock Photos)

Figure 2.35 Transport modes for raw materials (Arup, 2023c).

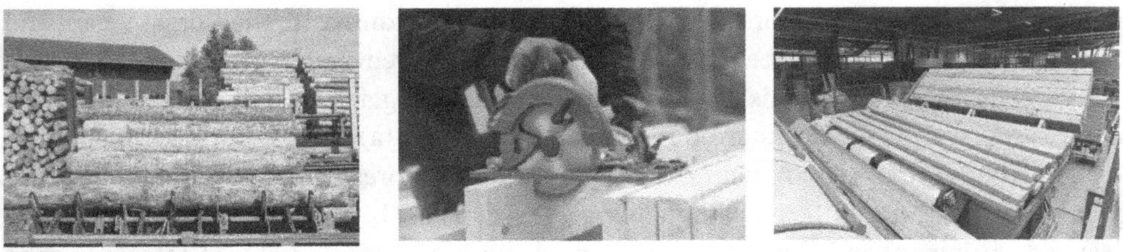

(a) Stripping bark (laurel/Shutterstock.com) **(b)** Sawing logs (Wiley Photo) **(c)** Board production (industryviews/Shutterstock.com)

Figure 2.36 Processing steps of timber (Arup, 2023c).

2.6.3.3 Manufacturing (Processing of the Wood, the Kiln-drying, Fabrication) (A3)

The manufacturing stage of timber products involves processing of wood, kiln-drying, and fabrication of the sawn timber into timber products such as cross-laminated timber (CLT) and glued laminated timber (glulam).

2.6.3.3.1 Processing

Processing timber involves multiple steps to ensure the wood is suitable for construction and other uses. Initially, the logs are stripped of their bark and sawed into the required section sizes. This process generates wood chips and shavings, which can be recycled and reused. The fresh wood is then kiln-dried to reduce its moisture content to approximately 12–18%. This drying process is usually fueled by natural gas or by using offcuts and forest waste, making it a crucial step in preparing the wood for further processing (Figure 2.36).

2.6.3.3.2 Kiln-drying

Kiln-drying is the part of the timber processing stage (Figure 2.37), where the fresh wood is dried to achieve the desired moisture content. This process ensures the wood is less susceptible to warping and shrinkage, making it more stable for construction purposes. The kilns are typically fueled by natural gas or renewable resources such as offcuts and forest waste. Proper drying is vital to ensure the wood's durability and longevity in construction applications. For glued structural timber products, kiln-drying is essential since excessive moisture prevents proper adhesion.

2.6.3.3.3 Sorting, Gluing, and Finishing (Post Processing)

Postprocessing includes various fabrication processes such as sorting, gluing, and finishing. Timber is transported to the manufacturing site, graded according to its strength properties, and then sorted for internal finishes or use in engineered timber. The fabrication processes involve manufacturing custom-sized engineered timber, including cross-laminated timber (CLT) and glulam. These processes consist of finger jointing, gluing and bonding, maintaining pressure, and CNC finishing. Since the gluing process often takes place far from sawmills or kiln-drying facilities, transportation can have a significant impact on sustainability (Figures 2.38 and 2.39).

2.6.3.3.4 Fire Protection and Moisture Control

Timber products may require fire protection and moisture control to ensure safety and durability. Fire protection (Figure 2.40a) can be achieved by oversizing members to maintain strength in case of fire or

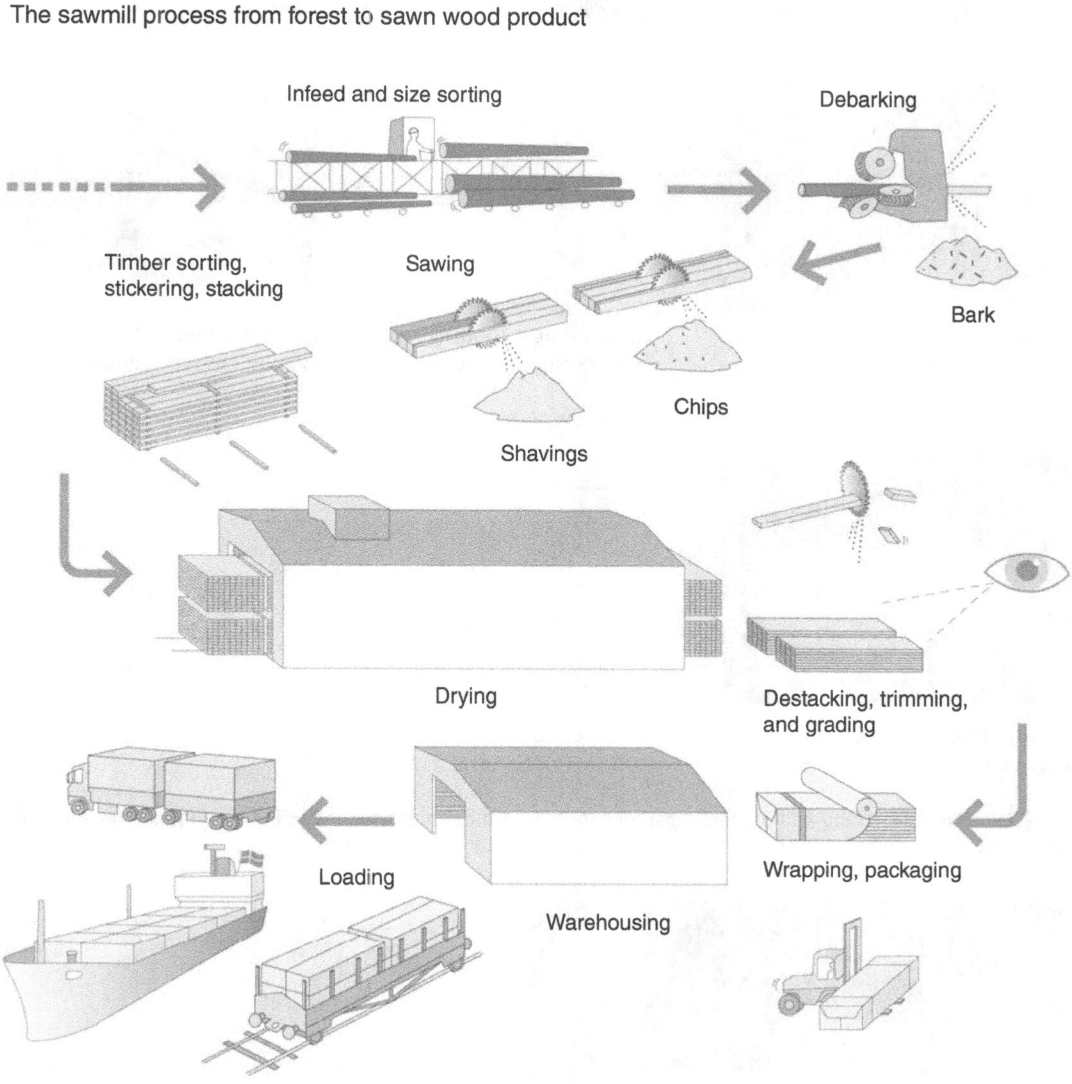

Figure 2.37 The sawmill process from raw timber to finished wood products, including sorting, debarking, sawing, drying, grading, packaging, and distribution (courtesy of Svenskt Trä).

Figure 2.38 Timber sorting and storing (Nomad_Soul/Adobe Stock Photos; Andrey VP/Adobe Stock Photos).

(a) Sorting and processing (SpockyPo/Shutterstok.com)

(b) Finger jointing (rawf8/Shutterstok.com)

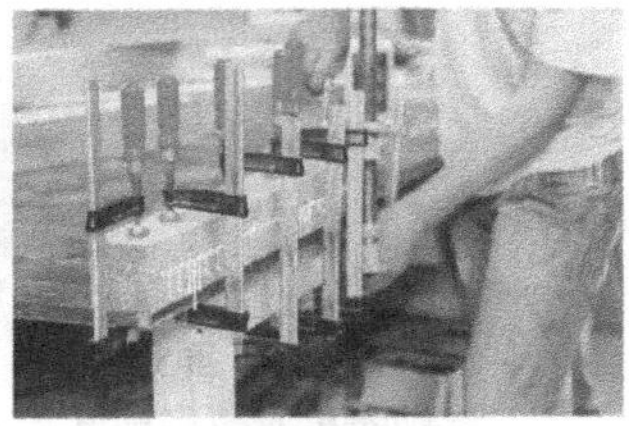

(c) Gluing and bonding (Vadim Orlov/Shutterstok.com)

(d) Maintaining pressure (kuzovkov/Adobe Stock Photos)

(e) CNC finishing (courtesy of Svenskt Trä)

Figure 2.39 Manufacturing and processing stages for mass timber elements including sorting, jointing, bonding, and finishing.

(a) Fire protection (il21/Shutterstock.com)

(b) Moisture control (courtesy of Svenskt Trä)

Figure 2.40 Examples of timber protection measures.

using fire-rated boards such as plasterboard. Moisture control (Figure 2.40b) involves designing structures to avoid standing water, hidden leaks, and reliance on roof membranes alone. These measures are crucial to prevent structural damage and degradation over time. To further enhance durability and prevent long-term structural damage, the installation of sensors for continuous moisture monitoring is always recommended (F. Piva, personal communication, 2025).

2.6.3.4 Opportunities to Reduce Upfront Carbon (A1–A3) Impact of Timber

Timber production emissions differ by region due to variations in energy grid mixes, forest types, and transportation distances. In areas powered by renewable energy, harvesting timber leads to lower emissions compared to regions dependent on fossil fuels. The use of electric or hybrid machinery further reduces the carbon footprint, making the production process more sustainable. Choosing smaller, fast-growing trees may help relieve pressure on old-growth forests while still offering high-strength materials suitable for modern construction. Substituting traditional synthetic chemicals with bio-based adhesives and finishes also cuts down on harmful emissions and reduces reliance on fossil fuels during production.

Looking ahead, European forecasts indicate that shortages of structural timber may occur (Grünig et al., 2026). The spread of wood-boring insects and rising temperatures have already diminished the availability of high-quality raw material, as evidenced by recent industry presentations. Although certification protocols remain important, the limited availability of timber requires a careful approach to design. It is essential to quantify CO_2 emissions accurately and optimize wood usage in construction. Today, some design offices intentionally oversize timber structures to increase a building's calculated carbon storage. While this practice artificially improves carbon metrics on paper, it wastes a limited resource and inflates project costs. Consequently, developers may abandon these designs for cheaper, carbon-intensive materials like concrete or steel. Research into hybrid construction systems can offer a promising way to reduce the cost of timber structures; a recent example is shown in Figure 2.41.

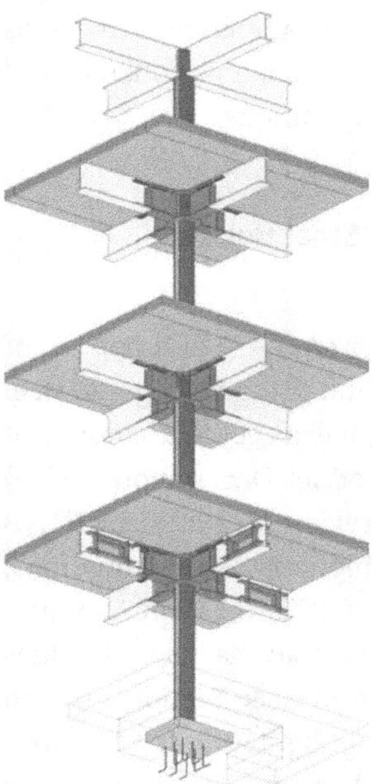

(a) Steel tubular columns, open profile beams, and CLT slab.

(b) A look inside the hybrid steel-timber building (courtesy of Franco Piva, Ergodomus)

Figure 2.41 An example to hybrid construction: SMU – Singapore Management University Extension, courtesy of Franco Piva, Ergodomus (Ergodomus, 2025). More details about this project will follow in Chapter 8.

2.6.4 Boundaries of A1, A2, and A3 in EPDs

A common source of confusion when interpreting Environmental Product Declarations arises from the fact that Module A1 (Raw Material Supply) is not limited to extraction activities alone, but it may include all upstream processes, including the manufacturing of intermediate products. The scope of A1, A2, and A3 varies through the supply chain. For example, in the context of certain Product Category Rules (PCRs) for steel construction products, the way that **modules** A1, A2, and A3 are defined can differ depending on how the product is made (American Institute of Steel Construction [AISC], n.d.). For **"mill products"** (e.g. unfabricated rebar, hot-rolled sections, plate, coil, wire), A1 usually covers the extraction and processing of feedstock materials, A2 accounts for transporting those materials to the mill, and A3 includes the mill's own operations. For **"manufactured products"** (e.g. welded steel sections, hollow structural sections, cold-formed steel framing, welded wire reinforcement), the definition of A1 is often broadened: it includes the sum of the mill's upstream impacts (A1 and A2) and the mill's production impacts (A3) for the steel inputs that go into the manufactured product, including scrap yield effects. Then the module labeled A2 represents transport of mill products to the manufacturer, and A3 captures the manufacturing activities that transform those mill products into the finished item. For fabricated products (e.g. fabricated structural steel beams and columns, cut and bent rebar, custom plate girders), A1 is calculated as the industry-average of material needed to produce a unit of fabricated steel, A2 covers transport from the mill or manufacturer to the fabricator, and A3 accounts for fabrication operations themselves. What this means in practice is that depending on the product option chosen in the PCR, the numeric value reported for A1 may already include upstream and mill production impacts that would otherwise be split across traditional A1 and A3 stages in a simple raw-material-only interpretation of A1. Similar variations in system boundaries and module definitions can also occur for other materials, such as concrete and timber products, depending on the applicable standards and supply chain configurations.

2.6.5 Carbon Emission Breakdown Examples for (A1 to A3) Stage (for a Steel Building Frame)

Figure 2.42 shows the carbon emissions associated with the production stages (A1–A3) of a portion of a 10-story steel building with a metal deck floor, measuring 15 × 7.5 meters, and using S275 grade steel for beams and S355 grade steel for columns. The steel used in this building has a recycled content of 13.7% post-consumer. In the calculation of emissions, the Environmental Product Declarations (EPD) for the materials are used (Nordec Oy, 2020; FEDBETON, 2021; Celsa Steel Service A/S, 2021; LLENTAB Steel, 2022). The A1 stage (Raw Material Supply and Processing) has the highest carbon emissions, amounting to 2982.85 kgCO_2e, while the A2 stage (Transport) contributes with 140.27 kgCO_2e, and the A3 stage (Manufacturing) accounts for 233.18 kgCO_2e in emissions. Please note that the A1 values reported in the relevant steel EPDs already include upstream steelmaking impacts (see Section 2.6.4) in this example. These values can be reduced by employing several strategies highlighted in the previous sections. Using steel with higher recycled content, including reclaimed steel in the design to replace some virgin steel production, and eco-friendly concrete formulations can lower the carbon emissions associated with their production (A1–A3 stages). To mitigate emissions during the A2 stage (Transport), optimizing logistics to reduce travel distances, implementing rail or sea transport instead of road transport where feasible, and switching to electric or low-emission vehicles can play a role in lowering related

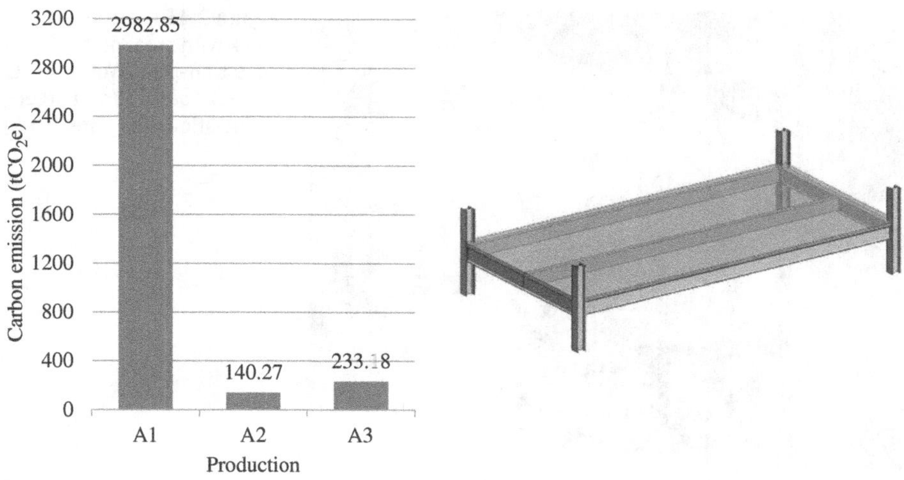

Figure 2.42 A1–A3 comparison for a portion of a 10-story steel building with a metal deck floor.

emissions. For the A3 stage (Manufacturing), enhancing energy efficiency in manufacturing plants by upgrading to more efficient machinery and adopting renewable energy sources can reduce carbon footprints (Figure 2.42).

2.7 Construction Stage Carbon for Building Structures (A4 to A5)

This section covers the carbon emissions produced during the construction life-cycle stages (A4 to A5) for building structures.

2.7.1 Steel Construction Stage Carbon (A4 to A5)

This subsection covers the construction life-cycle stages of steel during transport (A4) and construction installation (A5).

2.7.1.1 Transport (Road, Rail, Sea) of the Steel Product to the Construction Site (A4)

Transporting steel products (e.g. IPE, HEB profiles, tubes, plates) from the final manufacturing plant to the construction site includes all stages of the journey (Figures 2.43–2.45), covering both outbound and return trips. It is important to note that stage A4 differs from A2: while A2 involves the transportation of raw or intermediate steel materials earlier in the supply chain, A4 focuses solely on the final delivery of finished steel products to the construction site. Gibbons et al. (2025) report some average values regarding this step (Table 2.9). Reducing travel distances and optimizing transport modes can significantly contribute to lowering the embodied carbon of steel products. The electrification of railways and the development of hydrogen-powered heavy goods vehicles (HGVs) are promising changes that could significantly reduce carbon emissions from transport in the future. According to EN 15978 (2026), A4 is divided into two sub-categories: A4.1 (transport of construction products and materials) and A4.2 (transport of construction equipment).

Figure 2.43 Loading a truck with steel plates for road transport (World Steel, 2025c). *Source*: World Steel Association (worldsteel).

Figure 2.44 Loading a ship with steel plates for sea transport (World Steel, 2025j). *Source*: World Steel Association (worldsteel).

2.7.1.2 Construction and Installation of the Steel Product (A5)

The A5 phase focuses on the construction and installation activities associated with steel products (Figure 2.46). It covers on-site operations, including the use of electricity and fuel, equipment utilization, and waste management. A5 values can also vary slightly with transportation distance (for instance, longer transport routes might require additional or different packaging, leading to small shifts in A5 values). Module A5 is likely to account for a small, but not insignificant, percentage of structural embodied carbon over the life cycle of a building project; for heavy civil works, A5 may be more significant.

According to EN 15978:2026, the A5 module is divided into four sub-modules:

Pre-construction demolition (A5.1): Where a project requires demolition of an existing built asset, this should be accounted for in Module A5.1. The calculation process follows the same approach as for

Figure 2.45 Loading a train with steel plates for railway transport (World Steel, 2025h). *Source*: World Steel Association (worldsteel).

Table 2.9 Carbon emissions based on distance and means of transport some examples in UK and European context.

Transport scope	Module A4 carbon factor ($kgCO_2e/kg$)	Outbound distance
Local (between 20 and 50 km):	0.009	Road
National (between 80 and 120 km):	0.021	Road
European (<1500 km):	0.262	Road
Global (>1500 km):	0.284	10 000 km by sea and then 500 km by road

Source: Gibbons, O. P., & Orr, J. J./The Institution of Structural Engineers (a more detailed version of this table is available in the source).

end-of-life impact assessment (Modules C1–C4) of the existing asset being demolished. Any recovery of pre-construction demolition waste should be considered in Module D1.

Construction activities (A5.2): This sub-module covers site activities, temporary works, ground works, on-site assembly of products, and installation of formwork (along with their disposal, energy and water use for construction processes). Site energy-related emissions can be estimated from on-site electricity consumption and fuel use and should be monitored during construction to contribute to an accurate as-built embodied carbon calculation. The use of electrical equipment powered by renewable energy and minimizing diesel fuel use can lead to reduced greenhouse gas emissions. This phase can also benefit from off-site manufacturing techniques. Unless significant temporary works are required, their contribution to total structural embodied carbon is likely to be relatively small.

Waste and waste management (A5.3): This sub-module addresses material wastage on site. A significant portion of steel waste, typically in the form of scrap, is recovered. The carbon factor for

(a) Installation of a steel frame
(MaxSafaniuk/Adobe Stock Photos)

(b) Welding on site
(Christopher PB/Shutterstock)

(c) Bolted joints on site
(Vasiliy Ulyanov/Adobe Stock Photos)

(d) Bridge girder construction
(Randy Hergenrether/Shutterstock)

Figure 2.46 Example site activity for steel structures.

material wastage is calculated by multiplying a waste factor by the sum of the carbon factors associated with production (A1–A3), transportation to site (A4), transportation away from site for waste processing (C2), and waste processing or disposal (C3–C4). Average waste percentages for some structural elements are presented in Figure 2.47. Table 2.10 lists typical waste rates and corresponding waste factors for different structural materials. For instance, cast-in-place concrete and steel reinforcement each have a waste rate of 5% (waste factor of 0.053), while steel frames and precast concrete elements have a lower waste rate of 1% (waste factor of 0.010), indicating more efficient on-site material usage. Using off-site manufacturing techniques can help reduce waste, as prefabricated components typically have a maximum waste factor of 0.010.

Transport of construction workers (A5.4): This optional sub-module accounts for emissions from transporting workers to and from site.

Off-site manufacturing and using renewable energy-based electricity can reduce environmental impact, lower greenhouse gas emissions, and optimize material use at the A5 stage.

Details on calculation methodologies, default emission factors, and recommended waste rate data can be found in Gibbons et al. (2025) and EN 15978:2026.

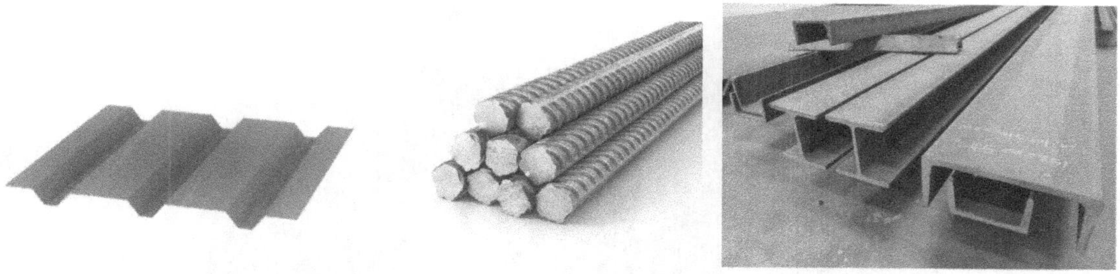

(a) Metal deck: 2% waste (SMD) **(b) Reinforcement steel:** 5–15% waste (urfinguss/Getty Images) **(c) Steel sections:** Near 0% waste (gooutsight/Shutterstock)

Figure 2.47 Average waste percentages for various structural elements.

Table 2.10 Waste factors for typical structural materials (Gibbons et al., 2025).

Material/product	Waste factor (WF)
Concrete in situ	0.053
Concrete precast (beams and frames)	0.010
Steel reinforcement	0.053
Steel frame (beams, columns, braces)	0.010
Timber frames (beams, columns, braces)	0.020
Timber floor (joists, boards)	0.111

2.7.2 Concrete Construction Stage Carbon (A4 to A5)

This subsection covers the construction life-cycle stages of concrete during transport (A4) and construction–installation (A5).

2.7.2.1 Transport of the Concrete Product to the Construction Site (A4)

While A2 addresses the transport of raw materials or intermediates (like aggregates and cement) to the batching plant, A4 specifically deals with delivering the finished or batched concrete to the site. After batching (A3 step), concrete is brought to the construction site by trucks (Figure 2.48). Minimal transport time is desired to maintain the integrity and workability of the concrete (an average transport distance being around 16 kilometers in the UK) (Arup, 2023a). This short distance helps ensure that the concrete remains in optimal condition upon arrival at the site. Therefore, the CO_2 emissions associated with this stage are relatively small compared to the total emissions in the A1–A5 life-cycle stages. However, it is essential to consider that even small contributions to CO_2 emissions accumulate over multiple projects and can have a significant cumulative effect.

Efforts to minimize transport distances and improving the efficiency of transport logistics are vital in reducing the carbon footprint of concrete construction. Also, like other cases mentioned above about the transport, exploring alternative fuel options for transport vehicles, such as electric

Figure 2.48 Concrete mixer bringing batched concrete to the site (XPhantom/Shutterstock.com).

or hydrogen-powered trucks, can further diminish the environmental impact of transporting concrete. According to EN 15978 (2026), A4 is divided into two sub-categories: A4.1 (transport of construction products and materials) and A4.2 (transport of construction equipment).

2.7.2.2 Construction and installation of the Concrete Products (A5)

The A5 phase focuses on the construction and installation activities associated with reinforced concrete (Figure 2.49). It covers on-site operations, including the use of electricity and fuel, equipment utilization, and waste management. Module A5 is likely to account for a small, but not insignificant, percentage of structural embodied carbon over the life cycle of a building project; for heavy civil works, A5 may be more significant.

According to EN 15978:2026, the A5 module is divided into four sub-modules:

Pre-construction demolition (A5.1): Where a project requires demolition of an existing built asset, this should be accounted for in Module A5.1. The calculation process follows the same approach as for end-of-life impact assessment (Modules C1–C4) of the existing asset being demolished. Any recovery of pre-construction demolition waste should be considered in Module D1.

Construction activities (A5.2): This sub-module covers site activities, temporary works, ground works, on-site assembly of products, and installation of formwork (along with their disposal, energy and water use for construction processes). For reinforced concrete, this stage includes concrete placing activities that involve fuel consumption for pumping and vibration. Formwork represents a significant component of construction emissions; it is commonly reused multiple times before being discarded or recycled. Efficient reuse and extending the lifespan of formwork can substantially reduce waste and associated emissions. Site energy-related emissions should be monitored during construction to contribute to an accurate as-built embodied carbon calculation.

Figure 2.49 Reinforced concrete construction (Unkas Photo/Adobe Stock Photos).

(a) Concrete in situ: 5% waste (ETAJOE/Adobe Stock Photos)

(b) Concrete precast (beams and frames): 5% waste (Wrangler/Adobe Stock Photos)

(c) Timber formwork: 10% waste (Oleg/Adobe Stock Photos)

Figure 2.50 Average waste percentages for concrete construction.

Waste and waste management (A5.3): This sub-module addresses material wastage on site. The amount of concrete waste generated on-site depends on the processes employed by the contractor. Managing material orders accurately and improving on-site practices can help minimize this waste. Precast concrete elements are typically manufactured off-site and transported to the construction site, where precise installation practices can reduce the generation of waste. Figure 2.50 presents average waste percentages during concrete construction.

Transport of construction workers (A5.4): This optional sub-module accounts for emissions from transporting workers to and from site.

The calculation method for A5 is provided in Gibbons et al. (2025).

2.7.3 Timber Construction Stage Carbon (A4 to A5)

This subsection covers the construction life-cycle stages of timber during transport (A4) and construction–installation (A5).

2.7.3.1 Transport (Road, Rail, Sea) of the Timber Product to the Construction Site (A4)

Transporting timber products (e.g. logs, beams, boards) from the final manufacturing plant to the construction site includes all stages of the journey, covering both outbound and return trips (Figure 2.51). The average values considered for steel in Table 2.9 can be considered valid also in this case. Reducing travel distances and optimizing transport modes can significantly contribute to lowering the embodied carbon of timber products. The electrification of railways and the development of hydrogen-powered heavy goods vehicles (HGVs) are promising changes that could significantly reduce carbon emissions from transport in the future. According to EN 15978 (2026), A4 is divided into two sub-categories: A4.1 (transport of construction products and materials) and A4.2 (transport of construction equipment).

2.7.3.2 Construction and Installation of Timber Products (A5)

This stage involves assembling prefabricated elements, such as cross-laminated timber (CLT) and glued laminated timber (glulam), on-site (Figure 2.52). Because these elements are typically precut

Figure 2.51 Transport of timber products (duckeesue/Shutterstock).

(a) (b)

Figure 2.52 Timber structures under construction. (a): Roberto Sorin/Adobe Stock Photos; (b): Jarama/Adobe Stock Photos.

and prepared, the need for on-site modifications is greatly reduced, minimizing waste. The prefabrication process also allows for quick and efficient installation by relatively small crews. Additionally, precise manufacturing methods contribute to improved construction efficiency and lower material waste.

According to EN 15978:2026, the A5 module is divided into four sub-modules:

Pre-construction demolition (A5.1): Where a project requires demolition of an existing built asset, this should be accounted for in Module A5.1. The calculation process follows the same approach as for end-of-life impact assessment (Modules C1–C4) of the existing asset being demolished. Any recovery of pre-construction demolition waste should be considered in Module D1.

Construction activities (A5.2): This sub-module covers site activities, temporary works, ground works, on-site assembly of products, and installation (along with their disposal, energy and water use for construction processes). For timber construction, this stage involves the installation of timber products such as CLT and glulam at the construction site. These activities include the use of machinery and equipment that consume fuel, particularly for lifting and placing elements. By optimizing equipment usage and fuel efficiency, emissions associated with these activities can be minimized. The use of renewable energy-powered machinery can further enhance the sustainability of timber construction projects. Unless significant temporary works are required, their contribution to total structural embodied carbon is likely to be relatively small.

Waste and waste management (A5.3): This sub-module addresses material wastage on site. CLT and glulam components generally arrive at the construction site ready for use, resulting in minimal waste. The amount of timber waste generated on-site can be minimized through proper material management and the use of prefabricated elements. By ensuring accurate cutting and assembly in controlled environments, timber waste can be significantly reduced.

Transport of construction workers (A5.4): This optional sub-module accounts for emissions from transporting workers to and from site.

The calculation method for A5 is provided in Gibbons et al. (2025).

2.7.4 Carbon Emission Breakdown Examples for (A1 to A5) Stage (for a Steel Building Frame)

Figure 2.53 shows the carbon emissions associated with the production and construction stages (A1–A5) of a portion of a 10-story steel building with a metal deck floor, measuring 15 × 7.5 meters, and using S275 grade steel for beams and S355 grade steel for columns. The structural steel frame incorporates 13.7% postconsumer recycled content. The calculations reference Environmental Product Declarations (EPDs) for the materials used (Nordec Oy, 2020; FEDBETON, 2021; Celsa Steel Service A/S, 2021,; LLENTAB Steel, 2022). The highest emissions occur during the A1 stage (Raw Material Supply and Processing), totaling 2982.85 kgCO_2e, while A2 (Transport) and A3 (Manufacturing) contribute 140.27 kgCO_2e and 233.18 kgCO_2e, respectively, following a pattern like previous studies. Please note that the A1 values reported in the relevant steel EPDs already include upstream steelmaking impacts (see Section 2.6.4). The graph also highlights emissions from the A4 (Transport to Site) and A5 (Construction Process) stages, which amount to 70.72 kgCO_2e and 75.22 kgCO_2e, respectively, indicating the additional impact of on-site construction activities and material transportation. Supply chain optimization, prefabrication techniques, lightweight yet high-strength materials, and electric or low-emission transport vehicles can be implemented to lower emissions in these phases.

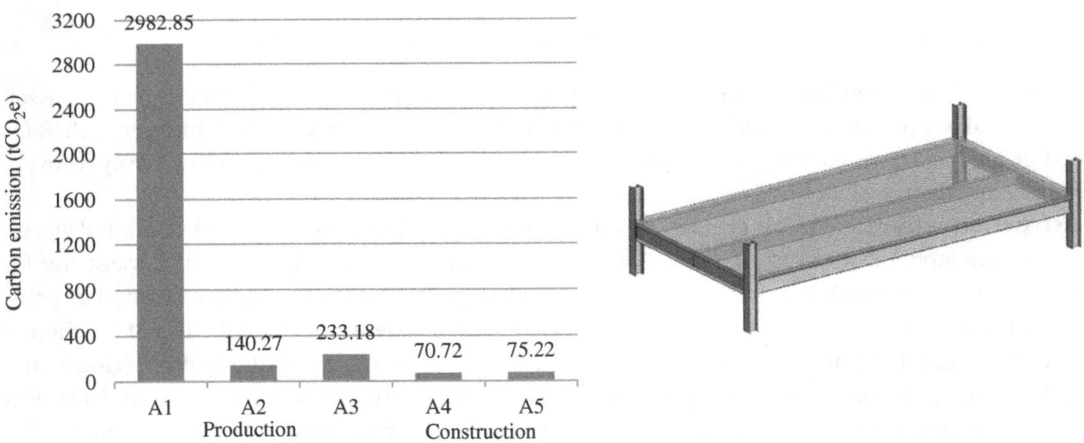

Figure 2.53 A1–A5 comparison for a portion of a 10-story steel building with a metal deck floor.

2.7.4.1 Modules B1–B5

In a building's life cycle, Modules B1–B5 address the use phase, maintenance, repair, replacement, and refurbishment. While B1 (use), B2 (maintenance), and B5 (refurbishment) can contribute to a building's overall carbon footprint, the most direct embodied carbon impacts in this phase typically occur in B3 (repair) and B4 (replacement) because they involve additional materials and processes.

Repair events can include partial or full replacement of damaged components, as well as extra materials (such as sealants or patches). Even though repairs may be smaller in scope than full replacements, they still add to the total embodied carbon over the building's service life.

Replacement events often have a more pronounced impact on embodied carbon than repairs, especially for components with shorter service lives. The number of replacements during the building's Reference Study Period (RSP) depends on each component's estimated lifespan (CL).

Components with shorter service lives will require more frequent replacements, which leads to a higher cumulative impact on embodied carbon. The calculation method for B1–B5 is provided in Gibbons et al. (2025).

2.8 End-of-life Stages

The end-of-life phase in a life-cycle assessment (LCA) includes the final stages of a product's life cycle, involving the demolition of structures, transportation of waste materials to disposal sites, processing, and ultimate disposal. This phase deals with the impacts associated with the end of a product's usability and its transition into waste (e.g. steel is highly recyclable, concrete can be crushed and reused, and timber may be repurposed or used as biomass based on its condition). The steps involved are categorized into four stages: Deconstruction (C1), Waste Transport (C2), Waste Processing (C3), and Waste Disposal (C4).

2.8.1 Deconstruction (C1)

Deconstruction refers to the dismantling or demolition of structures (Figure 2.54a). This process is about managing the waste generated at the end of a product's life. The methods employed can vary significantly, ranging from mechanical demolition using heavy machinery to more manual, labor-intensive approaches aimed at salvaging materials for reuse. The efficiency and environmental impact of deconstruction depend on the techniques used and the extent to which materials can be recovered and reused. The environmental impact of deconstruction is influenced by the energy consumption of machinery, the dust and noise generated, and the potential release of hazardous substances. Deconstruction practices must aim to minimize these impacts by adopting methods that enhance material recovery and reduce emissions. According to RICS (2023), the embodied carbon resulting from demolition and deconstruction in Module C1 can be assumed as an average value of 3.4 $kgCO_2e$ per square meter (to be multiplied by the Gross Internal Area of the building), in the absence of another information from the contractor.

2.8.2 Waste Transport (C2)

Once deconstruction is complete, the next step is the transportation of waste materials to processing or disposal facilities. This stage involves the logistics of moving large volumes of waste, which can include a variety of materials such as metals, concrete, wood, and hazardous substances. The transportation phase contributes to carbon emissions, fuel consumption, and traffic congestion. Optimizing transport routes, utilizing fuel-efficient vehicles, and consolidating loads can mitigate some of these impacts. The distance to disposal sites and the type of transport used (road, rail, sea) are critical factors in determining the overall environmental footprint. Table 2.11 presents values proposed by Gibbons et al. (2025) for the embodied carbon factors of Module C2.

Table 2.11 Default values of the embodied carbon factors of Module C2, some examples in UK and European context.

Scenario	Kilometer by road	$ECF_{C2,i}$ ($kgCO_2e/kg$)
Reuse/recycling on site	0	0.0
Reuse/recycling elsewhere	50	0.009
Landfill/incineration	Average between two closest landfill sites	0.009 (based on 50 km by road)

Source: Gibbons, O. P. and Orr, J. J./The Institution of Structural Engineers (a more detailed version of this table is available in the source).

Figure 2.54 Deconstruction (a) and waste disposal (b and c) (a: author; b, c: worradirek/Adobe Stock Photos).

2.8.3 Waste Processing (C3) and Disposal (C4)

Waste processing involves the treatment of waste materials to prepare them for recycling, reuse, or disposal (Figure 2.54b,c). This can include sorting, shredding, crushing, and other processes that convert waste into usable raw materials or reduce their volume for disposal. Processing activities consume energy and resources, contributing to the overall carbon footprint. An effective waste processing can significantly reduce the volume of waste sent to landfills and recover valuable materials, thus offsetting some of the environmental costs through material recovery and recycling. The final step in the end-of-life phase is waste disposal, where non-recyclable and non-reusable materials are disposed of in landfills, incineration facilities, or other disposal methods. This stage marks the end of the product's life cycle in terms of material utility. Waste disposal can have a substantial environmental impact, including land use for landfills, emissions from waste decomposition, and potential groundwater contamination. Incineration, while reducing waste volume, can release pollutants unless effectively managed. Advanced disposal technologies and practices aim to minimize these impacts through better containment, emission controls, and resource recovery from waste. C3 and C4 carbon values must be reported in EPDs, and when they are not available, Gibbons et al. (2025) suggest an average value of 0.013 $kgCO_2e$ per kilogram of waste.

For timber construction, another way of calculating C3 and C4 factors is available due to the sequestered biogenic carbon in the wood material (locking this sequestered biogenic carbon into a timber structure is beneficial for the climate the longer the carbon is kept within that structure – and so designing structures and components to be reused will keep the biogenic carbon locked away for as long as possible). For example, Table 2.12 shows the UK default values for timber end-of-life scenarios C3 and C4, along with their emission factors (Gibbons et al., 2025).

Table 2.12 UK timber end-of-life scenarios and associated approximations C3 and C4 carbon factors, some examples in UK and European context.

Scenario	UK default emission factors	UK default emission factors
Reuse	For future use in other assets. Biogenic carbon transferred to next use.	1.64 $kgCO_2e/kg$ (equal to default sequestered Carbon)
Recycling	For use in animal bedding/surfacing, panel board products. Biogenic carbon transferred to next use. Fossil carbon due to premanufacturing processing.	1.67 $kgCO_2e/kg$
Incineration for energy recovery	Biomass fuel (domestic or export). Biogenic carbon emitted to atmosphere.	1.65 $kgCO_2e/kg$ (equal to default sequestered Carbon)
Landfill (no gas recovery)	A proportion of biogenic carbon re-released as CO_2 and CH_4, with the remainder transferred to nature.	2.38 $kgCO_2e/kg$

Source: Gibbons, O. P. and Orr, J. J./The Institution of Structural Engineers (a more detailed version of this table is available in the source).

2.8.4 Examples, Life-cycle Stages A to C (for a Steel Building Frame)

Commonly, for steel and concrete, most emissions occur during the production and construction stages, with the end-of-life phase generally having a smaller impact, yet the absolute effect of emissions from deconstruction, waste transport, processing, and disposal can be substantial, contributing significantly to the overall carbon footprint. Timber is a bit different; its end-of-life phase can notably influence the overall emissions due to its potential for carbon sequestration and release. Figures 2.55–2.58 show four scenarios that highlight the impact of varying recycled content and structural components in the

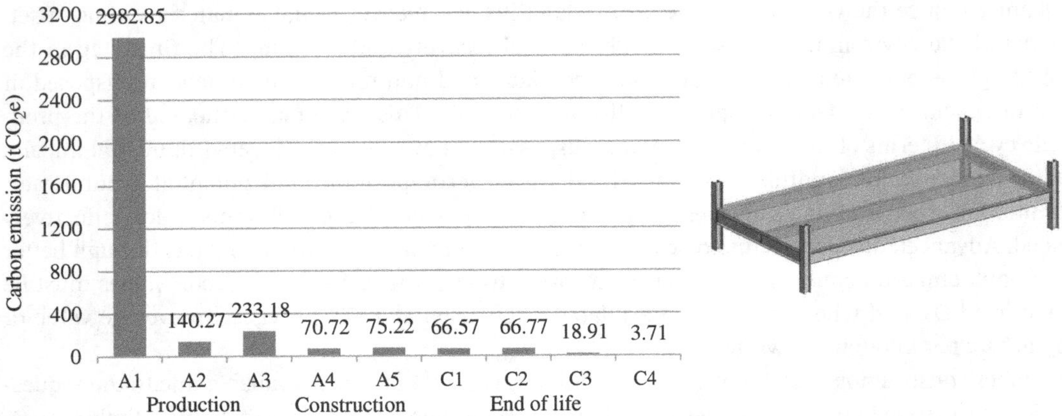

Figure 2.55 A + C comparison for the entire structural steel building (including RC slabs and foundation), steel recycled content of 13.7% postconsumer. The Environmental Product Declarations (EPDs) are used in the analysis (Nordec Oy, 2020; FEDBETON, 2021; Celsa Steel Service A/S, 2021; LLENTAB Steel, 2022; Kanyilmaz et al., 2025).

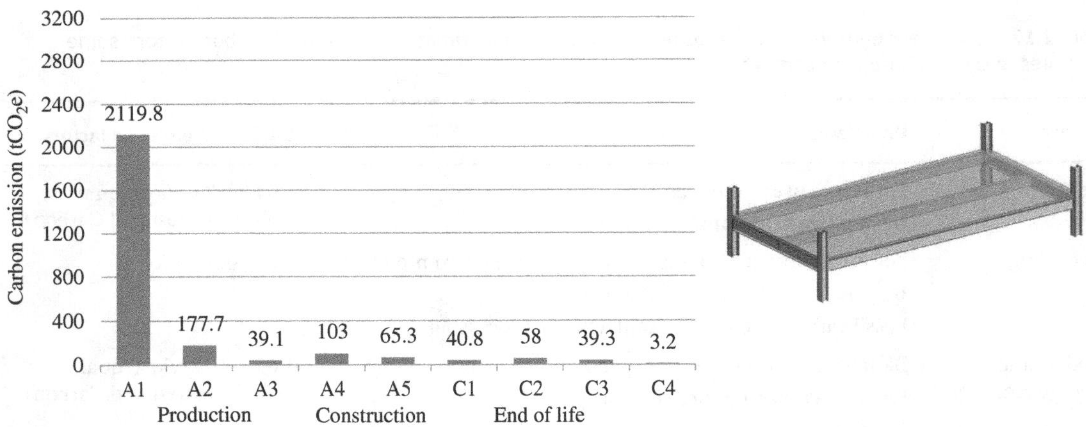

Figure 2.56 A + C comparison for the entire structural steel building (including RC slabs and foundation), steel recycled content of 86.7%. The Environmental Product Declarations (EPDs) are used in the analysis (Nordec Oy, 2020; FEDBETON, 2021; Celsa Steel Service A/S, 2021; LLENTAB Steel, 2022; Kanyilmaz et al., 2025).

A + C comparison. The reference EPDs are Nordec Oy (2020), FEDBETON (2021), Celsa Steel Service A/S (2021), and LLENTAB Steel (2022) emissions. Please note that the A1 values reported in the relevant steel EPDs already include upstream steelmaking impacts (see Section 2.6.4).

Figures 2.55 and 2.56 show the comparison considering both steel frame and the reinforced concrete slab, and the following observations can be made:

Structural Steel Building with 13.7% Recycled Content: The structural steel building consists of reinforced concrete (RC) slabs and foundations, with the steel recycled content at 13.7% postconsumer. The end-of-life stage adds 155.96 t CO_2e to the total emissions, which, although lower than production, still represents a significant impact.

Structural Steel Building with 86.7% Recycled Content: The steel recycled content is significantly higher at 86.7%. This comparison aims to show the reduction in carbon emissions achievable through higher recycling rates. The end-of-life stage sees a reduction to 141.3 t CO_2e, which reflects the decreased need for waste processing and disposal due to the higher recyclable content.

Figures 2.57 and 2.58 show the comparison considering only the steel frame. The total embodied carbon for the structural steel frame is calculated as 2042.24 t CO_2e, with an embodied carbon intensity of 91.68 $kgCO_2e/m^2$ when the recycled content is 13.7%. In contrast, the total embodied carbon for the

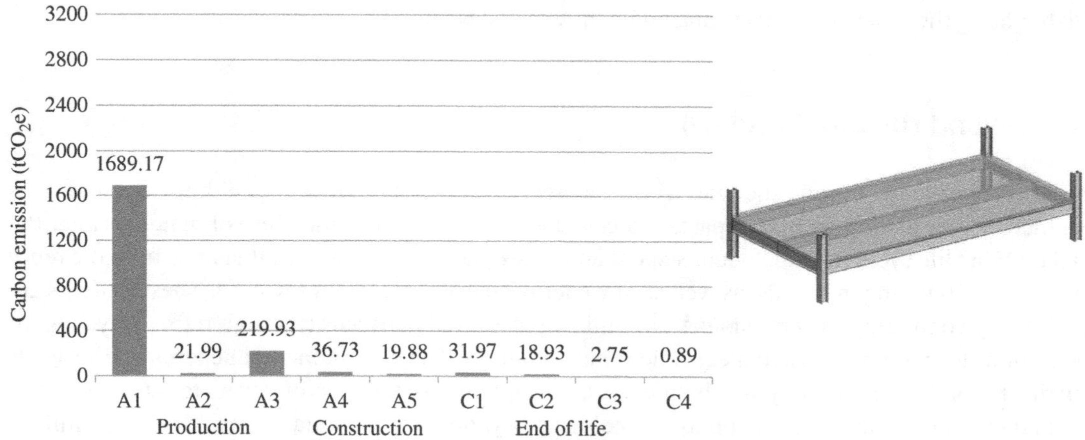

Figure 2.57 Structural steel frame, steel recycled content of 13.7% postconsumer. The calculations in reference to Environmental Product Declarations (EPDs) for structural steel (Kanyilmaz et al., 2025; Nordec Oy, 2020).

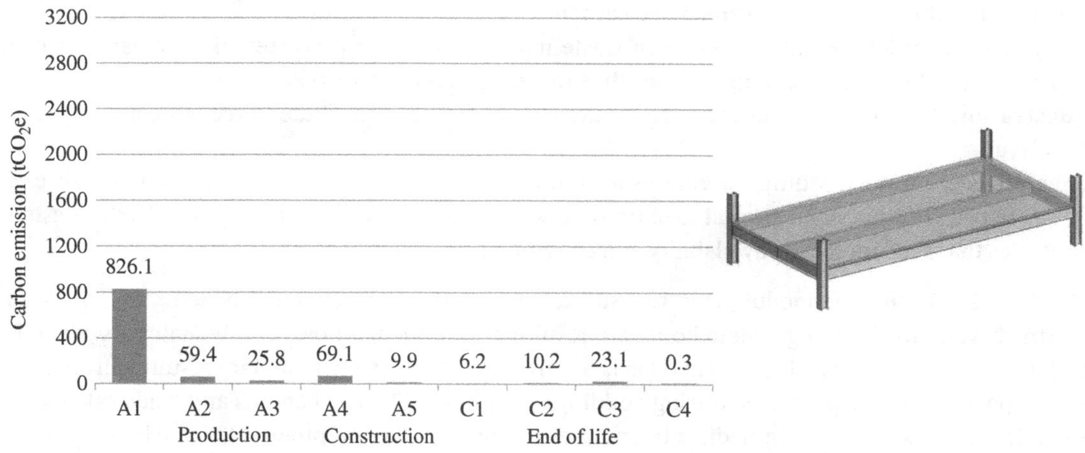

Figure 2.58 Structural steel frame steel recycled content of 86.7% postconsumer. The calculations in reference to Environmental Product Declarations (EPDs) for structural steel (BE Group Sverige AB, 2021; Kanyilmaz et al., 2025).

structural steel frame with 86.7% recycled content is calculated as 1030.18 t CO_2e, with an embodied carbon intensity of 46.25 kgCO_2e/m^2. The following observations can be made:

Structural Steel Frame with 13.7% Recycled Content: This scenario isolates the structural steel frame of the building, maintaining the 13.7% recycled content. The analysis focuses on the impact of removing RC slabs and foundations. The end-of-life emissions decrease to 54.54 t CO_2e, showing that the less material-intensive structure results in lower disposal emissions.

Structural Steel Frame with 86.7% Recycled Content: Finally, this scenario assesses the structural steel frame with the same high recycled content as in scenario 2, providing a comparison of the impact of high recycling rates on a lighter structure. The end-of-life stage emissions are minimal at 39.8 t CO_2e, highlighting the efficiency of recycling and lightweight design.

2.9 Beyond the Life Cycle (D)

Stage D, referred to as "Beyond the Life Cycle," focuses on the post-use phase of building materials. This stage includes the environmental impacts associated with the recycling, and reuse of materials after the initial building life cycle has concluded. Stage D addresses the potential benefits that arise from the reuse or recycling of building materials, as well as any energy recovery processes involved. These benefits are quantified by comparing the emissions from utilizing recovered materials (through reuse, recycling, or incineration) to those of producing equivalent new products. The environmental benefits and/or loads occurring beyond the building system boundary resulting from the net flows of raw materials or products (D1), and environmental benefits and loads occurring beyond system boundary resulting from utilities (e.g. power and heat generated during building operation) (D2). Key points in stage D include:

Reuse and Recycling: Implementing strategies for reusing materials in new construction projects can significantly reduce the carbon emissions associated with the production of new materials (A1–A3 emissions). For instance, reusing steel beams from a demolished building in a new construction can lead to substantial reductions in embodied carbon.

Energy Recovery: Recovering energy from waste materials through processes like incineration can offset the need for primary energy sources, thus reducing overall emissions.

Sequestration: Storing carbon in recovered materials can further contribute to reducing atmospheric CO_2 levels.

Scenario Development: Multiple scenarios for future reuse, recycling, and recovery of materials can be developed within a single project to optimize environmental benefits. This requires flexible design strategies that account for the availability and condition of reusable parts.

EN 15978:2026 splits the module D in two sub-categories: the environmental benefits and/or loads occurring beyond the building system boundary resulting from the net flows of raw materials or products (D1), and environmental benefits and loads occurring beyond system boundary resulting from utilities (e.g. power and heat generated during building operation) (D2). The benefits and loads calculated in stage D extend beyond the immediate boundaries of the construction project. This includes considering future uses and the life-cycle impacts of materials as they enter new systems or are repurposed

(a) Steel from the demolished building

(b) New building with reclaimed steel

Figure 2.59 Photos of (a) steel salvaged from a demolished building. (b) New construction using reclaimed steel (Kanyilmaz et al., 2023/Springer Nature/CC BY 4.0).

(Figure 2.59). These calculations can significantly influence the overall assessment of a project's environmental footprint by incorporating the long-term impacts of material reuse and recycling.

Module D gives a metric of circularity in the building. It represents the estimated benefits or burdens of building materials beyond the end of life of the building considered. The impacts are considered beneficial when the future use of the material, after the end of life, leads to less carbon emission than the business-as-usual use case. This impact is commonly reported separately to all other life-cycle stages (unless otherwise requested) and its calculation is described in detail in Gibbons et al. (2025).

Some typical Module D scenarios for the three most common construction materials are the following:

Steel:

- Steel sections for recycling are cut into shorter lengths and sent to a steel plant.
- The remelting process is energy-intensive and varies significantly depending on the fuel used, resulting in substantial carbon emissions.
- Steel scrap is often exported and processed in different regions, which can increase transportation-related carbon emissions depending on distances and logistics.
- Reusing steel sections avoids the emissions tied to the recycling process, making it a more carbon-efficient option.

Concrete:

- Recycled aggregates involve crushing concrete to reuse as aggregates. This approach is regulated, making sense only if locally sourced to avoid increased transportation emissions. However, it might increase cement and water demand.
- Reusing concrete elements may reduce natural gravel extraction by recycled concrete aggregates but still require equivalent cement quantities, maintaining a comparable level of global warming potential (GWP).
- When maintaining the structure is impossible, another solution is reusing concrete elements with minimal transformation, such as extracting elements from buildings planned for demolition and reassembling them into new structures. This practice is mainly at the research level.

Timber:

- Engineered timber combines timber and adhesives, making it difficult to recycle. It is typically recycled as nonstructural parts when split into smaller sections.
- Timber structures require a robust maintenance strategy to ensure a prolonged service life.
- Reusing timber elements is possible but less common in structural applications compared to materials like steel, due to variability in material properties, connection systems, and certification requirements; however, when feasible, reuse can significantly reduce environmental impacts.

2.10 Conclusion

Life-cycle assessment (LCA) plays a vital role in shaping more resource-conscious structural engineering. By evaluating building materials from raw resource extraction (A1) to beyond life (D), LCA offers clarity on where and how environmental impacts occur. This chapter presented the concepts and strategies of LCA for building structures made of reinforced concrete, steel, and timber at each phase: upfront carbon (A1–A3), construction (A4–A5), in-use (B1–B5), or end-of-life (C1–C4), and beyond life (D).

We introduced several LCA methodologies: Cradle to Gate, Cradle to Site, Cradle to Grave, Cradle to Cradle, Whole Life Carbon Assessment (WLCA), and Dynamic Life Cycle Assessment (DLCA). Cradle to Gate focuses solely on material production; Cradle to Site covers transport and construction; Cradle to Grave continues through end-of-life disposal; Cradle to Cradle broadens the boundary further, incorporating material reuse and recycling. WLCA includes both embodied and operational carbon, while DLCA accounts for emission timing and emphasizes near-term environmental damage.

We discussed how each construction material category (concrete, steel, and timber) contributes distinct environmental burdens and advantages. Moving forward, data transparency, standardization, and innovation in materials and methods will help further integrate LCA into mainstream structural practice. Through these steps, LCA aligns design decisions with sustainability targets, ensuring that resource conservation and emissions reduction remain integral elements of modern engineering.

2.11 Questions

Answer the following questions to enhance your understanding of the concepts covered in this chapter.

Short Answer Questions

1 What is life-cycle assessment (LCA) and how does it contribute to sustainable design?

2 Which LCA stages correspond to raw material extraction, manufacturing, and construction, and what do they represent?

3 How do the “Cradle to Grave” and “Cradle to Cradle” approaches differ in their assessment boundaries?

4 What is the significance of "system boundaries" in an LCA study?

5 What factors contribute most to the upfront carbon emissions (A1–A3) in steel, concrete, and timber production?

6 In what ways do Supplementary Cementitious Materials (SCMs) help reduce the carbon footprint of concrete?

7 How does recycling influence the end-of-life (C) stage emissions for materials like steel and timber?

8 What is the role of biogenic carbon in timber's LCA, particularly at the end-of-life stage?

9 How does the Beyond-Life Stage (D) enhance the overall carbon assessment of a building's life cycle?

10 Why is transportation efficiency critical in determining the emissions associated with the A2 stage?

True or False

1 Life-cycle assessment (LCA) evaluates the environmental impacts of a building across all stages, from raw material extraction to disposal.

2 The "Cradle to Gate" approach in LCA only considers the production phase and stops at the factory gate.

3 Upfront carbon emissions (A1–A3) are mainly associated with raw material extraction, processing, and manufacturing.

4 The end-of-life phase (C1–C4) in LCA includes demolition, waste transport, processing, and disposal.

5 Recycled aggregates typically increase the water demand of a concrete mix due to the fine materials generated during crushing.

6 Timber's LCA uniquely factors in biogenic carbon because it can sequester carbon during its service life.

7 The "Cradle to Cradle" approach extends the LCA boundary to include recycling and reuse in the beyond-life stage (D).

8 Dynamic life cycle assessment (DLCA) assigns a heavier weight to present emissions than to those in the future.

9 LCA methodologies are standardized under the ISO 14000 series.

10 Transportation emissions in the A2 stage can vary significantly based on distance and the mode of transport used.

11 Increasing the use of recycled steel in production generally lowers the overall carbon footprint of the steel product.

12 For steel and concrete, the production phase (A1–A3) typically contributes the highest proportion of carbon emissions.

13 LCA only focuses on carbon emissions and does not consider other environmental impacts such as resource depletion or water usage.

14 Life-cycle assessment also assesses energy demand, acidification, resource depletion, and other environmental impacts beyond carbon.

15 LC3 cement is an alternative binder that can reduce the clinker content by approximately half compared to traditional Portland cement.

16 The carbon sequestration properties of timber are factored into its LCA, affecting both production and end-of-life assessments.

17 Environmental Product Declarations (EPDs) provide data that help quantify the carbon emissions associated with different life-cycle stages.

18 Waste heat recovery systems in cement plants can reduce the overall energy consumption and associated emissions.

19 The structural carbon rating scheme (SCORS) considers emissions from both the production and construction phases of a building.

20 The LCA process is divided into multiple modules: preconstruction (A0), production (A1–A3), construction (A4–A5), use (B), end-of-life (C), and beyond-life (D).

Discussion Questions

1 How can life-cycle assessment (LCA) help policymakers and construction professionals make more informed decisions about building materials?

2 What role does balancing renewable and nonrenewable resource use play in creating a more sustainable construction industry?

3 How do Cradle to Gate, Cradle to Site, Cradle to Grave, Cradle to Cradle, and dynamic life cycle assessment (DLCA) differ in their treatment of a building's life-cycle phases?

4 What are the primary contributors to carbon emissions in the production of concrete, steel, and timber, and how do these differ among the three materials?

5 Why can the choice of transport method (road, rail, sea) significantly influence overall embodied carbon, and which key strategies help reduce emissions in these stages (A2, A4)?

6 Explain the roles of C1 (deconstruction), C2 (waste transport), C3 (waste processing), and C4 (disposal). Which strategies can lower emissions in each of these phases?

7 How does biogenic carbon storage affect the life-cycle impacts of timber, and what happens to that stored carbon during the end-of-life stage?

8 Identify three strategies for lowering A1–A3 emissions when using concrete or steel. Which of these do you think is most effective in typical building projects?

9 Describe how Module D can generate positive or negative emissions results. What design factors increase the likelihood of beneficial reuse or recycling?

10 During A5, what specific on-site activities generate notable carbon impacts, and how might off-site manufacturing reduce these?

References

American Institute of Steel Construction. (n.d.). *Product category rule*. https://www.aisc.org/aisc/sustainability/environmental-product-declarations/

ArcelorMittal. (n.d.). Hydrogen-based steelmaking to begin in Hamburg. Retrieved May 13, 2026, from https://corporate.arcelormittal.com/media/cases-studies/hydrogenbased-steelmaking-to-begin-in-hamburg

Arup. (2023a). Buildings & infrastructure priority actions for sustainability—embodied carbon: concrete (07762000-RP-SUS-0003). Retrieved from https://www.istructe.org/IStructE/media/Public/Resources/ARUP-Embodied-carbon-concrete_1.pdf

Arup. (2023b). Buildings & infrastructure priority actions for sustainability—embodied carbon: steel (07762000-RP-SUS-0001). Retrieved from https://www.istructe.org/IStructE/media/Public/Resources/ARUP-Embodied-carbon-steel_1.pdf

Arup. (2023c). Buildings & infrastructure priority actions for sustainability—embodied carbon: timber (07762000-RP-SUS-0002). Retrieved from https://www.shareyourgreendesign.com/research/buildings-infrastructure-priority-actions-for-sustainability-embodied-carbon-timber/

ASTM International. (2022). *ASTM E2921-22: Standard Practice for Minimum Criteria for Comparing Whole Building Life Cycle Assessments for Use in Green Building Rating Systems*. ASTM International. https://doi.org/10.1520/E2921-22

BE Group Sverige AB. (2021). "Environmental product declaration: steel beams." *The International EPD® System*. Available from https://api.environdec.com/api/v1/EPDLibrary/Files/6574f4e7-2e7c-4f70-39f2-08d99c9745fc/Data

Celsa Steel Service A/S. (2021). "Environmental product declaration: steel reinforcement products for concrete – Danish production." *The International EPD® System*. Available from https://api.environdec.com/api/v1/EPDLibrary/Files/b8f8bd47-8c39-480b-3a09-08d98fadb225/Data

Chehade, M. (2024). *Embodied Carbon Optioneering for Building Structures [MSc Thesis, supervised by Alper Kanyilmaz]*. Politecnico di Milano.

Circular Ecology. (2020). The inventory of carbon and energy (The ICE Database) [Dataset]. Retrieved from https://carbon.tips/ice3

EN 15643. (2021). EN 15643: 2021. Sustainability of construction works. Framework for assessment of buildings and civil engineering works. Available from https://knowledge.bsigroup.com/products/sustainability-of-construction-works-framework-for-assessment-of-buildings-and-civil-engineering-works

EN 15804:2012+A2:2019. Sustainability of construction works—Environmental product declarations—Core rules for the product category of construction products. Brussels, Belgium. Retrieved from https://standards.iteh.ai/

EN 15978. (2026). Sustainability of construction works – Assessment of environmental performance of buildings – Requirements and guidance, https://standards.cencenelec.eu/ords/f?p=CEN:110:::::FSP_PROJECT,FSP_ORG_ID:78444,481830&cs=12DCD4CE54ABF529EBAD126251E81609B

EN 17472. (2022). EN 17472: 2022. Sustainability of construction works. Sustainability assessment of civil engineering works. Calculation methods. Available from https://www.sis.se/en/produkter/construction-materials-and-building/buildings/general/ss-en-174722022/

EN 197-1. (2011). EN 197-1:2011—Cement—Part 1: Composition, specifications, and conformity criteria for common cements. Brussels, Belgium. Retrieved from https://standards.iteh.ai/catalog/standards/cen/64d327b1-d5ac-45e3-8b04-fafec9e0698e/en-197-1-2011

EN 1992-1-1. (2023). EN 1992-1-1:2023. Eurocode 2—Design of concrete structures. Available from https://knowledge.bsigroup.com/products/eurocode-2-design-of-concrete-structures-general-rules-and-rules-for-buildings-bridges-and-civil-engineering-structures/standard

Energiron. (2025). *Energiron* [Graphic]. Retrieved from https://www.energiron.com/solutions

Engman, M. (2025). *Lastbil med timmer* [Graphic]. Retrieved from https://skogsindustrierna.qbank.se/alla/media/145714

Ergodomus. (2025). *SMU - Singapore Management University Extension—Singapore, Singapore | Projects*. Ergodomus. Retrieved from https://www.ergodomus.it/projects/smu-singapore-management-university/

European Commission's Joint Research Centre (JRC). (2024, February 7). "Emerging bio-based products have nearly half the GHG footprint of fossil-based counterparts." *European Commission*. Available from https://joint-research-centre.ec.europa.eu/jrc-news-and-updates/emerging-bio-based-products-have-nearly-half-ghg-footprint-fossil-based-counterparts-2024-02-07_en

Fantilli, A. P., Mancinelli, O., & Chiaia, B. (2019). The carbon footprint of normal and high-strength concrete used in low-rise and high-rise buildings. *Case Studies in Construction Materials, 11*, e00296. https://doi.org/10.1016/j.cscm.2019.e00296

FEDBETON. (2021). Environmental product declaration: Typical Belgian ready-mixed concrete (B-EPD n° 21-0069-002-00-01-EN). Belgian: Federal Public Service of Health, Food Chain Safety and Environment. Available from https://www.google.com/search?q=https://www.federbeton.it/pubblicazioni/rapporti-di-sostenibilita

Geneva Environment Network. (2024, March 21). Importance of forests and the role of Geneva. Retrieved from https://www.genevaenvironmentnetwork.org/fr/ressources/nouvelles/importance-of-forests-and-the-role-of-geneva/

Gibbons, O. P., & Orr, J. J. (2022). *How to Calculate Embodied Carbon* (2nd edition). The Institution of Structural Engineers.

Gibbons, O., Orr, J., & Arnold, W. (2025). *How to Calculate Embodied Carbon* (Third Edition). The Institution of Structural Engineers. https://www.istructe.org/resources/guidance/how-to-calculate-embodied-carbon/.

Grünig, M., Rammer, W., Senf, C., et al. (2026). Climate change will increase forest disturbances in Europe throughout the 21st century. *Science*. https://doi.org/10.1126/science.adx6329

Gupta, S. K., & Kaul, S. K. (2011). Waste heat recovery power plants in cement industry. Retrieved from https://holtecnet.com/holtecdocs/TechnicalPapers/p_2011_6.pdf

HBI S.p.A. (n.d.). Group news. Retrieved May 13, 2026, from https://www.hbisco.com/en/news/group/t688/2699

Hewlett, P. C., & Liska, M. (2019). *Lea's Chemistry of Cement and Concrete*. Butterworth-Heinemann.

HYBRIT Development. (n.d.). HYBRIT. Retrieved May 13, 2026, from https://www.hybritdevelopment.se/en/

HYDRA Project IPCEI. (n.d.). HYDRA project. Retrieved May 13, 2026, from https://www.hydra-project-ipcei.eu

Hydrogen Europe. (2022). "Steel from solar energy: A techno-economic assessment of green steel manufacturing." *Hydrogen Europe*. Available from https://hydrogeneurope.eu/wp-content/uploads/2022/06/Steel_from_Solar_Energy_Report_05-2022_DIGITAL.pdf

ICE. (2022). Low carbon concrete routemap. Retrieved from https://www.ice.org.uk/media/q12jkljj/low-carbon-concrete-routemap.pdf

IEA. (2020). "Iron and steel technology roadmap." *International Energy Agency*. Available from https://www.iea.org/reports/iron-and-steel-technology-roadmap

IEA. (2023). "Steel – Breakthrough Agenda Report 2023 – Analysis." *IEA*. Available from https://www.iea.org/reports/breakthrough-agenda-report-2023/steel

International Energy Agency, "Iron and Steel Technology Roadmap," IEA, Paris, 2020. [Online]. Available: https://www.iea.org/reports/iron-and-steel-technology-roadmap

International Organization for Standardization. (2022). *Sustainability in Buildings and Civil Engineering Works – Framework for Methods of Assessment of the Environmental, Social and Economic Performance of Construction Works as a Basis for sustainability assessment – Part 1: Buildings*. ISO Standard No. 21931-1:2022.

ISO 14000. (2025). ISO 14000—Environmental management standards. Retrieved from https://www.iso.org/standards/popular/iso-14000-family

ISO 14044:2006. (2006). https://www.iso.org/standard/38498.html

Kanyilmaz, A., Birhane, M., Fishwick, R., & del Castillo, C. (2023). Reuse of steel in the construction industry: Challenges and opportunities. *International Journal of Steel Structures*, *23*(5), 1399–1416. https://doi.org/10.1007/s13296-023-00778-4

Kazama, V. S., Corte, A. P. D., Robert, R. C. G., et al. (2021). Global review on forest road optimization planning: Support for sustainable forest management in Amazonia. *Forest Ecology and Management*, *492*, 119159. https://doi.org/10.1016/j.foreco.2021.119159

Kibert, C. J. (2022). *Sustainable Construction: Green Building Design and Delivery* (5th edition). Wiley.

LC3. (2025). LC3—Limestone calcined clay cement. Retrieved from https://lc3.ch/
LLENTAB Steel. (2022). Environmental product declaration (EPD) for LL high profile sheet (EPD HUB, HUB-0056). Retrieved from https://www.llentabsteel.se/wp-content/uploads/EPD-Hogprofil-1.pdf
Mason, T. O., & Lea, F. M. (2026). "Cement." *Encyclopedia Britannica*. Available from https://www.britannica.com/technology/cement-building-material
Mission Possible Partnership. (2021). "Net-zero steel sector transition strategy." *Mission Possible Partnership*. Retrieved from https://www.energy-transitions.org/wp-content/uploads/2021/12/MPP-Steel_Transition-Strategy.pdf
Neville, A. M. (2011). *Properties of Concrete* (5th edition). Pearson: Harlow Munich.
Nordec Oy. (2020). "Environmental product declaration for welded and painted structural steel products—Based on European steel material (RTS-67-20; p. 14)." *The Building Information Foundation RTS*. Available from https://nordec.com/wp-content/uploads/2022/02/RTS-EPD_68-20_NordecOy_Welded-and-painted-structural-steel-products_SSAB-steel-material.pdf
Piva, F. (2025). *Interview with Franco Piva from Ergodomus* [Personal communication].
Repo, A., Eyvindson, K., Halme, P., & Mönkkönen, M. (2020). Forest bioenergy harvesting changes carbon balance and risks biodiversity in boreal forest landscapes. *Canadian Journal of Forest Research*, *50*(11), 1184–1193. https://doi.org/10.1139/cjfr-2019-0284
RICS. (2023). *Whole Life Carbon Assessment for the Built Environment* (2nd edition). Royal Institution of Chartered Surveyors (RICS).
Rocky Mountain Institute. (2019). *Pulling the Weight of Heavy Truck Decarbonization*. Rocky Mountain Institute. https://rmi.org/wp-content/uploads/2019/06/rmi-pulling-the-weight-of-heavy-truck-decarbonization.pdf
Santarossa, S. (2018). "A first for Turkey." *World Cement*. Available from https://www.turboden.com/upload/blocchi/X11799allegato1-2X_46935_WORLD_CEMENT-articolo-Cimko.pdf
Scrivener, K., John, V. M., & Gartner, E. (2018). Eco-efficient cements: Potential economically viable solutions for a low-CO_2 cement-based materials industry. *Cement and Concrete Research*, *114*, 2–26. https://doi.org/10.1016/j.cemconres.2018.03.015
Swedish Wood. (2022). *Moisture-Proof CLT Construction without a Full Temporary Shelter*. Swedish Wood. https://www.swedishwood.com/siteassets/5-publikationer/pdfer/moisture-proof-clt-construction-without-a-full-temporary-shelter.pdf
Swedish Wood. (2025a). *From Raw Material to Wood Product*. Swedish Wood. https://www.swedishwood.com/wood-facts/about-wood/from-log-to-plank/
Swedish Wood. (2025b). *Swedish Wood* [Graphic]. Swedish Wood. https://www.swedishwood.com/optimized/default/siteassets/2-bygg-med-tra/1-byggande/avt-sid99.jpg
The Concrete Center. (2020). *Specifying Sustainable Concrete*. The Concrete Center. https://www.concretecentre.com/Resources/Publications/Specifying-Sustainable-Concrete.aspx
Trade Map. (2023a). Trade Map—Concentration of exporting countries and average distance with their destination countries | for the selected product in 2023 | Product: 72 Iron and steel. Retrieved from https://www.trademap.org/Country_SelProduct_Graph.aspx?nvpm=1%7c%7c%7c%7c%7c72%7c%7c%7c2%7c1%7c1%7c2%7c1%7c1%7c2%7c1%7c1%7c2
Trade Map. (2023b). Trade Map—Concentration of importing countries and average distance with their supplying countries | for the selected product in 2023 | Product: 72 Iron and steel. Retrieved from https://

www.trademap.org/Country_SelProduct_Graph.aspx?nvpm=1%7c%7c%7c%7c%7c72%7c%7c%7c2%7c1%7c1%7c1%7c1%7c1%7c2%7c1%7c1%7c1%7c2

Turbines, T. (2024). *Leading the Way in Waste Heat Recovery in Cement Industry*. Triveni Turbine Ltd. https://www.triveniturbines.com/waste-heat-recovery-in-cement-industry/

UK Steel. (2025). *Electric Arc Furnaces in the UK*. UK Steel. https://www.uksteel.org/electric-arc-furnaces-1

World Economic Forum. (2023). *4 Breakthrough Actions to Decarbonize Steelmaking*. World Economic Forum. https://www.weforum.org/stories/2023/11/4-breakthrough-actions-to-decarbonize-steelmaking/

World Economic Forum. (2024). *Net-Zero Industry Tracker 2024 Edition*. World Economic Forum. https://www.weforum.org/publications/net-zero-industry-tracker-2024/

World Steel. (2025a). "Coating steel sheet with paint" [Graphic]. *Worldsteel.Org*. Available from https://worldsteel.org/media/images/worldsteelterniummonterrey1516jpg-622/

World Steel. (2025b). "Galvanising steel sheet in the pickling line." *Worldsteel.Org*. Available from https://worldsteel.org/media/images/worldsteelterniummonterrey1403jpg-603/

World Steel. (2025c). "Moving plates by crane." *Worldsteel.Org*. Available from https://worldsteel.org/media/images/worldsteelposcosouthkoreahotrolling51mp4-355/

World Steel. (2025d). "Raw materials." *Worldsteel.Org*. Available from https://worldsteel.org/other-topics/raw-materials/

World Steel (Director). (2025e). "Scrap metal and iron ore ready to be sorted and turned into new steel in the electric arc furnace" [Video recording]. *Worldsteel.Org*. Available from https://worldsteel.org/media/images/worldsteelnucornorthamericaironmaking7mp4-417/

World Steel. (2025f). "Stacks of steel beams in the shipping yard." *Worldsteel.Org*. Available from https://worldsteel.org/media/images/worldsteelnucornorthamericafinalproduct6jpg-361/

World Steel. (2025g). "Steelmaking." *Worldsteel.Org*. Available from https://worldsteel.org/about-steel/steelmaking-process/

World Steel. (2025h). "Transporting stacks of slab by train." *Worldsteel.Org*. Available from https://worldsteel.org/media/images/worldsteelusiminasbrazilfacility-exterior4jpg-1006/

World Steel. (2025i). "Unloading steel scrap to be recycled in an electric arc furnace" [Graphic]. *Worldsteel.Org*. Available from https://worldsteel.org/media/images/worldsteelbswgermanyscrap1jpg-554/

World Steel. (2025j). "Views of loading slabs onto ship for export at the facility port." *Worldsteel.Org*. Available from https://worldsteel.org/media/images/ternium-clip-3mp4-1030/

World Steel. (2025k). "What are the input materials to make steel in an EAF." *Worldsteel.Org*. Available from https://worldsteel.org/about-steel/facts/steelfacts/what-is-steel/what-are-the-input-materials-to-make-steel-in-an-eaf/

World Steel Association. (2021). *Scrap use in the steel industry* [Fact sheet]. https://www.worldsteel.org

World Steel Association, "World Steel in Figures 2025," Brussels, Belgium, 2025. [Online]. Available: https://worldsteel.org/data/world-steel-in-figures/world-steel-in-figures-2025/

3

Measuring and Reducing Embodied Carbon in Structures

LIST OF CONTENTS

With the descriptions of embodied carbon and life-cycle assessment provided in the previous chapters, this chapter will describe how to measure and reduce embodied carbon in structural systems.

The chapter begins with the fundamental calculation equation, exploring how Bill of Quantities, carbon factors, and Environmental Product Declarations (EPDs) form the backbone of accurate assessment. We then examine the tools, databases, and benchmarking approaches that enable practitioners to quantify embodied carbon with confidence and compare performance across projects. Then, we will explore ten key strategies for reducing embodied carbon through design decisions, material selection, and construction approaches. A comprehensive case study demonstrates these concepts. Walking through the complete embodied carbon assessment of a multi-story building.

To measure the Global Warming Potential (GWP) of structures, we first need to answer two critical questions:

- How much material is involved in the structural components?
- How much embodied carbon is stored in each structural component?

These two questions are brought together in the fundamental equation for measuring embodied carbon, which provides the simple mathematical framework for our analysis.

3.1 Embodied Carbon Equation

The embodied carbon is measured by multiplying the quantity of each material involved in a product (building's structural system for the scope of this book) by its carbon factor (normally measured in $kgCO_2e$ per kilogram of material) as shown in Equation (3.1).

$$\text{Embodied Carbon} = \text{Material Quantity} \times \text{Carbon Factor} \tag{3.1}$$

Here, material quantity is the mass of a material (kg) taken from a Bill of Quantities (BoQ), and the carbon factor represents the environmental impact of the material and process ($kgCO_2e/kg$) found in an Environmental Product Declaration (EPD). Both of these critical inputs (how to calculate quantities and where to find the appropriate carbon factors) are the primary focus of the detailed discussions that follow.

Table 3.1 BoQ example for a single-story residential building, simplified.

Material	Quantity	Unit	Description
Concrete	150	m^3	Concrete mix for foundation and structural elements
Steel	20	tons	Reinforcement bars
Timber	50	m^3	Timber beams
Insulation	200	m^2	Thermal insulation for walls and roof
Glass	100	m^2	Double-glazed windows

3.1.1 Bill of Quantities

Imagine you are baking a cake. To know the cake's total calorie content, you first need to list out all the ingredients and their respective amounts. Similarly, the Bill of Quantities (BoQ) details every construction material, from steel beams to concrete slabs, giving us a precise inventory to work with. This detailed breakdown allows us to quantify the amount of each material involved. Just as a BoQ is essential for calculating the total cost of a construction project, it is also necessary for determining the overall embodied carbon. An example of BoQ is shown in Table 3.1.

3.1.2 Carbon Factor (Embodied Carbon "Equivalent")

Carbon factor is defined as the amount of CO_2e emitted per kilogram of material over its life cycle ($kgCO_2e/kg$).

Since the carbon factor accounts for all greenhouse gases, not just CO_2 (as explained in Chapter 2), we determine a total CO_2e by summing the contributions of each gas, weighted by its Global Warming Potential (GWP):

$$CO_2e = CO_2 + GWP_{CH_4} \times CH_4 + GWP_{N_2O} \times N_2O + \ldots$$

This "CO_2 equivalent" metric allows us to capture the material's total climate impact by considering not only carbon dioxide but also other significant greenhouse gases. Therefore, in this book (and in general in documents related to the embodied carbon calculation), almost always you will see and use this metric.

Once we know the quantities of materials, the next step is to determine how much carbon each material stores.

Example, Steel Beam

Let us calculate the embodied carbon (A1–A5 stages) of an IPE 270 beam of 7 m length (example of an IPE steel beam is shown in Figure 3.1).

The dimensions and properties for the IPE 300 beam are typically provided in standard tables available from steel manufacturers:

Height (h):	270 mm
Flange width (b):	135 mm
Web thickness (t_w):	6.6 mm
Flange thickness (t_f):	10.2 mm
Average mass:	36.07 kg/m (this value is obtained from the standard tables)

Calculate the Total Mass:
We have a beam that is 7 m long. The total mass of the IPE 300 beam can be calculated as:

Mass per meter (from standard table):	36.07 kg/m
Length of the beam:	7 m
Total mass:	36.07 kg/m × 7 m
	= 252.49 kg

Obtain the Carbon Footprint for Steel:
In this example, let us assume an average carbon footprint for steel of 1.2 $kgCO_2e/kg$ of steel for A1–A5 LC steps (a generic value based on typical ranges found in the literature). As an alternative, EPDs can also be used to obtain a project-specific value.

Calculate the Embodied Carbon:
Using the carbon factor and the total mass, the embodied carbon for the IPE 300 beam can be calculated as follows:

Total mass of the IPE 300 beam:	253 kg
Carbon factor for steel:	1.2 $kgCO_2e/kg$ (representative average value adopted for this example)
Embodied carbon calculation	= 253 kg × 1.2 $kgCO_2e/kg$
	= 303.6 $kgCO_2e$

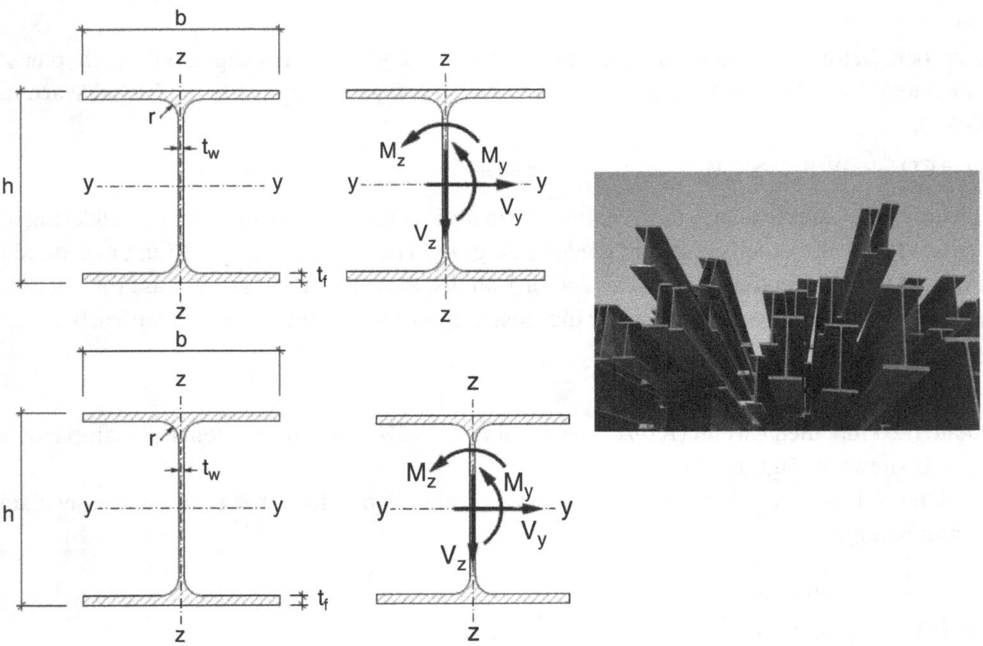

(a) (b)

Figure 3.1 I beam example. (a) Geometry of a rolled I section with main dimensions and reference axes (Eurocode Applied, 2025); (b) photograph of stacked steel I sections.

Example, Timber Column

Let us calculate the A1–A5 embodied carbon of a timber column with the following characteristics: (example of a timber column is shown in Figure 3.2).

Height (h):	3 m
Cross-sectional area (A):	$0.4\ m \times 0.4\ m = 0.16\ m^2$
Density (ρ):	$500\ kg/m^3$ (this value is obtained from the standard properties of the timber used)

Calculate the Total Mass:

Suppose we have a timber column with the given dimensions. The total mass of the column can be calculated as:

Volume of the column	= Height × Cross-sectional area $= 3\ m \times 0.16\ m^2 = 0.48\ m^3$
Density of timber:	$500\ kg/m^3$
Total mass	= Volume × Density $= 0.48\ m^3 \times 500\ kg/m^3 = 240\ kg$

Obtain the Carbon Footprint for Timber:

In this example, let us consider an average carbon footprint for timber of 0.4 $kgCO_2e/kg$ of timber for A1–A5 LC steps (GWP fossil). As an alternative, EPDs can also be used to obtain a project-specific value.

Calculate the Embodied Carbon:

Using the carbon factor and the total mass, the embodied carbon for the timber column can be calculated as follows:

Total mass of the timber column:	240 kg
Carbon factor for timber:	0.4 $kgCO_2e/kg$
Embodied carbon calculation:	240 kg × 0.4 $kgCO_2e/kg$ = 96 $kgCO_2e$

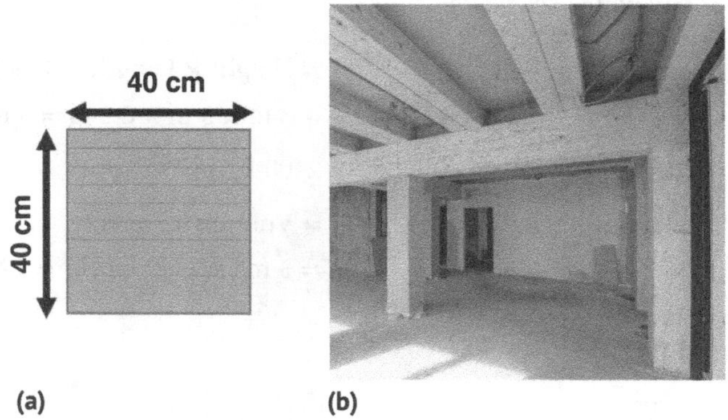

Figure 3.2 Timber column example. (a) Timber cross-section; (b) timber columns and beams (Swedish Wood, 2019/The Swedish Forest Industries Federation). Jarama/Adobe Stock Photos.

Example, Reinforced Concrete Wall

Let us calculate the A1–A5 embodied carbon of a reinforced concrete wall with the following characteristics: (example to a reinforced concrete wall is shown in Figure 3.3)

Height (h):	3 m
Length (L):	5 m
Thickness (t):	0.2 m
Density (ρ):	2400 kg/m^3

Figure 3.3 Reinforced concrete wall example (Unkas Photo/Adobe Stock Photos).

Calculate the Total Mass:

Suppose we have a reinforced concrete wall with the given dimensions. The total mass of the wall can be calculated as:

Volume of the wall	= Height × Length × Thickness
	$= 3\ \text{m} \times 5\ \text{m} \times 0.2\ \text{m} = 3\ \text{m}^3$
Density of reinforced concrete:	2400 kg/m^3
Total mass	= Volume × Density
	$= 3\ \text{m}^3 \times 2400\ \text{kg/m}^3 = 7200\ \text{kg}$

Obtain the Carbon Footprint for Concrete and Reinforcement Steel:
For the reinforced concrete wall, we need to consider both concrete and reinforcing steel, as each material has a distinct carbon footprint and must be accounted for separately. Let's assume the amount of reinforcing steel is 65 kg/m^3 of concrete (a typical reinforcement ratio for structural walls; Cobb, 2017).

Volume of concrete: 3 m^3

Amount of steel: 3 m^3 × 65 kg/m^3 = 195 kg

In this example, let us consider the following average carbon footprints:

Concrete: 0.13 kgCO_2e/kg for A1–A5 LC steps.

Reinforcing steel: 0.8 kgCO_2e/kg for A1–A5 LC steps.

Calculate the Embodied Carbon:
Using the carbon factors and the total mass, the embodied carbon for the reinforced concrete wall can be calculated as follows:

Total mass of the reinforced concrete: 7200 kg

Carbon factor for reinforced concrete: 0.13 kgCO_2e/kg

Embodied carbon for concrete: 7200 kg × 0.13 kgCO_2e/kg
= 936 kgCO_2e

Total mass of reinforcing steel: 195 kg

Carbon factor for reinforcing steel: 0.8 kgCO_2e/kg

Embodied carbon for steel: 195 kg × 0.8 kgCO_2e/kg
= 156 kgCO_2e

Total embodied carbon: 936 kgCO_2e (concrete)
+156 kgCO_2e (steel) = 1092 kgCO_2e.

In the remainder of this chapter, we will scale up this procedure from individual elements to entire buildings, demonstrating how to aggregate the embodied carbon contributions of all structural components into a comprehensive whole-building assessment.

3.2 Environmental Product Declaration (EPD)

An Environmental Product Declaration (EPD) is an independently verified and registered document that communicates the output of a life-cycle assessment of a product in a standardized way (it is compliant with the ISO14025 standard). Product Category Rules (PCR) define the rules and requirements for EPDs of a specific product category, enabling transparency and comparability between different EPDs.

Essentially, an EPD serves as a comprehensive "environmental footprint" for a product, detailing its impact on various aspects such as energy consumption, greenhouse gas emissions, water usage, and other relevant environmental indicators.

Usually, EPDs follow a scheme showcasing the following information:

- EPD general information
- Product information
- LCA calculation of product
- LCA results of product

Figure 3.4 shows the cover pages of 4 different EPDs used in the sector.

You can think of an EPD like a nutrition label on the packaging of the ingredients you buy for baking a cake. Just as the label tells you the calories, fats, and vitamins in your food, an EPD tells you the

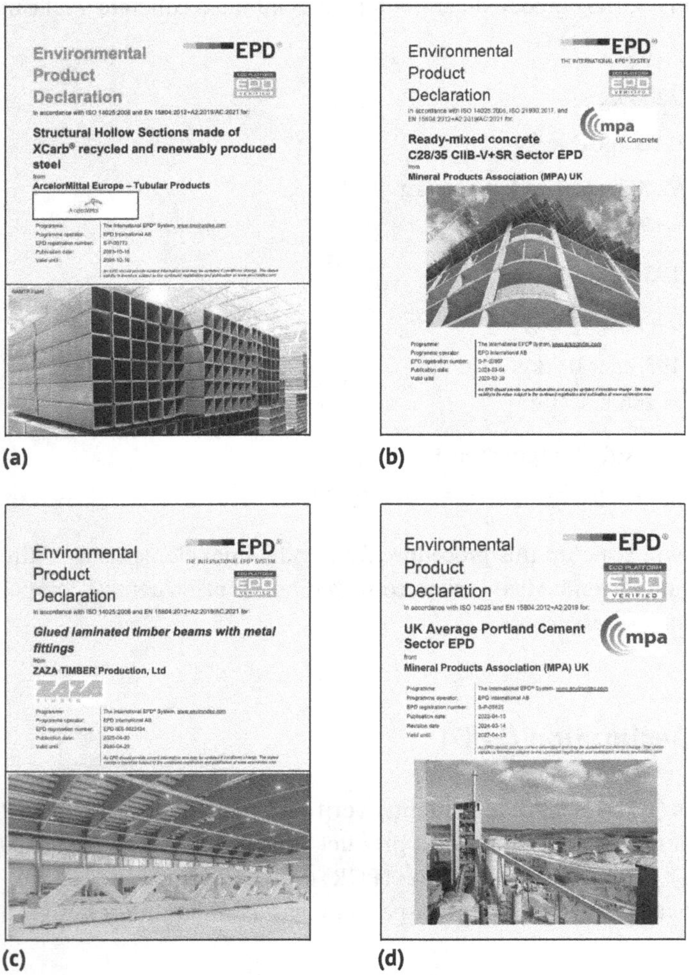

Figure 3.4 EPD examples. (a) Steel EPD example (Product EPD) (Arcelor Mittal Europe, 2022/EPD). With permission of ArcelorMittal; (b) concrete EPD example (Sector EPD) (UK Average Portland Cement Sector EPD from Mineral Products Association (MPA) UK). With permission of MPA, UK; (c) timber EPD example (Product EPD) (Swedish Wood, 2021/EPD). With permission of Swedish Wood; (d) cement EPD example (Sector EPD) (Ready-mixed concrete C28/35 CIIB-V+SR Sector EPD from Mineral Products Association (MPA) UK). With permission of MPA, UK.

embodied carbon and other environmental impacts of a building product. EPD goes a step further by revealing how the material was "prepared" – whether it was recycled, sustainably manufactured, or how it was transported (e.g. a steel beam with an EPD might indicate that it contains 50% recycled content and was produced using renewable energy, giving it a lower carbon footprint compared to conventional steel).

By referring to the EPDs, we can accurately assess the carbon footprint of each material listed in the BoQ. Depending on the industry or country, EPD results may vary due to differences in technology and equipment used throughout the product's life cycle.

3.2.1 Project-specific (EPD) vs. Generic Carbon Factor Data

Calculating the embodied carbon of a structure requires accurate environmental impact data for its materials. This data is available in two primary forms: project-specific data, derived from manufacturer-specific Environmental Product Declarations (EPDs), and generic data, which represents industry averages. The choice between them often depends on the project's design stage and the desired level of accuracy, with each having distinct advantages and applications in a life-cycle assessment.

3.2.2 Regional EPD Databases

EPD databases worldwide provide life-cycle assessment data. Table 3.2 highlights some of the most prominent EPD databases worldwide, including key national, regional, and global systems.

Table 3.2 Overview of major databases and programs providing environmental product declarations and embodied carbon data, including their regional coverage, applicable standards, and access links.

Database/Resource	Region/Country	Applicable standards	Website/URL
EC3 (Embodied Carbon in Construction Calculator)	USA, Canada, Global	Uses EPDs compliant with ISO 14025, EN 15804, ISO 21930	buildingtransparency.org/ec3
ASTM International EPD Program	USA (Globally recognized)	ISO 14025, ISO 21930	astm.org/epd
Ökobaudat	Germany (EU usage)	EN 15804, ISO 14025	oekobaudat.de/en
INIES	France	EN 15804 & French national complements	inies.fr/en
EPD Norge	Norway (Nordic)	EN 15804, ISO 14025, ISO 21930	epd-norge.no
International EPD System	Global (Sweden-based)	ISO 14025, EN 15804	environdec.com
EPD Danmark	Denmark	EN 15804, ISO 14025, ISO 21930	epddanmark.dk/uk
ICE Database	UK (Used globally)	A data source, not an EPD program. Data is derived from sources aligned with various standards.	circularecology.com

(Continued)

Table 3.2 (Continued)

Database/Resource	Region/Country	Applicable standards	Website/URL
AENOR Global EPD	Spain (International)	ISO 14025, EN 15804	globalepd.com
EPD Italy	Italy	EN 15804, ISO 14025	epditaly.it/en
EPD Ireland	Ireland	EN 15804, ISO 14025	igbc.ie/epd-ireland
EPD Turkey	Turkey	EN 15804, ISO 14025	epdturkey.org
Japan EPD Program (SuMPO)	Japan	ISO 14025, ISO 14040/14044	ecoleaf-label.jp/english
EPD Australasia	Australia & New Zealand	ISO 14025, EN 15804	epd-australasia.com
EPD Southeast Asia	Indonesia, Singapore, etc.	ISO 14025, EN 15804 (as a hub of the International EPD System)	epd-southeastasia.com
Korea EPD Program	South Korea	ISO 14025	epd.or.kr/eng
EPD Latin America	Latin America (Chile-based)	ISO 14025, EN 15804	epd-latinamerica.com
EPD Brasil	Brazil	ISO 14025, EN 15804	epdbrasil.com.br
EPD India	India	ISO 14025, EN 15804	epd-india.com

Driven by increasing regulations and the rising demands of green building standards, the construction sector has seen explosive growth in environmental reporting. Since 2019, many of these programs have reported increases of 200–500% in their published EPDs (Figure 3.5).

This rapid adoption is accelerated by digital transformation across all regions. Today, machine-readable formats (like XML or ILCD+EPD), API access, and seamless building information modeling (BIM) integration have become standard features.

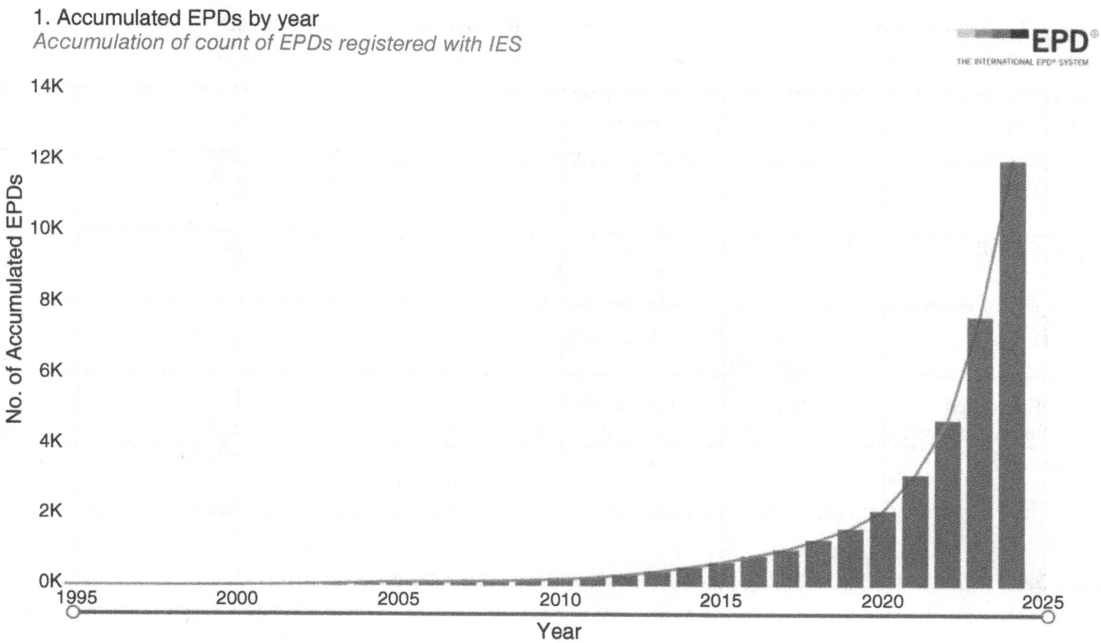

Figure 3.5 Cumulative number of Environmental Product Declarations registered with the International EPD System by year (image courtesy of EPD International AB, www.environdec.com).

3.2.3 Regional Generic Data Sources

While product-specific EPDs offer high accuracy, generic data sources can be useful for early-stage design and benchmarking before specific manufacturers are chosen. The environmental impact values of structural system elements vary between countries, not only due to different calculation methodologies but more critically due to variations in regional practices, energy sources, material supply chains, and the availability of data. For instance, steel produced in a country with a renewable-energy-intensive grid will have a lower product stage carbon factor compared to steel from a country relying heavily on fossil fuels.

Various databases have been developed to categorize generic embodied carbon values for construction materials. Some examples are shown in Tables 3.3 to 3.6. These databases enable direct identification of the carbon factors required to calculate the embodied carbon of structural systems, especially when benchmarking or performing early-stage assessments.

Table 3.3 Global foundational databases for life cycle inventory and embodied carbon data.

Global foundational databases	Managing organization(s)	Geographic coverage	Features
Ecoinvent	Ecoinvent Association	Global	Process-based LCI database.
Sphera (GaBi)	Sphera	Global (strong focus on EU, US, DE, JP, CN)	Process-based LCI database.
GLAD	UNEP/Life Cycle Initiative	Global (Directory)	A network/directory of independent LCA databases.
EC3 (Embodied carbon in construction calculator)	Building Transparency	Global (focus on North America & Europe)	Besides being a comprehensive database of EPDs (that allows users to find and compare low-carbon materials), provides some industry average values.

Table 3.4 Regional databases for life cycle inventory and embodied carbon data.

Regional databases	Managing organization(s)	Geographic coverage	Features
Ökobaudat	German Federal Ministry for Housing, Urban Development and Building (BMWSB)	Germany/Europe	German national database for building LCA. Contains generic datasets and specific EPDs. Compliant with EN 15804. Mandatory for BNB certification.
INIES Database	Alliance HQE-GBC (France)	France	French national reference database. Compliant with EN 15804 + French national supplements.
USLCI Database	National Renewable Energy Laboratory (NREL)/US Dept. of Energy (DOE)	United States/North America	Public repository of critically reviewed LCI data for the US context.
AusLCI	Australian Life Cycle Assessment Society (ALCAS)	Australia	National LCI database for Australia. Standardizes interpretation of ISO 14040 for the Australian context. Links to Ecoinvent for background data.
NABERS Embodied Carbon	Australian Government (Administered by NSW)	Australia	A national framework to measure, verify, and benchmark upfront embodied carbon (modules A1–A5). Uses EPDs and default data.

(Continued)

Table 3.4 (Continued)

Regional databases	Managing organization(s)	Geographic coverage	Features
NZ Construction Carbon Data	CIL Masterspec & BRANZ	New Zealand	New Zealand's national embodied carbon data repository. A centralized platform for EPDs and LCAs to replace and expand on previous datasets.
IDEA	National Institute of AIST & JEMAI (Japan)	Japan	Hybrid inventory database combining statistical and process-based data.
TianGong	TianGong Initiative	China	Research-driven, open LCI database for China.
CLCD	IKE Environmental Technology	China	One of the first major LCI databases for China. Provides foundational data for products manufactured or consumed in the region.

Table 3.5 Industry-specific databases for life cycle inventory and embodied carbon data.

Industry-specific databases	Managing organization(s)	Geographic coverage	Sectoral focus	Features
World Steel LCI Database	World Steel Association (worldsteel)	Global	Steel Products	Provides global average cradle-to-gate LCI data for steel products. Data is updated every few years based on inputs from member companies worldwide.
GCCA LCI Database	Global Cement and Concrete Association (GCCA)	Global	Cement & Concrete	Provides regional and global LCI and CO_2 emissions data for cement and clinker production.
AWC Industry-Wide EPDs	American Wood Council (AWC)	North America	Timber/Wood Products	Provides industry-average EPDs for major North American wood products (e.g. lumber, plywood, CLT).
ICE Database	Circular Ecology (originated at University of Bath)	UK/International	Construction Materials	Embodied carbon (cradle-to-gate) database created from extensive literature review.

Table 3.6 LCA software and data integration platforms for embodied carbon assessment.

LCA software & data integration platforms	Managing organization(s)	Geographic coverage	Features
SimaPro		Global	A leading professional LCA software, widely used in industry and academia. Integrates with Ecoinvent.
One Click LCA	One Click LCA Ltd.	Global (platform)	A commercial software platform that aggregates data from hundreds of sources, including EPDs and generic databases.
openLCA	GreenDelta	Global (platform)	Software for conducting LCAs. Connects to dozens of databases (free and commercial) via its Data Nexus.
Athena Impact Estimator	Athena Sustainable Materials Institute	North America	A specialized software for comparing the environmental impacts of whole buildings from cradle to grave using North American LCI data.
Brightway	Open Source Community	Global (framework)	A Python-based framework for advanced LCA. It is a developer's tool for custom, flexible, and large-scale calculations.

3.2.4 Handling Uncertainties in EPDs and Generic Data

Both project-specific and generic data come with inherent uncertainties that must be managed in any life-cycle assessment.

Project-specific EPDs, while more accurate for specific products, may not fully represent the actual manufacturing conditions or supply chain variations that occur during construction. Moreover, EPDs typically have validity periods of 3–5 years, during which manufacturing processes, energy grids, and supply chains may evolve significantly.

For generic data, uncertainty is typically higher, as values are averaged across a wide range of technologies, locations, and manufacturers, which can mask significant performance differences.

To address the uncertainties, practitioners can employ several strategies, as outlined by international standards for LCA (ISO 14044, 2006):

- In early design stages or for materials with high impact variability, a conservative or higher-impact generic value is often chosen to ensure the design meets its target even under a worst-case scenario.
- Conducting sensitivity analyses to identify which materials contribute most significantly to overall uncertainty.
- Clearly documenting all data sources, assumptions, and the justification for their use is critical for the credibility of the assessment.

To progressively reduce uncertainty, as a project moves from design to procurement, generic data should be replaced with manufacturer-specific EPDs wherever possible (this refinement helps ensure the final calculated embodied carbon is as close to the real-world value as possible). Ultimately, embodied carbon results should be presented with explicit acknowledgment of data limitations and uncertainty ranges. This transparency enables informed decision-making and proper interpretation of comparative assessments between design alternatives.

3.3 Embodied Carbon Variability Across Construction Products

The environmental impact of materials is far from uniform, with significant variations observed both between and within each material category. These variations are directly linked to fundamental differences in production technology, chemical composition, and manufacturing processes. In order to clarify these differences, this section presents A1–A3 cradle-to-gate impacts across different steel, concrete, and timber construction products according to the values provided in company-specific EPDs.

We explore how factors such as primary production technology, raw material selection, secondary processing, and material chemistry dictate the final GWP of products ranging from steel and concrete to mass timber. The analysis focuses on the A1–A3 (upfront) life-cycle stages, because this boundary represents critical information (the emissions already released before the building is in service) for architects, engineers, and procurement specialists making decisions about which materials to specify and purchase.

To build this comparison, we analyzed more than 1000 Environmental Product Declarations (EPDs) from multiple regions and manufacturers (worldwide) using EC3, EPD International, and IBU databases (EC3, 2025; EPD International, 2025). All environmental impact values presented in this study include an uncertainty factor as defined in the EC3 methodology (Waldman et al., 2020). This uncertainty factor reflects variability related to supply chains, product groupings, manufacturer coverage, methodological differences, and general measurement error (for instance, the GWP of EAF steel is highly sensitive to the carbon intensity of the regional electricity grid that powers the furnace: steel produced in a region with a coal-heavy grid will have a significantly higher GWP than the same product made in a region with a grid dominated by renewables or hydropower).

These comparisons are provided for study purposes and benchmarking to understand typical ranges and drivers and should not be applied directly to a project. Figure 3.6 shows the main statistical parameters shown in the box plot used in the presentation of results, and Table 3.7 describes their meaning.

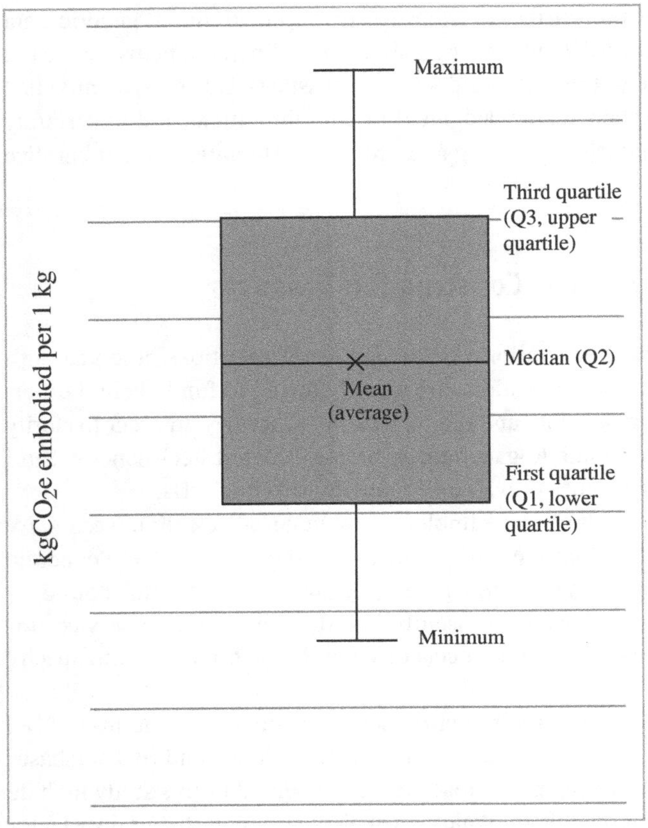

Figure 3.6 Statistical parameters used in the box plot graphs.

Table 3.7 Definition of the statistical parameters used in the box plot graphs.

Value	Description
Minimum (min)	The smallest observed value in the EPD dataset.
First quartile (Q1)	25% of the data falls below this value.
Median (Q2)	The middle value; half the data is below and half is above.
Mean (average)	The arithmetic average of all data points.
Third quartile (Q3)	75% of the data falls below this value.
Maximum (max)	The largest observed value in the EPD dataset.

3.3.1 Carbon Factor Variability Across Steel Product Types (A1–A3 Stages)

Different steel construction products show wide variability in cradle-to-gate GWP (A1–A3) as shown in Figure 3.7: Hot-rolled structural sections and steel rebars tend to have lower average embodied carbon (~1.2 kg CO_2e/kg) compared to flat-rolled products like steel plate or coil (~2.2–2.3 kg CO_2e/kg), which is in line with some other studies in the literature (Arup, 2023b; LCA eco-Profiles, 2021; Pinto et al., 2019). This difference occurs because products like rebar and hot-rolled sections are frequently produced in EAF mills, which specialize in recycling steel scrap and offer production flexibility. Flat products like plate steel and steel coil have historically been associated with the BOF route (Garvey et al., 2022; Suer et al., 2022), which produces virgin steel from iron ore, although EAF technology is increasingly capable of producing these products as well (Arens et al., 2017).

Cold-formed steel framing has a higher average GWP because its manufacturing involves additional steps after the initial hot-rolling stage: the production of a cold-formed steel product begins with a coil of hot-rolled steel, often referred to as a "hot band." This hot-rolled coil is the baseline material. To create a cold-formed product, this coil undergoes a series of processes at or near room temperature, where it is shaped into its final form (e.g. a C-section or tube) by being fed through a series of rollers and dies – a process that requires significant mechanical energy (Yu & LaBoube, 2010). Therefore, the GWP of a cold-formed product is cumulative, including the entire GWP of the initial hot-rolled coil plus the added environmental burden of the energy consumed during cold reduction, annealing, and roll forming (Qian et al., 2024). Despite this, cold-formed sections offer benefits such as precise dimensional tolerances, consistent quality, and the ability to produce thin-walled, lightweight profiles that can reduce overall steel consumption in a structure.

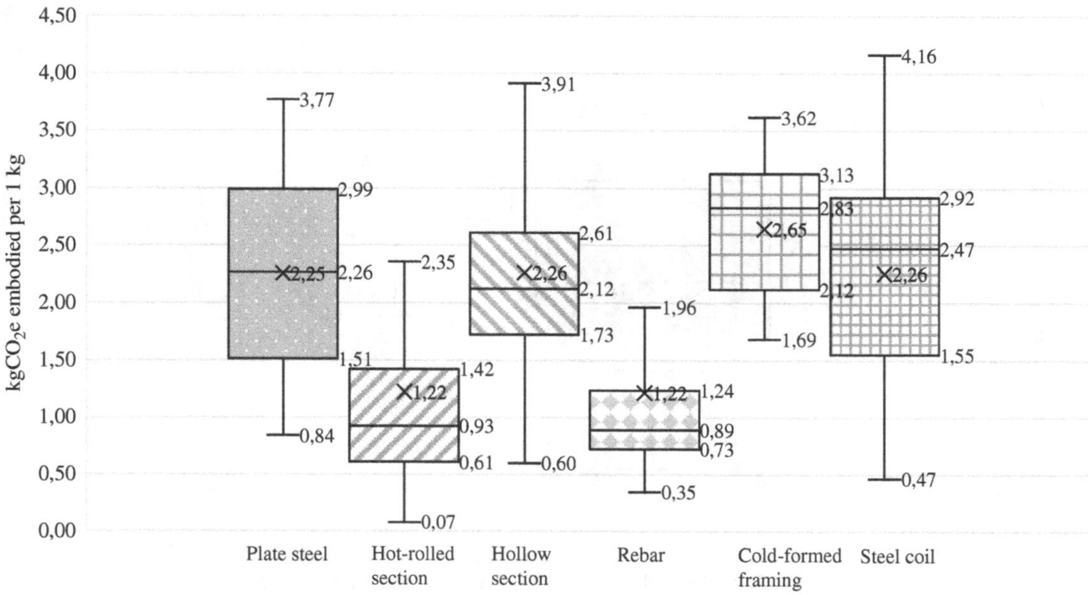

Figure 3.7 The GWP values (A1–A3 stages) for different types of steel products (EC3, n.d.; EPD International, n.d.; and IBU databases).

Hollow structural sections (HSS) (closed-profile steel members used extensively in construction for their high strength-to-weight ratio and excellent resistance to torsional forces) exhibit intermediate cradle-to-gate GWP values (A1–A3). This is slightly higher than open hot-rolled sections but lower than flat products, reflecting production routes: many HSS are made by cold-forming and welding of steel coil (thus inheriting the higher upstream emissions of BOF-produced coil). By contrast, HSS produced via scrap-fed EAF routes can achieve much lower GWP, approaching the levels of hot-rolled sections (arup.com). While our dataset indicates a world average of 2.26 $kgCO_2e/kg$, industry EPD data from an STI report (steeltubeinstitute.org) indicates a lower midpoint of around 1.7 $kgCO_2e/kg$, reflecting a more efficient, EAF-based regional production. Despite their intermediate embodied carbon, HSS can be favored for their structural efficiency, which allows for lighter designs and material savings at the system level, so the overall carbon footprint of a structure may ultimately be smaller.

3.3.2 Cold-formed vs. Hot-rolled Hollow Section

Cold-formed tubes (e.g. welded hollow sections made from cold-bent sheet/coil) and hot-rolled tubes (e.g. seamless pipe or hot-finished sections) show some differences in GWP due to how they're made (Figure 3.8). While cold-forming a tube at room temperature is a lower-energy forming process than hot rolling (no additional heat is needed to shape the steel, which avoids the substantial fuel use of

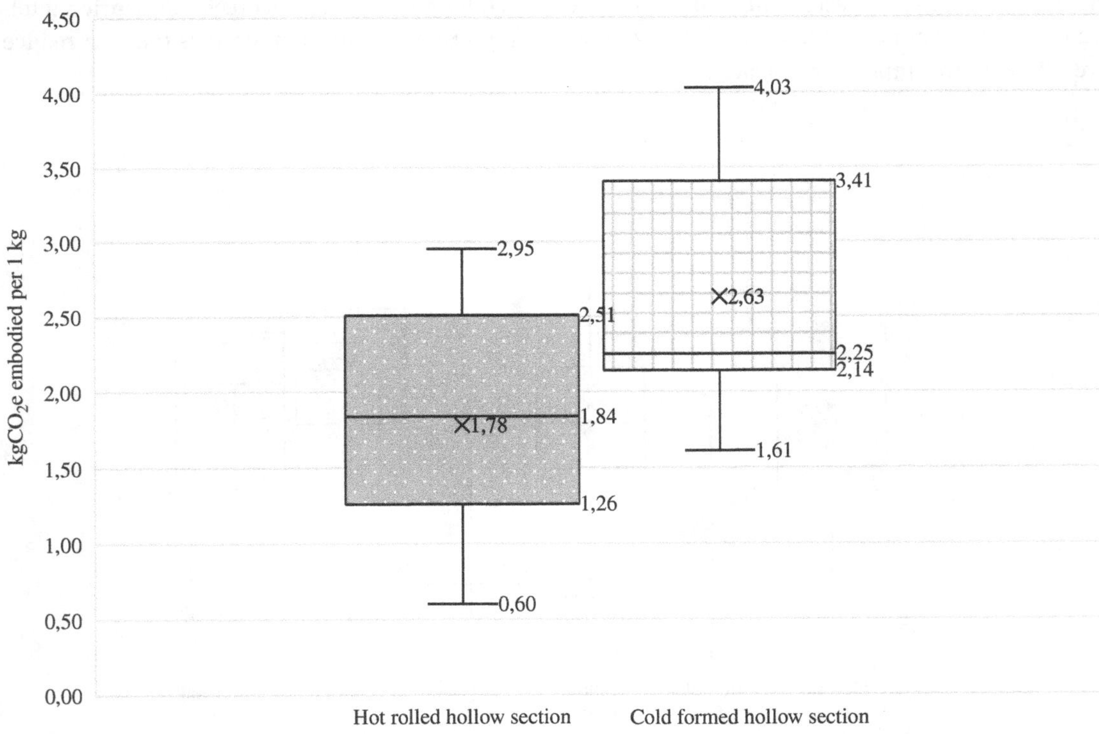

Figure 3.8 A comparison of the GWP values (A1–A3 stages) between hot-rolled and cold-formed hollow section steel products (EC3, n.d.; EPD International, n.d.; and IBU databases).

hot forming), cold-formed HSS typically starts from flat steel stock (like coil or plate) that often comes from the carbon-intensive BOF route (although in some cases EAF-produced flat products are used, and these are becoming increasingly common). This means the upstream emissions embedded in a cold-formed tube can be quite high. Some other figures mentioning the GWP values of cold-formed and hot-rolled hollow profiles can be found in Europe | Seamless Pipe – Other (2025), Europe | Welded Pipe – Construction (2025).

3.3.3 BF-BOF vs. EAF vs. DRI-EAF Steel Production Routes

The vast GWP disparity between steel production routes (shown in Figure 3.9) comes from fundamentally different chemistry and inputs, which are described in detail in Chapter 2. To recall, the BOF route is carbon-intensive, using iron ore and coke in blast furnaces where chemical reduction produces enormous CO_2 emissions as a direct process byproduct, plus emissions from fossil fuel combustion. The EAF route is electricity-based and circular (using recycled steel scrap as primary feedstock). By melting existing steel, EAF bypasses iron ore reduction and its associated process emissions entirely. Electric arcs between graphite electrodes melt scrap at high temperatures, making GWP primarily dependent on grid electricity emissions rather than chemical transformation (World Steel, 2025). Despite EAF's clear environmental advantage, a complete industry transition is not straightforward due to scrap steel

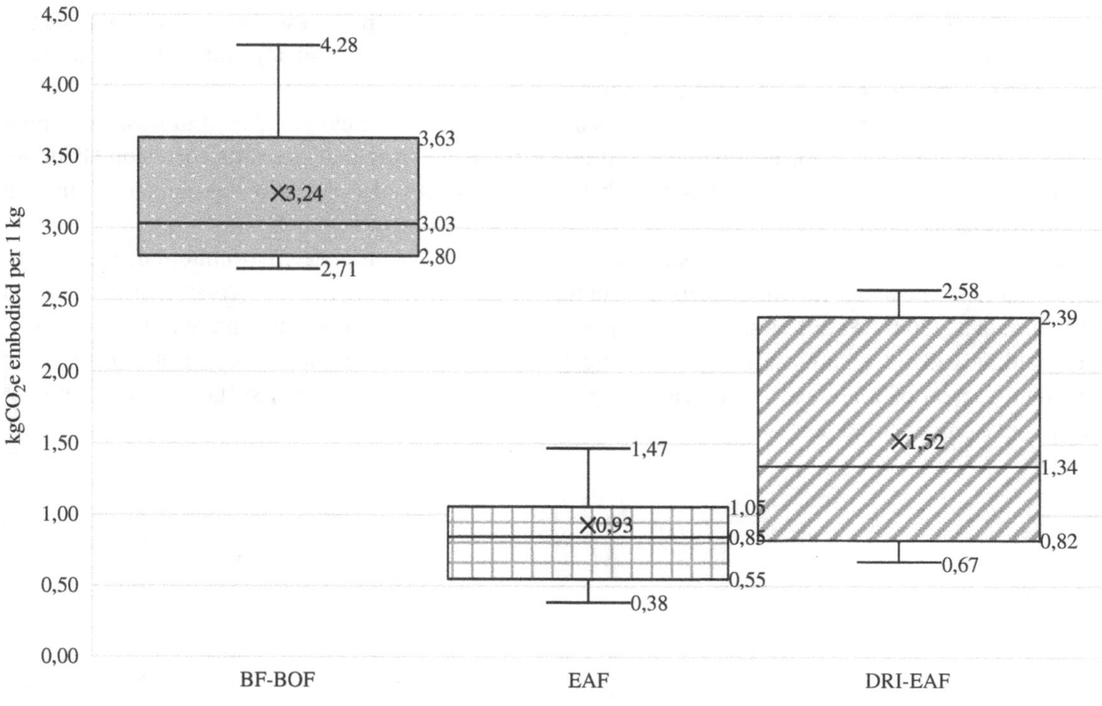

Figure 3.9 The GWP values (A1–A3 stages) for different types of steel production routes (EC3, n.d.; EPD International, n.d.; and IBU databases).

availability constraints. Current global steel demand significantly exceeds available scrap, necessitating continued BOF production of "virgin" steel. Regional ratios vary dramatically, while the share of EAF steel in global production is 28.6% (World Steel in Figures, 2024). In the literature, there are other studies mentioning the difference between the GWP values of different production routes (Guitar et al., 2025; LCA eco-Profiles, 2021; Pinto et al., 2019; Qian et al., 2024; Suer et al., 2022).

DRI-EAF represents an intermediate pathway between BOF and scrap-based EAF. In this route, natural gas (or hydrogen in emerging low-carbon variants) reduces iron ore pellets to direct reduced iron (DRI), which is then fed into an EAF for melting. Compared to BOF, the absence of blast furnaces reduces process emissions significantly, and the overall GWP depends strongly on the reducing agent used. The currently practical approach, natural gas-based DRI, still carries substantial CO_2 emissions, while hydrogen-based DRI powered by renewable electricity has the potential to approach near-zero GWP, although at present it remains at the pilot and demonstration stage (Arens et al., 2017; IEA, 2022). DRI-EAF is particularly relevant for regions with limited scrap availability but access to low-carbon electricity (and hydrogen in the future), providing a scalable option to decarbonize primary steelmaking.

3.3.4 Galvanized vs. Stainless Steel

A GWP disparity between galvanized and stainless steel (Figure 3.10) arises from fundamentally different methods used to achieve corrosion resistance. Galvanizing is a surface treatment, whereas stainless steel's resistance is integral to its material composition. A galvanized profile is standard carbon steel coated with a thin, protective layer of zinc, typically through a hot-dip process. Its total GWP equals the base carbon steel plus an additional impact from zinc mining, smelting, and application (Baddoo, 2008; *Galvanized Steel – an Overview | ScienceDirect Topics*, n.d.).

Stainless steel, in contrast, is not coated but is a distinct alloy. To be classified as stainless, steel must contain chromium, which forms a passive, self-repairing oxide layer that prevents corrosion (Baddoo, 2008). Many common grades also contain substantial quantities of other alloying elements, such as nickel, to enhance properties.

However, a cradle-to-gate GWP comparison is insufficient for a complete environmental assessment. Both materials' primary function is long-term performance in potentially corrosive environments (Ebrahimi et al., 2025). Stainless steel generally offers a much longer service life and requires less maintenance than galvanized alternatives. A comprehensive life-cycle assessment (LCA) must account for embodied carbon of maintenance activities (e.g. repainting failed galvanized surfaces) and manufacturing replacement components.

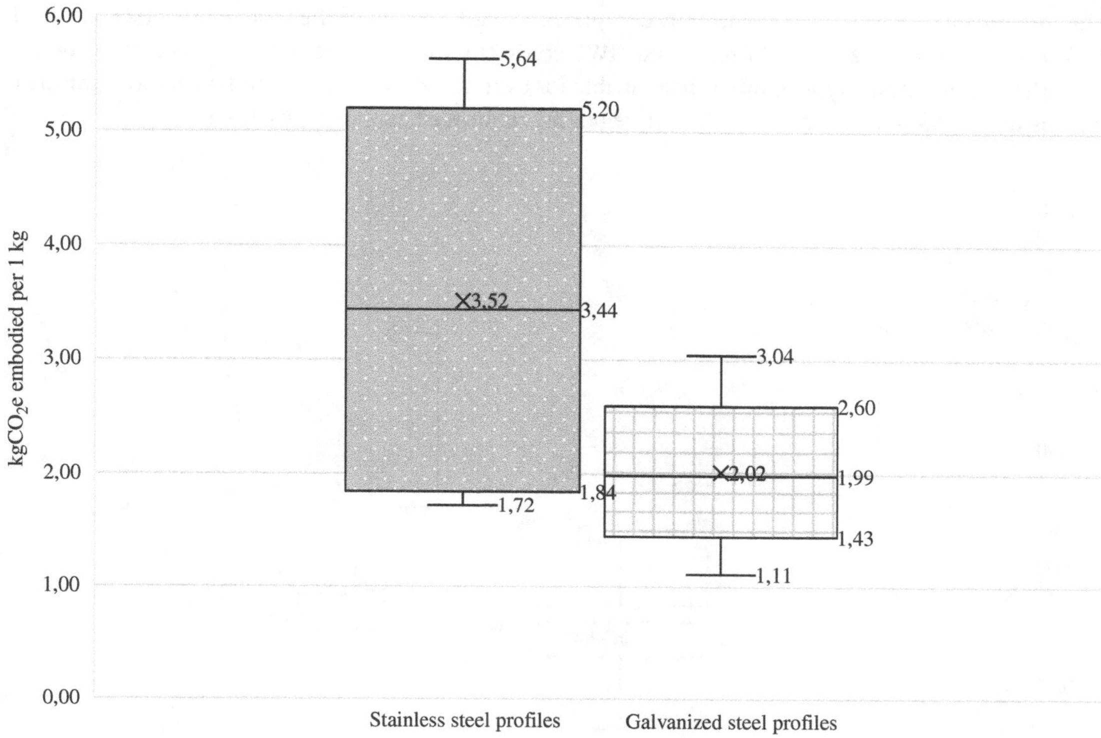

Figure 3.10 A comparison of the GWP values (A1–A3 stages) between stainless steel and galvanized steel products (EC3, n.d.; EPD International, n.d.; and IBU databases).

3.3.5 Comparison of Reinforced Concrete Components

In a typical reinforced concrete element, the vast majority of the embodied carbon comes from the cement in the concrete and the steel rebar, whereas aggregates contribute marginally (Figure 3.11). It's also noteworthy that aggregates (and even mixing water) are practically carbon-neutral in comparison – even transporting aggregates may produce more CO_2 than the aggregate itself (Driessen & Grönlund, 2024). This can mean that techniques like aggregate substitution or reduction don't impact GWP nearly as much as binder substitution. In the literature, similar values of GWPs for each product can be observed (Arup, 2023a; Driessen & Grönlund, 2024; Ige et al., 2024).

The comparison shows that cement is the GWP "hotspot" of the concrete mix on a percentage contribution basis. Although reinforcing steel has a higher GWP per kilogram than cement, the mass of cement used in a typical reinforced concrete element is much greater than the mass of steel. Consequently, cement's contribution to the total GWP of the system is dominant. This is confirmed by detailed LCA studies, which show that the Portland cement component is responsible for up to 90% of a concrete mix's total GWP, with coarse and fine aggregates contributing up to around 6 and 2%, respectively.

To complete the picture, we should mention the typical proportions of ingredients used in a typical reinforced concrete mix: it is composed of up to 70–80% aggregates (by mass), 10–15% cement, and the remainder water. Reinforcing steel usually comprises a small fraction of the total volume (see Table 3.8 for a representative breakdown of mass and GWP contributions). Despite rebar's higher per-kilogram GWP, the cement in this typical mix is responsible for over 80% of the final embodied carbon, confirming it as the primary target for reduction strategies for reinforced concrete structures.

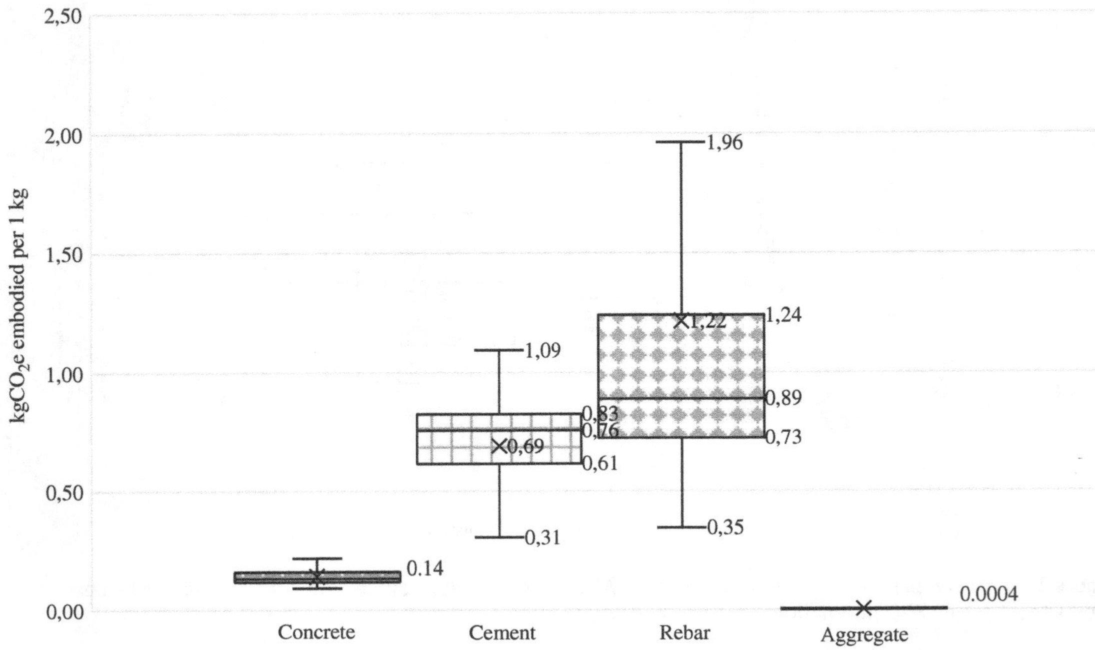

Figure 3.11 A comparison of the GWP values (A1–A3 stages) between the components of reinforced concrete products (EC3, n.d.; EPD International, n.d.; and IBU databases).

Table 3.8 A generic example of the constituents and their percentage in reinforced concrete products.

Component	Typical mass (kg)	GWP per kg ($kgCO_2e/kg$)	Total GWP ($kgCO_2e$)	% of Total GWP	% of Total mass
Cement	350	0.8	280	73.1	15.4
Fine aggregate	750	0.004	3	0.8	32.9
Coarse aggregate	1100	0.004	4.4	1.1	48.3
Reinforcing steel (1% vol)	78.5	1.22	95.77	25	3.40
Total	2278.5	-	383.17	100	100

Note: Water and admixtures are excluded for simplicity. Proportions are illustrative.

3.3.6 Precast vs. Ready-mix Concrete

Precast concrete elements generally show a higher GWP (A1–A3 per kg) than equivalent ready-mix (cast-in-place) concrete (Figure 3.12). This difference arises largely because precast manufacturers often use *richer cement mixes* and energy-intensive curing to achieve rapid strength gain (Arup, 2023a). Precast concrete is typically made with higher-strength concrete. Moreover, precast often uses less SCM (supplementary cementitious material) substitution, because fly ash or slag can slow down strength development. As a result, precast mixes tend to rely more on straight Portland clinker (Kelechi et al., 2022; Lee et al., 2016). Also, in many of the EPDs of precast concrete, reinforcement steel has been counted in the GWP.

Since precast is optimized for speed and quality control, precast units may enable material savings at the structural level (through prestressing, thinner sections, etc.), potentially offsetting their per-kg carbon penalty when looking at the whole building. Therefore, the potential for precast to achieve a lower GWP stems from the efficiencies inherent in a controlled, factory-based production environment.

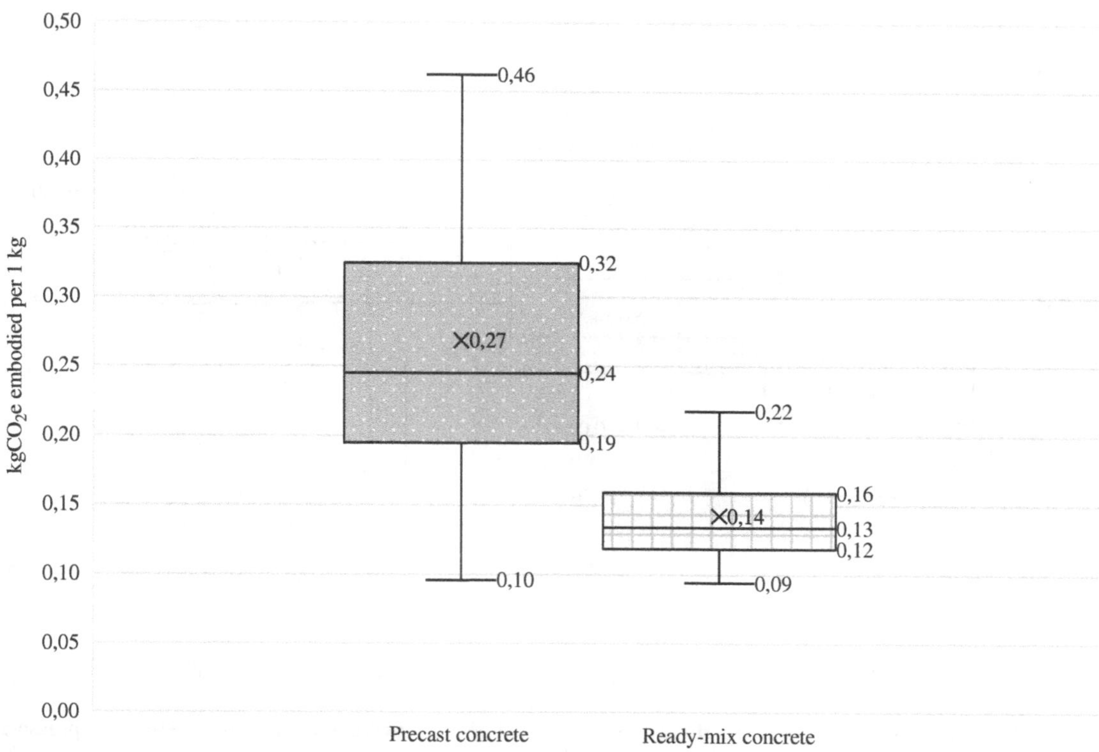

Figure 3.12 A comparison of the GWP values (A1–A3 stages) between precast and ready-mix concrete products (EC3, n.d.; EPD International, n.d.; and IBU databases).

3.3.7 Ready-mix Concrete with Different Compressive Strength

To achieve higher compressive strengths, a richer mix is required, which primarily means increasing the proportion of cement relative to aggregates (like sand and gravel) and water (Arup, 2023a; Munro et al., 2024). Since cement is the dominant source of CO_2 in concrete, as its proportion rises, so does the mix's embodied carbon (Figure 3.13). The variability seen in the GWP for each strength class likely reflects the varying degrees of SCM replacement in the different products included in the dataset (SCM content strongly lowers GWP) (Witte & Garg, 2024).

A key point is that structural design choices influence these outcomes: using higher-strength concrete can sometimes allow for smaller member sizes or longer spans, which reduces the total concrete volume required. This means that while the per-kg GWP is higher, the overall embodied carbon at the element or building level may not increase proportionally.

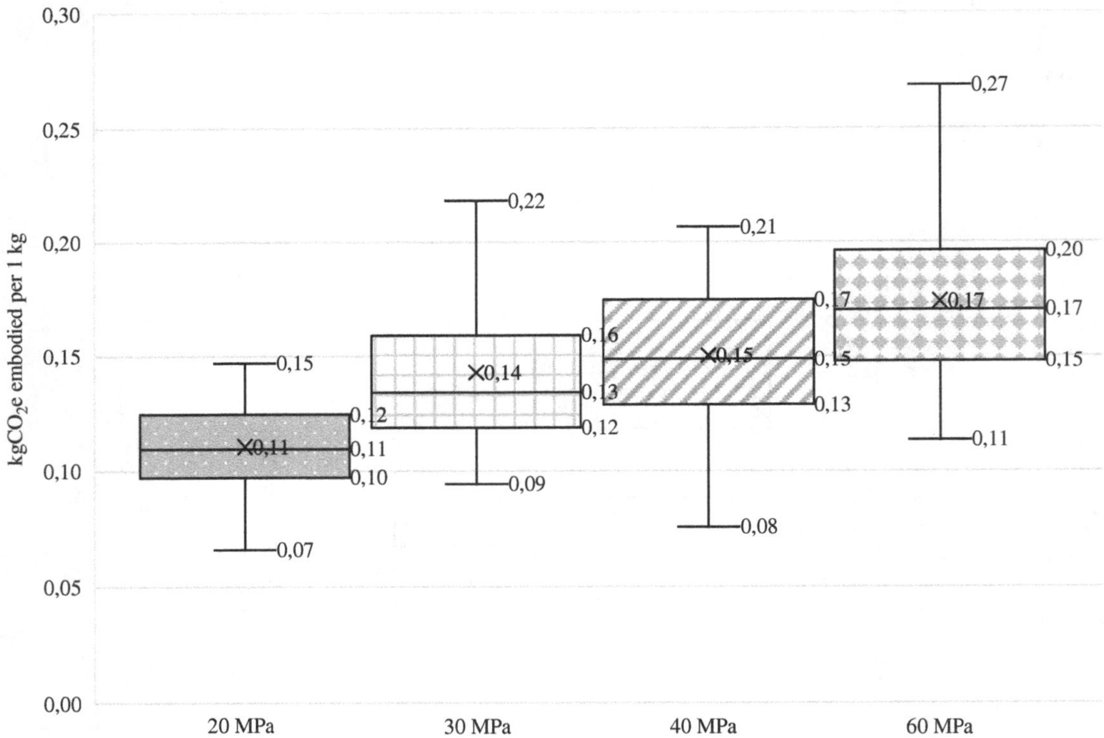

Figure 3.13 A comparison of the GWP values (A1–A3 stages) between concrete products of different compressive strength (EC3, n.d.; EPD International, n.d.; and IBU databases).

3.3.8 Timber Product Types (CLT, GLT, Sawn Timber)

The embodied carbon of timber products varies according to the degree of processing and adhesive use. As shown in Figure 3.14, sawn timber has the lowest cradle-to-gate GWP, while cross-laminated timber (CLT) and glued-laminated timber (GLT) show higher emissions, reflecting the additional processing and adhesive layers. While engineered products such as CLT and GLT provide superior structural performance and dimensional stability, this comes at a carbon cost compared to minimally processed sawn timber. Although sawn timber has lower impacts, its limitations in size and consistency make CLT and GLT more suitable for modern construction, explaining their wider use despite the higher carbon cost.

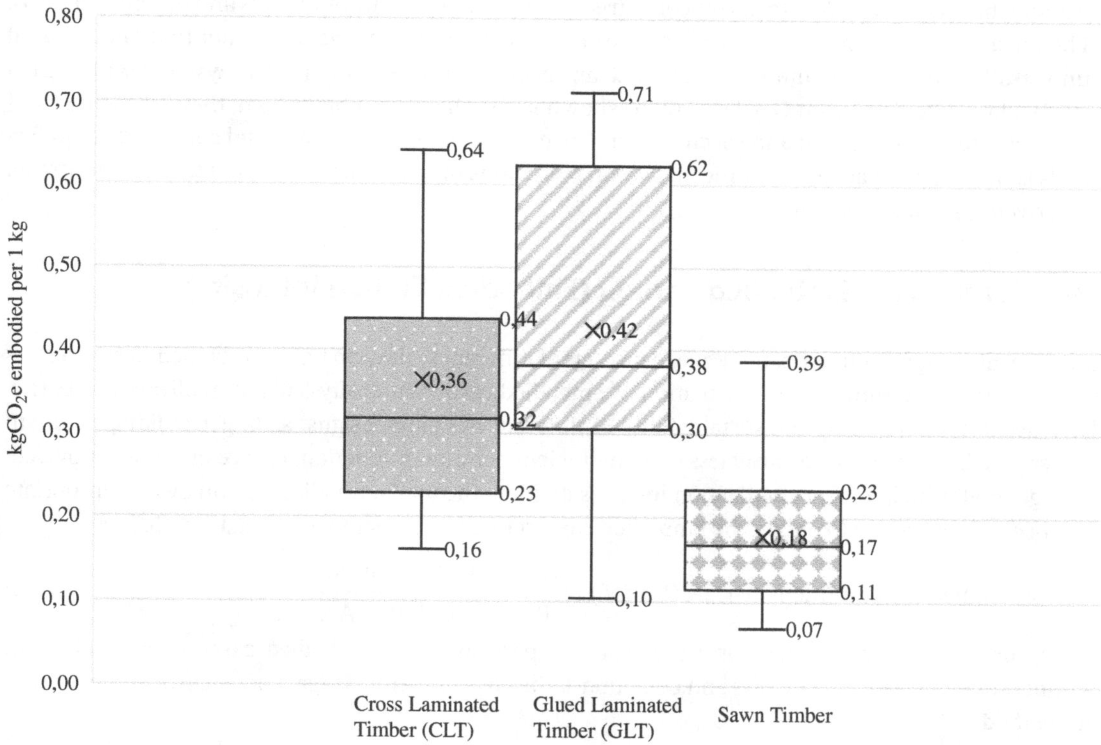

Figure 3.14 A comparison of the GWP-fossil values (A1–A3 stages) between different mass timber products (EC3, n.d.; EPD International, n.d.; and IBU databases).

3.3.9 Comparison Across All Products

Considering all the construction materials together, the overarching trend is that industrial products requiring high-temperature reduction of ores (metals, cement) have one to two orders of magnitude higher GWP per kg than materials that are basically modified natural resources (wood, aggregates, etc.).

While the above comparisons present variations observed both between and within each material category linked to differences in their production technology, chemical composition, and manufacturing, for building structures, comparing only materials on a per-mass basis would provide an incomplete picture. A valid environmental comparison between structural options can only be made using a "functional unit" (a common measure of performance, such as a beam designed to span a specific distance and support a specific load, or a square meter of floor area in a completed building). This reframes the question from "which material is best?" to "what is the lowest-carbon way to design this beam, floor, or building?" For example, mass timber (0.4 $kgCO_2/kg$) is low per kg, but you might need a larger mass of wood to replace a steel beam. Similarly, concrete has a very low GWP per kilogram but has a lower strength-to-weight ratio than steel or timber. Consequently, a concrete structure requires a much greater mass of material to achieve the same performance as a steel or mass timber alternative. Readers can apply this same reasoning to derive comparisons for other structural elements and design scenarios.

The most powerful conclusion to be drawn from this comprehensive dataset is not that one material is universally "better" than another, but that clear, specific decarbonization pathways exist within each material class. The significant GWP variations shown in the charts are not random; they are the result of specific choices in production and specification. The ultimate insight for design and construction professionals is that they can achieve significant carbon reductions in any structural system through informed, data-driven material selection.

3.4 Normalizing Embodied Carbon (Embodied Carbon Intensity)

To enable meaningful comparisons between projects of different scales and types, embodied carbon results must be normalized through the calculation of embodied carbon intensity. This normalization is particularly useful for comparing the sustainability of different building designs, as larger buildings contain more carbon but are not necessarily less carbon-efficient per unit of functional space or service provided.

The general principle of normalization involves dividing the total embodied carbon by an appropriate reference unit of area as to obtain the embodied carbon intensity, as shown in Equation (3.2).

$$\text{Embodied Carbon Intensity}\left(\text{kgCO}_2\text{e/m}^2\right) = \frac{\text{Embodied Carbon}\left(\text{kgCO}_2\text{e}\right)}{\text{Reference Unit of Area}\left(\text{m}^2\right)} \tag{3.2}$$

This normalized result allows for a practical comparison of the embodied carbon across different buildings, making it easier to assess and benchmark the environmental impact of construction materials and methods.

The choice of functional unit varies by structure type to ensure the most relevant and practical comparison basis. Different sectors have adopted specific normalization approaches that best reflect their operational characteristics and design priorities. The following subsections detail the primary normalization methods used across various infrastructure types.

3.4.1 Gross Internal Area (GIA) – Buildings

The Gross Internal Area (GIA) of a building is a measurement that covers the area within the internal face of the perimeter walls at each floor level. This measurement is crucial for calculating embodied carbon intensity, expressed as $kgCO_2e/m^2$. In principle, every level providing enclosed, weather-protected floor area contributes to the GIA calculation (RICS, 2015) as illustrated in Figure 3.15.

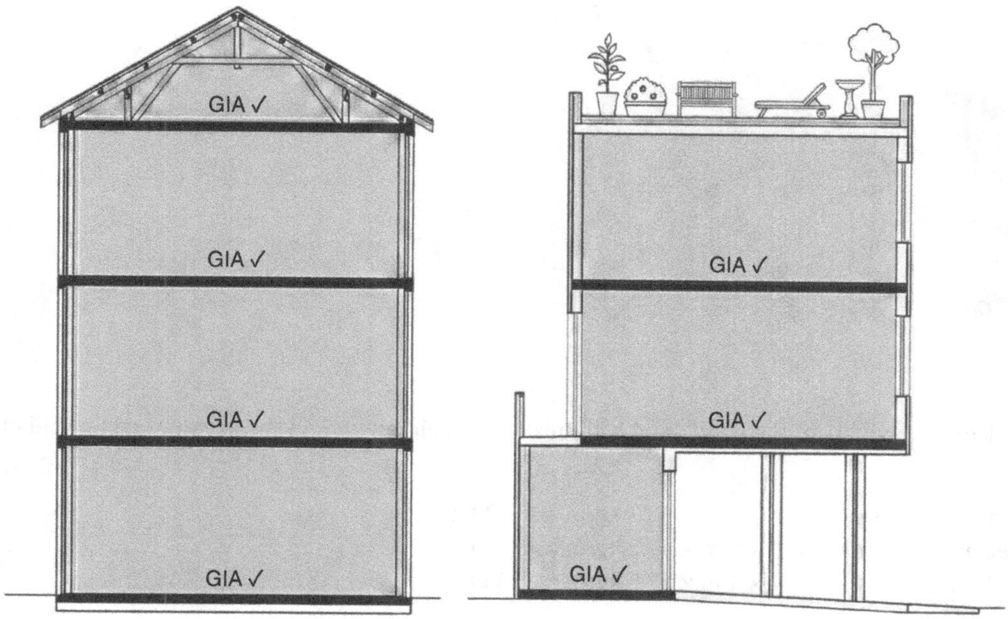

Figure 3.15 Schematic representation of Gross Internal Area (GIA) for buildings, showing the areas included in normalization of embodied carbon results.

The GIA Excludes:

- Open balconies, fire escapes, covered ways, and terraces (roof areas that are not permanently covered)
- Open undercrofts, where a building steps back at ground floor creating a covered but external area beneath a cantilevered upper floor.

For refurbishment projects, the total embodied carbon is typically normalized by the entire GIA of the refurbished building (including both new construction and retained existing floor areas), since this approach ensures fair comparison with new-build projects while acknowledging the carbon benefits of retaining existing structures.

Example

Let us calculate the embodied carbon intensity (A1–A5 stages) of the steel frame shown in Figure 3.16 (we will ignore the slab for this example) with the following characteristics:

Primary beams:	2 × IPE500
Secondary beams:	3 × IPE300
Columns:	4 × CHS 244.5 mm × 10.0 mm
A1–A5 carbon factor of steel:	1.2 $kgCO_2e/kg$ (represents the European average for structural steel based on industry EPD data, assuming a typical production mix of primary (BF-BOF) and recycled (EAF) steel).

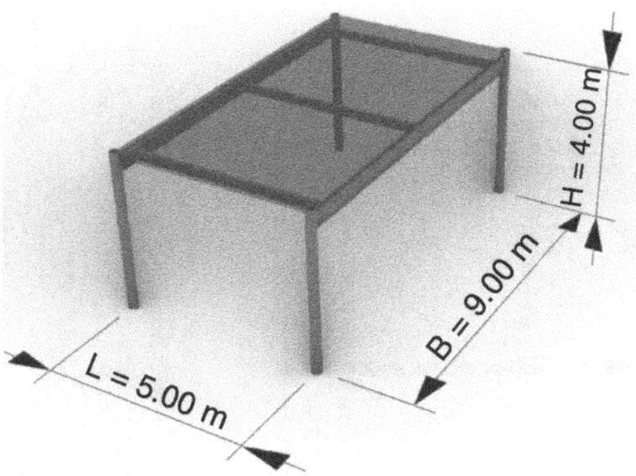

Figure 3.16 3D view of steel frame.

We first calculate the mass of the beams and columns using their length and the mass per meter (bill of quantities):

Mass of primary beams $= 2 \times 9\ \text{m} \times 90.7\ \text{kg/m} \approx 1632\ \text{kg}$
Mass of secondary beams $= 3 \times 5\ \text{m} \times 42.2\ \text{kg/m} \approx 633\ \text{kg}$
Mass of columns $= 4 \times 4\ \text{m} \times 57.8\ \text{kg/m} \approx 925\ \text{kg}$

Note: The mass values are obtained from standard steel section tables, which provide the mass per meter for different profiles based on their cross-sectional properties.

We then calculate the embodied carbon for the beams and columns by multiplying their total mass by the A1–A5 carbon factor of steel:

Embodied Carbon, Total Primary Beams $= (1632\ \text{kg}) \times (1.2\ \text{kgCO}_2\text{e/kg}) \approx 1958\ \text{kgCO}_2\text{e}$
Embodied Carbon, Total Secondary Beams $= (633\ \text{kg}) \times (1.2\ \text{kgCO}_2\text{e/kg}) \approx 760\ \text{kgCO}_2\text{e}$
Embodied Carbon, Total Columns $= (925\ \text{kg}) \times (1.2\ \text{kgCO}_2\text{e/kg}) \approx 1110\ \text{kgCO}_2\text{e}$
Embodied Carbon, Total $= 1958\ \text{kgCO}_2\text{e} + 760\ \text{kgCO}_2\text{e} + 1110\ \text{kgCO}_2\text{e} \approx 3828\ \text{kgCO}_2\text{e}$

Finally, we divide the total embodied carbon by the floor area to find the embodied carbon intensity:

Floor Area $= 5\ \text{m} \times 9\ \text{m} = 45\ \text{m}^2$
Embodied Carbon Intensity $= 3828\ \text{kgCO}_2\text{e}/45\ \text{m}^2 \approx 85.1\ \text{kgCO}_2\text{e/m}^2$

3.4.2 Functional Area (FA) – Bridges

The Functional Area (FA) for bridges is defined as the area measured in square meters (m^2). According to the RICS professional standards and guidance (RICS, 2015), this includes the width designated for various uses, such as rail and pedestrian traffic. The embodied carbon (EC) intensity for bridges is expressed as $\text{kgCO}_2\text{e/m}^2$. This metric allows for the normalization of carbon results, facilitating a more accurate comparison of the environmental impacts associated with different bridge designs and construction methods (Figure 3.17).

Figure 3.17 A bridge under construction (Wiley).

Example: Embodied Carbon Intensity for a Bridge

Let us calculate the upfront (A1–A5) embodied carbon intensity of the bridge with the following characteristics:

Bridge type	Simple beam bridge (steel and concrete deck)
Length of bridge:	35 m
Width of bridge:	6 m
Deck thickness:	0.4 m
Support type:	Two reinforced concrete piers, each 3 m wide, 9.3 m tall, and 2 m thick.

Main Beams:

Two steel I-beams, each 35 m long, with the following dimensions:

Depth:	1 m (1000 mm)
Flange width:	600 mm
Flange thickness:	30 mm
Web thickness:	20 mm
Mass per Beam:	15 000 kg

Embodied Carbon Assumptions, per Kilogram (assumptions selected for this example):

Embodied carbon per kilogram of concrete (A1–A5):	0.13 $kgCO_2e/kg$.
Embodied carbon per kilogram of steel:	1.2 $kgCO_2e/kg$.
Embodied carbon per kilogram of reinforcement steel:	0.8 $kgCO_2e/kg$.

Mass of Steel Reinforcement in Deck:
Assume 375 kg of steel per m^3 of concrete.

Mass of Reinforced Concrete Deck:

Volume of deck:	The deck is 0.4 m thick, with a length of 35 m and a width of 6 m.
Volume of deck	= Thickness × Length × Width = 0.4 m × 35 m × 6 m = 84 m^3.
Density:	2571 kg/m^3
Mass of deck	= 84 m^3 × 2571 kg/m^3 = 216 000 kg

Embodied Carbon of Reinforced Concrete Part:

Embodied carbon per kilogram of concrete (A1–A5):	0.13 $kgCO_2e/kg$
Embodied carbon of concrete in Deck:	216 000 kg × 0.13 $kgCO_2e/kg$ = 28 080 $kgCO_2e$
Steel reinforcement in deck:	Assume 375 kg of steel per m^3 of concrete
Total reinforcement steel	= 84 m^3 × 375 kg/m^3 = 31 500 kg
Embodied carbon per kilogram of reinforcement steel:	0.8 $kgCO_2e/kg$
Embodied carbon (reinforcement)	= 31 500 kg × 0.8 $kgCO_2e/kg$ = 25 200 $kgCO_2e$
Total embodied carbon (deck)	= 28 080 $kgCO_2e$ + 25 200 $kgCO_2e$ = 53 280 $kgCO_2e$

Embodied Carbon of Steel Profiles:

Mass of each beam:	15 000 kg
Total steel mass:	15 000 kg × 2 beams = 30 000 kg
Embodied carbon per kilogram of steel:	1.2 $kgCO_2e/kg$
Embodied carbon of steel beams	= 30 000 kg × 1.2 $kgCO_2e/kg$ = 36 000 $kgCO_2e$

Mass of Reinforced Concrete Piers:

Volume of each pier: Each pier is 3 m wide, 9.3 m tall, and 2 m thick

Volume of each pier $= \text{Width} \times \text{Height} \times \text{Thickness}$

$= 3\text{ m} \times 9.3\text{ m} \times 2\text{ m} = 56\text{ m}^3$.

Total volume of piers $= 56\text{ m}^3 \times 2 = 112\text{ m}^3$

Density: 2571 kg/m^3

Mass of piers $= 112\text{ m}^3 \times 2571\text{ kg/m}^3$

$= 288\,000\text{ kg}$.

Embodied Carbon of Concrete in Piers:

Embodied carbon (concrete piers) $= 288\,000\text{ kg} \times 0.13\text{ kgCO}_2\text{e/kg}$

$= 37\,440\text{ kgCO}_2\text{e}$.

Steel reinforcement in piers: 375 kg of steel per m^3 of concrete.

Total reinforcement steel $= 112\text{ m}^3 \times 375\text{ kg/m}^3 = 42\,000\text{ kg}$.

Embodied carbon (reinforcement) $= 42\,000\text{ kg} \times 0.8\text{ kgCO}_2\text{e/kg}$

$= 33\,600\text{ kgCO}_2\text{e}$.

Total Embodied Carbon of (Piers) $= 37\,440\text{ kgCO}_2\text{e} + 33\,600\text{ kgCO}_2\text{e} = 71\,040\text{ kgCO}_2\text{e}$.

The total embodied carbon is the sum of the embodied carbon for the deck, beams, and piers:

$$\text{Total EC} = \text{Embodied Carbon (Deck)} + \text{Embodied Carbon (Beams)} + \text{Embodied Carbon (Piers)}$$

$$\text{Total EC} = 53\,280\text{ kgCO}_2\text{e} + 36\,000\text{ kgCO}_2\text{e} + 71\,040\text{ kgCO}_2\text{e} = 160\,320\text{ kgCO}_2\text{e}$$

The Embodied Carbon Intensity is calculated by dividing the total embodied carbon by the functional area:

EC intensity $= \text{Total EC/FA}$

FA $= \text{Width of Deck} \times \text{Length of Bridge}$

FA $= 6\text{ m} \times 35\text{ m} = 210\text{ m}^2$

EC intensity $= 160\,320\text{ kgCO}_2\text{e}/210\text{ m}^2$

$= 763.4\text{ kgCO}_2\text{e/m}^2$.

3.4.3 Number of Spectator Seats (NSS) or Enclosed Area (EA) – Stadia

For stadia (Figure 3.18), the environmental impact can be normalized using either the number of spectator seats (NSS) or the enclosed area (EA). The embodied carbon intensity for stadia can be expressed in two ways:

Figure 3.18 Photo of a stadium (Blurra/iStock/Getty Images).

$kgCO_2e$/seat:	This metric calculates the carbon footprint per spectator seat, providing a view of the impact relative to the seating capacity.
$kgCO_2e/m^2$:	This metric measures the carbon footprint per square meter of the enclosed area, providing a view into the overall environmental impact of the stadium's internal space.

Example: Embodied Carbon Intensity for a Stadium

Let us calculate the upfront embodied carbon intensity (A1–A5) of a stadium with the following characteristics:

Total seating capacity (NSS):	50 000 seats
Enclosed area (EA):	40 000 m^2
Reinforced concrete superstructure:	25 000 m^3 of reinforced concrete
Steel roof:	6500 tons of structural steel
Steel reinforcement for concrete:	Assume 400 kg/m^3
Embodied carbon assumption per kilogram of concrete (A1–A5):	0.13 $kgCO_2e$/kg
Embodied carbon assumption per kilogram of steel:	1.2 $kgCO_2e$/kg
Embodied carbon assumption per kilogram of reinforcement steel:	0.8 $kgCO_2e$/kg

Mass of Reinforced Concrete:
The density of reinforced concrete is approximately 2400 kg/m^3.

Mass of Reinforced Concrete = 25 000 $m^3 \times 2400\ kg/m^3$ = 60 000 000 kg

Mass of steel reinforcement (assume that each cubic meter of concrete contains 400 kg of steel reinforcement):

Mass of Steel Reinforcement = 25 000 $m^3 \times 400\ kg/m^3$ = 10 000 000 kg

Mass of Steel Roof:
The roof contains 6500 tons of structural steel (6 500 000 kg):

Embodied Carbon Calculations:

Embodied Carbon of Reinforced Concrete (Excluding Reinforcement)

$$= 60\ 000\ 000\ kg \times 0.13\ kgCO_2e/kg = 7800\ 000\ kgCO_2e$$

Embodied Carbon of Steel Reinforcement = 10 000 000 kg × 0.8 $kgCO_2e/kg$ = 8 000 000 $kgCO_2e$

Embodied Carbon of Steel Roof = 6 500 000 kg × 1.2 $kgCO_2e/kg$ = 7 800 000 $kgCO_2e$

The total embodied carbon of the stadium is the sum of the contributions from all materials:

$$\begin{aligned}\text{Total Embodied Carbon} &= 7\ 800\ 000 + 8000\ 000 + 7\ 800\ 000\\ &= 23\ 600\ 000\ kgCO_2e\end{aligned}$$

Normalize the Total Embodied Carbon Using Two Metrics:
Embodied carbon intensity per seat ($kgCO_2e$/seat):

$$\begin{aligned}\text{Embodied Carbon Per Seat} &= \text{Total Embodied Carbon/ Total Seating Capacity}\\ \text{Embodied Carbon Per Seat} &= 23\ 600\ 000\ kgCO_2e/50\ 000\ \text{seats} = 472\ kgCO_2e/\text{seat}\end{aligned}$$

Embodied carbon intensity per square meter ($kgCO_2e/m^2$):

$$\begin{aligned}\text{Embodied Carbon Per } m^2 &= \text{Total embodied carbon/ Enclosed area}\\ \text{Embodied Carbon Per } m^2 &= 23\ 600\ 000\ kgCO_2e/40\ 000\ m^2 = 590\ kgCO_2e/m^2\end{aligned}$$

Embodied carbon intensity per seat: 472 $kgCO_2e$/seat
Embodied carbon intensity per square meter: 590 $kgCO_2e/m^2$

3.4.4 Energy Delivered (ED) – Energy Infrastructure

In the context of energy infrastructure, such as wind towers and offshore infrastructure (Figure 3.19), the embodied carbon intensity can be measured in terms of energy delivered. The metric used is $kgCO_2e$/kWh, which calculates the carbon footprint per kilowatt-hour of energy produced. This approach allows for the assessment of the environmental impact of constructing and maintaining energy infrastructure, normalized by the amount of energy it delivers.

Figure 3.19 Photo of an offshore wind turbine tower (Monty Rakusen/DigitalVision/Getty Images).

Example: Embodied Carbon Intensity for a Wind Farm

Let us calculate the upfront embodied carbon intensity (A1–A3) of an offshore wind turbine with the following characteristics:

Turbine capacity:	6 MW (megawatts)
Expected operational life:	25 years
Capacity factor:	Assume 40% (typical for offshore wind turbines)
Tower structure:	1200 tons of structural steel
Foundation structure:	2000 tons of concrete
Steel reinforcement for concrete:	400 kg/m^3
Embodied carbon assumption per kilogram of concrete (A1–A5):	0.15 kgCO_2e/kg.
Embodied carbon assumption per kilogram of steel:	1.80 kgCO_2e/kg.
Embodied carbon assumption per kilogram of reinforcement steel:	0.8 kgCO_2e/kg
Density of concrete:	2400 kg/m^3

The energy delivered by the turbine over its lifetime is calculated using the formula:

$$\text{Energy Delivered (kWh)} = \text{Turbine Capacity (kW)} \times \text{Capacity Factor} \times \text{Hours Per Year} \times \text{Operational Life (Years)}$$

Turbine capacity: 6 MW = 6000 kW

Capacity factor: 40% = 0.4

Hours per year: 365 days × 24 hours = 8760 hours

Operational life: 25 years

Energy delivered = 6000 × 0.4 × 8760 × 25 = 525 600 000 kWh

Mass of structural steel (tower) $= 1200 \text{ tons} \times 1000\frac{\text{kg}}{\text{ton}} = 1\,200\,000 \text{ kg}$

Mass of concrete $= 2000 \text{ tons} \times 1000\frac{\text{kg}}{\text{ton}} = 2\,000\,000 \text{ kg}$

Mass of steel reinforcement $= 833.33 \text{ m}^3 \times 400 \text{ kg/m}^3$
$= 333\,333 \text{ kg}$

Embodied carbon of structural steel (tower) $= 1\,200\,000 \text{ kg} \times 1.80 \text{ kgCO}_2\text{e/kg}$
$= 2\,160\,000 \text{ kgCO}_2\text{e}$

Embodied carbon of concrete (foundation) $= 2\,000\,000 \text{ kg} \times 0.15 \text{ kgCO}_2\text{e/kg}$
$= 300\,000 \text{ kgCO}_2\text{e}$

Embodied carbon of steel reinforcement (foundation) $= 333\,333 \text{ kg} \times 0.8 \text{ kgCO}_2\text{e/kg}$
$= 266\,667 \text{ kgCO}_2\text{e}$

The total embodied carbon of the wind turbine is the sum of the embodied carbon of the materials

$$= 2\,160\,000 + 300\,000 + 266\,667$$
$$= 2\,726\,667 \text{ kgCO}_2\text{e}$$

The embodied carbon intensity is calculated by dividing the total embodied carbon by the energy delivered over the turbine's lifetime:

Embodied Carbon Intensity ($\text{kgCO}_2\text{e/kWh}$) = Total Embodied Carbon (kgCO_2e)/Energy Delivered (kWh)

$$\text{Embodied Carbon Intensity} = 2\,726\,667 \text{ kgCO}_2\text{e}/525\,600\,000 \text{ kWh}$$
$$\approx 0.00519 \text{ kgCO}_2\text{e/kWh}$$

The embodied carbon of different types of structures varies significantly, as reported across multiple sources in the literature (Archer-Jones and Green, 2021; Gibbons et al., 2022; LETI, 2020b; RIBA, 2021; Collings, 2022). For buildings, the embodied carbon may range from 400 to 1200 $\text{kgCO}_2\text{e/m}^2$. For the structural systems within buildings, the range is often between 100 and 600 $kgCO_2e/m^2$. Bridges typically have a higher range of embodied carbon, between 800 and 6000 $kgCO_2e/m^2$. These values highlight the importance of material choices and construction methods in minimizing the carbon footprint of various types of infrastructure.

3.4.5 Embodied Carbon Rating System

The embodied carbon intensity in $kgCO_2e/m^2$ can be used to compare design decisions using the Structural Carbon Rating Scheme (SCORS) presented in Figure 3.20, provided by Arnold et al. (2020). This scheme puts in context the embodied carbon of the building and assesses it as high or low in embodied carbon compared to the typical range of industry practice. For bridges, Structural Carbon Rating Scheme for Bridgres (SCORBS) can be used (Archer-Jones and Green, 2021).

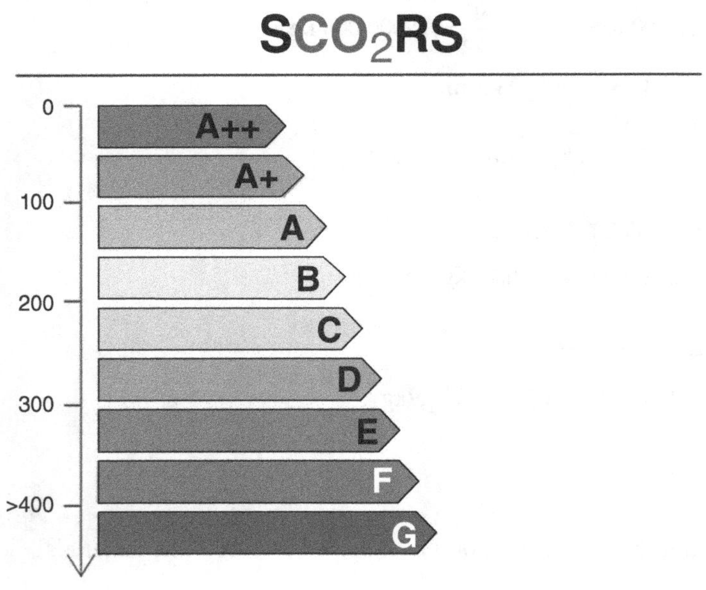

Figure 3.20 SCORS rating (based on total A1–A5 emissions for superstructure plus substructure, excluding biogenic carbon or offsetting, in accordance with the IStructE guide on how to calculate embodied carbon (Gibbons, O. P. and Orr, J. J./with permission of The Institution of Structural Engineers).

3.5 Ten Strategies for Reducing Embodied Carbon in Construction

Reducing the environmental impact of building construction requires interventions across its life cycle. In life-cycle assessment (LCA) terms, this involves cutting upfront embodied carbon (modules A1–A5, covering raw material production through construction), minimizing in-use impacts (module B, covering maintenance, repair, replacement, and associated transport and energy during the building's service life), facilitating low-impact *end-of-life* processes (modules C1–C4, covering demolition), and maximizing *beyond-life* benefits (module D, crediting reuse/recycling in future systems). At several sources in the literature, comprehensive strategies are listed (ARUP, 2025; LETI, 2020a; *Low Carbon*

Design Guidance – Hera, n.d.; SE2050, n.d.). This chapter summarizes the ten strategies presented in Figure 3.21 (putting into practice the key responsibilities for engineers and architects introduced in Chapter 1). We explain how each strategy relates to specific LCA stages, and detail them in the subsections that follow.

	Strategy	A1–A5	B	C1–C4	D
1	Design for and with circularity	✓	✓	✓	✓
2	Use high-strength materials where effective	✓	✓		
3	Explore conceptual options early	✓			
4	Adopt hybrid & composite systems	✓			
5	Employ prefabrication & modular construction	✓	✓	✓	✓
6	Source low-carbon materials	✓			
7	Engage stakeholders early in design	✓	✓	✓	✓
8	Apply digital, AI & automation	✓			
9	Design for durability & adaptability		✓		
10	Design for robustness		✓		

Figure 3.21 Relevance of each reduction strategy across LCA steps (A1–A5, B, C1–C4, D), with check marks showing the phase where each strategy has its greatest effect.

3.5.1 Design for and with Circularity

Design for and with circularity addresses both how we prepare buildings for future recovery of materials (design for circularity) and how we actively use reclaimed or recycled materials during construction (design with circularity). Together, these strategies aim to extend material life, reduce dependence on virgin (i.e., newly extracted, previously unused) resources, and lower embodied carbon across all stages of a building's life cycle (Figure 3.22). Importantly, design with circularity extends beyond the boundaries of a single project. It requires an understanding of the wider material ecosystem: decisions made on one project can have consequences for others. For instance, specifying a limited "low-carbon" resource (such as GGBS or reclaimed timber) in excessive quantities on a single project does not improve the industry overall if it simply depletes the available supply for other projects. True circular thinking therefore demands a systems-level perspective, balancing individual project optimization with the collective needs of the construction sector.

Design for circularity is relevant to the entire life cycle of a construction project, from material selection to end-of-life scenarios (Figure 3.22). One key principle is using mechanical connections (such as bolts or dry joints) instead of permanent methods like welding or adhesives. This allows structural components to be disassembled without being damaged, making future reuse or reassembly possible. Another strategy involves enabling component reuse by standardizing section profiles, limiting custom geometries, and documenting material specifications (such as steel grade or timber treatment). This facilitates easier classification, certification, and future repurposing of components. Efficient recycling should also be planned from the beginning, with clear separation of materials, clean demolition practices, and pathways for routing scrap back into production loops.

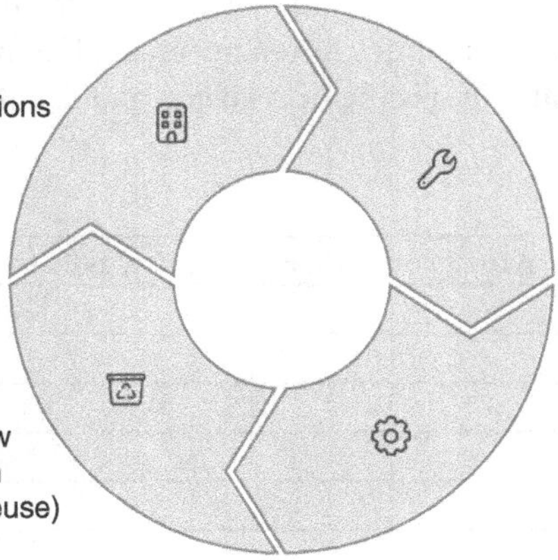

Figure 3.22 Relevance of circular construction to the whole life cycle of a building construction.

Design with circularity, on the other hand, focuses on the intentional use of reclaimed and recycled materials in new construction. Reclaimed structural steel is one of the most promising examples. When quality is verified and traceability is ensured (through testing, certification, or digital tracking), it can substitute virgin steel in many applications, dramatically reducing embodied carbon. Using reclaimed steel can also support a more robust secondary materials market and shift the role of demolition from waste generation to resource harvesting. The methods and challenges for implementing this specific circular approach are explored in detail in Kanyilmaz et al. (2023a).

All three major construction systems (steel, reinforced concrete, and timber) can benefit from circular strategies. Steel is inherently suited to reuse and recycling due to its durability, standardized components, and well-established collection and processing infrastructure. Reinforced concrete elements, while more challenging to reuse as whole components, can be designed for easy separation of concrete and rebars or for prefabricated forms that facilitate material recovery. Mass timber elements can be deconstructed and reused with minimal processing if their connections are carefully detailed and materials are tracked. Current end-of-life data illustrate the important differences between materials (SteelConstruction.info, n.d.): Steel has the strongest circularity profile, with 93% recycled and 6% directly reused, leaving only 1% sent to landfill. Concrete performs moderately, with 20% recycled and 75% downcycled (typically crushed for use as aggregate), though direct reuse of concrete elements remains negligible. Timber presents the greatest challenge: only 13% is recycled and 13% reused, while the majority is either incinerated (58%) or downcycled (10%), with 6% ending up in landfill. These figures highlight that, despite its biogenic carbon storage potential, timber's end-of-life reality currently undermines much of its circular promise, reinforcing the need for improved recovery infrastructure and design-for-disassembly practices.

However, there are practical limitations to this strategy. Designing for deconstruction may lead to more complex assemblies or slightly higher upfront material use. Using reclaimed materials often requires additional inspection, cleaning, or recertification efforts, which may not yet be used in all markets. Also, geographic variations (e.g. some regions have robust secondary steel markets but limited reclaimed timber supply) must be considered. Despite these challenges, integrating circular thinking (both in how we design for future disassembly and how we incorporate reused materials) offers a solid pathway to reduce environmental impact, close material loops, and build more resilient construction practices.

Before considering the recovery of individual materials or components, it is important to recognize that the most impactful circular strategy is often to retain and retrofit existing buildings in their entirety. Refurbishment and adaptive reuse of whole structures preserve the vast majority of the embodied carbon already invested in foundations, frames, and floors, avoiding the emissions associated with demolition, waste processing, and new construction altogether. Where a building's structural integrity and spatial configuration allow it, prioritizing retrofit over new-build should therefore be the first option considered within any circular design framework.

3.5.2 Use High-strength Materials Where Effective

High-strength materials have long played a central role in structural systems, enabling lighter, more efficient designs by improving performance per unit mass. In this progression, high-strength steel and high-performance concrete represent the most recent step, offering notable advantages when applied strategically. As shown in Figure 3.23, high-strength steel refers to grades that exceed the mechanical performance of standard structural steel (typically defined as having a yield strength higher than 460 MPa). These materials are produced through advanced metallurgy and postprocessing techniques, enabling higher load-bearing capacity with reduced structural weight. For instance, choosing S460 instead of S355 can allow for slimmer beam and column profiles, reducing both steel weight and the size of connections (Baddoo and Chen, 2020). While the alloying elements required for higher yield strength may marginally raise carbon intensity per kilogram, the net emissions per unit of load capacity usually drop when material savings are large. Higher-strength steels like S690 can make it possible to use very slender members in key locations (such as long-span roofs, minimizing cross-sections, and enabling more efficient designs). It is important to note that this benefit applies only when strength is the controlling limit (if deflection governs the design, section size remains tied to stiffness; not to strength). Because deflection depends on the elastic modulus (which doesn't increase with higher yield strength), using a high-strength grade won't reduce member dimensions in stiffness-controlled designs.

	15th century	18th century	19th century	20th century⁺
	Cast iron	**Wrought iron**	**Steel**	**High strength steel**
Better than	Better than **wood**	Better than **cast iron**	Better than **wrought iron**	Better* than **steel**
Step ahead	Higher strength	Higher ductility	Superior mechanical performance & higher production efficiency	Higher strength & lighter structures
Technology milestone	Hot furnaces	Steam engines & metallurgy	Bessemer process	Advanced metallurgy & post-processing

Figure 3.23 Evolution of metals (*high-strength steel can be considered "better" than steel for the applications where strength is the governing design parameter due to its lower mass-per-capacity).

A similar principle applies to concrete. Using higher-strength concrete (Fantilli et al., 2019) (which often has more cement per volume) can reduce total material quantity and thus net emissions. For example, using 60 MPa concrete for columns might allow significantly smaller cross-sections than 30 MPa concrete, which reduces the volume of concrete and steel reinforcement needed. The *embodied carbon per strength* can improve with high-strength mixes if the volume reduction is substantial. Ultra-high performance concrete (UHPC) (Farahzadi et al., 2024), while cement-rich, can enable very thin, long-lasting structures (its use in strategic locations, such as thin shell roofs) can minimize material. Additionally, high-performance concrete often has better durability, since its denser microstructure and lower permeability resist water ingress and chemical attacks (Neville, 2011). Which reduces future maintenance/replacement (an indirect benefit outside A1–A5, but still relevant to whole-life carbon footprint).

However, there are two main counterpoints to consider. First, high-strength materials can be more expensive and may require specialized fabrication and detailing, which can offset some environmental and economic gains. Second, the benefits are not universal (if used inappropriately or in stiffness-driven designs, the material savings may be negligible, while carbon intensity per unit mass remains higher). Having in mind these limitations, when applied in the right context, high-strength materials offer a clear path to reducing structural mass and embodied carbon, particularly in strength-driven designs and repetitive structural elements.

3.5.3 Explore Conceptual Options Early

Early-stage design decisions play a crucial role in the entire life-cycle performance of a structure. While choosing a scheme with minimal material use also impacts durability, adaptability, and maintenance during use (B), as well as future dismantling and reuse options (C–D), its greatest advantage is reducing the upfront carbon (A1–A5) because material quantities, structural layout, and system selection are largely determined at this stage, locking in the majority of embodied carbon before detailed design begins.

During conceptual design, multiple structural schemes (such as steel frame, reinforced concrete, or timber framing) can be explored to identify configurations that meet performance requirements with the least material use. Another important aspect is the exploration of building geometry and spatial configuration. Decisions such as the size of the structural grid, the span-to-depth ratio of floor systems, or whether to include a basement can have a carbon impact comparable to (or even exceeding) the choice of structural system itself. For example, a wider structural grid may reduce the number of columns and connections but demand heavier floor beams, while the inclusion of a basement introduces significant quantities of concrete, waterproofing, and excavation-related emissions. These geometric and configurational parameters should therefore be tested and compared alongside material system options from the earliest stages. Using computational optioneering and parametric design tools can allow rapid evaluation of multiple design schemes to identify the lowest-carbon option. Central to this process is the principle of data-driven design: rather than simply designing buildings and measuring their carbon after the fact, the goal is to ensure that real, project-specific data (such as verified carbon factors, regional supply chain information, and benchmark datasets) are available so that designers have proper metrics upon which to make genuinely informed decisions. As an example, Figure 3.24 – generated using a GA-based (genetic algorithm) preliminary design tool – shows how different structural systems (steel (circles), reinforced concrete (squares), and mass timber (triangles)) compare in terms of both cost and carbon footprint across varying floor space scenarios. The tool explored here a wide range of configurations to identify

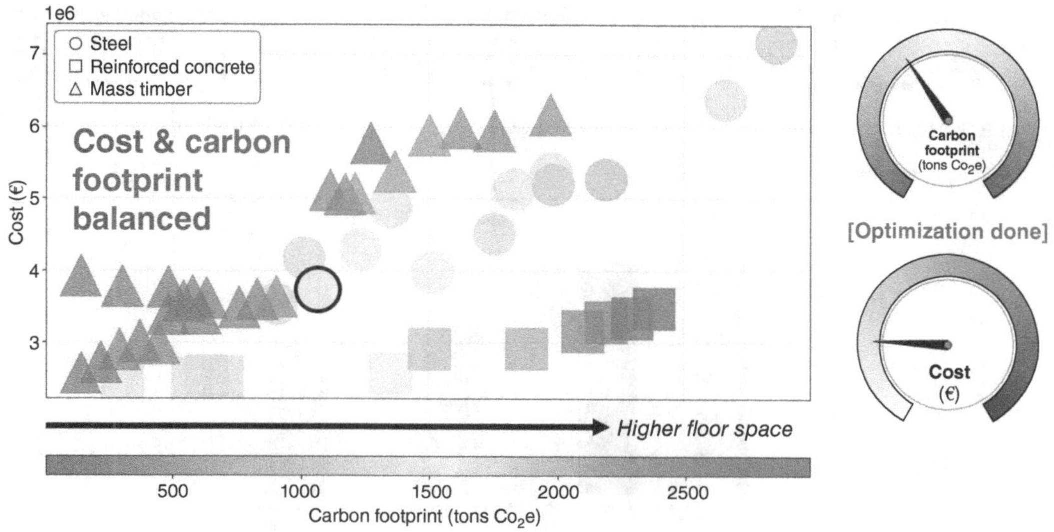

Figure 3.24 Optioneering example at conceptual design stage (Kanyilmaz et al., 2022, 2023b).

optimal trade-offs for a given design brief, with the black-circled option representing a balanced solution (Chapter 6 reports more examples related to this topic).

We should also note that the effectiveness of early-stage optimization depends on the quality of input assumptions, which are often based on rough estimates at this phase. In addition, design choices made too early without stakeholder input may reduce flexibility for architectural or functional changes later on. Having in mind these potential limitations, investing in concept-phase optioneering remains an effective strategy to reduce the environmental impact of the structural system.

3.5.4 Adopt Hybrid and Composite Systems

Combining materials in hybrid or composite systems allows each material to be used where it performs best, helping to reduce overall embodied carbon (e.g. composite reinforced concrete–steel floors, timber decks on steel beams, or steel-braced timber frames). This is especially relevant for floor systems, which often represent a significant portion of the total structural mass and carbon footprint. By enabling composite action (such as steel beams working together with a reinforced concrete slab), the structural depth and steel weight can be reduced without compromising performance. As an example, Figure 3.25 shows this with a comparison of three floor systems: two noncomposite steel beam options and one steel–concrete composite alternative. While the noncomposite options rely solely on heavy steel sections (HEB 360 or IPE 550) to carry loads, the composite system uses a lighter IPE 400 section in combination with a C25/30 reinforced concrete slab, exploiting the compressive strength of concrete and the tensile capacity of steel. As shown in the bar chart below, this results in a significantly lower A1–A3 embodied carbon value (167 $kgCO_2e/m^2$ compared to 294 $kgCO_2e/m^2$ and 230 $kgCO_2e/m^2$ for the noncomposite options).

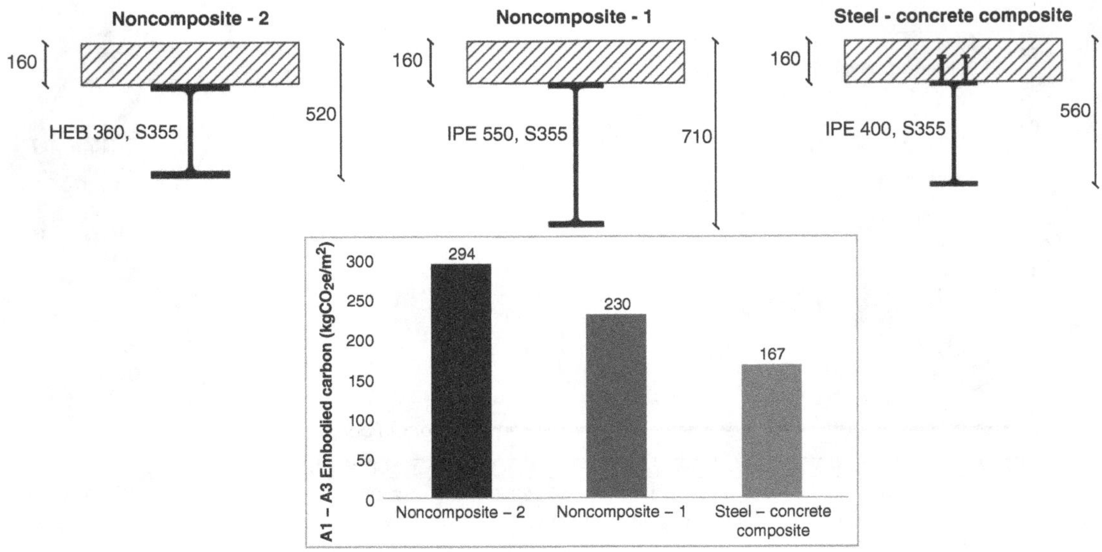

Figure 3.25 Composite construction can help save structural mass and eventually embodied carbon (image is produced thanks to a collaboration study between Florentia Kavoura and Alper Kanyilmaz).

However, hybrid and composite systems may involve more complex detailing and execution, requiring higher coordination during design and construction. In addition, end-of-life separation of materials can be more difficult, potentially limiting reuse and increasing recycling effort. Taking into account such limitations, composite systems offer a clear opportunity to reduce upfront carbon, especially in repetitive structural elements like floors.

3.5.5 Employ Prefabrication and Modular Construction

Prefabrication and modular construction can shift much of the work from the construction site to controlled factory environments (Figure 3.26). Off-site fabrication of modules, panels, or structural elements enables high-precision manufacturing, optimized material yields, and the use of reusable formwork, all of which can significantly reduce material waste. The factory-controlled setting also improves quality assurance and facilitates integration of building services into modules. Once transported to the site, the components can be rapidly assembled with minimal disruption, reducing on-site energy consumption, construction time, and labor requirements.

Prefabrication approach is applicable across all major structural materials: steel elements are often cut, welded, or bolted into frames or modules off-site; reinforced concrete components such as precast slabs, beams, and façade panels can be cast in reusable molds under controlled conditions; and mass timber products (e.g. CLT or glulam panels) are CNC-machined to precise dimensions for fast on-site assembly. Each material benefits from reduced waste, tighter tolerances, and improved assembly logistics.

This approach also leads to cleaner and safer construction sites, with lower noise and emissions, and can support future disassembly and relocation of components. Beyond construction, modular systems also support life cycle adaptability. Units designed for disassembly can be relocated or reused at end-of-life,

Figure 3.26 Example of modular construction.

reducing demolition waste and enabling material recovery (LCA phases C1–C4, module D). However, prefabrication is not without trade-offs. A notable challenge arises with precast concrete: factory production cycles prioritize rapid turnaround, which can limit the use of cement replacement materials (such as GGBS or fly ash) that slow down curing times. This tension between production speed and low-carbon mix design means that precast elements may, in some cases, carry a higher cement content (and therefore higher embodied carbon) than equivalent in situ concrete. There are solutions to this (such as steam curing, accelerators, or adjusted production scheduling), but they require designers to engage with precasters early in the design process to ensure that carbon reduction targets are not compromised by manufacturing constraints. While transport logistics and initial design coordination may require more planning, the long-term gains in efficiency, flexibility, and carbon reduction make prefabrication and modular approaches an increasingly attractive strategy in sustainable construction.

3.5.6 Source Low-carbon Materials

Low-carbon material sourcing focuses on reducing embodied emissions at the material production stage (A1–A3) and minimizing additional impacts during transport and site operations (A4–A5). This strategy involves selecting materials manufactured through cleaner processes, optimizing supply chain logistics, and incorporating reclaimed or recycled materials where possible. All three structural systems (steel, concrete, and timber) can benefit from low-carbon sourcing.

For steel, sourcing from electric arc furnace (EAF) routes using high scrap content and powered by renewable electricity can substantially lower emissions compared to traditional blast furnace steel. As shown in several industry studies, steel produced through EAF with a clean grid can cut A1–A3 emissions significantly (some examples are: ArcelorMittal, 2024; SSAB, 2025; Tata Steel UK, 2025). Another impactful strategy is using reclaimed steel (recovered from deconstructed buildings), where traceability and mechanical testing can confirm its suitability for structural applications. This not only avoids the need for remelting but also contributes to a circular resource loop with minimal added processing.

Reducing transport-related impacts (A4) can be achieved by sourcing materials and fabrications locally, while off-site prefabrication helps cut site waste (A5) by producing components to exact specifications. Timber offers a unique advantage through biogenic carbon storage: as trees grow, they absorb CO_2 from the atmosphere and sequester it within their wood fibers. When this timber is used in long-lived structural applications, the stored carbon remains locked in the building for the duration of its service life, effectively acting as a temporary carbon sink. This benefit is not typically captured within the A1–A3 or A1–A5 modules of a standard LCA (it is instead reported separately), but it should be an important consideration when comparing material options, particularly where the design also ensures that the timber will be recovered and reused or kept out of landfill and incineration at end of life. The climate advantage of timber holds only if the material comes from sustainably managed forests. For example, FSC certification (*Home | Forest Stewardship Council*, 2023) ensures responsible harvesting practices, maintaining carbon stocks, and promoting reforestation (without such certification, timber's environmental benefits are not guaranteed). Though today only around 9% of global forests are certified for sustainable management (PEFC, n.d.).

In concrete, the main source of carbon emissions is clinker in Portland cement (Yang et al., 2015). Reducing clinker content through supplementary cementitious materials (SCMs) (such as fly ash, ground granulated blast-furnace slag (GGBS), silica fume, or calcined clay (LC3)) is currently the most effective approach to cutting emissions. These substitutions can significantly lower A1–A3 impacts, provided the mix is engineered to maintain performance.

However, there are trade-offs. The carbon benefits of EAF steel depend heavily on the electricity mix (if powered by fossil fuels, the advantage is reduced to scrap use only). Reclaimed steel, while promising, requires careful quality control and is not always available at scale. FSC-certified timber may have higher costs and limited availability, and SCMs can vary in chemistry and performance, requiring rigorous validation. Despite these challenges, sourcing lower-carbon materials remains one of the most direct and scalable ways to cut upfront emissions, especially when integrated early in the design and procurement process.

3.5.7 Engage Stakeholders (e.g. Contractor) Early in Design

Engaging different stakeholders early in design (Figure 3.27), in particular contractors and structural fabricators, can significantly improve resource efficiency and reduce environmental impact during the early life-cycle stages (A1–A5). This is because involving key players from the start can bring practical views into buildability, fabrication, and logistics that design teams may otherwise overlook. For example, engaging structural steel fabricators early allows design choices to align with standard material dimensions and fabrication processes. This can lead to optimized bay spacing, reduced cutting and welding, and simpler connections (resulting in lower material waste and lighter structures). Similarly, early input can support decisions around modularization, transport logistics, and erection sequencing, improving both environmental and economic outcomes. While this approach requires coordination and may shift some design effort earlier in the process, the benefits include reduced rework, fewer site modifications, and a more streamlined construction phase, all contributing to lower embodied carbon and a more efficient project overall.

Figure 3.27 Building contractor with design team.

3.5.8 Apply Digital, AI-driven, and Automated Methods

Advanced digital, AI-driven, and automated methods are increasingly used across the construction sector to improve efficiency, reduce material waste, and lower environmental impacts, particularly during the early life-cycle stages (A1–A5). These approaches differ but are often integrated:

Digital methods refer to the use of data-rich models, such as BIM and parametric design, which allow for more accurate planning, coordination, and material quantification.

AI methods involve the use of machine learning, optimization algorithms, or generative design to support better decision-making – such as finding structurally efficient forms or identifying waste hotspots.

Automated methods include robotic fabrication, CNC machining, laser and plasma cutting, and automated assembly systems that increase precision and reduce manual errors in production.

As shown in the matrix in Figure 3.28, the combined effect of higher automation and targeted waste-reduction strategies can lead to significant reductions in material waste (up to 100% in cases where both are applied at high levels).

Each of the major structural material systems (steel, reinforced concrete, and timber) can benefit from these technologies. In steel construction, BIM-integrated workflows can generate automated shop drawings directly linked to robotic cutting, welding, and bolting systems, improving fabrication efficiency and reducing offcuts. In reinforced concrete, digital models can be used to generate precise formwork layouts and reinforcement schedules, while AI-based estimators can predict material demand more accurately. Automation in precast plants (such as robotic formwork handling or rebar bending) can enhance consistency and minimize waste. For mass timber, parametric design tools can be used to optimize panel layouts and jointing strategies. Automated CNC routers and nailing machines produce high-precision components, reducing tolerances and enabling rapid on-site installation with minimal adjustments. These technologies not only reduce scrap and rework but also enable better traceability of material inputs, supporting integration with life-cycle assessment tools.

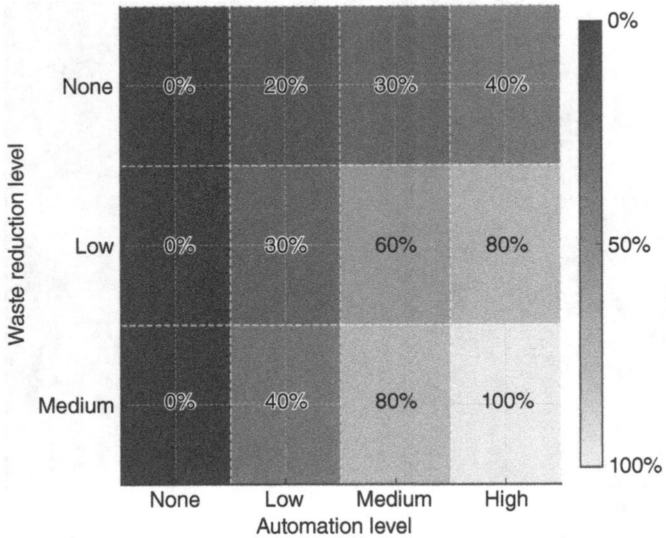

Figure 3.28 Correlation of automation level with waste reduction (% values are indicative).

Despite these benefits, some limitations remain. Initial investment costs in hardware, software, and training can be significant, especially for small- to medium-sized firms. Furthermore, integrating AI and automation into traditional workflows requires careful coordination and may not yield immediate savings in every project type. Over-reliance on digital tools can also lead to decisions made without a full understanding of construction realities if human oversight is reduced. Nevertheless, when thoughtfully applied, advanced digital, AI-based, and automated methods offer a powerful set of tools to improve material efficiency, reduce embodied carbon, and align construction practices with environmental and economic performance goals.

3.5.9 Design for Durability and Adaptability

Design for durability and adaptability is a strategy aimed at maximizing the service life of buildings while ensuring they can adjust to future demands of society (Figure 3.29). It responds to both environmental and functional goals by reducing the need for frequent interventions, replacements, or premature demolition – making it especially relevant to LCA stage B (use phase), with indirect effects on stages C1–C4.

Designing for durability means preventing material degradation through appropriate detailing, material selection, and protection strategies. For instance, minimizing moisture ingress is critical to avoid corrosion in steel, spalling in concrete, or biological decay in timber. Protective coatings, durable surface finishes, and effective drainage systems can greatly extend the lifespan of exposed components. Choosing materials based on their performance under site-specific environmental conditions (such as UV exposure, freeze–thaw cycles, or coastal corrosion) is essential for long-term resilience. Designing for adaptability focuses on making buildings flexible enough to accommodate future changes in use, occupancy, or layout without major structural modifications. This can involve providing generous

Figure 3.29 Structures needed for society, evolving over time.

floor-to-floor heights, moveable partitions, modular service zones, or oversized structural capacity to support potential load increases. By reducing the need for structural alterations during renovations, this approach limits material use, labor, and emissions associated with building upgrades.

All structural systems (steel, reinforced concrete, and timber) can incorporate durability and adaptability principles. Steel frames can be detailed with bolted connections and protective coatings to ease modifications and extend lifespan. Concrete structures can use durable mixes, minimize thermal bridging, and be designed for fire and corrosion resistance in harsh environments. Timber systems benefit from moisture management, replaceable components, and design for reversibility through dry jointing or modularity.

However, there are trade-offs. Designing for higher durability may increase material quantities (e.g. thicker coatings, protective cladding) and raise initial costs. Creating adaptable layouts can lead to underused space or oversizing of elements. Anticipating all future needs is rarely possible, so some adaptability features may remain unused. Nonetheless, prioritizing durability and adaptability from the outset contributes to more resilient, longer-lasting, and environmentally responsible buildings, supporting both performance and sustainability over time.

3.5.10 Design for Robustness

Design for robustness ensures that a structure can tolerate unexpected events (such as accidental loads or localized damage) without major loss of function. This reduces the need for repairs and interventions during use (LCA module B) and delays end-of-life activities (C1–C4), helping to lower the overall environmental impact. Robust design typically involves redundancy, fail-safe detailing (Figure 3.30),

Figure 3.30 Robust structures last more.

and the use of ductile materials. While robustness may increase material use, it helps preserve structural elements over time; especially valuable since these are often the longest-lasting parts of a building. Enhancing robustness increases the likelihood that core components will be retained through future adaptations, supporting long-term sustainability.

The strategies presented in this section highlight the multiple entry points available to reduce embodied carbon across a building's life cycle. While each approach targets different LCA phases (from early material selection to end-of-life planning), they are most effective when applied in combination. Whether through high-strength materials, circular design, early stakeholder involvement, or digital tools, these strategies provide a practical foundation for integrating life-cycle thinking into design and construction workflows.

3.6 Exercise: Embodied Carbon Calculation of Structural Elements Under Different Conditions and Assumptions

This chapter guides you through calculating and comparing the cradle-to-gate embodied carbon of three common structural systems, then exploring how key input assumptions influence the results. For the purposes of this exercise, the structural design is assumed to be complete since the focus is not on structural sizing or optimization, but on calculating the embodied carbon associated with the resulting design choices. In Part 1, we will quantify embodied carbon for a representative building model using fixed Global Warming Potential (GWP) factors. In Part 2, we will perform a parameter analysis by varying those same factors to see which material choices and section sizes have the greatest impact.

The benchmark model is a 30 m × 30 m, five-story building with 4-m-high floors, as shown in Figure 3.31, with both isometric and ground-floor plan views for each structural system. All three versions share identical geometry, column grid, and lateral-resisting elements; only material, member sizing, and connection details change.

Reinforced-concrete Frame:

- Primary beams span 6 m, secondary beams 5 m, all assumed as pinned at beam-to-column and column-to-foundation joints.
- Two reinforced-concrete shear wall systems to provide lateral stability.

Steel Frame:

- Universal-section columns and I-section beams fabricated from three sourcing scenarios: 100% recycled, some scrap/raw, and virgin steel.
- Primary beams span 6 m, secondary beams span 5 m at every 2 m; all connections are assumed as pinned joints.
- Reinforced-concrete shear walls used for lateral resistance.

Timber–CLT Hybrid Frame:

- CLT walls at perimeter and shear-wall locations carry lateral loads; Glulam columns and beams form the gravity frame.
- Primary glulam beams span 6 m, secondary members span 5 m at every 1.5 m.
- Pinned dowel and bracket connections throughout; timber specified as Service Class 1 (dry interior environment).

For each system, we will calculate embodied carbon for three member types (a secondary beam, a primary beam, and a column) using the GWP factors in Table 3.9. Comparing results across these identical geometries reveals how material choice, sourcing, and section sizing drive embodied carbon differences. Detailed sensitivity runs in Part 2 will then show how robust those conclusions are to changes in assumed GWP values (e.g. with increased structural demand).

Notes on Assumptions:

The chapter simplifies geometry by holding spans and grids constant across systems. In reality, design codes require stiffness and strength checks, leading to varied section sizes and details.

To isolate embodied-carbon differences, all three frames use the same spans, section layouts, and fully pinned joints. In practice, each material would require specific sizing, connection stiffness, and diaphragm detailing to satisfy deflection, buckling, and lateral-load requirements, and timber elements must meet service-class and moisture-control criteria. These simplifications do not affect our carbon comparison, but they would need to be replaced by code-compliant designs before any real-world application.

Embodied carbon calculations are based on Global Warming Potential (GWP) factors (focusing only on cradle-to-gate emissions, i.e. A1–A3), expressed in kilograms of CO_2 equivalent per kilogram of material ($kgCO_2e/kg$), provided in Table 3.9. The values chosen here fall within typical literature ranges but are used solely for consistency (whether they sit at the high or low end of reported GWPs won't affect our

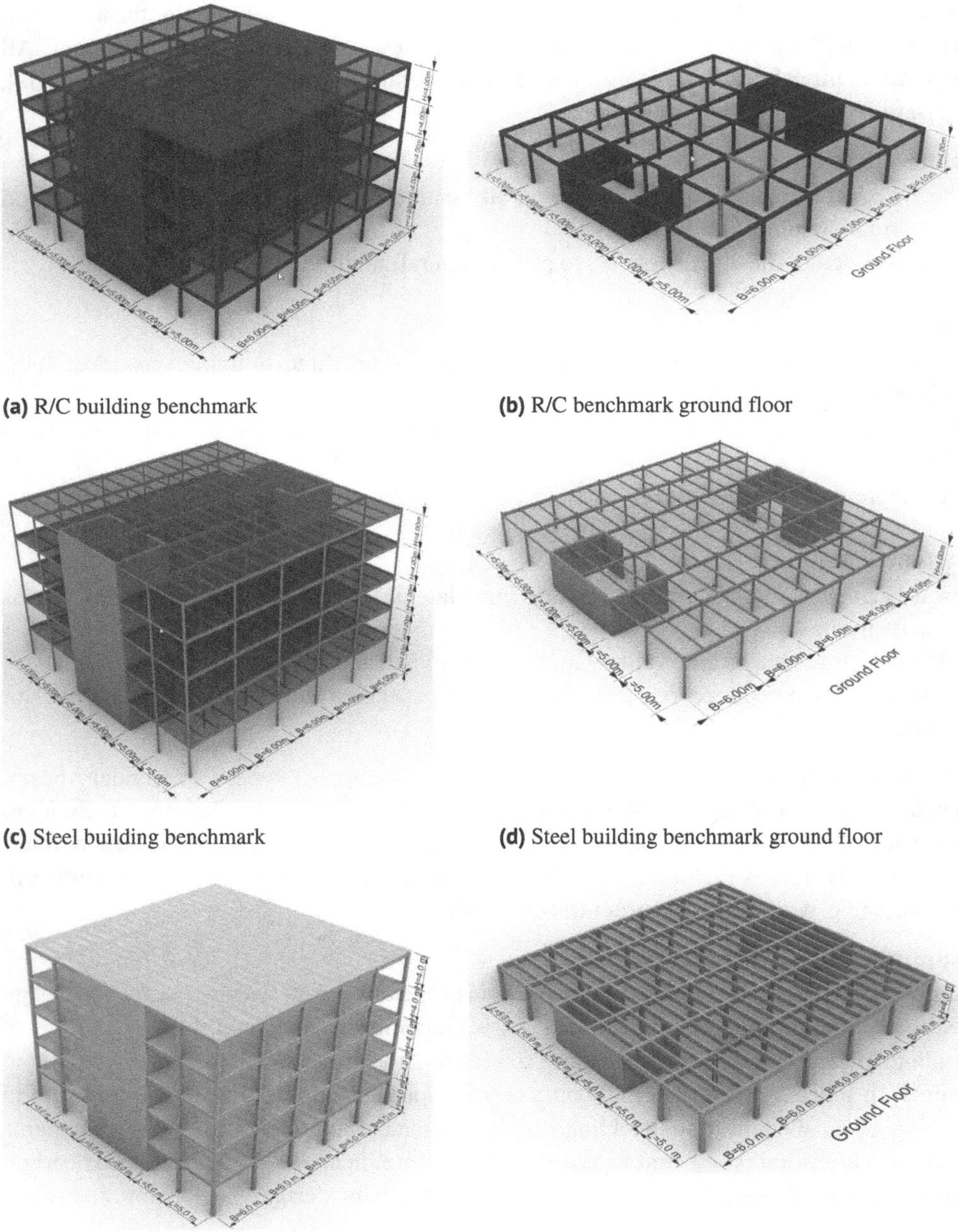

(a) R/C building benchmark

(b) R/C benchmark ground floor

(c) Steel building benchmark

(d) Steel building benchmark ground floor

(e) Timber building benchmark

(f) Timber building benchmark ground floor

Figure 3.31 Isometric and ground-floor plan views for each benchmark structural system.

Table 3.9 Global Warming Potential (GWP) values for selected construction materials ($kgCO_2e/kg$, modules A1–A3). Values are representative and illustrate the influence of production method and sourcing on embodied carbon: steel ranges from EAF-route (high scrap content) to BF-BOF route (virgin ore); timber values reflect the carbon accounting difference between sustainably managed forests (where regrowth offsets harvest) and unsustainable sources (where deforestation releases stored carbon). Concrete values correspond to a standard C25/30 mix.

Material	GWP ($kgCO_2e/m^2$)
Steel (produced with high scrap content)	0.500
Mix of recycled and raw steel	1.800
Steel from raw material	2.500
Concrete C25/30	0.133
Rebars	0.760
Timber from sustainable source	0.474
Timber from unsustainable source	2.100

comparison). Global GWP values differ widely due to local energy grids, production methods, and supply chains. By holding these factors constant, we can see how different structural materials respond in terms of embodied-carbon sensitivity. Normally, the values can vary depending on supply chain specifics.

We performed calculations for a "secondary beam," a "primary beam," and "a column" (for a ground floor) indicated in Figure 3.34.

The following design cases are considered:

Timber structure: Two sets of calculations for each element, using both sustainable and unsustainable timber values.

Steel structure: Based on steel sourced from high, some, and low scrap content material.

Reinforced concrete structure: Taking into account the emissions from concrete and reinforcement steel (rebar).

Once the values are determined, they are compared across all materials to illustrate how sourcing and material choice influence embodied carbon emissions. Assumed values for calculations include a timber density of 470 kg/m^3, concrete density of 2400 kg/m^3, a rebar ratio of 2% by concrete volume, and a rebar density of 7850 kg/m^3. In the calculations, connection details and other elements than stated (such as fire protection) are not taken into account.

3.6.1 Solution Part 1: Benchmark Structural Demand

For this exercise, the structural design is assumed to be complete (our focus here is not on structural sizing but on calculating the embodied carbon of the design outcomes). The cross-sectional properties of all structural elements are summarized in Table 3.10. These form the basis for the embodied carbon calculations detailed in the following sections.

Table 3.10 Cross-sectional dimensions and shapes of structural elements for timber, reinforced concrete, and steel systems.

	Timber	Reinforced Concrete	Steel
Secondary Beam Section L = 5 m	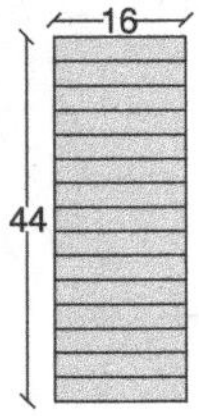	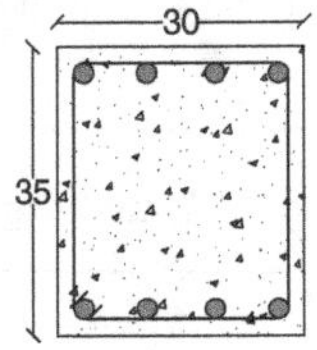	
	GL44 cm × 16 cm	**B35 cm × 30 cm**	**IPE220**
Primary Beam Section L = 6 m	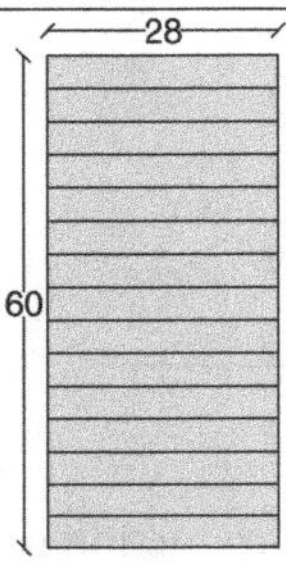	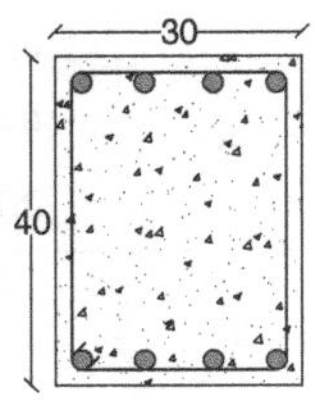	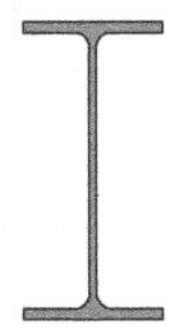
	GL60 cm × 28 cm	**B40 cm × 30 cm**	**IPE360**
Column Section L = 4 m	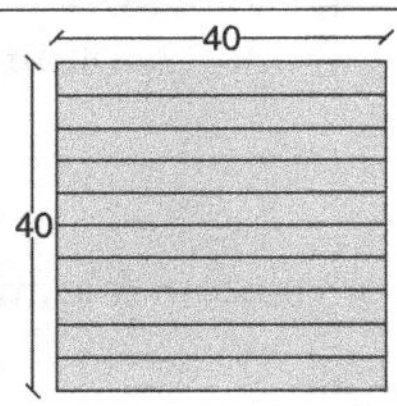	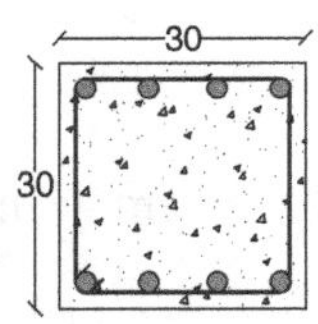	
	GL40 cm × 40 cm	**C30 cm × 30 cm**	**CHS 219.1 mm × 10.0 mm**

3.6.1.1 Timber Beams, Option 1 (GWP = 0.474 $kgCO_2e/kg$)

Secondary Beam Section (L = 5 m):

Volume	$= 5\ m \times 0.44\ m \times 0.16\ m$	$= 0.352\ m^3$
Mass	$= 0.352\ m^3 \times 470\ kg/m^3$	$= 166\ kg$
Embodied Carbon	$= 166\ kg \times 0.474\ kgCO_2e/kg$	$\approx 79\ kgCO_2e$

Primary Beam Section (L = 6 m):

Volume	$= 6\ m \times 0.60\ m \times 0.28\ m$	$= 1.008\ m^3$
Mass	$= 1.008\ m^3 \times 470\ kg/m^3$	$= 474\ kg$
Embodied Carbon	$= 474\ kg \times 0.474\ kgCO_2e/kg$	$\approx 225\ kgCO_2e$

Column Section (L = 4 m):

Volume	$= 4\ \text{m} \times 0.40\ \text{m} \times 0.40\ \text{m}$	$= 0.640\ \text{m}^3$
Mass	$= 0.640\ \text{m}^3 \times 470\ \text{kg/m}^3$	$= 301\ \text{kg}$
Embodied Carbon	$= 301\ \text{kg} \times 0.474\ \text{kgCO}_2\text{e/kg}$	$\approx 143\ \text{kgCO}_2\text{e}$

3.6.1.2 Timber Beams, Option 2 (GWP = 2.100 $kgCO_2e/kg$)

Since the dimensions, volume, and mass of the elements remain the same, only the embodied carbon values need to be recalculated using the higher GWP value.

Secondary Beam Section (L = 5 m):

Embodied Carbon $= 166\ \text{kg} \times 2.100\ \text{kgCO}_2\text{e/kg} \approx 349\ \text{kgCO}_2\text{e}$

Primary Beam Section (L = 6 m):

Embodied Carbon $= 474\ \text{kg} \times 2.100\ \text{kgCO}_2\text{e/kg} \approx 995\ \text{kgCO}_2\text{e}$

Column Section (L = 4 m):

Embodied Carbon $= 301\ \text{kg} \times 2.100\ \text{kgCO}_2\text{e/kg} \approx 632\ \text{kgCO}_2\text{e}$

3.6.1.3 Reinforced Concrete Embodied Carbon Calculation

Reinforcement is assumed at 2% of concrete volume, applied consistently across beams and columns.

Secondary Beam Section (L = 5 m)

Volume:

Concrete Volume $= 5\ \text{m} \times 0.35\ \text{m} \times 0.30\ \text{m} = 0.525\ \text{m}^3$
Rebar Volume $= 0.525\ \text{m}^3 \times 0.02 \approx 0.011\ \text{m}^3$

Mass:

Concrete Mass $= 0.525\ \text{m}^3 \times 2400\ \text{kg/m}^3 \approx 1260\ \text{kg}$
Rebar Mass $= 0.011\ \text{m}^3 \times 7850\ \text{kg/m}^3 \approx 86\ \text{kg}$

Embodied Carbon, Total:

$(1260\ \text{kg} \times 0.133\ \text{kgCO}_2\text{e/kg}) + (86\ \text{kg} \times 0.760\ \text{kgCO}_2\text{e/kg}) \approx 232\ \text{kgCO}_2\text{e}$

Primary Beam Section (L = 6 m)

Volume:

Concrete Volume $= 6\ \text{m} \times 0.40\ \text{m} \times 0.30\ \text{m} = 0.720\ \text{m}^3$
Rebar Volume $= 0.720\ \text{m}^3 \times 0.02 \approx 0.014\ \text{m}^3$

Mass:

Concrete Mass = 0.720 m^3 × 2400 kg/m^3 ≈ 1728 kg
Rebar Mass = 0.014 m^3 × 7850 kg/m^3 ≈ 110 kg

Embodied Carbon, Total:

(1728 kg × 0.133 kgCO_2e/kg) + (110 kg × 0.760 kgCO_2e/kg) ≈ 313 kgCO_2e

Column Section (L = 4 m)

Volume:

Concrete Volume = 4 m × 0.30 m × 0.30 m = 0.360 m^3
Rebar Volume = 0.360 m^3 × 0.02 ≈ 0.007 m^3

Mass:

Concrete Mass = 0.360 m^3 × 2400 kg/m^3 ≈ 864 kg
Rebar mass = 0.007 m^3 × 7850 kg/m^3 ≈ 55 kg

Embodied Carbon, Total:

(864 kg × 0.133 kgCO_2e/kg) + (55 kg × 0.760 kgCO_2e/kg) ≈ 156 kgCO_2e

3.6.1.4 Steel Embodied Carbon Calculation

Steel (Produced with High Scrap Content) (GWP = 0.500 kgCO_2e/kg)

Secondary Beam Section (L = 5 m):

Cross-section = IPE220 (mass per meter = 26.2 kg/m)
Mass = 5 m × 26.2 kg/m = 131 kg
Embodied Carbon = 131 kg × 0.500 kgCO_2e/kg ≈ 66 kgCO_2e

Primary Beam Section (L = 6 m):

Cross-section = IPE360 (mass per meter = 57.1 kg/m)
Mass = 6 m × 57.1 kg/m ≈ 343 kg
Embodied Carbon = 343 kg × 0.500 kgCO_2e/kg ≈ 172 kgCO_2e

Column Section (L = 4 m):

Cross-section = CHS 219.1 mm × 10.0 mm (mass per meter = 51.6 kg/m)
Mass = 4 m × 51.6 kg/m ≈ 207 kg
Embodied Carbon = 207 kg × 0.500 kgCO_2e/kg ≈ 104 kgCO_2e

Steel (Produced with Some Scrap Content) (GWP = 1.800 kgCO$_2$e/kg)
Since the dimensions, volume, and mass of the elements remain the same, only the embodied carbon values need to be recalculated using the higher GWP value.

Secondary Beam Section (L = 5 m):

Embodied Carbon = 131 kg × 1.800 kgCO$_2$e/kg ≈ 236 kgCO$_2$e

Primary Beam Section (L = 6 m):

Embodied Carbon = 343 kg × 1.800 kgCO$_2$e/kg ≈ 617 kgCO$_2$e

Column Section (L = 4 m):

Embodied Carbon = 207 kg × 1.800 kgCO$_2$e/kg ≈ 373 kgCO$_2$e

Steel (Produced with Low Scrap Content) (GWP = 2.500 kgCO$_2$e/kg)
Again, since the dimensions, volume, and mass of the elements remain the same, only the embodied carbon values need to be recalculated using the higher GWP value.

Secondary Beam Section (L = 5 m):

Embodied Carbon = 131 kg × 2.500 kgCO$_2$e/kg ≈ 328 kgCO$_2$e

Primary Beam Section (L = 6 m):

Embodied Carbon = 343 kg × 2.500 kgCO$_2$e/kg ≈ 858 kgCO$_2$e

Column Section (L = 4 m):

Embodied Carbon = 207 kg × 2.500 kgCO$_2$e/kg ≈ 518 kgCO$_2$e

Now that the embodied carbon and mass of each structural element have been calculated, we can proceed to compare the results across different materials. The following sections visually compare the secondary beams, primary beams, and columns made from timber, reinforced concrete, and steel. These comparisons highlight not only the differences in mass but also their environmental impact based on embodied carbon emissions.

3.6.2 Component-wise Comparison

This section compares the mass and embodied carbon emissions of secondary beams, primary beams, and columns constructed from timber, reinforced concrete, and steel. The analysis demonstrates how material selection influences both the physical weight and the environmental impact of structural elements. Particular attention is given to how sourcing strategies, especially for steel and timber, significantly affect embodied carbon outcomes.

3.6.2.1 Beams

As shown in Figures 3.32 and 3.33, reinforced concrete is by far the heaviest. Greater mass typically leads to increased transportation demands, more challenging handling on site, and the need for more substantial foundations.

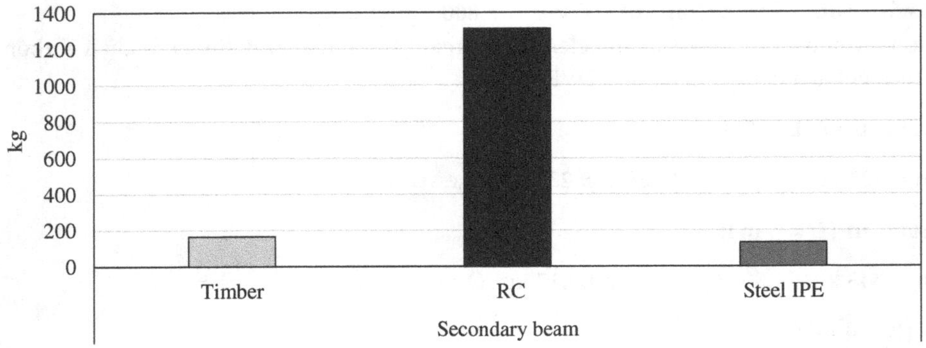

Figure 3.32 Mass comparison of secondary beams made of timber, reinforced concrete, and steel.

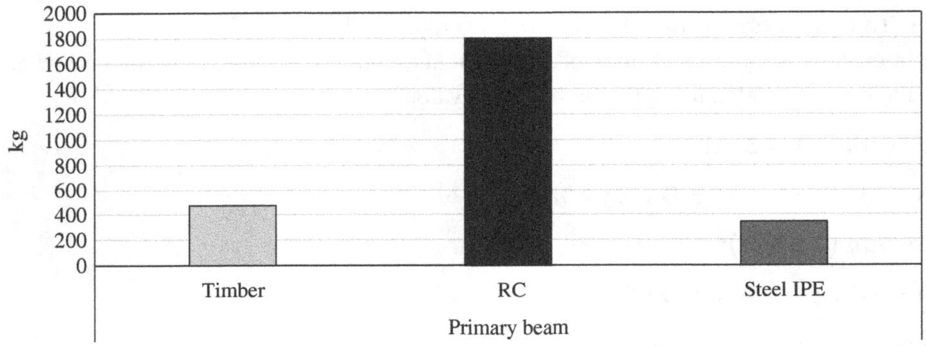

Figure 3.33 Mass comparison of primary beams made of timber, reinforced concrete, and steel.

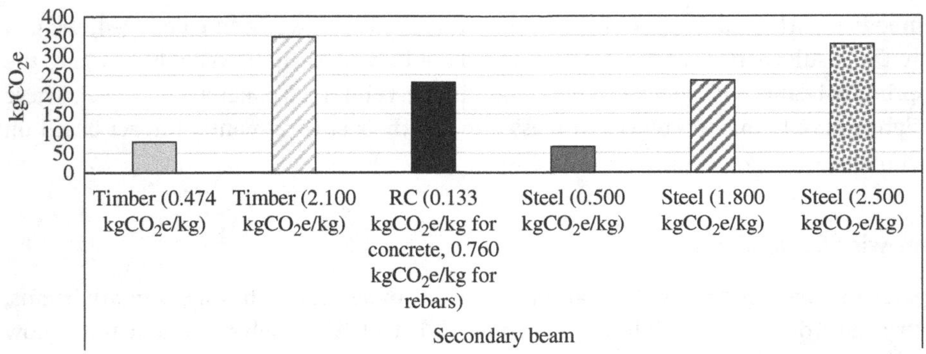

Figure 3.34 Embodied carbon comparison of secondary beams.

The embodied carbon emissions associated with each beam are shown in Figures 3.34 and 3.35. Reinforced concrete beams, despite their large mass, maintain a moderate level of embodied carbon. Timber beams demonstrate the critical role of sourcing: when sustainably sourced, timber achieves very low emissions, whereas the use of unsustainable timber greatly increases the carbon footprint, even higher than raw steel. Beams manufactured from steel (produced with high scrap content) result in relatively low emissions, while reliance on partial scrap or raw materials leads to an increase in the embodied carbon.

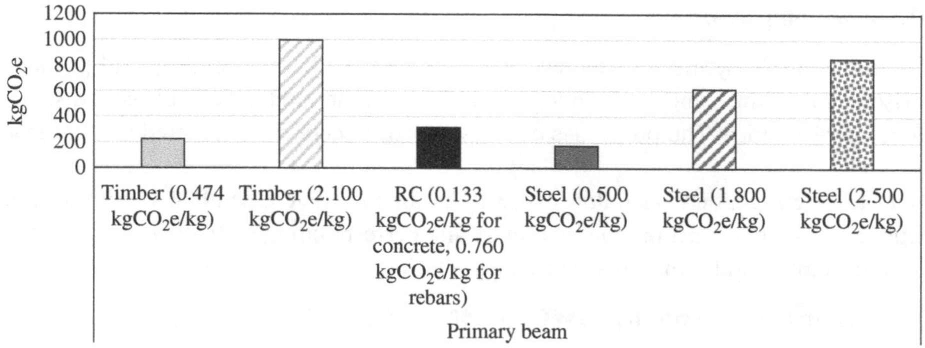

Figure 3.35 Embodied carbon comparison of primary beams.

3.6.2.2 Columns

The mass and embodied carbon comparison for columns, shown in Figures 3.36 and 3.37, follows a similar pattern with the beams.

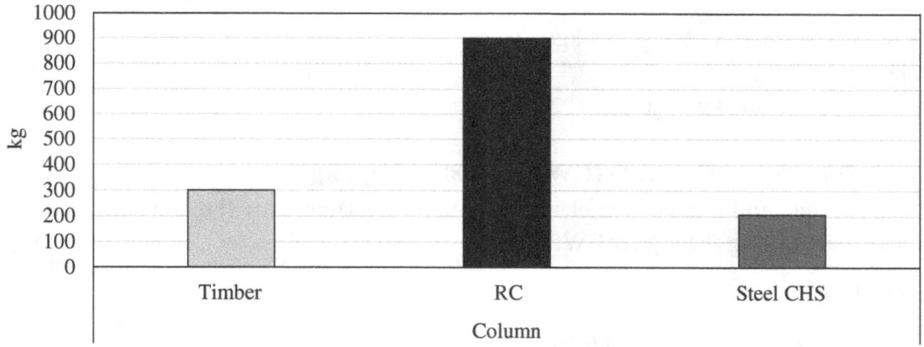

Figure 3.36 Mass comparison of columns made of timber, reinforced concrete, and steel.

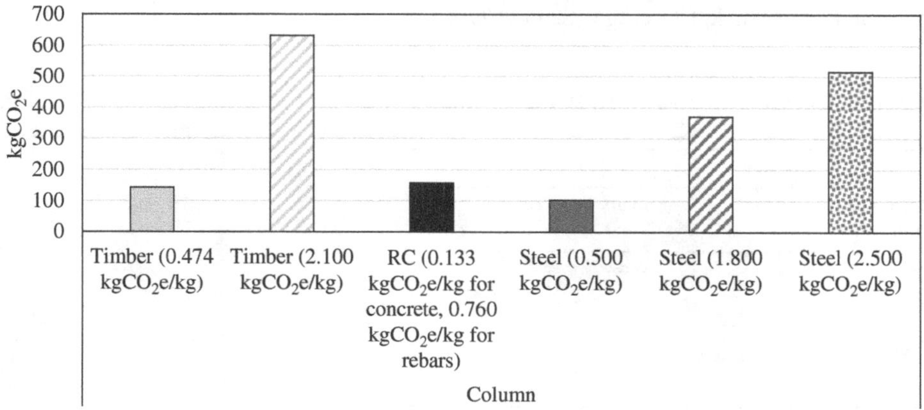

Figure 3.37 Embodied carbon comparison of columns.

3.6.3 Steel Profile Type Comparison

This section extends the analysis by introducing steel HEA profiles for both secondary and primary beams. The comparisons follow the same methodology and highlight the implications of section selection on environmental performance. The properties of the steel HEA sections considered are summarized in Table 3.11.

The mass per meter for each section is as follows: 42.3 kg/m for the HEA200 and 76.4 kg/m for the HEA280. These values are used to determine the total mass and corresponding embodied carbon emissions for each structural element under the three sourcing scenarios.

Steel (Produced with High Scrap Content) (GWP = 0.500 kgCO$_2$e/kg)

Secondary Beam Section (L = 5 m):

Cross-section = HEA200 (Mass per Meter = 42.3 kg/m)
Mass = 5 m × 42.3 kg/m ≈ 212 kg
Embodied Carbon = 212 kg × 0.500 kgCO$_2$e/kg ≈ 106 kgCO$_2$e

Primary Beam Section (L = 6 m):

Cross-section = HEA280 (Mass per Meter = 76.4 kg/m)
Mass = 6 m × 76.4 kg/m ≈ 458 kg
Embodied Carbon = 458 kg × 0.500 kgCO$_2$e/kg ≈ 229 kgCO$_2$e

Steel (Produced with Some Scrap Content) (GWP = 1.800 kgCO$_2$e/kg)
Since the dimensions, volume, and mass of the elements remain the same, only the embodied carbon values need to be recalculated using the higher GWP value.

Secondary Beam Section (L = 5 m):

Embodied Carbon = 212 kg × 1.800 kgCO$_2$e/kg ≈ 382 kgCO$_2$e

Primary Beam Section (L = 6 m):

Embodied Carbon = 458 kg × 1.800 kgCO$_2$e/kg ≈ 824 kgCO$_2$e

Table 3.11 Cross-sectional types and dimensions for HEA steel profiles.

Steel	
Secondary beam section L = 5 m	Primary beam section L = 6 m
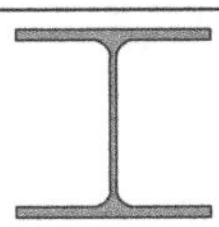	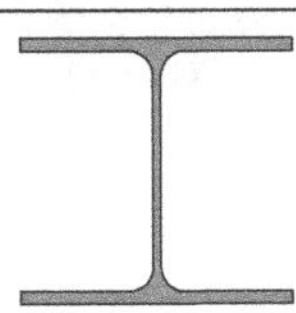
HEA200	HEA280

Steel (Produced Low Scrap Content) Calculations (GWP = 2.500 $kgCO_2e/kg$)

Secondary Beam Section (L = 5 m):

Embodied Carbon = 212 kg × 2.500 $kgCO_2e/kg \approx 530\ kgCO_2e$

Primary Beam Section (L = 6 m):

Embodied Carbon = 458 kg × 2.500 $kgCO_2e/kg \approx 1145\ kgCO_2e$

3.6.3.1 Beams

The mass comparison for secondary beams, now including HEA sections, is presented in Figure 3.38. The embodied carbon emissions for these secondary beams are shown in Figure 3.39. While the mass of HEA beams is higher than IPE beams, their embodied carbon still strongly depends on the material sourcing. As with IPE beams, the embodied carbon impact for HEA sections rises steeply when the proportion

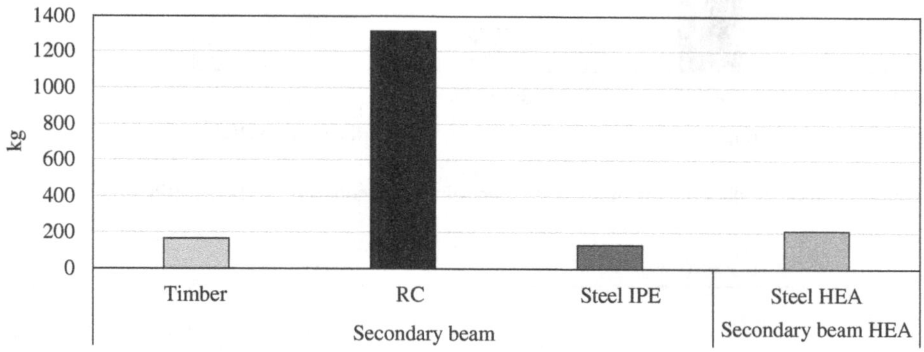

Figure 3.38 Mass comparison of secondary beams made of timber, reinforced concrete, and steel (IPE and HEA profiles).

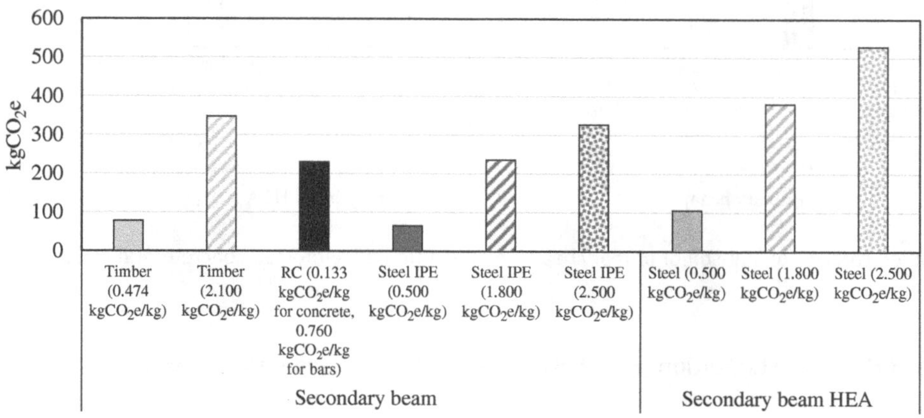

Figure 3.39 Embodied carbon comparison of secondary beams made of timber, reinforced concrete, and steel (IPE and HEA profiles).

of raw material increases. Figures 3.40 and 3.41 present the mass and embodied carbon comparison for primary beams including HEA profiles. The pattern is similar to that observed in secondary beams. Although HEA profiles are heavier than IPE sections, the environmental penalty remains relatively small if steel (produced with high scrap content) is specified. Furthermore, HEA sections have a lower depth than IPE profiles of comparable capacity, reducing floor zone thickness. In high-rise construction, these savings can accumulate to permit an additional floor within the same building height—an outcome with clear economic and functional benefits.

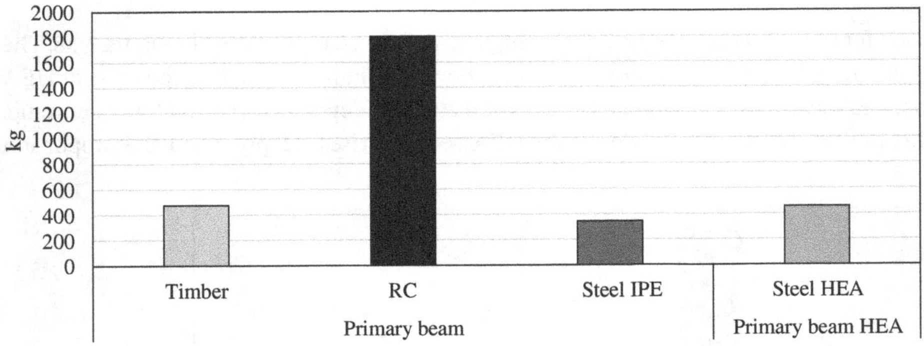

Figure 3.40 Mass comparison of primary beams made of timber, reinforced concrete, and steel (IPE and HEA profiles).

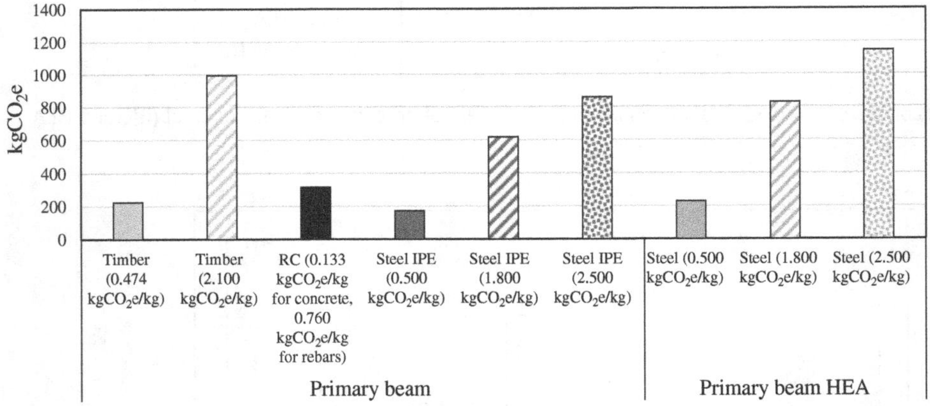

Figure 3.41 Embodied carbon comparison of primary beams made of timber, reinforced concrete, and steel (IPE and HEA profiles).

3.6.4 Influence of Boundary Condition (Fixed vs. Pinned) on Embodied Carbon Results of Steel Beams

In previous comparisons, steel beams were considered with pinned connections (Figure 3.42a). This section examines how adopting rigid (fixed-end) connections (Figure 3.42b) can influence the embodied

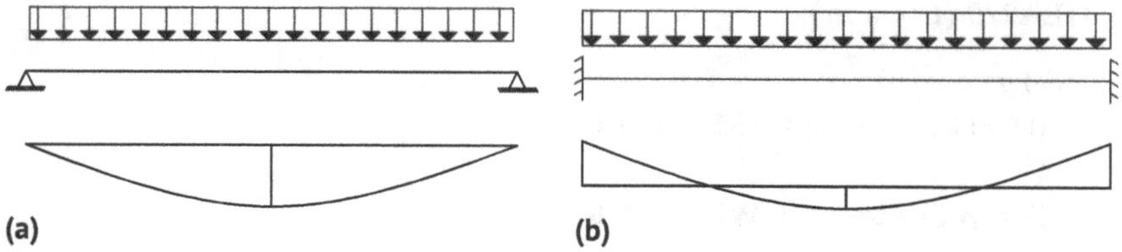

Figure 3.42 Typical bending moment diagram of beams with distributed load under different boundary conditions. (a) Simply supported pinned beam; (b) rigidly supported (fixed-end) beam.

Table 3.12 Cross-sectional types and dimensions for fixed-end steel primary beams.

Steel	
Primary beam IPE section L = 6 m	Primary beam HEA section L = 6 m
	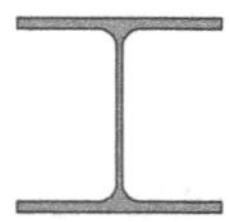
IPE300	HEA220

carbon emissions of primary beams. In beams with rigid end connections, higher bending moments are transferred between the beam and the supporting columns, but the internal moment distribution becomes more favorable, with lower bending moments at midspan. This reduction in midspan bending can allow for smaller section sizes to be used while satisfying the same load requirements. As a result, lighter beams can sometimes be adopted compared to simply supported designs, which may reduce total mass and embodied carbon.

The analysis focuses on the primary steel beams studied in the previous section (HEA280 with pinned-end assumption). When they are redesigned with a rigid-end assumption, the new sizes become: an IPE300 section and an HEA220 section (Table 3.12). These two alternatives are presented to compare different structural profiles that achieve similar performance while offering distinct trade-offs in weight, stiffness, and embodied carbon. The embodied carbon is calculated for three different steel sourcing scenarios, using the same Global Warming Potential (GWP) values as presented in Table 3.9.

The mass per meter for each section is as follows: 42.2 kg/m for the IPE300 and 50.5 kg/m for the HEA220. These values are used to calculate total mass and embodied carbon under each sourcing condition.

Steel (Produced with High Scrap Content) (GWP = 0.500 kgCO$_2$e/kg)

Primary Beam Section – IPE300 (L = 6 m):

Mass = 6 m × 42.2 kg/m = 253 kg
Embodied Carbon = 253 kg × 0.500 kgCO$_2$e/kg ≈ 127 kgCO$_2$e

Primary Beam Section – HEA220 (L = 6 m):

Mass = 6 m × 50.5 kg/m = 303 kg
Embodied Carbon = 303 kg × 0.500 kgCO_2e/kg ≈ **152 kgCO_2e**

Steel (Produced with Some Scrap Content) (**GWP = 1.800 kgCO_2e/kg**)

Primary Beam Section – IPE300 (L = 6 m):

Embodied Carbon = 253 kg × 1.800 kgCO_2e/kg ≈ 455 kgCO_2e

Primary Beam Section – HEA220 (L = 6 m):

Embodied Carbon = 303 kg × 1.800 kgCO_2e/kg ≈ **545 kgCO_2e**

Steel (Produced Low Scrap Content) Calculations (**GWP = 2.500 kgCO_2e/kg**)

Primary Beam Section – IPE300 (L = 6 m):

Embodied Carbon = 253 kg × 2.500 kgCO_2e/kg ≈ 633 kgCO_2e

Primary Beam Section – HEA220 (L = 6 m):

Embodied Carbon = 303 kg × 2.500 kgCO_2e/kg ≈ 758 kgCO_2e

These calculations show that fixed-end steel connections, by reducing midspan bending, can allow the use of more compact sections with improved structural efficiency (see Figure 3.43: simply supported IPE 360 (6 m) = 343 kg and HEA 280 (6 m) = 458 kg; rigid-end IPE 300 (6 m) = 253 kg and HEA 220 (6 m) = 303 kg).

The embodied carbon values are presented in Figure 3.44. Rigidly connected steel beams have lower upfront embodied carbon than simply supported steel beams across all steel sourcing conditions.

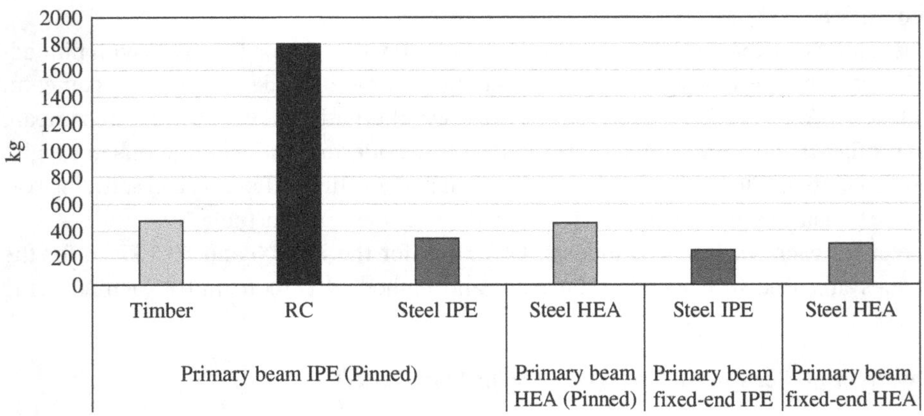

Figure 3.43 Mass of primary beams under simple-pinned (Steel IPE 360, Steel HEA 280) vs. rigid-connection conditions (Steel rigid IPE, Steel rigid HEA), with timber and RC for context.

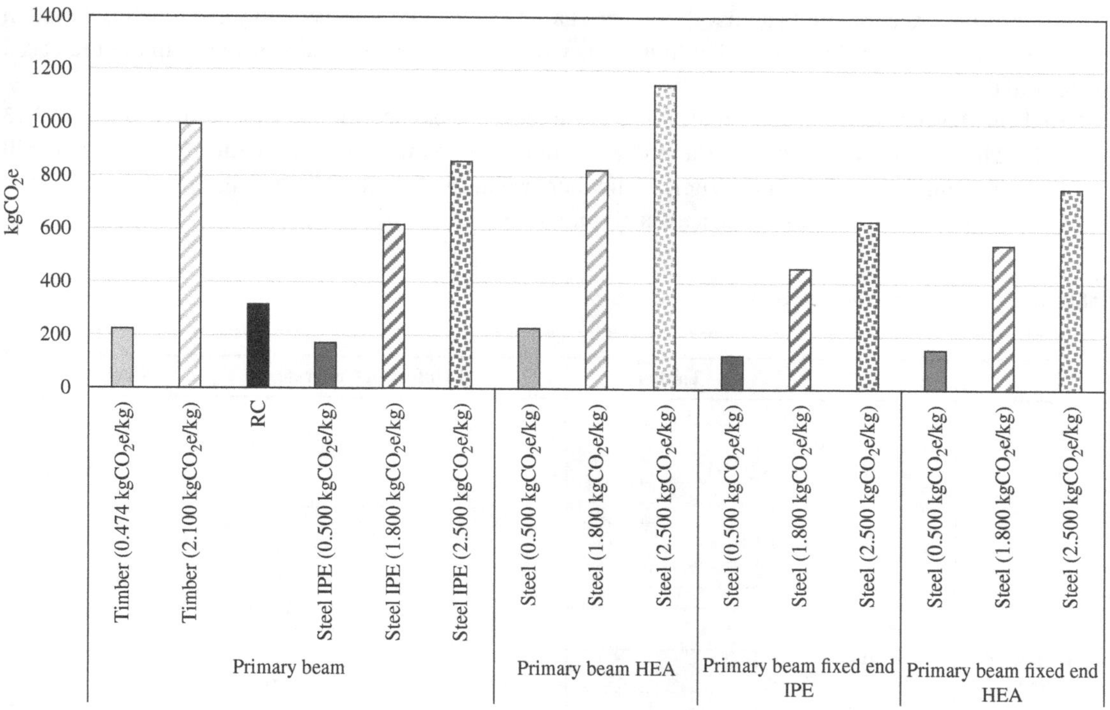

Figure 3.44 Embodied carbon comparison of primary beams made of timber, reinforced concrete, and steel (including rigid connections).

Please note also that, while rigid end connections can allow smaller beam sections by providing greater moment resistance, they also introduce greater fabrication complexity (requiring tighter tolerances, additional stiffeners or flanges, precise welding or bolting sequences, and more rigorous quality control). Therefore, the balance between size savings and manufacturing effort must be judged on a case-by-case basis. Our analysis here focuses solely on the section-reduction benefits offered by fixed connections.

3.6.5 Solution Part 2: Increased Structural Demand

Part 2 examines how an increased structural demand on the benchmark structure affects the embodied carbon results. The following updates are applied to the baseline design:

Primary beams: Increase the span from 6 to 9 m, keeping the same loading conditions.
Columns: Increase the number of stories from five to six, recalculating the vertical load accordingly.

With respect to Part 1, each structural element is resized to meet revised strength and serviceability criteria. Volumes and masses are recalculated based on updated dimensions, and embodied carbon is determined using the same GWP values from Table 3.9.

These results are compared against those from Part 1 to evaluate how sensitive each material system is to structural changes. The discussion highlights which materials are most affected from the increased structural demand.

Based on the modified design conditions, the updated cross-sections are presented in Table 3.13 and 3.14. These new sections reflect the higher structural demands imposed by the changes while still meeting the required performance criteria. The increase in the primary beam span to 9 m had a particularly strong effect on section sizing across all materials.

Table 3.13 Part II structural sections.

		Timber	Reinforced Concrete	Steel
Increased Span	**Primary Beam Section L = 9 m**	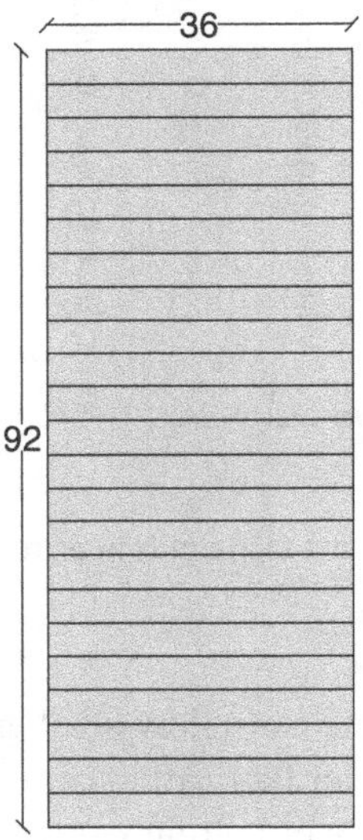	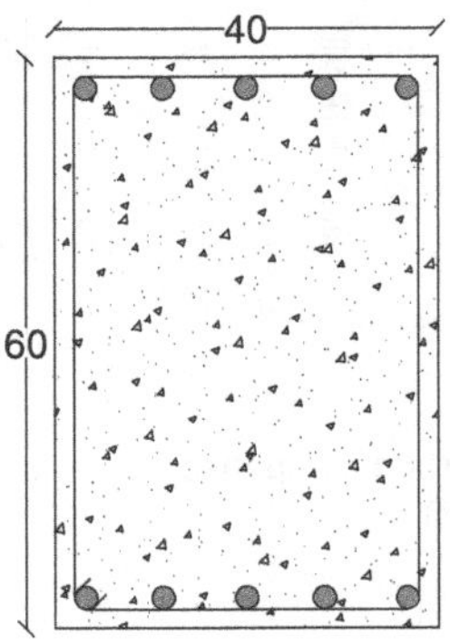	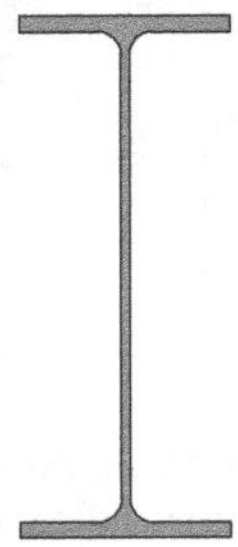
		GL92 cm × 36 cm	**B60 cm × 40 cm**	**IPE500**
Additional Story	**Column Section L = 4 m**	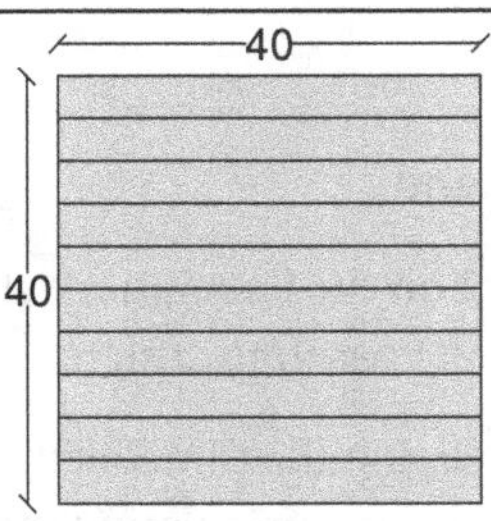	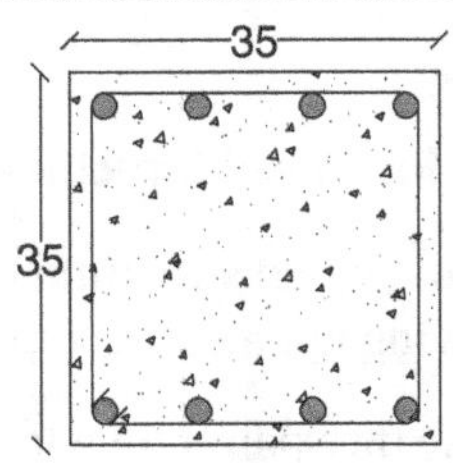	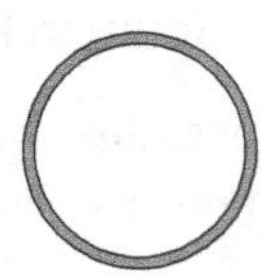
		GL40 cm × 40 cm	**C35 cm × 35 cm**	**CHS 244.5 mm × 10.0 mm**

Table 3.14 Part II steel structural sections.

		Steel	Steel HEA	Steel Fixed-End IPE	Steel Fixed-End HEA
Increased Span	**Primary Beam Section L = 9 m**		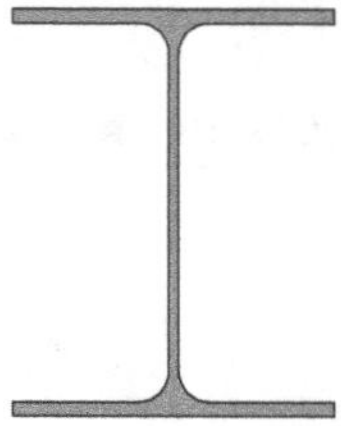		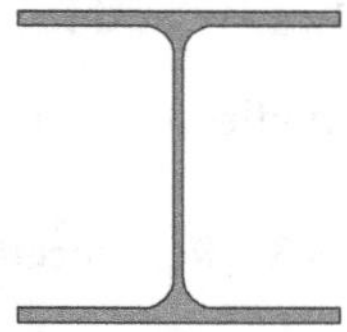
		IPE500	**HEA400**	**IPE400**	**HEA300**
Additional Story	**Column Section L = 4 m**		–	–	–
		CHS 244.5 mm x 10.0 mm			

3.6.5.1 Timber Beams, Option 1 (GWP = 0.474 $kgCO_2e/kg$)

Secondary Beam Section (L = 5 m):

Volume = 5 m × 0.52 m × 0.20 m = 0.52 m^3
Mass = 0.52 m^3 × 470 kg/m^3 = 244 kg
Embodied Carbon = 244 kg × 0.474 $kgCO_2e/kg$ ≈ 116 $kgCO_2e$

Primary Beam Section (L = 9 m):

Volume = 9 m × 0.92 m × 0.36 m = 2.98 m^3
Mass = 2.98 m^3 × 470 kg/m^3 ≈ 1401 kg
Embodied Carbon = 1401 kg × 0.474 $kgCO_2e/kg$ ≈ 664 $kgCO_2e$

Column Section (L = 4 m):

Volume = 4 m × 0.40 m × 0.40 m = 0.64 m^3
Mass = 0.64 m^3 × 470 kg/m^3 ≈ 301 kg
Embodied Carbon = 301 kg × 0.474 $kgCO_2e/kg$ ≈ 143 $kgCO_2e$

3.6.5.2 Timber Beams, Option 2 (GWP = 2.1 $kgCO_2e/kg$)

Secondary Beam Section:

Embodied Carbon = 244 kg × 2.100 $kgCO_2e/kg \approx 512\ kgCO_2e$

Primary Beam Section:

Embodied Carbon = 1401 kg × 2.100 $kgCO_2e/kg \approx 2942\ kgCO_2e$

Column Section:

Embodied Carbon = 301 kg × 2.100 $kgCO_2e/kg \approx 632\ kgCO_2e$

3.6.5.3 Reinforced Concrete

Primary Beam Section (L = 9 m):

Volume = 9 m × 0.60 m × 0.40 m = 2.16 m^3
Rebar Volume = 2.16 m^3 × 0.02 ≈ 0.043 m^3

Mass:

Concrete = 2.16 m^3 × 2400 kg/m^3 ≈ 5184 kg
Rebar = 0.043 m^3 × 7850 kg/m^3 ≈ 338 kg
Embodied Carbon = (5184 × 0.133) + (338 × 0.760) ≈ 946 $kgCO_2e$

Column Section (L = 4 m):

Volume = 4 m × 0.35 m × 0.35 m = 0.490 m^3
Rebar Volume = 0.490 m^3 × 0.02 = 0.010 m^3

Mass:

Concrete = 0.490 m^3 × 2400 kg/m^3 ≈ 1176 kg
Rebar = 0.010 m^3 × 7850 kg/m^3 ≈ 79 kg
Embodied Carbon = (1176 × 0.133) + (79 × 0.760) ≈ 216 $kgCO_2e$

3.6.5.4 Steel – IPE Sections

Steel (Produced with High Scrap Content) (GWP = 0.500 $kgCO_2e/kg$)

Primary Beam: IPE500 (90.7 kg/m):

Mass = 9 m × 90.7 kg/m ≈ 816 kg
Embodied Carbon = 816 × 0.500 ≈ 408 $kgCO_2e$

Column: CHS 244.5 mm × 10 mm (57.8 kg/m):

Mass = 4 m × 57.8 kg/m ≈ 231 kg
Embodied Carbon = 231 × 0.500 ≈ 116 $kgCO_2e$

Steel (Produced Some Scrap Content) (GWP = 1.800 $kgCO_2e/kg$)
Primary Beam = 816 kg × 1.800 ≈ 1469 $kgCO_2e$
Column: 231 kg × 1.800 ≈ 416 $kgCO_2e$

Steel (Produced Low Scrap Content) (GWP = 2.500 $kgCO_2e/kg$)
Primary Beam = 816 kg × 2.500 ≈ 2040 $kgCO_2e$
Column = 231 kg × 2.500 ≈ 578 $kgCO_2e$

3.6.5.5 Steel – HEA Sections

Steel (Produced with High Scrap Content) (GWP = 0.500 $kgCO_2e/kg$)

Primary Beam: HEA400 (124.8 kg/m):

Mass = 9 m × 124.8 kg/m = 1123 kg
Embodied Carbon = 1123 × 0.500 ≈ 562 $kgCO_2e$

Steel (Produced Some Scrap Content) (GWP = 1.800 $kgCO_2e/kg$)

Primary Beam = 1123 kg × 1.800 ≈ 2021 $kgCO_2e$

Steel (Produced Low Scrap Content) (GWP = 2.500 $kgCO_2e/kg$)

Primary Beam = 1123 kg × 2.500 ≈ 2808 $kgCO_2e$

3.6.5.6 Fixed-end Steel Calculations, Primary Beams

Steel (Produced with High Scrap Content) (GWP = 0.500 $kgCO_2e/kg$)

Primary Beam: IPE400 (66.3 kg/m):

Mass = 9 m × 66.3 kg/m = 597 kg
Embodied Carbon = 597 kg × 0.500 $kgCO_2e/kg$ ≈ 299 $kgCO_2e$

Primary Beam: HEA300 (88.3 kg/m):

Mass = 9 m × 88.3 kg/m = 795 kg
Embodied Carbon = 795 kg × 0.500 $kgCO_2e/kg$ ≈ 398 $kgCO_2e$

Steel (Produced Some Scrap Content) (GWP = 1.800 $kgCO_2e/kg$)

Primary Beam: IPE400:

Embodied Carbon = 597 kg × 1.800 $kgCO_2e/kg$ ≈ 1075 $kgCO_2e$

Primary Beam: HEA300:

Embodied Carbon = 795 kg × 1.800 $kgCO_2e/kg$ ≈ 1431 $kgCO_2e$

Steel (Produced Low Scrap Content) (GWP = 2.500 kgCO$_2$e/kg)

Primary Beam: IPE400:

Embodied Carbon = 597 kg × 2.500 kgCO$_2$e/kg ≈ 1493 kgCO$_2$e

Primary Beam: HEA300:

Embodied Carbon = 795 kg × 2.500 kgCO$_2$e/kg ≈ 1988 kgCO$_2$e

Figure 3.45 shows the mass comparison for beams. The increased span led to a much larger reinforced concrete section, resulting in a significant increase in mass. Timber and steel beams also grew in size, though they remained lighter than reinforced concrete.

The embodied carbon results are shown in Figure 3.46.

Figure 3.47 shows the mass comparison for columns. The mass of the reinforced concrete columns increases significantly due to the added height and resulting increase in design load. Timber and steel columns also show a moderate increase in weight, but both remain lighter than reinforced concrete options.

The embodied carbon comparison is presented in Figure 3.48.

This chapter demonstrated how material selection, sourcing, and design assumptions shape embodied carbon outcomes for timber, concrete, and steel structural elements. The sensitivity analysis highlights that longer spans or additional stories can shift the material, which is optimal. Including both sustainable and unsustainable GWP values emphasizes the critical role of responsible sourcing. While the simplified geometry clarifies mass-driven carbon differences, real-world designs require detailed code checks for stiffness and strength. Acknowledging fabrication and transport penalties (e.g. extra welding

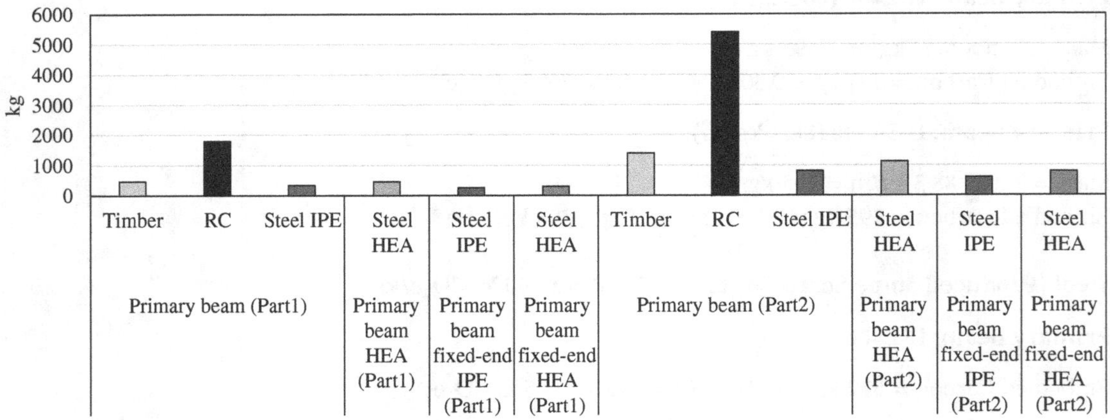

Figure 3.45 Mass comparison of primary beams made of timber, reinforced concrete, and steel (including fixed-end connections).

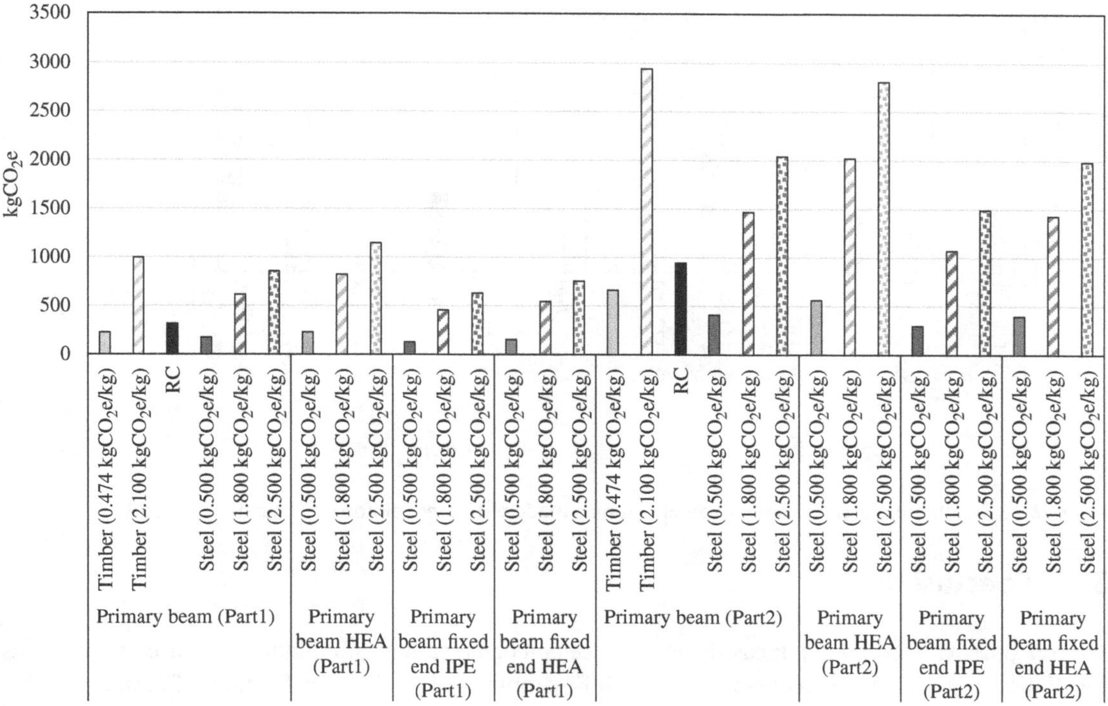

Figure 3.46 Embodied carbon comparison of primary beams made of timber, reinforced concrete, and steel (including rigid connections).

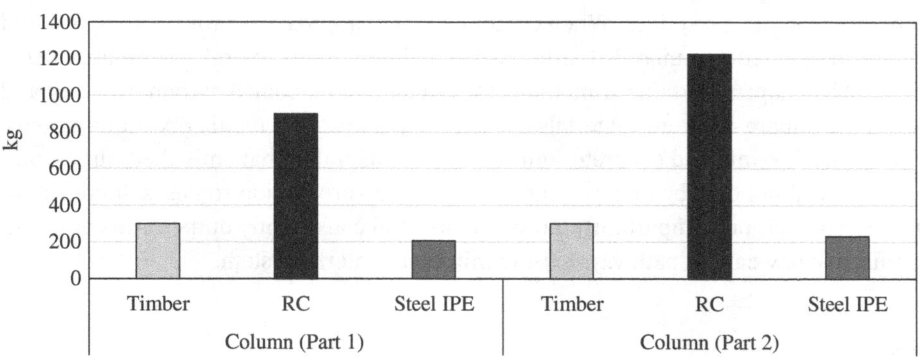

Figure 3.47 Mass comparison of columns made of timber, reinforced concrete, and steel.

carbon) would further refine results. Overall, this exercise aims to equip readers with a clear methodology for cradle-to-gate carbon calculation and shows the importance of contextualizing GWP data when evaluating structural alternatives.

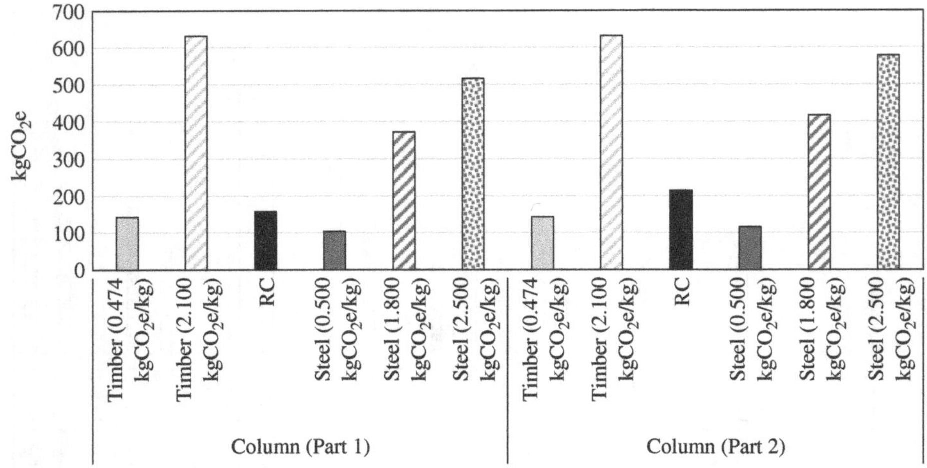

Figure 3.48 Embodied carbon comparison of columns made of timber, reinforced concrete, and steel.

3.7 Conclusion

This chapter explored how to measure and reduce embodied carbon in building structures, covering its calculation, regulation, and assessment through Environmental Product Declarations (EPDs) and rating systems. The discussion of EPDs and databases highlighted both the opportunities and limitations of available data, emphasizing the importance of understanding variability, regional context, and uncertainty when interpreting results. The ten strategies presented in this chapter underline that meaningful carbon reductions are achievable through early design decisions, informed material sourcing, system-level optimization, and collaboration across disciplines. When combined, these approaches provide designers and engineers with practical tools to reduce embodied carbon while maintaining structural performance, resilience, and adaptability. The chapter introduced methods for calculating embodied carbon intensity and presented examples that compare different materials, structural systems, and regulatory requirements. Comparative analyses of steel, reinforced concrete, and timber demonstrated that embodied carbon outcomes are strongly influenced not only by material choice, but also by production routes, sourcing strategies, and structural design decisions. Importantly, no single material consistently outperforms others in all contexts; rather, multiple low-carbon pathways exist within each material system.

3.8 Questions

Answer the following questions to enhance your understanding of the concepts covered in this chapter.

Short Answer Questions

1 Define embodied carbon and explain how it differs from operational carbon.

2 What is the purpose of the Bill of Quantities (BoQ) in calculating embodied carbon?

3 Describe what an Environmental Product Declaration (EPD) is and why it is important in embodied carbon assessments.

4 What is the equation used to calculate embodied carbon using material quantity and the carbon factor?

5 Explain the role of Global Warming Potential (GWP) in converting various greenhouse gases to a common CO_2e metric.

6 What is meant by embodied carbon intensity, and how is it calculated?

7 How does the use of recycled content in structural steel affect the embodied carbon of a building?

8 Identify one strategy for reducing the embodied carbon of concrete and explain its impact.

9 When comparing different EPDs in a case study, what key factors should be considered?

10 Why is it useful to normalize embodied carbon values (e.g. by using the Gross Internal Area), and how does this aid in sustainable design comparisons?

11 What are the different methods of calculation of normalized values for different structure typologies (buildings, bridges, stadia, energy infrastructure)? How do they differ from each other?

True or False

1 Embodied carbon includes greenhouse gas emissions from material extraction, manufacturing, transportation, construction, maintenance, and disposal.

2 The Bill of Quantities (BoQ) is used to determine the specific quantities of materials in a building project for embodied carbon calculations.

3 An Environmental Product Declaration (EPD) provides standardized and verified data on the environmental impacts, including embodied carbon, of a construction material.

4 The carbon factor for a material is typically expressed in kgCO_2e per kilogram of material.

5 Global Warming Potential (GWP) is used to convert the impacts of different greenhouse gases into a common unit (CO_2e).

6 Embodied carbon intensity is calculated by dividing the total embodied carbon by the building's Gross Internal Area (GIA).

7 Increasing the recycled content in structural steel generally increases its embodied carbon due to the extra processing required.

8 EPDs must adhere to Product Category Rules (PCRs) to ensure that the data provided are transparent and comparable across similar products.

9 Transportation of materials to the construction site is not considered in embodied carbon calculations.

10 In life-cycle assessment terminology, stages A1–A3 represent the processes of raw material extraction, processing, and manufacturing.

11 Using a higher percentage of postconsumer recycled steel in a building's structural elements generally reduces the overall embodied carbon.

12 A higher carbon factor of a material always indicates a lower environmental impact.

13 Replacing a portion of Portland cement in concrete with supplementary cementitious materials (SCMs) such as GGBS can reduce the embodied carbon of the concrete mix.

14 Embodied carbon assessments only consider the production and construction phases and do not include end-of-life impacts.

15 The Structural Carbon Rating Scheme (SCORS) is used to compare the embodied carbon intensity of different structural systems within a building.

Discussion Questions

1 How do Environmental Product Declarations (EPDs) and the Bill of Quantities (BoQ) work together to support accurate embodied carbon assessments, and what challenges can arise when gathering this data?

2 In what ways do the different life-cycle stages (A1–A5, C, and D) influence design decisions, and how can engineers prioritize reductions in embodied carbon across these stages?

3 How does the percentage of recycled content in structural steel affect the embodied carbon of a building, and what practical challenges exist for increasing recycled content?

4 How does normalizing embodied carbon (e.g. per square meter, per seat, or per kWh) improve comparisons between different projects, and what are the benefits and limitations of this approach?

5 How can regulatory frameworks and rating systems, such as the Structural Carbon Rating Scheme (SCORS), drive innovation in low-carbon design, and what potential trade-offs might designers face when aiming to reduce embodied carbon?

6 Why is it important to distinguish between product-specific EPDs and generic data when calculating embodied carbon? In what project phases might each be more appropriate, and what are the risks of relying solely on generic values?

7 HEA sections are heavier than IPE sections of equivalent bending capacity. Under what conditions might specifying a heavier section be justified from both structural and environmental perspectives?

8 Considering the wide variability in GWP values shown across different steel production routes, concrete mixes, and timber sources, how should designers communicate uncertainty in embodied carbon calculations to clients and decision-makers?

9 In what ways do steel production routes (BF-BOF, EAF, DRI-EAF) affect embodied carbon outcomes, and how should designers respond to these differences in practice?

10 Timber is often described as a low-carbon material. Under what conditions can timber result in higher embodied carbon than steel or concrete, based on the chapter's examples?

11 Why does cement dominate the embodied carbon of reinforced concrete elements, and which design or material strategies are most effective in reducing this impact?

12 How can structural boundary conditions, such as pinned versus fixed-end connections, influence embodied carbon outcomes without changing the material type?

13 What are the advantages and limitations of using a representative structural portion instead of a full-building model when estimating embodied carbon during early design?

14 How do increased structural demands, such as longer spans or additional stories, change the relative embodied carbon performance of timber, steel, and reinforced concrete systems?

15 How can early structural layout decisions influence material efficiency and long-term environmental performance before any material grades or suppliers are selected?

16 In what situations might increasing material performance per unit mass lead to lower overall embodied carbon, and when might it fail to deliver meaningful benefits?

17 How can design choices made to simplify fabrication and construction workflows reduce environmental impacts beyond material production alone?

18 What design features can enable a structure to remain useful under changing functional requirements, and how do these features affect embodied carbon over the asset's lifetime?

19 How can digital design tools support lower-carbon outcomes while still requiring critical engineering judgment to avoid unintended consequences?

Calculation Exercises

Question 1. A multistory office building has the following characteristics:

Gross Internal Area (GIA): 5000 m^2
Structural System Information:

- **Reinforced Concrete**: 2000 m^3(density = 2400 kg/m^3, A1–A5 embodied carbon = 0.11 $kgCO_2e/kg$)
- **Steel Reinforcement**: 350 kg/m^3 of concrete (A1–A5 embodied carbon = 0.8 $kgCO_2e/kg$)
- **Structural Steel Beams and Columns**: 500 tons (500 000 kg, A1–A5 embodied carbon = 1.2 $kgCO_2e/kg$)

Tasks:

A. Calculate the total embodied carbon for the building (in $kgCO_2e$) for life-cycle stages A1–A5.
B. Determine the embodied carbon intensity ($kgCO_2e/m^2$).

Question 2. A pedestrian bridge has the following specifications:

Length: 40 m, **Width**: 5 m
Deck Information:

- **Reinforced Concrete Volume:** 120 m^3(density = 2400 kg/m^3, A1–A5 embodied carbon = 0.11 $kgCO_2e/kg$)
- **Steel Reinforcement**: 350 kg/m^3(A1–A5 embodied carbon = 0.8 $kgCO_2e/kg$)
- **Steel Beams**: 10 tons (10 000 kg, A1–A5 embodied carbon = 1.2 $kgCO_2e/kg$)

Tasks:

A. Calculate the total embodied carbon for the bridge (in $kgCO_2e$) for life-cycle stages A1–A5.
B. Determine the embodied carbon intensity ($kgCO_2e/m^2$) using the **Functional Area (FA)** of the bridge.

Question 3. A stadium has the following specifications:

Seating Capacity: 30 000 seats
Enclosed Area (EA): 25 000 m^2
Structural System Information:

- **Reinforced Concrete Superstructure:** 10 000 m^3 (density = 2400 kg/m^3, A1–A5 embodied carbon = 0.11 $kgCO_2e/kg$)
- **Steel Reinforcement**: 350 kg/m^3 of concrete (A1–A5 embodied carbon = 0.8 $kgCO_2e/kg$)
- **Steel Roof**: 2000 tons (2000 000 kg, A1–A5 embodied carbon = 1.2 $kgCO_2e/kg$)

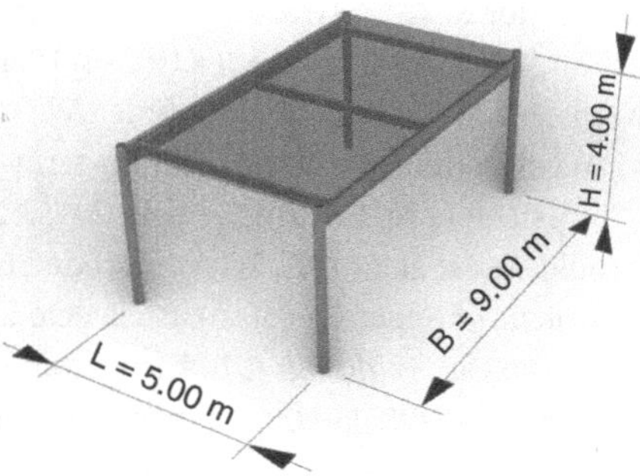

Figure 3.49 Steel structural frame with dimensions and member specifications used for embodied carbon calculation.

Tasks:

A. Calculate the total embodied carbon for the stadium (in $kgCO_2e$) for life-cycle stages A1–A5.

B. Determine the embodied carbon intensity in $kgCO_2e$/seat and $kgCO_2e/m^2$.

Question 4. Simple structural frame:

This exercise gives readers the opportunity to apply the calculation techniques introduced in earlier sections. The objective is to determine the embodied carbon of a simple steel frame using given section properties, dimensions, and a specified GWP value.

The frame includes the following elements:

Secondary beams: 3 members, IPE300, 42.2 kg/m
Primary beams: 2 members, IPE500, 90.7 kg/m
Columns: 4 members, CHS 244.5 mm × 10.0 mm, 57.8 kg/m

The structural layout is illustrated in Figure 3.49. The reinforced concrete slab is excluded to isolate the carbon contribution of the steel elements. The frame spans in one direction and 9 m in the other, with a story height of 4 m. All calculations should use the provided GWP value (A1–A3) of (i) 1.800 $kgCO_2e$/kg, which reflects a low-scrap content and (ii) 0.500 $kgCO_2e$/kg, which reflects a high-scrap content. As a final step, redo the calculation by using two different EPDs suitable for the steel profiles shown in this exercise, from a producer within your geographical zone. To complete the exercise, the reader should:

- Calculate the total length of each type of structural member.
- Multiply the length by the mass per meter to obtain the total mass of each member group.
- Apply the GWP factor to determine the total embodied carbon in $kgCO_2e$ for the full frame.

References

ArcelorMittal Europe. (2022). Environmental Product Declaration – Structural Hollow Sections Made of XCarb® Recycled and Renewably Produced Steel. Retrieved from https://tubular.arcelormittal.com/sites/default/files/2023-12/structural_hollow_sections_made_of_xcarbr_recycled_and_renewably_produced_steel.pdf

ArcelorMittal. (2024). *Environmental product declaration: XCarb recycled and renewably produced heavy plates* (Registration No. S-P-10991). EPD International AB. https://www.environdec.com/library/epd10991

Archer-Jones, C., & Green, D. (2021). Carbon targets for bridges: A proposed SCORS-style rating scheme. *The Structural Engineer, 99*(10), 14–18. istructe.info/htcec3-scorbs, accessed January 2026.

Arens, M., Worrell, E., Eichhammer, W., Hasanbeigi, A., & Zhang, Q. (2017). Pathways to a low-carbon iron and steel industry in the medium-term—The case of Germany. *Journal of Cleaner Production, 163*, 84–98. https://doi.org/10.1016/j.jclepro.2015.12.097

Arnold, W., et al. (2020). Setting carbon targets: An introduction to the proposed SCORS rating scheme. *The Structural Engineer, 98*(10), 5. istructe.info/htcec3-scors, accessed January 2026.

Arup. (2023a). *Buildings & Infrastructure Priority Actions for Sustainability—Embodied Carbon: Concrete* (Nos 07762000-RP-SUS-0003). https://www.istructe.org/IStructE/media/Public/Resources/ARUP-Embodied-carbon-concrete_1.pdf

Arup. (2023b). *Buildings & Infrastructure Priority Actions for Sustainability—Embodied Carbon: Steel* (Nos 07762000-RP-SUS-0001). https://www.istructe.org/IStructE/media/Public/Resources/ARUP-Embodied-carbon-steel_1.pdf

ARUP. (2025). Circular Buildings Toolkit. Retrieved from https://ce-toolkit.dhub.arup.com/

Baddoo, N. R. (2008). Stainless steel in construction: A review of research, applications, challenges and opportunities. *Journal of Constructional Steel Research, 64*(11), 1199–1206. https://doi.org/10.1016/j.jcsr.2008.07.011

Baddoo, N., & Chen, A. (2020). SCI P432: High Strength Steel Design and Execution Guide.

Cobb, F. (2017). *Structural Engineer's Pocket Book: Eurocodes* (3rd edition). CRC Press.

Collings, D. (2022). The carbon footprint of bridges. *Structural Engineering International, 32*(4). https://doi.org/10.1080/10168664.2021.1917326

Driessen, E., & Grönlund, E. (2024). Circular concrete scenarios and their environmental impacts: A life cycle assessment modelled after a Swedish city. *Journal of Cleaner Production, 485*, 144348. https://doi.org/10.1016/j.jclepro.2024.144348

Ebrahimi, N., Khanlarian, M., Momeni, M., Ahmadi, L., & Gallant, D. (2025). Toward sustainable steel bridge maintenance: Anti-corrosion coating systems. *Journal of Coatings Technology and Research*. 1–20. https://doi.org/10.1007/s11998-025-01126-3

EC3. (2025). *Building Transparency*. Retrieved 10 September 2025, from https://buildingtransparency.org/ec3

EPD International. (2025). *EPD International*. Retrieved 10 September 2025, from https://www.environdec.com/home

EurocodeApplied. (2025). Eurocode Design & Calculation Tools for Structural Engineers. EurocodeApplied.Com. Retrieved from https://eurocodeapplied.com/

Europe | Seamless pipe—Other. (2025). *Worldsteel.Org*. Retrieved 9 September 2025, from https://worldsteel.org/wider-sustainability/life-cycle-thinking/lca-eco-profiles-2022/europe-seamless-pipe-other/

Europe | Welded pipe—Construction. (2025). *Worldsteel.Org*. Retrieved 9 September 2025, from https://worldsteel.org/wider-sustainability/life-cycle-thinking/lca-eco-profiles-2022/europe-welded-pipe-construction/

Fantilli, A. P., Mancinelli, O., & Chiaia, B. (2019). The carbon footprint of normal and high-strength concrete used in low-rise and high-rise buildings. *Case Studies in Construction Materials, 11*, e00296. https://doi.org/10.1016/j.cscm.2019.e00296

Farahzadi, L., Tellnes, L. G. F., Shafei, B., & Kioumarsi, M. (2024). Life-cycle environmental assessment of ultra-high-performance concrete with sustainable materials and fiber substitutions. *Cleaner Engineering and Technology, 23*, 100846. https://doi.org/10.1016/j.clet.2024.100846

Galvanized Steel—An overview | ScienceDirect Topics. (n.d.). Retrieved 9 September 2025, from https://www.sciencedirect.com/topics/materials-science/galvanized-steel

Garvey, A., Norman, J. B., & Barrett, J. (2022). Technology and material efficiency scenarios for net zero emissions in the UK steel sector. *Journal of Cleaner Production, 333*, 130216. https://doi.org/10.1016/j.jclepro.2021.130216

Gibbons, O. P., Orr, J. J., Archer-Jones, C., Arnold, W., & Green, D. (2022). *How To Calculate Embodied Carbon* (2nd edition). The Institution of Structural Engineers.

Greenhouse Gas Protocol. (2011). *Corporate Value Chain (Scope 3) Accounting and Reporting Standard: Supplement to the GHG Protocol Corporate Accounting and Reporting Standard*. World Resources Institute and World Business Council for Sustainable Development. https://ghgprotocol.org/sites/default/files/standards/Corporate-Value-Chain-Accounting-Reporing-Standard_041613_2.pdf

Guitar, M. A., Thome, A., & Britz, D. (2025). How green will the green-steel production be? *Discover Sustainability, 6*(1), 405. https://doi.org/10.1007/s43621-025-01254-1

Hera. (n.d.). Low Carbon Design Guidance. Retrieved May 30, 2025, from https://hera.org.nz/technical-support/sustainability/low-carbon-design-guidance/

Home | Forest Stewardship Council. (2023, September 7). The Future of Forests Is in Our Hands. Retrieved from https://fsc.org/en

Ige, O. E., Von Kallon, D. V., & Desai, D. (2024). Carbon emissions mitigation methods for cement industry using a systems dynamics model. *Clean Technologies and Environmental Policy, 26*(3), 579–597. https://doi.org/10.1007/s10098-023-02683-0

Kanyilmaz, A., Birhane, M., Fishwick, R. et al. (2023a). Reuse of steel in the construction industry: challenges and opportunities. *International Journal of Steel Structures, 23*, 1399–1416. https://doi.org/10.1007/s13296-023-00778-4

Kanyilmaz, A., Dang, V. H., & Kondratenko, A. (2023b). How does conceptual design impact the cost and carbon footprint of structures? *Structures, 58*, 105102. https://doi.org/10.1016/j.istruc.2023.105102

Kanyilmaz, A., Tichell, P. R. N., & Loiacono, D. (2022). A genetic algorithm tool for conceptual structural design with cost and embodied carbon optimization. *Engineering Applications of Artificial Intelligence, 112*, 104711. https://doi.org/10.1016/j.engappai.2022.104711

Kelechi, S. E., Adamu, M., Uche, O. A. U., Okokpujie, I. P., Ibrahim, Y. E., & Obianyo, I. I. (2022). A comprehensive review on coal fly ash and its application in the construction industry. *Cogent Engineering, 9*(1), 2114201. https://doi.org/10.1080/23311916.2022.2114201

Lafarge Beton. (2021). Environmental Product Declaration—Ready Mix Concrete (S-P-04982). Retrieved from https://api.environdec.com/api/v1/EPDLibrary/Files/1aa5fb57-21eb-4126-252a-08dc176fe0d5/Data

LCA eco-profiles. (2021). *Worldsteel.Org*. Retrieved 9 September 2025, from https://worldsteel.org/wider-sustainability/life-cycle-thinking/lca-eco-profiles-2021/

Lee, C., Lee, S., & Nguyen, N. (2016). Modeling of compressive strength development of high-early-strength-concrete at different curing temperatures. *International Journal of Concrete Structures and Materials, 10*(2), 205–219. https://doi.org/10.1007/s40069-016-0147-6

LETI. (2020a). LETI Embodied Carbon Primer—Supplementary guidance to the Climate Emergency Design Guide. Retrieved from https://www.leti.uk/_files/ugd/252d09_8ceffcbcafdb43cf8a19ab9af5073b92.pdf

LETI. (2020b). Embodied Carbon Primer. London Energy Transformation Initiative. www.leti.uk/ecp

Mineral Products Association (MPA) UK. (2022). *Environmental Product Declaration: UK Average Portland Cement Sector EPD* (S-P-05825). The International EPD System, EPD International AB. https://cement.

mineralproducts.org/MPACement/media/Cement/Publications/2017/EPD_UK_Average_Portland_Cement_04-2022.pdf

Munro, M., Kanavaris, F., & Orr, J. (2024). A critical review of embodied carbon classification schemes for concrete. *Case Studies in Construction Materials, 21*, e04021. https://doi.org/10.1016/j.cscm.2024.e04021

Neville, A. M. (2011). *Properties of Concrete* (5th ed.). Pearson.

PEFC, Facts and figures (Forest Management). (n.d.). Retrieved August 30, 2025, from https://www.pefc.org/discover-pefc/facts-and-figures

Pinto, J. T. M., Sverdrup, H. U., & Diemer, A. (2019). Integrating life cycle analysis into system dynamics: The case of steel in Europe. *Environmental Systems Research, 8*(1), 15. https://doi.org/10.1186/s40068-019-0144-2

Qian, Y., Li, Y., Hao, Y., Yu, T., & Hu, H. (2024). Greenhouse gas control in steel manufacturing: Inventory, assurance, and strategic reduction review. *Carbon Research, 3*(1), 27. https://doi.org/10.1007/s44246-024-00118-z

Ready-mixed concrete C28/35 CIIB-V+SR Sector EPD from Mineral Products Association (MPA) UK. The International EPD System, valid until 28-02-2029.

RIBA. (2021). RIBA 2030 *Climate Challenge – Version 2*. Royal Institute of British Architects.

RICS. (2015). *Code of Measuring Practice* (6th edition). Royal Institution of Chartered Surveyors. https://www.rics.org/content/dam/ricsglobal/documents/standards/May_2015_Code_Of_Measuring_Practice_6th_Edition.pdf

SE2050. (n.d.). Design Guidance for Reducing Embodied Carbon in Structural Systems. Retrieved May 30, 2025, from https://se2050.org/resources-overview/structural-materials/lean-design-guidance/

SSAB. (2025). *Environmental product declaration: Hot rolled SSAB Zero steel sheets and coils* (Registration No. S-P-18717). EPD International AB. https://www.ssab.com/en/fossil-free-steel/ssab-zero

SteelConstruction.info. (n.d.). The recycling and reuse survey. Retrieved April 1, 2026, from https://www.steelconstruction.info/The_recycling_and_reuse_survey

Suer, J., Ahrenhold, F., & Traverso, M. (2022). Carbon footprint and energy transformation analysis of steel produced via a direct reduction plant with an integrated electric melting unit. *Journal of Sustainable Metallurgy, 8*(4), 1532–1545. https://doi.org/10.1007/s40831-022-00585-x

Swedish Wood. (2019). "Open spaces in modern materials—Wood Magazine." *Swedish Wood*. Retrieved from https://www.swedishwood.com/publications/wood-magazine/2019-1/open-spaces-in-modern-materials/

Swedish Wood. (2021). *Environmental Product Declaration: Swedish Sawn Dried Timber of Spruce or Pine* (S-P-02537). The International EPD System, EPD International AB. https://api.environdec.com/api/v1/EPDLibrary/Files/8d0a16a5-41dd-49e4-9fcf-08d8f8b9d146/Data

Tata Steel UK. (2025). *Green steel future: The transition to electric arc furnace steelmaking*. Tata Steel. https://www.tatasteeluk.com/green-steel-future

UK Average Portland Cement Sector EPD from Mineral Products Association (MPA) UK. The International EPD System, valid until 13-04-2027.

Waldman, B., Huang, M., & Simonen, K. (2020). Embodied carbon in construction materials: A framework for quantifying data quality in EPDs. *Buildings and Cities, 1*(1), 625–636. https://doi.org/10.5334/bc.31

Witte, A., & Garg, N. (2024). Quantifying the global warming potential of low carbon concrete mixes: Comparison of existing life cycle analysis tools. *Case Studies in Construction Materials, 20*, e02832. https://doi.org/10.1016/j.cscm.2023.e02832

World Steel in Figures. (2024). *Worldsteel.Org*. Retrieved 9 September 2025, from https://worldsteel.org/data/world-steel-in-figures/world-steel-in-figures-2024/

World Steel. (2025). Raw materials. *Worldsteel.Org*. https://worldsteel.org/other-topics/raw-materials/

Yang, K. H., Jung, Y. B., Cho, M. S., & Tae, S. H. (2015). Effect of supplementary cementitious materials on reduction of CO_2 emissions from concrete. *Journal of Cleaner Production, 103*, 774–783. https://doi.org/10.1016/j.jclepro.2014.03.018

Yu, W., & LaBoube, R. A. (2010). *Cold-Formed Steel Design*. 1st ed. Wiley. https://doi.org/10.1002/9780470949825

4

Life-cycle Parameter Analysis (LCPA) at Component Level

LIST OF CONTENTS

In the previous chapters, you learned how to use life-cycle assessment (LCA) to calculate a structure's total environmental impact. The LCA gives us a final score (e.g. global warming potential [GWP] value), but looking only at that result, we cannot tell which of our many design choices had the biggest influence or where the most effective opportunities for reducing embodied carbon exist. Was it the choice of steel over concrete? The height of the column? The shape of the beam?

This lack of clarity may lead to several critical issues. First, it prevents us from prioritizing our optimization efforts (we might spend considerable time refining minor details while overlooking major contributors to environmental impact), and we cannot identify which parameters truly matter. Second, it obstructs our ability to make trade-offs between environmental performance, cost, and structural requirements during the design process. Therefore, without this understanding, we cannot effectively communicate to clients and stakeholders which design modifications would result in the greatest environmental benefits, potentially missing opportunities for significant carbon reduction.

Beyond design parameters, a similar and equally critical uncertainty exists within the LCA data itself. The industry is often confused about which data to use. The current debates are: should we rely on broad, industry-average data or specific environmental product declarations (EPDs)? If using averages, which one is most representative? If using EPDs, how do we handle the wide performance variations between different suppliers for the same material? This data variability means that the same design can yield significantly different environmental impact scores, create confusion, and make it difficult to draw firm conclusions (Figure 4.1).

To answer these questions and overcome the limitations (both in design and data), we need to understand the sensitivity of each selection we make. This chapter introduces life-cycle parameter analysis (LCPA) as a method to evaluate how individual choices impact the embodied carbon of structural components. By systematically varying both design parameters (like material choice) and data inputs (like the source of GWP values), LCPA reveals which factors have the most influence on the final result. Besides, it can allow us to quantify the impact of switching from a generic dataset to a specific EPD, thereby managing data uncertainty effectively and identifying which data points are most critical to refine. This

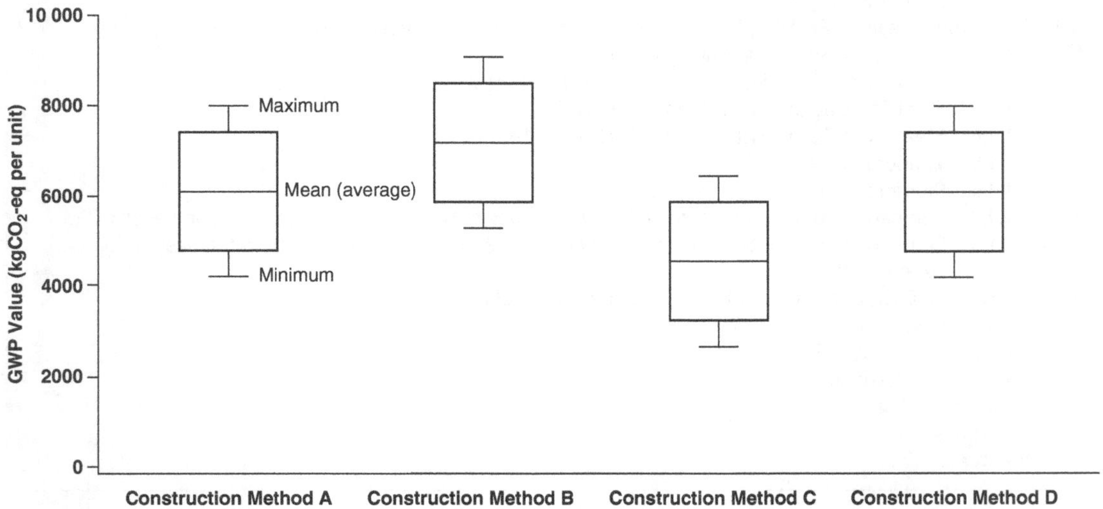

Figure 4.1 A generic image showing the high variability between GWP values vs. selected construction methods.

helps us to make defensible design decisions, even when faced with data variability. Chapter 5 extends these concepts to the building scale, demonstrating how parameter sensitivity can guide decisions across entire structural systems. Before moving forward, we first introduce the concept of "parameter analysis."

4.1 Principles of Parameter Analysis

As a structural engineer or architect, every decision you make (from the span of a beam to the material selection of a column) has an effect on a building's cost, performance, and environmental footprint. The earliest stages of design are the most critical, offering the greatest opportunity to create structures that are not only safe but also sustainable, an argument explored in detail in Chapters 5 and 6 (Figure 4.2).

Parameter analysis is a powerful "what-if" method used to determine how changes in a set of independent input variables impact a specific output. For example, it's widely used in finance to see how a stock's price might change with fluctuations in interest rates and in climate modeling to understand how variations in greenhouse gas emissions affect global temperature rise. We can apply the same logic to a building's structural design.

In structural design, our input variables can include:

- Material choice (e.g. steel, timber, concrete)
- Component geometry (e.g. I-beam vs. hollow tube vs. truss)
- Dimensions (e.g. column height or beam span)
- Material properties (e.g. standard vs. high-strength concrete)

Figure 4.2 Parameter analysis of an engineer at work (Thapana_Studio/Adobe Stock Photo).

By systematically changing one variable while keeping others constant, we can identify the design decisions that have significant effects on our desired outcome, whether that's mass, cost, or, for this chapter, embodied carbon. This can be done both at the component and building levels.

ISO 14044:2006 (EN ISO 14044:2006/A2:2020) requires sensitivity analysis at multiple stages (LCI, LCIA, Interpretation) within LCA methodology, to test assumptions and validate conclusions. It integrates this process throughout the LCA framework to ensure results are robust:

Inventory Phase (LCI): Sensitivity analysis is used to refine the system boundary by determining the significance of various inputs and outputs. It's also used to test the consequences of different allocation procedures when a process creates multiple products.

Impact Assessment Phase (LCIA): The analysis helps determine how methodological choices and data variations affect the final environmental impact results.

Interpretation Phase: The standard requires a final sensitivity check to "assess the reliability of the final results and conclusions." This step ensures your recommendations are dependable and not merely an artifact of a single assumption. For studies making comparative public claims, a detailed sensitivity analysis is mandatory.

Today, conducting such an analysis became fast thanks to digital tools such as building information modeling (BIM), and in the near future, the integration of artificial intelligence (AI) promises to make it even more efficient.

Building on the same principles outlined in ISO 14044, this book introduces LCPA. While ISO 14044 uses sensitivity analysis mainly to test assumptions within the LCA framework, LCPA extends this idea to the design phase itself. In this sense, LCPA differs from sensitivity analysis: rather than being used primarily as a validation step after results are obtained, it is applied proactively during design to guide decision-making and explore alternative solutions. It treats the structure as a system of interdependent parameters (geometry, material, and performance) evaluated not only for environmental impact but also for efficiency and constructability. In other words, LCPA transforms sensitivity analysis from a verification tool into a design-driving method, guiding engineers and architects toward solutions that balance strength, cost, and embodied carbon from the earliest project stages. This chapter focuses on LCPA at the component level (columns and beams), while the next chapters expand it to entire structural systems.

4.2 Combining Life-cycle Assessment (LCA) and Parameter Analysis: Life-cycle Parameter Analysis (LCPA)

When we merge the environmental accounting of LCA with the parameter analysis, we get Life-cycle Parameter Analysis (LCPA). Rather than delivering a single carbon value, LCPA aims to reveal the underlying drivers of that result, helping us understand how different design choices influence environmental performance. It will help designers identify the most critical "levers" we can pull to make meaningful reductions in a structure's environmental impact. This is a fundamental skill for any designer committed to sustainability.

In the case studies that follow, we will apply this LCPA methodology to two of the most common structural components: columns and beams. For each, we will set up a benchmark model with a fixed layout and load. Then, we will vary one design parameter at a time to see how it affects the component's total mass and its GWP. This one-at-a-time variation approach, as recognized in ISO 14044:2006, is particularly effective for isolating and understanding the influence of individual parameters.

As you work through the analyses, focus on understanding the underlying reasons for the results. You will see firsthand how structural phenomena like flexural buckling, composite construction, prestressing, trusses, types of beams, and column shapes can dramatically alter the environmental outcome. Each component is analyzed under varying conditions:

Columns are studied at heights of 4, 6, and 9 m, with cross-sections made from steel (HEA, square hollow section [SHS], circular hollow section [CHS]), composite (partially and fully encased), reinforced concrete (square and circular), and glulam.

Beams are studied for spans of 6, 12, and 18 m, comparing options like steel IPE and HEA profiles, steel trusses with open or tubular sections, glulam, and reinforced concrete.

4.3 Case Study: Columns (Steel, Timber, Reinforced Concrete, Composite)

This section presents a comparative analysis of structural column profiles, examining their mass and carbon emissions for the heights of 4, 6, and 9 m. The study evaluates various column types, including steel HEA, steel hollow sections (such as tubular and square profiles), encased composite columns (both partially and fully encased), glulam (glued laminated timber), and reinforced concrete (both circular and square sections).

4.3.1 Benchmark Structural Configuration

Three column heights were analyzed based on the framing plan of 8 m span (Figure 4.3) in a 10-story building (Figure 4.4). The columns are designed for heights of 4, 6, and 9 m, which would accommodate various structural needs in a real building (for example, the 4-m columns could be used for residential or small commercial buildings, the 6-meter columns for medium-sized office buildings, and the 9-m columns for large industrial structures or high-rise buildings).

4.3.2 Column Types Analyzed (Steel HEA, SHS, CHS, Partially and Fully Encased Composite, Square and Circular Reinforced Concrete, Timber)

Eight types of column sections were analyzed, as shown in Figure 4.5; steel HEA (European wide flange column), steel SHS, steel CHS, partially encased composite columns, fully encased composite columns, reinforced concrete (including circular and square sections) and glulam (glued laminated timber) columns. Each of these column types offers distinct characteristics in terms of strength, weight, and material efficiency, providing a range of options suitable for various structural applications: for example, steel HEA and SHS columns might be preferred for their high strength-to-weight ratio, while glulam columns could be chosen for their environmental benefits and aesthetic appeal; reinforced concrete columns offer robust fire resistance and can support heavy loads effectively. Composite columns such as partially and fully encased, combine steel and concrete providing a balance of strength,

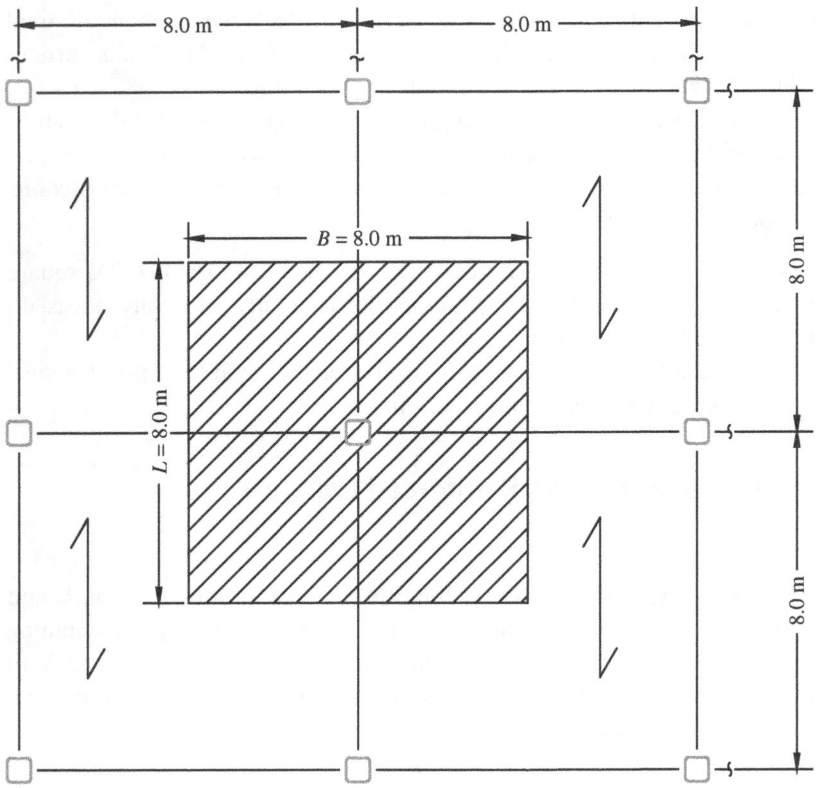

Figure 4.3 Layout of benchmark frame studied in this chapter.

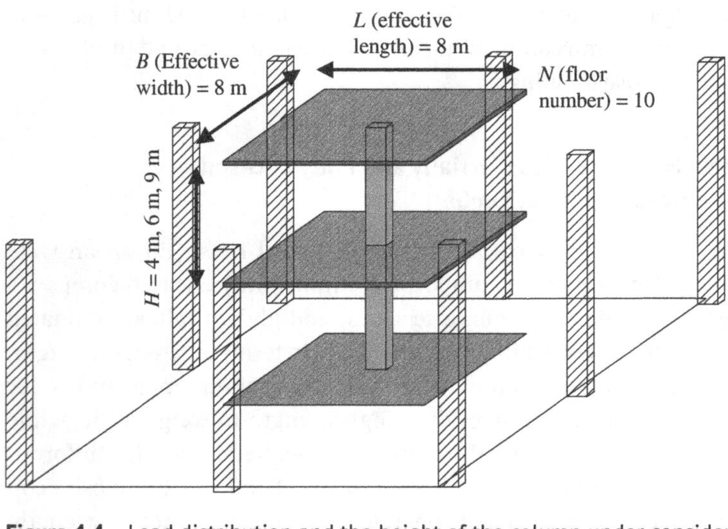

Figure 4.4 Load distribution and the height of the column under consideration.

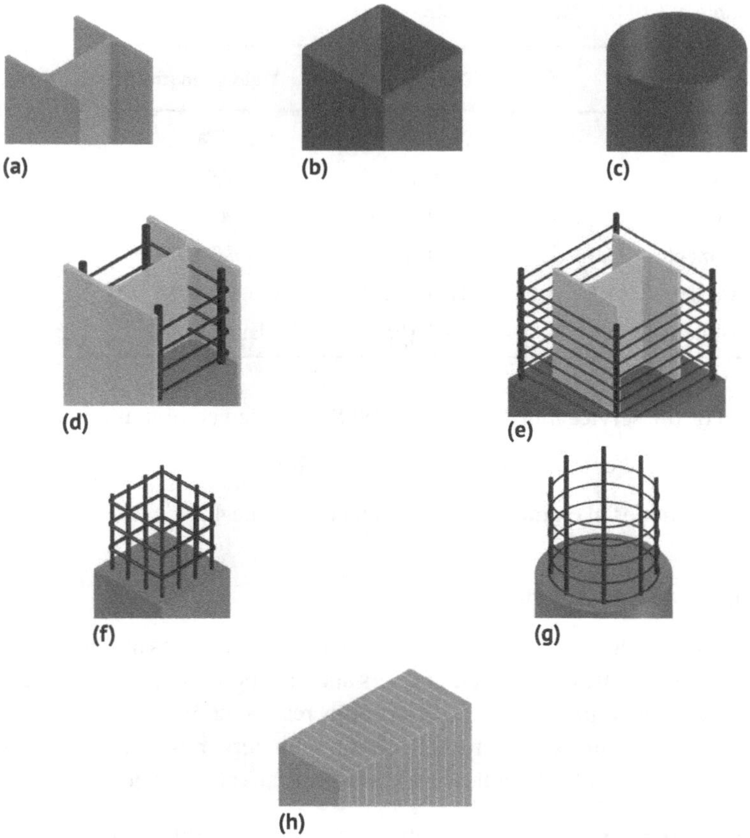

Figure 4.5 Column types analyzed in this chapter: (a) steel HEA, (b) steel SHS, (c) steel CHS, (d) partially encased composite, (e) fully encased composite, (f) square reinforced concrete, (g) circular reinforced concrete, and (h) glulam.

Table 4.1 Load cases.

Load cases	Load [kN/m^2]
Self-weight (G1)	3
Permanent nonstructural (G2)	2
Live load office: category B (Q)	2

stiffness, and fire resistance, making them ideal for high-rise buildings and structures requiring enhanced durability.

The loading for an office building with metal deck system flooring has been considered. The loads are categorized as self-weight (G1), permanent nonstructural (G2), and live load (Q). The self-weight load (G1) is 3 kN/m^2, while the permanent nonstructural load (G2) and the live load office load (Q) are both 2 kN/m^2 (Table 4.1).

For the ultimate limit state (ULS), the load combination is calculated as:

$$\text{Load combination (ULS)} = 1.35 \times (G1 + G2) + 1.5 \times Q$$

Table 4.2 Material characteristics.

Material type	Density	Yield strength (f_y)
Steel S355	7850 kg/m^3	355 MPa
Steel S275	7850 kg/m^3	275 MPa
Reinforcing steel B450	7850 kg/m^3	450 MPa
Concrete C25/30	2400 kg/m^3	Class C25/30
Concrete C45/55	2400 kg/m^3	Class C45/55
Glulam GL28h	470 kg/m^3	Class GL28h

For the serviceability limit state (SLS), the load combination is:

Load combination (SLS) = G1 + G2 + Q

The material characteristics are listed in Table 4.2.

4.3.3 Steel Columns

Table 4.3 details the performance of steel HEA columns at heights of 4, 6, and 9 m. For each height, the table lists critical design parameters such as the maximum axial design force, selected column section, flexural buckling resistance, maximum resistance ratio, governing criteria, mass per length, and total mass. Since the design of steel column members must account for flexural buckling to ensure stability under axial loads, the following key aspects are also considered for their design:

Buckling Length: For fully supported members, the buckling length is taken as equal to the column height (this length is crucial in determining the member's susceptibility to buckling).

Flexural Buckling Criteria: These criteria are applied to each segment of the column, defined by two consecutive restrained sections. These segments are analyzed to determine the risk of buckling along both the major and minor axes of the column.

The following equations are used during design:

$$\text{Ratio}_y = \frac{N_{\text{Ed}}}{N_{\text{by,Rd}}} \leq 1$$

$$\text{Ratio}_z = \frac{N_{\text{Ed}}}{N_{\text{bz,Rd}}} \leq 1$$

$N_{\text{by,Rd}}$ and $N_{\text{bz,Rd}}$ are the design buckling resistances calculated according to Section 6.3.1 of EN 1993-1-1 (2022).

This method is used to ensure that the applied axial forces do not exceed the buckling resistances in both major (y-axis) and minor (z-axis) directions. Table 4.4 provides the design data for steel SHS columns, while Table 4.5 presents the design checks for steel CHS (circular hollow section) columns. Upon comparing the three tables, it can be observed that for the same column height (in all three cases), SHS provides the lightest design solution across all heights, mainly due to its high buckling resistance in both axes.

Table 4.3 Steel HEA column design checks.

Height [m]	P (design force) [kN]	Selected section	Flexural buckling resistance [kN]	Maximum resistance ratio	Governing criteria	Mass [kg/m]	Total mass [kg]
4	6240	HEA700	6550	0.96	Flexural buckling	204	816
6	6240	HE 1000 × 393	7274	0.86	Flexural buckling	393	2356
9	6240	HE 1000 × 584	6345	0.99	Flexural buckling	584	5256

Table 4.4 Steel SHS column design checks.

Height [m]	P (design force) [kN]	Selected section	Flexural buckling resistance [kN]	Maximum resistance ratio	Governing criteria	Mass [kg/m]	Total mass [kg]
4	6240	SHS 350 × 350 × 14.2	6420	0.98	Flexural buckling	148	592
6	6240	SHS 400 × 400 × 12.5	6310	0.99	Flexural buckling	151	906
9	6240	SHS 400 × 400 × 14.2	6340	0.99	Flexural buckling	170	1530

Table 4.5 Steel CHS column design checks.

Height [m]	P (design force) [kN]	Selected section	Flexural buckling resistance [kN]	Maximum resistance ratio	Governing criteria	Mass [kg/m]	Total mass [kg]
4	6240	CHS 406.4 × 16	6670	0.94	Flexural buckling	154	616
6	6240	CHS 457 × 14.2	6500	0.96	Flexural buckling	155	930
9	6240	CHS 457 × 16	6460	0.97	Flexural buckling	174	1586

4.3.4 Timber Columns

Glued Laminated (Glulam) timber columns with the material class GL28h were designed for three different heights: 4, 6, and 9 m. Glulam is a structural timber product made by bonding multiple layers of wood laminates together with adhesive. The size of a glulam column is often limited by the dimensions of the original lumber from which the laminates are sourced. Due to these size limitations (which are governed by available lamella thickness, maximum manufacturable cross-sections, and production and

Table 4.6 Glulam column design checks.

Height [m]	P (design force) [kN]	Number of columns	Timber column dimension [m]	Timber class	Axial compression resistance [kN]	Maximum resistance ratio	Governing criteria	Total timber mass [kg]
4	6240	4	0.40 × 0.40	GL28h	7600	0.82	Flexural buckling	1178
6	6240	4	0.40 × 0.40	GL28h	6320	0.98	Flexural buckling	1766
9	6240	7	0.40 × 0.40	GL28h	6860	0.91	Flexural buckling	4637

transport constraints), the load-bearing capacity of a single glulam column is restricted. Therefore, in some cases, since a single glulam column cannot meet the required design load, the total force is distributed across multiple identical glulam columns.

For the design of timber column members, the buckling length is considered equal to the system length for supported members. Flexural buckling criteria are calculated for each segment, which is defined by two consecutive restrained sections along the column:

$$\text{Ratio} = \frac{N_{\text{Ed}}}{N_{\text{Rd}}} \leq 1$$

where N_{Ed} is the axial design force and N_{Rd} is the buckling resistance, calculated according to EN 1995-1-1 (2014).

Table 4.6 provides a detailed overview of the design checks for the glulam columns, including key parameters such as the maximum axial design force, the number of glulam columns required, column dimensions, timber class, axial compression resistance, maximum resistance ratio, governing criteria, and mass per length.

4.3.5 Reinforced Concrete Columns

Twenty four different reinforced concrete columns were designed to examine the impact of various factors (such as column length, section shape, concrete strength, and reinforcement density) on the final reinforced concrete column dimension and material usage. The analysis considers three column lengths (4, 6, and 9 m) across two section shapes: square and circular. Concrete strength varies between two classes, C25/30 and C45/55. Reinforcement density is categorized as low density ($\rho \downarrow$, less than 2%) and high density ($\rho \uparrow$, more than 2%).

For design methodology for reinforced concrete column members, the buckling length is considered equal to the column height. Flexural buckling criteria are calculated for each segment delimited by two consecutive restrained sections. The relationship is expressed as:

$$\text{Ratio} = \frac{N_{\text{Ed}}}{N_{\text{Rd}}} \leq 1$$

where N_{Rd} is the design buckling resistances calculated according to (EN 1992-1-1, 2023).

Tables 4.7 and 4.8 present the design checks for square and circular reinforced concrete columns. For each height, the table lists design criteria, including the maximum axial design force, column dimensions,

Table 4.7 Square reinforced concrete column design checks.

Height [m]	P (design force) [kN]	Concrete column dimension [m]	Concrete class	Rebar ratio (high >2% or low <2%)	Axial compression resistance [kN]	Maximum resistance ratio	Governing criteria	Rebars mass [kg/m]	Total rebars mass [kg]	Concrete volume [m^3]	Total concrete mass [kg]
4	6240	0.50 × 0.50	C25/30	3.4% (p ↑)	6159	0.99	Axial compression	80	320	1.00	2400
6	6240	0.55 × 0.55	C25/30	2.6% (p ↑)	5753	0.92	Flexural buckling	76	456	1.82	4356
9	6240	0.55 × 0.55	C25/30	4% (p ↑)	6196	0.99	Flexural buckling	114	1022	2.72	6534
4	6240	0.60 × 0.60	C25/30	1.1% (p ↑)	5865	0.94	Axial compression	45	182	1.44	3456
6	6240	0.60 × 0.60	C25/30	1.4% (p ↑)	5678	0.91	Flexural buckling	55	330	2.16	5184
9	6240	0.60 × 0.60	C25/30	1.8% (p ↑)	6152	0.99	Flexural buckling	70	629	3.24	7776
4	6240	0.45 × 0.45	C45/55	1.9% (p ↑)	6047	0.97	Axial compression	42	167	0.81	1944
6	6240	0.45 × 0.45	C45/55	3.1% (p ↑)	6215	1.00	Flexural buckling	60	358	1.22	2916
9	6240	0.55 × 0.55	C45/55	3.9% (p ↑)	6153	0.99	Flexural buckling	111	1002	2.72	6534
4	6240	0.50 × 0.50	C45/55	1.1% (p ↑)	5291	0.85	Flexural buckling	29	115	1.00	2400
6	6240	0.50 × 0.50	C45/55	1.1% (p ↑)	5803	0.93	Flexural buckling	29	173	1.50	3600
9	6240	0.60 × 0.60	C45/55	1.4% (p ↑)	6183	0.99	Flexural buckling	55	495	3.24	7776

Table 4.8 Circular reinforced concrete column design checks.

Height [m]	P (design force) [kN]	Concrete column dimension [m]	Concrete class	Rebar ratio (high >2% or low <2%)	Axial compression resistance [kN]	Maximum resistance ratio	Governing criteria	Rebars mass [kg/m]	Total rebars mass [kg]	Concrete volume [m³]	Total concrete mass [kg]
4	6240	0.55	C25/30	3.9% (p ↑)	6215	1.00	Axial compression	80	318	0.85	2280
6	6240	0.60	C25/30	2.8% (p ↑)	5971	0.96	Flexural buckling	70	421	1.70	4069
9	6240	0.60	C25/30	3.4% (p ↑)	6159	0.99	Flexural buckling	83	743	2.54	6104
4	6240	0.65	C25/30	1.4% (p ↓)	6233	1.00	Axial compression	44	175	1.33	3184
6	6240	0.65	C25/30	1.6% (p ↓)	6096	0.98	Flexural buckling	49	292	1.99	4776
9	6240	0.70	C25/30	1.4% (p ↓)	6084	0.98	Flexural buckling	49	444	3.46	8308
4	6240	0.50	C45/55	2.1% (p ↑)	6228	1.00	Axial compression	39	158	0.79	1884
6	6240	0.50	C45/55	3.9% (p ↑)	6077	0.97	Flexural buckling	66	399	1.18	2826
9	6240	0.60	C45/55	3.9% (p ↑)	5997	0.96	Flexural buckling	95	853	2.54	6104
4	6240	0.55	C45/55	1.1% (p ↓)	5616	0.90	Flexural buckling	28	111	0.85	2280
6	6240	0.55	C45/55	1.4% (p ↓)	5916	0.95	Flexural buckling	33	195	1.42	3419
9	6240	0.65	C45/55	1.2% (p ↓)	6240	1.00	Flexural buckling	39	349	2.98	7164

concrete class, ratio of rebar used, axial compression resistance, maximum resistance ratio, governing criteria, mass of rebars, concrete volume, and the total concrete mass.

Overall, both geometries satisfy design requirements under comparable loads, yet their performance and material efficiency differ. Circular columns show slightly higher axial resistance for similar or lower material quantities, reflecting a more uniform stress distribution and reduced susceptibility to buckling. Square columns, however, could be easier to construct and connect to orthogonal framing systems.

Building on the principle that high-strength materials can improve structural efficiency by reducing required material quantities, when comparing concrete classes, the use of high-strength concrete (C45/55) enhances axial resistance while reducing required cross-sectional dimensions and total concrete volume. The reinforcement ratio has a direct influence on both strength and material efficiency. Columns with higher reinforcement ratios ($\rho > 2\%$) display improved axial and flexural capacity. Such improvements related to high strength concrete and reinforcement ratio become less pronounced at higher column heights, where slenderness effects are critical.

4.3.6 Composite Columns (Steel–concrete)

Composite columns (Figure 4.6) combine the compressive strength of concrete with the tensile capacity of steel. The steel core increases the column's load-bearing capacity, while the reinforced concrete provides additional compressive strength, fire resistance, and protection for the steel and improves column's resistance to buckling. This combination can result in columns, potentially allowing for slimmer profiles compared to using steel or concrete alone. From an embodied carbon perspective, this efficiency can lead to reduced material quantities; however, the overall impact depends on the balance between the high carbon intensity of steel and the cement content of concrete. In this section, two types of steel–concrete composite columns (partially encased and fully encased) are examined, and the concrete class used in the composite columns is Class C25/30.

In the design of composite column members, special consideration must be given to flexural buckling, ensuring the stability of the column under axial loads. Since the design of steel column members must account for flexural buckling to ensure stability under axial loads, the following key aspects are also considered for their design:

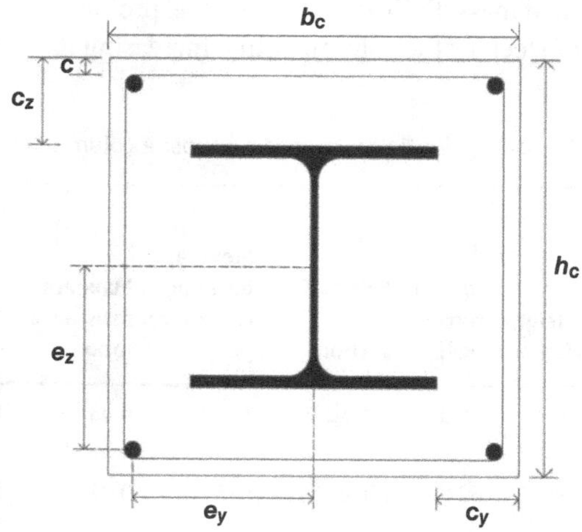

Figure 4.6 Fully encased composite column design dimensioning limits (adapted from EN 1994-1-1, 2004).

Buckling Length: For supported members, the buckling length is taken to be equal to the column height (this length determines the column's effective buckling behavior under loading).

Flexural Buckling Criteria: Flexural buckling is assessed for each segment of the column, which is defined by two consecutive restrained sections. This ensures that both major and minor axes are analyzed for potential buckling.

The following equations are used during design:

$$\text{Major axis buckling: Ratio}_y = \frac{N_{\text{Ed}}}{\chi \times N_{\text{Ply,Rd}}} \leq 1$$

$$\text{Minor axis buckling: Ratio}_z = \frac{N_{\text{Ed}}}{\chi \times N_{\text{Plz,Rd}}} \leq 1$$

where

N_{Ed} is the applied axial design force.

$N_{\text{Ply,Rd}}$ and $N_{\text{Plz,Rd}}$ are the design plastic buckling resistances in the major (y-axis) and minor (z-axis) directions, respectively.

χ is the buckling reduction factor, which accounts for the slenderness of the column. These values are determined according to Section 6.7.3.2(1), Eq. (6.30), and Section 6.7.3.3(2) of EN 1994-1-1 (2004).

Slenderness Condition:
According to Section 6.7.3.1(1) of EN 1994-1-1 (2004), the reduced slenderness λ of a composite column must satisfy the following conditions: $\bar{\lambda}_y \leq 2$; $\bar{\lambda}_z \leq 2$

Concrete Cover Thickness:
As per Section 6.7.3.1(1) of EN 1994-1-1 (2004), the thickness of the concrete that encases the steel profile should not exceed values shown in Figure 4.6.

Tables 4.9 and 4.10 detail the performance of partially and fully encased composite columns, respectively, at heights of 4, 6, and 9 m. For each height, the table lists critical design parameters such as the maximum axial design force, selected column section, flexural buckling resistance, maximum resistance ratio, governing criteria, mass per length of steel section, and rebars as well as the concrete volume and total mass. Fully incasing the steel profile in reinforced concrete leads to a reduction in the required mass of steel for the column (while increasing the volume of concrete).

Table 4.9 Partially encased composite column design checks.

Height [m]	P (design force) [kN]	Selected steel section	Flexural buckling resistance [kN]	Maximum resistance ratio	Governing criteria	Steel section mass [kg/m]	Total steel section mass [kg]	Total rebars mass [kg]	Concrete volume [m^3]	Total concrete mass [kg]	Total column mass [kg]
4	6240	HEA450	6698	0.94	Flexural buckling	140	560	83	0.58	1392	2035
6	6240	HEA650	6454	0.97	Flexural buckling	190	1140	214	1.15	2760	4114
9	6240	HE 1000 × 438	7080	0.89	Flexural buckling	438	3942	919	2.82	6768	11 269

Table 4.10 Fully encased composite column design checks.

Height [m]	P (design force) [kN]	Selected steel section	Concrete column dimension [m]	Flexural buckling resistance [kN]	Maximum resistance ratio	Governing criteria	Total steel section mass [kg]	Total rebars mass [kg]	Concrete volume [m³]	Total concrete mass [kg]	Total column mass [kg]
4	6240	HEA280	0.40 × 0.45	6638	0.94	Flexural buckling	308	118	0.72	1728	2154
6	6240	HEA300	0.45 × 0.50	6500	0.96	Flexural buckling	534	177	1.35	3240	3951
9	6240	HEA360	0.55 × 0.50	6303	0.99	Flexural buckling	1009	355	2.47	5940	7304

4.3.7 Comparison of Masses Between Different Types of Columns with Increasing Height

Figure 4.7 compares the total mass (in kg) of eight different types of structural columns designed for a similar load-bearing capacity, across three different heights: 4, 6, and 9 m. The choice of column material and type has a significant impact on its total weight, and this effect becomes much more pronounced as the column height increases. For every column type, the mass increases as the height increases from 4 to 6 to 9 m. This is expected not only because the column is longer, but also because taller, more slender columns often require larger cross-sections to resist buckling, further increasing their mass. The following specific observations can be made:

- Slenderness is a primary driver of mass in columns. As columns become taller, stability (rather than strength) becomes the critical design factor. This often requires a disproportionate increase in material to achieve the necessary stiffness, highlighting a fundamental trade-off between height and material efficiency.
- For columns with nonsymmetrical cross-sections (such as HEA and partially encased HEA), resistance to buckling along the weaker axis frequently governs the design, especially as height increases. This single factor can control the overall sizing of the column, requiring a significant increase in cross-sectional dimensions or the selection of heavier profiles, even when the material strength is sufficient.
- The way materials are combined in a composite system is critical. A design that fully encases a steel core profile might seem to use more material, but it can create a more efficient cross-section for resisting buckling, ultimately resulting in a lighter element than a partially encased alternative under certain conditions.
- Hollow sections often provide a better stiffness-to-weight ratio than solid or open profiles because they distribute material more effectively away from the element's central axis.

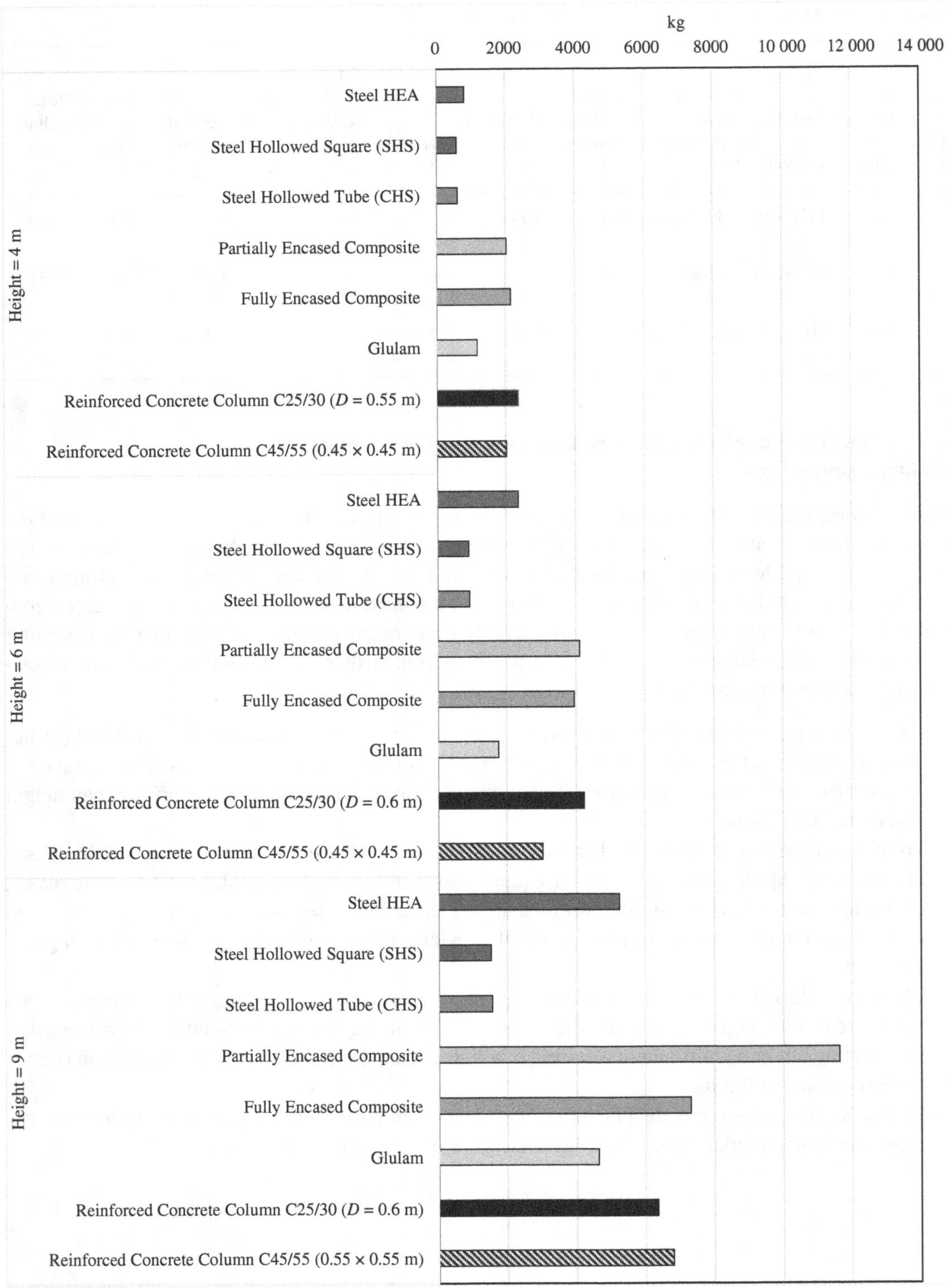

Figure 4.7 The masses of the steel, composite, and timber columns for different column heights.

4.3.8 Influence of Concrete Strength and Reinforcement Ratio on Masses of Columns with Different Heights

This section explores how two key design parameters, concrete strength and reinforcement ratio, influence the mass of reinforced concrete columns of different heights, helping to identify where material optimization is most effective. In Figure 4.8, the masses of all reinforced concrete columns with low reinforcement ratio are compared. Increasing the concrete class results in a reduction of the total column mass. This reduction is more pronounced in shorter columns, where material strength has a greater impact. For longer columns, however, flexural buckling becomes the dominant factor, and column size plays a more critical role than concrete strength. This is because higher-strength concrete does not significantly increase the elastic modulus, meaning that stiffness, rather than strength, governs the design of slender columns, limiting the benefits of using higher-strength materials. Circular section columns exhibit a lower overall mass compared to square sections, emphasizing the material efficiency of circular geometries when using lower rebar densities. This comparison highlights the importance of concrete class and section shape in optimizing structural mass and material usage in reinforced concrete column design.

Figure 4.9 presents the masses of reinforced concrete columns with high reinforcement ratios. The conclusions align with those drawn for the low reinforcement ratio columns. As the concrete class increases, the total mass of the columns can be reduced, with this effect being more pronounced in shorter columns. In longer columns, where flexural buckling is a more significant concern, the column dimensions take precedence over material strength. Circular sections maintain lower overall mass compared to square sections, further reinforcing the material efficiency of circular geometries in designs with higher rebar densities.

Figure 4.10 shows that increasing the reinforcement ratio significantly reduces the overall mass of columns with the same height, regardless of whether they are circular or square in shape. This trend highlights the efficiency gained by using more reinforcement, which compensates for the required strength, allowing for a reduction in the overall concrete volume.

Although this comparison gives a rapid indication of how material choice influences mass, other factors such as structural demands, local supply chains, and installation methods also shape overall

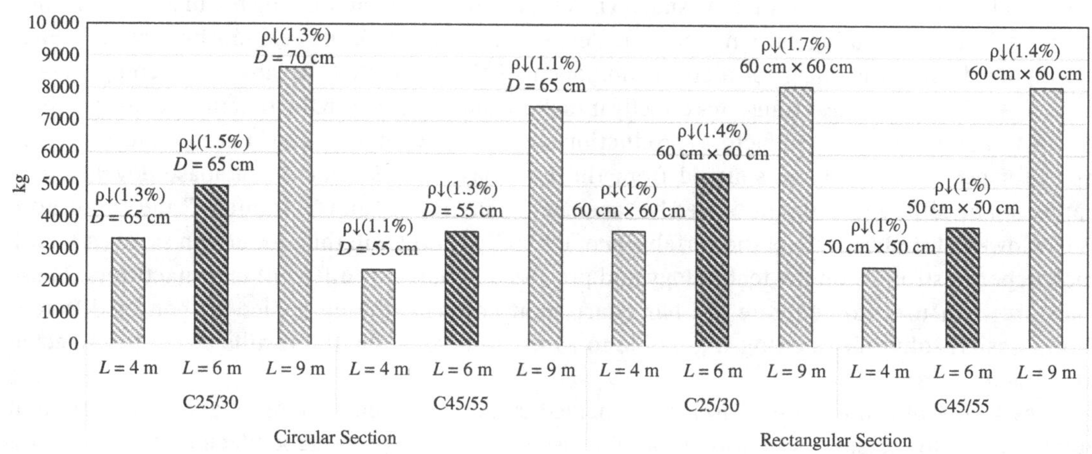

Figure 4.8 The masses of columns with low reinforcement ratio.

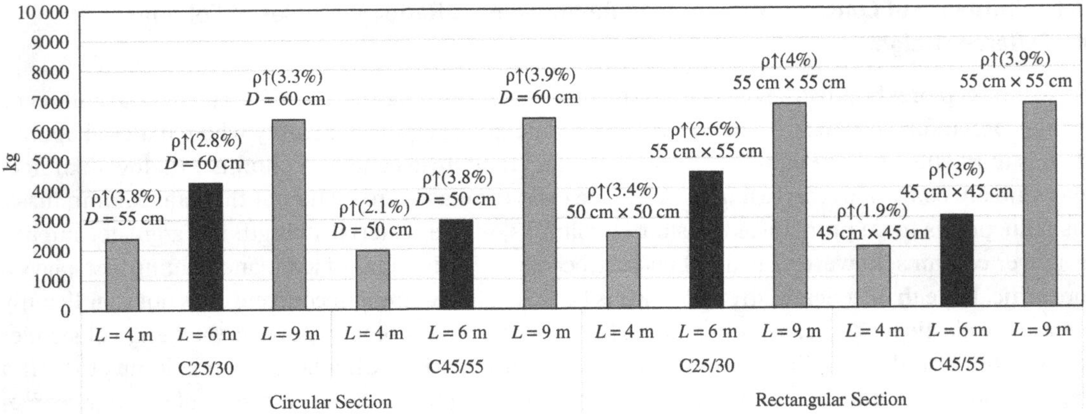

Figure 4.9 The masses of columns with high reinforcement ratio.

performance and sustainability outcomes. Although mass is a good preliminary indicator, please note that it does not fully define performance or environmental outcomes. The complete design context (structural needs, local resources, and life cycle considerations) will be fundamental in choosing the right column type. Overall, this analysis shows that optimizing column design requires understanding whether strength or stability governs the behavior, and that effective material use depends on balancing concrete class, reinforcement ratio, and geometry within the broader design context.

4.3.9 Comparison of Embodied Carbon Between Different Types of Columns with Increasing Height

In this subsection, the embodied carbon of all column types is compared. The calculations are based on a 4-m column height. Material quantities are taken from Tables 4.3 to 4.10, and carbon factors (production stage A1–A3 [EN 15804, n.d.; EN 15978, 2026]) are listed in Table 4.11.

Since embodied carbon in production stage A1–A3 is calculated by multiplying the mass of each component by its corresponding carbon factor (as detailed in Chapter 3), the relationship between mass and embodied carbon is linear for a given component. This means that, for any single component, a percentage reduction in mass translates directly into the same percentage reduction in embodied carbon.

Table 4.11 shows the carbon factors (production stage A1–A3) that are used to calculate embodied carbon for each column type, sourced from the EPD-based carbon factor database developed in Chapter 3. The different values indicated by "average," "maximum," and "minimum," and "minimum GWP & low-sust timber" reflect the variability of embodied carbon in the production stage (A1–A3), which depends strongly on the technology and processes adopted by different manufacturers. These values are then used to generate the bar charts that compare the embodied carbon of different column design solutions, making it possible to assess how sensitive the results are to the selected carbon factors.

Figures 4.11–4.14 show a comparison of embodied carbon for steel, composite, and glulam column solutions across four distinct scenarios: a baseline using average industry carbon factors (Figure 4.11), a worst-case scenario using maximum factors (Figure 4.12), a best-case scenario using minimum factors

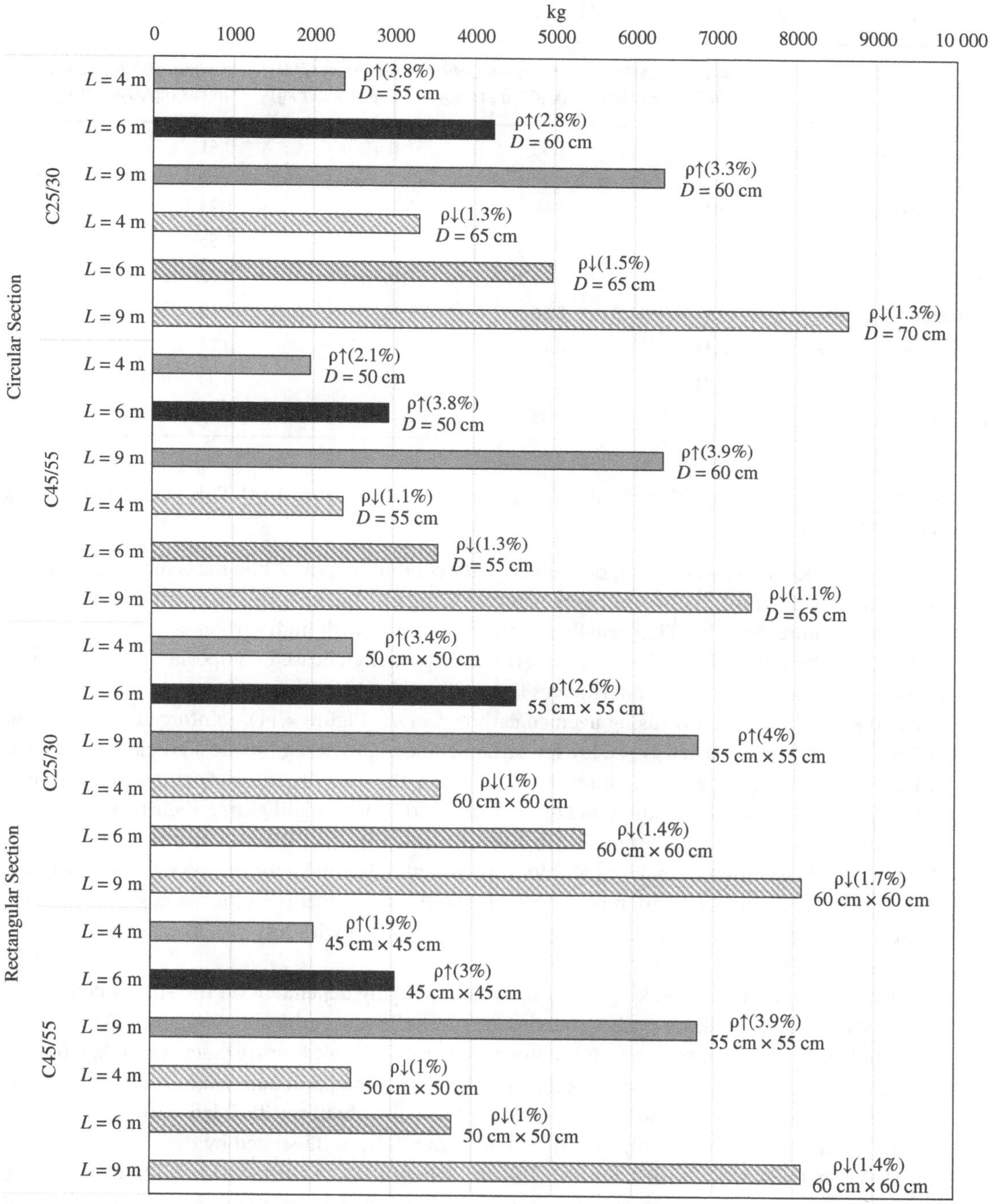

Figure 4.10 The masses reinforced concrete columns.

Table 4.11 Carbon factors (production stage A1–A3).

	Average GWP ($kgCO_2e$ per kg)	Maximum GWP ($kgCO_2e$ per kg)	Minimum GWP ($kgCO_2e$ per kg)	Minimum GWP& low-sust timber ($kgCO_2e$ per kg)
Structural hot-rolled profiles	1.22	1.42	0.61	0.61
Tubular sections	2.26	2.60	1.73	1.73
Plate girders	2.28	3.02	1.55	1.55
Rebars	1.21	1.24	0.73	0.73
Concrete (C25/30)	0.14	0.16	0.12	0.12
Concrete (C45/55)	0.15	0.17	0.13	0.13
Timber	0.41	0.55	0.25	0.83
Precast concrete	0.26	0.33	0.19	0.19

(Figure 4.13), and a specific sensitivity analysis on timber sourcing (Figure 4.14). This parameter analysis reveals several key findings:

- Embodied carbon increases disproportionately with column height. While mass increases predictably (as seen in Figure 4.7), the need to prevent buckling in taller columns (6 m and 9 m) demands significantly more material. This heavily penalizes materials with high carbon factors, causing the embodied carbon of options like the Steel HEA and partially encased composite columns to rise dramatically compared to other alternatives less susceptible to buckling.
- Under the baseline scenario (using average carbon factors, Figure 4.11), reinforced concrete and glulam columns are consistently the lowest-carbon choices, providing the most efficient balance of structural capacity and low carbon intensity. In contrast, the partially encased composite column at 9 m and the steel HEA at 6 m and 9 m are the worst performers, highlighting their material inefficiency when designed for slenderness.
- The strategic distribution of material within a cross-section is crucial for carbon efficiency. Hollow steel sections (SHS and CHS) are more effective than the solid steel HEA profile at greater heights because they achieve the necessary stiffness to resist buckling with less overall mass, leading to a lower carbon footprint.
- A material's environmental ranking is not fixed and is highly dependent on the supply chain. The specific sensitivity analysis on timber sourcing provides the most critical observation (Figure 4.14): when modeled with a high carbon factor representing unsustainable forestry practices, glulam (one of the best performers in other scenarios) becomes a higher-carbon option than reinforced concrete. This proves that a material's perceived green credentials are void without certified, low-carbon sourcing.
- The study's conclusions are highly sensitive to the input data, as illustrated by the significant divergence between the worst-case (maximum emissions, Figure 4.12) and best-case (minimum emissions, Figure 4.13) scenarios. Therefore, designing for low-carbon outcomes requires moving beyond default material choices. Engineers and designers must prioritize materials with transparent and certified supply chains, as the data shows that sourcing and manufacturing practices are as influential as the choice of the material itself, making verified LCA data essential for meaningful carbon reduction.

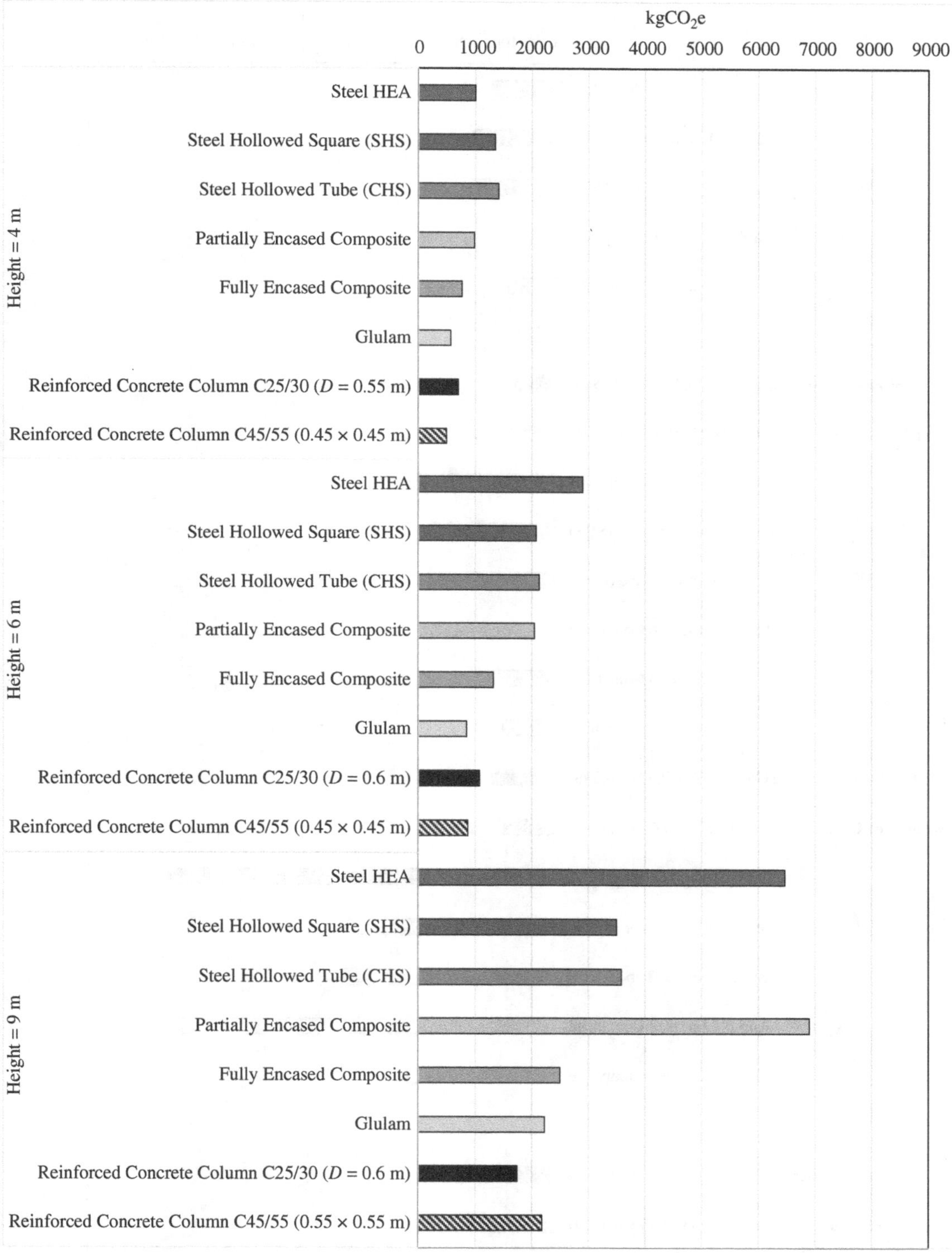

Figure 4.11 The comparison of GWP values with respect to various column types (using average industry carbon factors).

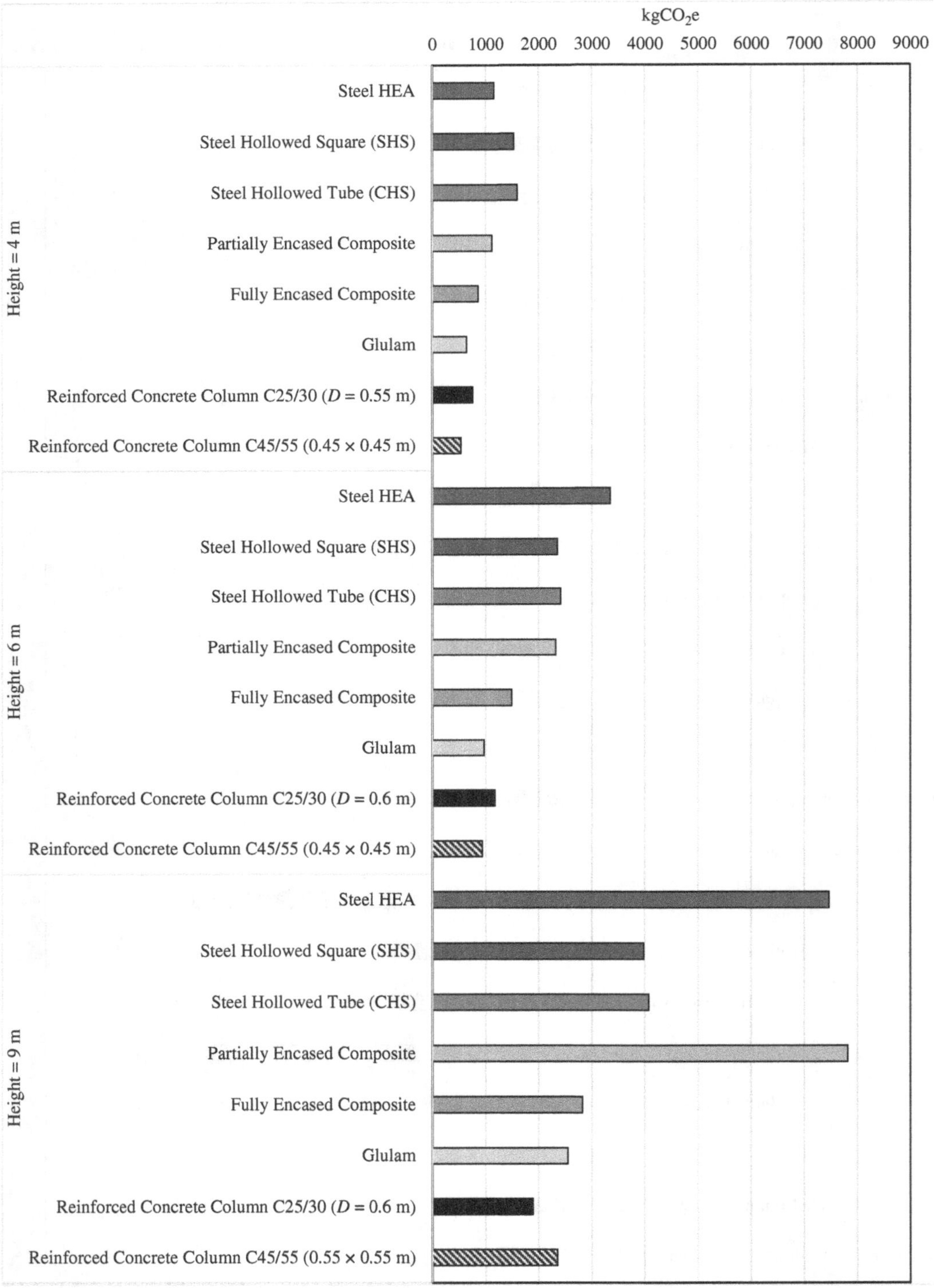

Figure 4.12 The comparison of GWP values with respect to various column types (using "worst-case" scenario using maximum industry carbon factors).

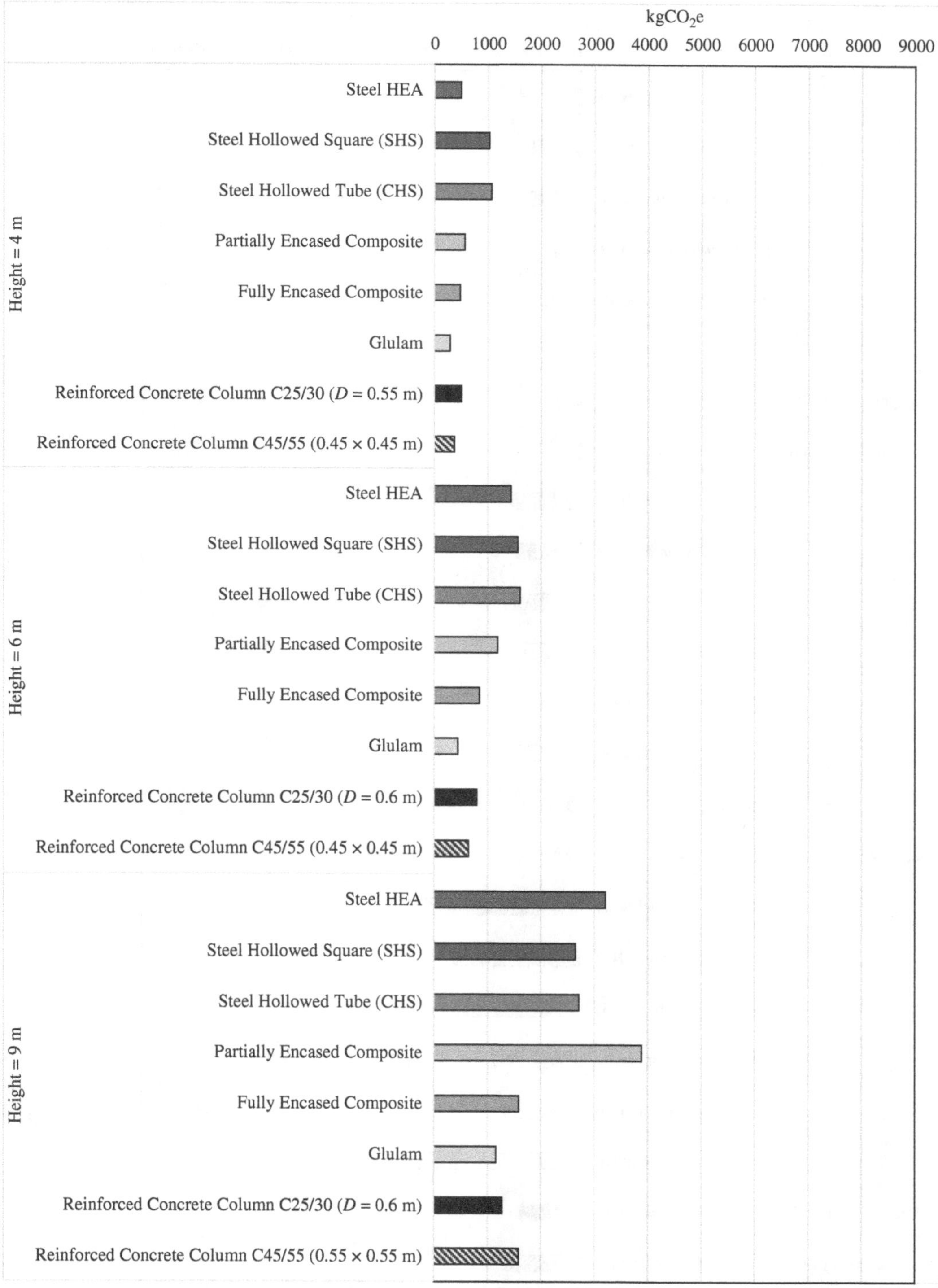

Figure 4.13 The comparison of GWP values with respect to various column types (using "best-case" scenario using minimum industry carbon factors).

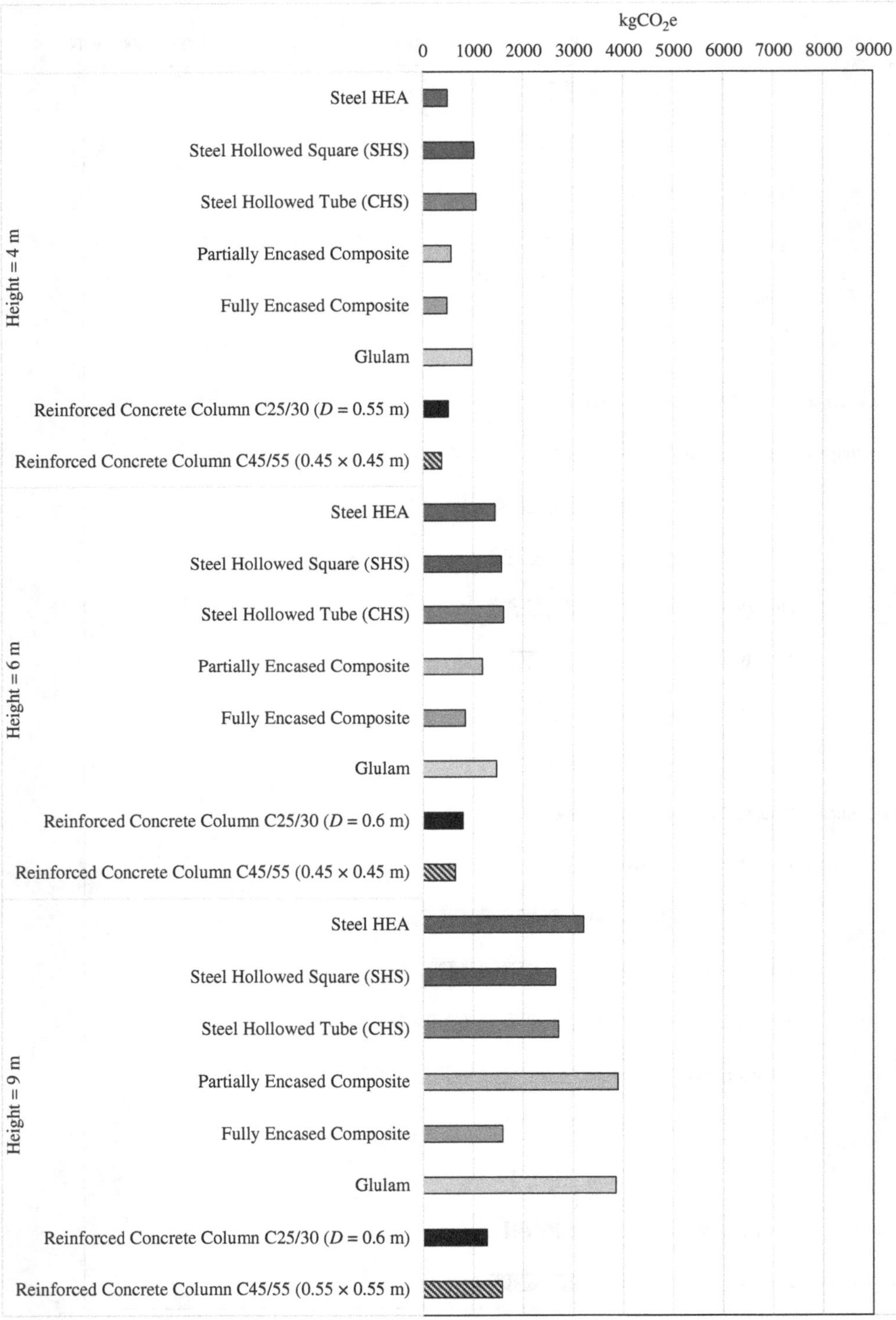

Figure 4.14 The comparison of GWP values with respect to various column types (using "best-case" scenario using minimum industry carbon factors, except for timber, which assumes low-sustainability sourcing).

It is important to recognize the limitations of this analysis. The results are confined to embodied carbon from the production stage (A1–A3) and do not account for whole life-cycle impacts, such as transportation, construction, maintenance, or end-of-life scenarios. Furthermore, this study does not consider other important design factors, including material cost, construction speed, or fire performance. However, the analysis remains valuable because the production stage typically accounts for the largest portion of a structural element's total embodied carbon, making it a critical phase for optimization, and the area where designers can exert the most influence at early design stages.

4.3.10 Influence of Concrete Strength and Reinforcement Ratio on Embodied Carbon of Columns with Different Heights

For the reinforced concrete columns, as shown in Figure 4.15, the GWP of circular sections is lower than that of square sections of the same height. In shorter columns, increasing the concrete class can lower the GWP: although higher-strength concrete carries a higher carbon factor per unit volume, the reduced quantity of material required for the same load capacity offsets this and leads to an overall reduction. In taller columns, where flexural buckling governs, increasing the concrete class no longer brings this advantage and instead results in a slight rise in GWP. Additionally, while increasing the ratio of reinforcement bars helps reduce the size of the columns, it can also result in higher embodied carbon. This highlights the trade-offs between material optimization and environmental impact, especially relevant for taller structures.

Please note that the entire environmental comparison is critically dependent on the chosen embodied carbon factors for each material. These figures can vary significantly based on regional data, manufacturing methods, or the age of the database. A different, yet equally valid, set of GWP values could completely alter the results. For example, using a lower GWP factor for steel (reflecting a higher percentage of recycled content) could make a steel hollow section the most sustainable option, even if it ranked lower in this specific comparison. Therefore, while such parameter analysis is a valuable preliminary tool, its conclusions depend also on the input (carbon factor) data, and final design decisions should be based on project-specific GWP values.

This section highlighted the importance of balancing material efficiency and structural stability, particularly in the design of taller columns where slenderness becomes a critical factor. As column height increases, the counter effects of slenderness on performance become more pronounced, making it essential to carefully consider section dimensions and material selection. Optimizing both weight and stability is crucial in such cases, especially for structures that demand longer spans or greater heights. Achieving this balance ensures not only structural integrity but also material sustainability, which is increasingly important in modern architectural and engineering practice. By addressing these design challenges, engineers can create more efficient, stable, and environmentally responsible structures.

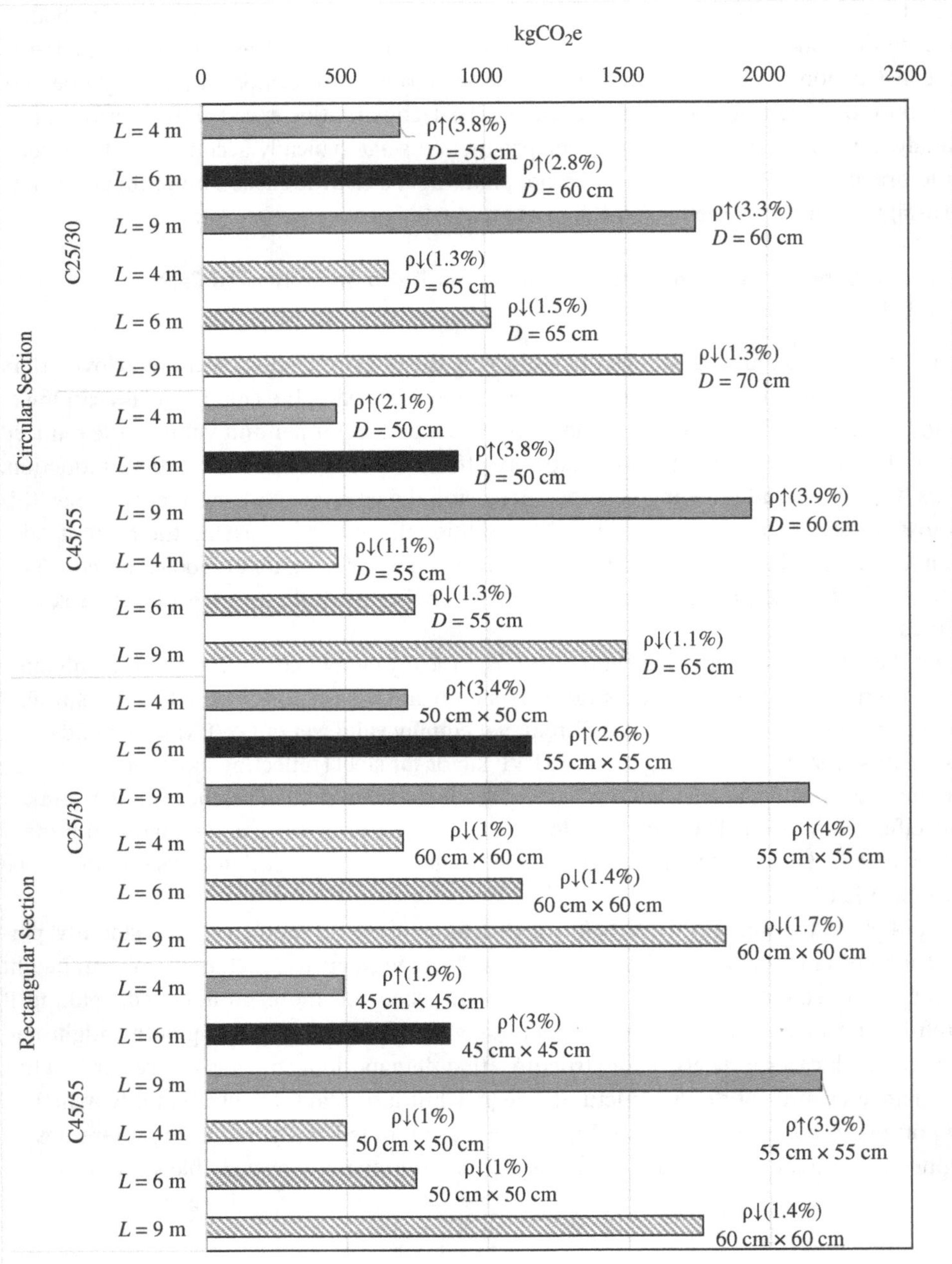

Figure 4.15 The GWP of reinforced concrete columns.

4.4 Case Study: Beams (IPE, HEA, Truss, Steel, Timber, Reinforced Concrete)

This section presents a comparative study of structural beam profiles, focusing on their mass and carbon emissions across spans of 6, 12, and 18 m. The analysis covers beam types such as steel IPE, steel HEA, steel trusses with open and tubular sections, steel plate girder, glulam (glued laminated timber), and reinforced concrete, including prestressed sections.

4.4.1 Benchmark Structural Configuration

Three lengths were considered to calculate different beam section types based on the framing plan. The framing plan consists of two sections, each with a width of 6.0 m (also considered as the effective width for load distribution) (Figures 4.16 and 4.17). The beams are designed for three different span

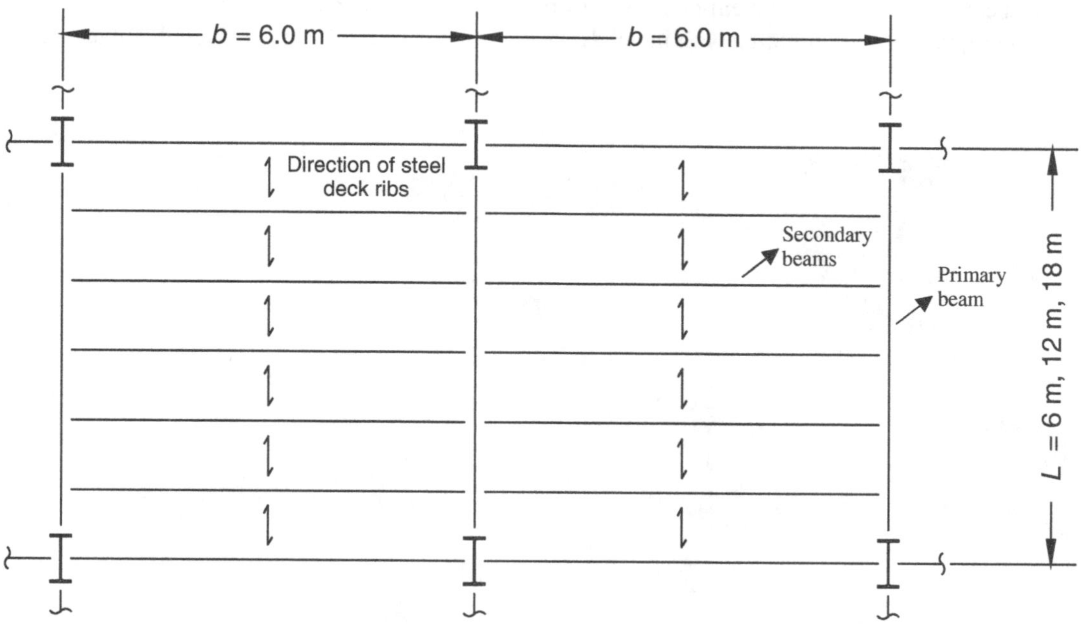

Figure 4.16 Layout of benchmark frame studied in this chapter.

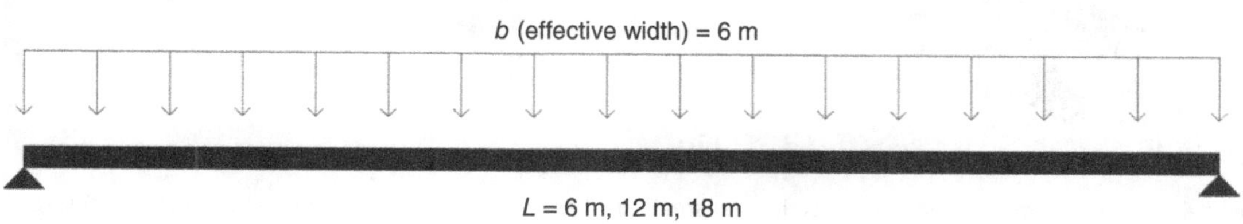

Figure 4.17 Load distribution and effective length of the beam under consideration.

lengths (6, 12, and 18 m), representing a practical range commonly used in building construction (from short spans typical of residential and office buildings to longer spans used in commercial and industrial facilities). This variation allows a systematic comparison of how span length influences the required cross-section size and, consequently, the embodied carbon of the structural element. Each beam is subjected to a uniformly distributed load across its entire length, which is a common loading condition for floor beams supporting a concrete slab, as the slab transfers the floor loads to the beam in a continuous manner along its span. The direction of the steel deck ribs (reinforced concrete slab with corrugated deck) is assumed to run perpendicular to the span of the secondary beams.

4.4.2 Beams with Open Section Girders (IPE, HEA)

Seven types of beam sections have been analyzed, which are shown in Figure 4.18: steel IPE (European I-beam), steel HEA (European wide flange beam), steel truss beams (i) using either tubular sections [a] circular and [b] rectangular tubes), (ii) open sections (such as I-sections and angle profiles), glulam, and reinforced concrete beams (i) conventional (ii) prestressed reinforced concrete beams.

The loading for an office building with metal deck system flooring has been considered (Table 4.12).

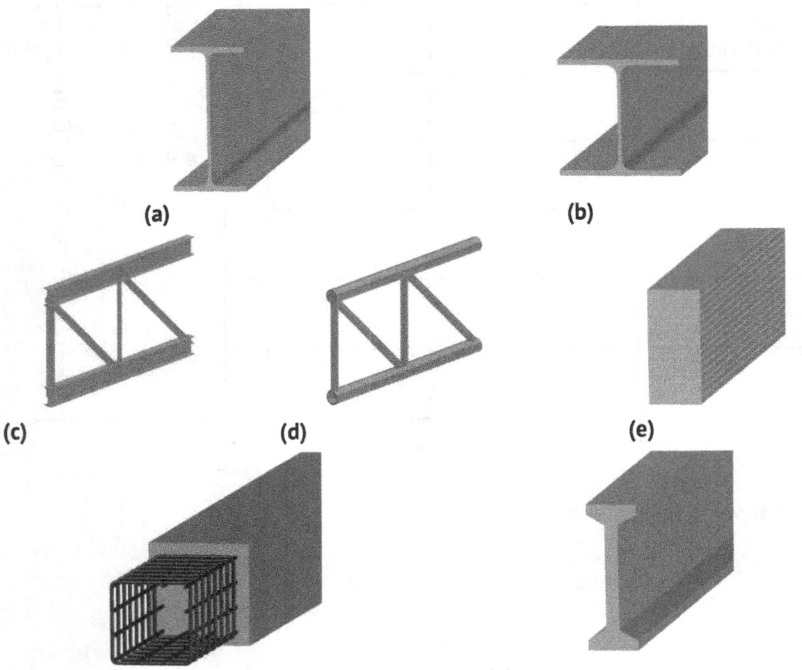

Figure 4.18 Beam type is used in this section: (a) steel IPE, (b) steel HEA, (c) steel open section trusses, (d) steel tubular section trusses, (e) glulam beam, (f) reinforced concrete, and (g) prestressed reinforced concrete.

Table 4.12 Load cases.

Load cases	Load [kN/m^2]
Self-weight (G1)	3
Permanent nonstructural (G2)	2
Live load office: category B (Q)	2

Table 4.13 Material characteristics.

Material type	Density	Yield strength (f_y)
Steel S355	7850 kg/m^3	355 MPa
Steel S275	7850 kg/m^3	275 MPa
Reinforcing steel B450	7850 kg/m^3	450 MPa
Concrete C25/30	2400 kg/m^3	Class C25/30
Concrete C45/55	2400 kg/m^3	Class C45/55
Glulam GL28h	470 kg/m^3	Class GL28h

Table 4.14 Steel IPE beam design checks.

Span [m]	Maximum bending moment [kN·m]	Selected cross-section	Cross-section class	Design resistance for bending [kN·m]	Maximum resistance ratio (ULS)	Deformation (SLS)	Governing criteria	Mass [kg/m]	Total mass [kg]
6	287	IPE360	Class 1	280	0.95	21 mm (L/285 < L/250)	ULS	57	342
12	1085	IPE750 × 134	Class 1	1286	0.85	36 mm (L/372 < L/250)	ULS	134	1608
18	2470	PG115 × 1.25 × 2.5	Class 4	2544	0.98	51 mm (L/352 < L/250)	ULS	188	3391

For the ULS, the load combination is calculated as

$$\text{Load combination (ULS)} = 1.35 \times (G1 + G2) + 1.5\ (Q)$$

For SLS, the load combination is:

$$\text{Load combination (SLS)} = G1 + G2 + Q$$

The material characteristics are listed in Table 4.13.

Table 4.14 details the performance of steel open profiles (IPE, welded sections) at spans of 6, 12, and 18 m. Each span lists the maximum bending moment, selected beam section, cross-section class,

Table 4.15 Steel HEA beam design checks.

Span [m]	Maximum bending moment [kN·m]	Selected cross-section	Cross-section class	Design resistance for bending [kN·m]	Maximum resistance ratio (ULS)	Deformation (SLS)	Governing criteria	Mass [kg/m]	Total mass [kg]
6	276	HEA300	Class 2	380	0.73	19 mm (L/310 < L/250)	SLS	88	529
12	1095	HEA600	Class 1	1417	0.78	39 mm (L/303 < L/250)	SLS	178	2136
18	2504	HEA900	Class 1	2864	0.88	68 mm (L/262 < L/250)	SLS	252	4536

design resistance for bending, maximum resistance ratio for ULS, deformation under SLS, governing criteria, mass per length, and total mass. Table 4.15 presents similar information for steel HEA beams. A comparison of the two tables reveals a fundamental difference in the governing design criterion between the two profile families. For IPE sections (which have a relatively narrow flange and a deeper, more slender web), the design is primarily governed by the Ultimate Limit State (ULS), meaning that the cross-section is selected based on its capacity to resist the applied bending moment and shear forces without exceeding the material strength. In contrast, for HEA sections (which feature wider flanges and a more compact, stockier geometry), the Serviceability Limit State (SLS) becomes the governing criterion. This means that, even though the HEA section has sufficient strength to carry the applied loads, its selection is dictated by the need to limit the vertical deflection of the beam to acceptable values (typically span/250 or span/300, depending on the code and the type of floor finish). The fact that SLS governs for HEA sections indicates that these profiles are inherently stiffer than what would be strictly required by strength alone, resulting in heavier cross-sections being selected. In practical terms, the use of HEA beams offers a clear advantage in situations where the available structural depth is limited (for example, when the floor-to-floor height must be minimized), since HEA profiles are shallower than their IPE counterparts for equivalent load-bearing capacity. However, this reduction in beam height comes at the cost of increased total mass, which in turn leads to higher embodied carbon for the same structural function.

4.4.3 Steel Truss Beams with Open Sections

The steel truss beams are designed using open sections, with chords made from IPE sections and braces composed of double angles (Figure 4.19).

Two truss design options have been developed: a shallow truss and a deep truss. The rationale behind exploring two different depths lies in the fundamental structural behavior of trusses: as the depth of the truss increases, the internal lever arm between the top and bottom chords also

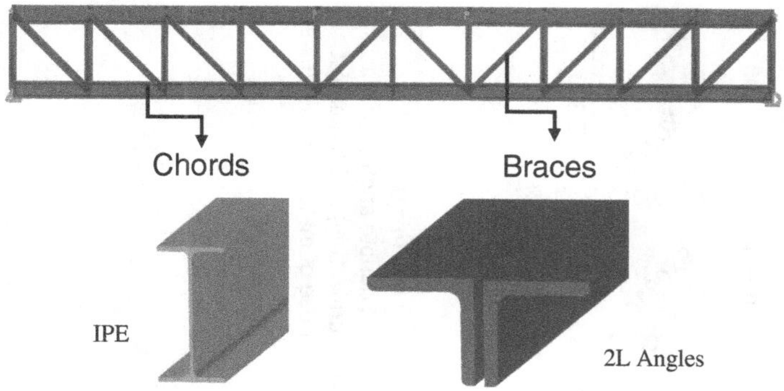

Figure 4.19 Steel truss beam layout showing the general configuration of the truss with Warren-type bracing pattern. The chords (top and bottom horizontal members) are made from IPE open sections, while the diagonal braces are composed of double angles (2L angles) welded or bolted back-to-back, forming a symmetric cross-section that provides equal buckling resistance about both axes.

increases. Since the bending moment carried by the truss is resisted by a pair of axial forces (compression in the top chord and tension in the bottom chord) acting at a distance equal to the truss depth, a larger lever arm means that smaller axial forces are needed to resist the same applied bending moment. Consequently, the chord members can be selected with lighter cross-sections. In addition, the reduced chord forces also lead to lower forces in the diagonal braces, further contributing to an overall reduction in material. The deep truss therefore requires less steel mass to carry the same load over the same span, resulting in lower embodied carbon. However, this structural efficiency comes with a practical trade-off: a deeper truss occupies more vertical space within the building, which may conflict with architectural constraints such as limited floor-to-floor height or the need to accommodate building services within the structural depth. Figures 4.20 and 4.21 show respectively the maximum resistance ratio of each truss, highlighting their optimization in terms of structural performance. The maximum resistance ratio indicate that the design achieves a balanced distribution of forces, reducing the likelihood of overloading any single component and guaranteeing the safety of trusses.

Tables 4.16 and 4.17 summarize the design outcomes for the shallow and deep truss configurations, respectively. For each span length, the tables report the selected IPE sections for the upper and lower chords, which may differ since the upper chord is subjected to compression (and is therefore more susceptible to buckling) while the lower chord acts in tension. The maximum resistance ratio under the ULS indicates how efficiently the truss utilizes its load-bearing capacity, with values closer to 1.0 representing a more optimal design. The deformation column reports the maximum midspan deflection under SLS loading, compared against the standard limit of L/250. Notably, for both truss configurations, the ULS governs across all span lengths, which contrasts with the solid-web HEA beams discussed earlier where the SLS was the controlling criterion. The total mass column confirms that, for each span length, the deep truss achieves a consistently lower mass than the shallow truss, validating the structural benefit of the increased internal lever arm.

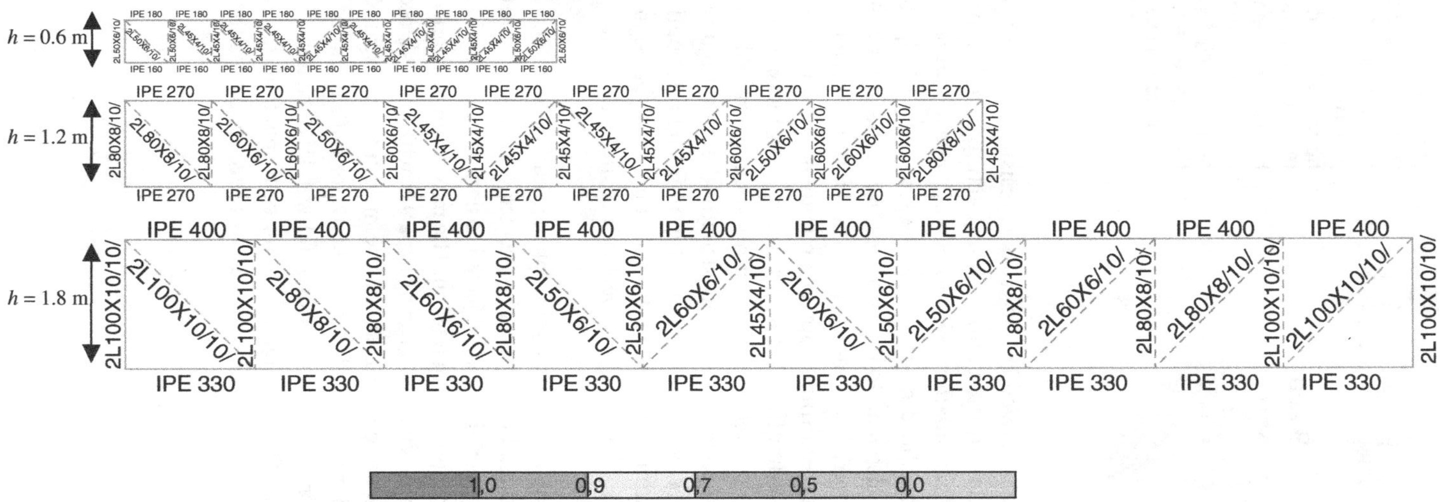

Figure 4.20 Maximum resistance ratio factor of shallow steel truss with open sections for three different span lengths: 6 m (h = 0.6 m), 12 m (h = 1.2 m), and 18 m (h = 1.8 m). Each member is labeled with its cross-section designation (IPE sections for chords and double angles for braces). The colored-scale shading of each member indicates its resistance utilization ratio, ranging from 0.0 (lightest, lowest utilization) to 1.0 (darkest, full utilization). A well-optimized design features utilization ratios close to 1.0 across all members, indicating efficient use of material, while lower values suggest that the member has reserve capacity beyond what is strictly required by the applied loading.

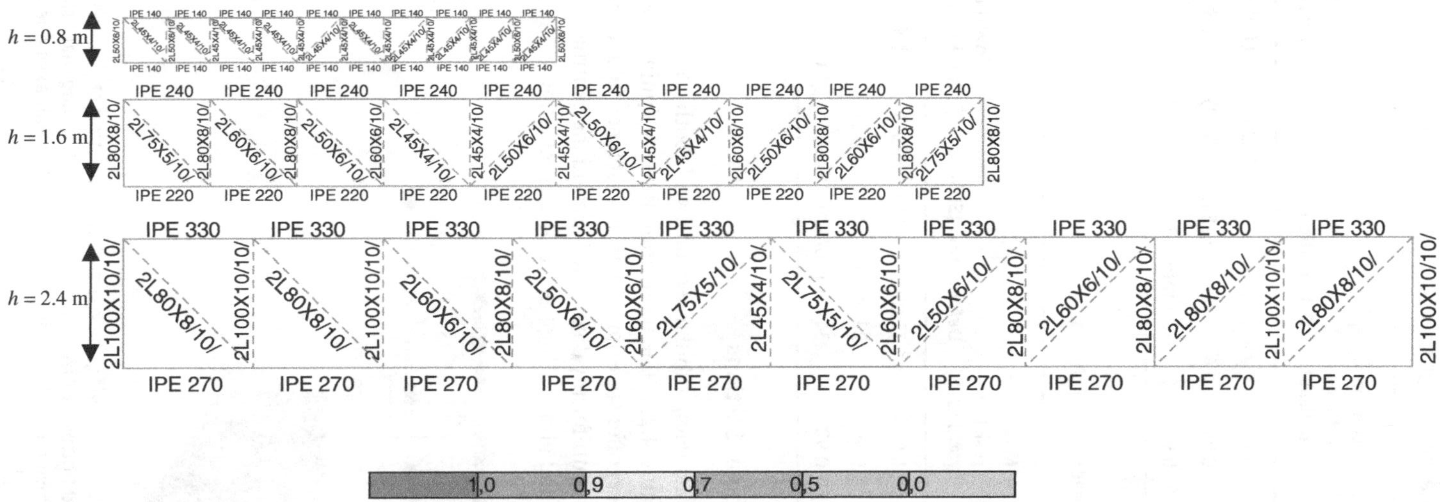

Figure 4.21 Maximum resistance ratio factor of deep steel truss with open sections for three different span lengths: 6 m (h = 0.8 m), 12 m (h = 1.6 m), and 18 m (h = 2.4 m). Each member is labeled with its cross-section designation (IPE sections for chords and double angles for braces). The colored-scale shading of each member indicates its resistance utilization ratio, ranging from 0.0 (lightest, lowest utilization) to 1.0 (darkest, full utilization). Compared to the shallow truss (Figure 4.20), the increased structural depth results in smaller chord sections for the same span length, confirming that a larger internal lever arm reduces the axial forces in the chords and allows for lighter member selection.

Table 4.16 Open-section steel truss (shallow) design checks.

Span [m]	Upper cord section	Lower chord section	Maximum resistance ratio (ULS)	Deformation (SLS)	Governing criteria	Total mass [kg]
6	IPE180	IPE160	0.88	12 mm (L/500 < L/250)	ULS	334
12	IPE270	IPE270	0.98	22 mm (L/545 < L/250)	ULS	1317
18	IPE400	IPE330	0.92	37 mm (L/486 < L/250)	ULS	3170

Table 4.17 Open-section steel truss (deep) design checks.

Span [m]	Upper cord section	Lower chord section	Maximum resistance ratio (ULS)	Deformation (SLS)	Governing criteria	Total mass [kg]
6	IPE140	IPE140	0.99	11 mm (L/545 < L/250)	ULS	269
12	IPE240	IPE220	0.99	23 mm (L/521 < L/250)	ULS	1111
18	IPE330	IPE270	0.92	34 mm (L/529 < L/250)	ULS	2542

4.4.4 Steel Truss Beams with Tubular Sections

The steel truss beams are also designed using tubular sections, with both chords and braces made from Circular Hollow Sections (CHS) (Figure 4.22). The motivation for redesigning the trusses with tubular profiles is twofold. First, CHS members offer superior structural efficiency under compression compared to open sections, since their material is uniformly distributed around the centroidal axis, providing equal buckling resistance in all directions and eliminating the weak-axis buckling vulnerability typical of IPE

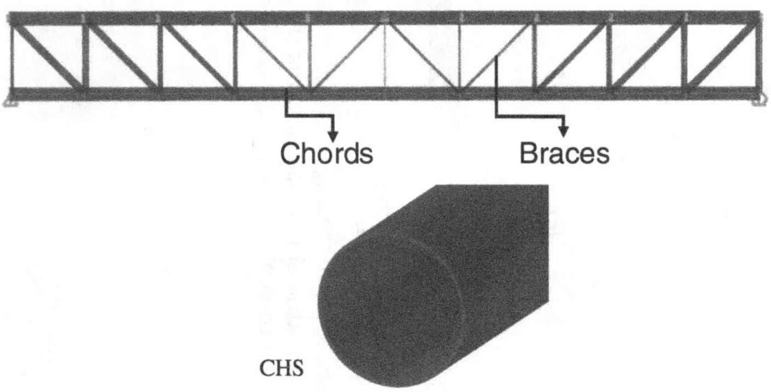

Figure 4.22 Tubular steel truss beam layout showing the general configuration of the truss with Warren-type bracing pattern. Both the chords (top and bottom horizontal members) and the diagonal braces are made from Circular Hollow Sections (CHS), which provide equal buckling resistance about all axes due to their axially symmetric cross-section.

profiles and angles. Second, CHS-based trusses benefit from a more favorable surface-area-to-volume ratio, which reduces the amount of material exposed to corrosion and can lower maintenance-related costs and environmental impacts over the structure's service life. Moreover, the aesthetic quality and cleaner geometry of tubular trusses make them a preferred choice in architecturally exposed applications. By comparing the tubular truss designs with the open-section trusses presented in Section 4.4.3, it becomes possible to evaluate how the choice of profile type influences the total mass and, consequently, the embodied carbon of the structural element for the same span and loading conditions.

Two truss design options have been developed: a shallow truss and a deep truss (featuring greater structural depth that required less material, resulting in a lower total mass). Figures 4.23 and 4.24 show respectively the maximum resistance ratio of each truss, highlighting their optimization in terms of structural performance. The maximum resistance ratio indicates that the design achieves a balanced distribution of forces, reducing the likelihood of overloading any single component and guaranteeing the safety of trusses.

Tables 4.18 and 4.19 summarize the design outcomes for the shallow and deep CHS truss configurations, respectively. For each span length, the tables report the selected CHS sections for the upper and lower chords. Unlike the open-section trusses (Tables 4.16 and 4.17), where the upper and lower chords were assigned different IPE profiles, the CHS trusses use the same section for both chords at each span length. This is because the axially symmetric geometry of CHS provides equal buckling resistance in all directions, making the section equally efficient in compression (upper chord) and tension (lower chord). The maximum resistance ratio under the ULS shows that all CHS truss designs achieve utilization values above 0.90, indicating a highly efficient use of the material's load-bearing capacity. The deformation column reports the maximum midspan deflection under SLS loading, which remains well within the L/250 limit for all configurations. As with the open-section trusses, the ULS governs in all cases, and the deep truss consistently achieves a lower total mass than the shallow truss for the same span length.

4.4.5 Timber Beams

Having explored the design space for steel beams (both solid-web and truss configurations with open and tubular sections), the analysis now turns to timber as a structural material. Timber behaves fundamentally differently from steel in several respects that will influence the design outcomes. First, timber has a significantly lower strength and stiffness compared to steel, which means that much larger cross-sections are required to carry the same loads over the same spans. This results in deeper and bulkier members, making the available structural depth a more critical constraint. Second, unlike steel profiles that are industrially produced in standardized catalogues with a wide range of section sizes, glulam beams are manufactured by bonding timber laminates together, and their dimensions are therefore limited by the size of the source lumber and by practical manufacturing constraints (particularly the maximum achievable width). Third, timber can have a substantially lower carbon factor per kilogram compared to steel (as shown in Table 4.11), which means that even though the total mass of a timber beam may be higher, its embodied carbon could potentially be comparable or even lower. This section therefore investigates whether the mass penalty of timber is offset by its environmental advantage in terms of carbon intensity.

A glulam beam using the material class GL28h was designed for three-span lengths: 6, 12, and 18 m, respectively (Glulam-Glued Laminated Timber is a structural timber beam constructed from multiple

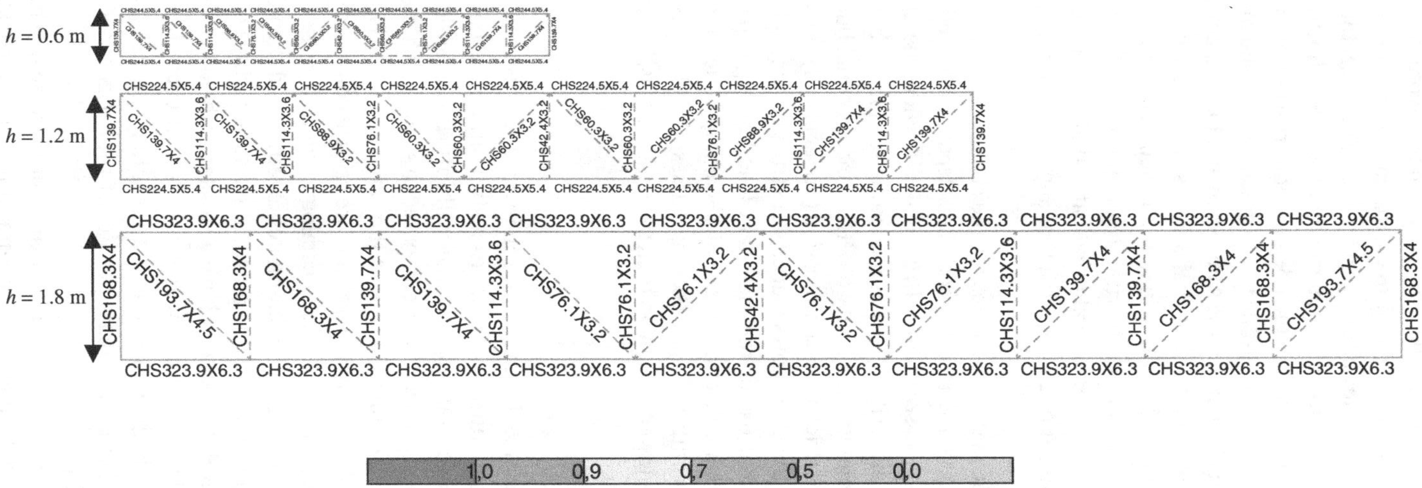

Figure 4.23 Maximum resistance ratio of shallow truss with tubular sections for three different span lengths: 6 m (h = 0.6 m), 12 m (h = 1.2 m), and 18 m (h = 1.8 m). Each member is labeled with its CHS designation (outer diameter × wall thickness). The grayscale shading indicates the resistance utilization ratio, ranging from 0.0 (lightest, lowest utilization) to 1.0 (darkest, full utilization). Note that, unlike the open-section trusses (Figure 4.20), the same CHS section is used for both the upper and lower chords at each span length, since the symmetric cross-section provides comparable performance in both compression and tension.

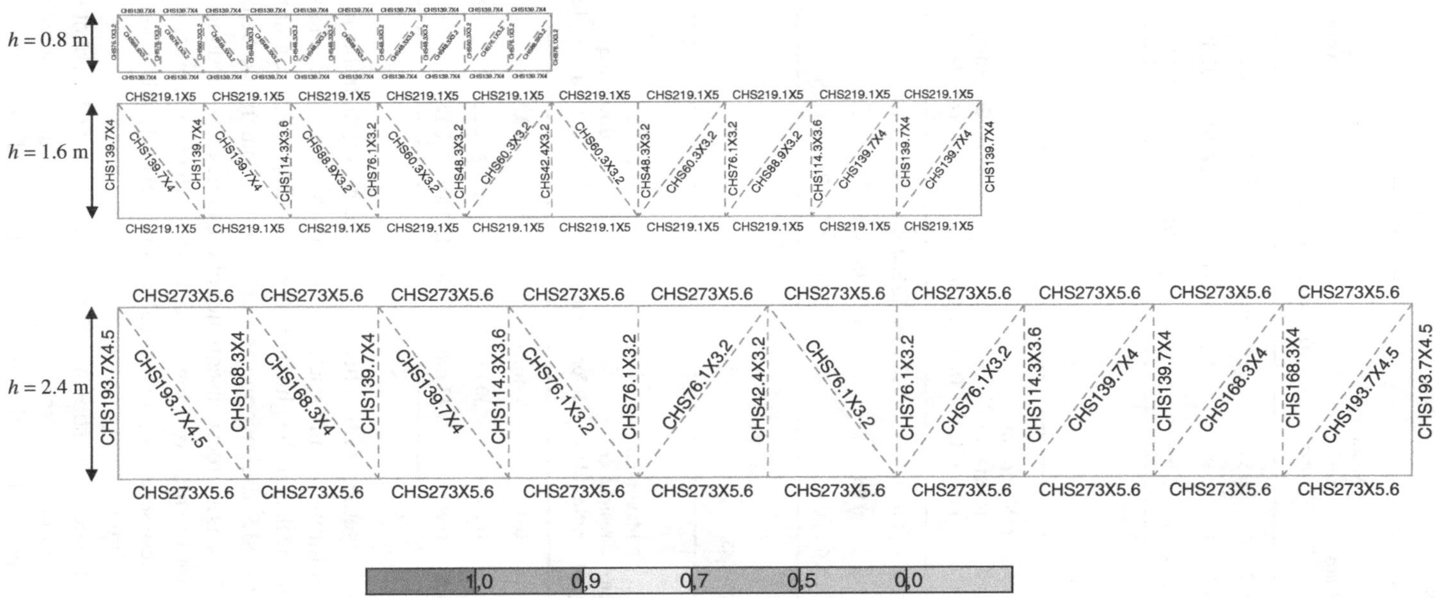

Figure 4.24 Maximum resistance ratio of deep truss with tubular sections for three different span lengths: 6 m (h = 0.8 m), 12 m (h = 1.6 m), and 18 m (h = 2.4 m). Each member is labeled with its CHS designation (outer diameter × wall thickness). The grayscale shading indicates the resistance utilization ratio, ranging from 0.0 (lightest, lowest utilization) to 1.0 (darkest, full utilization). Compared to the shallow CHS truss (Figure 4.23), the increased structural depth results in smaller chord sections for the same span length, while the brace sections also decrease due to the reduced axial forces transmitted through the truss.

Table 4.18 CHS truss design checks.

Span [m]	Upper cord section	Lower chord section	Maximum resistance ratio (ULS)	Deformation (SLS)	Governing criteria	Total mass [kg]
6	CHS168.3 × 4	CHS168.3 × 4	0.92	15 mm (L/400 < L/250)	ULS	264
12	CHS244.5 × 5.4	CHS244.5 × 5.4	0.99	31 mm (L/387 < L/250)	ULS	1016
18	CHS323.9 × 6.3	CHS323.9 × 6.3	0.98	46 mm (L/391 < L/250)	ULS	2318

Table 4.19 CHS truss design checks.

Span [m]	Upper cord section	Lower chord section	Maximum resistance ratio (ULS)	Deformation (SLS)	Governing criteria	Total mass [kg]
6	CHS139.7 × 4	CHS139.7 × 4	0.91	12 mm (L/500 < L/250)	ULS	247
12	CHS219.1 × 5	CHS219.1 × 5	0.96	25 mm (L/480 < L/250)	ULS	956
18	CHS273 × 5.6	CHS273 × 5.6	0.96	40 mm (L/450 < L/250)	ULS	2027

Table 4.20 Glulam beam design checks.

Span [m]	Maximum bending moment [kN·m]	Selected section GL h (cm) × b (cm)	Maximum resistance ratio (ULS)	Deformation (SLS)	Governing criteria	Mass [kg/m]	Total mass [kg]
6	269	GL 68 × 28 cm	0.94	9 mm (L/615 < L/250)	ULS	88	526
12	1097	GL 104 × 38 cm	0.93	30 mm (L/400 < L/250)	ULS	182	2182
18	2522	GL 160 × 38 cm	0.96	43 mm (L/418 < L/250)	ULS	280	5040

layers of wood laminates bonded together with adhesive. The size of a glulam beam is often limited by the dimensions of the original lumber from which the laminates are sourced). Table 4.20 provides a detailed overview of the design checks for the glulam beam, including the maximum resistance ratio for the ULS, deformation under the SLS, the governing criteria, and the total mass. Across all three span lengths, ULS governs the design. This is a direct consequence of the large cross-sectional depths required by glulam beams to compensate for the relatively low bending strength of timber compared to steel. At the same time, the large depth of these sections provides substantial flexural rigidity, keeping deflections well below the L/250 serviceability limit.

The dimensions and sections of the beam are shown in Figure 4.25. The total height of the glulam beam (h) ranges from 68 to 160 cm, depending on the span length, while the width of the glulam beam (w) is constrained to 38 cm. These dimensions reflect the design limitations and considerations for structural

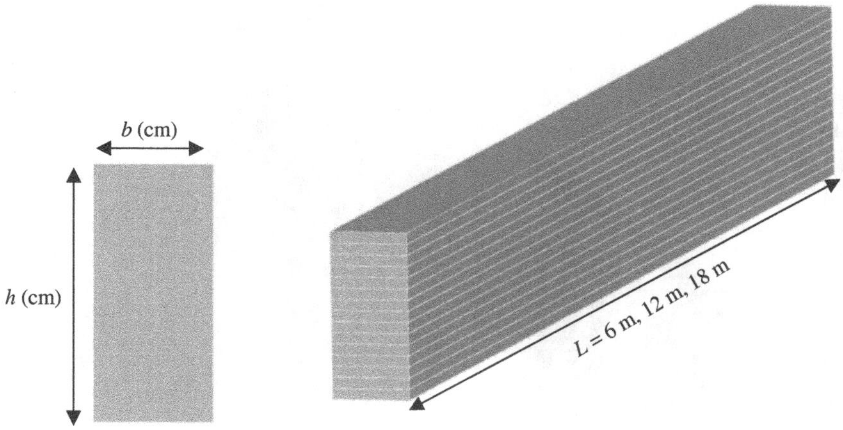

Figure 4.25 Glulam beam geometry: rectangular cross-section with depth h and width b (left), and three-dimensional view of the laminated layup (right). The beam is designed for three representative span lengths (L = 6, 12, and 18 m) to evaluate how cross-sectional dimensions and total mass scale with increasing span under the same loading conditions.

performance across different spans (in practice, solid timber beams of 18 m span are rarely adopted, as such lengths are neither economical nor practical; long-span timber systems are typically achieved using trusses or hybrid configurations. However, for the sake of this exercise, the 18 m beam case is intentionally included to demonstrate this noneconomic condition, and to quantify its material implications).

4.4.6 Reinforced Concrete Beams

This analysis includes various reinforced concrete beams, including two different concrete classes and a prestressed beam. Table 4.21 provides details on the ULS resistance ratio, governing criteria, total rebar mass, and concrete volume for these beams. Notably, Table 4.21 demonstrates that higher concrete

Table 4.21 Reinforced concrete beam design checks.

Span [m]	Concrete class	Maximum bending moment [kN·m]	Selected Section B h (cm) × b (cm)	Governing criteria	Maximum resistance ratio	Mass of rebar [kg/m]	Total rebars mass [kg]	Concrete volume [m³]	Total concrete mass [kg]
6	C25/30	288	B55 × 30 cm	ULS	0.91	25	155	1.0	2378
6	C45/55	288	B50 × 30 cm	ULS	0.95	23	141	0.9	2160
12	C25/30	1295	B100 × 40 cm	ULS	0.91	63	753	4.8	11520
12	C45/55	1259	B 80 × 40 cm	ULS	0.94	53	640	4.0	9720
18	C25/30	3394	B150 × 50 cm	ULS	0.92	118	2120	13.5	32400
18	C45/55	3291	B135 × 5 cm	ULS	0.95	106	1907	12.1	29160

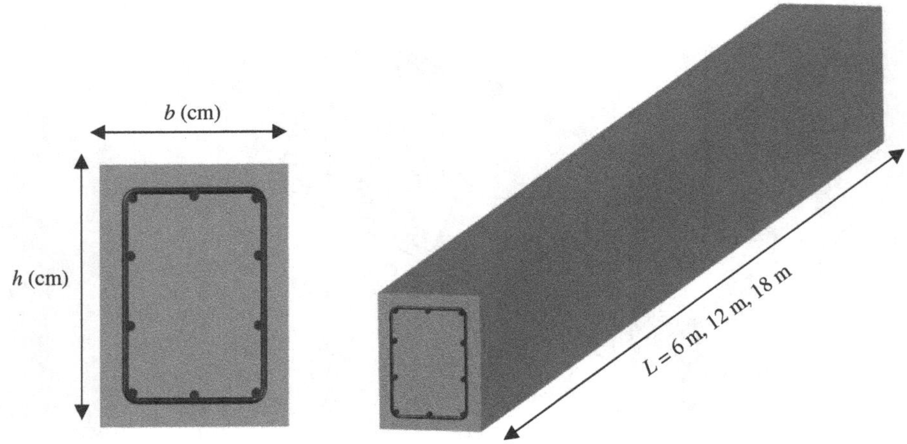

Figure 4.26 Reinforced concrete beam geometry: rectangular cross-section with depth h, width b, and longitudinal and transverse reinforcement layout (left), and three-dimensional view (right). The beam is designed for three representative span lengths ($L = 6, 12$, and 18 m) to assess how cross-sectional dimensions, reinforcement demands, and total mass evolve with increasing span under the same loading conditions.

classes enable the design of beams with a smaller height, highlighting an advantage in terms of beam dimensions.

Figure 4.26 shows the schematics of the section and the length of the reinforced concrete beams.

In the case of the 18 m span, the reinforced concrete beam represents an unrealistic solution from a construction and economic standpoint (casting in situ such a long span monolithically would require extensive formwork, heavy reinforcement, and significant on-site labor, making it inefficient compared to alternative systems such as prestressed beams or composite girders). Nevertheless, this case is intentionally included to show the limitations of conventional reinforced concrete for long spans and to quantify how rapidly material demand increases when approaching such spans. To address this inefficiency, a prestressed reinforced concrete beam is also analyzed, offering a more feasible and economical solution for the same span range while maintaining structural performance and reducing material use. Prestressing works by introducing a compressive force into the beam (typically through high-strength steel strands tensioned before or after casting) that counteracts the tensile stresses generated by external loading, effectively allowing the concrete to participate more fully in resisting bending. This delays cracking, reduces deflections, and permits the use of shallower cross-sections for a given span and load level. The prestressed reinforced concrete beam material is the concrete class C45/55. By prestressing the reinforced concrete beam, the total height of the beam and, eventually, the volume of the concrete can be reduced significantly. Table 4.22 shows the maximum resistance ratio for the ULS, governing criteria, the total mass of rebars and strands, and the volume of the concrete.

Figure 4.27 shows the span lengths and schematic of the section of the prestressed reinforced concrete.

Table 4.22 Prestressed reinforced concrete beam design checks.

Span [m]	Max bending moment [kN·m]	Selected section PCB *h* (cm) × *b* (cm)	Governing criteria	Maximum resistance ratio	Rebars total mass [kg]	Strands total mass [kg]	Volume of concrete [m³]	Total prestressed beam mass [kg]
6	271	PCB45 × 20 cm	ULS	0.91	83	53	0.34	843
12	1130	PCB87 × 30 cm	ULS	0.98	83	258	1.64	4110
18	2760	PCB117 × 50 cm	ULS	0.98	83	809	5.15	12879

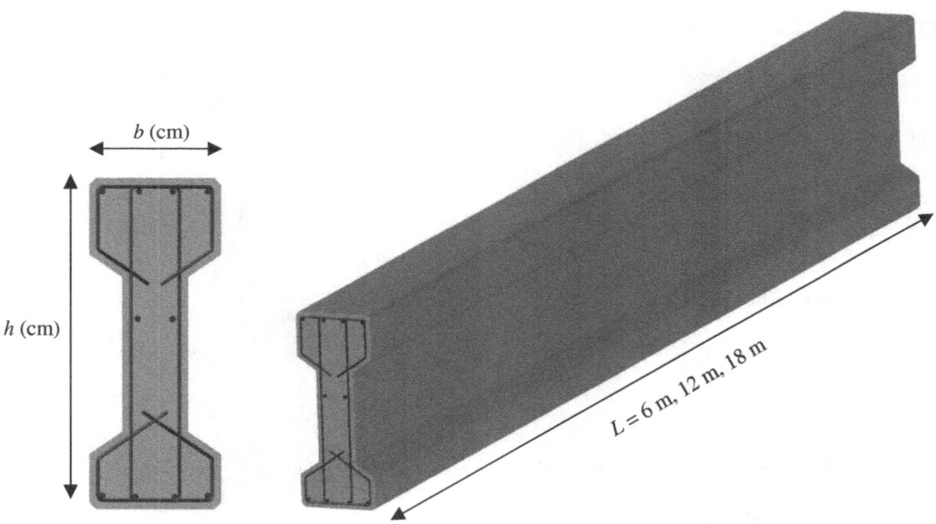

Figure 4.27 Prestressed reinforced concrete beam (left: cross-section; right: span lengths).

4.4.7 Comparison of Masses Between Different Types of Beams with Increasing Span Length

Figure 4.28 provides a comprehensive mass comparison of different structural beam types, including steel IPE, steel HEA, steel trusses with open and tubular sections, steel plate girders, glulam beams, and reinforced concrete beams with varying material classes and prestressing, across spans of 6, 12, and 18 m.

- For a span of 6 m, IPE and HEA beams demonstrate similar masses, approximately 500 kg, while both the open and tubular truss sections are slightly lighter. As the span extends to 12 m, the masses of the truss sections remain relatively low, but the mass of HEA beams increases slightly compared to the 6 m span. At a span of 18 m, a notable rise in mass is observed in the plate girder, reaching approximately 1500 kg, while the mass of HEA beams also increases significantly.
- Reinforced concrete beams, regardless of material class or prestressing, are consistently the heaviest for all span lengths.

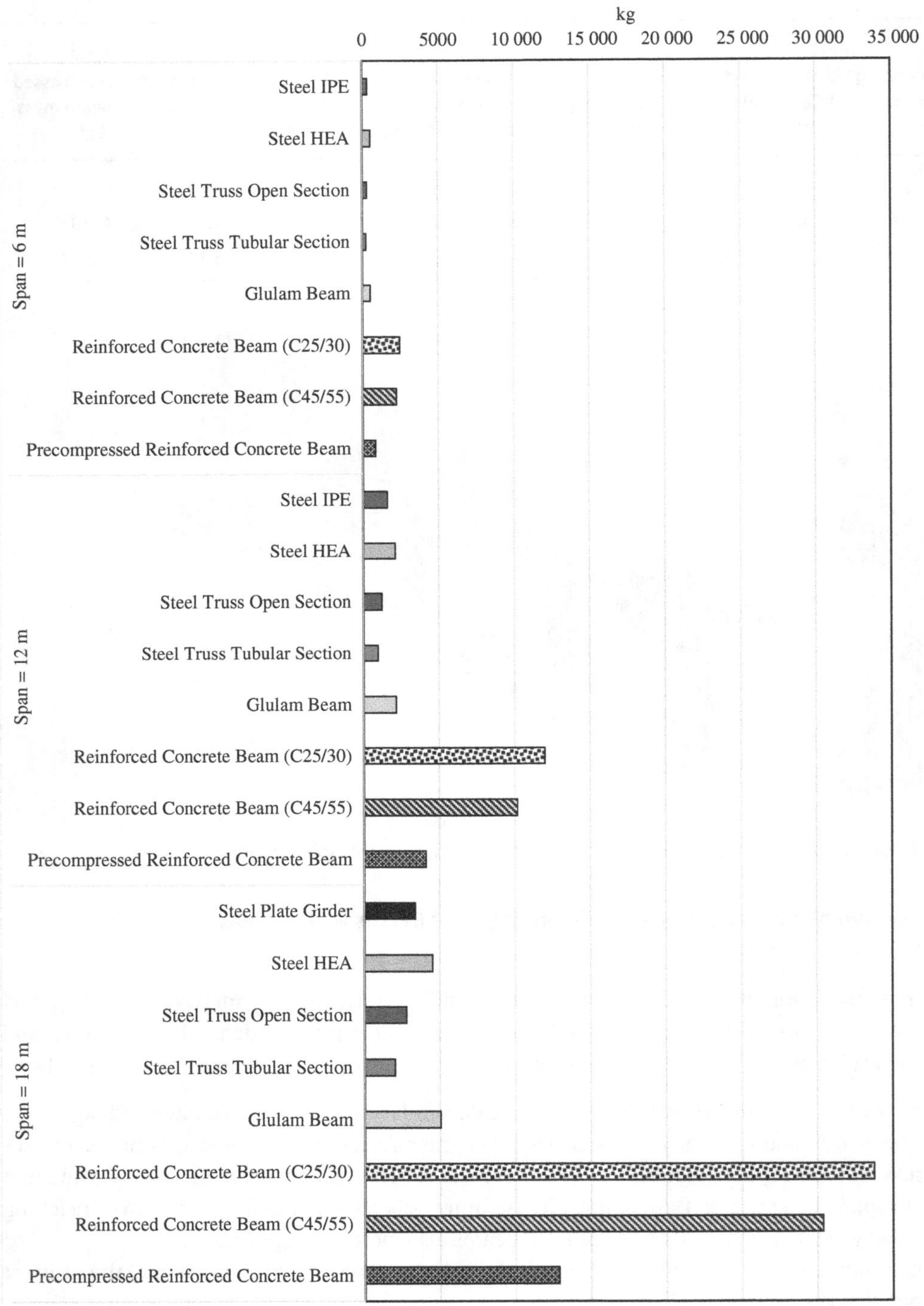

Figure 4.28 The masses of various options in with changing span length of the beams.

The data clearly show that as span length increases, the mass of certain beam types (particularly HEA, glulam, and reinforced concrete beams) grows substantially, highlighting the relationship between span length and material weight in structural design. This analysis emphasizes the importance of selecting appropriate beam types based on span requirements, considering factors such as material efficiency, mass, and structural performance.

4.4.8 Comparison of Embodied Carbon Between Different Types of Beams with Increasing Span Length

In this subsection, the embodied carbon of all beam types is compared. While the following calculations are based on a 6-m beam length, the approach can be applied similarly to other lengths. Quantities are taken from Tables 4.14–4.22, and the carbon factors used (production stage A1–A3 (EN 15804, 2012; EN 15978, 2026) are listed in Table 4.23.

Since embodied carbon in production stage A1–A3 is calculated by multiplying the mass of each component by its corresponding carbon factor (as detailed in Chapter 3), the relationship between mass and embodied carbon is linear for a given component. This means that, for any single component, a percentage reduction in mass translates directly into the same percentage reduction in embodied carbon.

Table 4.23 shows the carbon factors (production stage A1–A3) that are used to calculate embodied carbon for each beam type, sourced from the EPD-based carbon factor database developed in Chapter 3. The different values indicated by "average," "maximum," "minimum," and "minimum GWP & low-sust timber" reflect the variability of embodied carbon in the production stage (A1–A3), which depends strongly on the technology and processes adopted by different manufacturers. These values are then used to generate bar charts that compare the embodied carbon of different beam design solutions, making it possible to assess how sensitive the results are to the selected carbon factors.

Figures 4.29–4.32 show a comparison of embodied carbon for steel, composite, and glulam beam solutions across four distinct scenarios: a baseline using average industry carbon factors (Figure 4.29),

Table 4.23 Carbon factors (production stage A1–A3) used in the analysis (derived from the Environmental Product Declaration (EPD) database analysis presented in Chapter 3).

	Average GWP ($kgCO_2e$ per kg)	Maximum GWP ($kgCO_2e$ per kg)	Minimum GWP ($kgCO_2e$ per kg)	Minimum GWP & low-sust timber ($kgCO_2e$ per kg)
Structural Hot-rolled profiles	1.22	1.42	0.61	0.61
Tubular sections	2.26	2.60	1.73	1.73
Plate girders	2.28	3.02	1.55	1.55
Rebars	1.21	1.24	0.73	0.73
Concrete (C25/30)	0.14	0.16	0.12	0.12
Concrete (C45/55)	0.15	0.17	0.13	0.13
Timber	0.41	0.55	0.25	0.83
Precast concrete	0.26	0.33	0.19	0.19

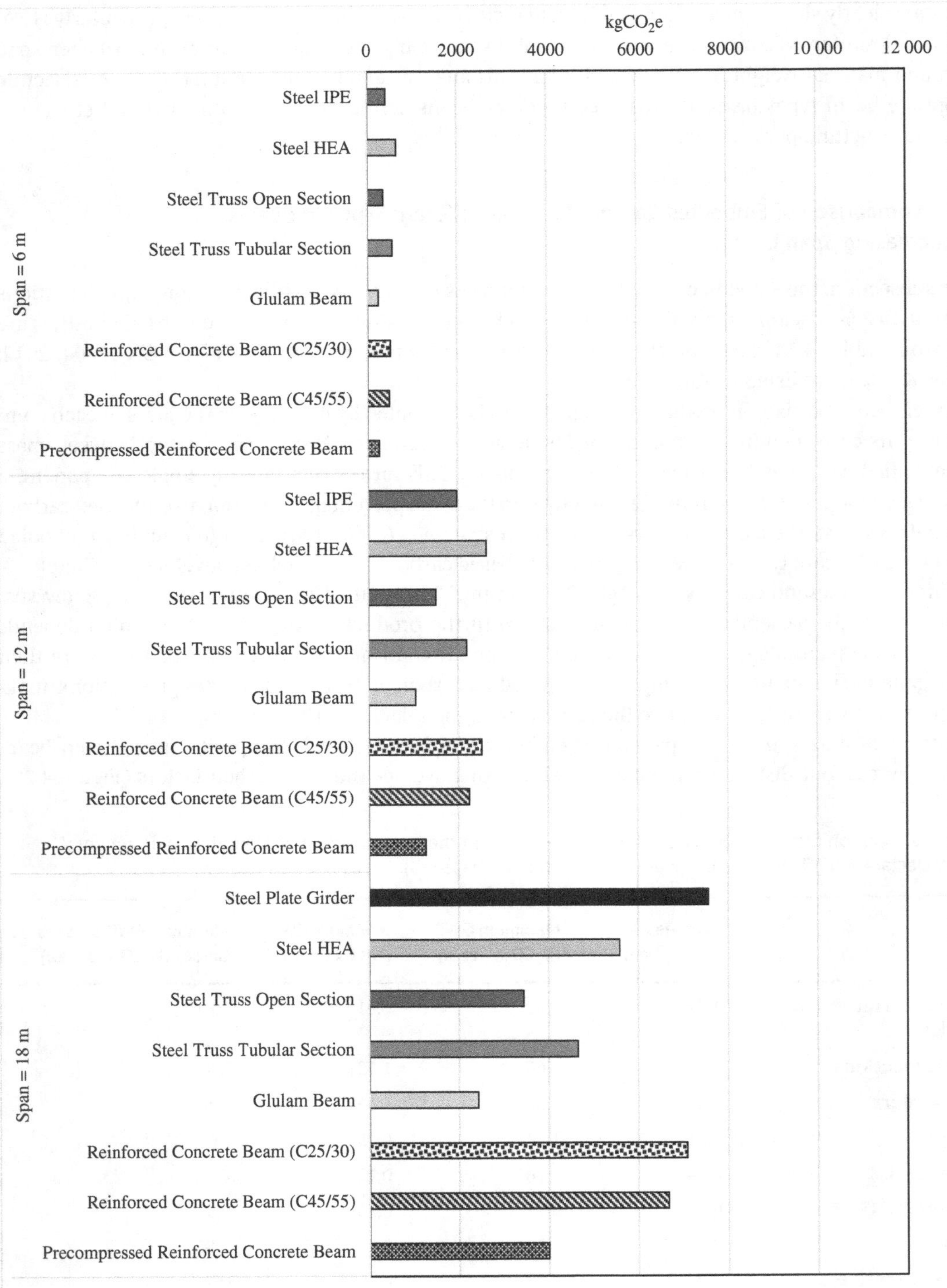

Figure 4.29 The comparison of GWP values with respect to various beam sections (using average industry carbon factors).

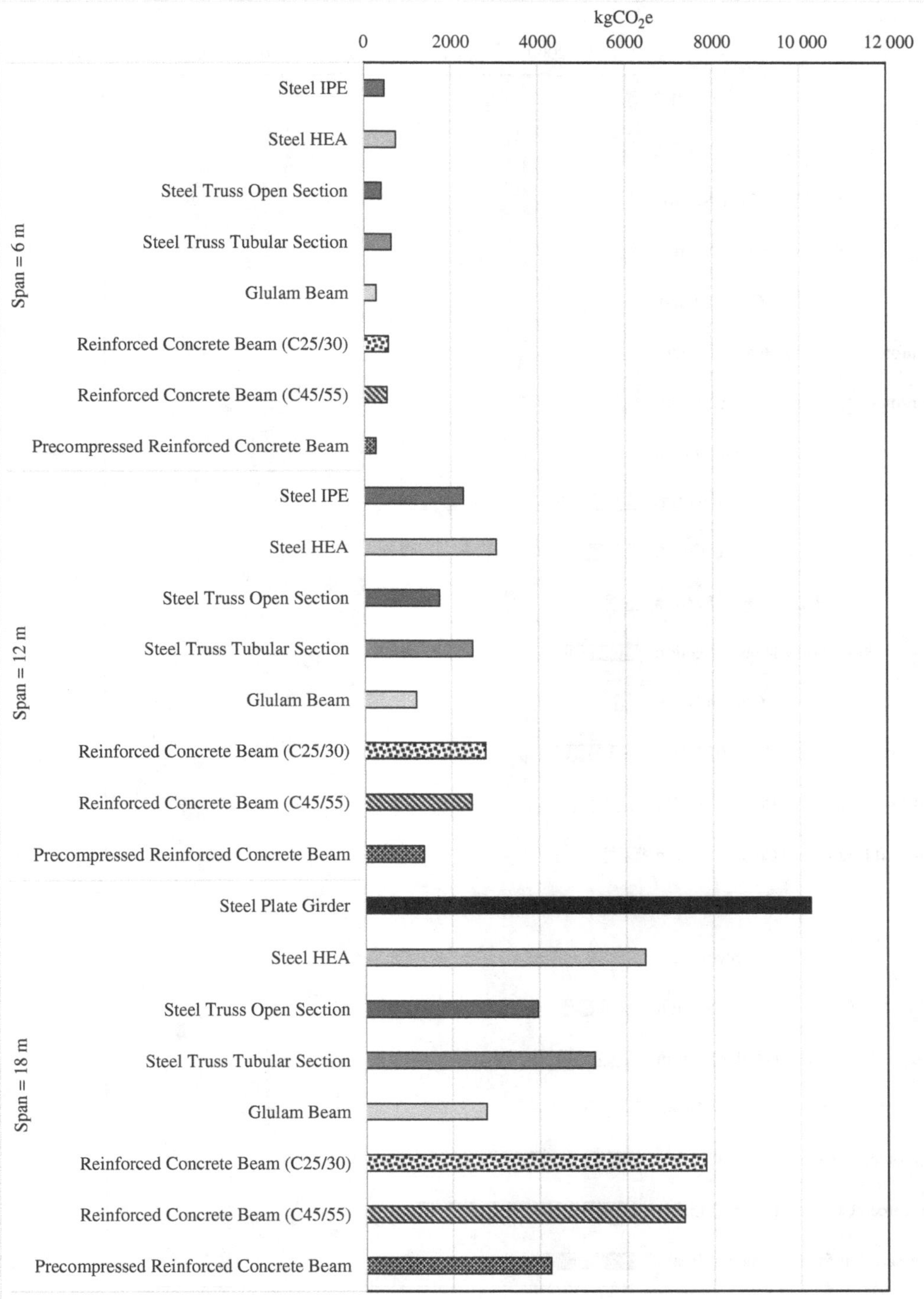

Figure 4.30 The comparison of GWP values with respect to various beam sections (using "worst-case" scenario using maximum industry carbon factors).

Figure 4.31 The comparison of GWP values with respect to various beam sections (using "best-case" scenario using minimum industry carbon factors).

Figure 4.32 The comparison of GWP values with respect to various beam sections (using "best-case" scenario using minimum industry carbon factors, except for timber, which assumes low-sustainability sourcing).

a worst-case scenario using maximum factors (Figure 4.30), a best-case scenario using minimum factors (Figure 4.31), and a specific sensitivity analysis on timber sourcing (Figure 4.32). This parameter analysis reveals several key findings:

- Embodied carbon increases disproportionately with beam span. While mass increases predictably (as seen in the previous Figure 4.28), the need to resist bending moments and deflection in longer beams (12 m and 18 m) demands significantly more material. This heavily penalizes solutions requiring large mass for stiffness, causing the embodied carbon of options like reinforced concrete beams to rise dramatically compared to more structurally efficient alternatives.
- Under the baseline scenario (using average carbon factors, Figure 4.29), glulam beams and steel trusses are consistently the lowest carbon choices, providing the most efficient balance of structural capacity and low carbon intensity, especially at longer spans.
- The strategic distribution of material through structural form is crucial for carbon efficiency. Steel trusses (both open and tubular section) are more effective than reinforced concrete and glulam profiles at longer spans because they achieve the necessary depth and stiffness to resist bending with less overall mass, leading to a significantly lower carbon footprint.
- For the 18 m span, a steel plate girder was analyzed as a structurally optimized alternative to standard profiles, yet it emerges as the highest carbon solution. Although a custom-fabricated plate girder can be more mass-efficient than a standard steel HEA beam at long spans, this benefit is negated by the high carbon intensity of its constituent material. The plates used have a carbon factor (2.28 $kgCO_2e/kg$) nearly double that of hot-rolled profiles (1.22 $kgCO_2e/kg$), a difference typically attributed to plates being produced via the more carbon-intensive BF-BOF route, while profiles often utilize more recycled steel in electric arc furnaces (however, EAF-produced plates are available, highlighting why a detailed analysis must rely on project-specific environmental product declarations rather than generic industry averages).
- A material's environmental ranking is not fixed and is highly dependent on the supply chain. The specific parameter analysis on timber sourcing provides a critical observation (Figure 4.32): when modeled with a high carbon factor representing unsustainable forestry practices, glulam (one of the best performers in other scenarios) becomes a higher carbon option than the steel trusses. This proves that a material's perceived green credentials are void without certified, low-carbon sourcing.
- The study's conclusions are highly sensitive to the input data, as illustrated by the significant divergence between the worst-case (maximum emissions, Figure 4.30) and best-case (minimum emissions, Figure 4.31) scenarios. Therefore, designing for low-carbon outcomes requires moving beyond default material choices. Designers must prioritize materials with transparent and certified supply chains, as the data shows that sourcing and manufacturing practices are as influential as the choice of the material itself, making verified LCA data essential for meaningful carbon reduction.

It is important to recognize the limitations of this analysis. The results are confined to embodied carbon from the production stage (A1–A3) and do not account for whole life cycle impacts, such as transportation, construction, maintenance, or end-of-life scenarios. Furthermore, this study does not consider other important design factors, including material cost, construction speed, or fire performance. However, the analysis remains valuable because the production stage typically accounts for the largest portion of a structural element's total embodied carbon, making it the most critical phase for optimization and the area where designers can exert the most influence at early design stages.

4.5 Integrating Cost, Durability, and Fire Resistance into LCPA

While this chapter focused on comparing structural components based on their mass and embodied carbon, a real-world design process is equally, if not more, driven by other factors such as cost, fire resistance, and durability. The LCPA framework is inherently designed to accommodate these additional dimensions, and integrating them strengthens its ability to capture the full spectrum of trade-offs involved in structural design decisions. The lowest-mass or lowest-carbon solution may not always be the most affordable, and cost can significantly influence the final design decision.

4.5.1 Cost

The lowest-mass or lowest-carbon solution may not always be the most affordable, and cost can significantly influence the final design decision. Several key observations from our case studies highlight this interaction:

- a steel truss beam is a prime example of this trade-off. As shown in the analysis, trusses are highly efficient in terms of mass and, consequently, have a lower embodied carbon footprint compared to solid steel beams over long spans. However, their fabrication requires cutting, welding, and assembly operations. This translates to labor costs, which may outweigh, in some cases, the savings from using less material. In contrast, a standard rolled steel IPE or HEA beam, while heavier, is mass-produced and requires minimal fabrication, making it a more cost-effective choice in many scenarios.
- reinforced concrete components can be cost-effective from a raw material perspective. However, their construction can be slower due to the need for formwork and curing times, potentially increasing overall project timelines and labor costs. Furthermore, a comprehensive cost analysis must extend beyond the initial construction to consider whole-life costs. For designers seeking to formally integrate this long-term perspective, established standards provide a clear methodology. Key normative references include ISO 15686-5:2017 (ISO 15686-5, n.d.) and EN 16627:2015 (EN 16627, n.d.). These standards offer a structured framework for calculating costs over a building's entire life, including maintenance, durability, and end-of-life value.
- the cost of materials is not static. Steel prices, for example, can be volatile and subject to global market fluctuations. The cost of timber depends heavily on the species, treatment, and certification (sustainably sourced timber is often more expensive). These economic uncertainties must be factored into the decision-making process.

A holistic design approach requires integrating cost into the parameter analysis. A more complete LCPA would treat cost as a key output variable alongside embodied carbon and mass, following the principles outlined in the aforementioned standards. This would empower designers to visualize the trade-offs more clearly, allowing them to present clients with a range of options that balance environmental responsibility, structural performance, and budget constraints. While this chapter has focused on the environmental aspect, it is crucial to remember that in practice, cost remains a decisive factor that shapes every stage of a project.

4.5.2 Durability

Durability affects both the service life of structural components and their life cycle environmental performance. A material with a higher upfront carbon footprint may still provide a lower whole-life impact if it requires fewer replacements or repairs over the building lifespan. For instance, reinforced concrete with carefully designed cover depth and proper curing can ensure long-term resistance against carbonation and chloride ingress, reducing maintenance needs. Timber, while relatively low in embodied carbon, often requires protective treatments and strict detailing to control moisture and biological decay. Weathering steel sections can resist corrosion in aggressive environments, extending service life compared to standard steel. When durability is incorporated into LCPA, it highlights how replacement cycles (B4 stage) and maintenance (B2–B3 stages) can significantly shift the balance between alternatives. Design choices that optimize durability often reduce both environmental and financial burdens across the structure's lifetime.

4.5.3 Fire Resistance

Fire resistance is another critical factor influencing structural solutions. Steel, despite its strength, loses stiffness and strength rapidly at elevated temperatures and often requires passive protection such as intumescent coatings or fireproof cladding. These additional layers increase both embodied carbon and cost. Concrete generally provides inherent fire resistance due to its mass and low thermal conductivity, but spalling in high-intensity fires remains a risk. Mass timber performs differently: although combustible, it chars predictably, and thick sections can maintain load-bearing capacity for extended durations, reducing the need for additional protection in some cases. Including fire resistance in LCPA can ensure a balanced comparison, as the carbon and cost of protective systems can alter the perceived advantage of each material.

4.6 Conclusion

This chapter has demonstrated that LCPA can be a tool for the modern structural designer to make decisions, moving beyond a single carbon score to reveal the critical decisions that drive a structure's environmental impact. Through the detailed analysis of columns and beams, we have seen that structural challenges such as slenderness, buckling, and span length are often the primary drivers of embodied carbon. By asking "what-if" and systematically testing your assumptions, you can identify the most effective strategies to create structures that are not only safe and efficient but also genuinely sustainable. While embodied carbon is a critical metric, this chapter also highlights that a real-world decision is a complex trade-off. A truly optimal solution must balance environmental performance against other crucial factors like cost, durability, and fire resistance. In the next chapters, we will see how this strategy is applied at the building level. Some additional studies of this kind can be found in the literature (Abushama et al. 2025; Ching and Carstensen 2022; Kaddoura et al. 2025; Morris et al. 2021; Rossi et al. 2011; Su et al. 2016), presenting case analyses of individual structural components such as beams and columns; portal frame LCA comparing timber, steel, and concrete spans; glulam vs. steel beam LCA with sensitivity scenarios; truss topology optimization exploring hybrid timber and steel; pile optimization minimizing carbon for timber, concrete, and steel; column LCA comparing steel, composite, concrete, and timber impacts, CFST, and RC columns.

4.7 Questions

Answer the following questions to enhance your understanding of the concepts covered in this chapter.

Discussion Questions

1 Explain why a designer might choose a composite column over a purely steel or reinforced concrete column for a high-rise building.

2 Describe the concept of "flexural buckling." Explain how this phenomenon can lead to a significant increase in a tall column's embodied carbon compared to a shorter column carrying the same load.

3 Discuss the trade-offs between using a steel truss beam vs. a steel I-beam in terms of both structural efficiency and embodied carbon.

4 Discuss how high strength concrete can impact a reinforced concrete element's embodied carbon.

5 How can the governing design criteria (ULS vs. SLS) lead to different material quantities and, consequently, different embodied carbon outcomes for two beams that span the same distance?

6 If you wanted to reduce the total mass of a steel or reinforced concrete beam, what are two different design changes you could make? Explain the potential impact of each.

7 Why can hollow structural sections (like SHS or CHS) offer more mass efficiency for columns than open profiles like HEA?

8 Describe the purpose of prestressing a reinforced concrete beam. How does this process allow for more efficient designs?

9 Beyond material type (e.g. steel vs. timber), what other factors related to a material's origin and manufacturing process can significantly influence its carbon factor and overall GWP?

10 An engineer needs to solve a buckling problem in a tall, slender steel column. They could either (a) switch to a larger steel section, (b) use a composite design (e.g. encase it in concrete), or (c) use a higher strength steel. Discuss the potential embodied carbon implications of each choice.

11 Discuss the benefits of LCPA at component level in reducing the environmental impact of the structural design decisions.

12 Explain why the structural component with the lowest embodied carbon might not be the most cost-effective choice for a project.

13 Beyond the initial price of raw materials, what other cost-related factors can influence the selection of reinforced concrete for a project?

14 How does durability challenge the idea of simply choosing materials with the lowest upfront embodied carbon? Provide an example of how a material with a higher initial carbon footprint could be more sustainable over the building's entire life.

15 Discuss the trade-offs between steel, concrete, and mass timber regarding fire resistance. How can the need for fire protection systems alter the overall embodied carbon and cost comparison between these materials?

16 What is the main argument for expanding the sustainability topic to include metrics like cost, durability, and fire resistance, rather than focusing solely on mass and embodied carbon?

17 Explain why a column's embodied carbon often increases disproportionately with height, rather than linearly. What structural phenomenon causes this, and at what point does it typically become the dominant design consideration?

18 Two columns of identical material must carry the same axial load. Column A is 4 m tall; column B is 8 m tall. Why might column B require more than twice the material of column A? Relate your answer to the governing design criteria.

19 Explain the relationship between the Euler buckling formula and embodied carbon for slender columns. Why does the exponent in the buckling relationship (involving length squared) have profound implications for the environmental efficiency of tall structures?

20 Why do hollow structural sections (tubes) often outperform solid or open sections for tall columns in terms of material efficiency?

21 A designer is choosing between an H-shaped steel section and a square hollow section for a slender column. The H-section has different stiffness about its two principal axes. How does this asymmetry affect the design, and why might it lead to inefficient material use?

22 Explain the structural rationale behind composite columns (steel encased in concrete). How does this combination address the weaknesses of each material individually, and what are the trade-offs in terms of embodied carbon?

23 Explain why truss beams can achieve longer spans with less embodied carbon than solid beams of the same depth. What is the fundamental structural principle that makes trusses efficient?

24 Concrete has a much lower carbon factor per kilogram than steel, yet concrete structures are not commonly the lowest carbon option. Explain the circumstances under which steel might have lower embodied carbon than concrete for the same structural application.

25 Increasing the reinforcement ratio in a concrete column reduces the required column size. However, this doesn't always reduce embodied carbon. Explain the trade-off between concrete volume and steel reinforcement from an environmental perspective.

26 Timber is often considered an inherently "green" material. Under what circumstances might a timber structural element have higher embodied carbon than a steel or concrete alternative? What factors beyond the material itself determine timber's environmental performance?

27 Explain why the environmental ranking of structural materials is not absolute but depends heavily on the specific application (span length, load magnitude, height, etc.). What does this imply for simplified material selection guidelines?

28 The same material (e.g. structural steel) can have significantly different carbon factors depending on the manufacturer and production process. Explain the main sources of this variability and discuss how a designer should handle this uncertainty when making material selection decisions.

29 Why is specifying "steel" or "concrete" carbon factors insufficient for accurate embodied carbon calculations? What additional information about the supply chain is needed, and how can designers obtain this information?

30 Explain why the production stage (A1–A3) is often the focus of structural LCPA, despite not representing the complete life cycle. What are the limitations of this scope, and when might it lead to misleading conclusions?

31 Fire resistance requirements can significantly alter the embodied carbon comparison between materials. Explain how protective measures for steel, concrete's inherent fire resistance, and timber's charring behavior affect the overall environmental assessment.

32 Durability affects life cycle environmental performance through maintenance and replacement requirements. Explain how a material with higher initial embodied carbon might still be the most sustainable choice over a building's lifetime.

33 Construction speed and labor costs significantly influence material selection. Discuss how fabrication complexity (e.g. trusses vs. rolled sections) creates a disconnect between structural efficiency and economy.

34 How would you explain the concept of LCPA to a client who has no engineering background? Develop an analogy that communicates both the method and its value.

35 Design codes typically optimize for safety and serviceability, not environmental performance. Discuss whether and how LCPA insights should influence future code development. What barriers exist to incorporating environmental criteria into structural design standards?

36 Early design stages offer the greatest opportunity for carbon reduction, but also have the greatest uncertainty. Explain the paradox this creates for LCPA and discuss strategies for making robust decisions with incomplete information.

Scenario Questions

1 A client asks, "Which is more sustainable: steel or concrete?" Explain why this question cannot be answered without additional context, and list the key parameters that would determine the answer for a specific project.

2 An architect requests a 9-m tall column for a lobby. They prefer the aesthetic of an open H-section (HEA profile). You, the engineer, suggest a hollow section (SHS) or a reinforced concrete column instead. Question: referencing Figures 4.7 (Mass comparison) and 4.11 (GWP comparison), explain the structural phenomenon that causes the HEA profile to perform poorly in terms of mass and embodied carbon at this specific height compared to the SHS. Why does the geometry of the HEA make it inefficient for long columns?

3 A client wants to achieve a "Net Zero" rating and demands glulam columns for the entire structure, assuming wood is always the most sustainable choice. You perform a parameter analysis using the data in Figures 4.11–4.14. Question: under what specific supply chain condition does the glulam column perform worse than a standard reinforced concrete column? Construct an argument to present to the client explaining why "sustainable timber" is a supply chain decision, not just a material decision.

4 A colleague argues: "we can just choose low-carbon materials and the building will be sustainable." Construct a counterargument using the principles learned in this chapter. What critical information does simple material selection overlook?

5 The chapter demonstrates that slenderness is "a primary driver of mass in columns." Derive a conceptual relationship between the Euler buckling formula and embodied carbon. At what slenderness ratio would you expect the environmental "penalty" of buckling resistance to become dominant over pure compression strength requirements?

6 A client has specified that a 12-m span beam must have a maximum depth of 800 mm due to floor-to-floor height constraints. Calculate the embodied carbon for the options that meet the depth constraint. If the Glulam option were "pushed" to fit within 800 mm by increasing width, estimate the new cross-sectional dimensions and recalculate the GWP. Is a constrained glulam still competitive?

7 A beam's span-to-depth ratio has significant implications for material efficiency. Explain why very shallow beams (high span-to-depth ratio) require disproportionately more material, and how this affects the environmental comparison between materials with different stiffness properties.

8 Why might a beam designed for a 6-m span have fundamentally different optimal material choices than a beam designed for an 18-m span? What changes in the relative importance of different structural behaviors?

9 Higher-strength concrete (e.g. C45/55 vs. C25/30) has a higher carbon factor per unit volume but can sometimes result in lower total embodied carbon. Explain under what structural conditions this occurs, and when using higher strength concrete might actually increase embodied carbon.

10 A structural solution with the lowest embodied carbon may not be the most cost-effective. Explain the common scenarios where carbon optimization and cost optimization lead to different design choices. How should a designer balance these competing objectives?

Calculation Questions

Question 1: An engineering team is designing a 9-m-tall column for the ground floor of a high-rise office building. The architectural plans have a strict limit on the column's footprint, making a slender design preferable. The team is comparing a steel hollow section (SHS) with a fully encased composite column.

Option A (Steel SHS): A SHS with a total mass of 1530 kg. The steel used for this tubular section has a high carbon factor of 2.50 $kgCO_2e/kg$.

Option B (Composite): A fully encased composite column. It uses a smaller steel section weighing 1010 kg (carbon factor = 1.79 $kgCO_2e/kg$), 7 300 kg of concrete (carbon factor = 0.133 $kgCO_2e/kg$), and 355 kg of steel rebars (carbon factor = 0.76 $kgCO_2e/kg$).

Instructions

- Calculate the total GWP in $kgCO_2e$ for both Options A and B.
- Based strictly on the GWP values, which option is the more sustainable choice?
- Considering the architectural requirement for a slender design, what is the primary trade-off between these two designs?

Question 2: For a long 18-m span bridge component, engineers are comparing a custom-fabricated steel plate girder with a steel truss made from open sections.

- **Option A (Plate Girder):** A plate girder beam with a total mass of 3390 kg. The steel plates are from a process with a carbon factor of 1.5 $kgCO_2e/kg$.
- **Option B (Steel Truss):** A lighter steel truss made of IPE and angle sections. Its total mass is 2570 kg. The steel used also has a carbon factor of 1.5 $kgCO_2e/kg$.

Instructions

- Calculate the total GWP for both the plate girder and the steel truss.
- While the truss in this case is lighter, its fabrication is much more complex, requiring more cutting, welding, and labor than the simple plate girder. How might these "hidden" factors, which are not included in the production stage (A1–A3) carbon factors, influence the true, life cycle environmental impact of the truss compared to the plate girder?

Question 3: An engineer is designing a 6-m-tall interior column that is susceptible to flexural buckling by comparing a standard HEA (wide flange) section with an SHS (square hollow) section.

Option A (HEA Section): An HEA1000 used as a column with a carbon factor of 1.5 $kgCO_2e/kg$.
Option B (SHS Section): An SHS 400 × 400 × 12.5 section with a carbon factor of 2.50 $kgCO_2e/kg$.

Instructions

- Calculate the total GWP for both the HEA and SHS column options.
- Compute the percentage difference in GWP between Options A and B to quantify the relative reduction in carbon emissions.

Exercise: Life-cycle Parameter Analysis (LCPA) for Structural Frame Components

You are a structural engineer tasked with designing the structural frame for a 5-story office building. The client has expressed interest in understanding how different design choices affect the building's embodied carbon. You will perform a LCPA to identify the most influential design decisions.

The structural plan (Figure 4.33) shows a typical floor layout (two alternatives of different slab directions) with the following characteristics:

Grid spacing: 10.0 m × 6.0 m
Number of bays: 2 × 2
Floor system: non-composite steel deck with reinforced concrete
Building use: office (Category B)

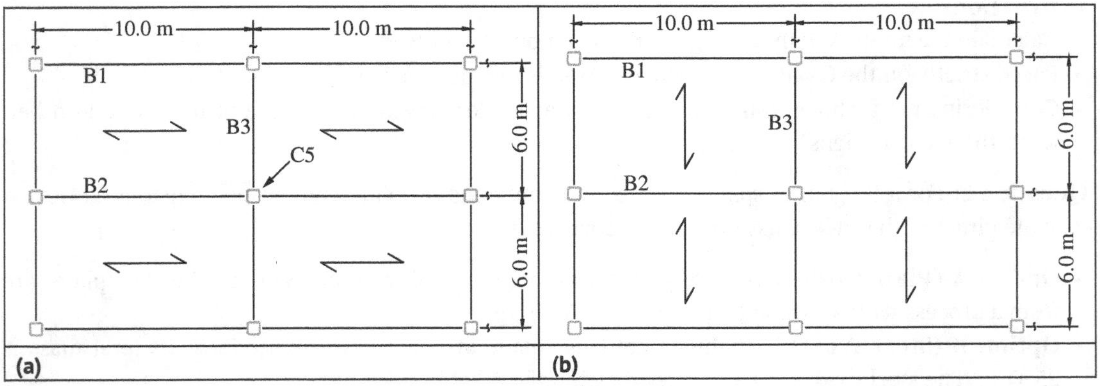

Figure 4.33 Two alternative typical floor framing plans showing different slab span directions for the 10.0 m × 6.0 m grid. (a) Slab plan direction a; (b) slab plan direction b.

Self-weight of floor system (G_1): 3.5 kN/m^2
Permanent non-structural loads (G_2): 3.5 kN/m^2
Imposed load – office Category B (Q): 2.0 kN/m^2

Load combinations:

ULS: $1.35 \times (G_1 + G_2) + 1.5 \times Q$
SLS: $G_1 + G_2 + Q$

Analysis Assumptions

- Column effective length factor: K = 1.0 (pinned-pinned, i.e. no moment transfer at connections, which is conservative for carbon estimation purposes).
- Beam support conditions: simply supported, assuming that lateral resistance is guaranteed by reinforced concrete cores that are not considered in this exercise for simplicity.
- For embodied carbon estimation, assume a steel reinforcement density of 100 kg/m^3 for slabs and beams, and 150 kg/m^3 for columns.
- Assume steel beams are non-composite.
- Ground floor column load: cumulative from all floors above.
- You can use the carbon factors provided in this chapter for life-cycle stages A1–A3. Where necessary, you may consult the databases indicated in this book.
- The load G1 (3.5 kN/m^2) includes the reinforced concrete slab and metal deck but excludes the self-weight of the beams/columns. You must add the self-weight of the structural frame members you design.
- For physical and mechanical properties (e.g. yield strength, elastic modulus, bending strength), you must research and select appropriate values from standard engineering sources (e.g. Eurocodes EC2/EC3/EC5, structural engineering handbooks, or reputable material databases).

Part 1: Column Analysis (Interior Column C5), Span Direction a (Figure 4.33)

The interior column C5 supports the tributary area shown (shaded region). Analyze this column for three different heights representing different floor configurations:

Case A: Story height = 3.5 m
Case B: Story height = 5.0 m
Case C: Story height = 7.5 m

1. Calculate the design axial load on column C5 from a single floor.
2. For each column height, determine appropriate sections for:
   - Steel HEA section, class S355
   - Steel SHS (Square Hollow Section), S355
   - Steel HEA section, class S460
   - Steel SHS (Square Hollow Section), class S460
   - Reinforced concrete (square section, class C25/30)
   - Reinforced concrete (square section, class C45/55)
3. Calculate the total mass and A1–A3 stage embodied carbon (GWP) for each option using average carbon factors.

4. Create a comparison chart showing how embodied carbon varies with column height for each material type.
5. Identify which design parameter (material choice or column height) has the greater influence on embodied carbon. Justify your answer with calculations.

Part 2: Beam Analysis (Beams B1, B2, B3), Span Direction a (Figure 4.33)

1. Calculate the design loads (ULS and SLS) on beams B1, B2, and B3.
2. For each beam, determine appropriate sections for:
   - Steel IPE section, class S355
   - Steel HEA section, class S355
   - Steel IPE section, class S460
   - Steel HEA section, class S460
   - Glulam beam
   - Reinforced concrete beam (C25/30)
   - Reinforced concrete beam (C45/55)
3. Calculate the total mass and embodied carbon for each option.
4. Identify the governing design criterion (ULS or SLS) for each beam type and explain how this affects material efficiency.
5. At what span length does the relative ranking of materials (by embodied carbon) change? Explain the structural reason for this shift.

Part 3: Life-cycle Parameter Analysis (LCPA)

1. Recalculate the embodied carbon for your column designs using:
   - Minimum carbon factors (best-case scenario)
   - Maximum carbon factors (worst-case scenario)
   - Average carbon factors
2. For which material does the choice of carbon factor have the greatest impact on the final result? What does this imply for design decisions?
3. If a timber supplier provides an EPD showing a carbon factor of 0.80 $kgCO_2e/kg$ (indicating poor forestry practices), how does this change the ranking of materials for the 6 m column?

Part 4: System-level Thinking

1. For the entire floor plate shown, estimate the total embodied carbon for:
   Option 1: All steel frame (HEA columns, IPE beams), assuming S355 class
   Option 2: All steel frame (HEA columns, IPE beams), assuming S460 class
   Option 3: All reinforced concrete frame, assuming C25/30 class concrete
   Option 4: All reinforced concrete frame, assuming C45/55 class concrete
   Option 5: A hybrid system (reinforced concrete columns, steel beams) with material classes of your choice
2. Which specific design changes would have the greatest impacts on reducing the total embodied carbon of the structural frame? Support your answer with calculations.

3. The architect proposes increasing the grid spacing from 10.0 m × 6.0 m to 10.0 m × 10.0 m to have more free space without column interruption. Discuss, by showing calculations, how this change would affect:
   - Individual member sizes
   - Total number of members
   - Overall embodied carbon

Part 5: Critical Reflection

1. What are the limitations of this LCPA? What factors have been excluded that might influence the final design decision?
2. If you could only communicate three key findings from your analysis to the client, what would they be?
3. How would your recommendations change if the client's primary concern shifted from minimizing embodied carbon to minimizing construction cost?

Part 6: Solve Parts 1–5 Considering the Span Direction b (Figure 4.33)

How does the choice of span direction affect the tributary areas and loads on beams B1, B2, and B3 in each case? Based on this, explain which configuration results in a lower total embodied carbon for the floor system, and why.

References

Abushama, K., Hawkins, W., Pelecanos, L., & Ibell, T. (2025). Optimizing the embodied carbon of concrete, timber, and steel piles with a case study. In: M. Kioumarsi and B. Shafei (Eds.), *The 1st International Conference on Net-Zero Built Environment* (pp. 899–911). Springer Nature Switzerland. https://doi.org/10.1007/978-3-031-69626-8_76

Ching, E., & Carstensen, J. V. (2022). Truss topology optimization of timber-steel structures for reduced embodied carbon design. *Engineering Structures*, *252*, 113540. https://doi.org/10.1016/j.engstruct.2021.113540

EN 1992-1-1. (2023). EN 1992-1-1: 2023. Eurocode 2: Design of concrete structures.

EN 1993-1-1. (2022). EN 1993-1-1: 2022. Eurocode 3: Design of steel structures—Part 1-1: General rules and rules for buildings.

EN 1994-1-1. (2004). EN 1994-1-1: 2004. Eurocode 4: Design of Composite Steel and Concrete Structures.

EN 1995-1-1. (2014). EN 1995-1-1: 2004+A2:2014. Eurocode 5: Design of Timber Structures – Part 1-1.

EN 15804: 2012+A2:2019. (n.d.). Sustainability of construction works—Environmental product declarations—Core rules for the product category of construction products. iTeh Standards. Retrieved September 20, 2025, from https://standards.iteh.ai/catalog/standards/cen/c98127b4-8dc2-48a4-9338-3e1366b16669/en-15804-2012a2-2019

EN 15978. (2026). EN 15978: 2011. Sustainability of construction works. Assessment of environmental performance of buildings. Requirements and guidance.

EN 16627: 2015. (n.d.). Sustainability of construction works—Assessment of economic performance of buildings—Calculation methods. iTeh Standards. Retrieved September 8, 2025, from https://standards.iteh.ai/catalog/standards/cen/873a2524-808c-4843-b1db-35520c48146a/en-16627-2015

EN ISO 14044: 2006/A2:2020. (n.d.). Retrieved September 5, 2025, from https://store.uni.com/en-iso-14044-2006-a2-2020

ISO 15686-5: 2017. (n.d.). ISO. Retrieved September 8, 2025, from https://www.iso.org/standard/61148.html

Kaddoura, M., Majeau-Bettez, G., Amor, B., & Margni, M. (2025). Global sensitivity analysis reduces data collection efforts in LCA: A comparison between two additive manufacturing technologies. *Science of the Total Environment*, *975*, 179269. https://doi.org/10.1016/j.scitotenv.2025.179269

Morris, F., Allen, S., & Hawkins, W. (2021). On the embodied carbon of structural timber versus steel, and the influence of LCA methodology. *Building and Environment*, *206*(108285). https://doi.org/10.1016/j.buildenv.2021.108285

Rossi, B., Lukic, I., Du, G., Cregg, D., Borg, R. P., & Haller, P. (2011). Life-cycle impacts assessment of steel, composite, concrete and wooden columns: *Final International Conference of COST Action C25. Final International Conference of COST Action C25 - Sustainability of Constructions*, 277–285.

Su, S., Li, X., Wang, T., & Zhu, Y. (2016). A comparative study of environmental performance between CFST and RC columns under combinations of compression and bending. *Journal of Cleaner Production*, *137*, 10–20. https://doi.org/10.1016/j.jclepro.2016.07.043

5

Life-cycle Parameter Analysis (LCPA) at Building Level

LIST OF CONTENTS

In this chapter, we will apply the principles studied in the earlier chapters to assess and compare the embodied carbon of various structural configurations at the building level. We will analyze a case study that has been designed in multiple ways, allowing a comparison of embodied carbon across different materials (concrete, steel, and timber) and varied structural configurations. Each design scenario is analyzed under changing assumptions – for instance, adjustments to carbon factors, transportation distances, and complexity of connections – to demonstrate how these factors can affect a project's overall impact. This systematic evaluation of how varying parameters influence the final results is a form of life-cycle parameter analysis (LCPA) at the building level, which we can use to identify the most critical factors in a project's overall carbon footprint. While the case study focuses on a specific building type, the methodology presented here is general: the same step-by-step approach (defining structural options, assigning carbon factors from EPDs or databases, varying key parameters, and interpreting results as ranges rather than fixed values) can be applied to any material system, building typology, or regional context. The goal is not only to present results for a specific project, but to equip the reader with a repeatable framework that can be adapted to their own design challenges, whether working with reinforced concrete, engineered timber, hybrid systems, or emerging materials.

5.1 Why Does Optioneering Matter During the Structural Conceptual Design?

Through a series of parameter analyses (i.e. optioneering), we can see how a building's embodied carbon is influenced not only by the choice of materials, but also by specific life-cycle considerations such as material sourcing, design decisions and construction methods. By examining these influences, the chapter demonstrates how even small changes in assumptions can lead to different conclusions about sustainability and efficiency. The study is split into several parts, focusing on two specific targets:

i. Correlation of embodied carbon intensity of buildings and their representative portion, allowing an approximate evaluation of the entire building based only on a portion of it.
ii. Evaluation of the embodied carbon intensity of several options with the sake of making early-stage decisions related to the materials, structural system, and construction methods.

During early stage (conceptual) design, calculating the total embodied carbon intensity of a building by summing up the contributions of each structural element can be time-consuming and complex. Instead, analyzing a single representative bay (Figure 5.1) can provide a useful early-stage approximation to embodied carbon intensity, enabling rapid early-stage design comparisons. Although such a rapid approach may introduce some under- or overestimations, this simplification is valuable because it allows designers to quickly explore multiple design alternatives before committing significant resources to detailed analysis.

The case study in this chapter also quantifies the difference between using a representative portion and assessing the entire building through a percentage comparison. Several factors contribute to this variation. In full-building designs, beam and column sections vary within a single slab, leading to differences in structural element weights. In contrast, the portion-based approach assumes a single section for beams and columns, typically based on internal elements, introducing some redundancy. Moreover, since internal sections are often larger than edge or corner elements, this method may slightly overestimate embodied carbon. By accounting for these differences, this methodology provides a basic approach for early stage assessments, enabling us to evaluate the impact of material choices, span lengths, and structural configurations on embodied carbon.

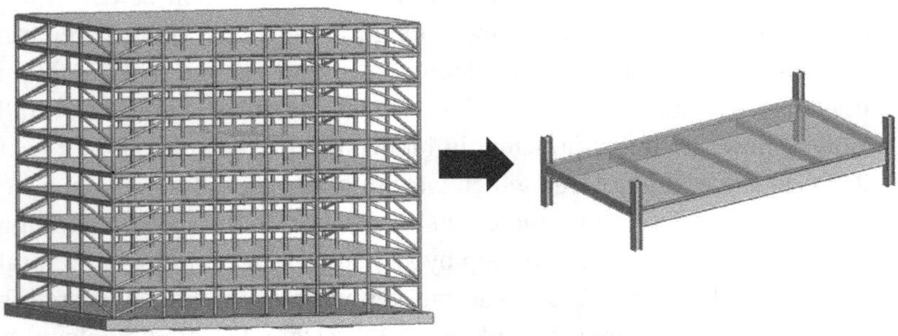

Figure 5.1 Representative portion of a building structure allowing a rapid embodied carbon optioneering during conceptual design.

5.2 Benchmark Case Study: Two Multistory Building Configurations (Reinforced Concrete and Steel)

In their book, Norman et al. (2020) describe the conceptual structural design for a 10-story office building using reinforced concrete with flat slabs and structural steel with composite metal deck floors. This building is used for the parameter analysis of the embodied carbon study in this chapter.

The main elements implemented within the buildings under study, regardless of the material used, are designed respecting the structural safety rules according to the UK building standards (NHBC, 2024; UK Government, 2010), within a MSc thesis study (Chehade, 2024). The superstructure of the studied systems consists mainly of slabs, beams, columns, and the lateral load resisting systems. The substructure comprises the ground floor slab, the ground beams, and the footings (strips and pads). Composite action (i.e. steel beams working together with concrete slabs to increase load capacity and reduce material use) has been assigned to the secondary beams in some cases in order to observe its effect on the building's embodied carbon, with the study focusing on changes in material usage and structural efficiency. Strip footings are used along the full perimeter of the building to support cladding, in addition to the internal pad footings to also suit the ground conditions. It is assumed that a competent natural ground is not present at shallow depth (that is why reinforced concrete ground beams are needed).

Benchmark Case Study 1: Ten-story Reinforced Concrete Office Building with Open-plan Layout

Spanning 45 by 45 m in plan, the building is laid out in a 5 × 5 grid, with each bay measuring 9 m by 9 m. Table 5.1 presents the structural elements of this case study and their dimensions. Figures 5.2 and 5.3 present a visual representation of the building's typical floor plan and its 3D model. The columns are reinforced concrete sections, sized at 55 × 55 cm for edges, 42.5 × 42.5 cm at the corners, and 75 × 75 cm for interior positions, each extending 4 m in height per story. Footings include perimeter strip foundations and internal pads. The floors are solid reinforced concrete slabs 350 mm thick, reinforced with 90 kg/m^3 of steel bars. A 200 mm ground-floor slab covers the entire footprint, while beams and connections tie the structure together for stability.

Benchmark Case Study 2: Ten-story Structural Steel Office Building with Open-plan Layout

Spanning 45 by 45 m in plan, this ten-story steel-framed building is arranged in a 3 × 6 bay grid, with each bay measuring 15 by 7.5 m. Table 5.2 presents the structural elements of this case study and their dimensions. Figures 5.4 and 5.5 present a visual representation of the building's typical floor plan and its 3D model. Columns are open steel sections placed at the edges, corners, and interior positions with different steel profiles, supporting primary and secondary beams. The floors are made of metal steel decks with concrete slab. Lateral stability is provided by diagonal bracing in each elevation, ensuring resistance to wind and seismic forces. Foundation support comes from perimeter strip footings and internal pad footings in reinforced concrete.

The different profile sections of the beams and columns in the case study are represented in Table 5.3.

Table 5.1 Parameters of the different structural elements included in the benchmark case study 1 (reinforced concrete building).

Element	Component description		Dimensions
Slabs	Concrete C30/37		45 × 45 m, 350 mm depth
	Reinforcing bars (90 kg/m^3)		
Columns	Concrete C50/60	Edge	55 × 55 cm, 4 m height
	Rebar (250 kg/m^3)	Corner	42.5 × 42.5 cm, 4 m height
		Internal	75 × 75 cm, 4 m height
Footings	Concrete GEN1	Perimeter (strip)	3.2 m width, 1.6 m depth
	Rebar (80 kg/m^3)	Internal (pads)	4.5 × 4.5 m, 2.25 m depth
Ground floor slab	Concrete C30/37		45 × 45 m, 200 mm depth
	Rebar A193 (20 kg/m^3)		
Ground floor beams	Concrete C30/37		30 × 100 cm, 9 m span
	Rebar (230 kg/m^3)		
Connections	Concrete rebar		10% of steel reinforcement rebar mass
Structural	Concrete C40/50		9.0 × 4.5 m, 250 mm thickness
	Reinforcing bars (178 kg/m^3)		

Note: The connection allowance of 10% of rebar mass accounts for the additional reinforcement concentrated in beam-column joint regions, where increased steel quantities are typically required to ensure adequate force transfer and confinement.

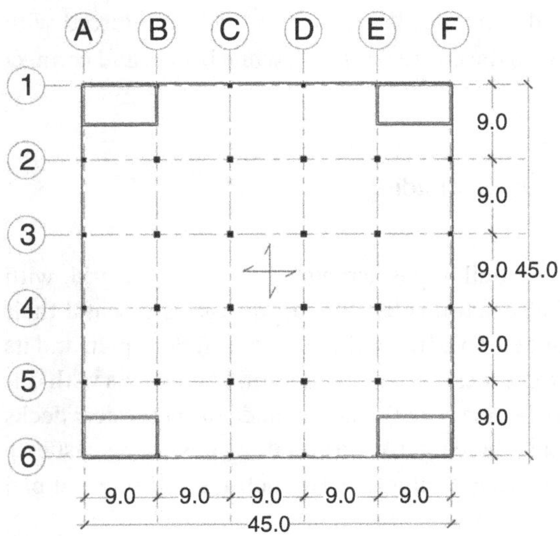

Figure 5.2 Conceptual plan of the case study 1 (Norman et al./with permission of The Institution of Structural Engineers).

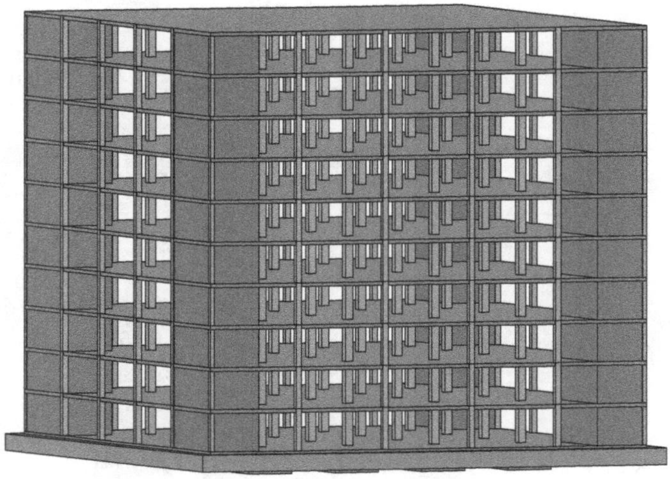

Figure 5.3 3D modeling of the case study 1.

Table 5.2 Parameters of the different structural elements included in the benchmark case study 2 (steel building).

Element	Component description		Dimensions
Slabs	Concrete C30/37 (0.105 m^3/m^2)		45 × 45 m, 150 mm depth
	Rebar A252 (3.95 kg/m^2)		
	Metal deck (1382 mm^2/m)		80 mm deep × 0.9 mm thick
Beams	Primary	Edge, S355	533 × 210 × 92UB
	(noncomposite)	Internal, S355	610 × 229 × 125UB
	Secondary, S275 (composite)		533 × 210 × 101UB
Columns	Edge, S355		356 × 368 × 202UC, 4 m height
	Corner, S355		356 × 368 × 129UC, 4 m height
	Internal, S355		356 × 406 × 287UC, 4 m height
Footings	Concrete GEN1	Perimeter (strip)	3 m width; 1.5 m depth
	Rebar (80 kg/m^3)	Internal (pads)	3.65 × 3.65 m; 1.825 m depth
Ground floor slab	Concrete C30/37		45 × 45 m, 200 mm depth
	Rebar A193 (20 kg/m^3)		
Ground floor beams	Concrete C30/37		30 × 60; 30 × 75; 30 × 45 cm, 7.5 m span
	Rebar (230 kg/m^3)		
Connections	Structural steel		10% of structural steel mass
	Concrete rebar		10% of rebar mass
Lateral load resisting system	Structural steel		10% of structural steel mass

Note: The 10% allowances for connections and LLRS are simplified estimates to account for steel used in beam-column joints and bracing elements, respectively. The same percentage applies to rebar for additional reinforcement in concrete connection zones.

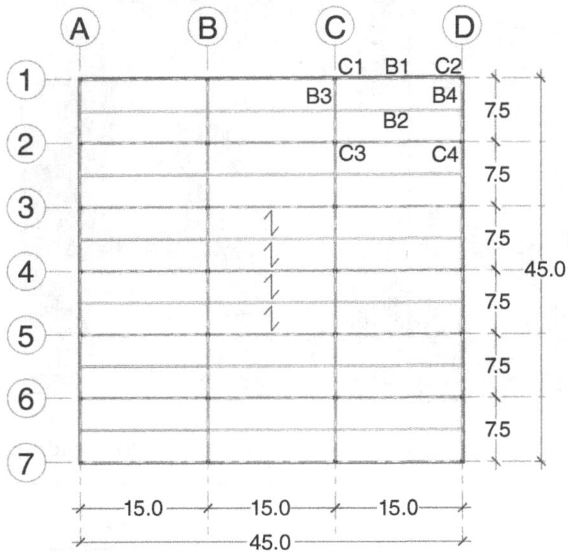

Figure 5.4 Conceptual plan of the case study 2.

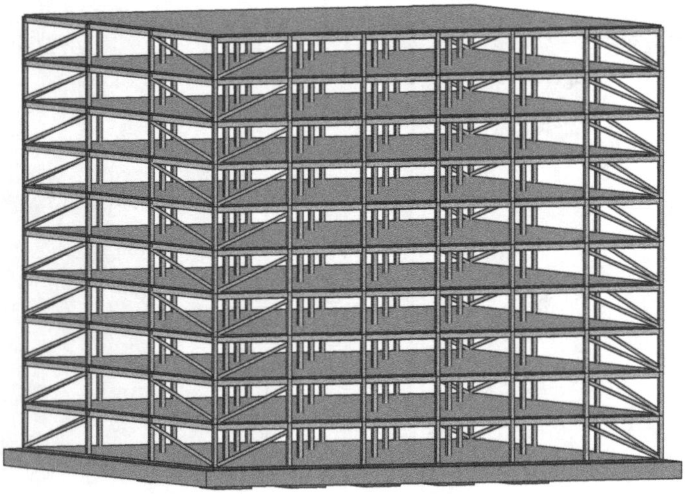

Figure 5.5 3D modeling of the case study 2.

5.2.1 Gross Internal Area (GIA) Calculation

To obtain the embodied carbon intensity ($kgCO_2e/m^2$), the Gross Internal Area (GIA) can be used (this concept is explained in the Chapter 3). Since both case studies have the same typical floor dimension and number of floors, the reinforced concrete building is taken into consideration to calculate the GIA, used for both entire building cases (Table 5.4 and Figure 5.6). The GIA (m^2) involves all the floors of the buildings, except the roof of that is not accessible except for maintenance. The

Table 5.3 Visual representation of the structural steel profiles used in the case study.

Beams			Columns		
610 × 229 × 125UB	533 × 210 × 92UB	533 × 210 × 101UB	356 × 368 × 202UC	356 × 368 × 129UC	356 × 406 × 287UC

Primary internal beams	Primary edge beams	Secondary beams	Edge columns	Corner columns	Internal columns

Table 5.4 Calculation of Gross Internal Area for a building with 10 floors.

Parameter	Calculation	Result
Internal floor area	45 × 45 m	2025 m^2 per floor
Number of floors	9 (number of upper floors except roof) +1 (ground floor)	10 floors
Gross Internal Area	2025 m^2 per floor × 10 floors	20 250 m^2

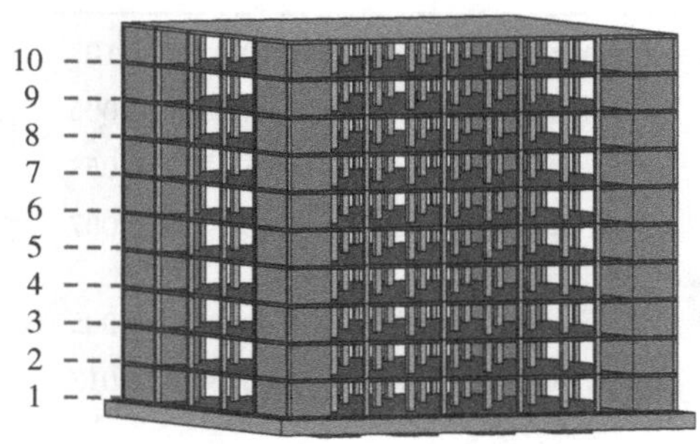

Figure 5.6 Number of floors in buildings for GIA calculation.

GIA is calculated as follows, considering the structural grid (with center-to-center distances and dimensions) of 45 × 45 m (Table 5.4):

$$\text{GIA} = (\text{Internal floor area}) \times (n^\circ \text{ of floors except roof}) \tag{5.1}$$

This value will be used in the next sections to determine the embodied carbon intensity of the entire building calculation and the total embodied carbon of the building, derived from the intensity.

5.2.2 Selection of the Embodied Carbon Factors

To ensure comparability across different structural systems, a common source of carbon factors has been used in this study (Tables 5.5 and 5.6). The values are derived from the ICE database (Circular

Table 5.5 Carbon factors ($kgCO_2e/kg$, average) of the different materials and elements used in reinforced concrete building cases.

	Concrete				Steel
Stages	**C30/37**	**C40/50**	**C50/60**	**GEN1 (C10)**	**Rebar**
A1–A3 (production)	0.10	0.138	0.165	0.103	0.76
A4 (transport)	0.0055	0.005	0.0055	0.005	0.032
WF (waste factor)	0.053	0.053	0.053	0.053	0.053
A5w (material wastage on site)	0.006	0.009	0.009	0.007	0.043
C2 (waste transport)	0.003	0.005	0.003	0.005	0.005
C3/C4 (waste processing & disposal)	0.0003	0.013	0.00035	0.013	0.013

Table 5.6 Carbon factors ($kgCO_2e/kg$, average) of the different materials and elements used in structural steel building cases.

	Concrete		Steel			
Stages	**C30/37**	**GEN1 (C10)**	**Open sections**	**Closed sections**	**Rebar**	**Galvanized sheet**
A1–A3 (production)	0.10	0.103	1.74	2.5	0.76	2.76
A4 (transport)	0.0055	0.005	0.032	0.032	0.032	0.183
WF (waste factor)	0.053	0.053	0.01	0.01	0.053	0.01
A5w (material wastage on site)	0.006	0.007	0.0179	0.026	0.043	0.03
C2 (waste transport)	0.003	0.005	0.005	0.005	0.005	0.005
C3/C4 (waste processing and disposal)	0.0003	0.013	0.013	0.013	0.013	0.013
D (end of life)	−0.0228	0.000	−0.920	−1.53	0.351	−1.320

Note: These are industry-average values; for greater accuracy, product-specific EPDs should be the reference.

Ecology, 2020), which provides average carbon emission factors for concrete, steel, and other construction materials.

It's important to recognize the limitations of relying on generic databases for carbon factors. While providing valuable benchmarks, they contain industry-average data, which may not accurately reflect the specific materials chosen for a project. The environmental impact of a material can vary significantly depending on the manufacturer, production process, and supply chain. For more accurate assessments, product-specific Environmental Product Declarations (EPDs) should be used. EPDs provide verified and transparent data about a product's life cycle environmental impact. However, in the early stages of design, when specific products have not yet been selected, using established databases is a common and practical approach. These values should be used cautiously as placeholders and replaced with specific EPD data as the project progresses to ensure a more accurate final carbon calculation.

Carbon Factors Used in Reinforced Concrete Calculations:
Table 5.5 presents the carbon factors for different grades of concrete (C30/37, C40/50, C50/60) and steel rebar used in reinforced concrete structures, derived from the ICE database (Circular Ecology 2020) as average values. The production stage (A1–A3) shows the highest carbon impact, with rebar having a carbon factor of 0.76 $kgCO_2e/kg$, significantly higher than the values for concrete, which range from 0.1 to 0.165 $kgCO_2e/kg$ depending on the mix. To put this in perspective: steel has a higher carbon footprint per kilogram, but its high strength means that less material is needed overall (e.g. despite its higher carbon intensity per unit weight, the total amount of steel required in a reinforced concrete beam is significantly lower than concrete). Even though rebar is just a small part of the concrete mix (in terms of volume), it's crucial since it handles the tensile forces that concrete alone cannot. The transportation stage (A4) contributes relatively low emissions, as does waste transport (C2) and waste processing (C3/C4). The material wastage on-site (A5w) also adds a minor contribution to the overall embodied carbon.

The assumptions for concrete mixes include Ground Granulated Blast Furnace Slag (GGBS) replacement, with 30% GGBS in C30/37, 25% in C40/50, and no replacement in C50/60, reflecting standard sustainability practices in concrete mix design. Steel reinforcement is assumed to be sourced from the UK market, where 97% of rebars are produced via Electric Arc Furnace (EAF) route (Gibbons et al., 2022), reducing carbon intensity compared to BF-BOF route (see Chapter 2 for details).

Carbon Factors Used in Structural Steel Calculations:
Table 5.6 provides carbon factors for different structural steel elements, including open and closed sections, rebar, and galvanized sheets, derived from the ICE database (Circular Ecology 2020) as generic values. The production stage (A1–A3) has the most significant impact, with closed steel sections emitting 2.5 $kgCO_2e/kg$, while open sections emit 1.74 $kgCO_2e/kg$. This variation is due to differences in material processing and fabrication techniques (these values are on the higher end of the spectrum, as they include a significant share of BF-BOF route production, which is considerably more carbon-intensive than EAF route steel made from recycled scrap). Rebar has a production carbon factor of 0.76 $kgCO_2e/kg$, same with that seen in the reinforced concrete case, while galvanized steel sheets have the highest impact at 2.76 $kgCO_2e/kg$, reflecting the additional processes involved in their coating.

For steel rebars, the analysis assumes 97% recycled content via Electric Arc Furnace (EAF) production, following the sector average for the UK market (please remember from Chapter 2, that compared to primary steel production via the blast furnace route (BF-BOF), EAF relies more on scrap steel recycling and electricity, rather than raw iron ore and coking coal).

Concrete C30/37 is assumed with a 30% Ground Granulated Blast Furnace Slag (GGBS) replacement, while GEN1 (C10) includes a 25% GGBS content (please remember from Chapter 2 that these supplementary cementitious materials help reduce embodied carbon by partially replacing Portland cement, which is one of the largest contributors to emissions in concrete production). Transportation emissions (A4) remain relatively low. Waste factors (WF), on-site material wastage (A5w), and disposal (C3/C4) have minimal emissions.

A notable aspect of steel's carbon profile is the negative value in Module D (end-of-life), where open sections (−0.920 $kgCO_2e/kg$) and closed sections (−1.53 $kgCO_2e/kg$) achieve substantial carbon credits because, at end-of-life, these structural elements can be recovered and recycled into new steel products, thereby avoiding the carbon emissions that would otherwise be required to produce virgin steel. This credit reflects the "avoided burden" principle: when steel is recycled, it displaces the need for primary production, and the associated emission savings are assigned back to the original product. In contrast, rebars have a positive Module D value, because the input material is already assumed as recycled (scrap), and the system boundary yields no net benefit from further recycling. In addition, the energy required to recover the rebar outweighs the marginal credit for replacing scrap in future cycles.

5.2.3 Challenges in Selecting the Embodied Carbon Factors

The carbon factor assumptions presented in the previous section are highly dependent on specific regional, material, and industry-related conditions, meaning they can vary significantly in different contexts. Several key factors influence these values, including material sourcing, manufacturing processes, transportation distances, and end-of-life scenarios:

a. **Material Sourcing**
As explained in detail in the Chapter 2, the carbon intensity of concrete and steel depends heavily on the local availability of raw materials and the use of supplementary cementitious materials (SCMs) such as Ground Granulated Blast Furnace Slag (GGBS) or Fly Ash. In some regions, higher percentages of SCMs may be used, significantly lowering the embodied carbon of concrete. Similarly, the recycled content of steel varies, with Electric Arc Furnace (EAF) steel typically having a lower carbon footprint than the steel obtained by BF-BOF route, which dominates in some markets. However, while increasing the use of SCMs and recycled steel can significantly reduce embodied carbon, their availability is not always guaranteed. In some regions, the supply of GGBS or Fly Ash may be limited due to production constraints or competing industrial demands, leading to a reliance on traditional Portland cement with higher emissions. Similarly, while EAF steel has a lower carbon footprint, its adoption depends on regional recycling infrastructure and electricity sources – if the grid is heavily dependent on fossil fuels, the benefits of EAF production may be diminished. These are some of the reasons why project-specific EPDs, which account for the local factors, are essential for final accurate calculations.

b. **Manufacturing Processes**
Variations in energy sources used in production facilities impact carbon factors. In regions where electricity grids are dominated by renewable energy, steel and concrete production tend to have lower embodied carbon compared to areas reliant on coal or fossil-fuel-based power. The efficiency of production plants, kiln technology in cement manufacturing, and rolling processes in steel mills also contribute to differences in emissions.

c. **Transportation Distances and Modes**
The projects may involve longer or shorter transport routes depending on local supply chain networks. Imported materials, whether steel or cement, may introduce higher transport emissions, particularly for long-distance shipping. Besides, the mode of transport (road, rail, or sea) significantly affects emissions, with sea transport often being more carbon-intensive per unit distance than rail or road transport.

d. **Construction and Material Wastage**
The assumptions for material wastage on-site also depend on construction practices, precision of material orders, and fabrication methods. For example, precast concrete or modular steel fabrication

often reduces wastage compared to in situ methods. In some cases, higher on-site losses can lead to increased material demand, which raises embodied carbon beyond the assumed values.

e. **End-of-life Scenarios**
The recycling and reuse potential of materials plays a major role in the overall carbon balance. In regions with well-established steel recycling infrastructure, Module D benefits can be higher than assumed in this study. Similarly, concrete recycling rates vary, with some locations having advanced crushing and reuse techniques, while others rely more on landfilling, leading to higher disposal emissions.

Given these variables, the carbon factor assumptions used in this chapter are specific to the study at hand and should not be generalized beyond this analysis. The results are bound by these assumptions, meaning that while the trends and comparisons remain relevant, absolute values may shift significantly in different geographic, economic, and industrial contexts.

5.3 Early Stage Design Alternatives Using Representative Portions

To estimate the embodied carbon intensity of the entire building rapidly, a smaller representative portion is analyzed instead of evaluating every structural element of the entire building. Figure 5.7 demonstrates this concept by extracting a single portion from the full reinforced concrete building. This method allows for a quicker assessment of different design options while maintaining reasonable accuracy. In this study, one portion of the reinforced concrete building (shown in Figure 5.8) and six selected steel building alternatives (Figures 5.9–5.14) are used to calculate the embodied carbon for the whole structure.

5.3.1 Reinforced Concrete Structure

The studied reinforced concrete portion is presented in Figure 5.8 and described in Table 5.7. It consists of a 9 × 9 m slab supported by four columns. The slab is 350 mm thick and made of C30/37 concrete with an estimated 90 kg/m^3 of steel reinforcement. Each column measures 750 × 750 mm in cross-section, stands 4.0 m high, and is composed of C50/60 concrete with around 250 kg/m^3 of steel reinforcement.

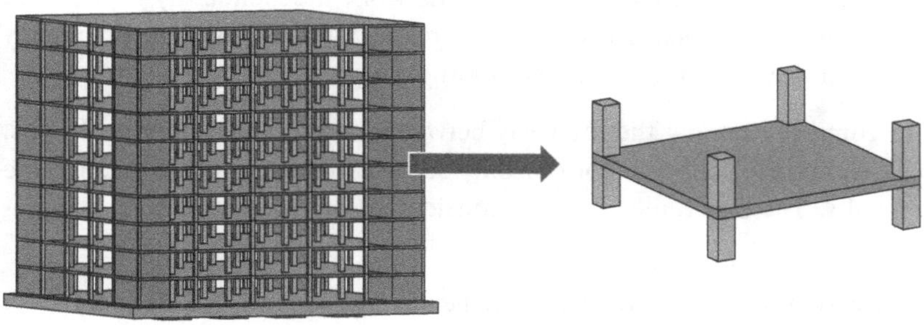

Figure 5.7 Representation of entire building as a representative portion (reinforced concrete case).

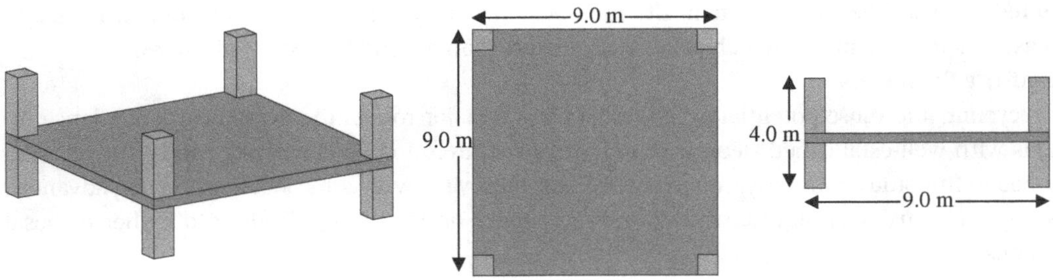

Figure 5.8 Visual representation of the reinforced concrete portion considered for the study.

Table 5.7 Description of the structural elements and dimensions of the RC portion considered.

Element	Material	Dimensions
Slab	Concrete C30/37 Reinforcing bars (90 kg/m^3)	9 × 9 m, 350 mm depth
Columns	Concrete C50/60 Reinforcing bars (250 kg/m^3)	750 × 750 mm, 4 m height, ¼ at each corner*

* Since the representative bay shares its corner columns with adjacent bays, only one quarter of each column is attributed to this portion. This allocation reflects the tributary area principle: each corner column serves four bays, so its embodied carbon contribution is divided equally among them.

5.3.2 Steel Structure with Metal Decking

This section presents six design variations for the steel segment of the same building. Each option specifies a different arrangement of spans (15 × 7.5 m or 7.5 × 7.5 m), composite or noncomposite action, and steel grades (S275 for beams and S355 for columns). At this point, an explanation is needed about composite or noncomposite action (for more information, you can refer to this source (SteelConstruction, 2025):

- In composite action, different structural materials – typically a steel beam and a concrete slab – are connected (usually with shear connectors) so they act together as a single unit. This integrated system distributes loads between the materials, increasing the overall strength and stiffness of the element.
- With noncomposite action, the steel and concrete components work independently rather than together. Without the connection to share loads, each material resists forces on its own, which may lead to a less efficient structural system compared to a composite design.

Using these options, we are going to see how the interplay between long or short primary beams, along with the choice to apply composite action to both or only secondary beams will influence the mass and embodied carbon results. The six studied options considered for the calculations are presented as follows:

Option 1. 15 × 7.5 m, noncomp. action (both beams), long prim. beams, S275 (beams), S355 (col.)
Option 2. 15 × 7.5 m, comp. action (sec. beams), long prim. beams, S275 (beams), S355 (col.)

Option 3. 15 × 7.5 m, comp. action (sec. beams), short prim. beams, S275 (beams), S355 (col.)
Option 4. 15 × 7.5 m, comp. action (sec. beams), short prim. beams, S275 (sec. beams), S355 (prim. beams and col.)
Option 5. 7.5 × 7.5 m, noncomp. action (both beams), open sec. col., S275 (beams), S355 (col.)
Option 6. 7.5 × 7.5 m, noncomp. action (both beams), open sec. col., S355 (beams and col.)

Visual representation of six of the options considered for the study is shown in Figures 5.9–5.14. Table 5.8 presents the characteristics of the studied options.

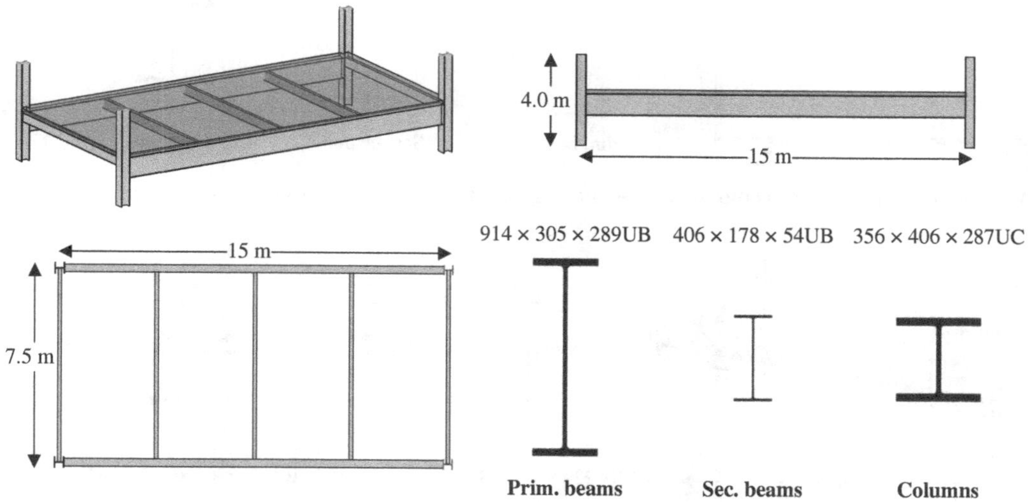

Figure 5.9 Option 1. 15 × 7.5 m, noncomp. action (both beams), long prim. beams, S275 (beams), S355 (col.).

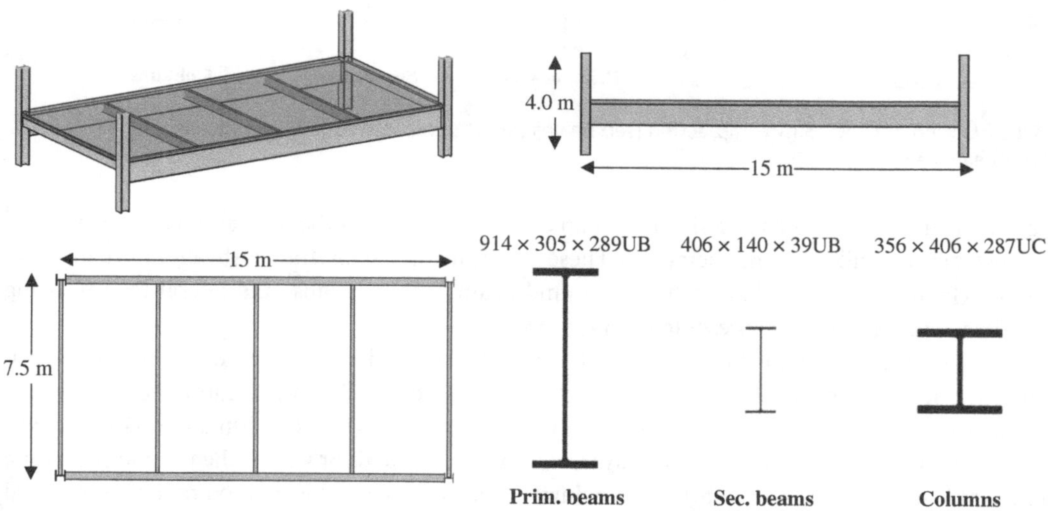

Figure 5.10 Option 2. 15 × 7.5 m, comp. action (sec. beams), long prim. beams, S275 (beams), S355 (col.).

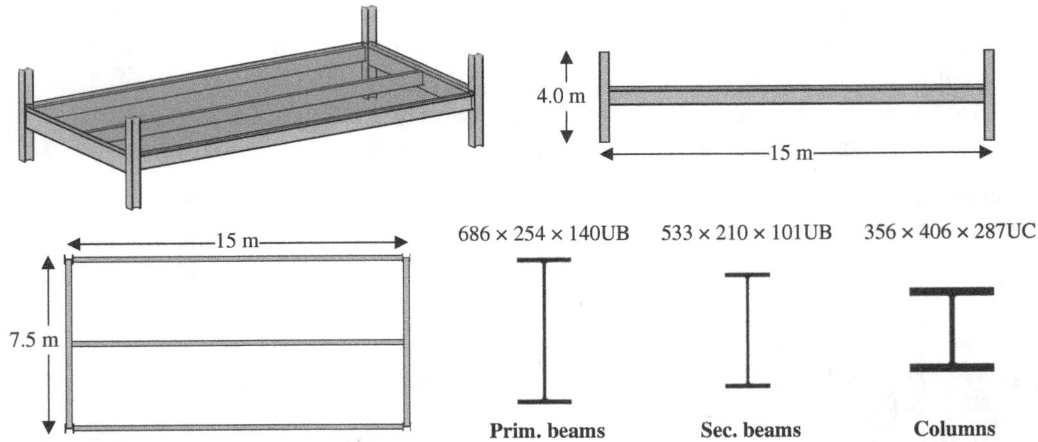

Figure 5.11 Option 3. 15 × 7.5 m, comp. action (sec. beams), short prim. beams, S275 (beams), S355 (col.).

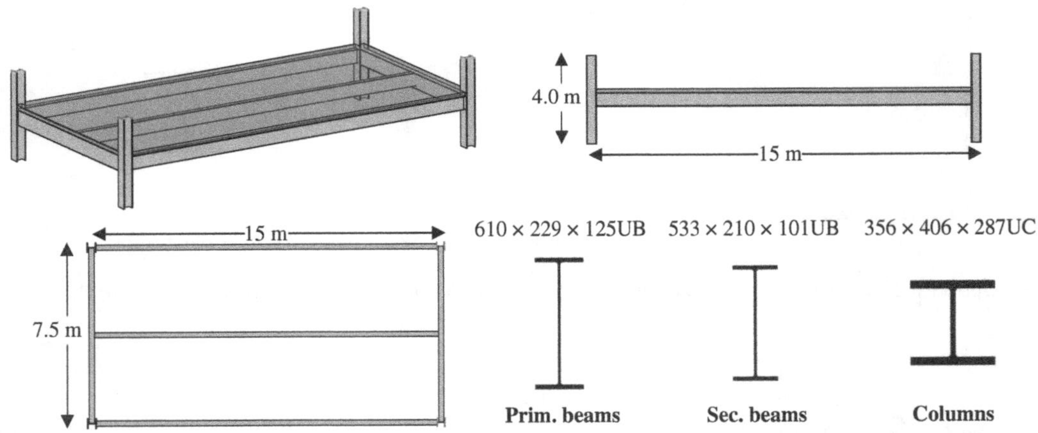

Figure 5.12 Option 4. 15 × 7.5 m, comp. action (sec. beams), short prim. beams, S275 (sec. beams), S355 (prim. beams and col.).

Using the calculation methods described in Chapter 3, we determined the total embodied carbon and its intensity for entire buildings and their parts. These results include slabs, frames (beams and columns), substructure (footings, ground floor slab, and ground beams), connections, and lateral load resisting systems. The results obtained are presented in Table 5.9.

The slight variance in embodied carbon intensity between the entire buildings and their representative portions, not exceeding 10%, can be attributed to several factors. When evaluating only a portion, common structural elements like columns and beams are calculated without redundancy. For instance, a beam or column shared between multiple bays is partially accounted for within the portion (i.e. half a primary beam from each side or a quarter of a column at each corner). This method reduces the actual number of beams and columns, as well as structural steel connections (expressed as a percentage of the structural steel weight), leading to an underestimation of the total mass and, consequently, the embodied

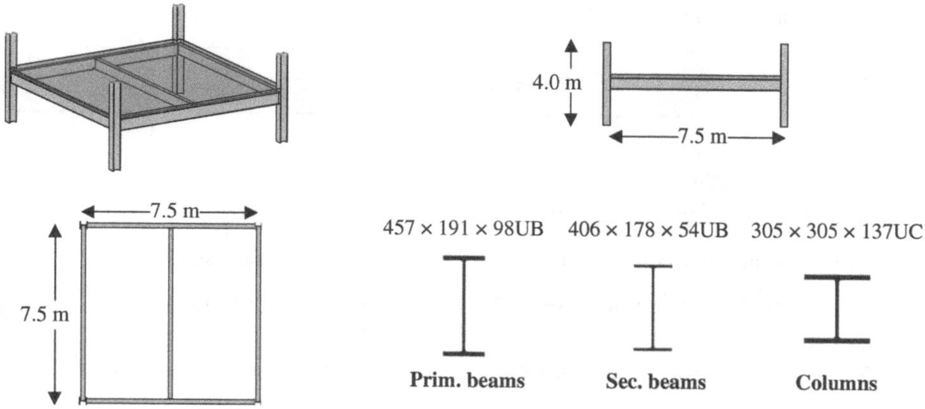

Figure 5.13 Option 5. 7.5 × 7.5 m, noncomp. action (both beams), open sec. col., S275 (beams), S355 (col.).

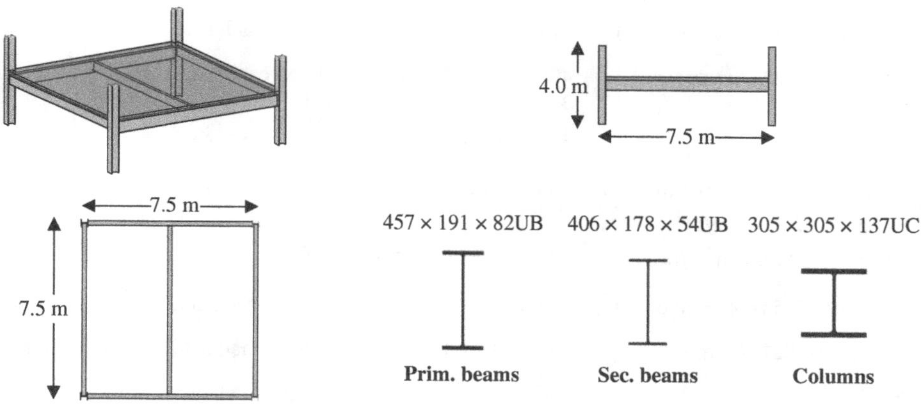

Figure 5.14 Option 6. 7.5 × 7.5 m, noncomp. action (both beams), open sec. col., S355 (beams and col.).

carbon intensity of the portion compared to the entire building. As a result, the frame's embodied carbon intensity in the portion appears lower than that of the whole building it represents. To account for such an underestimation, it is recommended to introduce an uncertainty factor when extrapolating from portion-based calculations to full-building estimates. In this case, the uncertainty factor is assumed as 1.1, which compensates for the shared elements that are only partially captured in the representative bay analysis, providing a more accurate approximation of the building's total embodied carbon intensity.

It has been shown that during the life cycle of the building, the Production stage (A1–A3) is the one with the highest amount of embodied carbon (total or intensity). To illustrate this, the total embodied carbon at each life-cycle stage is shown for the reinforced concrete option (Figure 5.15) and the structural steel system Option 1 (Figure 5.16), both excluding and including Module D.

Figures 5.17 and 5.18 show that when Module D is considered, embodied carbon and the embodied carbon intensity of the buildings are reduced because Module D accounts for the benefits and loads beyond the system boundary, which includes the reuse, recovery, and recycling potential of the materials.

Table 5.8 Description of the structural elements and dimensions of the six options considered.

Option	Element	Component	Dimensions
1	Slab	Concrete C30/37 (0.105 m^3/m^2)	15 × 7.5 m, 150 mm depth
		Rebar A252 (3.95 kg/m^2)	
		Metal deck (1382 mm^2/m)	80 mm deep, 0.9 mm thick
	Beams	Primary, S275 (noncomposite)	914 × 305 × 289UB, 15 m, ½ on each side.
		Secondary, S275 (noncomposite)	406 × 178 × 54UB, 7.5 m, ½ on each side and 3 internal at 3.75 m interval
	Columns	S355	356 × 406 × 287UC, 4 m height, ¼ at each corner
2	Slab	Concrete C30/37 (0.105 m^3/m^2)	15 × 7.5 m, 150 mm depth
		Rebar A252 (3.95 kg/m^2)	
		Metal deck (1382 mm^2/m)	80 mm deep, 0.9 mm gauge
	Beams	Primary, S275 (noncomposite)	914 × 305 × 289UB, 15 m, ½ on each side.
		Secondary, S275 (composite)	406 × 178 × 54UB, 7.5 m, ½ on each side and 3 internal at 3.75 m interval
	Columns	S355	356 × 406 × 287UC, 4 m height, ¼ at each corner
3	Slab	Concrete C30/37 (0.105 m^3/m^2)	15 × 7.5 m, 150 mm depth
		Rebar A252 (3.95 kg/m^2)	
		Metal deck (1382 mm^2/m)	80 mm deep, 0.9 mm gauge
	Beams	Primary, S275 (noncomposite)	686 × 254 × 140UB, 7.5 m, ½ on each side
		Secondary, S275 (composite)	533 × 210 × 101UB, 15 m, ½ on each side and 1 in center
	Columns	S355	356 × 406 × 287UC, 4 m height, ¼ at each corner
4	Slab	Concrete C30/37 (0.105 m^3/m^2)	15 × 7.5 m, 150 mm depth
		Rebar A252 (3.95 kg/m^2)	
		Metal deck (1382 mm^2/m)	80 mm deep, 0.9 mm gauge
	Beams	Primary, S355 (noncomposite)	610 × 229 × 125UB, 7.5 m, ½ on each side
		Secondary, S275 (composite)	533 × 210 × 101UB, 15 m, ½ on each side and 1 in center
	Columns	S355	356 × 406 × 287UC, 4 m height, ¼ at each corner
5	Slab	Concrete C30/37 (0.105 m^3/m^2)	7.5 × 7.5 m, 150 mm depth
		Rebar A252 (3.95 kg/m^2)	
		Metal deck (1382 mm^2/m)	80 mm deep, 0.9 mm gauge
	Beams	Primary, S275 (noncomposite)	457 × 191 × 98UB, 7.5 m, ½ on each side.
		Secondary, S275 (noncomposite)	406 × 178 × 54UB, 7.5 m, ½ on each side and 1 internal
	Columns	S355	305 × 305 × 137UC, 4 m height, ¼ at each corner

Option	Element	Component	Dimensions
6	Slab	Concrete C30/37 (0.105 m^3/m^2)	7.5 × 7.5 m, 150 mm depth
		Rebar A252 (3.95 kg/m^2)	
		Metal deck (1382 mm^2/m)	80 mm deep, 0.9 mm gauge
	Beams	Primary, S355 (noncomposite)	457 × 191 × 82UB, 7.5 m, ½ on each side.
		Secondary, S355 (noncomposite)	406 × 178 × 54UB, 7.5 m, ½ on each side and 1 internal
	Columns	S355	305 × 305 × 137UC, 4 m height, ¼ at each corner

Table 5.9 Calculation results of embodied carbon intensity and total embodied carbon of the entire buildings and their representative portions.

	EC intensity (A–C) ($kgCO_2e/m^2$)		Building total EC (A–C) (tCO_2e)		
Option	**Building**	**Portion**	**From building**	**From portion**	**% Difference between building and portion**
Reinforced concrete (9 × 9 m)	223.46	218.17	4525.07	4417.94	3%
1. 15 × 7.5 m, noncomp. action (both beams), long prim. beams, S275 (beams), S355 (col.)	251.04	229.26	5083.52	4642.52	9%
2. 15 × 7.5 m, comp. action (sec. beams), long prim. beams, S275 (beams), S355 (col.)	239.10	220.73	4841.68	4469.84	8%
3. 15 × 7.5 m, comp. action (sec. beams), short prim. beams, S275 (beams), S355 (col.)	204.30	193.89	4137.00	3926.27	5%
4. 15 × 7.5 m, comp. action (sec. beams), short prim. beams, S275 (sec. beams), S355 (prim. beams and col.)	202.09	191.78	4092.28	3883.50	5%
5. 7.5 × 7.5 m, noncomp. action (both beams), open sec. col., S275 (beams), S355 (col.)	176.50	162.73	3574.10	3295.28	8%
6. 7.5 × 7.5 m, noncomp. action (both beams), open sec. col., S355 (beams and col.)	171.23	158.13	3467.35	3202.13	8%

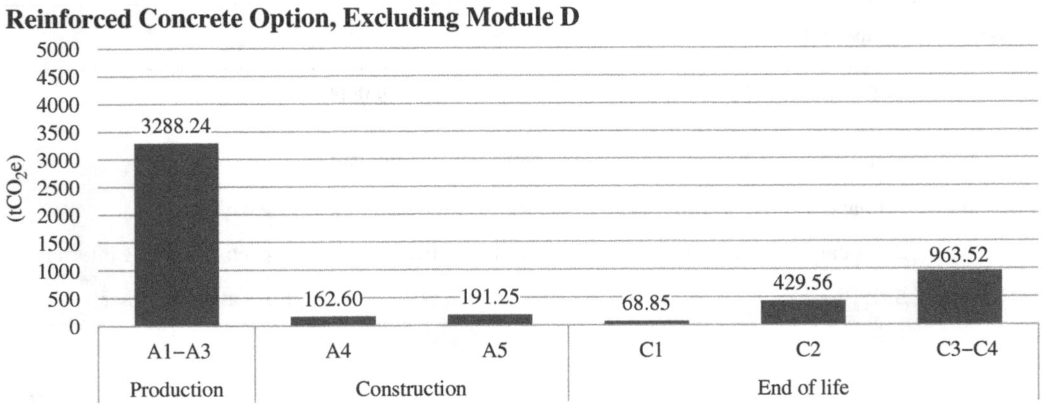

Figure 5.15 Embodied carbon emission per stage (A–D) of reinforced concrete option without Stage D.

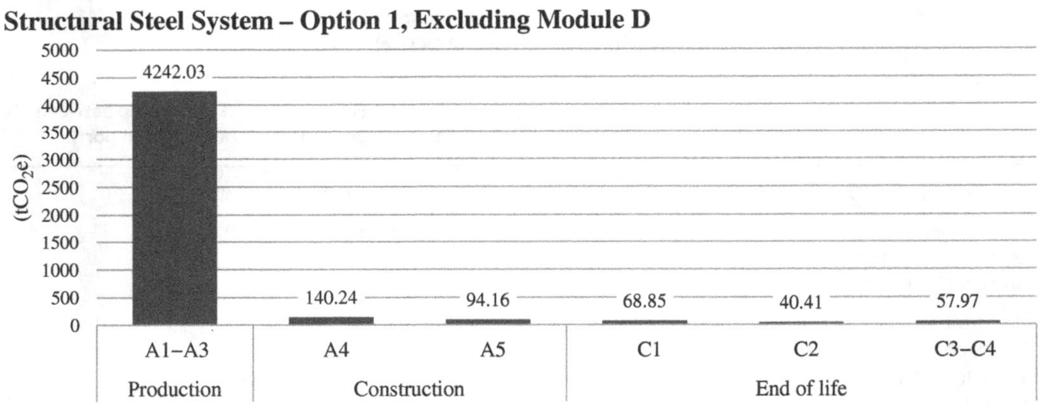

Figure 5.16 Embodied carbon emission per stage (A–D) of Option 1 without Stage D.

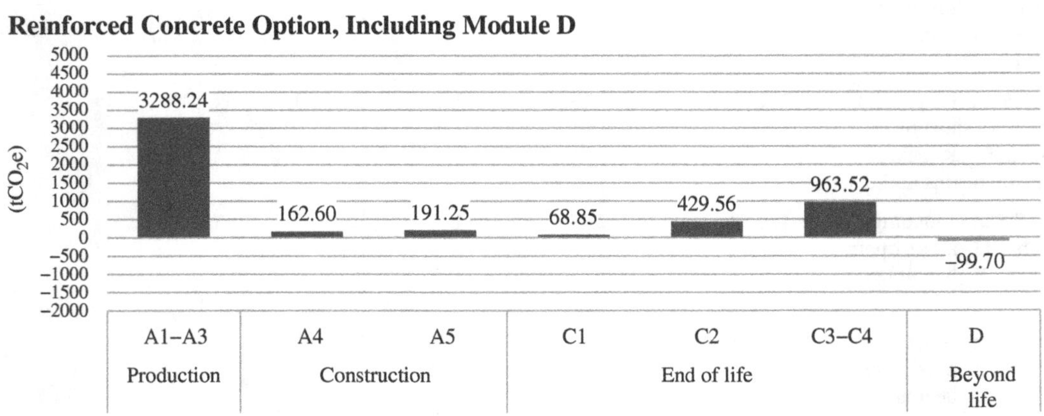

Figure 5.17 Embodied carbon emission per stage (A–D) of reinforced concrete option with stage D.

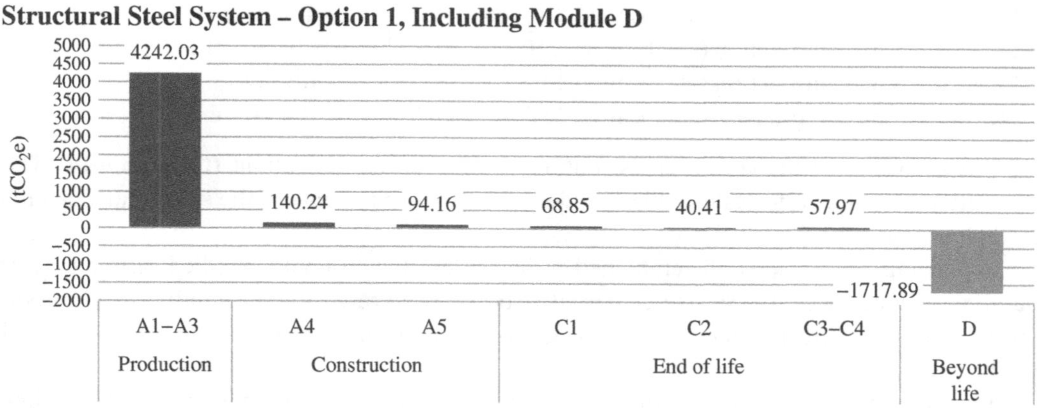

Figure 5.18 Embodied carbon emission per stage (A–D) of Option 1 with stage D.

In summary, the representative portion approach provides a practical method for rapid early-stage embodied carbon estimation, with results within 10% of full-building calculations in this specific case. The analysis demonstrates that design decisions (such as span configuration, composite versus noncomposite action, and material selection) significantly influence embodied carbon outcomes. Furthermore, including Module D in the assessment reveals the substantial end-of-life benefits of steel structures, highlighting the importance of considering the complete life cycle when comparing design alternatives.

5.4 The Impact of Tubular Profiles and Higher-strength Steel on the Embodied Carbon

This section explores the parameter analysis (i.e. an investigation into how different design changes affect outcomes) of 14 different structural configurations, evaluating their impact on embodied carbon intensity. The study expands on previous findings by introducing eight additional options, each varying in beam action (composite or noncomposite), column type, and material grade. The configurations include long-span and short-span beam arrangements, open-section and closed-section columns, and material selections ranging from standard steel grades (S275, S355) to high-strength steel (S460):

Option 1. 15 × 7.5 m, noncomp. action (both beams), long prim. beams, S275 (beams), S355 (col.)
Option 2. 15 × 7.5 m, comp. action (sec. beams), long prim. beams, S275 (beams), S355 (col.)
Option 3. 15 × 7.5 m, comp. action (sec. beams), short prim. beams, S275 (beams), S355 (col.)
Option 4. 15 × 7.5 m, comp. action (sec. beams), short prim. beams, S275 (sec. beams), S355 (prim. beams and col.)
Option 5. 7.5 × 7.5 m, noncomp. action (both beams), open sec. col., S275 (beams), S355 (col.)
Option 6. 7.5 × 7.5 m, noncomp. action (both beams), open sec. col., S355 (beams and col.)
Option 7. 7.5 × 7.5 m, noncomp. action (both beams), open sec. col., S275 (beams and col.)
Option 8. 7.5 × 7.5 m, comp. action (sec. beams), open sec. col., S275 (beams), S355 (col.)
Option 9. 7.5 × 7.5 m, comp. action (sec. beams), open sec. col., S355 (beams and col.)
Option 10. 7.5 × 7.5 m, comp. action (sec. beams), open sec. col., S355 (beams), S460 (col.)

Option 11. 7.5 × 7.5 m, comp. action (sec. beams), CHS col., S355 (beams and col.)
Option 12. 7.5 × 7.5 m, comp. action (sec. beams), CHS col., S355 (beams), S460 (col.)
Option 13. 7.5 × 7.5 m, comp. action (sec. beams), SHS col., S355 (beams and col.)
Option 14. 7.5 × 7.5 m, comp. action (sec. beams), SHS col., S355 (beams), S460 (col.)

Figures 5.19–5.26 present the visual representation of the 8 additional options (Options 7–14). The elements' sections are shown to scale. Table 5.10 describes the components of the structural elements and their dimensions.

Similar to previous calculations, the embodied carbon intensity and total embodied carbon for different building options are computed using alternative portions representing the buildings and carbon factors from Table 5.6, with parameter analysis results for all 14 options shown in Table 5.11 and Figures 5.27–5.30.

Reducing the span length from 15 to 7.5 m significantly lowers embodied carbon intensity. The baseline configuration (Option 1: 15 × 7.5 m, noncomposite action, long-span primary beams, S355 columns) exhibits

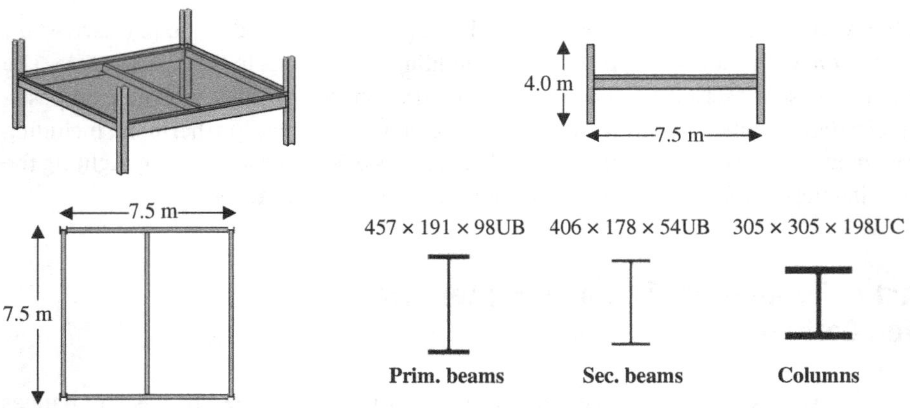

Figure 5.19 [Option 7] 7.5 × 7.5 m, noncomp. action (both beams), open sec. col., S275 (beams and col.).

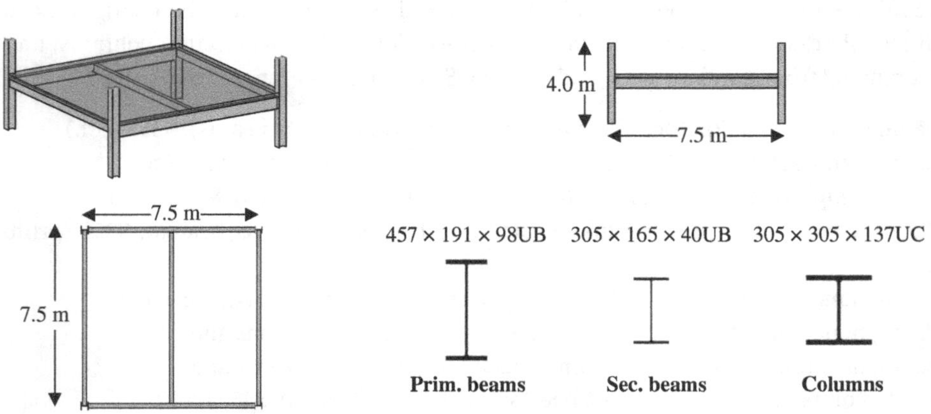

Figure 5.20 [Option 8] 7.5 × 7.5 m, comp. action (sec. beams), open sec. col., S275 (beams), S355 (col.).

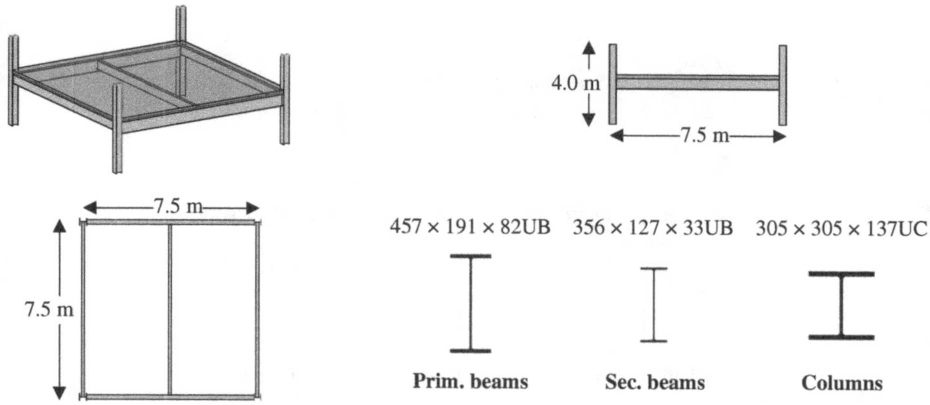

Figure 5.21 [Option 9] 7.5 × 7.5 m, comp. action (sec. beams), open sec. col., S355 (beams and col.).

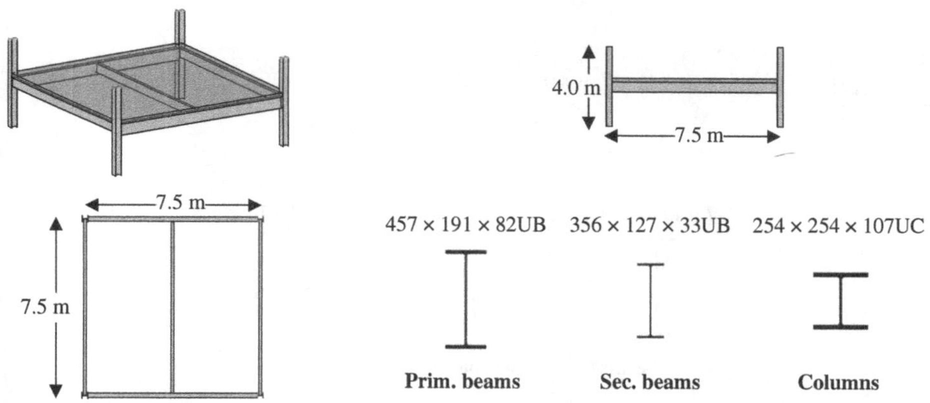

Figure 5.22 [Option 10] 7.5 × 7.5 m, comp. action (sec. beams), open sec. col., S355 (beams), S460 (col.).

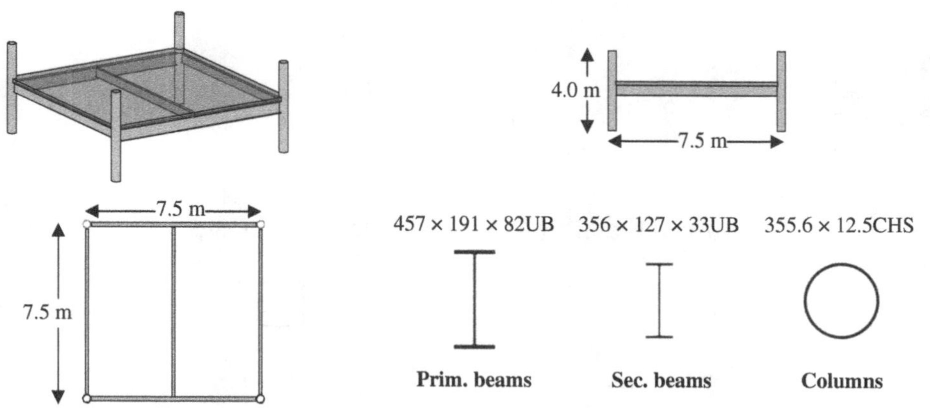

Figure 5.23 [Option 11] 7.5 × 7.5 m, comp. action (sec. beams), CHS col., S355 (beams and col.).

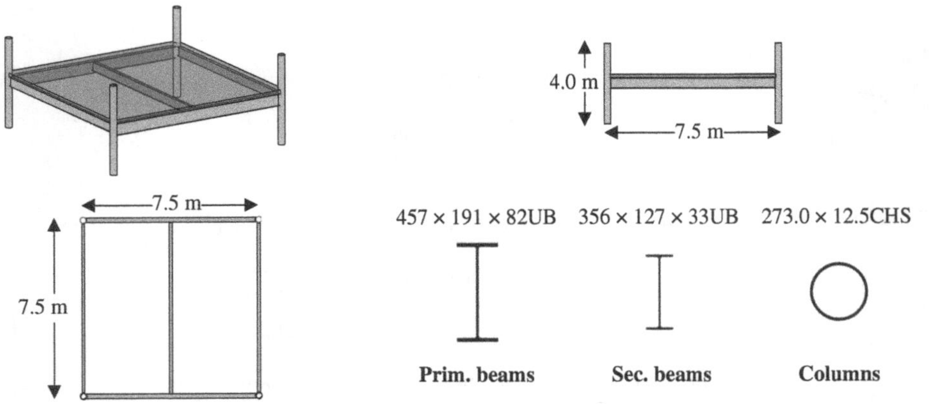

Figure 5.24 [Option 12] 7.5 × 7.5 m, comp. action (sec. beams), CHS col., S355 (beams), S460 (col.).

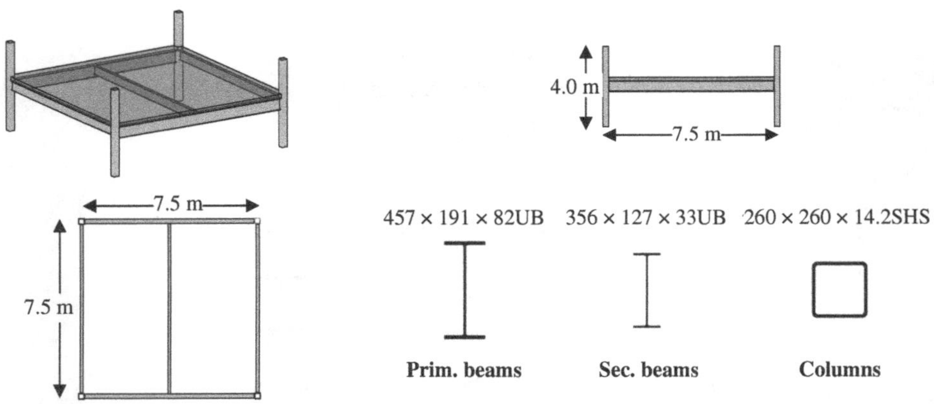

Figure 5.25 [Option 13] 7.5 × 7.5 m, comp. action (sec. beams), SHS col., S355 (beams and col.).

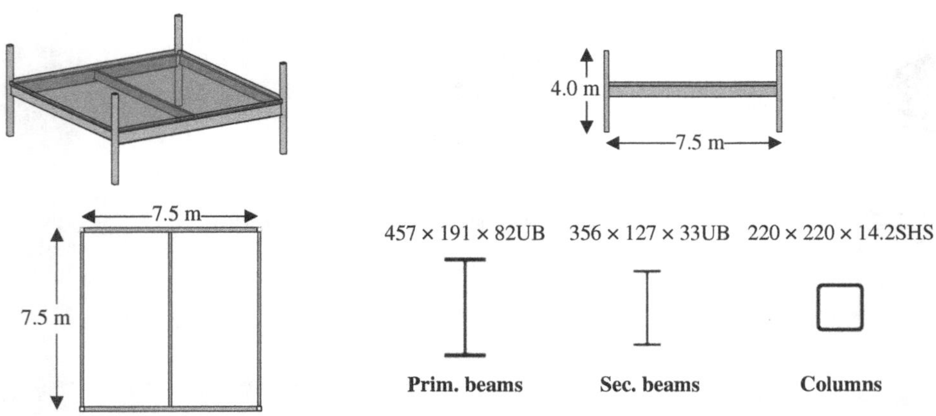

Figure 5.26 [Option 14] 7.5 × 7.5 m, comp. action (sec. beams), SHS col., S355 (beams), S460 (col.).

Table 5.10 Description of the structural elements and dimensions of the eight additional options considered.

Option	Element	Component	Dimensions
7	Slab	Concrete C30/37 (0.105 m^3/m^2)	7.5 × 7.5 m, 150 mm depth
		Rebar A252 (3.95 kg/m^2)	80 mm deep, 0.9 mm gauge
		Metal deck (1382 mm^2/m)	
	Beams	Primary, S275 (noncomposite)	457 × 191 × 98UB, 7.5 m, ½ on each side
		Secondary, S275 (noncomposite)	406 × 178 × 54UB, 7.5 m, ½ on each side and 1 internal
	Columns	S275	305 × 305 × 198UC, 4 m height, ¼ at each corner
8	Slab	Concrete C30/37 (0.105 m^3/m^2)	7.5 × 7.5 m, 150 mm depth
		Rebar A252 (3.95 kg/m^2)	80 mm deep, 0.9 mm gauge
		Metal deck (1382 mm^2/m)	
	Beams	Primary, S275 (noncomposite)	457 × 191 × 98UB, 7.5 m, ½ on each side.
		Secondary, S275 (composite)	305 × 165 × 40UB, 7.5 m, ½ on each side and 1 internal
	Columns	S355	305 × 305 × 137UC, 4 m height, ¼ at each corner
9	Slab	Concrete C30/37 (0.105 m^3/m^2)	7.5 × 7.5 m, 150 mm depth
		Rebar A252 (3.95 kg/m^2)	80 mm deep, 0.9 mm gauge
		Metal deck (1382 mm^2/m)	
	Beams	Primary, S355 (noncomposite)	457 × 191 × 82UB, 7.5 m, ½ on each side.
		Secondary, S355 (Composite)	356 × 127 × 33UB, 7.5 m, ½ on each side and 1 internal
	Columns	S355	305 × 305 × 137UC, 4 m height, ¼ at each corner
10	Slab	Concrete C30/37 (0.105 m^3/m^2)	7.5 × 7.5 m, 150 mm depth
		Rebar A252 (3.95 kg/m^2)	80 mm deep, 0.9 mm gauge
		Metal deck (1382 mm^2/m)	
	Beams	Primary, S355 (noncomposite)	457 × 191 × 82UB, 7.5 m, ½ on each side.
		Secondary, S355 (Composite)	356 × 127 × 33UB, 7.5 m, ½ on each side and 1 internal
	Columns	S460	254 × 254 × 107UC, 4 m height, ¼ at each corner
11	Slab	Concrete C30/37 (0.105 m^3/m^2)	7.5 × 7.5 m, 150 mm depth
		Rebar A252 (3.95 kg/m^2)	80 mm deep, 0.9 mm gauge
		Metal deck (1382 mm^2/m)	
	Beams	Primary, S355 (noncomposite)	457 × 191 × 82UB, 7.5 m, ½ on each side.
		Secondary, S355 (Composite)	356 × 127 × 33UB, 7.5 m, ½ on each side and 1 internal
	Columns	S355	355.6 × 12.5CHS, 4 m height, ¼ at each corner

(Continued)

Table 5.10 (Continued)

Option	Element	Component	Dimensions
12	Slab	Concrete C30/37 (0.105 m^3/m^2) Rebar A252 (3.95 kg/m^2) Metal deck (1382 mm^2/m)	7.5 × 7.5 m, 150 mm depth 80 mm deep, 0.9 mm gauge
	Beams	Primary, S355 (noncomposite)	457 × 191 × 82UB, 7.5 m, ½ on each side.
		Secondary, S355 (Composite)	356 × 127 × 33UB, 7.5 m, ½ on each side and 1 internal
	Columns	S355	273.0 × 12.5CHS, 4 m height, ¼ at each corner
13	Slab	Concrete C30/37 (0.105 m^3/m^2) Rebar A252 (3.95 kg/m^2) Metal deck (1382 mm^2/m)	7.5 × 7.5 m, 150 mm depth 80 mm deep, 0.9 mm gauge
	Beams	Primary, S355 (noncomposite)	457 × 191 × 82UB, 7.5 m, ½ on each side.
		Secondary, S355 (Composite)	356 × 127 × 33UB, 7.5 m, ½ on each side and 1 internal
	Columns	S355	260 × 260 × 14.2SHS, 4 m height, ¼ at each corner
14	Slab	Concrete C30/37 (0.105 m^3/m^2) Rebar A252 (3.95 kg/m^2) Metal deck (1382 mm^2/m)	7.5 × 7.5 m, 150 mm depth 80 mm deep, 0.9 mm gauge
	Beams	Primary, S355 (noncomposite)	457 × 191 × 82UB, 7.5 m, ½ on each side.
		Secondary, S355 (Composite)	356 × 127 × 33UB, 7.5 m, ½ on each side and 1 internal
	Columns	S460	220 × 220 × 14.2SHS, 4 m height, ¼ at each corner

the highest EC intensity at 229.26 $kgCO_2e/m^2$. In contrast, the best-performing configuration (Option 12: 7.5 × 7.5 m, composite secondary beams, high-strength steel columns) achieves a reduction of 38%, bringing the EC intensity down to 142.14 $kgCO_2e/m^2$. However, this represents a drastic change in span, which may not always be feasible in practice. Instead, a moderate reduction in span (e.g. from 15 to 12.5 m) can provide a balanced approach for optimizing embodied carbon while maintaining spatial efficiency.

The composite actions in beams lead to a reduction of embodied carbon, as observed in comparisons between similar noncomposite and composite configurations. Exploiting composite action can allow for larger-span buildings while still achieving lower embodied carbon, making it a valuable strategy for sustainable structural design.

Using high-strength steel (S460) in columns, as seen in Options 10, 12, and 14, provided reductions in total embodied carbon. Closed-section columns (CHS and SHS) perform comparably to open-section UC columns but offer potential benefits in stability and material efficiency.

Figure 5.27 presents the estimated embodied carbon for different structural configurations, considering both upfront emissions (Modules A1–A5) and total emissions up to end-of-life (Modules A–C). The results show a clear trend where shorter spans and optimized material selection significantly reduce embodied carbon. Figures 5.28 and 5.29 present a comparative analysis of the embodied carbon intensity and total embodied carbon across different structural configurations.

Table 5.11 Sensitivity analysis results of structural steel buildings.

Opt.	Span (m)	Comp. action	Prim. beams span	Column grade	Prim. beams grade	Sec. beams grade	Columns section type	EC intensity ($kgCO_2e/m^2$)		Building total EC (tCO_2e) (A–C)	EC intensity % change from Opt. 1 (A–C)
								(A–C)	(A1–A5)		
1.	15 × 7.5	NO	Long	S355	S275	S275	UC	229.26	221.00	4642.52	—
2.	15 × 7.5	Sec. beams	Long	S355	S275	S275	UC	220.73	212.56	4469.84	−3.72%
3.	15 × 7.5	Sec. beams	Short	S355	S275	S275	UC	193.89	185.98	3926.27	−15.43%
4.	15 × 7.5	Sec. beams	Short	S355	S355	S275	UC	191.78	183.89	3883.50	−16.35%
5.	7.5 × 7.5	NO	—	S355	S275	S275	UC	162.73	156.36	3295.28	−29.02%
6.	7.5 × 7.5	NO	—	S355	S355	S355	UC	158.13	151.81	3202.13	−31.03%
7.	7.5 × 7.5	NO	—	S275	S275	S275	UC	171.95	165.49	3481.95	−25.00%
8.	7.5 × 7.5	Sec. beams	—	S355	S275	S275	UC	154.94	148.65	3137.54	−32.42%
9.	7.5 × 7.5	Sec. beams	—	S355	S355	S355	UC	146.27	140.06	2962.02	−36.20%
10.	7.5 × 7.5	Sec. beams	—	S460	S355	S355	UC	141.78	135.62	2871.05	−38.16%
11.	7.5 × 7.5	Sec. beams	—	S355	S355	S355	CHS	147.26	141.25	2985.04	−35.77%
12.	7.5 × 7.5	Sec. beams	—	S460	S355	S355	CHS	142.14	136.01	2878.34	−38.00%
13.	7.5 × 7.5	Sec. beams	—	S355	S355	S355	SHS	147.82	141.65	2993.36	−35.52%
14.	7.5 × 7.5	Sec. beams	—	S460	S355	S355	SHS	144.15	138.01	2919.04	−37.12%

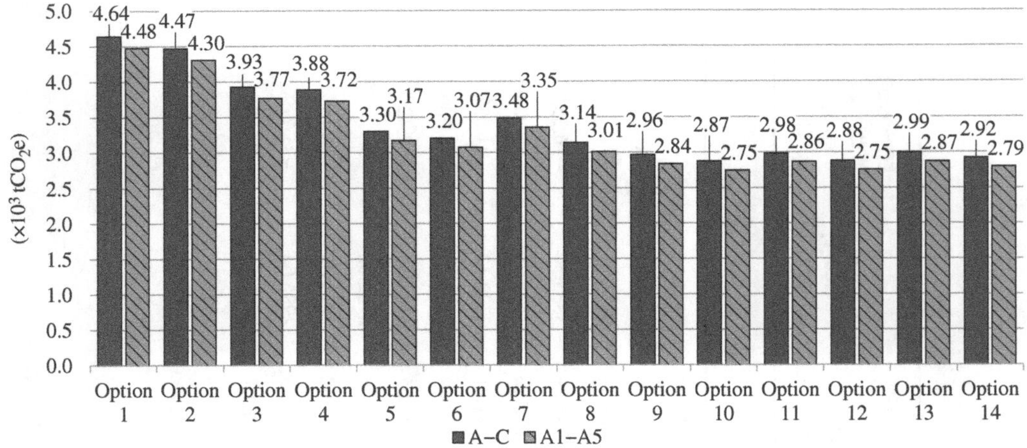

Figure 5.27 Modules A1–A5 and A–C estimated embodied carbon of the different options.

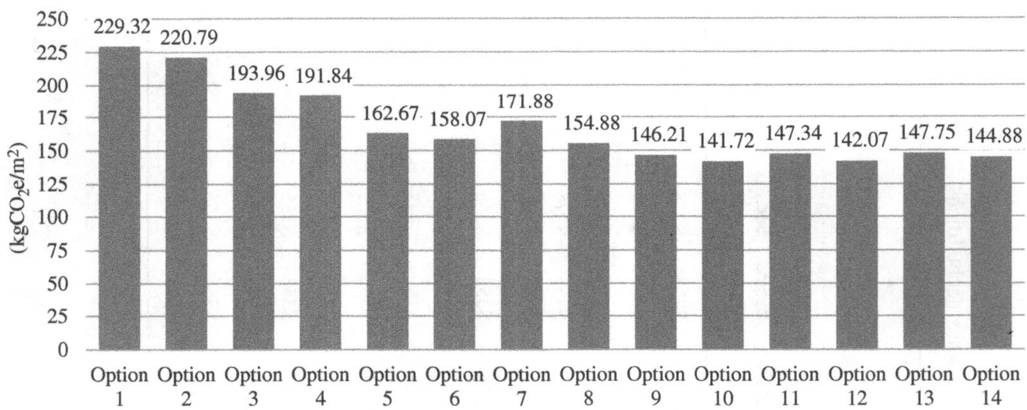

Figure 5.28 Estimated Modules A–C embodied carbon intensity of different options.

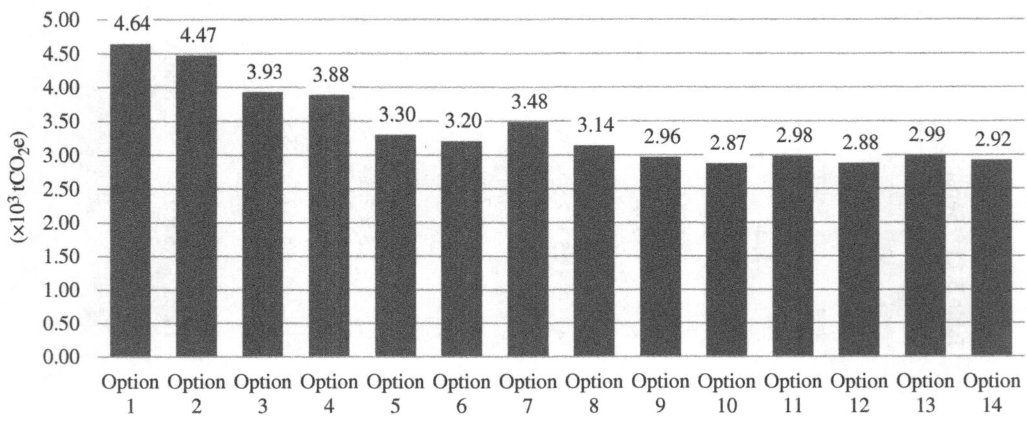

Figure 5.29 Estimated Modules A–C total embodied carbon of different options 1–14.

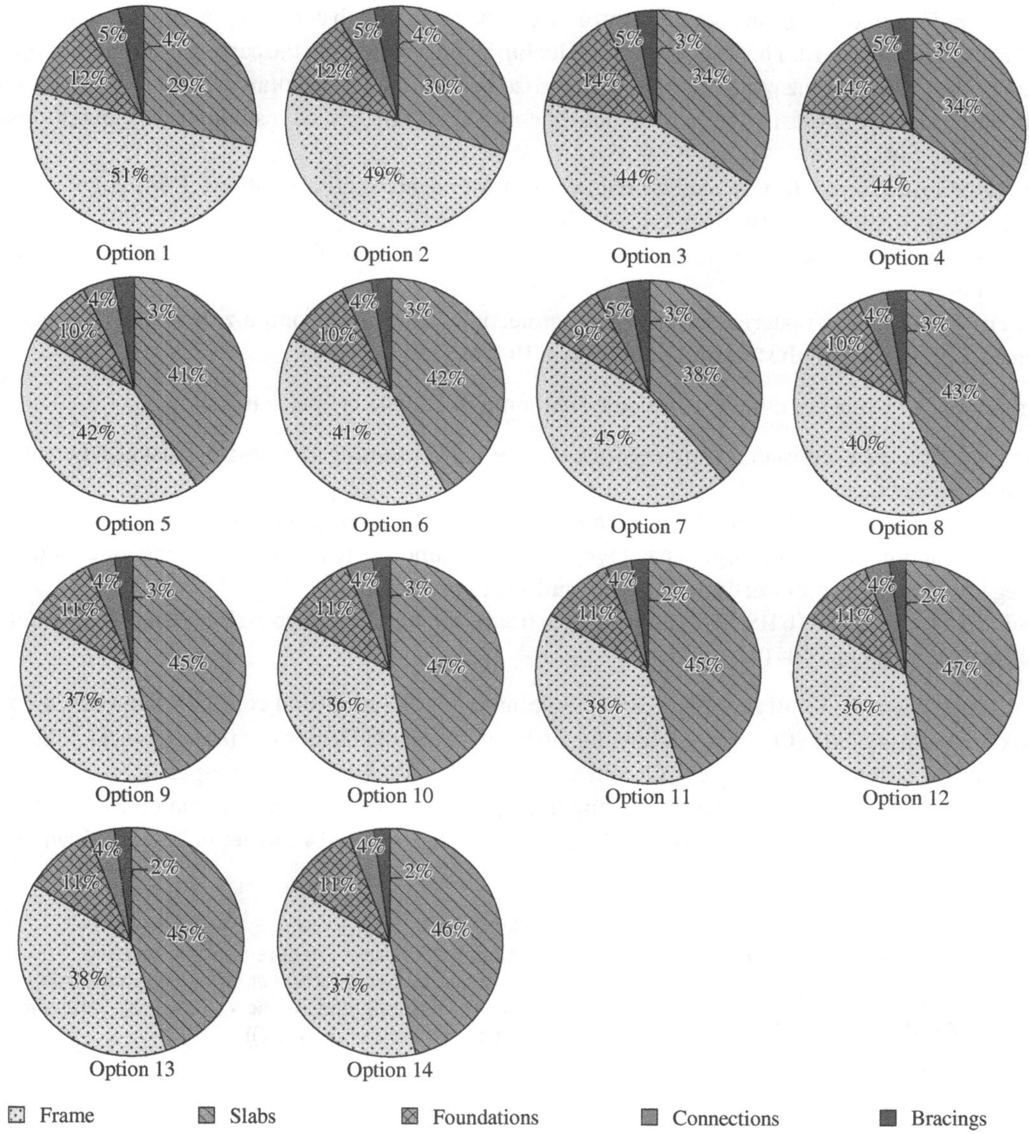

Figure 5.30 Element emission contribution (%) of the different options (A–C), options 1–14.

Figure 5.30 presents the breakdown of embodied carbon contributions from different structural elements across the 14 design options (the elements considered include frames, slabs, foundations, connections, and bracings). We can state some key points as follows:

- The structural frame accounts for the largest share of embodied carbon for the options 1, 2, 3, 4, 5, and 7, ranging from 36% to 51%.
- Options 1 and 2 (15 m span, noncomposite and composite beams, respectively) have the highest frame-related emissions (49–51%) due to their longer span requiring heavier structural members.

- As spans reduce and composite action is introduced, the relative slab contribution increases, reaching 47% in Options 10–12. This suggests that reducing beam spans shifts the carbon burden from the frame to the slabs, emphasizing the need to optimize both elements simultaneously.
- Foundations contribute 9–14% of total embodied carbon, with higher values in longer-span options (e.g. Options 1–4).
- The reduction in foundation-related emissions in shorter-span options (Options 9–14) aligns with the overall decrease in structural loads due to lighter frames.
- Connections and bracings represent small yet consistent contributions (~4–5%) across all configurations.
- Efficient detailing and material selection for connections can further optimize embodied carbon, especially in high-strength steel designs (Options 10–14).

In terms of design strategies aimed at embodied carbon reductions, the pie charts show:

- Span reduction shifts emissions from the frame to the slabs, meaning slab design must also be optimized alongside beam spans.
- There is an opportunity of embodied carbon reduction if the span length can be reduced.
- Composite action allows for larger spans with reduced frame-related carbon, offering an effective strategy for balancing architectural flexibility and sustainability.
- Closed-section columns (CHS, SHS) help balance frame and foundation emissions, leading to better distribution in Options 10–14.

Table 5.12 and Figure 5.31 present the SCORS ranking for the 14 structural configurations based on Modules A1–A5 emissions ($kgCO_2e/m^2$), following the IStructE SCORS framework (Gibbons et al., 2022). The top-rated (A) options, including 10, 12, and 14, feature 7.5 × 7.5 m spans, composite beams, and high-strength steel (S460) columns, demonstrating that shorter spans and optimized materials improve carbon performance. Closed-section columns (CHS, SHS) in Options 12–14 further enhance efficiency.

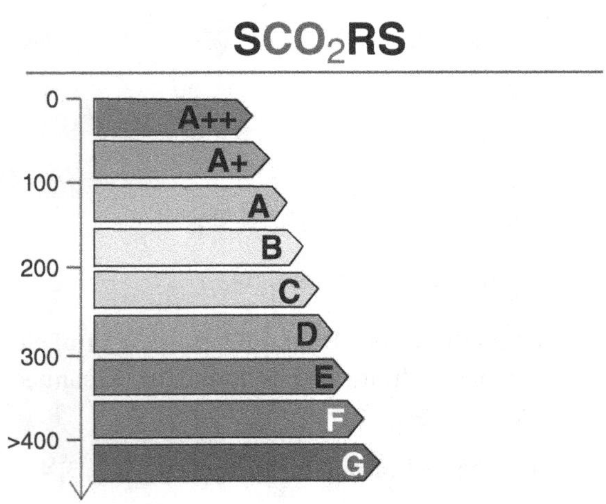

Figure 5.31 SCORS rating (based on total A1–A5 emissions for superstructure plus substructure, excluding biogenic carbon or offsetting, in accordance with the IStructE guide on how to calculate embodied carbon (Gibbons et al., 2022)).

Table 5.12 Classification of the 14 studied options according to SCORS ranking.

Scheme rankings		
Rank	**Scheme**	**SCORS rating & A1–A5 Emissions ($kgCO_2e/m^2$)**
1st	Option 10	A (135.62)
2nd	Option 12	A (136.01)
3rd	Option 14	A (138.01)
4th	Option 9	A (140.06)
5th	Option 11	A (141.25)
6th	Option 13	A (141.65)
7th	Option 8	A (148.65)
8th	Option 6	B (151.81)
9th	Option 5	B (156.36)
10th	Option 7	B (165.49)
11th	Option 4	B (183.89)
12th	Option 3	B (185.98)
13th	Option 2	C (212.56)
14th	Option 1	C (221.00)

The B-rated options, such as 5–7 and 3–4, could improve with composite action or high-strength steel. The C-rated designs (Options 1 and 2) rely on long-span, noncomposite beams, leading to higher embodied carbon. This ranking shows that composite action, span optimization, and high-strength materials can be key to achieving better SCORS ratings, reinforcing the importance of strategic structural choices for sustainable construction.

In summary, the results show the importance of structural optimization in minimizing embodied carbon. By adjusting span lengths, material grades, and beam-column configurations, designers can significantly reduce the environmental footprint of steel buildings while maintaining structural integrity.

5.5 Influence of the Carbon Factors on the Final Results

This section inspects several aspects that, when modified, contribute to a significant change in the embodied carbon of the building. While the analysis focuses on structural steel as the primary variable, the methodology is equally applicable to other structural materials. For concrete, similar sensitivity studies could examine how variations in cement type, supplementary cementitious material content, or aggregate sourcing affect embodied carbon outcomes. For timber, the carbon factor can vary substantially depending on species, forest management practices, transportation distances, and the treatment or engineering processes involved (e.g. glulam versus CLT versus sawn timber). The principles demonstrated here (isolating a single parameter and observing its effect on total building emissions) provide a transferable framework that designers can adapt to any material system, enabling informed decisions regardless of the structural typology chosen. The studied aspects are:

- The effect of the carbon factor of the material (structural steel).
- Sourcing of the steel (related to the manufacturing stage A1–A3).
- The transportation distance (stage A4).
- The complexity of connections (stage A5).

We will consider the Option 9 (7.5 × 7.5 m, comp. action (sec. beams), open sec. col., S355 (beams and col.)) as benchmark.

5.5.1 Range of Generic Carbon Factors

This section examines the variation in embodied carbon intensity and total embodied carbon of the "Option 9" by explicitly assigning different carbon factors to structural steel: 0.5, 1.0, 1.5, 2.0, and 2.5 $kgCO_2e/kg$ (Modules A–C). These values were deliberately chosen as round, generic figures rather than product-specific data to interpolate and compare results. Such a range can reflect potential differences in production methods, energy sources, and processing efficiencies, which can alter the steel's carbon footprint (in practice such values are defined in databases or EPDs as explained in Chapters 2 and 3). The outputs are presented in Table 5.13 and Figures 5.32–5.34.

As the steel carbon factor increases from 0.5 to 2.5 $kgCO_2e/kg$, EC intensity rises from 100.71 to 168.28 $kgCO_2e/m^2$, highlighting the sensitivity of structural steel's footprint.

Figure 5.32 shows how the distribution of embodied carbon among structural elements shifts with increasing steel carbon factors. At 0.5 $kgCO_2e/kg$, slabs dominate emissions at 67%, while the frame contributes only 15%. As the steel carbon factor rises, the frame's share increases to 45% (from 15%) in the 2.5 $kgCO_2e/kg$ case, while the slab contribution drops to 39% (from 67%). Foundations contribute less as the steel factor rises, while connections and bracings maintain a small share. These results show the

Table 5.13 Building embodied carbon intensity and total embodied carbon as a factor structural steel carbon factor.

Option	EC intensity ($kgCO_2e/m^2$)	Building total EC (tCO_2e)
Steel 0.5 A–C	100.71	2039.45
Steel 1.0 A–C	117.57	2380.74
Steel 1.5 A–C	134.47	2722.98
Steel 2.0 A–C	151.37	3065.30
Steel 2.5 A–C	168.28	3407.62

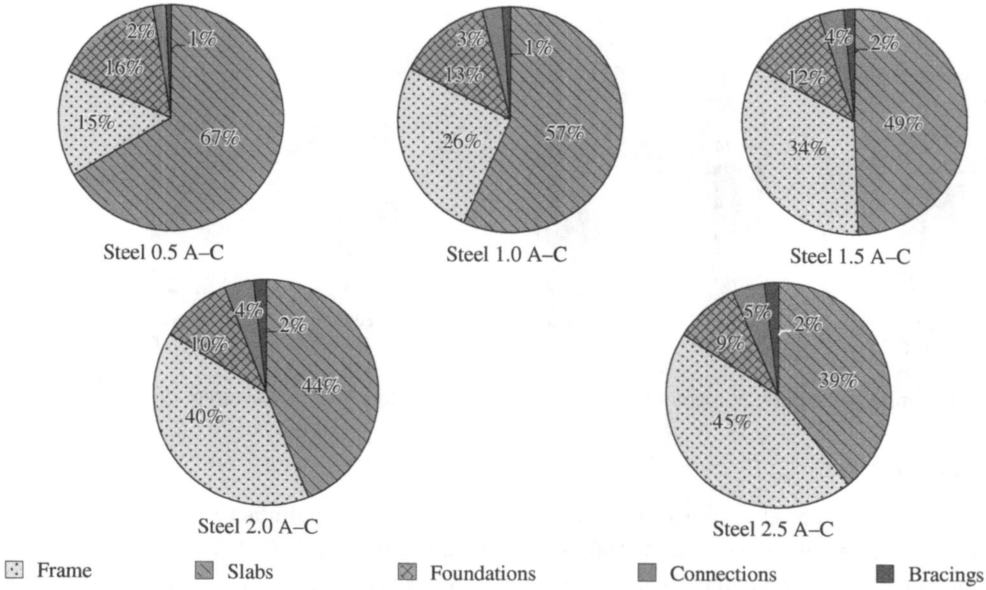

Figure 5.32 Element emission contribution (%) of the different options (A–C) with different steel carbon factor assumptions.

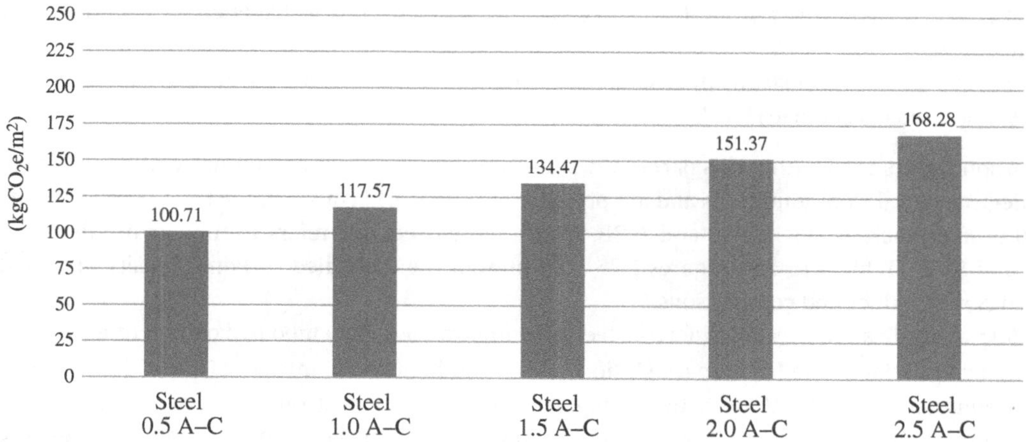

Figure 5.33 Estimated Modules A–C embodied carbon intensity of different options with changing steel carbon factors.

importance of sourcing lower-carbon steel, and of optimizing structural design to use steel efficiently, as both strategies can significantly reduce the structure's contribution to overall building emissions.

Presenting these results quantifies the direct impact of varying carbon factors on the overall emissions, reinforcing the sensitivity of the system to this parameter.

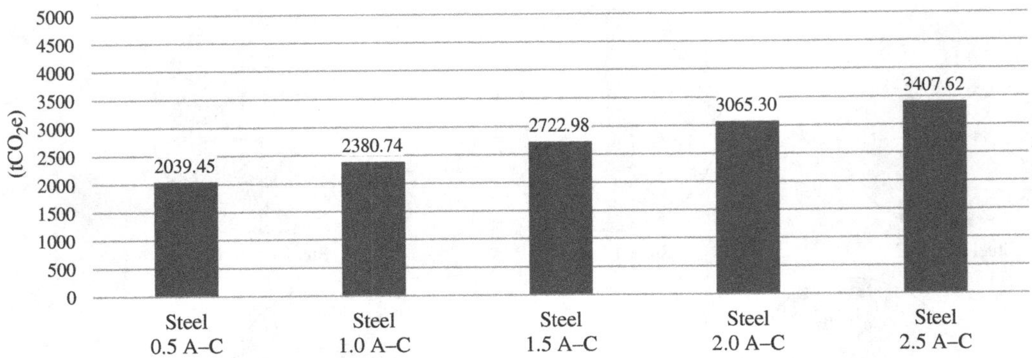

Figure 5.34 Total embodied carbon of Modules A–C with different steel carbon factor assumptions.

5.5.2 Material Sourcing

In Chapter 2, we have observed how variations in material properties, manufacturing processes, and sourcing strategies influence the environmental performance of structural steel. To show the impact of such parameters on the embodied carbon results, four conditions are put in evidence here, each one leading to the change in steel carbon factors:

Option 9.1. Benchmark (known as Option 9 of the previous studies and representing a recommended default value averaging the UK production).

Option 9.2. 72% Postconsumer scrap, 12.4% preconsumer scrap, 0% renewable energy used during the manufacturing

Option 9.3. 100% Pre- and postconsumer scrap, 100% renewable energy used during manufacturing.

Option 9.4. Reused (Reclaimed) Steel.

The carbon factors used within this part of the study are extracted from real industry EPDs reflecting the studied structural steel conditions and are presented in Table 5.14 (in table, the term "open section" refers to profiles such as IPE, HEA, and HEB, while "closed section" refers to tubular profiles such as CHS and SHS). Table 5.15 and Figures 5.35–5.38 present the embodied carbon intensity and total embodied carbon values and comparisons.

Table 5.15 quantifies the effect of different steel sourcing strategies on embodied carbon intensity and total building emissions. The benchmark Option 9.1 has the highest EC intensity (146.27 $kgCO_2e/m^2$) and total emissions (2961.97 tCO_2e). Increasing the proportion of postconsumer scrap (Option 9.2) reduces emissions by approximately 16%, while adopting 100% recycled content with renewable energy (Option 9.3) leads to a 33% reduction. The lowest emissions occur in Option 9.4 (reclaimed steel), with EC intensity dropping to 90.50 $kgCO_2e/m^2$, reflecting a 38% reduction compared to the benchmark. These results highlight the carbon savings associated with recycled and reclaimed steel, underlining the importance of material circularity in structural design.

Figure 5.35 shows the impact of different steel sourcing strategies on the distribution of embodied carbon across structural elements. In Option 9.1 (benchmark steel mix), the frame accounts for 37% of emissions, while slabs contribute 45%. As the proportion of recycled content and renewable energy use increases, the structural frame's contribution decreases, shifting more emissions toward slabs. In

Table 5.14 Carbon factors used ($kgCO_2e/kg$).

	Concrete		Steel							
			Open sections				Closed sections		Rebar	Galvanized sheet
Stages	C30/37	GEN1	Option 9.1	Option 9.2	Option 9.3	Option 9.4	Options 9.1 & 9.4	Options 9.2 & 9.3		
A1–A3	0.10	0.103	1.74	1.13	0.333	0.0466	2.5	0.646	0.76	2.76
A4	0.0055	0.005	0.032	0.032	0.032	0.032	0.032	0.032	0.032	0.183
WF	0.053	0.053	0.01	0.01	0.01	0.01	0.01	0.01	0.053	0.01
A5w	0.006	0.007	0.0179	0.012	0.004	0.001	0.026	0.007	0.043	0.03
C2	0.003	0.005	0.005	0.005	0.005	0.0043	0.005	0.0263	0.005	0.005
C3/C4	0.0003	0.013	0.013	0.0018	0.0017	0.0050	0.013	0.0014	0.013	0.013
D	−0.0228	0.000	−0.920	−0.413	0.214	0.000	−1.53	−0.108	0.351	−1.320

Table 5.15 Building embodied carbon intensity and total embodied carbon as a factor structural steel initial condition (A–C).

Option	EC intensity ($kgCO_2e/m^2$)	Building total EC (tCO_2e)
Option 9.1	146.27	2961.97
Option 9.2	123.18	2494.40
Option 9.3	97.06	1965.41
Option 9.4	90.50	1832.63

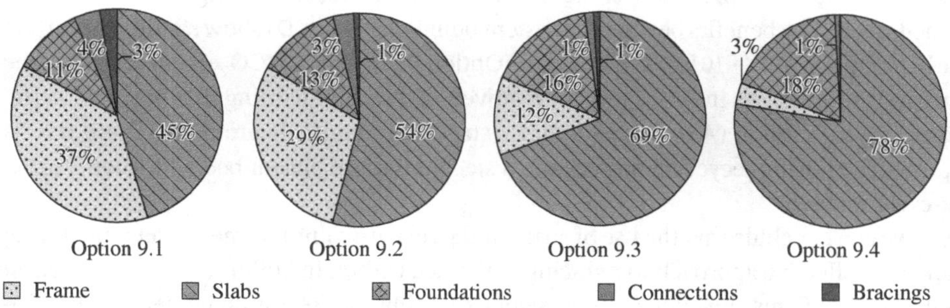

Figure 5.35 Element emission contribution (%) of the different options (A–C) options 9.1–9.4.

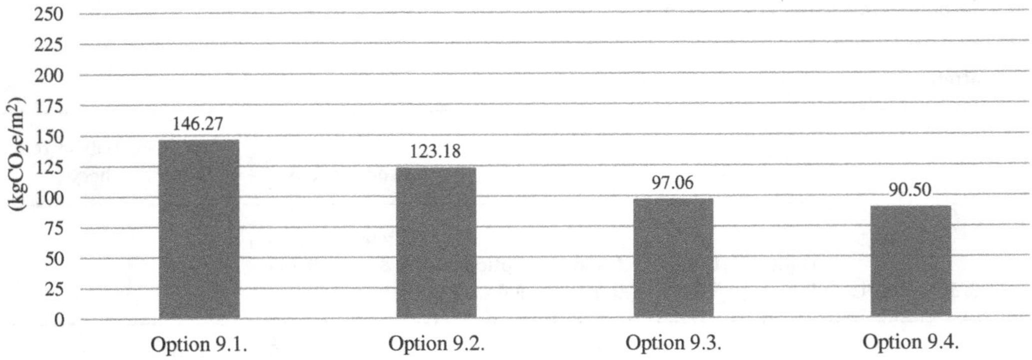

Figure 5.36 Estimated Modules A–C Embodied carbon intensity of different options 9.1–9.4.

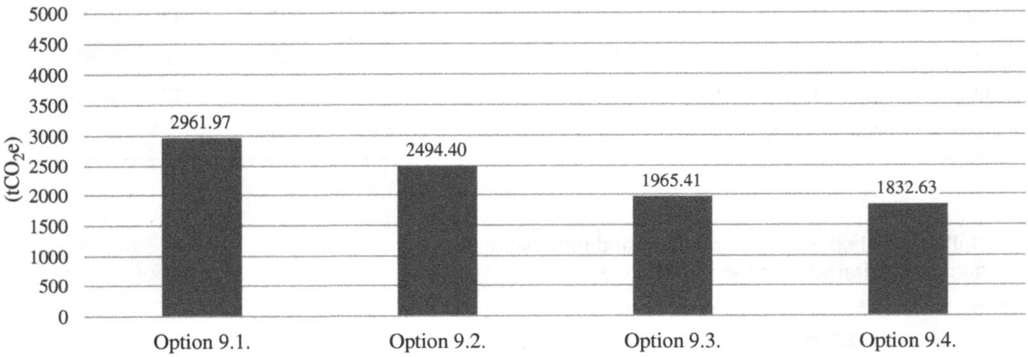

Figure 5.37 Total embodied carbon of Modules A–C options 9.1–9.4.

Option 9.3 (100% recycled steel, renewable energy), slab emissions increase to 69%, while the frame's share drops to 12%. The effect is even more pronounced in Option 9.4 (reclaimed steel), where slabs dominate at 78%, and the frame accounts for only 3%.

Figure 5.38 presents the emissions distribution across life-cycle stages for different steel sourcing scenarios. Option 9.1 (benchmark) has the highest emissions in A1–A3 (2661.10 tCO_2e) due to BF-BOF steel production. As the recycled content increases, emissions in A1–A3 decrease, reaching 1475.27 tCO_2e in Option 9.4 (reclaimed steel). The benefits beyond the system boundary (Stage D) show significant carbon savings, particularly in Option 9.2 (−1013.72 tCO_2e) and Option 9.3 (−656.37 tCO_2e), emphasizing the impact of steel circularity. Emissions in A4–C remain relatively stable, highlighting that material sourcing primarily affects upfront carbon (A1–A3), while later stages see minimal variations. These results underline the importance of using recycled and reclaimed steel in minimizing embodied carbon, particularly in early life-cycle stages.

From this section, we can conclude that the use of structural steel with minimal energy demand during manufacturing can be an effective approach to reducing embodied carbon in buildings. The comparison of the four studied cases confirms this, with reused steel exhibiting the lowest embodied carbon, followed by 100% recycled steel produced with renewable energy, then partially recycled steel, and finally the benchmark case with conventional production methods.

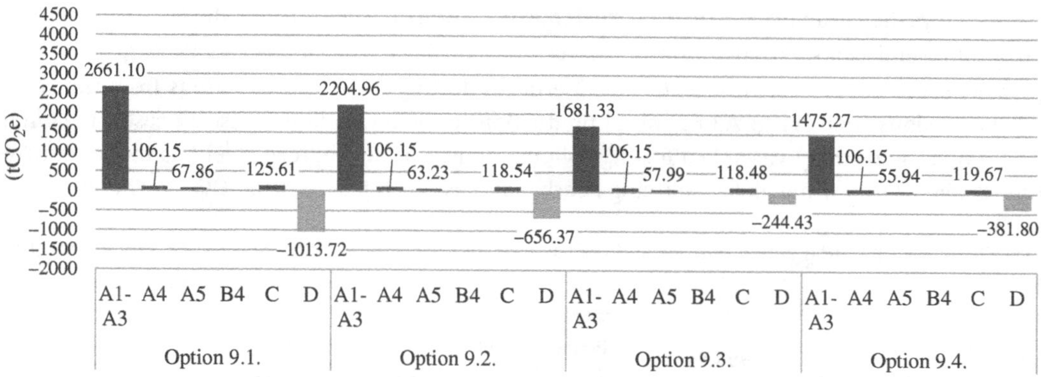

Figure 5.38 Emission by life-cycle stages (tCO_2e).

Reused steel almost completely eliminates the need for new material processing, significantly reducing energy consumption. Similarly, using 100% recycled steel within a renewable-energy-powered manufacturing process minimizes emissions by avoiding raw material extraction and reducing the impact of stage A1.

In contrast, lower recycled content increases demand for virgin steel, raising energy consumption and emissions. However, even partial recycling reduces the carbon footprint compared to conventional production. Since the manufacturing phase (A1–A3) is the most carbon-intensive, optimizing steel sourcing is key to lowering a building's overall embodied carbon.

It is also important to be aware of the challenges associated with these approaches. While reused steel offers the lowest embodied carbon, its widespread adoption depends on the availability of reclaimed structural elements that meet design and regulatory requirements. Material traceability, mechanical properties, and connection detailing may introduce extra complexities, requiring careful assessment before reuse. Similarly, although fully recycled steel produced with renewable energy significantly reduces emissions, its feasibility at scale is constrained by the availability of scrap material and the capacity of renewable energy infrastructure. High demand for scrap steel could drive up costs or limit supply, affecting the economic viability of this option.

Therefore, while optimizing steel sourcing is an effective strategy for reducing embodied carbon, it should be considered alongside other design choices, such as minimizing material quantities through efficient structural design, integrating hybrid material solutions, and exploring alternative construction methods.

5.5.3 Transportation Distances

Following the previous case study (Option 9), the transportation distance from the manufacturing mill to the construction site has been changed, affecting the carbon factor of stage A4. The evaluated options are:

- Locally (50 km) with a carbon factor of 0.005 $kgCO_2e/kg$,
- Nationally (300 km) with a carbon factor of 0.032 $kgCO_2e/kg$,
- European (1500 km) by road with a carbon factor of 0.161 $kgCO_2e/kg$,
- Globally (10 000 km) by sea with a carbon factor of 0.183 $kgCO_2e/kg$.

The outputs are presented in Table 5.16 and Figures 5.39–5.41.

Table 5.16 shows the influence of structural steel transportation distance on embodied carbon intensity and total building emissions. Locally sourced steel results in the lowest EC intensity (145.08 $kgCO_2e/m^2$), while globally sourced steel reaches 150.64 $kgCO_2e/m^2$, reflecting the added emissions from transportation. The emissions from A1 to A3 (material production) remain constant across all cases, meaning transportation is the primary variable affecting A4 emissions. Locally sourced steel has the lowest A4 emissions (83.66 tCO_2e), while global sourcing nearly doubles this figure (195.03 tCO_2e). Therefore, we

Table 5.16 Results of building embodied carbon (A–C) as a factor of transportation distance of structural steel to the site.

Transportation distance	EC intensity ($kgCO_2e/m^2$)	Building Total EC (tCO_2e)	Emission by life-cycle stages (tCO_2e)				
			A1–A3	A4	A5	C	D
Locally	145.08	2937.84	2661.10	83.66	67.46	125.61	−1013.72
Nationally	146.21	2960.73	2661.10	106.15	67.86	125.61	−1013.72
European	150.63	3050.22	2661.10	194.76	68.74	125.61	−1013.72
Globally	150.64	3050.50	2661.10	195.03	68.75	125.61	−1013.72

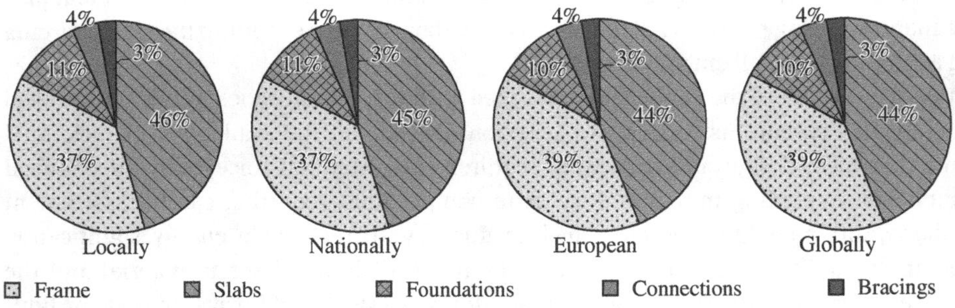

Figure 5.39 Element emission contribution (%) of the different options (A–C).

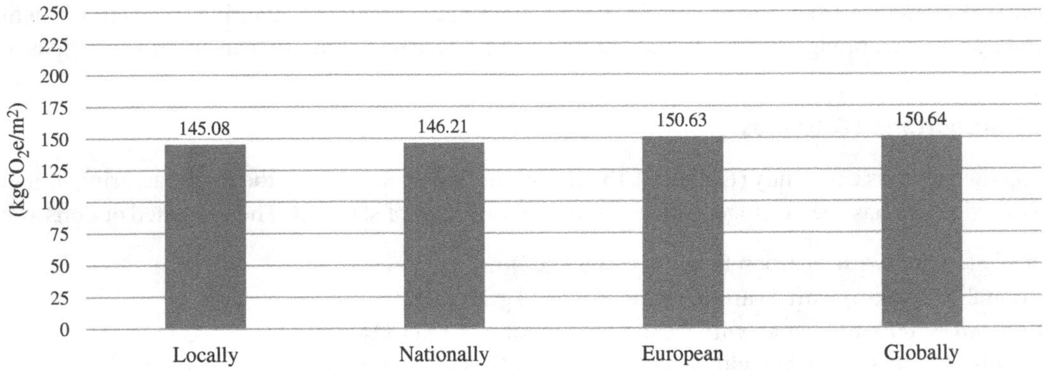

Figure 5.40 Estimated Modules A–C embodied carbon intensity of different options with different transport assumptions.

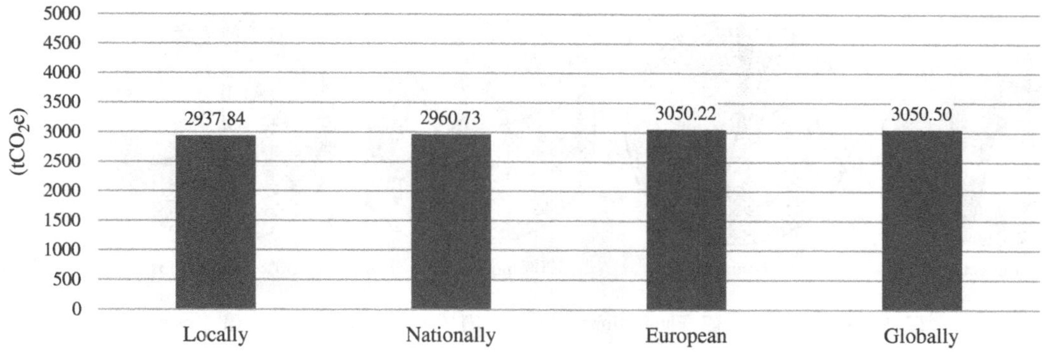

Figure 5.41 Total embodied carbon of Modules A–C with different transport assumptions.

can conclude that minimizing transportation distances can effectively reduce embodied carbon, reinforcing the importance of prioritizing locally or nationally sourced steel in sustainable construction.

Figure 5.40 shows how transportation distance affects the distribution of embodied carbon across structural components. Across all cases, slabs remain the dominant contributor (44–46%), followed by the frame (37–39%), foundations, and smaller contributions from connections and bracings. These results show that while transportation emissions influence total embodied carbon, they have a limited effect on the proportional distribution among structural elements. Prioritizing local or national steel sourcing remains beneficial for minimizing emissions, especially in transportation-heavy phases (A4).

5.5.4 Connection Complexity

In structural steel construction, connections (bolts, welds, plates, and other joining elements) are often estimated as a percentage of the total structural steel weight rather than being calculated individually. To conduct this analysis, this connection percentage is varied from 5% to 30%, representing a range from simple, standardized connections to highly complex, custom-fabricated joints, while all other components are kept constant. The outputs are shown in Table 5.17 and Figures 5.42–5.44, demonstrating how the complexity of structural steel connections influences embodied carbon intensity and total emissions. As the percentage of connections increases from 5% to 30%, EC intensity rises from 143.54 to 156.87 $kgCO_2e/m^2$, with total building emissions following the same trend. The A1–A3 phase (material production) shows the most significant increase, indicating that connection complexity

Table 5.17 Results of building embodied carbon as a factor of the complexity of the structural steel connection.

Option	EC intensity ($kgCO_2e/m^2$)	Building total EC (tCO_2e)	Emission by life-cycle stages (tCO_2e)				
			A1–A3	A4	A5	C	D
5% Connections	143.54	2906.74	2609.14	105.20	67.32	125.08	−986.25
10% Connections	146.21	2960.73	2661.10	106.15	67.86	125.61	−1013.72
20% Connections	151.54	3068.71	2765.03	108.06	68.93	126.69	−1068.67
30% Connections	156.87	3176.69	2868.96	109.98	70.00	127.76	−1123.62

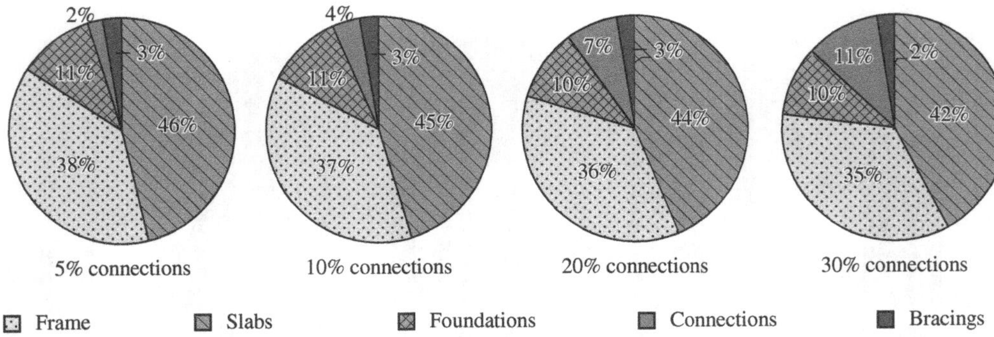

Figure 5.42 Element emission contribution (%) of the different options (A–C) with different connection complexity.

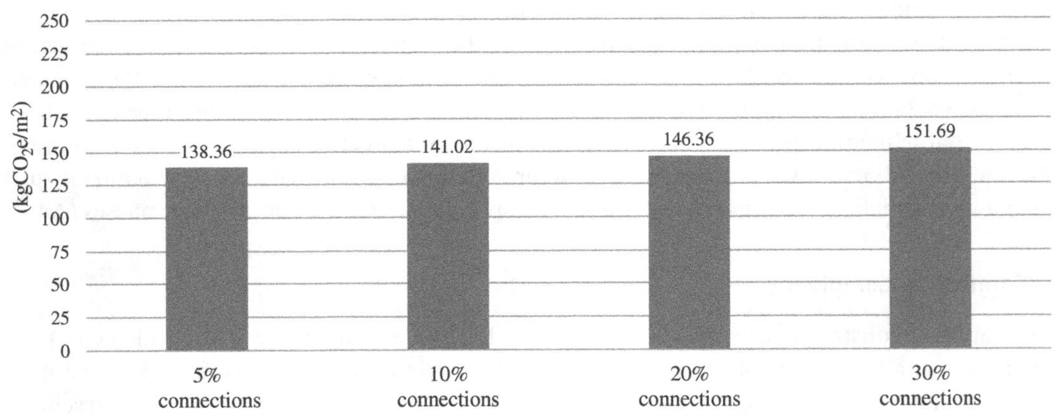

Figure 5.43 Estimated Modules A–C embodied carbon intensity of different options with different connection complexity.

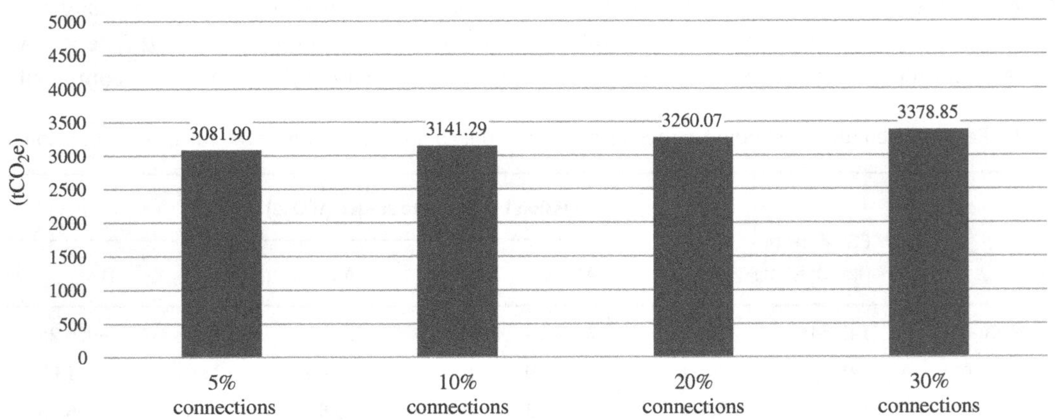

Figure 5.44 Total embodied carbon of Modules A–C.

directly affects upfront carbon emissions. Similarly, Stage D (benefits beyond the system boundary) varies, with higher connection complexity leading to larger carbon savings, likely due to greater material recovery potential. These results highlight the importance of optimizing steel connections to minimize embodied carbon while maintaining structural integrity.

Figure 5.43 shows the impact of connection complexity on the distribution of embodied carbon among structural elements. As the percentage of connections increases from 5% to 30%, the frame's share decreases from 38% to 35%, while the slab contribution drops from 46% to 42%. In contrast, the connection-related emissions increased significantly, from 2% in the 5% scenario, and to 11% in the 30% scenario, confirming the direct carbon impact of more complex connection designs. The foundation and bracing contributions remain relatively stable, showing that the main shifts occur between the frame, slabs, and connections.

5.6 What If We Use a Hybrid Approach Combining CLT Slabs with a Steel Frame?

This section explores the potential of hybrid construction, where cross-laminated timber (CLT) slabs replace conventional metal deck slabs in steel-framed structures. But why should we consider such a shift? By combining the strength and durability of steel with the lower weight and carbon-sequestering capacity of timber (although this benefit is highly contingent on sustainable forestry practices and effective end-of-life management to prevent the stored carbon from being re-released into the atmosphere), this approach can present an opportunity to reduce embodied carbon while maintaining the structural integrity. In Chapters 2 and 3, we have discussed the potential benefits of hybrid construction involving steel, timber, and reinforced concrete materials.

The study builds upon the previously analyzed 14 structural steel configurations by introducing 11 additional options that incorporate CLT slabs. These options fall into two categories: one with structural steel beams and the other with Glulam (GL28H) beams paired with steel columns. Variations in column section types (open, CHS, SHS) and steel grades (S355, S460) are considered to assess their impact on structural efficiency and embodied carbon performance. The 11 additional options, considered for the sensitivity analysis, are presented as follows:

Option 15. 7.5 × 7.5 m, open sec. col., S355 (beams and col.), CLT slabs
Option 16. 7.5 × 7.5 m, open sec. col., S355 (beams and optimal col.), CLT slab
Option 17. 7.5 × 7.5 m, open sec. col., S355 (beams), S460 (col.), CLT slab
Option 18. 7.5 × 7.5 m, CHS. col., S355 (beams and col.), CLT slab
Option 19. 7.5 × 7.5 m, CHS. col., S355 (beams), S460 (col.), CLT slab
Option 20. 7.5 × 7.5 m, SHS. col., S355 (beams and col.), CLT slab
Option 21. 7.5 × 7.5 m, open sec. col., Glulam GL28H (beams), S355 (col.), CLT slab
Option 22. 7.5 × 7.5 m, open sec. col., Glulam GL28H (beams), S460 (col.), CLT slab
Option 23. 7.5 × 7.5 m, CHS col., Glulam GL28H (beams), S355 (col.), CLT slab
Option 24. 7.5 × 7.5 m, CHS col., Glulam GL28H (beams), S460 (col.), CLT slab
Option 25. 7.5 × 7.5 m, SHS col., Glulam GL28H (beams), S355 (col.), CLT slab

Figures 5.45–5.55 and Tables 5.18 and 5.19 present the visual representation and structural elements description of the 11 additional options.

CLT slabs, structural steel beams, and structural steel columns

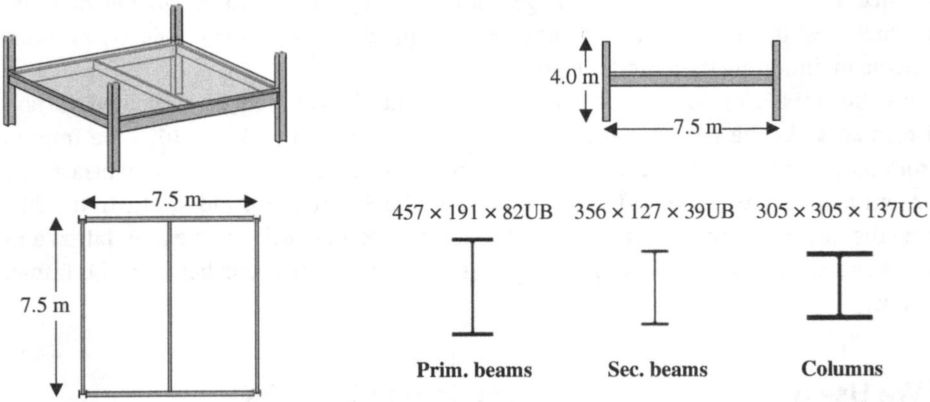

Figure 5.45 **Option 15.** 7.5 × 7.5 m, open sec. col., S355 (beams and col.), CLT slabs.

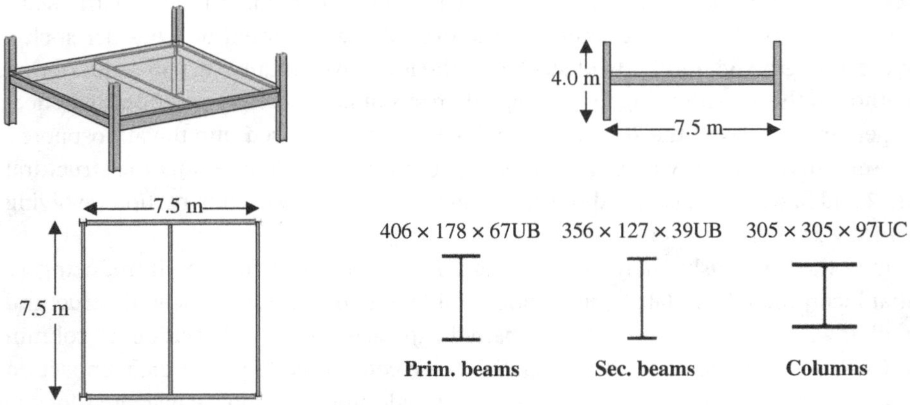

Figure 5.46 **Option 16.** 7.5 × 7.5 m, open sec. col., S355 (beams and optimal col.), CLT slab.

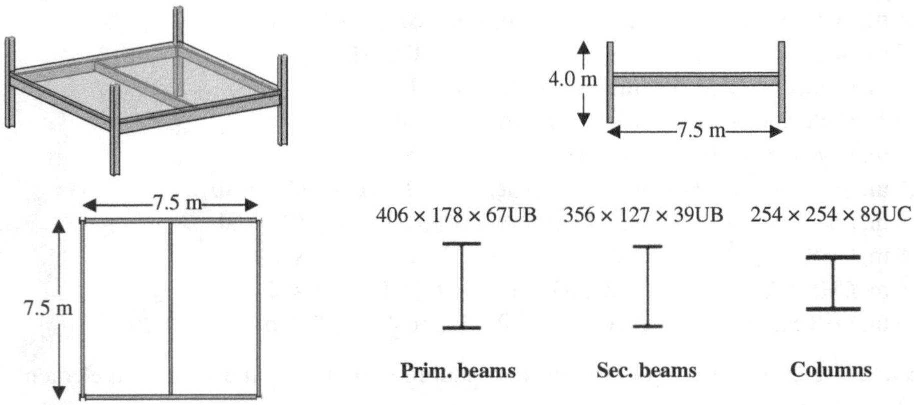

Figure 5.47 **Option 17.** 7.5 × 7.5 m, open sec. col., S355 (beams), S460 (col.), CLT slab.

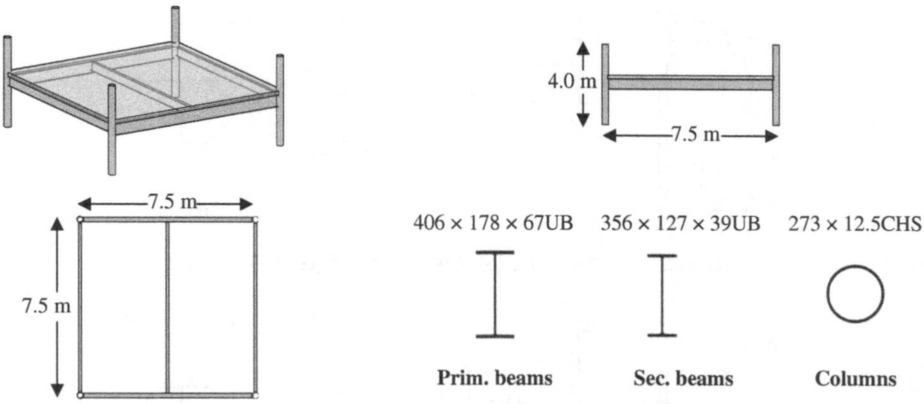

Figure 5.48 **Option 18.** 7.5 × 7.5 m, CHS. col., S355 (beams and col.), CLT slab.

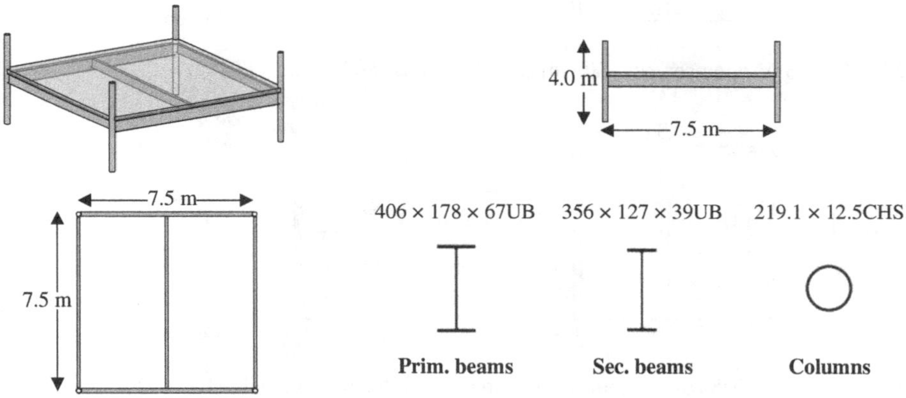

Figure 5.49 **Option 19.** 7.5 × 7.5 m, CHS. col., S355 (beams), S460 (col.), CLT slab.

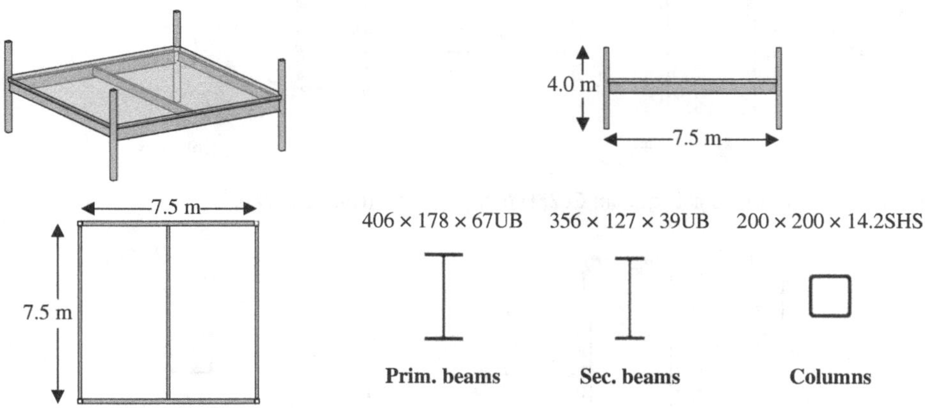

Figure 5.50 **Option 20.** 7.5 × 7.5 m, SHS. col., S355 (beams and col.), CLT slab.

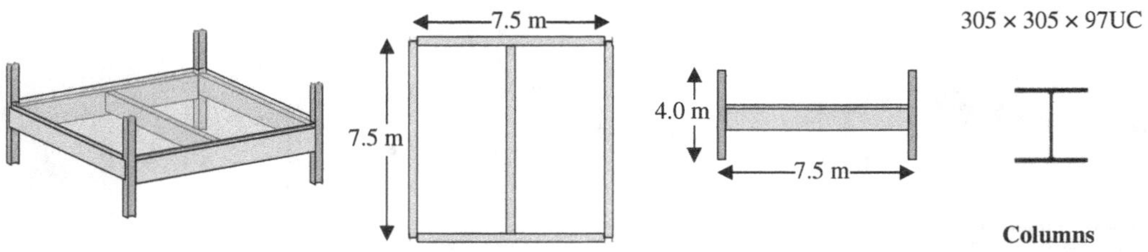

Figure 5.51 **Option 21.** 7.5 × 7.5 m, open sec. col., Glulam GL28H (beams), S355 (col.), CLT slab.

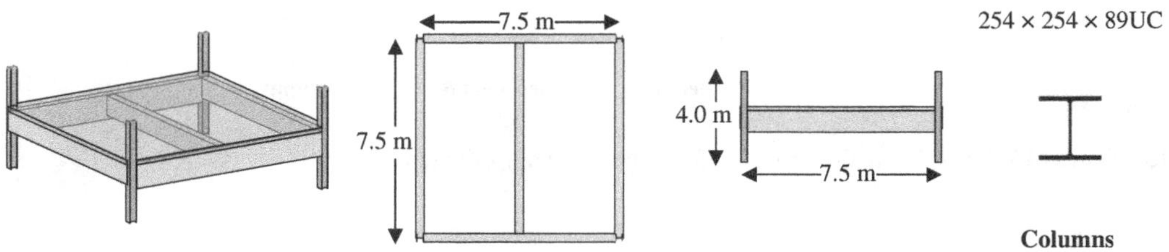

Figure 5.52 **Option 22.** 7.5 × 7.5 m, open sec. col., Glulam GL28H (beams), S460 (col.), CLT slab.

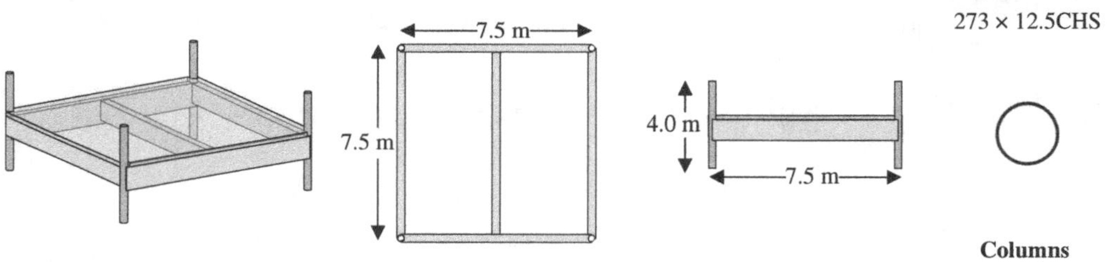

Figure 5.53 **Option 23.** 7.5 × 7.5 m, CHS col., Glulam GL28H (beams), S355 (col.), CLT slab.

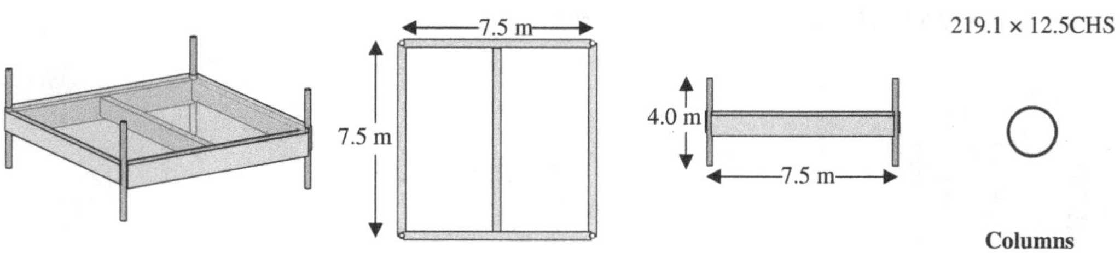

Figure 5.54 **Option 24.** 7.5 × 7.5 m, CHS col., glulam GL28H (beams), S460 (col.), CLT slab.

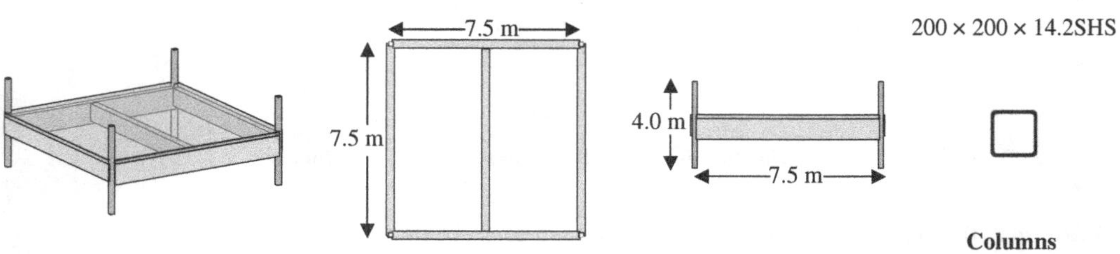

Figure 5.55 **Option 25.** 7.5 × 7.5 m, SHS col., glulam GL28H (beams), S355 (col.), CLT slab.

Table 5.18 Description of the structural elements and dimensions of the CLT slabs, structural steel beams, and steel columns options.

Option	Element	Component	Dimensions
15	Slab	Cross-laminated timber Class C24	7.5 × 7.5 m, 160 mm depth
	Beams	Primary, S355	457 × 191 × 82UB, 7.5 m, ½ on each side.
		Secondary, S355	356 × 127 × 39UB, 7.5 m, ½ on each side and 1 internal
	Columns	S355	305 × 305 × 137UC, 4 m height, ¼ at each corner
16	Slab	Cross-laminated timber Class C24	7.5 × 7.5 m, 160 mm depth
	Beams	Primary, S355	406 × 178 × 67UB, 7.5 m, ½ on each side.
		Secondary, S355	356 × 127 × 39UB, 7.5 m, ½ on each side and 1 internal
	Columns	S355	305 × 305 × 97UC, 4 m height, ¼ at each corner
17	Slab	Cross-laminated timber Class C24	7.5 × 7.5 m, 160 mm depth
	Beams	Primary, S355	406 × 178 × 67UB, 7.5 m, ½ on each side.
		Secondary, S355	356 × 127 × 39UB, 7.5 m, ½ on each side and 1 internal
	Columns	S460	254 × 254 × 89UC, 4 m height, ¼ at each corner
18	Slab	Cross-laminated timber Class C24	7.5 × 7.5 m, 160 mm depth
	Beams	Primary, S355	406 × 178 × 67UB, 7.5 m, ½ on each side.
		Secondary, S355	356 × 127 × 39UB, 7.5 m, ½ on each side and 1 internal
	Columns	S355	273 × 12.5CHS, 4 m height, ¼ at each corner
19	Slab	Cross-laminated timber Class C24	7.5 × 7.5 m, 160 mm depth
	Beams	Primary, S355	406 × 178 × 67UB, 7.5 m, ½ on each side.
		Secondary, S355	356 × 127 × 39UB, 7.5 m, ½ on each side and 1 internal
	Columns	S460	219.1 × 12.5CHS, 4 m height, ¼ at each corner
20	Slab	Cross-laminated timber Class C24	7.5 × 7.5 m, 160 mm depth
	Beams	Primary, S355	406 × 178 × 67UB, 7.5 m, ½ on each side.
		Secondary, S355	356 × 127 × 39UB, 7.5 m, ½ on each side and 1 internal
	Columns	S355	200 × 200 × 14.2SHS, 4 m height, ¼ at each corner

Table 5.19 Description of the structural elements and dimensions of the CLT slabs, glulam beams, and steel columns options.

Option	Element	Component	Dimensions
21	Slab	Cross-laminated timber Class C24	7.5 × 7.5 m, 160 mm depth
	Beams	Primary, Glulam Class GL28h	840 × 320 cm, 7.5 m, ½ on each side.
		Secondary, Glulam Class GL28h	680 × 320 cm, 7.5 m, ½ on each side and 1 internal
	Columns	S355	305 × 305 × 97UC, 4 m height, ¼ at each corner
22	Slab	Cross-laminated timber Class C24	7.5 × 7.5 m, 160 mm depth
	Beams	Primary, Glulam Class GL28h	840 × 320 cm, 7.5 m, ½ on each side.
		Secondary, Glulam Class GL28h	680 × 320 cm, 7.5 m, ½ on each side and 1 internal
	Columns	S460	254 × 254 × 89UC, 4 m height, ¼ at each corner
23	Slab	Cross-laminated timber Class C24	7.5 × 7.5 m, 160 mm depth
	Beams	Primary, Glulam Class GL28h	840 × 320 cm, 7.5 m, ½ on each side.
		Secondary, Glulam Class GL28h	680 × 320 cm, 7.5 m, ½ on each side and 1 internal
	Columns	S355	273 × 12.5CHS, 4 m height, ¼ at each corner
24	Slab	Cross-laminated timber Class C24	7.5 × 7.5 m, 160 mm depth
	Beams	Primary, Glulam Class C28h	840 × 320 cm, 7.5 m, ½ on each side.
		Secondary, Glulam Class GL28h	680 × 320 cm, 7.5 m, ½ on each side and 1 internal
	Columns	S460	219.1 × 12.5CHS, 4 m height, ¼ at each corner
25	Slab	Cross-laminated timber Class C24	7.5 × 7.5 m, 160 mm depth
	Beams	Primary, Glulam Class GL28h	840 × 320 cm, 7.5 m, ½ on each side.
		Secondary, Glulam Class GL28h	680 × 320 cm, 7.5 m, ½ on each side and 1 internal
	Columns	S355	200 × 200 × 14.2SHS, 4 m height, ¼ at each corner

Table 5.20 Carbon factors ($kgCO_2e/kg$) of the different materials and elements used in structural composite steel cases (timber slabs).

	Concrete	Steel			Timber	
Stages	**GEN1 (C10)**	**Open sections**	**Closed sections**	**Rebar**	**CLT**	**Glulam**
A1–A3 (production)	0.103	1.74	2.5	0.76	0.25	0.28
A4 (transport)	0.005	0.032	0.032	0.032	0.161	0.161
WF (waste factor)	0.053	0.01	0.01	0.053	0.010	0.01
A5w (material wastage on site)	0.007	0.0179	0.026	0.043	0.021	0.021
C2 (waste transport)	0.005	0.005	0.005	0.005	0.005	0.005
C3/C4 (waste processing and disposal)	0.013	0.013	0.013	0.013	1.662	1.662
D (end of life)	0.000	−0.920	−1.53	0.351	−0.524	−0.524
Biogenic carbon	—	—	—	—	−1.64	−1.64

Source: Adapted from Gibbons et al. (2022) and Circular Ecology (2020).

The carbon factors for timber, including Glulam and CLT materials, are shown in Table 5.20. These factors apply to the 11 additional timber slab options studied here.

The assumptions in this case are as follows:

- Concrete GEN1(C10): 25% GGBS content in cement
- Steel open sections: Average UK rolled sections
- Steel Rebar: UK 97% recycled EAF production

Figures 5.56 and 5.57 show the contribution of each life-cycle stage in the whole life-cycle assessment of Option 15, taking into consideration the biogenic carbon, excluding, and including Module D, respectively.

Without Module D, the total embodied carbon reaches 2962.02 tCO_2e, with an intensity of 146.27 $kgCO_2e/m^2$. When Module D is included, these values drop to 1488.16 tCO_2e, reducing the intensity to 73.49 $kgCO_2e/m^2$ due to the carbon benefits associated with material reuse and recycling.

Unlike what is observed in reinforced concrete and structural steel options, A1–A3 emissions are negative (−613.93 tCO_2e) due to biogenic carbon storage in CLT slabs, which offsets emissions from material production (as explained in the Chapter 2, "biogenic carbon" refers to the carbon absorbed by trees during their growth, which remains stored in timber products as long as they are in use). This reduces the upfront carbon footprint of the structure, making timber a lower-emission alternative to conventional materials.

However, C3–C4 (end-of-life stages) exhibit the highest emissions (2549.61 tCO_2e), as disposal significantly impacts total embodied carbon. In this case, these emissions are particularly high because CLT slabs, unlike steel, are subject to decomposition or incineration at the end of life, potentially releasing stored biogenic carbon back into the atmosphere as CO_2. Without proper reuse or recycling strategies,

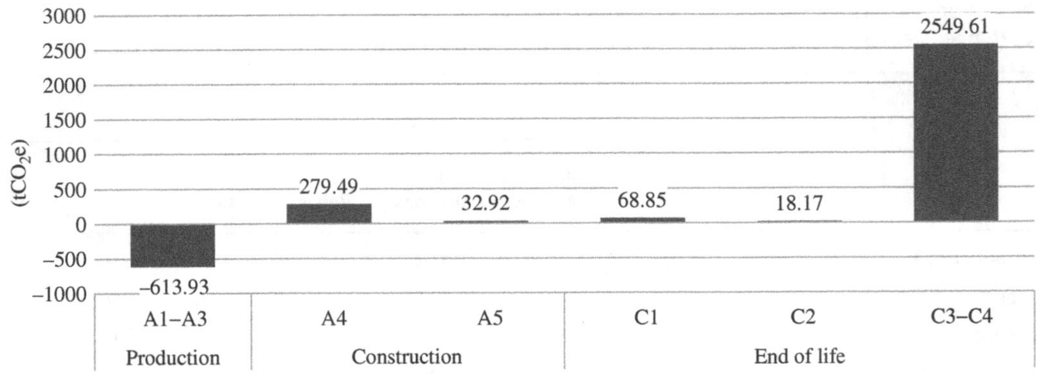

Figure 5.56 Embodied carbon emission per stage (A–D) of Option 15 without Module D.

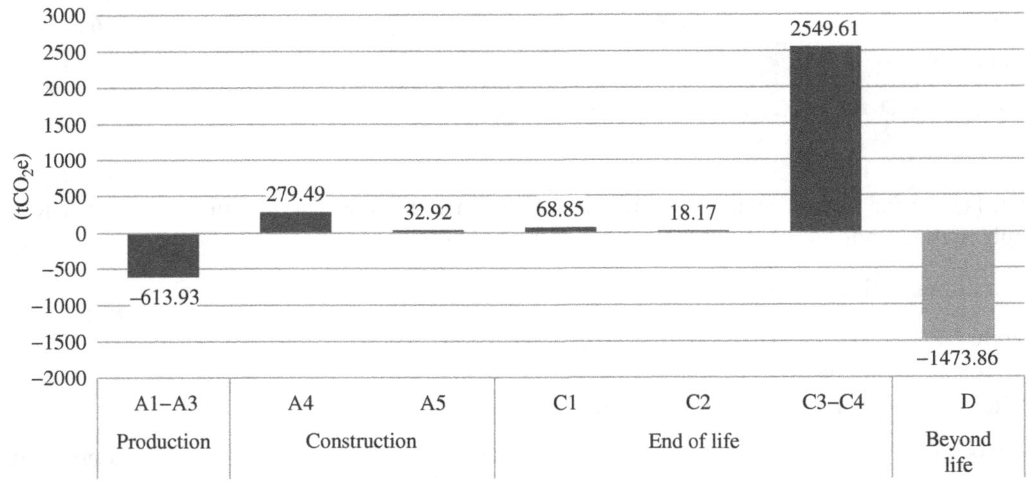

Figure 5.57 Embodied carbon emission per stage (A–D) of Option 15 with Module D.

biogenic carbon benefits achieved in A1–A3 can be offset by emissions from disposal. The incorporation of Module D (−1473.86 tCO_2e) balances these impacts, demonstrating the long-term environmental benefits of material recovery and reuse. This highlights the importance of considering the full life cycle when evaluating carbon performance in hybrid construction.

The sensitivity analysis results of all 14 options are presented in Table 5.21 and are graphically shown in Figures 5.58–5.60. Table 5.21 presents the sensitivity analysis results for composite structures combining structural steel with timber elements, focusing on embodied carbon intensity and total building emissions. Compared to Option 9 (fully steel-based structure with composite secondary beams), which has an EC intensity of 146.27 $kgCO_2e/m^2$, all hybrid options with CLT slabs show significant reductions in embodied carbon, with savings ranging from 21.19% to 43.16%.

Hybrid configurations featuring steel beams with CLT slabs (Options 15–20) achieve reductions between 21.19% and 29.22%, depending on the column section type and steel grade used. The introduction

Table 5.21 Parameter analysis results of composite structure building (structural steel–timber).

								EC intensity ($kgCO_2e/m^2$)			Building total EC (tCO_2e) (A–C)	EC intensity % change from Opt. 1 (A–C)
Opt.	Span (m)	Comp. action	Prim. beams span	Column grade	Prim. beams grade	Sec. beams grade	Columns section type	(A–C)	Slab (A–C)	(A1–A5)		
9.	7.5 × 7.5	Sec. beams	—	S355	S355	S355	UC	146.27	180.05	140.07	2962.02	—
15.	7.5 × 7.5	NO	—	S355	S355	S355	UC	115.28	92.33	108.41	2334.42	−21.19%
16.	7.5 × 7.5	NO	—	S355	S355	S355	UC	105.05	92.33	98.28	2127.26	−28.18%
17.	7.5 × 7.5	NO	—	S460	S355	S355	UC	103.85	92.33	97.09	2102.96	−29.00%
18.	7.5 × 7.5	NO	—	S355	S355	S355	CHS	106.92	92.33	100.17	2165.05	−26.90%
19.	7.5 × 7.5	NO	—	S460	S355	S355	CHS	103.53	92.33	96.81	2096.48	−29.22%
20.	7.5 × 7.5	NO	—	S355	S355	S355	SHS	107.10	92.33	100.35	2168.84	−26.78%
21.	7.5 × 7.5	NO	—	S355	GL28h	GL28h	UC	84.54	92.33	76.98	1711.91	−42.20%
22.	7.5 × 7.5	NO	—	S460	GL28h	GL28h	UC	83.36	92.33	75.82	1688.11	−43.01%
23.	7.5 × 7.5	NO	—	S355	GL28h	GL28h	CHS	86.49	92.33	81.95	1751.37	−40.87%
24.	7.5 × 7.5	NO	—	S460	GL28h	GL28h	CHS	83.14	92.33	75.63	1683.59	−43.16%
25.	7.5 × 7.5	NO	—	S355	GL28h	GL28h	SHS	86.65	92.33	79.11	1764.66	−40.76%

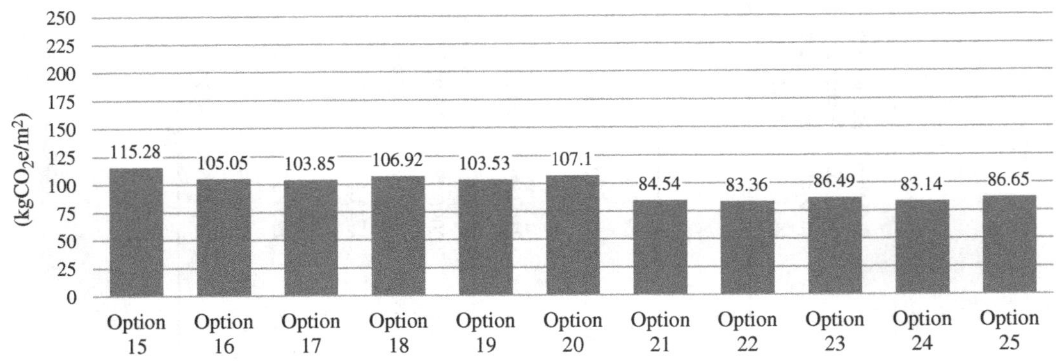

Figure 5.58 Estimated Modules A–C embodied carbon intensity of different options 15–25.

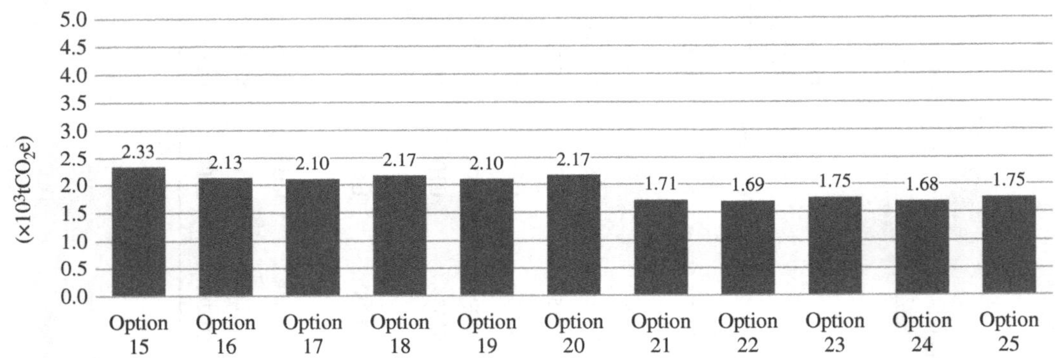

Figure 5.59 Estimated Modules A–C total embodied carbon of different options 15–25.

of Glulam beams (Options 21–25) further reduces EC intensity, with savings exceeding 40% compared to the steel-only benchmark. The lowest embodied carbon is observed in Option 22 (S460 columns with Glulam beams and CLT slabs), achieving a 43.01% reduction. This highlights that integrating timber into both slabs and beams provides the most effective carbon reduction strategy.

The results emphasize that replacing steel slabs with CLT significantly reduces embodied carbon, while the substitution of steel primary beams with Glulam enhances carbon savings further. However, these reductions must be balanced with considerations related to structural performance, fire safety, and long-term durability to ensure the feasibility of hybrid systems in practice. While the analysis shows that replacing steel beams with Glulam reduces embodied carbon in most cases, there are scenarios where keeping steel beams could be more favorable. If the structural spans increase significantly, the required cross-section of Glulam beams would need to be much larger than their steel counterparts to achieve the same load-bearing capacity. This would increase material use, offsetting the carbon savings from using timber. Besides, if high-strength steel (S460) is used, it allows for more material efficiency compared to Glulam, leading to lower embodied carbon when optimizing for long spans. Another scenario where steel beams might perform better is when recycling and reuse potential is maximized. Steel has higher recyclability at the end of life, whereas Glulam beams are often downcycled, incinerated, or landfilled,

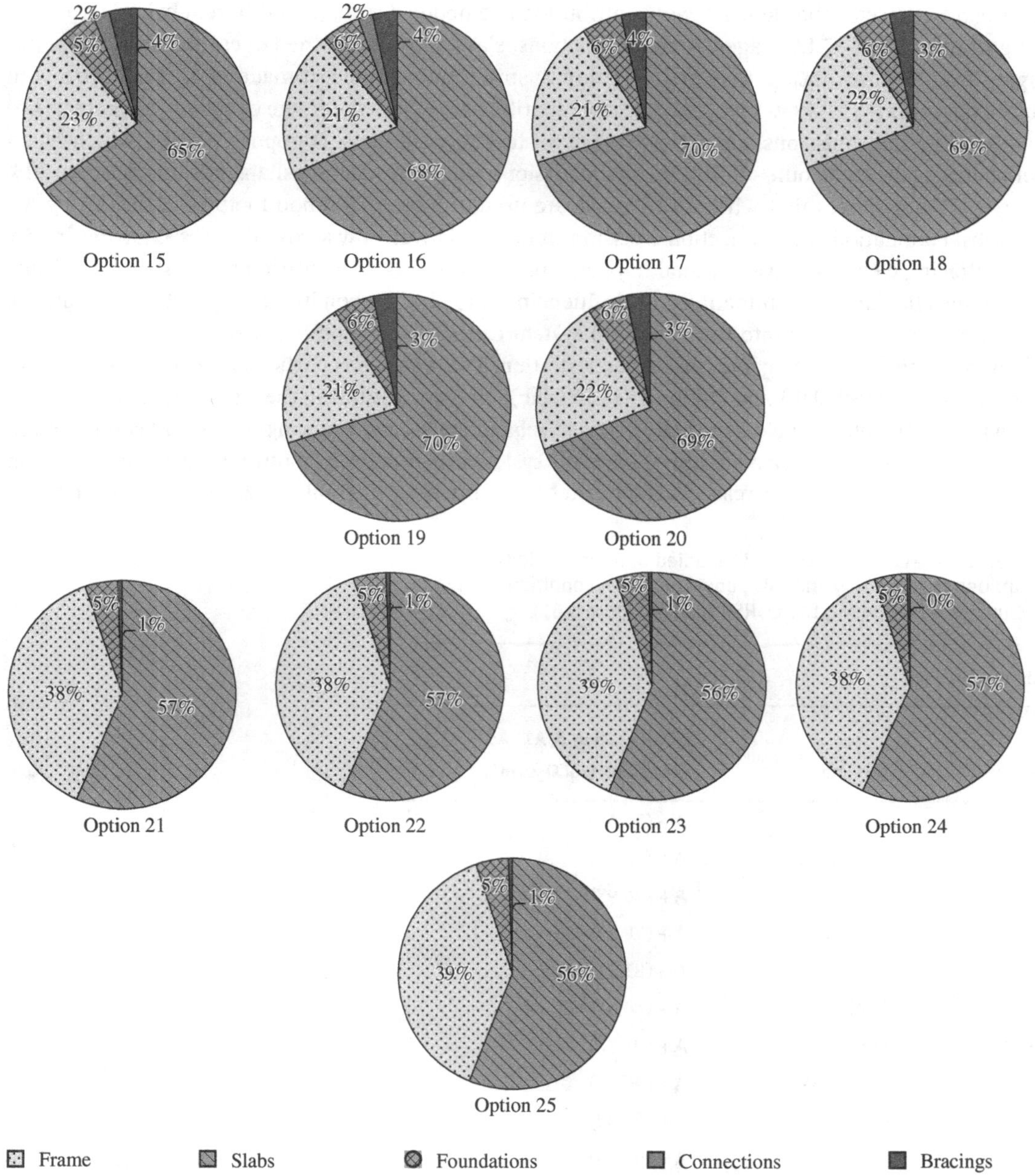

Figure 5.60 Element emission contribution (%) of the different options (A–C) options 15–25.

potentially releasing stored biogenic carbon. In cases where Module D (reuse benefits) plays a major role, steel beams could offer a better life-cycle carbon balance than Glulam. In conclusion, a case-by-case approach that considers both embodied carbon and practical constraints is necessary to determine the most effective strategy for a given project.

Figure 5.60 presents the element-wise contribution to embodied carbon for different hybrid structural configurations for A–C LC stages. Across all options, slabs account for the largest share of emissions, ranging from 56% to 70%, as they are the primary structural component made of CLT. Options with Glulam beams (21–25) show a shift in emission distribution, where the frame contribution increases to 38–39%, while slab emissions remain lower than in steel-beam options. For options 15–20, which retain steel beams, slabs contribute 65–70% of total emissions, while the frame's share remains between 21% and 23%. This indicates that when steel beams are used, the overall carbon footprint is driven by the CLT slabs. Foundation and connection emissions remain relatively low across all cases, contributing 5% or less. Bracing elements have minimal impact, typically below 1–2%, confirming that structural frame and slab selection are the dominant factors influencing embodied carbon in these hybrid configurations.

Integrating mass timber into the structural system can lead to a significant reduction in embodied carbon, primarily due to biogenic carbon sequestration. This effect was particularly evident when comparing Option 9 (steel slab) and Option 15 (CLT slab), where substituting the steel and concrete-based deck with CLT results in a clear reduction in slab-related emissions, lowering the overall carbon intensity of the structure. Further analysis of hybrid systems incorporating both CLT slabs and Glulam beams shows a continued decrease in embodied carbon intensity. In Table 5.22, Glulam-based options

Table 5.22 Classification of the 11 studied options in addition to the options of sensitivity analysis about the initial condition of the structural steel, according to SCORS ranking (Figure 5.61).

Scheme rankings		
Rank	**Scheme**	**SCORS rating & A1–A5 Emissions ($kgCO_2e/m^2$)**
1st	Option 24	A+ (75.63)
2nd	Option 22	A+ (75.82)
3rd	Option 21	A+ (76.98)
4th	Option 25	A+ (79.11)
5th	Option 23	A+ (81.95)
6th	Option 9.4	A+ (84.31)
7th	Option 9.3	A+ (91.20)
8th	Option 19	A+ (96.81)
9th	Option 17	A+ (97.09)
10th	Option 16	A+ (98.28)
11th	Option 18	A (100.17)
12th	Option 20	A (100.35)
13th	Option 15	A (108.41)
14th	Option 9.2	A (117.32)
15th	Option 9.1	A (140.06)

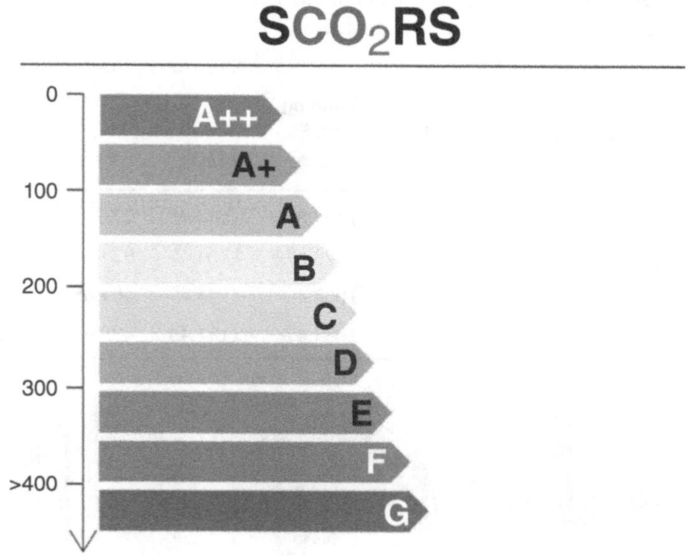

Figure 5.61 SCORS rating (based on total A1–A5 emissions for superstructure plus substructure, excluding biogenic carbon or offsetting, in accordance with the IStructE guide on how to calculate embodied carbon (Gibbons, O. P. and Orr, J. J./ with permission of The Institution of Structural Engineers).

consistently rank lower in carbon emissions than configurations retaining steel beams. Among the studied options, Option 24 achieves the lowest embodied carbon intensity, registering 75.63 $kgCO_2e/m^2$ for A1–A5 and 83.14 $kgCO_2e/m^2$ for A–C.

Now, we will explore how the transportation distance of timber affects the embodied carbon of the structure (Figures 5.63 and 5.64), specifically focusing on Option 21. The analysis modifies the carbon factor of Stage A4 (transportation) based on different distances: locally (50 km), nationally (300 km), European supply (1500 km), and global shipping (10 000 km by sea). The results in Table 5.23 show a progressive increase in total embodied carbon as transportation distances grow. Locally sourced timber results in the lowest embodied carbon intensity (64.58 $kgCO_2e/m^2$), while globally transported timber nearly adds 20 $kgCO_2e/m^2$, reaching 83.43 $kgCO_2e/m^2$. This increase is mainly attributed to Stage A4 emissions, which rise from 27.73 tCO_2e (local) to 405.62 tCO_2e (global). Sourcing timber locally or regionally provides an environmental advantage by minimizing transport-related emissions.

Figure 5.62 presents the element-wise contribution to embodied carbon across different transportation distances for timber. The frame and slab components remain the dominant contributors to total emissions, with their shares shifting as transportation distances increase.

In conclusion, hybrid steel-timber systems can help reduce the environmental impact of buildings. CLT slabs contribute to carbon storage, offsetting some of the emissions from steel production. Steel provides high load-bearing capacity, allowing for longer spans. Prefabricated CLT elements also enable faster construction, which can lower labor costs and improve project timelines. The aesthetic, biophilic, and thermal properties of exposed timber can enhance indoor living and environmental quality, offering additional benefits beyond structural performance. While hybrid construction presents advantages, certain challenges must be addressed. Fire resistance remains a critical factor, requiring careful design to meet safety standards. Moisture protection is necessary to prevent timber degradation, particularly in humid environments. Steel and timber exhibit different behaviors under load, temperature

Table 5.23 Results of building embodied carbon as a factor of transportation distance of timber to the site.

Transportation distance	EC intensity ($kgCO_2e/m^2$)	Building total EC (tCO_2e)	Emission by life-cycle stages (tCO_2e)					
			A1–A3	A4	A5	C	D	Biogenic carbon
Locally	64.58	1307.80	1104.45	27.73	22.75	4115.40	−1405.35	−3962.53
Nationally	67.84	1373.69	1104.45	92.97	23.40	4115.40	−1405.35	−3962.53
European	83.38	1688.49	1104.45	404.65	26.52	4115.40	−1405.35	−3962.53
Globally	83.43	1689.47	1104.45	405.62	26.53	4115.40	−1405.35	−3962.53

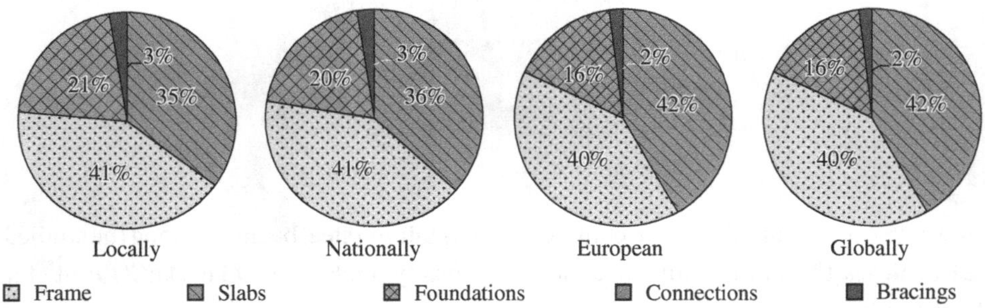

Figure 5.62 Element emission contribution (%) of the different options (A–C) (timber).

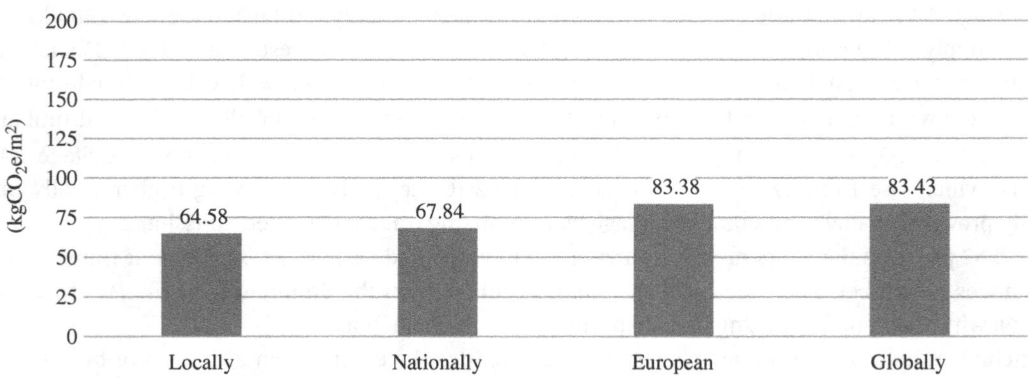

Figure 5.63 Estimated Modules A–C embodied carbon intensity of different options with different transport assumptions (timber).

variations, and humidity changes, requiring well-designed connections to accommodate these differences. The availability and cost of high-quality CLT and Glulam also vary by region, which may influence feasibility. This analysis quantifies the embodied carbon savings and structural implications of hybrid steel-CLT systems, providing a basis for evaluating their potential in sustainable building design. The

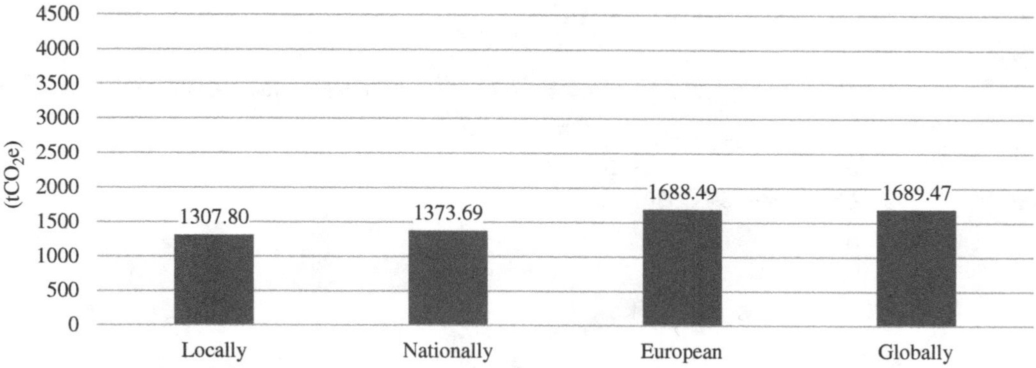

Figure 5.64 Total embodied carbon of Modules A–C, with different transport assumptions.

literature presents recent research on hybrid timber–steel composite slabs, where the combined action of both materials is engineered to optimize floor system performance (Safarik et al., 2024; Romero and Odenbreit, 2024; Trabucco and Perrucci, 2025).

5.7 How to Account for Uncertainty of Carbon Factors?

In any embodied carbon assessment, it is critical to recognize that the data inputs – such as the carbon emission factors for concrete, steel, and other construction materials – are estimates that can vary widely. As also demonstrated in this chapter, even small shifts in these input values can lead to notable differences in the overall embodied carbon of a building.

Variability can arise from differences in manufacturing processes, regional production technologies, and the use of supplementary materials. For instance, as explained in detail in Chapter 2, steel produced via Electric Arc Furnace (EAF) route has significantly lower carbon emissions than that produced by traditional BF-BOF route. Similarly, the concrete mix design – especially the level of Ground Granulated Blast Furnace Slag (GGBS) replacement – can result in carbon factors that vary from 0.1 to 0.165 $kgCO_2e/kg$. Transportation distances, on-site material wastage, and end-of-life recycling potentials (often captured in Module D of LCA studies) further add to the uncertainty. Omitting uncertainty analysis in embodied carbon assessments can lead to overly confident conclusions based on single-value estimates, rather than capturing the full range of possible outcomes, as also underlined by several researchers including Mendoza Beltran et al. (2018).

In this chapter, we tried to deal with this variability by making a parameter analysis. By systematically varying each input parameter across its plausible range, we could determine which factors most strongly influence the overall embodied carbon. We observed that altering parameters resulted in embodied carbon intensity differences of up to 38% among various design options.

While parameter analysis is useful for identifying key parameters, Monte Carlo simulation (MCS) can offer a robust framework for quantifying overall uncertainty. In an MCS, each uncertain input is assigned a probability distribution rather than a single fixed value. By running thousands of iterations – each time randomly sampling from these distributions – the simulation generates a probability distribution of the

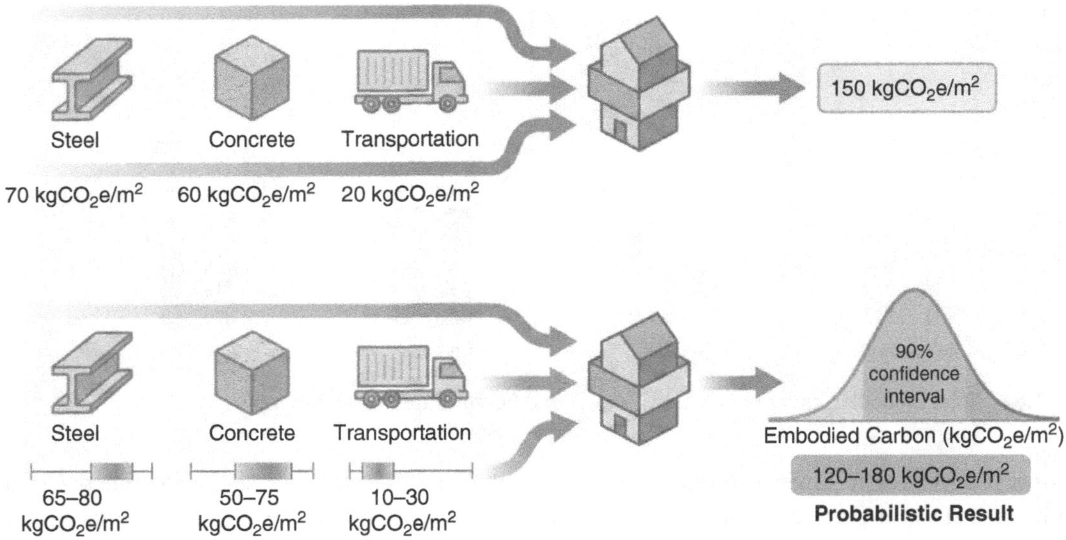

Figure 5.65 Deterministic versus probabilistic estimation of embodied carbon. Fixed input values for steel, concrete, and transportation yield a single result (e.g. 150 $kgCO_2e/m^2$), whereas representing inputs as ranges produces a distribution of outcomes, expressed as a 90% confidence interval (120–180 $kgCO_2e/m^2$). This approach better reflects real-world variability and provides a more informative basis for decision-making under uncertainty.

total embodied carbon outcome. This probabilistic approach provides decision-makers with a statistical range (e.g. a 90% confidence interval) rather than a single point estimate (Figure 5.65). Bin Thaneya et al. (2024) applied a scenario-aware Monte Carlo simulation to assess uncertainty in building material emissions. Their work demonstrated that when the carbon factor for steel production was allowed to vary between its lower and upper bounds, the resulting embodied carbon intensity for a building could be reliably described by a distribution rather than a fixed value.

Uncertainty in one input parameter does not exist in isolation – it propagates through the entire calculation, potentially affecting multiple aspects of the final result. For instance, the representative portion method used in Chapter 5 assumes that shared structural elements (like columns and beams) are fractionally accounted for in a single bay. This assumption introduces a slight underestimation compared to a full-building analysis. By propagating the uncertainty of each input through the model, one can obtain an overall uncertainty measure (often expressed as a confidence interval) around the total embodied carbon value. Recent literature, such as the review by Marsh et al. (2023), discusses techniques for combining individual uncertainties into a total uncertainty metric in LCA studies. These methods ensure that uncertainties are neither underestimated nor ignored, leading to more reliable environmental assessments.

5.8 Conclusion

This chapter has demonstrated how Life-cycle Parameter Analysis (LCPA) can play a decisive role in reducing embodied carbon at the building scale. By applying a representative portion approach, it was shown that reliable comparisons between structural systems can be achieved rapidly, with limited deviation from full building assessments. The case studies confirm that embodied carbon is highly sensitive to span length, material grade, composite action, and section type selection. Shorter spans, optimized member sizing, and the use of higher-strength steel consistently lead to lower embodied carbon intensities. The analysis also highlights the dominant influence of production stage emissions and the strong impact of carbon factor assumptions, steel sourcing, and end-of-life scenarios on final results. Hybrid solutions combining steel with timber elements show further potential for reduction, though practical constraints remain. Overall, the findings reinforce the value of embedding carbon focused decision making at the conceptual design stage, where design flexibility is greatest and meaningful reductions can be achieved with minimal redesign effort. In addition, it has been recommended that the embodied carbon results should be interpreted as ranges rather than fixed values, and that systematic parameter analysis and probabilistic approaches provide a more robust basis for early design decisions under variable material data and assumptions. It is important to recognize that this chapter has presented one detailed example to illustrate a broader methodology. The approach itself (systematic optioneering, parameter analysis across carbon factors, sourcing scenarios, and life-cycle stages) is not specific to steel or to any single building typology. The same framework can and should be applied to reinforced concrete structures, timber buildings, hybrid systems, and any other material combination a designer may consider. Whether the variable under investigation is the cement replacement ratio in concrete, the species and origin of engineered timber, or the recycled content in aluminum cladding, the underlying logic remains consistent: isolate key parameters, quantify their influence on total embodied carbon, and use the results to guide design decisions. Students are encouraged to adapt this methodology to their own projects, treating it as a transferable tool rather than a case-specific exercise.

5.9 Questions

Answer the following questions to enhance your understanding of the concepts covered in this chapter.

Short Answer Questions

1. Define "life-cycle parameter analysis (LCPA) at building level."

2. Define Gross Internal Area (GIA) and explain its importance in determining a building's embodied carbon intensity.

3. What is a "representative portion" in the context of early stage embodied carbon assessments, and why is this method used?

4. Describe the key factors that influence the carbon factor assumptions for materials such as reinforced concrete, steel, and timber.

5. Outline the main differences between embodied carbon assessments for reinforced concrete buildings and those for structural steel buildings as discussed in the chapter.

6 Explain the purpose of using benchmark buildings in the life-cycle parameter analysis (LCPA) at building level.

7 Describe how the representative portion method approximates the embodied carbon of an entire building.

8 Explain the role of transportation emissions (Module A4) in the overall embodied carbon calculation.

9 Define the concept of a SCORS rating and its significance in evaluating structural options.

10 How do supplementary cementitious materials (SCMs) in concrete mixes contribute to lower embodied carbon?

11 Describe the impact of applying composite action in beams on both structural efficiency and embodied carbon savings.

12 Why might designers use a conservative extrapolation factor when estimating whole-building embodied carbon from representative portions?

13 Explain why uncertainty analysis is essential in an LCA.

14 List three common sources of uncertainty in calculating the embodied carbon of building materials.

True or False

1 The Gross Internal Area (GIA) of a building is determined by multiplying the internal floor area of one level by the number of floors that are in use, excluding the roof (if it is not accessible except maintenance).

2 The representative portion method can lead to an underestimation of the total embodied carbon because shared elements (like beams and columns) are not fully accounted for.

3 Sensitivity analysis shows that design changes such as reducing span lengths or employing composite beam action generally result in lower embodied carbon intensity.

4 Regional variations, including differences in material sourcing and transportation distances, are negligible when selecting carbon factor assumptions for LCA.

5 The benchmarking process in the chapter involves comparing a reinforced concrete building with a structural steel building built to similar dimensions.

6 The calculation of Gross Internal Area (GIA) excludes the roof area because it is not considered usable space.

7 Sensitivity analysis in the chapter shows that reducing beam span lengths generally leads to a reduction in embodied carbon intensity.

8 Production stage emissions (A1–A3) are typically the dominant contributor to a building's embodied carbon, whereas transportation emissions (A4) contribute only a minor portion.

9 Composite action in secondary beams helps optimize material usage, thereby reducing the embodied carbon of a structure.

10 The SCORS rating evaluates structural configurations solely on economic factors without considering embodied carbon intensity.

11 Uncertainty analysis helps identify which design parameters have the greatest impact on the overall embodied carbon.

12 Sensitivity analysis involves varying one input parameter at a time to observe its effect on the embodied carbon outcome.

13 Scenario analysis always provides a single, definitive value for a building's embodied carbon, eliminating all uncertainty.

14 Incorporating uncertainty analysis in early design stages can lead to more resilient and sustainable design decisions.

Discussion Questions

1 Discuss the advantages and potential limitations of using a representative portion (e.g. a single bay) to estimate the overall embodied carbon of an entire building. How might this approach influence decision-making during early design stages?

2 Evaluate how different structural design choices – such as variations in span length, beam action (composite versus noncomposite), and column configuration – can impact the embodied carbon of a building. In your discussion, consider how these choices might be optimized to balance sustainability with structural integrity.

3 Consider the challenges in selecting carbon factors for materials used in construction. How do factors like material sourcing, manufacturing processes, transportation, and end-of-life scenarios contribute to uncertainties in these factors, and what strategies can designers use to address them?

4 Analyze how the selection of reinforced concrete versus structural steel influences the embodied carbon of a building, considering factors such as material sourcing and manufacturing processes.

5 Discuss the benefits and limitations of using a representative portion (such as a single bay) for estimating a building's overall embodied carbon in the early design phase.

6 How should designers balance the trade-offs between longer spans and the use of composite action in beams when aiming to reduce embodied carbon while maintaining architectural flexibility?

7 In what ways can transportation emissions influence the embodied carbon of a building, and how might changes in supply chain logistics affect design decisions?

8 Evaluate the challenges that regional differences in material production and energy sourcing pose for standardizing carbon factor assumptions in embodied carbon assessments.

9 Reflect on how the expanded sensitivity analysis of various structural options can guide designers toward optimizing both embodied carbon and structural performance.

10 How does integrating embodied carbon assessments into early stage design processes contribute to long-term sustainability, and what potential challenges might arise in communicating these findings to nontechnical stakeholders?

11 How can integrating uncertainty analysis into the building design process improve decision-making and contribute to long-term sustainability outcomes?

12 Life-cycle Parameter Analysis (LCPA) reveals that embodied carbon is not a fixed property of a structure, but a function of the assumptions we choose to make. If the same building can appear sustainable or unsustainable depending on input parameters, what does this imply about the objectivity of carbon assessments?

13 LCPA demonstrates that small changes in assumptions – carbon factors, transport distances, end-of-life scenarios – can shift rankings between design alternatives. If the "best" option depends on uncertain inputs, how should engineers justify their final design choice?

14 LCPA encourages questioning default values and testing alternative scenarios. However, this flexibility also opens the door to selective parameter choices that favor predetermined conclusions. How can the methodology guard against misuse while preserving its exploratory power?

15 Traditional design optimization seeks a single best solution, while LCPA suggests that "best" is conditional and context-dependent. Does this undermine the pursuit of optimal design, or does it offer a more honest representation of the decisions engineers actually face?

16 Early-stage embodied carbon assessments often rely on simplified methods and representative portions. What are the risks of making irreversible design decisions based on approximations, and how can engineers manage this uncertainty?

17 Hybrid timber–steel systems can achieve substantial upfront carbon savings through biogenic carbon storage, but end-of-life emissions may offset these benefits. How should designers account for the timing of emissions across a building's life cycle?

18 Reducing span lengths consistently lowers embodied carbon, but also limits architectural flexibility and usable floor space. How should design teams negotiate the trade-off between environmental performance and functional requirements?

19 Carbon factors vary significantly depending on data sources, regional conditions, and methodological assumptions. How should engineers communicate this uncertainty to clients and decision-makers without undermining confidence in the results?

20 Transportation emissions represent a relatively small share of total embodied carbon compared to material production. Does this mean local sourcing is overemphasized in sustainable design discussions, or are there other reasons to prioritize it?

21 Composite action between steel beams and concrete slabs improves structural efficiency and reduces material use, but increases construction complexity and limits future disassembly. How should designers weigh immediate carbon savings against long-term adaptability and circularity?

22 High-strength steel grades allow for smaller cross-sections and reduced material quantities, yet their production often involves higher energy inputs. Under what conditions does the trade-off between material efficiency and manufacturing intensity favor one approach over the other?

23 Embodied carbon assessments typically focus on structural elements, while connections, bracings, and secondary components are estimated using simplified allowances. What are the implications of this approach, and when might a more detailed analysis be warranted?

24 Optioneering enables comparison of multiple design alternatives, but time and budget constraints often limit how many options can realistically be explored. How should design teams prioritize which variables to test when resources are limited?

References

Bin Thaneya, A., Gursel, A. P., Kane, S., Miller, S. A., & Horvath, A. (2024). Assessing uncertainty in building material emissions using scenario-aware Monte Carlo simulation. *Environmental Research: Infrastructure and Sustainability*, *4*(2), 025003. https://doi.org/10.1088/2634-4505/ad40ce

Circular Ecology. (2020). The Inventory of Carbon and Energy (The ICE Database) [Dataset]. Retrieved from https://carbon.tips/ice3

Gibbons, O. P., Orr, J. J., Archer-Jones, C., Arnold, W., & Green, D. (2022). *How To Calculate Embodied Carbon* (2nd edition). The Institution of Structural Engineers.

Chehade, M. (2024). Embodied Carbon Optioneering for Building Structures [MSc Thesis, supervised by Alper Kanyilmaz]. Politecnico di Milano.

Marsh, E., Allen, S., & Hattam, L. (2023). Tackling uncertainty in life cycle assessments for the built environment: A review. *Building and Environment*, *231*, 109941. https://doi.org/10.1016/j.buildenv.2022.109941

Mendoza Beltran, M. A., Pomponi, F., Guinée, J. B., & Heijungs, R. (2018). Uncertainty analysis in embodied carbon assessments: What are the implications of its omission? In *Embodied Carbon in Buildings* (3–21). Springer.

NHBC. (2024). National House Building Council (NHBC) Standards 2024. Retrieved from https://nhbc-standards.co.uk/downloads/NHBC-Standards-2024-Complete.pdf

Norman, J., Broadbent, O., Carr, J. F., De'Ath, R., Harpin, R., Knowles, G., & Lloyd, I. (2020). *Conceptual Design of Buildings*. The Institution of Structural Engineers.

Romero, A., & Odenbreit, C. (2024). Experimental investigation on novel shear connections for demountable steel-timber composite (STC) beams and flooring systems. *Engineering Structures*, 304. https://doi.org/10.1016/j.engstruct.2024.117620

Safarik, D., Miranda, W., & Anthony. S. (2024). *Steel-timber Hybrid Buildings: Case Studies*. Chicago: Council on Tall Buildings and Urban Habitat.

SteelConstruction. (2025). "Composite construction." *Steelconstruction.Info*. Available from https://steelconstruction.info/Composite_construction

Trabucco, D., & Perrucci, G. (2025). Steel–timber hybrid buildings: A comparative life cycle assessment study of global warming potential impacts. *Sustainability*, *17*(2), 718. https://doi.org/10.3390/su17020718

UK Government. (2010). Approved Document A: Structure. Retrieved from https://assets.publishing.service.gov.uk/media/5a80437640f0b623026927b2/BR_PDF_AD_A_2013.pdf

6

Life-cycle Optimization (LCO)

LIST OF CONTENTS

The seven steps of a building project realization (RIBA, 2020) can be summarized with the following three main stages:

Conceptual design (Stages 0–2): In this phase, the client requirements are collected, the project brief is developed and approved, and the design concept and main building system are decided.

Detailed design (Stages 3–4): The architectural and engineering information is spatially coordinated. At this point the structural analyses are performed and all the design details are specified. The information required for manufacturing and construction is also created.

Construction and use (Stages 5–7): This stage includes the fabrication of the structural elements, the construction of the building, and the posterior performance feedback and usage.

Early design decisions (i.e. concept design) have a significant impact on a project's long-term (life-cycle) performance. Under tight time constraints, these choices affect the overall quality, cost, and embodied carbon of a building's construction (even more so compared to the later stages). This level of influence on the project cost has been historically represented in the cost influence curves first introduced by Paulson (1976) (Figure 6.1). Similar versions of this curve are presented in many articles, which show that C. Paulson's idea is still valid and up-to-date.

This chapter introduces Life-cycle Optimization (LCO): a multi-objective conceptual design method that simultaneously evaluates cost, embodied carbon, and structural efficiency across a building's life cycle. By rapidly generating multiple design alternatives, LCO enables engineers and architects to compare trade-offs and select promising solutions for further development during the early stages when design influence is greatest (Kanyilmaz et al., 2023b).

The construction sector is known for being very fragmented and multidisciplinary. Architects, structural engineers, and building physics engineers, among others (e.g. MEP – Mechanical, Electrical, Plumbing – Engineers, Clients, Developers, Construction Managers, Contractors, Sustainability Consultants, Geotechnical Engineers, Façade Engineers, Regulatory Authorities), are all involved in the same project and each one of them has its own specific objective (Figure 6.2). Besides, the communication among them is commonly low. Such lack of communication leads to mistakes and misunderstandings that rebound later in the design process, increasing the overall cost and decreasing the optimal performance of the building. While building information modeling (BIM) technique is tackling with this issue at the detailed design stage, currently the conceptual design stage still suffers from such problem.

Many configurations exist for a given design brief and, in the face of the impossibility of evaluating all of them, it is normally the intuition and previous experiences of engineers and architects what guide the decision-making process. Due to the complexity of the process, the conceptual design is assigned to senior engineers who have the necessary expertise, knowledge and intuition to select the best preliminary designs.

High influence levels during early design represent opportunities for impactful changes (Rafiq and Beck, 2008). Therefore, making the right decisions at this stage is crucial for project success and cost reduction. Therefore, the integration of the majority of the construction industry parties should be considered earlier

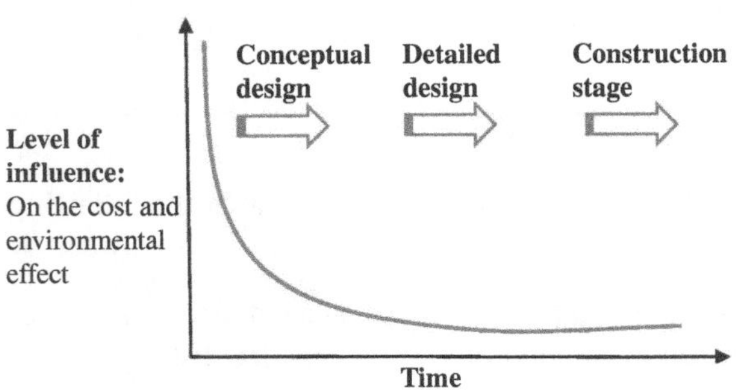

Figure 6.1 Cost influence curve introduced by Paulson (adapted from Kanyilmaz, Dang, et al., and Paulson).

Figure 6.2 A representation of how engineering, design, and sustainability experts join forces to deliver an eco-friendly project (alphaspirit/Adobe Stock Photos).

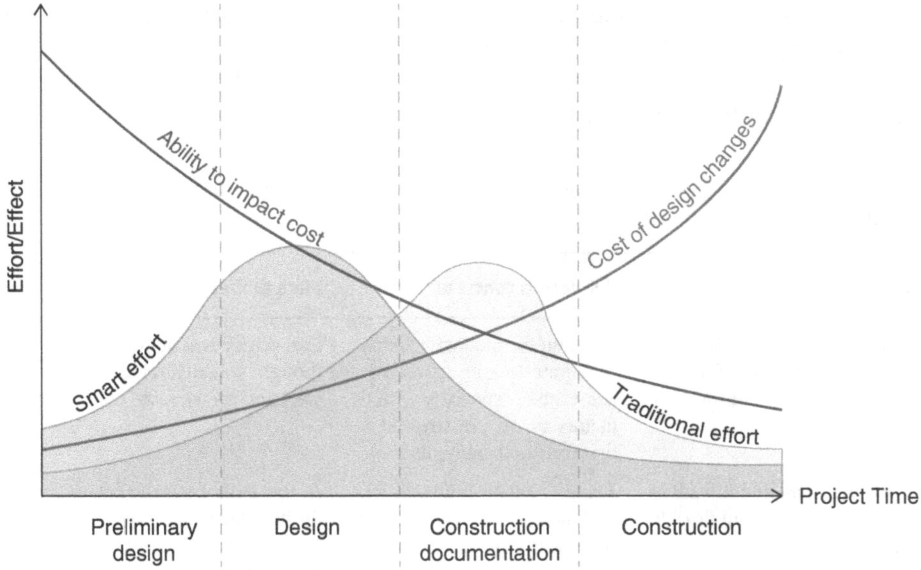

Figure 6.3 MacLeamy curve representation (adapted from MacLeamy, 2020).

in the process when the ability to impact the project is higher. This idea has been represented by the MacLeamy curve shown in Figure 6.3 (MacLeamy, 2020).

Nevertheless, the conceptual design is often an "ill-defined" process. This, together with the limitation in time and resources normally associated with this stage, results in decisions having to be taken quickly and under uncertain conditions, which leads to poor decision-making. This is where LCO becomes essential: by rapidly generating and evaluating multiple design alternatives, it transforms the traditional conceptual design into a structured, data-driven process.

6.1 Key Decisions to be Given at a Conceptual Design of Building Structures

The structural design process involves a series of interconnected decisions that collectively define the solution space for life-cycle optimization. To generate a structural layout, some key decisions have to be taken about material, lateral load system, floor system, and lateral load resisting system (LLRS). Once the decisions are taken, a preliminary sizing is performed to ensure the feasibility of the solution and to make a cost and embodied carbon estimation. Since most of those decisions are normally highly dependent between them, they have to be taken almost simultaneously.

6.1.1 Material Selection

There are principally three types of materials that are commonly used in building construction today: Steel, reinforced concrete, and mass timber (besides masonry can be also widespread, but often used as non-structural infill in modern frames and as load-bearing walls in low-rise or heritage buildings). Material selection in construction projects is governed by several parameters that must be carefully balanced to achieve optimal project outcomes. These range from foundational structural requirements such as desired span, building height, and site-specific ground conditions to critical performance criteria such as fire resistance, acoustic properties, and long-term durability (Table 6.1 shows a detailed list). These technical factors must then be weighed against overarching project drivers, including cost, construction speed, and sustainability goals, particularly the reduction of embodied carbon.

Table 6.1 Comparison of the characteristics of steel, reinforced concrete, and mass timber in relation to structural requirements and performance criteria.

Parameter	Steel	Reinforced concrete	Mass timber
Span length	Its high strength-to-weight ratio makes it the best choice for achieving very long, column-free spans.	Efficient for moderate spans. Moderate-to-long spans are achievable, especially with factory-made, pre-stressed prefabricated elements.	Can achieve moderate spans, though typically not as long as steel and pre-stressed concrete.
Building height	High strength and ductility allow for very tall, slender, and flexible structures that can safely handle wind and seismic forces when combined with efficient lateral load resisting systems (LLRS).	Thanks to their high stiffness and mass, reinforced concrete shear walls and cores are often used in combination with steel and mass timber buildings to resist wind and seismic loads.	Its low stiffness is the current limiting factor for going tall. Commonly combined with steel bracings or reinforced concrete shear walls.
Durability & exposure	Prone to corrosion (rust) if not properly protected with coatings.	Cast-in-situ is inherently watertight due to its monolithic form, while the long-term durability of prefabricated systems depends on the performance of their joint seals.	Susceptible to moisture-related decay and insect attack if not properly detailed.

Table 6.1 (Continued)

Parameter	Steel	Reinforced concrete	Mass timber
Geotechnical impact	Its high strength-to-weight ratio reduces the load on foundations, making it ideal for sites with poor soil conditions.	Requires substantial foundations regardless of whether it is cast-in-situ or prefabricated.	Its moderate strength-to-weight ratio reduces the load on foundations.
Fire safety	Non-combustible, but loses strength quickly at high temperatures. Requires applied fireproofing.	Inherently non-combustible and a poor conductor of heat, it provides superior fire resistance regardless of construction method.	Very combustible but chars at a slow rate. The char layer insulates the inner wood, maintaining structural integrity for a moderately long period.
Acoustics	Its low mass transmits sound easily. Requires detailed insulation assemblies.	Its high surface density is effective at sound insulation. Joints between prefabricated elements must be properly sealed to prevent sound leaks.	Requires carefully detailed floor and wall assemblies to meet acoustic standards.
Vibration control	(Steel) Its low mass and high flexibility make steel floors susceptible to footfall-induced vibrations, particularly in long-span configurations. Careful design of composite floor systems and, in some cases, tuned mass dampers may be required to satisfy serviceability limits.	(Reinforced Concrete) Its high mass and stiffness provide excellent vibration performance, inherently limiting footfall-induced floor vibrations with minimal need for additional measures.	(Mass Timber) Its relatively low mass and stiffness can make timber floors sensitive to footfall-induced vibrations, often requiring supplementary measures such as added mass, increased floor depth, or tuned damping systems.
Labor & availability	Requires a skilled workforce.	Universally available.	Availability can be limited and requires a specialized workforce for design and assembly.
Aesthetic vision	Allows for a slender, precise, and modern look with large open spaces and extensive glazing.	Can be fluid and monolithic when cast-in-situ, or modular and panelized with high-quality factory finishes when prefabricated.	Provides a warm, natural, and biophilic aesthetic, where the structure itself often serves as the final interior finish.
Construction speed	Prefabricated components are assembled quickly on-site.	Cast-in-situ is the slowest method due to on-site work and curing, while prefabricated systems allow for fast on-site assembly.	Prefabricated panels allow for rapid assembly.
Reuse potential	Standardized components connected with bolts can be easily disassembled and reused in new structures (if they are in good condition)	Very difficult to obtain reusable elements from demolished cast-in-situ structures. Prefabricated elements can be reused (if they are in good condition)	Standardized components connected with bolts can be easily disassembled and reused in new structures (if they are in good condition), but the range of standardized sections and connection types is much more limited than in steel construction.
Recycling potential	Can be melted down and reformed into new high-quality steel with no loss of properties.	Can be crushed and downcycled into aggregate for fill or road base, but it cannot be reformed into new concrete.	Can be downcycled into wood chips for particleboard or used as biomass for energy. It cannot be reformed into new structural timber.

6.1.2 Structural Grid

The structural grid establishes the geometric framework for positioning columns and other vertical elements within a building (Figure 6.4). The distances between columns in the two orthogonal directions depend on several factors. The first concern is the spanning capability of the beams and floor system used. This is influenced by the structural material and the loads applied. The second concern is the spatial coordination. In this case, the influencing factors are the room layouts and the module sizes of floors and cladding systems. Other factors such as the utility of the building (e.g. office, residential, car park, data-center) also influence the need for column-free spaces or layout flexibility which will result in larger span distances. Table 6.2 compares the impact of three materials on the structural grid selection.

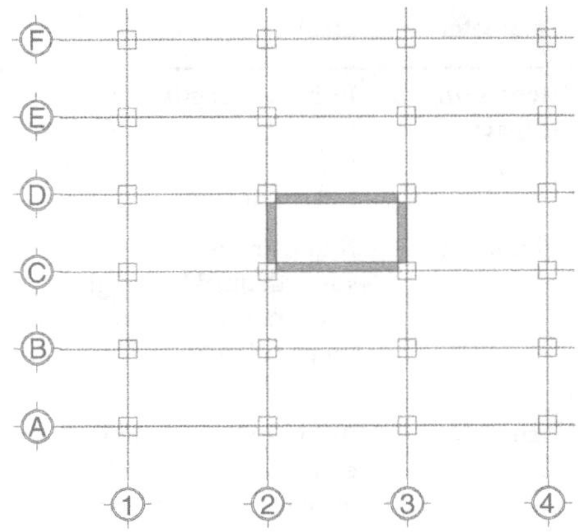

Figure 6.4 Structural grid example.

Table 6.2 Impact of materials on structural grid selection.

Parameter	Steel	Reinforced concrete	Mass timber
Structural layout	Its high strength-to-weight ratio allows for the long spans, giving architects maximum freedom in setting column grids. Useful also for future destination change (versatility)	Effective for moderate, repetitive grids. Longer spans are achievable but require deeper, heavier elements or post-tensioning, which adds complexity. Complex shapes are feasible due to the fluid nature of the concrete material.	Enables moderately long spans for open layouts, though typically not as long as steel does.
Spatial coordination & system integration	As a prefabricated system, it integrates well with modular facades and interior systems. Slender profiles also maximize usable floor space.	Cast-in-situ concrete can be formed to any layout. Prefabricated concrete offers speed but involves larger, bulkier elements than steel.	CNC-machined for high precision, making it ideal for coordinating with other prefabricated systems. The grid layout is planned by the factory.
Building use & function	The top choice for functions requiring maximum column-free space and flexibility, such as convention centers, industrial warehouses, data centers, and large open-plan offices.	Ideal for functions with regular, cellular layouts like apartments, hotels, and car parks, where its durability and moderate span capability are a match.	Preferred for offices, schools, and public buildings where both moderate spans and a biophilic qualities are desired.

6.1.3 Floor Systems

The choice of a floor system is determinant for the overall weight of the structure and, consequently, for its cost and carbon footprint. The floor thickness is also relevant to the overall height of the structure and the possibility of MEP systems' integration. The common floor system families are shown in Figure 6.5:

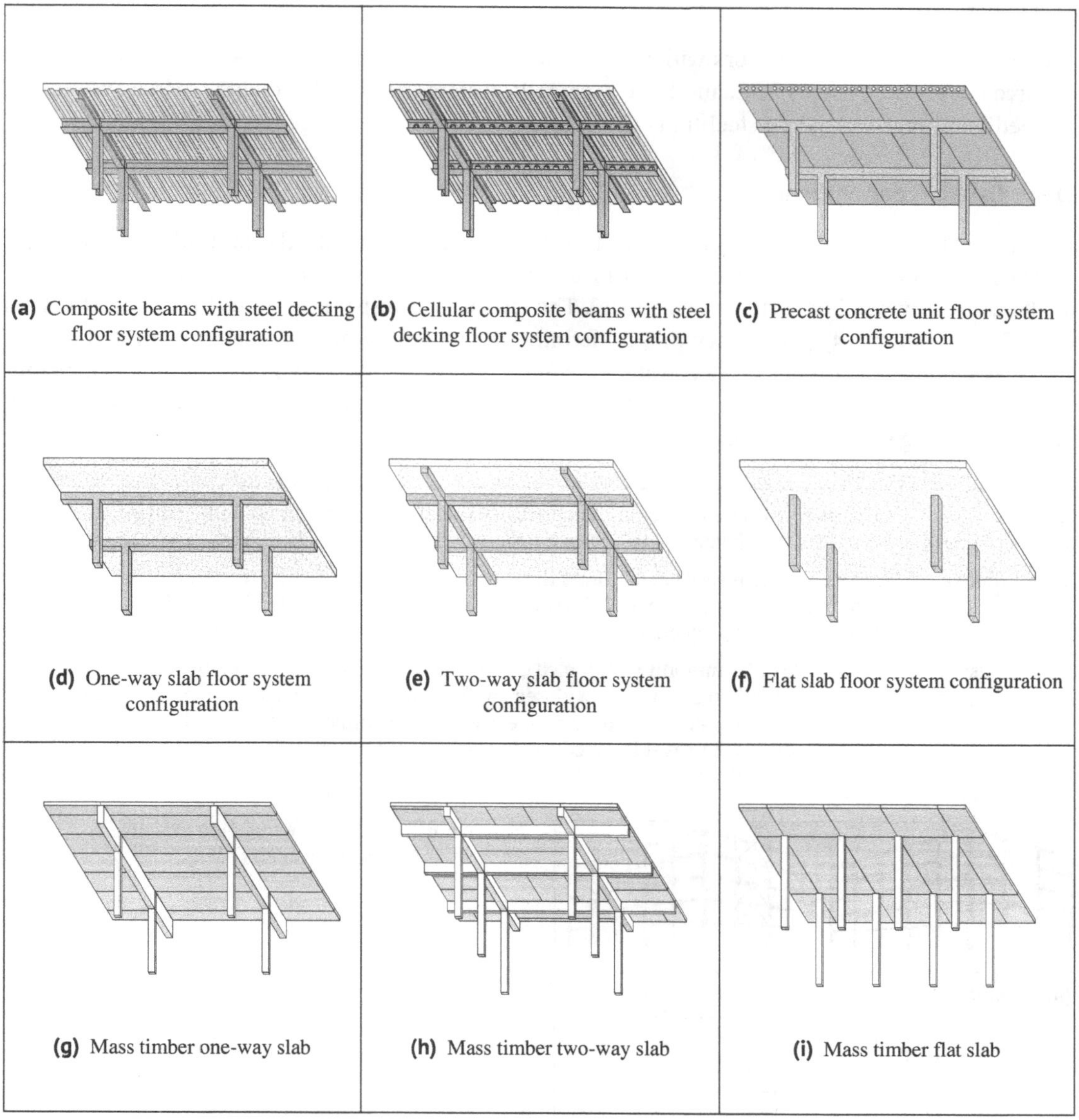

(a) Composite beams with steel decking floor system configuration
(b) Cellular composite beams with steel decking floor system configuration
(c) Precast concrete unit floor system configuration
(d) One-way slab floor system configuration
(e) Two-way slab floor system configuration
(f) Flat slab floor system configuration
(g) Mass timber one-way slab
(h) Mass timber two-way slab
(i) Mass timber flat slab

Figure 6.5 Different floor systems, commonly used in the construction practice.

A. Composite beams and composite slabs with steel decking
B. Cellular composite beams with composite slabs and steel decking
C. Precast concrete units
D. One-way slabs
E. Two-way slabs
F. Flat slab
G. Timber slabs (e.g. CLT)

Systems A and B commonly work with steel structures, while System C is applicable to both steel and reinforced concrete. Systems D, E, and F are used with steel and reinforced concrete, while System G is designed for timber and hybrid steel-timber structures.

6.1.4 Lateral Load System

Apart from the main structural system that transfers vertical loads into the ground, a building has to provide stability for lateral loads (lateral loads are horizontal forces acting on a structure and are usually caused by wind or earthquake actions). The lateral load-resisting system (LLRS) (e.g. the bold blue color in Figure 6.4) plays a key role in defining the structural grid. The placement and type of LLRS – whether braced frames, shear walls, or cores – affect the distribution of columns and beams, influencing both architectural flexibility and structural efficiency. There exist three main types of lateral load resisting systems (Figure 6.6):

Moment resisting frames:	Composed of beams and columns that provide lateral stability by means of the stiffness and strength of their joints. Allows for open floor plans and clear facades with no diagonal obstructions.
Braced frames:	A vertical truss resists lateral forces through axial tension and compression in its diagonal members. The diagonal braces obstruct bays and must be integrated into the facade design, limiting window placement.
Shear walls:	Walls made of (commonly) reinforced concrete that provide lateral stability through shear and bending. When arranged in a closed configuration (typically rectangular or C-shaped), they form a reinforced concrete core. Shear walls are commonly placed in the perimeter walls of the building, whereas reinforced concrete cores are normally placed around stairs and lifts.

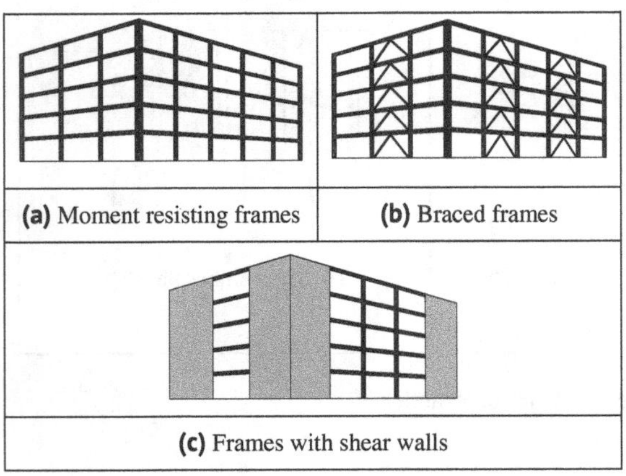

Figure 6.6 Lateral force resisting systems. Reproduced from Narayanan (2018).

Table 6.3 The impact of three materials on the LLRS selection.

Lateral resistant system	Steel	Reinforced concrete	Mass timber
Moment-resisting frames	The ductility, high strength, and versatility of steel profiles and connection elements allow for reliable, rigid and high-strength connections.	Cast-in-situ reinforced concrete is inherently monolithic, made of rigid connections.	Creating rigid, moment-resisting connections in timber is challenging, therefore not typically used as a primary lateral system.
Braced frames	Steel members are good in both tension and compression resistance, making them ideal for creating structurally efficient and lightweight vertical trusses.	Impractical for concrete, which performs poorly as a slender tension member in a truss.	Timber elements can be used as diagonals to create an efficient and aesthetically pleasing braced frame system. Not as common as steel bracings.
Shear walls	Specialized systems like steel plate shear walls are available, where a thin steel plate acts in tension to resist loads. Not as common as reinforced concrete shear walls.	This is the classic application for reinforced concrete. Its high stiffness creates a strong and economical system, often used as a central core of tall buildings.	Cross-laminated timber (CLT) panels have good in-plane strength and stiffness, making them a primary and highly effective shear wall system for timber buildings. Not as common as reinforced concrete shear walls.

The decision of which is the best lateral system depends on several other design parameters of the building. Table 6.3 compares the impact of three materials on the LLRS selection.

In summary, given the numerous variables and conflicting objectives in material, grid, and system selection, a systematic approach is needed to move beyond intuition. This is where a life-cycle optimization (LCO) method will be of help.

6.2 From Life-cycle Parameter Analysis (LCPA) to Life-cycle Optimization (LCO)

The conceptual design stage is a difficult phase of a construction project because designers must navigate a complex web of interconnected and often conflicting parameters (from structural efficiency and cost to construction speed and sustainability). The traditional approach to this challenge relies heavily on an engineer's or architect's past experience, intuition, and simplified rules of thumb, often resulting in a sequential, iterative process. This approach, however, carries a significant risk: by committing to a design path early with limited information, truly sustainable or higher-performing alternatives may be unintentionally discarded.

In Chapter 4, we proposed a method tackling this, called life-cycle parameter analysis (LCPA). While powerful, the LCPA method is most effective when refining an already-chosen design concept. Its limitation, however, emerges when faced with multiple, competing project goals like minimizing cost, maximizing the free spaces (or other parameters such as construction speed), *and* reducing embodied carbon, all at the same time. While LCPA is well suited to evaluating a limited set of predefined solutions, LCO supports the identification of balanced design strategies across a wide range of alternative structural configurations.

LCO is a data-driven, multi-objective process that uses the principles of LCPA not just to analyze, but to score and rank hundreds of computer-generated design variations. By doing so, it moves beyond simply optimizing a single design and instead identifies a whole family of optimal solutions giving stakeholders a choice between several "best" designs, each offering a different trade-off – for instance, one might be the cheapest, another the fastest, and a third the most sustainable (this is often visualized with a Pareto front, as will be explained later in the section).

In this chapter, we will detail the methodology behind LCO, focusing on the Genetic Algorithms (GAs) commonly used to drive the optimization. We will then introduce a conceptual design tool (Kanyilmaz et al., 2022) that implements this method and, finally, demonstrate the effectiveness of LCO with several examples.

6.2.1 Genetic Algorithms for Design Space Exploration

Genetic Algorithms (GAs) are heuristic adaptive search techniques based on the mechanisms of natural selection and natural genetics (Holland, 1975). They are classified as a type of evolutionary computing algorithms. The Darwinian strategy of "survival of the fittest" is used to explore a space of possible solutions in an optimization problem. In their application, a *population* of solutions undergoes a process of gradual changes. Each solution is considered an *individual*. These solutions are usually represented as a set of parameter values (also called *genes*) that together comprise the *chromosome* of each individual.

The algorithm sequence is as follows. An initial population of individuals is generated randomly and the *fitness level* of each solution, understood as an assessment of its performance, is calculated. Fitness function is a numerical measure used to evaluate how well each solution meets the objectives of the optimization problem (in our case, the fitness function calculation step will comprise the "cost," "span," and "embodied carbon" evaluation substeps). The fittest individuals among the population are selected by means of a *selection process* to create new solutions. Then, two methods are used for generating an offspring population: *Crossover* and *mutation*. Once the new generation is created, the process is repeated until meeting the convergence criteria (Varghese and Raj 2016).

GAs have emerged as an important tool to solve complex engineering optimization problems with wide search spaces. The main characteristics and advantages that make this method different from other optimization techniques are that GAs (Pezeshk and Camp, 2002):

- operate on a population of potential solutions rather than improving a single solution.
- do not require gradient information (i.e. the mathematical derivatives of the objective function with respect to each design variable, which indicate the direction of steepest improvement and are

essential for classical gradient-based optimization methods but are often difficult or impossible to obtain for complex, nonlinear, or discontinuous problems) and consequently, they can be used in ill-defined problems or unsmooth data.
- can operate with discrete design variables to solve discrete optimization problems.
- offer an open format for constraint statements, which can be applied as a penalty in the objective function.

As explained in the previous section, one of the main characteristics of conceptual design is its ill-defined nature (due to this characteristic, some optimization methods that need gradient information cannot be applied). GAs do not use gradients and consequently can be applied in this kind of problem. Moreover, GAs can be used to optimize nonlinear problems with continuous or discrete variables. These features broaden the modeling possibilities of the design problems.

Also, GAs are compatible with multi-objective optimization. Due to the fragmentation of the building sector, this is an interesting capability. The requirements of the different parts involved can be modeled as different objective functions in order to obtain more integrated designs.

Finally, it is important to note that GA is a stochastic algorithm, and it generates high-quality solutions but cannot guarantee optimality. However, this drawback is compensated with a greater exploration ability. This is the reason why this algorithm is presented as the best option for conceptual design problems where the main focus is set on the design space exploration.

6.2.2 Formulating a Multi-objective Optimization Problem

In this chapter, we approach the "life-cycle-driven" conceptual design of building structures as a multi-objective optimization (MOO) problem, which can be represented with the following generic form:

$$\max F(x) = [f_1(x), f_2(x), \ldots, f_l(x)]$$

$$\text{Find } x = [x_1, x_2, \ldots, x_t]$$

$$\begin{aligned} \text{subject to: } & h_j(x) = 0 \quad (j = 1, 2, \ldots, J) \\ & g_k(x) \le 0 \quad (k = 1, 2, \ldots, K) \end{aligned}$$

where

$f_i(x)$ $(i = 1, 2, \ldots, l)$ is the objective function,

$x = [x_1, x_2, x_3, \ldots, x_t]$ is the design variable, and

$h_j(x)$ and $g_k(x)$ are the constraints (conditions that the design must satisfy, such as structural requirements, material limitations).

$t, l, J,$ and K are the number of design variables, objective functions (the number of different goals we're optimizing for, like cost, sustainability, or safety), equality constraints (conditions that must exactly be met, like a specific architectural dimension), and inequality constraints (conditions setting limits, such as a maximum allowable weight or cost).

$F(x)$ in this context represents a collection (or set) of multiple goals we are trying to achieve simultaneously in the structural conceptual design process. Instead of having just one objective – such as cost – we have several objectives to balance at the same time. Each goal or objective (represented by $f1(x), f2(x), \ldots$) could be something like reducing cost, improving sustainability, enhancing fire safety, or improving comfort and accessibility. The multi-objective optimization then helps find the best combination of design variables that gives an optimal balance of these competing goals.

The strategy chosen will be closely related to the objective function. Different project objectives include: cost, fire safety, durability, acoustics, construction speed, reuse potential, or sustainability. It is clear that if the project objective is focused, for example, in cost, the objective function of the equivalent optimization problem would be minimizing the cost function.

Apart from project objective, the client brief normally includes other information about requirements and preferences over the project. These specifications can be translated into problem constraints or new optimization criteria. For example, a specification of the maximum height of the building would be an inequality constraint, whereas a preference of larger span distances could be translated into an objective function (maximizing span distance). Nevertheless, the project brief normally uses fuzzy language to describe the project requirements, which makes the translation into a modelized optimization problem difficult. Moreover, sometimes the difficult part relies on establishing a direct relationship between the candidate building system and the objective function (e.g. if the objective function is an aesthetic preference). In conclusion, conceptual design can be treated as an optimization process, but its inherent characteristics (ill-defined, fuzzy language, etc.) result in a big effort having to be done to create a mathematical model of the problem.

6.3 Description of an LCO Conceptual Design Method and Its Applications

In this section, the methodology of an LCO conceptual design method based on NSGA II (developed by Kanyilmaz et al. (2022)) is described to enable adoption within other design ecosystems, followed by application examples that demonstrate the sensitivity of conceptual design decisions on both building cost and embodied carbon based on the theory described in the previous sections.

Figures 6.7–6.9 describe the algorithm workflow of the method describing the iterative process and the convergence criteria. The model consists of a genetic algorithm (GA) engine including the material selection, the floor system, grid size, building dimension selections, and a lateral load-resisting system (LLRS) module that determines the best LLRS to use, sizes, and locates it in the building floor plan.

To begin with, a random parent population of size N is generated, where each parent solution represents a potential conceptual design for a building. Each created parent solution is defined by a specific combination of design variables. These design variables serve as inputs for the genetic algorithm and encompass various conceptual design decisions, such as material selection, grid and structural layouts, and the floor system. The fitness function calculation step comprises the "cost," "span," and "embodied carbon" evaluation substeps. New individuals from previous populations are created by using GA operators (Selection, Crossover, Mutation) and Repair modules to eliminate some unfeasible and conflicted options. The termination conditions occur when a certain number of generations (predefined by users) are reached. Table 6.4 describes the material and structural compatibilities and the material properties considered, respectively. The detailed information about the tool and its GA methodology can be read in a previous article of the author (Kanyilmaz et al., 2022).

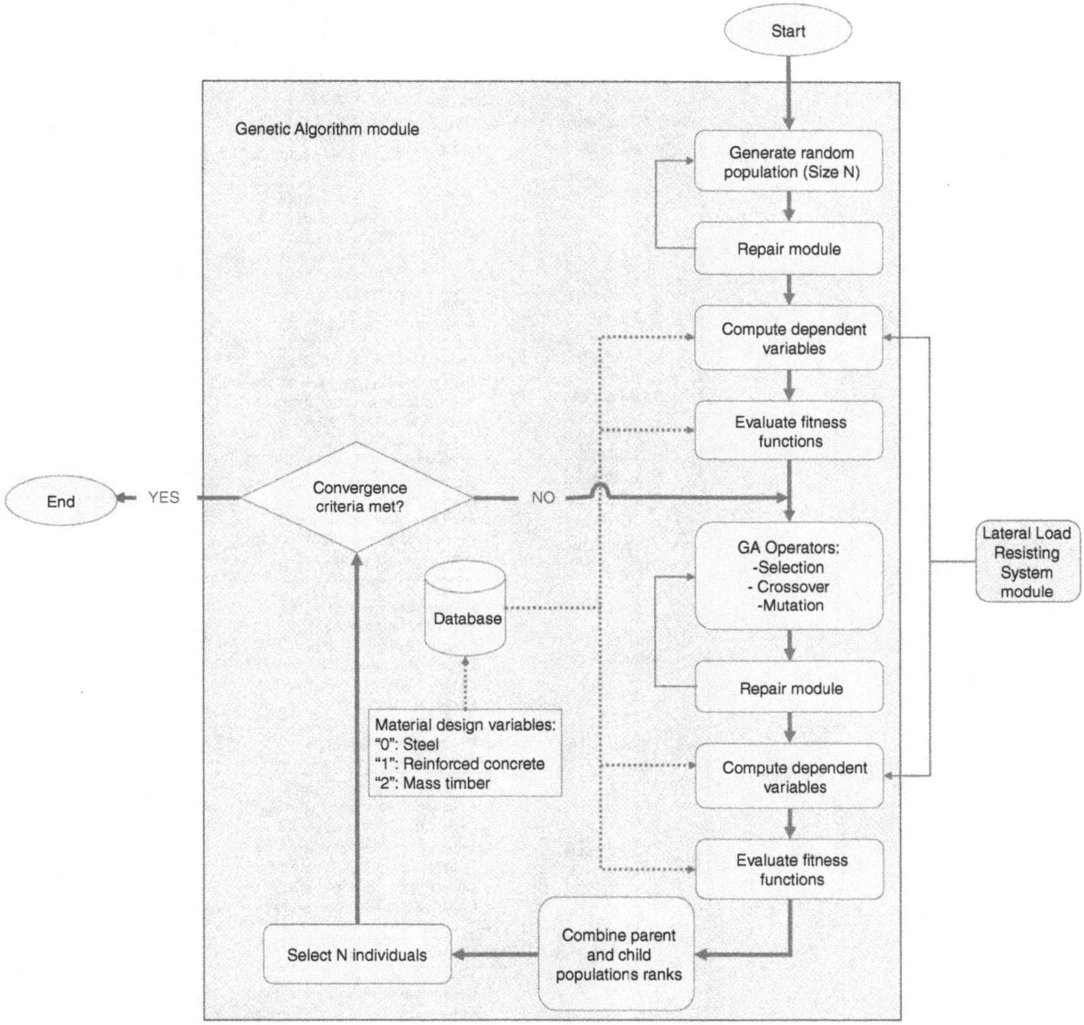

Figure 6.7 General algorithm workflow.

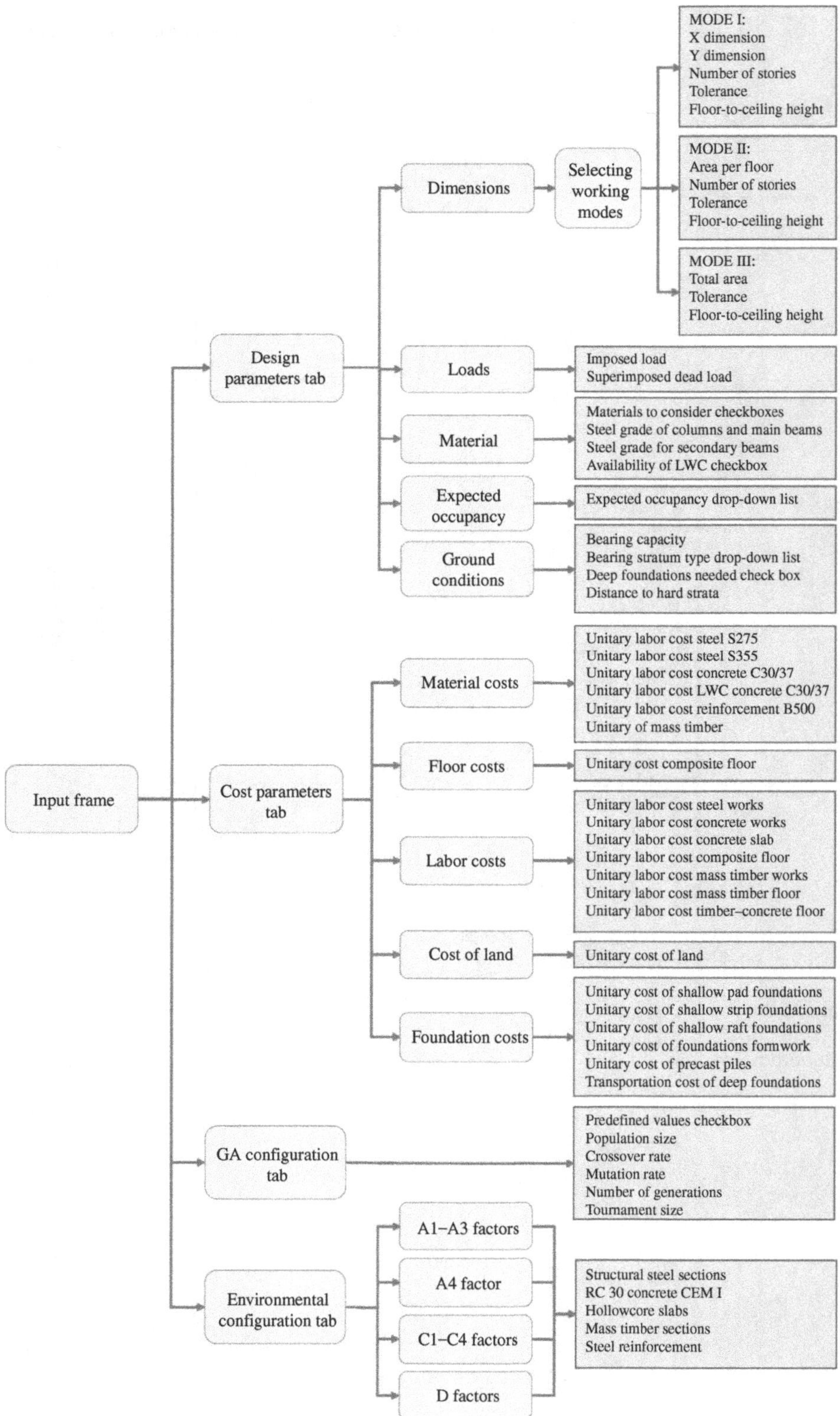

Figure 6.8 The main structure of the input frame.

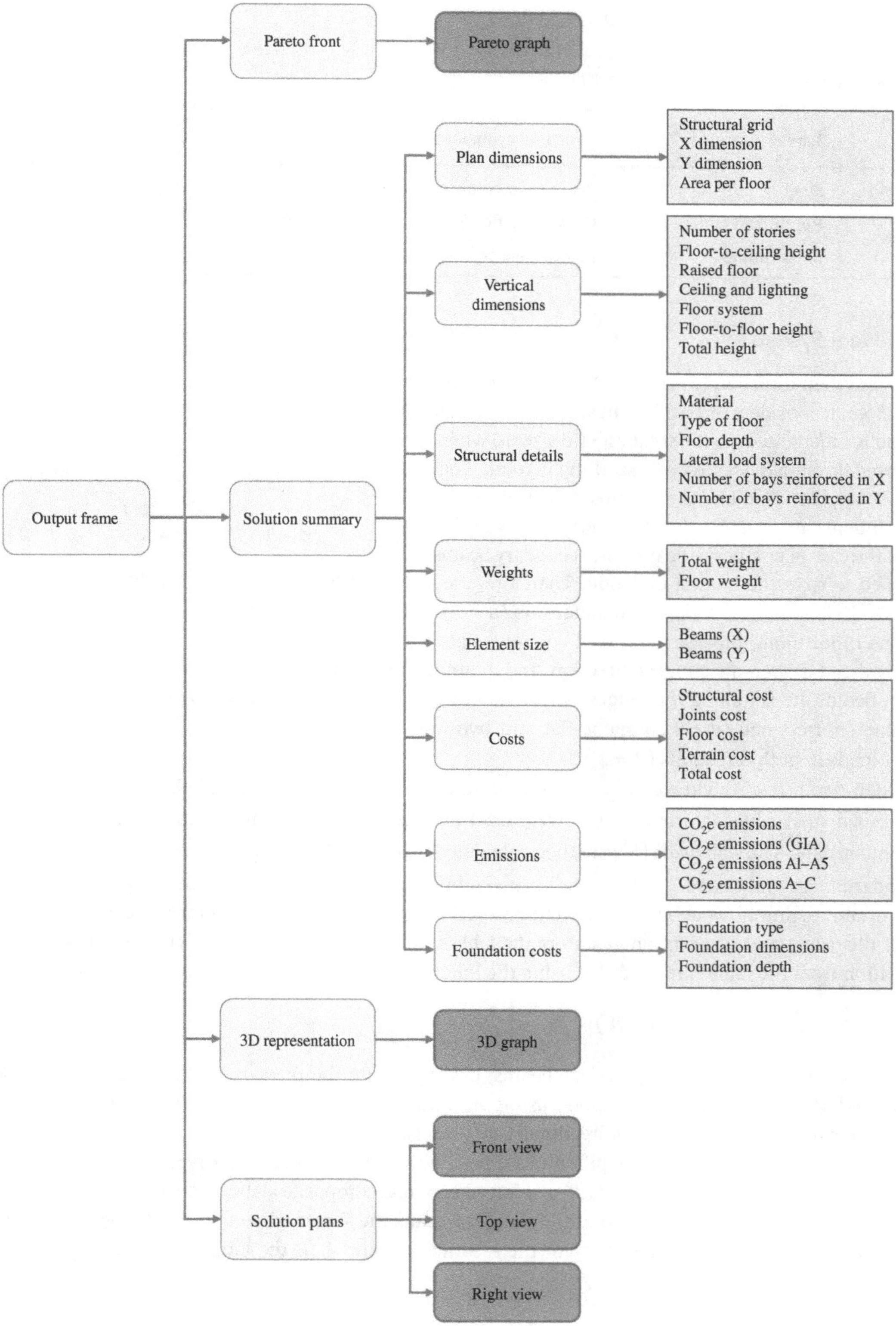

Figure 6.9 The main structure of the output frame.

Table 6.4 Materials and structural compatibilities.

No.	Type of materials	Structural components
1	Steel	Beams and columns
2	Reinforced concrete	Composite floors, building core, and foundations
3	Mass timber	Floors, beams, columns, and braces

6.3.1 Floor System

The floor system direction (d) (Figure 6.10) is defined by the type of floor. It considers three different scenarios: i) when the slab load is transmitted via the smallest grid size, ii) when it is transferred through the largest grid size, and iii) both. The tool uses this variable "d" to identify the direction of the loads and which dimensions to utilize for calculating the contributing area. In the case of composite beams, the secondary beams are positioned to span the longest direction. Therefore, the slab transmits the load in the smallest grid dimension ($d = 0$).

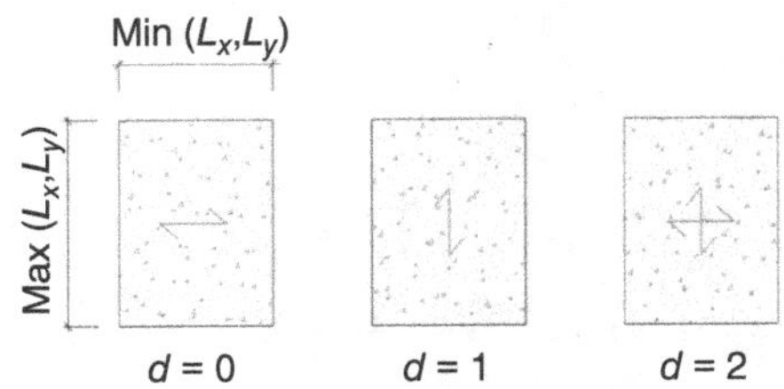

Figure 6.10 Types of the different floor system directions.

On the other hand, precast reinforced concrete slabs are often arranged to span the greatest direction, and therefore, the load is transferred in the long dimension ($d = 1$). Beams are set along the longest dimension in reinforced concrete one-way slabs, and the slab spans the shortest one ($d = 0$). Finally, flat and two-way reinforced concrete and mass timber slabs transfer loads in both directions ($d = 2$).

The slab depth (e.g. thickness) (S_D) is required to calculate the slab load and the actions over the beams. Load-span tables (Norman, 2018) were used to establish the minimum slab depths (S_D) for all mass timber slabs. The slab loads (S_L) are then calculated using the slab depth and a density of 4.8 kN/m^3 for standard mass timber and 25 kN/m^3 for normal reinforced concrete. The load-span tables commonly make certain assumptions about the estimated superimposed dead loads (SDL) for finishes and partitions. If the user's specifications change from the table's values, the Imposed Loads (IL) used to retrieve information from the tables are modified using the following formula:

$$IL = IL + \frac{1.35}{1.5} \cdot (\text{SDL} - \text{target SDL}) \tag{6.1}$$

The slab's load is received by secondary beams, passing it into the primary beams. The direction of primary and secondary beams is identified using the values of the floor direction (d). No primary and secondary beams are allocated to a bidirectional slab ($d = 2$).

Additional secondary beams are required to support the slab load in particular types of floor systems, for example, composite decking slabs, CLT-reinforced concrete composite slabs, and CLT slabs with steel frames. In these cases, it is assumed that the slab load spans the shorter direction, while the secondary beams must span the longest. The length of the secondary beams L_S is the longest between L_X and L_y,

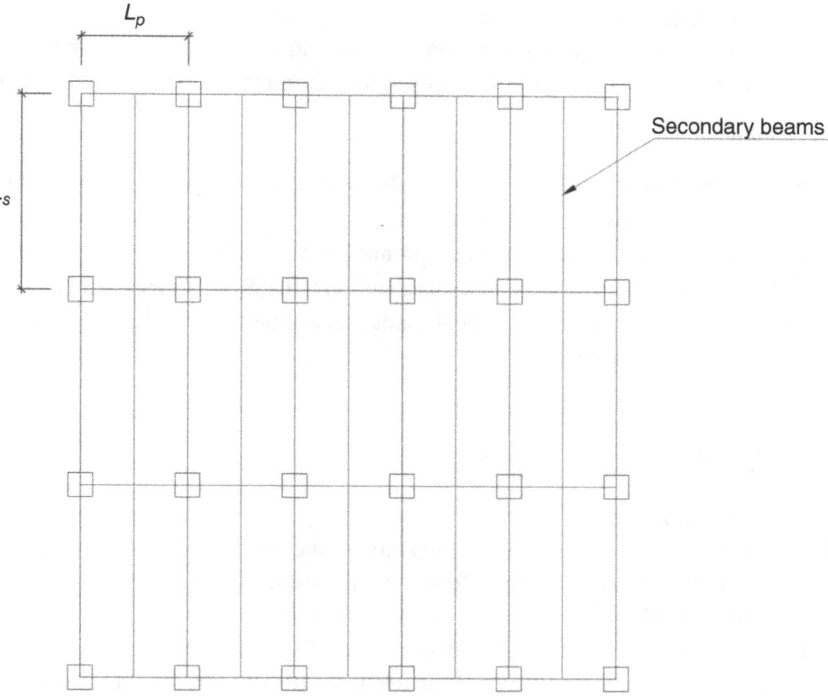

Figure 6.11 Secondary beams distribution.

whereas the length of the primary beams L_p is the shortest. Figure 6.11 shows a typical distribution. The ideal spacing between secondary beams set between 3 and 4 m. The number of secondary beams (NSB) is computed by dividing the length of the primary beams by three and rounding to the following integer number to obtain a constant spacing between beams.

$$\text{NSB} = \frac{L_p}{3} \tag{6.2}$$

After that, the distance between secondary beams (SB_{dis}) is computed by dividing the distance between primary bays by the number of secondary beams required.

$$SB_{\text{dis}} = \frac{L_p}{\text{NSB}} \tag{6.3}$$

If no further secondary beams are required, NSB = 1 and the distance between secondary beams is equal to the primary beam length, $SB_{\text{dis}} = L_p$.

6.3.2 Beam and Column Design

A preliminary member sizing should be performed to guarantee the structural feasibility of the building system. This phase is also crucial for later estimating material quantities.

6.3.2.1 Actions and Considerations

The first step is to compute the actions over the structure. The following equation is used to determine actions in accordance with EN 1990 (European Committee for Standardization, 2002a):

$$q = \gamma_G G + \gamma_Q Q \tag{6.4}$$

where G is permanent actions, Q is variable actions, and γ_G and γ_Q are the partial safety factors for load combinations as shown in Table 6.5.

Only vertical loads have been incorporated in the preliminary sizing. The loads should be uniformly distributed throughout the beams. Slab self-weight (S_L), imposed loads (IL), and superimposed dead loads (SDL) are examples of vertical loads. As a result, the following formulas are used to determine permanent and variable loads:

$$G = S_L + \mathrm{SDL} \tag{6.5}$$

$$Q = IL \tag{6.6}$$

6.3.2.2 Beam Sizing

The loads across a single beam are determined once the combination of actions has been computed. The tributary area is defined by the different floor system directions (d), as shown in Figure 6.12.

Another consideration made for the sake of simplicity is that the section chosen for the beams and columns would be the same for each of the building's stories.

In the case of steel solutions, constraints on strength in ULS conditions and displacement in SLS conditions are applied after the loads are determined. Due to these constraints, the required bending moment and inertia are minimized. For the beams assumed as simply supported (for the gravity frame), the profile must bear the following minimum required bending moment under ULS conditions:

$$M_{\min} = \frac{\mathrm{UDL}_{uls} \cdot L^2}{8} \tag{6.7}$$

Table 6.5 Partial safety factors for load combinations.

	Dead loads (γ_G)	Live loads (γ_Q)
Ultimate limit state (ULS)	1.35	1.50
Serviceability limit state (SLS)	1	1

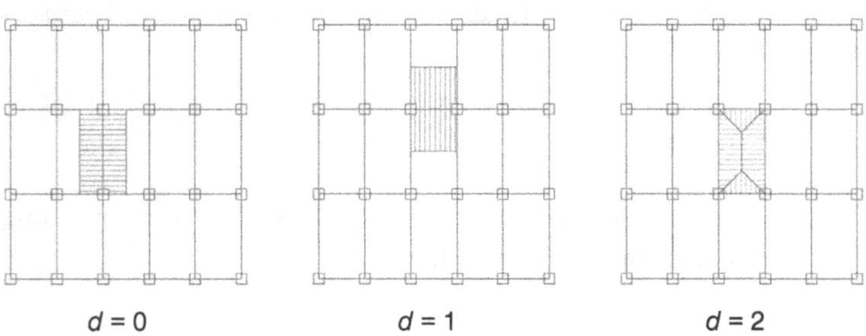

Figure 6.12 Tributary areas for each type of floors.

For SLS conditions, the minimum inertia requested for the profile (simply supported) is:

$$I_{\min} = \frac{5 \cdot \mathrm{UDL}_{sls} \cdot L^4}{384 \cdot E \cdot \delta_{\max}} \tag{6.8}$$

The maximum displacement in beams for the serviceability limit state is represented by $\delta_{\max}$, which has been set at $\delta_{\max} = \frac{L}{250}$

For beams, a uniform distributed load (UDL) is supposed. In the case of steel structures, the following formula is used to compute the UDL over secondary beams.

$$\mathrm{UDL}_s = q \times SB_{\mathrm{dis}} \tag{6.9}$$

The load over primary beams is computed in the same way but adding the weight of the secondary beams:

$$\mathrm{UDL}_p = (q + \gamma_G \times \text{weight of secondary beams}) \times L_s \tag{6.10}$$

In the case of reinforced concrete beams, the width of all beams is assumed to be 300 mm, with a reinforcement ratio of 2% of the total concrete volume. A span-length ratio of $\frac{L}{12}$ has been established to compute the required depth.

In the case of mass timber beams, glulam beams are used. The dimension of the beams is retrieved after the loads are determined. In terms of the floor type "0," which is the primary beam span in the longest direction, the ultimate distributed load is computed by multiplying the tributary area by the width of the smaller dimension. For floor type "1," the ultimate load may be calculated by multiplying the tributary area by the width of the greater dimension. To evaluate the load on the beams for floor type "2" in both directions, the yield line theory is utilized, as shown in Figure 6.13:

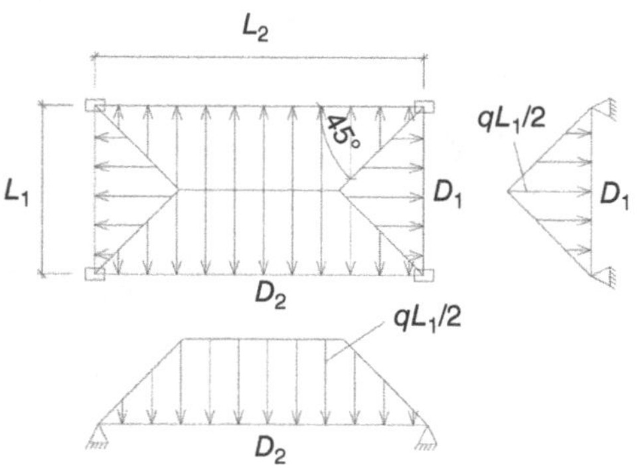

Figure 6.13 The load path for two-way slab.

The equivalent load approach (and Figure 6.14) is used to acquire the equivalent uniform distributed loads from the diagonal and triangular distributed loads.

Figure 6.14 Equivalent distributed loads.

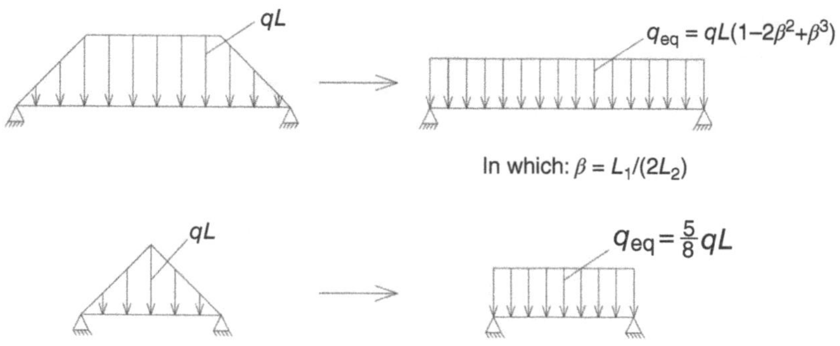

After finding the uniform distributed load acting on the slabs, the span length is considered to find the most appropriate sections according to the span-length table taken from Norman (2018).

6.3.2.3 Column Sizing

As seen in Figure 6.15, the internal columns and carry a heavier load than the external ones. The load over the columns is thus calculated as follows:

$$N = q \times L_x \times L_y \times NS \tag{6.11}$$

N: Axial load acting on a column (the vertical force the column must support).
q: Applied load per unit area (load intensity, typically given as weight per square meter, e.g. kN/m^2).
L_x: Length of the tributary area in the *x*-direction (the horizontal dimension of the area contributing load to the column along the x-axis).
L_y: Length of the tributary area in the *y*-direction (the horizontal dimension of the area contributing load to the column along the y-axis).
NS: Number of stories or floors supported by the column (how many levels above transfer their load to this column).

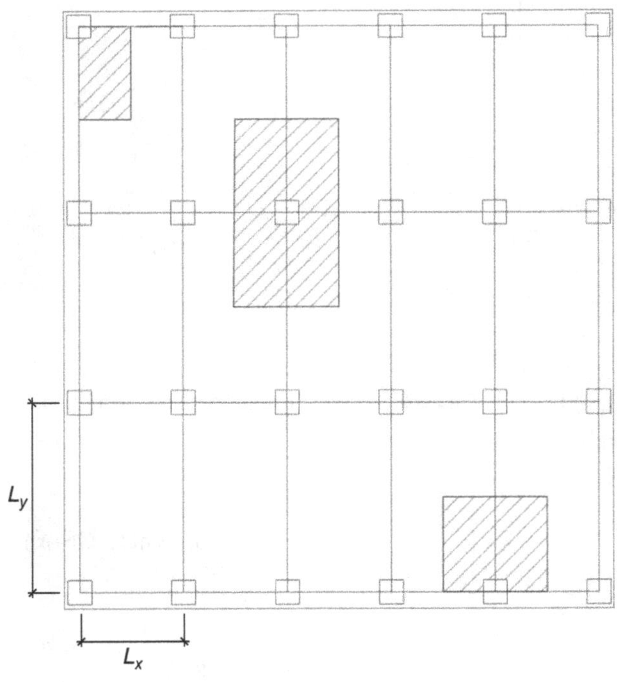

Figure 6.15 Tributary area of columns.

Mass timber columns are sized using the load-height tables (Norman, J / Exova BM TRADA) to get the column dimensions.

In the case of tall buildings and following EN 1991 (European Committee for Standardization, 2002b), the imposed loads on the columns can be lowered by applying the reduction factor α_n, which is computed using the number of stories as follows:

$$\alpha_n = \frac{2 + (n - 2) \cdot \psi_0}{\mathrm{n}} \tag{6.12}$$

where n is the total stories (> 2) and the factor ψ_0 is 0.7.

6.3.3 The Floor-to-floor Height

The *floor-to-floor height (h)* has been determined depending on the floor system used and its possible integration, which is explained in Figure 6.16.

In all the configurations, a raised floor (a secondary elevated floor structure built above a structural slab, creating a space typically used to house services such as electrical wiring, HVAC ducts, and communication cables) has been assumed.

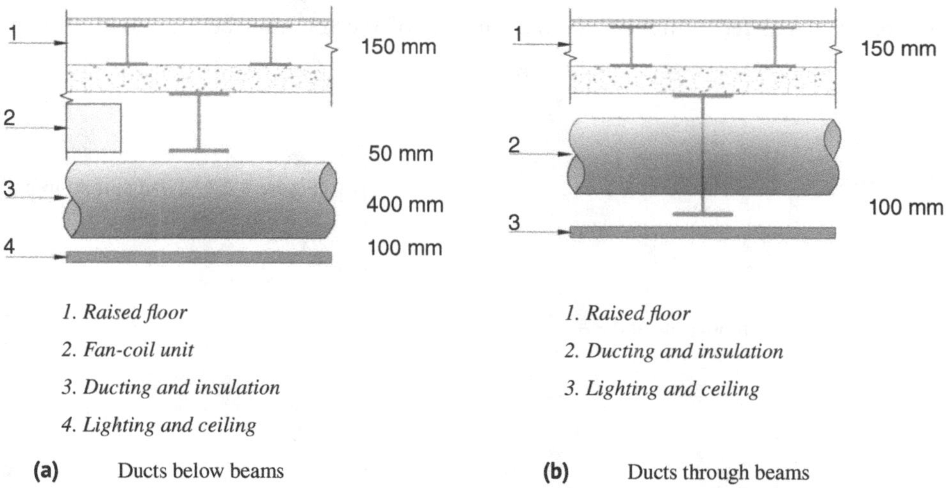

Figure 6.16 Floor zone cross-section of two different systems' integration strategies.

6.3.4 Foundation Systems

The foundation type and size have been included in the model as dependent variables. The types of foundations considered are the following:

Pad shallow foundations: Isolated footings have been chosen when the area of foundation needed is small (Figure 6.17).

Strip shallow foundations: Continuous foundations in one orthogonal direction have been chosen when the area of foundations needed results in individual footings too close to each other in one direction. The criterion utilized to decide strip foundations is:

$$B_f > D_c \cdot \frac{2}{3} \tag{6.13}$$

Where B_f is the length of the needed foundation per column (supposing it is squared) and D_C is the distance between columns (Figure 6.18).

Raft shallow foundations: Used when the area of foundation needed is high. The criteria used are the same as the one used for strip foundations, but in the case that this criterion is met in both orthogonal directions.

Deep foundations: Deep foundations are considered when the foundation area needed is higher than the building footprint, and the upper soil layers are thick but not strong enough to bear the superstructure loads. Precast deep foundations have been considered in this case (CYPE Ingenieros S.A., 2020).

The foundation area is calculated using the bearing capacity of the soil and the total load over the foundations.

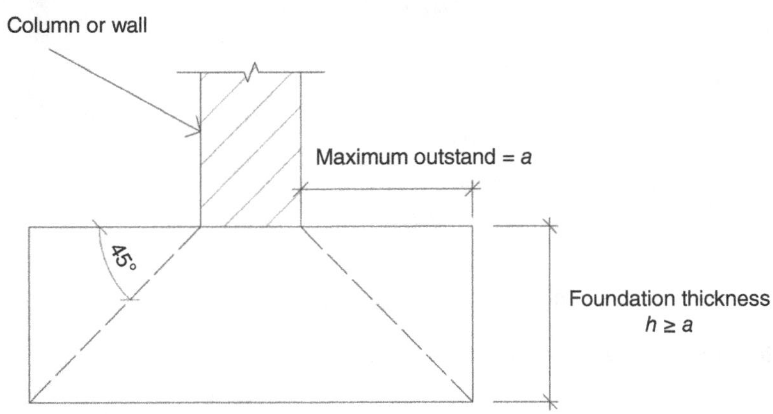

Figure 6.17 Load spread in pad shallow foundation.

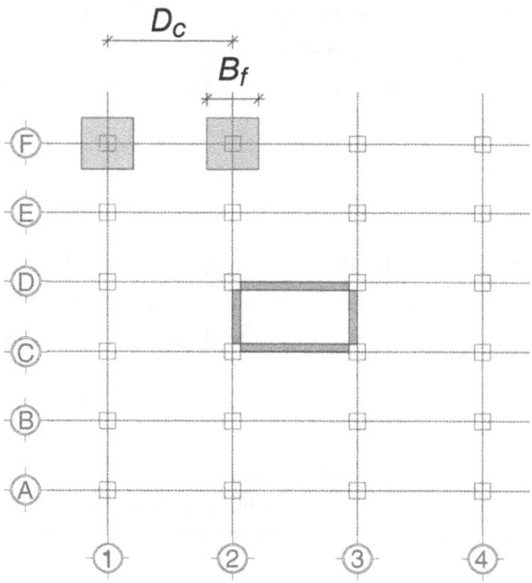

Figure 6.18 Foundation scheme.

6.3.5 Fitness Function to Address Different Parameters in an Optimization Problem

Optimization involves a fitness function that integrates total cost, open space, and embodied carbon objectives, while penalizing infeasible designs. The overall cost is divided into several parts:

Land cost: Computed by multiplying project area by a unit price per square meter.

Structural cost: Derived from material and labor expenses for elements such as columns, beams, and lateral systems.

Joint costs: Estimated based on connection type, with fixed joints costing more than pinned ones in steel.

Foundation costs: Use unit prices from established sources and consider both shallow and deep types, factoring in concrete volume, formwork area, and pile installation.

Floor costs: Depend on slab area and specific unit costs for various systems, including reinforced concrete and mass timber variants, calculated according to slab thickness and reinforcement ratio.

Optimal free space is evaluated using the diagonal span to balance directional dimensions. Embodied carbon is determined by multiplying material quantities by carbon factors for each life-cycle module, with analysis covering key construction stages and expressed per Gross Internal Area. This approach supports the multi-objective optimization of economic and environmental performance overall. Detailed indications about the cost and embodied carbon calculation parameters used in this chapter are already explained in this open access paper (Kanyilmaz et al., 2023b).

6.3.6 Repair Module to Handle Non-feasible Solutions

The incompatibilities between material and floor type, and between the floor type and span lengths are summarized in Tables 6.6 and 6.7, respectively. The span length capacity of RC and steel buildings is taken from the previous research of Kanyilmaz et al. (2022). The references for the mass timber floors are from Norman (2018) and Stora Enso's design guide (Stora Enso, 2021).

Table 6.6 Compatibility matrix between structural materials (steel, reinforced concrete, and mass timber) and floor system types. A check mark (✓) indicates that the floor system is structurally compatible with the given material, while a cross (×) denotes an incompatible combination excluded by the optimization algorithm's repair module.

Steel	RC	Mass timber	
✓	×	×	Composite beams
✓	×	×	Cellular composite beams
✓	×	×	CLT-reinforced concrete composite slab (for steel frames)
✓	×	×	CLT slab (for steel frames)
×	✓	×	Precast-reinforced concrete units
×	✓	×	One-way-reinforced concrete slab
×	✓	×	Two-way-reinforced concrete slab
×	✓	×	Flat-reinforced concrete slab
×	×	✓	CLT-reinforced concrete composite slab
×	×	✓	Mass timber ribbed slab
×	×	✓	One-way CLT slab
×	×	✓	Two-way CLT slab
×	×	✓	Flat CLT slab

Table 6.7 Compatibilities between floor systems and span lengths (m).

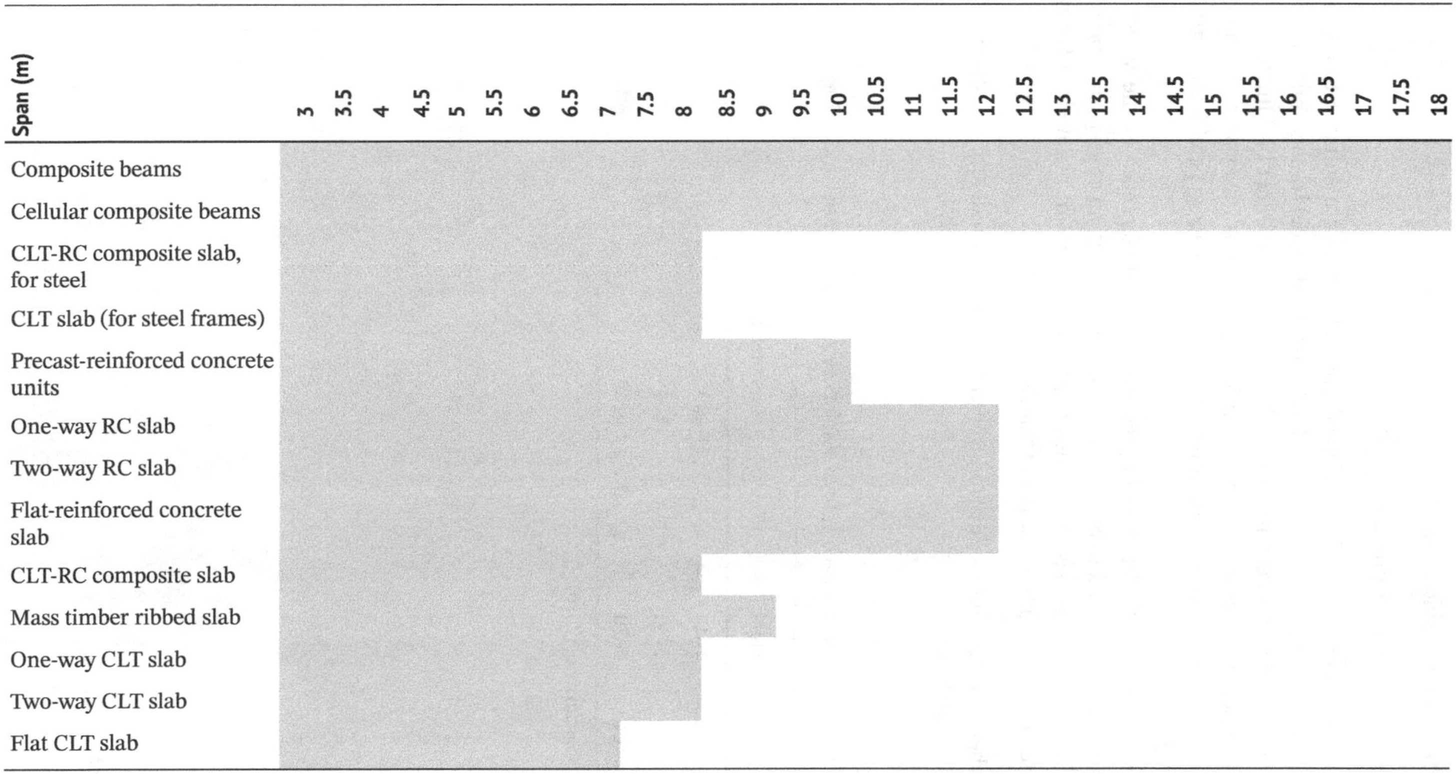

6.3.7 Selection of Lateral Stability Systems

The following factors have been considered:

Material: Moment-resisting frames and reinforced concrete cores are suitable to all material types. On the other hand, bracing systems are assumed to be ideal with steel frames and mass timber frames. In addition, the reinforced concrete core also can be used in mass timber buildings.

Number of floors: When a structure reaches a particular height, some systems become insufficient. According to Norman (2018), the standard number of floors for each type of lateral structural system is listed in Table 6.8.

Building occupancy: The use of the building requires some preferences in the structural grid. For example, if the building is planned to be used as an office building, an open plan is preferred.

Beams size: The use of moment-resisting frames requires a minimum beam size (400 mm depth) to install rigid connections for moment-resisting frames.

The factors have been merged into a decision tree created in the model for the stability system selection, as illustrated in Figure 6.19.

Table 6.8 Typical number of floors for each type of stability systems (residential and office).

	Steel	RC	Mass timber
Moment-resisting frames	1–4	1–2	1–2
Braced frames	1–20	NA	1–4
RC shear walls/RC cores	4–20	2–20	4–20

Source: Adapted from Norman et al. (2020).

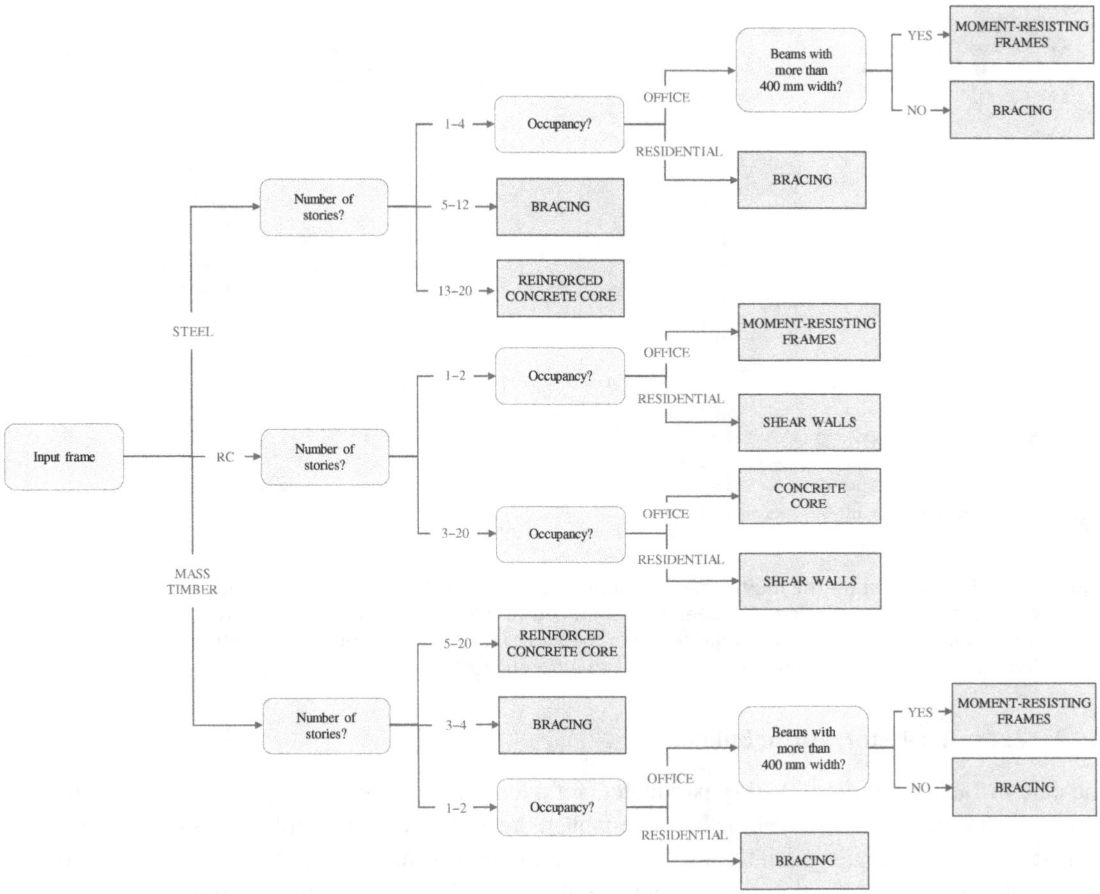

Figure 6.19 Decision tree to determine the most appropriate LLRS.

6.3.8 Pareto Front Representation

The Pareto front includes all solutions that cannot be improved in one objective without worsening another. Graphically, each solution is a point in objective space, and the Pareto front is formed by those points that define the boundary of feasible outcomes. Selecting one point on the front means trading off one objective for another. In engineering design, this can help decision makers explore various trade-offs and choose an option that best matches their priorities. A point on the Pareto front might reduce weight but increase material cost, or lower energy consumption at the expense of system reliability. Instead of focusing on a single best solution, one can select from a set of optimal trade-offs (Reddy and Kumar, 2014; Vermeulen, 2018). Figure 6.20 shows an example.

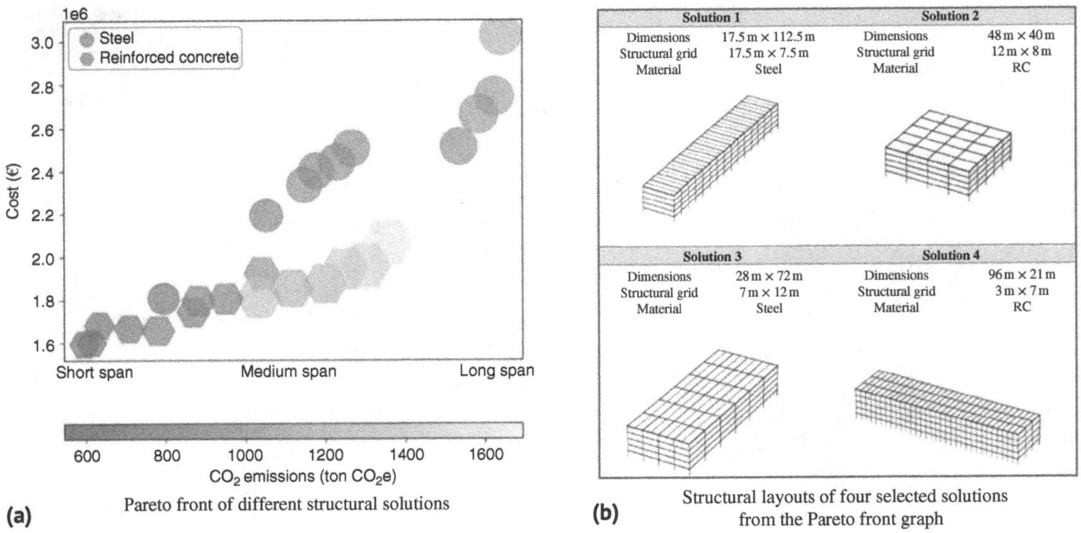

Figure 6.20 Pareto front for the multi-objective optimization of a building structure (example): (a) trade-off between cost and CO_2 emissions for steel and reinforced concrete solutions across different span lengths; (b) structural layouts of four selected solutions from the Pareto front, illustrating how varying the building proportions, grid spacing, and material choice leads to different cost–carbon combinations.

6.3.9 Example: 4-story Office Building

The data in Table 6.9 present the key parameters for a four-story office building. Its overall dimensions, number of floors, and floor-to-ceiling height establish the basic geometry, while the usage category confirms standard office loading conditions. The footprint is approximately 45 m by 45 m, with four stories. Each level features a 3 m floor-to-ceiling height, giving a total building height of nearly 12 m. The design load includes a 2.5 kN/m^2 imposed load and 1.5 kN/m^2 superimposed dead load. Soil-bearing capacity is 50 kN/m^2. Both primary and secondary steel members are S355 grade. The table also outlines the lateral load resisting system (LLRS) options, which might be braced frames, shear walls, or other stabilizing elements. In addition, it indicates possible floor types, such as composite steel decking, reinforced concrete slabs, or mass timber panels.

Lower cost and lower carbon solutions generally appear toward the bottom-left portion of the graph (Figure 6.22), while higher values lie toward the upper-right. Several solutions are highlighted (Figure 6.21), including a 9 m × 7.5 m mass timber scheme with an estimated cost of 3.07 million euros and approximately 93 kgCO_2e/m^2 GIA, a 7.5 m × 7.5 m steel arrangement at 2.18 million euros and around 124 kgCO_2e/m^2 GIA, and a 9 m × 7.5 m reinforced concrete option at 1.88 million euros and roughly 140 kgCO_2e/m^2 GIA.

The results demonstrate how changes in span length and material choice influence both cost and embodied carbon outcomes. Mass timber solutions often yield lower emissions, while steel variants can be cost-effective while emitting moderately low embodied carbon. Reinforced concrete typically presents moderate to high embodied carbon, but it comes with lower costs.

These outcomes are based on specific cost and unit carbon footprint assumptions. If these assumptions change, the relative performance of each solution on the Pareto front could shift significantly.

Table 6.9 Input parameters.

X dimension	45 m	Slab type-steel frame	RC composite slab
Y dimension	45 m	Slab type-RC frame	Precast slab
Number of stories	4	Slab type-timber frame	CLT ribbed slab
Floor-to-ceiling height	3 m	LLRS type-steel frame	Steel bracing
Imposed load	2.5 kN/m^2	LLRS type-RC frame	RC shear wall
Superimposed dead load	1.5 kN/m^2	LLRS type-timber frame	Timber bracing
Steel grade secondary	S355		
Steel grade primary	S355		
Occupancy	Office		
Soil-bearing capacity	50 kN/m^2		

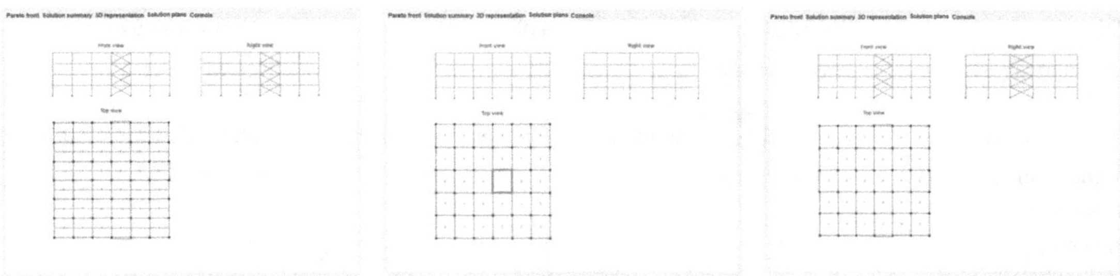

7.5 m × 7.5 m (steel) 7.5 m × 9 m (RC) 7.5 m × 9.0 m (Mass timber)

Figure 6.21 Selected configurations.

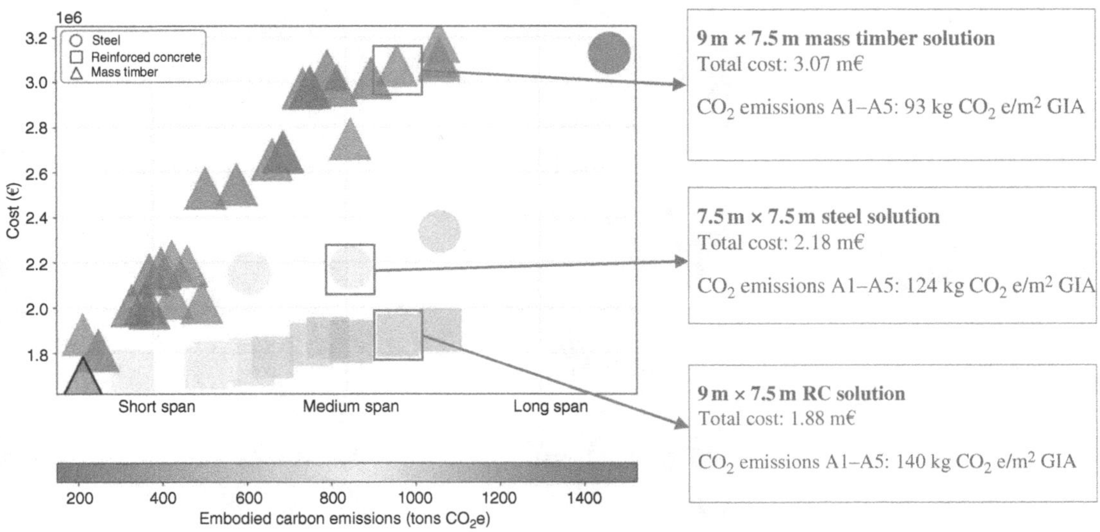

Figure 6.22 Pareto front, 4-story office building.

Table 6.10 details overall building dimensions, structural grid, floor type and depth, member sizes, lateral load resisting system, foundation specifics, total cost, and embodied CO_2 emissions of the three selected options from the Pareto front.

Table 6.10 Comparison between selected solutions.

	a) [Steel-optimized solution]	**b) [RC-optimized solution]**	**c) [Mass timber-optimized solution]**
Total building dimension:	45 m × 45 m	45 m × 45 m	45 m × 45 m
Structural grid:	7.5 m × 7.5 m	7.5 m × 9 m	7.5 m × 9 m
Type of floor:	Decking composite slab	Precast slab	Mass timber ribbed slab
Floor depth:	130 mm	250 mm	360 mm
Secondary beams:	IPE360	300 mm × 630 mm	160 mm × 360 mm
Primary beams:	IPE450	300 mm × 630 mm	320 mm × 840 mm
Columns:	HEA220	410 mm × 410 mm	480 mm × 480 mm
Lateral load resisting system:	Braced frame	RC core	Braced frame
Foundation type	Raft shallow foundations	Raft shallow foundations	Raft shallow foundations
Foundation dimension (w × l × t)(m)	45 × 45 × 0.3	45 × 45 × 0.3	45 × 45 × 0.3
Total cost:	2.18 m€	1.88 m€	3.07 m€
Embodied CO_2 A1 – A5	124 kg CO_2e/m^2 GIA	140 kg CO_2e/m^2 GIA	93 kg CO_2e/m^2 GIA
Embodied CO_2 A + C	133 kg CO_2e/m^2 GIA	153 kg CO_2e/m^2 GIA	100 kg CO_2e/m^2 GIA

6.3.10 Example: 8-story Residential Building

Table 6.11 presents the key input parameters for an eight-story residential building designed for residential occupancy. The footprint remains approximately 45 m by 45 m, but with eight stories, each featuring a 3 m floor-to-ceiling height, resulting in an overall height near 24 m. The design load includes a 2.5 kN/m^2 imposed load and a 1.5 kN/m^2 superimposed dead load, while the soil-bearing capacity is 50 kN/m^2. Both primary and secondary steel members are specified as S355 grade. The table also details options for the floor system and lateral load resisting system (LLRS): the steel frame option features an RC composite slab paired with steel bracing, the RC frame option uses a precast slab with RC shear walls, and the timber frame option employs a CLT ribbed slab in combination with a concrete core.

Pareto front (Figure 6.24) displays the trade-offs between total cost and embodied carbon emissions for the three solutions (Figure 6.23). It highlights a 9 m × 9 m mass timber solution with a total cost

Table 6.11 Input parameters.

X dimension	45 m	Slab type-steel frame	RC composite slab
Y dimension	45 m	Slab type-RC frame	Precast slab
Number of stories	8	Slab type-timber frame	CLT ribbed slab
Floor-to-ceiling height	3 m	LLRS type-steel frame	Steel bracing
Imposed load	2.5 kN/m^2	LLRS type-RC frame	RC shear wall
Superimposed dead load	1.5 kN/m^2	LLRS type-timber frame	Concrete core
Steel grade secondary	S355		
Steel grade primary	S355		
Occupancy	Residential		
Soil-bearing capacity	50 kN/m^2		

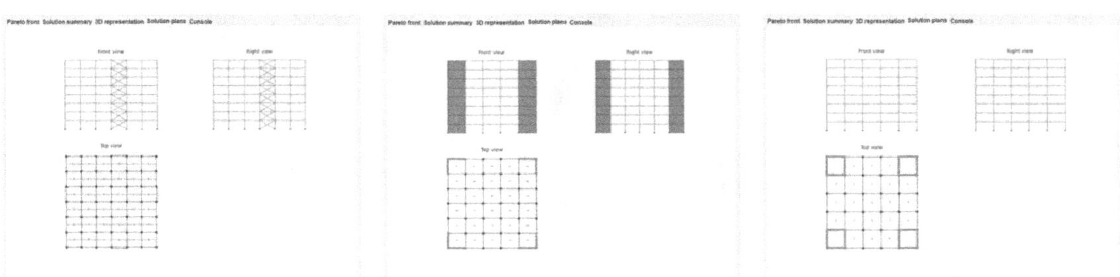

7.5 m × 7.5 m (steel) 9 m × 7.5 m (RC) 9 m × 9 m (Mass timber)

Figure 6.23 Selected configurations.

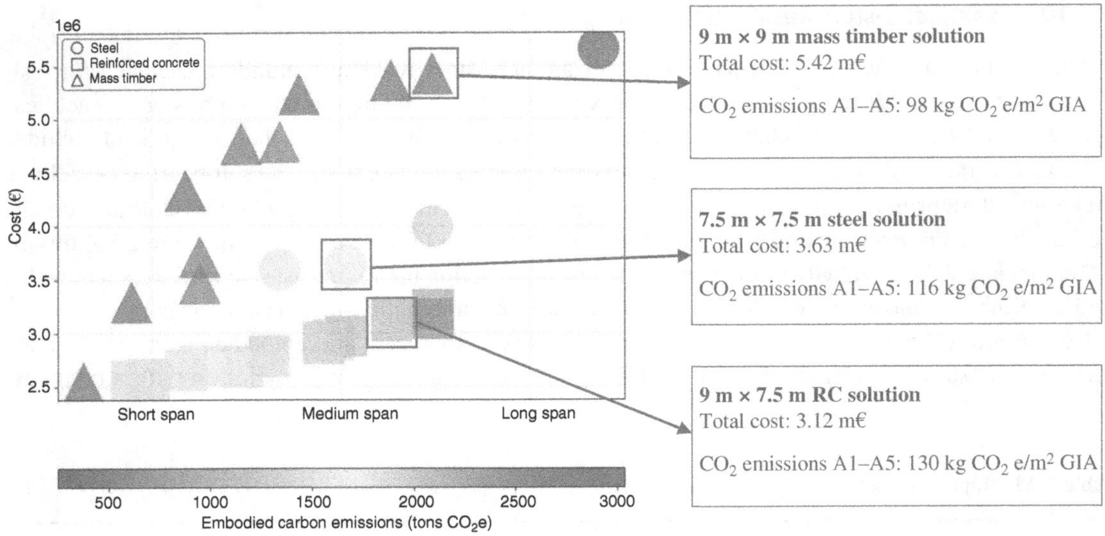

Figure 6.24 Pareto front, 8-story residential building.

of 5.42 m€ and embodied carbon of 98 $kgCO_2e/m^2$GIA; a 7.5 m × 7.5 m steel solution at 3.63 m€ and 116 $kgCO_2e/m^2$ GIA; and a 9 m × 7.5 m RC solution at 3.12 m€ and 130 $kgCO_2e/m^2$ GIA. These points show how variations in structural grid, material selection, and system configuration impact both financial and environmental performance. Stakeholders can use this information to balance cost efficiency with sustainability objectives, selecting the solution that best meets their project requirements.

Table 6.12 details overall building dimensions, structural grid, floor type and depth, member sizes, lateral load resisting system, foundation specifics, total cost, and embodied carbon emissions of the three selected solutions.

Table 6.12 Comparison between selected solutions.

	a) [Steel-optimized solution]	b) [RC-optimized solution]	c) [Mass timber-optimized solution]
Total building dimension:	45 m × 45 m	45 m × 45 m	45 m × 45 m
Structural grid:	7.5 m × 7.5 m	9 m × 7.5 m	9 m × 9 m
Type of floor:	Decking composite slab	Precast slab	Mass timber ribbed slab
Floor depth:	130 mm	250 mm	360 mm
Secondary beams:	IPE360	300 mm × 630 mm	160 mm × 360 mm
Primary beams:	IPE450	300 mm × 630 mm	320 mm × 1000 mm
Columns:	HEA320	570 mm × 570 mm	600 mm × 600 mm
Lateral load resisting system:	Braced frame	Shear walls	Concrete core
Foundation type	Deep piles foundations	Deep piles foundations	Deep piles foundations
Foundation dimension	6 piles, $D = 27.5$ cm, $Q = 75$ (ton)	10 piles, $D = 27.5$ cm, $Q = 75$ (ton)	7 piles, $D = 27.5$ cm, $Q = 75$ (ton)
Total cost:	3.63 m€	3.12 m€	5.42 m€
Embodied CO_2 A1–A5	116 $kgCO_2e/m^2$ GIA	130 $kgCO_2e/m^2$ GIA	98 $kgCO_2e/m^2$ GIA
Embodied CO_2 A + C	124 $kgCO_2e/m^2$ GIA	143 $kgCO_2e/m^2$ GIA	106 $kgCO_2e/m^2$ GIA

6.3.11 Sensitivity of Results to Cost and Carbon Factor Assumptions

The outcomes of section 6.3.10 are based on specific cost and unit carbon footprint assumptions. However, if these assumptions change, the relative performance of each solution on the Pareto front could shift significantly. For instance, if the production process for one material becomes more efficient, its embodied carbon may be reduced, making that option more attractive from an environmental perspective. Conversely, if market dynamics alter the cost structure – through fluctuations in raw material prices or changes in labor expenses – the cost-effectiveness of a given solution might improve or deteriorate. In one scenario, advancements in manufacturing could lower a material's carbon intensity without significantly affecting its cost, shifting its position toward a more sustainable outcome. In another scenario, market changes might drive up the cost of a material while its carbon footprint remains constant, rendering alternative solutions more competitive overall. These variations emphasize the need to continuously review and update the underlying assumptions to keep the design tool aligned with current economic and environmental realities.

6.4 A Parametric Study on Key Design Variables Using Life-cycle Optimization

In this section, using the methodology described previously, we are going to check the sensitivity of carbon factor choice on the results (an important topic we touch at several parts of the book). Table 6.13 presents the input parameters for this study, analyzing a 10-story residential building with a fixed 60 m × 60 m dimension in both directions and a 9 m × 9 m span length. Steel-framed structures incorporate RC composite slabs, RC-framed buildings use precast slabs, and timber-framed options employ CLT ribbed slabs. The lateral load-resisting system varies based on material choice, with steel frames utilizing steel bracing, RC buildings incorporating shear walls, and timber structures relying on timber bracing or a reinforced concrete core. These standardized parameters allow for a direct comparison of different structural solutions in terms of cost and embodied carbon efficiency.

Table 6.13 Input parameters.

X dimension	60 m	Slab type-steel frame	RC composite slab
Y dimension	60 m	Slab type-RC frame	Precast slab
Number of stories	10	Slab type-timber frame	CLT ribbed slab
Span length	9 m × 9 m	LLRS type-steel frame	Steel bracing
Floor-to-ceiling height	3 m	LLRS type-RC frame	RC shear wall
Imposed load	2.5 kN/m^2	LLRS type-timber frame	Timber bracing/Concrete core
Superimposed dead load	1.5 kN/m^2		
Steel grade secondary	S355		
Steel grade primary	S355		
Occupancy	Residential		
Soil-bearing capacity	50 kN/m^2		

6.4.1 Virgin, Recycled, and Reclaimed Steel

Figure 6.25 shows a significant change in the solution set composition when shifting from typical (having an industry-average carbon factor reflecting a market mix of different production routes: includes a significant portion of virgin steel) to recycled, and then to reused steel: the tool favored steel frames over timber and RC solutions, with the former offering better cost efficiency and enhanced embodied carbon metrics. The importance of reused steel indeed has been recently also highlighted in a recent paper (Kanyilmaz et al., 2023a).

With typical steel, the embodied carbon is elevated due to virgin material content in the production mix, reducing the competitiveness of steel solutions and favoring RC and timber alternatives in the optimal set. When recycled steel is introduced, steel solutions start to appear more frequently. With reused steel, this effect becomes even stronger. Because reused steel has the lowest embodied carbon

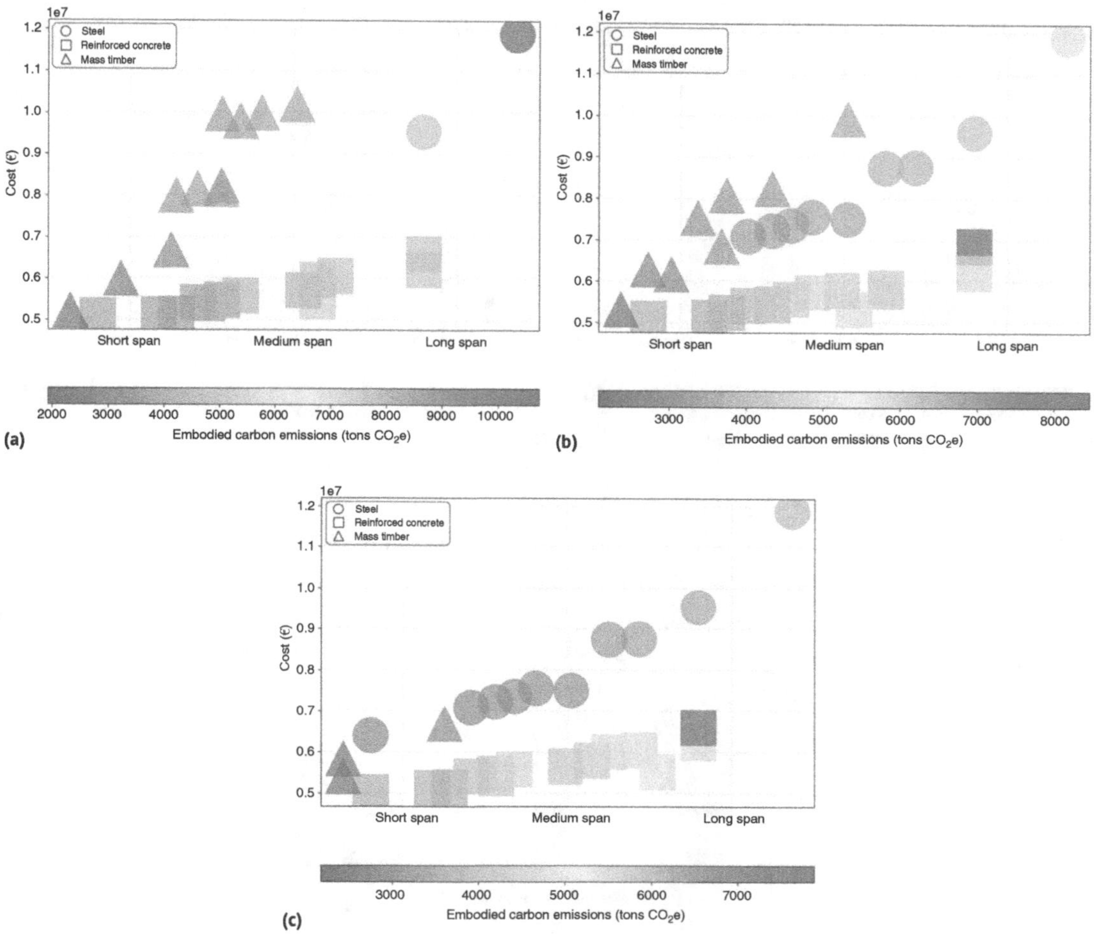

Figure 6.25 Comparison of solution sets with changing steel type: (a) typical steel, (b) recycled steel, and (c) reused steel.

among all steel options, steel-framed buildings become the most favored choice. The reduction in environmental impact is so significant that steel solutions now outperform RC and timber in most cases. This demonstrates that reusing structural steel is a highly effective way to lower carbon emissions while keeping costs under control, making it one of the most sustainable design choices.

6.4.2 Percentage of Ground Granulated Blast Furnace Slag

Figure 6.26 shows a clear shift in the solution set composition as the GGBS content in concrete increases. As GGBS percentage increases, RC solutions become significantly favorable in the optimization process, primarily due to their significantly reduced embodied carbon emissions. This reduction enhances the competitiveness of RC structures against steel and timber alternatives, especially in scenarios where minimizing embodied carbon is a priority.

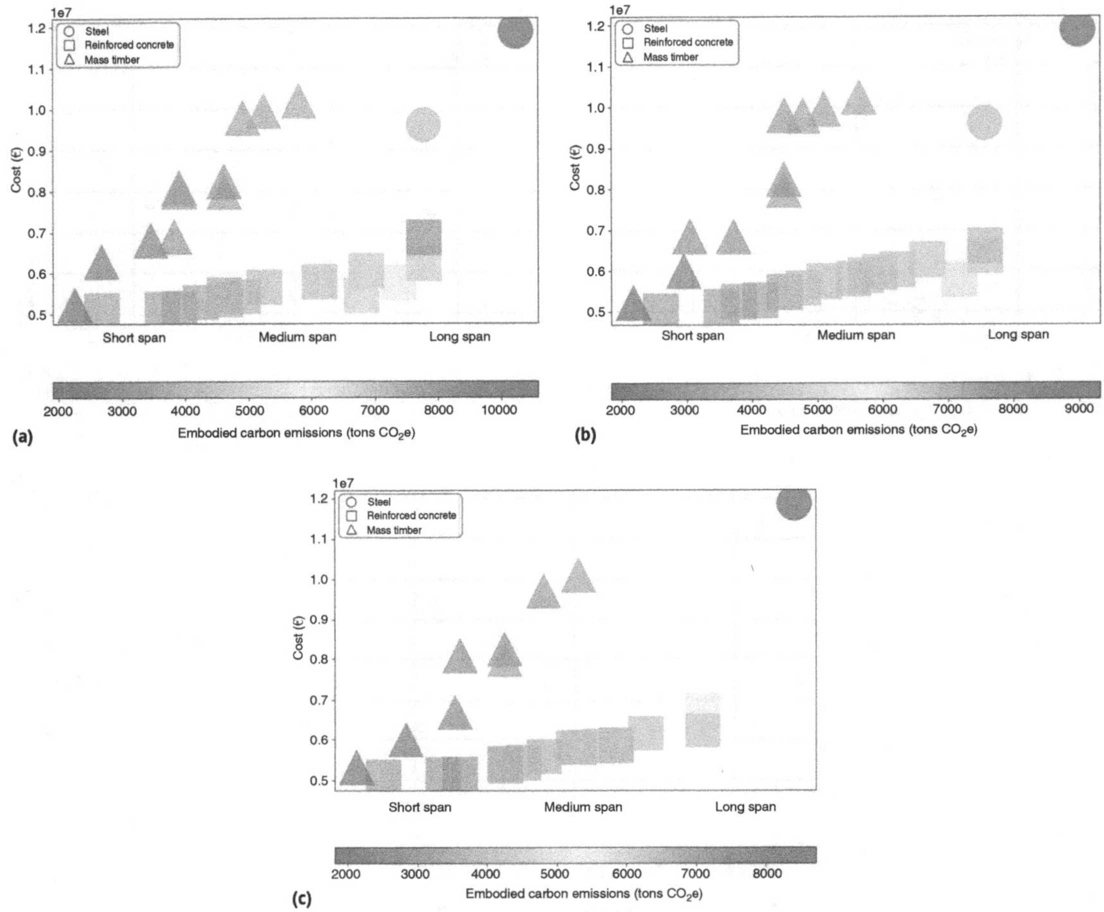

Figure 6.26 Comparison of solution sets with changing concrete type: (a) concrete 0% GGBS; (b) concrete 25% GGBS; (c) concrete 70% GGBS.

As the GGBS percentage in concrete increases, a noticeable shift occurs in the composition of the solution set. At 0% GGBS, RC solutions exhibit higher embodied carbon emissions, making them less competitive against steel and timber alternatives. When GGBS content reaches 25%, RC solutions start to become more viable. The partial replacement of cement with GGBS significantly lowers embodied carbon, allowing RC structures to better compete with steel and timber. At this stage, the optimization tool identifies a more balanced selection of solutions, where RC buildings appear more frequently in the optimal set. Then, with 70% GGBS, this trend becomes even more pronounced. The embodied carbon of RC structures is substantially reduced, making them one of the most favorable options in the optimization process. This allows RC solutions to outperform other materials in scenarios where carbon reduction is prioritized while maintaining cost efficiency. However, it is important to remember that GGBS availability is geographically limited and dependent on steel production, meaning this approach cannot serve as a universal solution for achieving low-carbon concrete globally, as explained in detail in Chapter 2.

In this comparison, please note that essentially, every run of the model balances a range of objectives, and even a small change in one parameter can cause other solutions to be reevaluated. This may change the best cross-section sizes, spans, or other design details in all materials, because the model is seeking optimal solutions in a multicriteria space.

In the following sections, using the methodology described above, we are going to check the impact of the geometric parameters on building cost and embodied carbon. Table 6.14 summarizes the carbon factors used in the analysis.

Table 6.14 Carbon factors used in this section, derived from industry-average data for each life-cycle stage (see Chapters 2 and 3 for a detailed discussion of carbon factor sources, variability, and selection criteria).

Material	A1–A3: PRODUCT stage ($kgCO_2e/kg$)	A4: CONSTRUCTION stage ($kgCO_2e/kg$)	C1–C4: USE stage ($kgCO_2e/kg$)	D: BEYOND LIFE ($kgCO_2e/kg$)
Structural steel sections	1.2	0.183	0.018	−0.34
RC30 concrete (CEMI)	0.129	0.005	0.018	0.004
Hollowcore slab	0.166	0.032	0.018	0.004
Mass timber sections	0.437	0.032	0.027	−0.524
Steel reinforcement	0.8	0.183	0.018	−0.79

6.4.3 Span Length

Table 6.15 shows the input parameters for this analysis aiming to isolate and examine the particular effect of changes in span length on costs and embodied carbon emissions, within a fixed total length of X and Y dimensions which are 54 m and 60 m respectively. Buildings consist of three types of frames (Figure 6.27). These are steel, RC, and timber frames. RC composite slabs, precast slabs, and CLT ribbed or flat slabs are adopted for steel, RC, and timber frames, respectively.

Table 6.15 Input parameters.

X dimension	54 m	Slab type-steel frame	RC composite slab
Y dimension	60 m	Slab type-RC frame	Precast slab
Number of stories	8	Slab type-timber frame	CLT ribbed/flat slab
Floor-to-ceiling height	3 m	LLRS type-steel frame	Steel bracing
Imposed load	2.5 kN/m^2	LLRS type-RC frame	RC shear wall
Superimposed dead load	1.5 kN/m^2	LLRS type-timber frame	Concrete core
Steel grade secondary	S355		
Steel grade primary	S355		
Occupancy	Residential		
Soil-bearing capacity	50 kN/m^2		

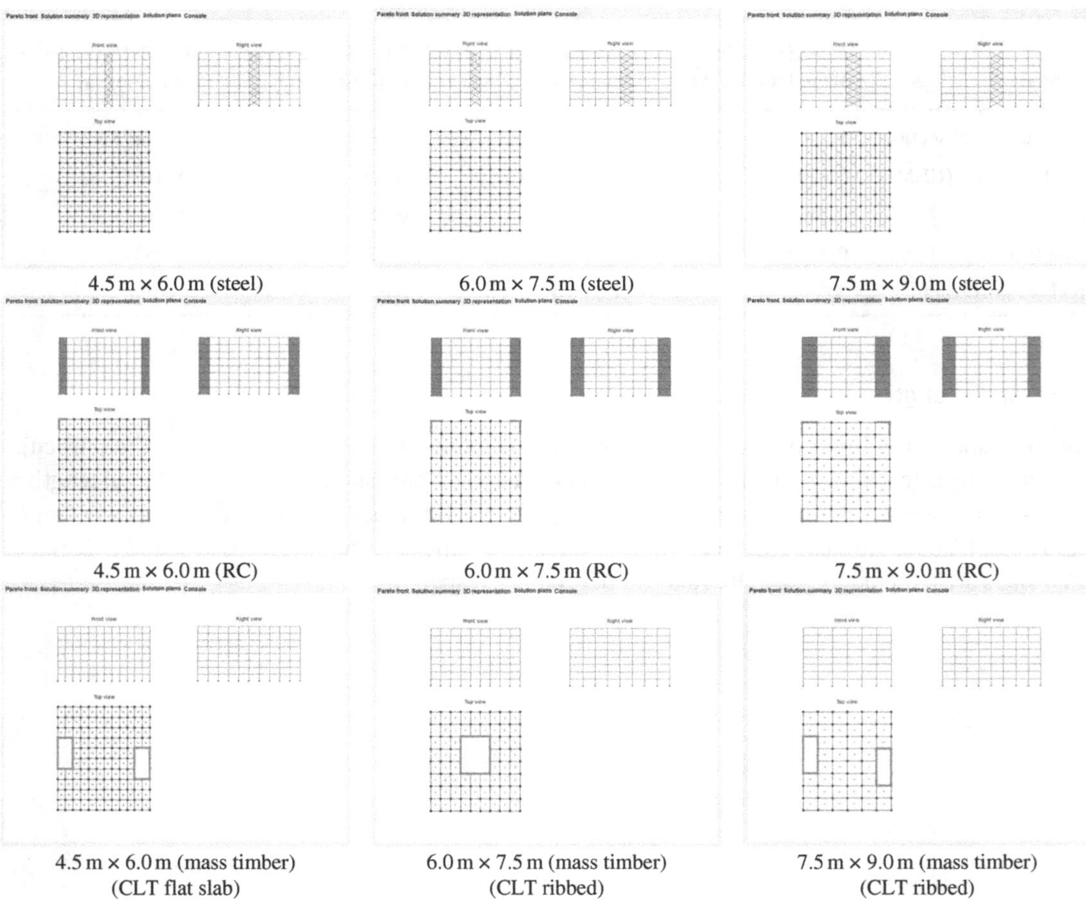

4.5 m × 6.0 m (steel) | 6.0 m × 7.5 m (steel) | 7.5 m × 9.0 m (steel)

4.5 m × 6.0 m (RC) | 6.0 m × 7.5 m (RC) | 7.5 m × 9.0 m (RC)

4.5 m × 6.0 m (mass timber) (CLT flat slab) | 6.0 m × 7.5 m (mass timber) (CLT ribbed) | 7.5 m × 9.0 m (mass timber) (CLT ribbed)

Figure 6.27 Output plan views.

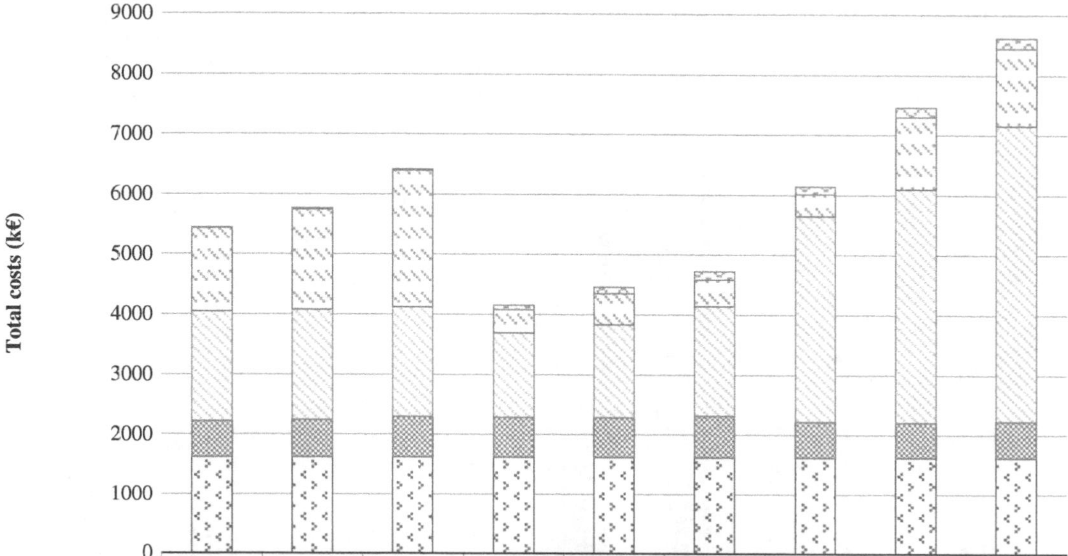

	4.5 m × 6 m (steel)	6 m × 7.5 m (steel)	7.5 m × 9 m (steel)	4.5 m × 6 m (RC)	6.0 m × 7.5 m (RC)	7.5 m × 9 m (RC)	4.5 m × 6 m (mass timber)	6 m × 7.5 m (mass timber)	7.5 m × 9 m (mass timber)
Lateral systems	7	12	21	80	107	137	128	159	162
Beams and columns	1387	1675	2272	380	525	450	357	1206	1293
Floors	1829	1829	1829	1412	1548	1824	3434	3888	4925
Foundations	596	623	668	660	664	696	596	583	611
Terrain	1620	1620	1620	1620	1620	1620	1620	1620	1620

Figure 6.28 Total cost comparison between different span lengths.

Figures 6.28 and 6.29 demonstrate the impact of the span length on the cost and embodied carbon values. The bigger span length options require higher lateral bracing and foundation mass for all frame types. Since the total area on which the building sits is fixed, the total terrain cost (i.e. the cost of the land) remains unchanged even if the span length changes. Figure 6.29 shows the embodied carbon intensity for steel, reinforced concrete (RC), and timber buildings for different span lengths which are 4.5 m × 6 m, 6 m × 7.5 m, and 7.5 m × 9 m.

Floors have the largest influence on carbon emission. Lateral load resistant systems, beams, and columns have a smaller impact but, their contributions go up as span length increases.

In all span configurations, increasing the span length results in higher total cost and embodied carbon per square meter due to the increased material requirements for larger structural components to increase the demanded resistance of the building (Figure 6.29). Please note that these results can differ according to the carbon factor assumptions related to the aspects such as manufacturing processes of materials.

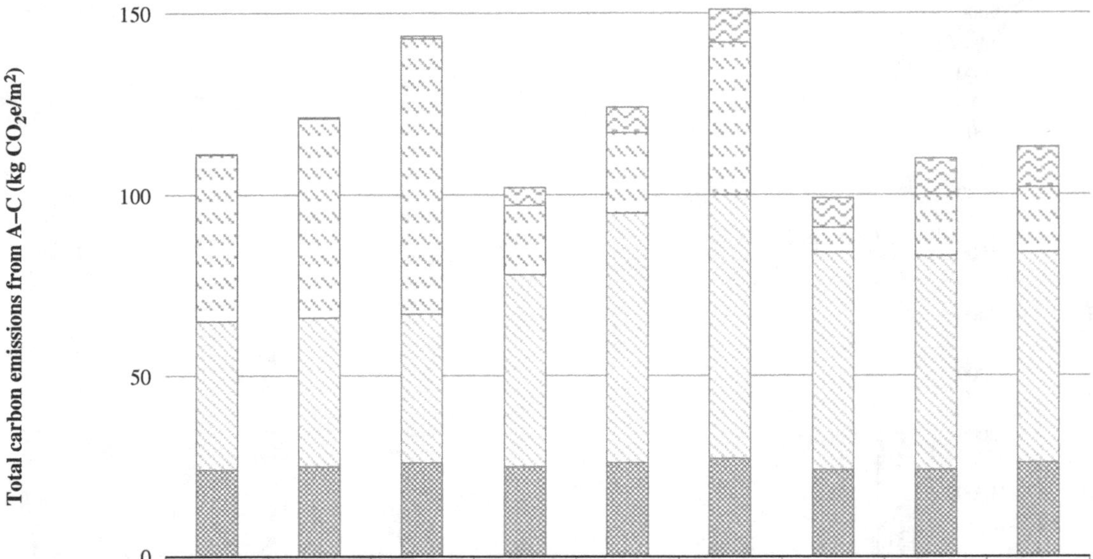

	4.5 m × 6 m (steel)	6 m × 7.5 m (steel)	7.5 m × 9 m (steel)	4.5 m × 6 m (RC)	6.0 m × 7.5 m (RC)	7.5 m × 9 m (RC)	4.5 m × 6 m (mass timber)	6 m × 7.5 m (mass timber)	7.5 m × 9 m (mass timber)
Lateral systems	0.2	0.4	0.7	5	7	9	8.0	10.0	11.0
Beams and columns	46	55	76	19	22	42	7	17	18
Floors	41	41	41	53	69	73	60	59	58
Foundations	24	25	26	25	26	27	24	24	26

Figure 6.29 Embodied carbon intensity comparison between different span lengths.

6.4.4 Building Height

Table 6.16 shows the input parameters for this analysis. The number of stories examined in this study varies between 4, 8, and 12, while the building footprint remains constant with X and Y dimensions set at 60 m × 60 m (Figure 6.30). A span length of 7.5 m × 7.5 m is maintained across all structures. Different structural materials are analyzed: steel-framed buildings utilize RC composite slabs, RC-framed structures incorporate precast slabs, and timber buildings employ CLT ribbed slabs. The lateral load-resisting system varies depending on the structural material, with steel frames relying on steel bracing, RC buildings incorporating both shear walls, and timber structures using either timber bracing or a reinforced concrete core.

Figures 6.31 and 6.32 illustrate the impact of building height on total costs and embodied carbon emissions. As the number of stories increases, structural elements such as lateral systems, beams, and columns become the dominant cost drivers. Figure 6.32 presents the embodied carbon intensity across steel, reinforced concrete (RC), and mass timber buildings at 4, 8, and 12 stories. In all cases, the plan dimensions (60 m × 60 m) remain constant, allowing a direct comparison of how increasing building height affects material consumption and embodied carbon emissions.

Table 6.16 Input parameters.

X dimension	60 m	Slab type-steel frame	RC composite slab
Y dimension	60 m	Slab type-RC frame	Precast slab
Number of stories	4/8/12	Slab type-timber frame	CLT ribbed slab
Span length	7.5 m × 7.5 m	LLRS type-steel frame	Steel bracing
Floor-to-ceiling height	3 m	LLRS type-RC frame	RC shear wall
Imposed load	2.5 kN/m^2	LLRS type-timber frame	Timber bracing/Concrete core
Superimposed dead load	1.5 kN/m^2		
Steel grade secondary	S355		
Steel grade primary	S355		
Occupancy	Residential		
Soil-bearing capacity	50 kN/m^2		

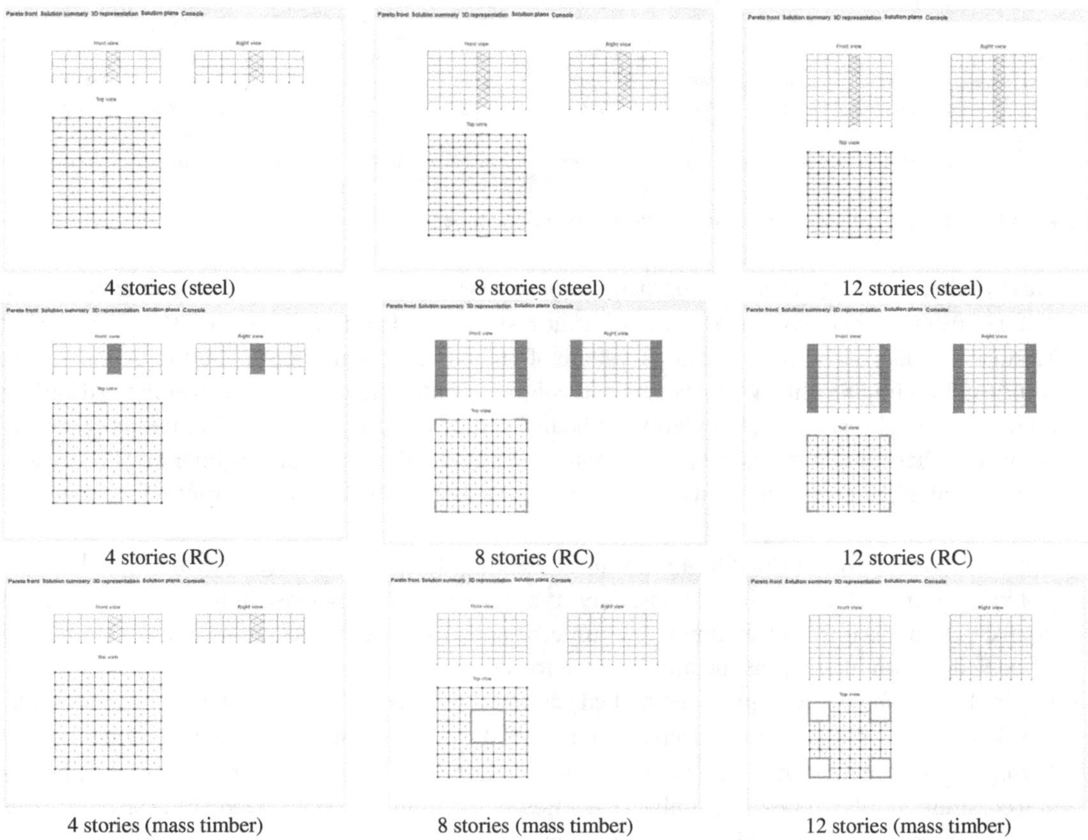

4 stories (steel) 8 stories (steel) 12 stories (steel)

4 stories (RC) 8 stories (RC) 12 stories (RC)

4 stories (mass timber) 8 stories (mass timber) 12 stories (mass timber)

Figure 6.30 Output plan views.

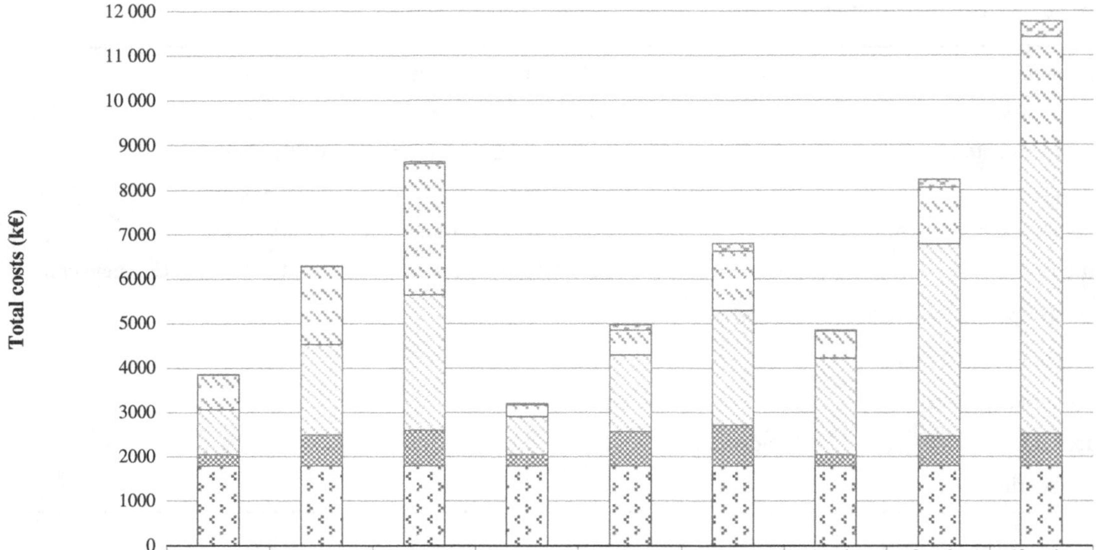

	4 stories (steel)	8 stories (steel)	12 stories (steel)	4 stories (RC)	8 stories (RC)	12 stories (RC)	4 stories (mass timber)	8 stories (mass timber)	12 stories (mass timber)
Lateral systems	4.4	17.2	36.9	30.7	122.9	184.4	6.5	176.7	353.3
Beams and columns	778.2	1748.4	2943.8	259.2	560.9	1319.0	626.3	1281.9	2406.5
Floors	1015.9	2031.8	3047.8	860.1	1720.2	2580.3	2160.0	4320.0	6480.0
Foundations	248.5	692.4	801.8	248.5	765.3	911.1	248.5	656.0	728.9
Terrain	1800.0	1800.0	1800.0	1800.0	1800.0	1800.0	1800.0	1800.0	1800.0

Figure 6.31 Total cost comparison between different building stories.

- Mass timber buildings begin at approximately 101 $kgCO_2e/m^2$ embodied carbon intensity for a 4-story structure, decrease to 96 $kgCO_2e/m^2$ at 8 stories, and further reduce to 94 $kgCO_2e/m^2$ at 12 stories. Floors account for the largest share of embodied carbon intensity and remain constant across all building heights. While beams and columns contribute increasingly to embodied carbon intensity as height increases, foundation embodied carbon intensity decrease. This reduction occurs because higher buildings require proportionally less foundation material per floor area, leading to a lower embodied carbon intensity from foundations despite the increased structural demands of taller buildings.
- For steel-framed structures, the 4-story option starts at around 129 $kgCO_2e/m^2$. It reduces to 119 $kgCO_2e/m^2$ at 8 stories and 118 $kgCO_2e/m^2$ at 12 stories. The increase in the embodied carbon intensity from beams and columns is counterbalanced by the reduction in foundation-related embodied carbon intensity as the number of stories rises.
- RC buildings show the highest embodied carbon intensity values, starting at approximately 136 $kgCO_2e/m^2$ for 4 stories, decreasing to 127 $kgCO_2e/m^2$ at 8 stories, and remaining at 127 $kgCO_2e/m^2$ at 12 stories. Floors account for the largest share of emissions, followed by beams and columns. Similar to steel and timber structures, foundation embodied carbon intensity decreases with height, offsetting the additional material requirements for taller buildings.

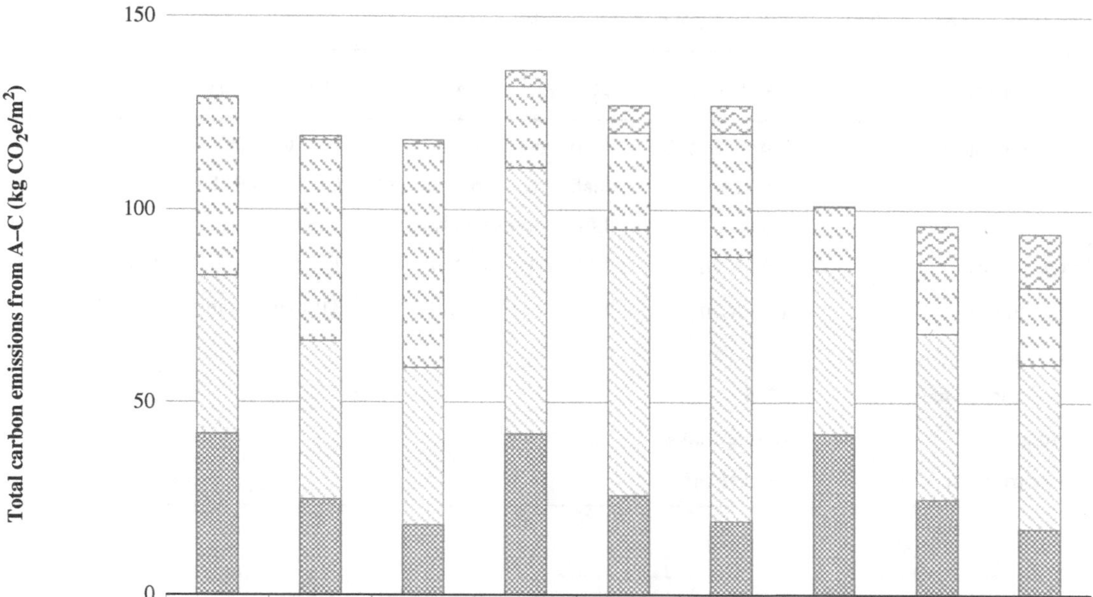

	4 stories (steel)	8 stories (steel)	12 stories (steel)	4 stories (RC)	8 stories (RC)	12 stories (RC)	4 stories (mass timber)	8 stories (mass timber)	12 stories (mass timber)
Lateral systems	0.3	1	1	4	7	7	0.2	10	14
Beams and columns	46	52	58	21	25	32	16	18	20
Floors	41	41	41	69	69	69	43	43	43
Foundations	42	25	18	42	26	19	42	25	17

Figure 6.32 Embodied carbon emissions (kg CO_2e/m^2) comparison between different building stories (Modules A + C).

In all three material types, the embodied carbon intensity decreases as building height rises because the large carbon footprint of the foundation is spread across more floors. Although taller structures require more materials for their structural frame (beams, columns, etc.), this increase can be smaller than the per-floor reduction gained by "amortizing" the foundation's impact. While this trend of decreasing carbon intensity per m^2 is observed in this 4-to-12-story analysis, it may not hold for very tall structures where the material demands of the lateral load-resisting system become the dominant factor.

6.4.5 Building Shape

Table 6.17 shows the input parameters for this analysis. The *Gross Internal Area (GIA)* for all building types is set at 5400 m^2, regardless of the changes in building shape (Figure 6.33). This approach made it possible to isolate and measure the specific impact that variations in building shape and height have on costs and embodied carbon intensity, all within a fixed total floor area of 5400 m^2. The number of stories varies between 2, 4, and 6. For buildings with a steel frame, an RC composite slab is used, while the RC frame buildings utilize a precast slab. Please note that in the taller options,

Table 6.17 Input parameters for building shape comparison.

Total floor area	**5400 m^2**	**Slab type-steel frame**	**RC composite slab**
Number of stories	2/4/6	Slab type-RC frame	Precast slab
Span length	7.5 m × 7.5 m	Slab type-timber frame	CLT ribbed slab
Floor-to-ceiling height	3.0 m	LLRS type-steel frame	Steel bracing
Imposed load	2.5 kN/m^2	LLRS type-RC frame	RC shear wall/RC core
Superimposed dead load	1.5 kN/m^2	LLRS type-timber frame	Timber bracing/Concrete core
Steel grade secondary	S355		
Steel grade primary	S355		
Occupancy	Residential		
Soil-bearing capacity	450 kN/m^2		

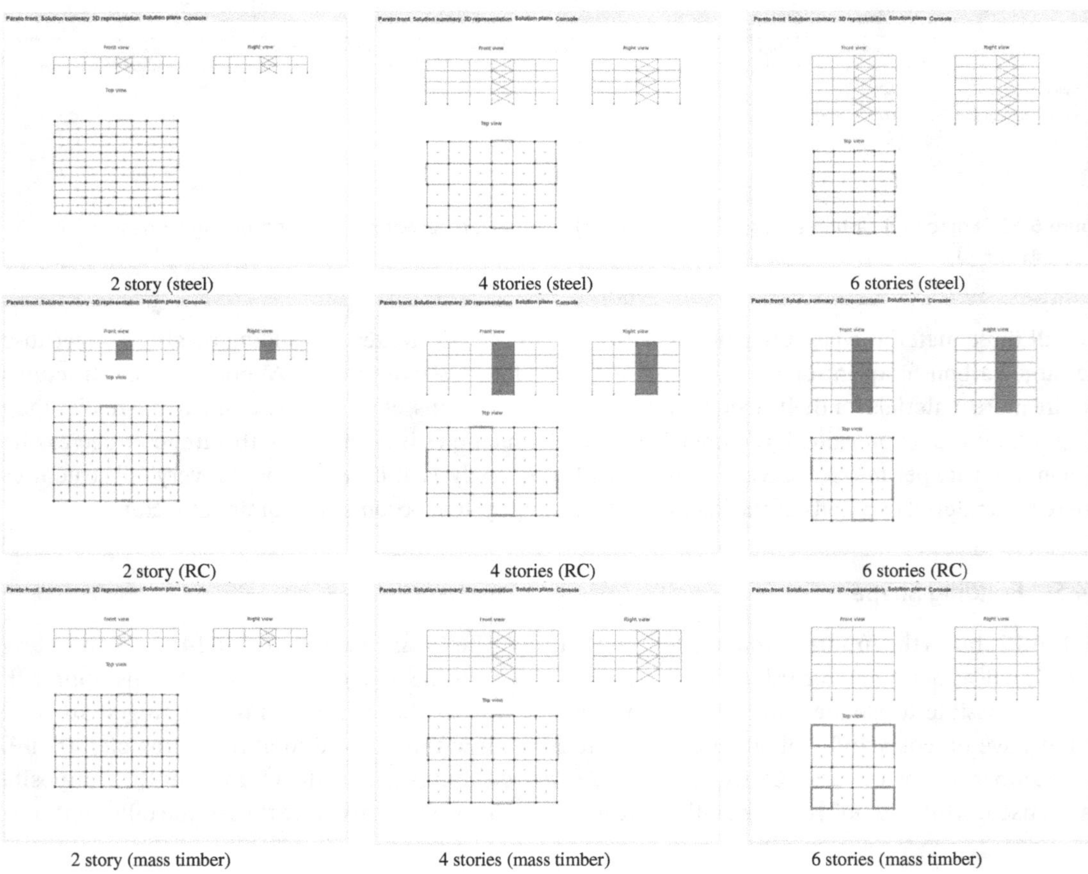

Figure 6.33 Output plan views (same GIA value for all structures).

additional demands on the lateral load-resisting systems has led to higher material usage, especially for RC and timber frames.

Figures 6.34 and 6.35 show the influence of a building's shape on the costs and embodied carbon intensity. The tallest options involve higher spending on lateral bracing, frames, and foundations. In situations where the building has more floors, structural elements drive the overall cost more than land. Figure 6.35 presents the embodied carbon intensity for steel, reinforced concrete (RC), and timber buildings at 2, 4, and 6 stories.

- Timber structures begin at approximately 68 $kgCO_2e/m^2$ for a 2-story design, increase to about 86 $kgCO_2e/m^2$ for a 4-story option, and reach around 107 $kgCO_2e/m^2$ at 6 stories. Floors account for the largest share of the emissions, followed by beams and columns. Lateral systems and foundations contribute smaller portions but become more significant as building height increases.
- For steel-framed buildings, the 2-story option is around 93 $kgCO_2e/m^2$. The embodied carbon intensity increases 109 $kgCO_2e/m^2$ for 4 stories and about 123 $kgCO_2e/m^2$ for 6 stories. Floors remain the main contributor. Lateral systems and foundations contribution rise as more floors are added.
- At 2 stories, RC buildings begin at approximately 70 $kgCO_2e/m^2$, increasing to about 90 $kgCO_2e/m^2$ by 4 stories and around 106 $kgCO_2e/m^2$ at 6 stories. Beams and columns make up a large portion, with floors also playing a major role. Lateral systems and foundations grow in proportion to height, though their contribution remain less than the primary structural elements.

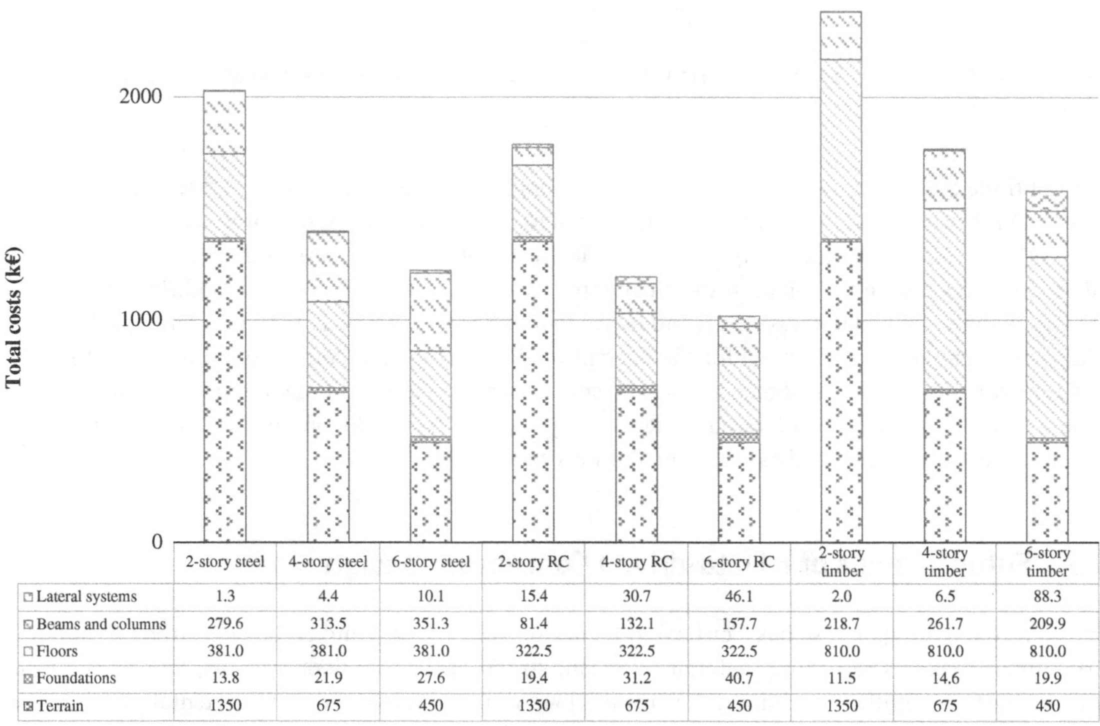

	2-story steel	4-story steel	6-story steel	2-story RC	4-story RC	6-story RC	2-story timber	4-story timber	6-story timber
Lateral systems	1.3	4.4	10.1	15.4	30.7	46.1	2.0	6.5	88.3
Beams and columns	279.6	313.5	351.3	81.4	132.1	157.7	218.7	261.7	209.9
Floors	381.0	381.0	381.0	322.5	322.5	322.5	810.0	810.0	810.0
Foundations	13.8	21.9	27.6	19.4	31.2	40.7	11.5	14.6	19.9
Terrain	1350	675	450	1350	675	450	1350	675	450

Figure 6.34 Total cost comparison between different building shapes.

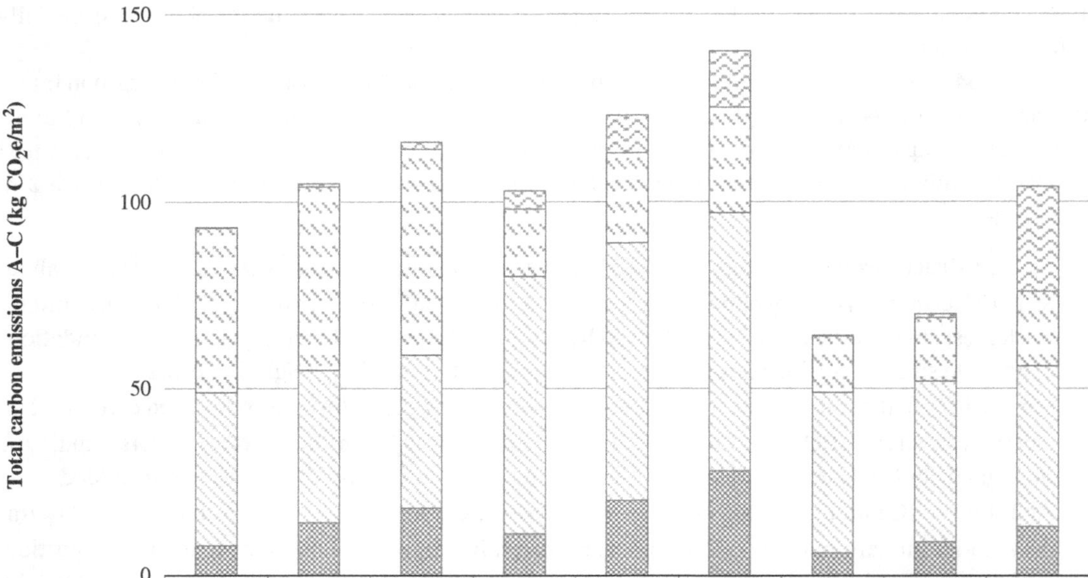

	2-story steel	4-story steel	6-story steel	2-story RC	4-story RC	6-story RC	2-story timber	4-story timber	6-story timber
Lateral systems	0.2	1	2	5	10	15	0.1	1	28
Beams and columns	44	49	55	18	24	28	15	17	20
Floors	41	41	41	69	69	69	43	43	43
Foundations	8	14	18	11	20	28	6	9	13

Figure 6.35 Embodied carbon emission (kg CO_2e/m^2) comparison between different building shapes (Modules A + C).

In all three materials, adding stories raises the overall embodied carbon per m², due to the extra lateral resistant system and foundation volume needed for taller configurations. Although lower buildings spread their area over a wider footprint, they typically require less lateral resistant systems. In contrast, taller structures need extra lateral resistant systems requiring as heavier columns, additional reinforcement, and more robust lateral systems, and hence foundations all of which add to the embodied carbon. While these patterns are observed for the current analysis parameters, other conditions might lead to different conclusions. For instance, if local seismic or wind load conditions require enhanced lateral systems for all materials, or if alternative design approaches change the proportions of structural components, the overall embodied carbon contributions could shift.

6.5 Future Trends of a Data-driven Conceptual Design

The conceptual design phase has a critical influence on both the cost and embodied carbon of building structures. Decisions regarding material selection, structural layout, floor systems, and lateral load-resisting systems significantly impact the project's long-term economic and environmental performance. Given the inherent complexity and time constraints of early-stage design, intuitive decision-making often dominates, leading to suboptimal solutions. Life-cycle Optimization (LCO), as introduced in this chapter, addresses this gap by providing a systematic framework for evaluating multiple design alternatives against competing objectives early in the design process.

Applying GAs for multi-objective optimization can offer a data-driven approach to balancing cost and embodied carbon, enabling a rapid exploration of a wide solution space, considering multiple trade-offs simultaneously.

It should be considered that the optimization results can be highly sensitive to evolving material technologies, economic conditions, and sustainability targets. The integration of alternative material compositions (e.g. higher GGBS content in concrete) or reuse strategies (e.g. reclaimed steel) can substantially alter the trade-offs, shifting the Pareto front toward more sustainable solutions.

Future trends in the optimization of building structures are moving toward increased sophistication and accessibility as emerging technologies continue to transform the design process. In the literature, there are several valuable studies in this direction (Fang et al., 2023; Feickert and Mueller, 2025; Hart et al., 2021; Kathiravel and Feng, 2024; Smith et al., 2024; De Wolf et al., 2017; Xue et al., 2022; Zhang and Zhang, 2021). Future work can likely introduce additional design variables such as connection capacity, acoustics, and fire protection systems, allowing these parameters to interact with current objectives and improve performance outcomes. Moreover, integrating new optimization criteria such as energy efficiency, constructability, and project scheduling will provide deeper conclusions into cost management and environmental impact. The adoption of artificial neural networks (an artificial intelligence method) shows significant promise for the evaluation of lateral stability and seismic performance while reducing computational costs. Enhancing user experience through the development of interactive web applications and ensuring interoperability with popular design software will make these advanced tools more accessible to industry professionals. Collectively, such innovations can be expected to drive the evolution of design methodologies toward more agile, efficient, and sustainable practices, ultimately supporting the creation of building systems that balance cost and embodied carbon effectively. These advancements can pave the way for smarter and more sustainable construction practices.

6.6 Questions

Answer the following questions to enhance your understanding of the concepts covered in this chapter.

Short Answer Questions

1. Describe how early design decisions affect a building's life cycle cost and embodied carbon.
2. What are the three generalized design stages rearranged from the RIBA plan of work?
3. How does material selection influence the possible choices for floor systems and lateral load systems?
4. What role does the genetic algorithm play in the multi-objective optimization process?

5 Define the Pareto front in the context of balancing cost and embodied carbon.

6 How can changes in span length affect the embodied carbon emissions of a building?

7 How does the use of recycled or reused steel influence the embodied carbon metrics of steel-framed buildings?

8 What changes occur in the solution set composition when shifting from typical to recycled steel?

9 How does increasing the GGBS content in concrete affect the competitiveness of RC solutions in sustainable design?

10 In what way does a small update in the steel carbon factor influence the optimal design details of different materials?

11 What role do carbon factors play in determining the balance between cost and embodied carbon in the optimization process?

12 How does an increase in span length affect the required material usage for lateral bracing and foundations in steel-framed buildings?

13 Describe the effect of increasing building height on the embodied carbon intensity for RC-framed structures.

14 How can building shape, particularly the footprint and number of stories, alter land-related costs and overall embodied carbon emissions?

15 What is the difference between Life-cycle Parameter Analysis (LCPA) and Life-cycle Optimization (LCO)?

True/False Questions

1 Design decisions made during the conceptual phase can influence up to 80% of the project's resource allocation.

2 In the cost influence curve, early design stages have a high influence on the final cost while incurring low expenditures.

3 A lack of communication among multidisciplinary professionals in construction can lead to increased costs and decreased performance.

4 A solution on the Pareto front can be improved in one objective without worsening any other objective.

5 Increasing the GGBS content in concrete reduces the embodied carbon emissions of RC structures.

6 The introduction of recycled steel makes steel-framed solutions more competitive by lowering their embodied carbon.

7 A change in the steel carbon factor has no impact on the optimal cross-section sizes or span lengths for other materials.

8 Timber frames show the same trend in embodied carbon increase with span length as steel and RC frames.

9 Increasing building height always results in higher embodied carbon emissions per square meter for all material types.

10 A single change in one parameter, like the steel carbon factor, can slightly adjust the entire solution set in a multi-objective optimization.

11 Reused steel, due to its lower embodied carbon, typically leads to the most favored solutions in the optimization process.

Discussion Questions

1 How can the use of genetic algorithms improve decision-making during the conceptual design phase? Consider both the potential benefits and limitations.

2 In what ways does the Pareto front assist stakeholders when balancing cost and environmental impact? Provide examples based on the chapter.

3 Discuss the influence of material selection on both economic outcomes and environmental performance. What factors should be prioritized in this choice?

4 How do geometric parameters such as span length and building height affect structural performance, cost, and embodied carbon? Use examples from the text to support your points.

5 Reflect on the challenges posed by low communication among different professionals in construction projects. What strategies might enhance collaboration and lead to better design outcomes?

6 Evaluate the trade-offs between cost efficiency and embodied carbon emissions when selecting different types of steel (typical, recycled, reused) for structural frames.

7 How might changes in GGBS content influence decision-making in RC structures, and what are the potential impacts on a project's sustainability performance?

8 Discuss how geometric factors such as span length and building height interact with material choices to affect both cost and embodied carbon outcomes.

9 Given that a small parameter change can alter the entire solution set, what measures can designers take to achieve robust optimization results under fluctuating material properties?

10 In the realm of sustainable construction, how can multi-objective optimization tools be further developed to effectively integrate both material-specific factors and geometric parameters?

References

CYPE Ingenieros S.A. (2020). Generador de precios de la construcción. https://www.generadordeprecios.info/#gsc.tab=0

Dang, V. H. (2022). *Conceptual Design of Mass Timber Structures Using Genetic Algorithms* [Master's thesis, supervised by Alper Kanyilmaz]. Politecnico di Milano.

De Wolf, C. C. E. L. (2017). *Low carbon pathways for structural design: Embodied life cycle impacts of building structures* [Thesis, Massachusetts Institute of Technology]. https://dspace.mit.edu/handle/1721.1/111491

European Committee for Standardization. (2002a). EN 1990: Eurocode—Basis of Structural Design. https://eurocodes.jrc.ec.europa.eu/EN-Eurocodes/eurocode-basis-structural-design

European Committee for Standardization. (2002b). Eurocode 1: Actions on Structures—Part 1-1: General Actions—Densities, Self-weight, Imposed Loads for Buildings. https://eurocodes.jrc.ec.europa.eu/EN-Eurocodes/eurocode-1-actions-structures

Fang, D., Brown, N., De Wolf, C., & Mueller, C. (2023). Reducing embodied carbon in structural systems: A review of early-stage design strategies. *Journal of Building Engineering*, *76*, 107054. https://doi.org/10.1016/j.jobe.2023.107054

Feickert, K., & Mueller, C. T. (2025). Policy and design levers for minimizing embodied carbon in United States buildings: A quantitative comparison of current and proposed strategies. *Building and Environment*, *270*, 112485. https://doi.org/10.1016/j.buildenv.2024.112485

FEMA. (2004). *Primer for Design Professionals: Communicating with Owners and Managers of New Buildings on Earthquake Risk* (Risk Management Series FEMA 389). Federal Emergency Management Agency.

Hart, J., D'Amico, B., & Pomponi, F. (2021). Whole-life embodied carbon in multistory buildings: Steel, concrete and timber structures. *Journal of Industrial Ecology*, *25*(2), 403–418. https://doi.org/10.1111/jiec.13139

Holland, J. H. (1975). *Adaptation in Natural and Artificial Systems: An Introductory Analysis with Applications to Biology, Control, and Artificial Intelligence*. The University of Michigan.

Kanyilmaz, A., Birhane, M., Fishwick, R., & del Castillo, C. (2023a). Reuse of steel in the construction industry: Challenges and opportunities. *International Journal of Steel Structures*, *23*(5), 1399–1416. https://doi.org/10.1007/s13296-023-00778-4

Kanyilmaz, A., Dang, V. H., & Kondratenko, A. (2023b). How does conceptual design impact the cost and carbon footprint of structures? *Structures*, *58*, 105102. https://doi.org/10.1016/j.istruc.2023.105102

Kanyilmaz, A., Tichell, P. R. N., & Loiacono, D. (2022). A genetic algorithm tool for conceptual structural design with cost and embodied carbon optimization. *Engineering Applications of Artificial Intelligence*, *112*, 104711. https://doi.org/10.1016/j.engappai.2022.104711

Kathiravel, R., & Feng, H. (2024). Structural and embodied carbon performance optimization for low carbon buildings through BIM-based integrated design. *Journal of Building Engineering*, *98*, 111008. https://doi.org/10.1016/j.jobe.2024.111008

MacLeamy, P. (2020). *Designing a World-Class Architecture Firm: The People, Stories, and Strategies Behind HOK*. Wiley. https://www.wiley.com/en-us/Designing+a+World-Class+Architecture+Firm%3A+The+People%2C+Stories%2C+and+Strategies+Behind+HOK-p-9781119685302

Narayanan, S. (2018). Chapter 17—Seismic resistant design. In *Design of Steel Structures Limit States Method* (709–761). Oxford University Press.

Norman, J. (2018). *Structural Timber Elements—A Pre-Scheme Design Guide*. TRADA.

Norman, J., Broadbent, O., De'Ath, R., Lloyd, I., Carr, J., & Harpin, R. (2020). *Conceptual Design of Buildings*. The Institution of Structural Engineers.

Paulson, B. C. (1976). Designing to reduce construction costs. *Journal of the Construction Division*, *102*(4), 587–592.

Pezeshk, S., & Camp, C. V. (2002). State of the art on the use of genetic algorithms in design of steel structures. In *Recent Advances in Optimal Structural Design* (55–79). American Society of Civil Engineers.

Rafiq, Y., & Beck, M. (2008). A decision support tool for multi-disciplinary conceptual design. *Structural Engineer*, *86*, 37–42.

Raquel Navarro Tichell, P., Kanyilmaz, A., & Loiacono, D. (2020). *Conceptual Design of Buildings with Artificial Intelligence [Master's Thesis]*.

Reddy, M. J., & Kumar, D. N. (2014). Elitist-mutated multi-objective particle swarm optimization for engineering design. In *Encyclopedia of Information Science and Technology* (3rd edition, 3534–3545). IGI Global Scientific Publishing.

RIBA. (2019). What's behind the updates to the 2020 Plan of Work. The *RIBA Journal*. https://www.ribaj.com/intelligence/updates-to-the-riba-plan-of-work-2019-dale-sinclair-gary-clark

RIBA. (2020). http://www.ribaplanofwork.com, accessed on February 2026.

Smith, M. S. I., Fang, D., Mueller, C., & Carstensen, J. V. (2024). Reducing embodied carbon with material optimization in structural engineering practice: Perceived barriers and opportunities. *Journal of Building Engineering*, *95*, 109943. https://doi.org/10.1016/j.jobe.2024.109943

Stora Ens. (2021). *CLT rib panel by Stora Enso*. https://www.storaenso.com/en/products/mass-timber-construction/building-products/rib-panels

Varghese, B. M., & Raj, R. J. S. (2016). A survey on variants of genetic algorithm for scheduling workflow of tasks. *2016 Second International Conference on Science Technology Engineering and Management (ICONSTEM)*, 489–492. https://doi.org/10.1109/ICONSTEM.2016.7560870

Vermeulen, D. (2018). *Structural Dynam(o)ite: Optimized Design and Fabrication Workflows with Dynamo*. Autodesk University. https://www.autodesk.com/autodesk-university/article/Structural-Dynamoite-Optimized-Design-and-Fabrication-Workflows-Dynamo-2019

Xue, Q., Wang, Z., & Chen, Q. (2022). Multi-objective optimization of building design for life cycle cost and CO_2 emissions: A case study of a low-energy residential building in a severe cold climate. *Building Simulation*, *15*(1), 83–98. https://doi.org/10.1007/s12273-021-0796-5

Zhang, X., & Zhang, X. (2021). Sustainable design of reinforced concrete structural members using embodied carbon emission and cost optimization. *Journal of Building Engineering*, *44*, 102940. https://doi.org/10.1016/j.jobe.2021.102940

7

Life-cycle-driven Seismic Design, Construction, Retrofitting, and Assessment

LIST OF CONTENTS

Earthquake (seismic) resilience of structures and their environmental impact are usually discussed in separate rooms. This approach is no longer sustainable. Figure 7.1 shows Earth's major tectonic plates (the intense geological activity along their boundaries is the primary source of seismic risk). Since many of the world's most populated regions are situated on or near these volatile zones (Figure 7.2), this fragmented approach will probably change soon.

Roughly one-third of the world's population lives in moderate to high seismic hazard zones, and the population living in seismic areas is expected to grow due to urbanization (Pesaresi et al., 2017).

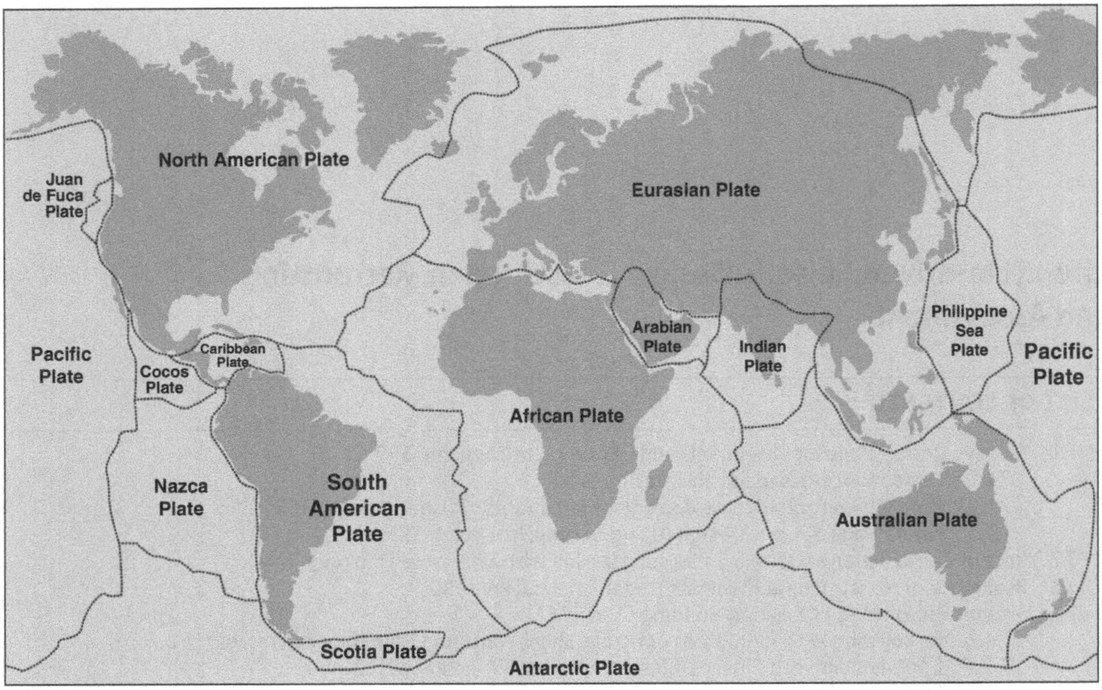

Figure 7.1 Tectonic plates of earth (Peter Hermes Furian/Shutterstock).

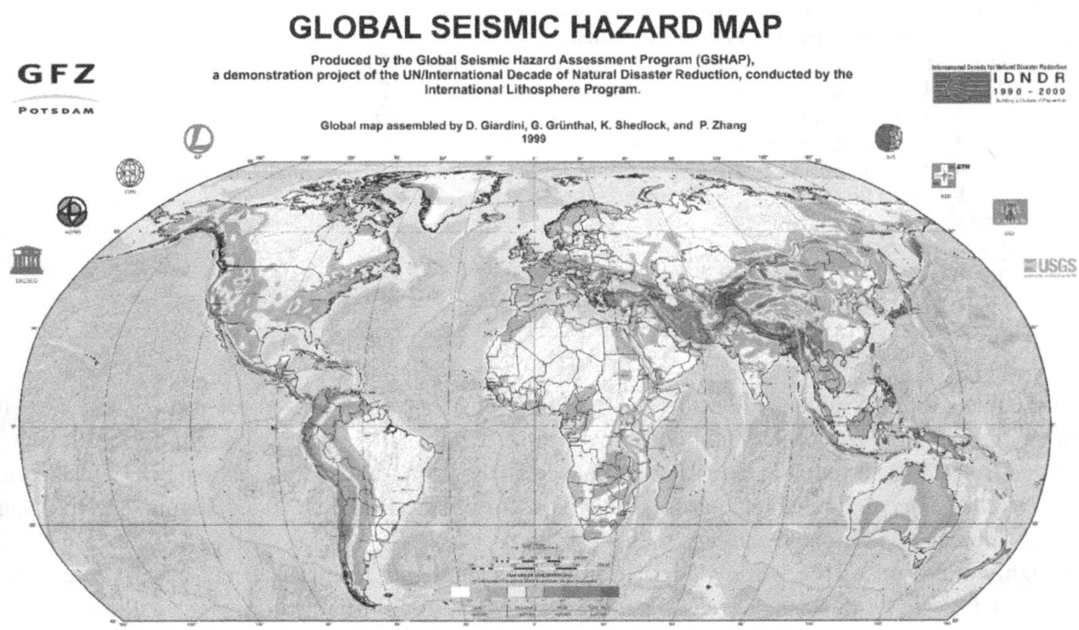

Figure 7.2 Global seismic hazard map (GSHAP / https://www.gfz.de/en/gshap / last accessed on 7 November 2025 / GFZ Helmholtz Centre for Geosciences).

Figure 7.3 Two highly populated cities in high seismic areas (left: Istanbul, Yavuz Sariyildiz/Shutterstock; right: Tokyo, leungchopan/Shutterstock).

Figure 7.4 Two examples of the consequence of an improper design in a seismic region. Left: a "short column" collapse (Stockbyte/Getty Images). Right: a global (soft story) collapse (courtesy of Bahadir Sadan), resulting in large, avoidable carbon footprints.

Such seismic regions also host the densest concentrations of built environment (Figure 7.3), meaning the bulk of construction-related CO_2 emissions occur where seismic risk is highest. Decisions made in these areas, therefore, shape not only structural safety but also global embodied carbon outcomes.

Beyond the tragic loss of lives from structural collapses, neglecting seismic design rules in earthquake-prone regions is also a very large source of avoidable embodied carbon. Structures not engineered to withstand seismic forces often face demolition after earthquakes, wasting resources and time (Figure 7.4). Although seismic codes in most countries help prevent this outcome, the engineer's responsibility to perform a proper seismic design remains a top priority to protect both people and the environment.

Furthermore, the seismic event itself can trigger subsequent disasters, creating complex multi-hazard scenarios. A frequent and destructive example is fire following an earthquake, often ignited by damaged gas lines and electrical systems. This catastrophic combination amplifies structural damage and complicates emergency response efforts, as shown by the fire-ravaged buildings (after an earthquake in Turkey) shown in Figure 7.5.

Figure 7.5 Two examples of multi-hazard (earthquake followed by fire) damage (courtesy of Tunc Deniz Uludag).

Table 7.1 Trade-offs between safety-first and sustainable seismic design.

Design strategy	Resilience level	Sustainability level	Outcome	LCA focus	Seismic design assumptions and carbon footprint
Safety-first	High	Low	Durable against earthquakes but high material use	LCA is not applied	Extra ductility, overstrength, and redundancy drive embodied carbon up
Sustainability-first	Low	High	Low-carbon footprint but lacks resilience and robustness potentially vulnerable to collapse or large damage	Reduced upfront carbon (A1–A5)	Lack of seismic resilience and robustness can lead to premature repair/replacement, hence extra carbon footprint
Integrated resilience + sustainability	High	High	Safe, long-lasting, and low-carbon structures	Balanced across all life stages (A–D)	Optimized ductility/overstrength/redundancy to minimize embodied and operational carbon

A new challenge for modern engineering is to create structures that are simultaneously resilient to earthquakes and environmentally sustainable. Table 7.1 summarizes the trade-offs between resilience and sustainability in seismic design, highlighting how different strategies can influence life-cycle phases, resource use, and carbon footprint. Resilient structures (durable against exceptional actions such as earthquakes) designed without environmental considerations can lead to increased resource consumption and a higher embodied carbon footprint. On the other hand, sustainable designs lacking resilience may cause poor robustness and premature failure, resulting in wasted resources. By merging structural resilience with environmental sustainability, we can create buildings that are safe, durable, and sustainable.

This chapter explores how life-cycle thinking can be applied to earthquake-resistant (seismic) design, ensuring that structures are both earthquake-resistant and sustainable. We will introduce three core seismic design philosophies – conventional (ductility-based), damage-controlled (repairable), and no-damage (base isolated or vibration controlled) – and find their positions within life-cycle thinking. Finally, the chapter addresses a question that is often overlooked in seismic design textbooks: how do we verify structural safety after an earthquake has occurred? Section 7.7 introduces the role of structural health monitoring (SHM) as an enabling technology for life-cycle-driven seismic design, and explores an emerging concept in which existing telecommunications fiber-optic networks could be dual-purposed (without interrupting internet service) to provide city-scale, continuous structural monitoring. This forward-looking topic connects the design philosophies discussed earlier in the chapter to the practical challenge of post-earthquake decision-making, where the ability to rapidly distinguish safe buildings from damaged ones has direct consequences for embodied carbon, community displacement, and recovery time.

7.1 Seismic Design Philosophies in Relation to Life-cycle Thinking

Modern seismic design can be categorized into three distinct philosophies (Figure 7.6), each with different implications for embodied carbon across the building life cycle:

Performance-based Design: Targets specific outcomes, such as immediate occupancy, life safety, or collapse prevention. A conventional ductility-based design is a subset of performance-based design, often targeting only life safety, which accepts controlled damage. This choice results in high post-earthquake embodied carbon from extensive repairs or complete reconstruction.

Resilient (Damage-controlled) Design: Focuses on minimizing damage and ensuring rapid recovery. By integrating systems like self-centering mechanisms and easily replaceable energy-dissipating components, it reduces the need for resource-intensive repairs, thereby lowering the building's total life-cycle carbon.

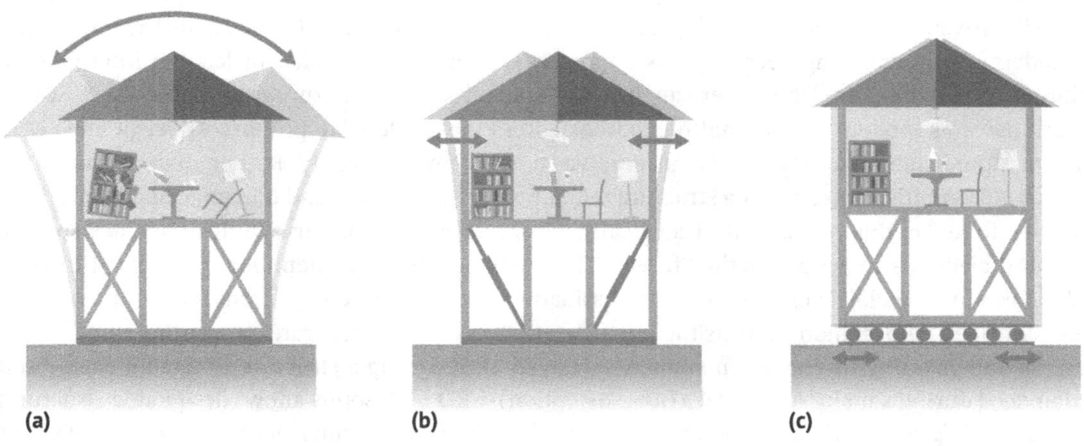

Figure 7.6 Three main seismic design philosophies: (a) conventional, (b) damage-controlled, and (c) no-damage (metamorworks/Adobe Stock Photos).

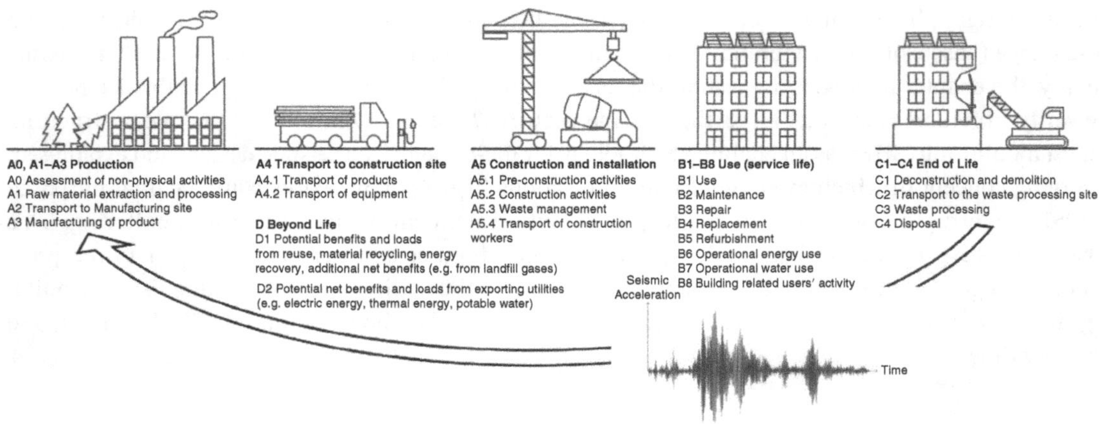

Figure 7.7 Life-cycle stages flowchart.

No-damage Design: Decouples the structure from the ground's motion using special bearings (e.g. elastomeric bearings, friction pendulum systems). By allowing the building to move independently from ground, it remains largely undamaged, drastically reducing the post-earthquake embodied carbon associated with structural repair or replacement and offering the highest level of protection.

Each philosophy represents a different balance between initial material investment, expected damage levels, and long-term sustainability outcomes. This section will demonstrate how each of these three seismic philosophies impacts embodied carbon at every stage of the life cycle (Figure 7.7), linking structural resilience directly to life-cycle sustainability under seismic demands.

7.1.1 Performance-based Seismic Design

Performance-based seismic design (PBD), the most common earthquake-resistant approach today, is primarily driven by the goal of creating economical and safe structures. It represents the foundation of modern earthquake engineering. This approach is formally embedded in leading international building codes, including the American ASCE 7 standards and the European Eurocode 8, which in turn serve as models for national regulations across the world. The primary driver of PBD is to achieve a predictable and reliable balance between economy and safety. Rather than pursuing the often cost-prohibitive objective of a structure remaining entirely elastic and undamaged during a rare, major earthquake, this philosophy strategically allows for controlled damage. It designates specific structural elements to act as ductile "fuses." These components are intentionally designed to yield and deform in an inelastic manner, thereby dissipating the immense energy from the ground motion. This controlled deformation protects the integrity of the rest of the structure, preventing catastrophic collapse and ensuring life safety. In summary, instead of designing a structure to remain elastic and undamaged during a major earthquake (a costly option), PBD philosophy allows designated structural elements (or "fuses") to yield and deform in a controlled, inelastic manner, absorbing the initial shock and enabling a safe evacuation.

Figure 7.8 shows this principle with two distinct examples. On the left, the nonstructural masonry infill walls of a reinforced concrete building have been sacrificed while the ductile behavior of the frame

Figure 7.8 Comparison of earthquake-induced damage in reinforced concrete and steel structures. The left image shows a reinforced concrete frame building where the non-structural masonry infill walls have suffered severe cracking and collapse, while the structural frame itself appears to remain standing. The right image illustrates the interior of a steel-framed building with concentric bracing, where the diagonal tube braces have undergone significant buckling deformation to dissipate seismic energy, leaving the main columns and beams largely intact (images generated by Gemini).

prevented a complete collapse, meeting the primary goal of life safety. On the right, a steel-framed structure shows a more repairable form of damage; the slender steel braces have buckled as designed.

The life-cycle implications of this seismic design philosophy are as follows:

Production Stage (A1–A3): Since ductile behavior helps dissipate the seismic energy (similar to how a car's bumper absorbs impact), it allows the designer to reduce the design forces by means of a "behavior factor" (a number that allows engineers to use smaller design forces because the building can bend without breaking: detailed provisions on these reduction factors are outlined in international seismic design standards such as Eurocode 8 or ASCE 7), leading to smaller structural members and, consequently, lower upfront embodied carbon.

Construction Stage (A4–A5): The reduction of primary structural material explained above comes with a hurdle in the construction site. Ductile design requires specific detailing to ensure controlled failure, such as additional confining reinforcement (steel ties) in reinforced concrete columns and robust connections in steel frames. Besides adding extra material that partially offsets the savings from smaller main members, the ductile detailing also adds complexity to the construction site. For example, a cast-in-place reinforced concrete frame with dense rebar confining reinforcement (common in ductile seismic design) requires careful assembly and quality control on-site (Figure 7.9a). This can slightly extend construction time and equipment usage, impacting fuel consumption and emissions during construction. Likewise, steel structures designed for high seismic performance may require specialized welding (Figure 7.9b) or bolting techniques, which involve additional fabrication steps on-site or in the shop. These requirements can increase construction-stage emissions due to extra welding (electricity usage) or the need for heavier cranes to lift stronger, heavier components. Similarly, mass timber seismic systems rely on the precise installation of numerous high-performance metal connectors (in addition to the reinforced concrete shear walls or steel bracings), which adds a layer of complexity and quality control to the assembly process.

However, some ductile details can actually simplify fabrication and assembly instead of complicating them. A good example is the Reduced Beam Section (RBS) connection, shown in Figure 7.10. In this solution, a small portion of the beam flanges is intentionally trimmed near the column to create a predictable plastic hinge away from the welds. This detail allows large inelastic rotations to occur in a

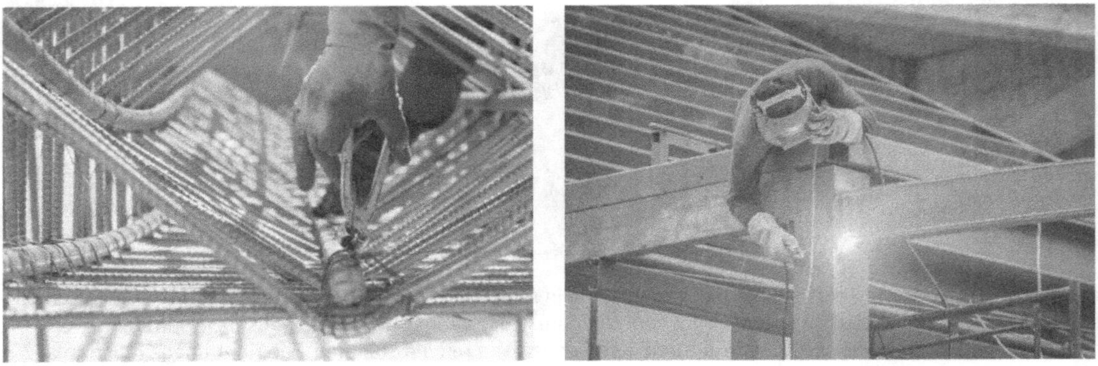

(a) (b)

Figure 7.9 Examples to site works: (a) reinforcement bars assembly (bannafarsai/Adobe Stock Photo) and (b) on-site welding (Christopher PB/Shutterstock).

Figure 7.10 Example of a "ductile" detail for steel structural systems: Reduced Beam Section (RBS), tested under controlled laboratory conditions to monitor seismic behavior; left: undamaged, right: damaged (courtesy of Massimo Latour).

controlled manner, protecting the column and connection zone from damage. The test in Figure 7.10 shows the undamaged beam (left) and the same beam after cyclic loading (right), where deformation is concentrated in the reduced section region while the rest of the structure remains intact. RBS connections therefore combine high seismic performance with fabrication simplicity, as they can be produced with standard cutting or milling operations without requiring additional reinforcement or welding.

(a) (b)

Figure 7.11 Examples of building damages following an earthquake event: (a) light damage (Claudia Weinmann/Alamy Stock Photo); (b) serious damage (NigelSpiers/Shutterstock).

Use and Repair Stage (B): This stage reveals the critical trade-off of the ductile design. A building designed according to the PBD is expected to be damaged after a design-level earthquake. While lives are saved, the structure becomes a damaged asset (Figure 7.11). Repairs can range from patching concrete cracks to replacing entire yielded steel members, all of which are carbon-intensive activities. Studies have shown this impact is significant. For example, Menna et al. (2013) found that post-earthquake repairs could add 25% or more to a building's initial environmental impact. The 2011 Christchurch earthquake in New Zealand is another real-world example: over 140 reinforced concrete buildings designed with ductile principles were demolished, resulting in an estimated 0.3 million tons of "wasted" embodied carbon (Gonzalez et al., 2021). Neglecting these future repair emissions drastically underestimates a building's true life-cycle carbon footprint.

End-of-life and Reuse Potential (C and D): Because significant damage is an expected outcome, a ductile structure that experiences a major earthquake may be irreparable, leading to premature demolition. The damaged and often monolithic nature of the construction (e.g. welded connections, cast concrete) makes deconstruction and high-value material reuse nearly impossible. Steel rebar can be recycled after concrete crushing, but the components themselves cannot be salvaged.

7.1.2 Damage-controlled (Low-damage or Repairable) Seismic Design

Damage-controlled design aims to confine inelastic behavior to specific, replaceable components, preserving the integrity of the primary structure and enabling rapid post-earthquake recovery. This approach can follow two main strategies:

i. A self-centering or minimal-repair design, where energy dissipation occurs through devices or materials that limit residual deformations and require little to no repair.

ii. A dissipative and repairable design, where designated "fuse" components such as replaceable steel links, friction dampers, or dissipative bracing connections absorb seismic energy and are replaced after an event.

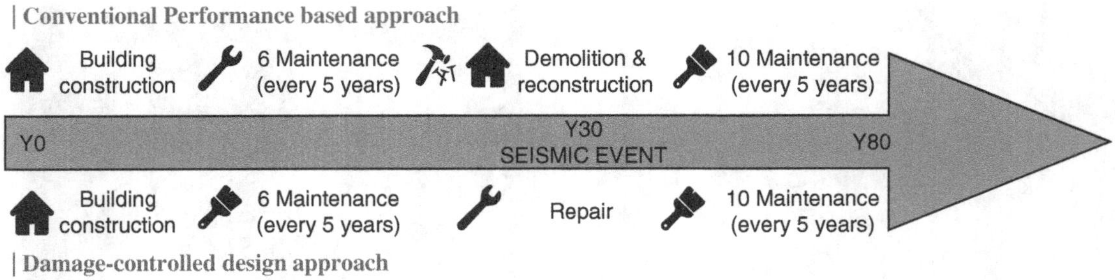

Figure 7.12 Life-cycle comparison of a conventional performance-based approach vs. a damage-controlled design approach, illustrating the outcome after a seismic event at year 30.

Their core principle is repairability, ensuring the building can be quickly and economically brought back to service. Figure 7.12 visually explains this in comparison with a conventional performance-based approach over an 80-year lifespan: following a major seismic event at year 30, the conventional structure requires complete demolition and reconstruction. In contrast, a damage-controlled structure only needs repair, demonstrating its resilience and significant long-term economic advantage.

Figure 7.13 provides an example of a "minimal-repair" steel beam–column connection. The diagram shows how components like T-stubs, internal and external angles, and friction pads are joined in order to form the connection. These elements are designed to absorb and dissipate seismic energy, protecting the main beam and column from significant damage, allowing for efficient (or minimal) repair and continued functionality of the structure.

Figure 7.14 shows a minimal-repair bracing connection concept based on frictional energy dissipation. In conventional concentrically braced frames, seismic energy is absorbed through the inelastic buckling and yielding of diagonal braces, which typically require replacement after a major earthquake. In contrast, this friction-based connection detail allows controlled sliding between steel plates, dissipating seismic energy through friction rather than material yielding. As a result, the main brace and gusset plate remain elastic and undamaged. After an earthquake, the system can be retightened or adjusted without component replacement, restoring full structural performance with minimal intervention. This design philosophy prioritizes not only life safety but also rapid functional recovery and preservation of the building's economic value.

Figure 7.15 shows an example of a dissipative and repairable design: dissipative and replaceable bracing connection (DrBRC) developed and tested within the DISSIPABLE project, funded by Research Fund for Coal and Steel (Menghini, Kanyilmaz, et al., 2025; Menghini, Mouzakis, et al., 2025). Unlike buckling-restrained braces or self-centering devices, the DrBRC is not a proprietary or factory-built product but a structural component composed of standard steel plates and pins, which allows fabrication by any qualified steel manufacturer without the need for specialized equipment, making it a cost-competitive option for dissipative seismic design. The system functions by concentrating plastic deformation in pins within replaceable plates, ensuring that energy dissipation occurs in a predictable and localized manner. Once the pins yield under cyclic loading, they can be easily replaced without affecting the surrounding frame or bracing. This strategy enables a damage-controlled response similar to more complex devices while maintaining compatibility with conventional steel fabrication workflows.

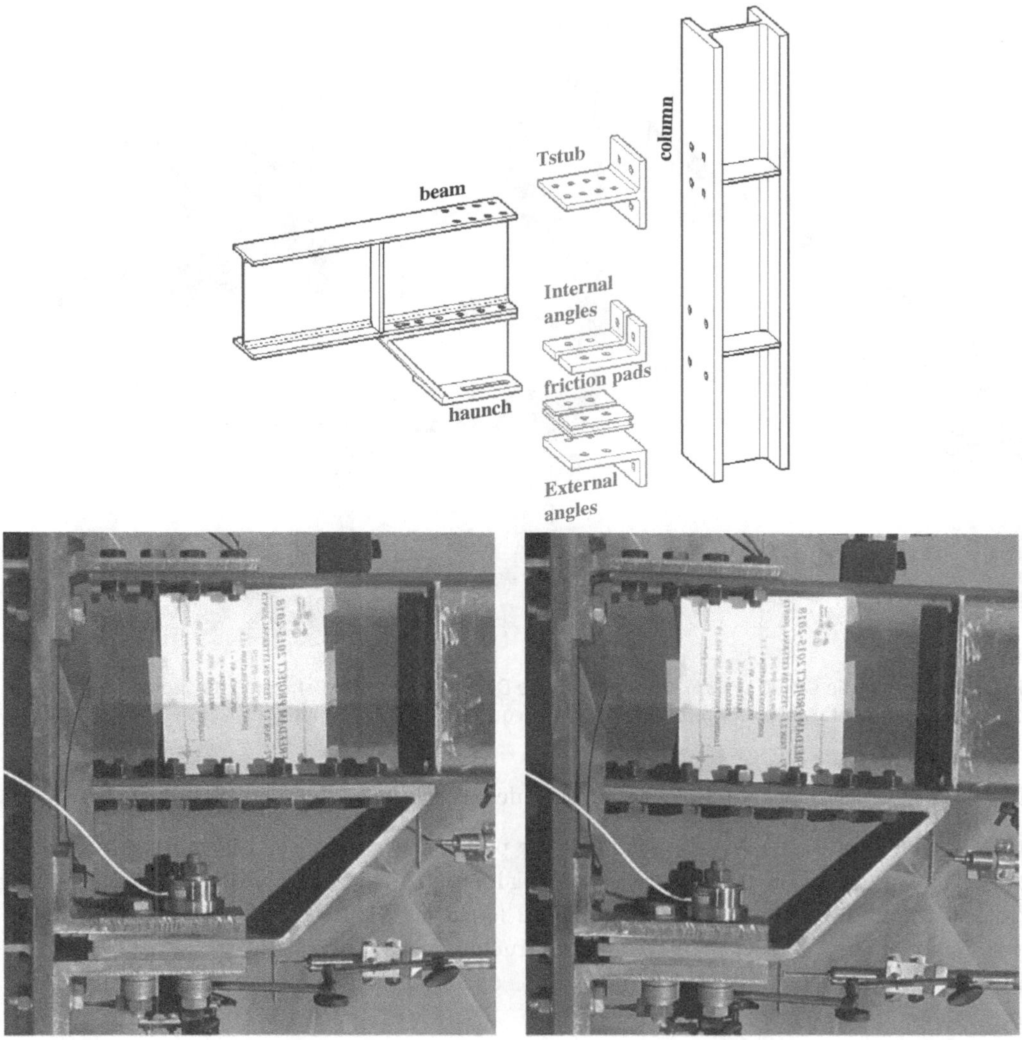

Figure 7.13 Example of a "minimal repair" repairable steel beam–column connection, tested under controlled laboratory conditions to monitor seismic behavior (modified from Latour et al., 2018).

Figure 7.14 Example of a "minimal repair" bracing connection, tested under controlled laboratory conditions to monitor seismic behavior (courtesy of Massimo Latour).

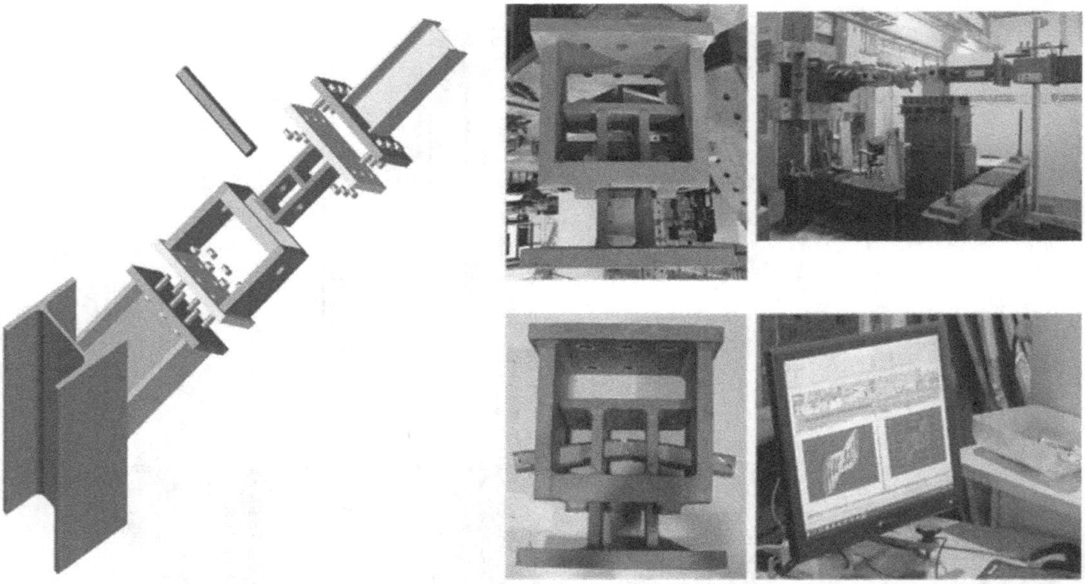

Figure 7.15 Overview of the DrBRC system developed in the DISSIPABLE project, showing its geometry, components, and experimental setup used for cyclic testing (courtesy of Prof. Luis Calado).

The life-cycle implications of the damage-controlled design philosophy are:

Production Stage (A1–A3): While this seismic design philosophy requires manufacturing specialized devices (e.g. buckling-restrained braces, self-centering links, dissipative and replaceable connections), the primary structural frame can often be designed for lower forces because the fuses handle the energy dissipation. This can lead to a lighter main structure, potentially resulting in a neutral or even slightly lower upfront carbon footprint compared to a conventional ductile design, despite the added devices.

As an example, Figures 7.16–7.19 show the sequential stages in the fabrication and preparation of a buckling-restrained brace (BRB) system. The process begins with welding the steel components that form the buckling-restrained core (Figure 7.16), continues with surface protection (Figure 7.17) and assembly (Figure 7.18) through coating and painting, and concludes with laboratory testing and storage before being transported to the site installation (Figure 7.19).

Figure 7.16 Welding of a steel component forming part of a buckling-restrained frame, showing fabrication precision required for reliable seismic energy dissipation (courtesy of Arfen BRB).

Figure 7.17 Completed buckling-restrained brace after assembly and coating, ready for quality inspection and shipment (courtesy of Arfen BRB).

Figure 7.18 Surface treatment and painting of assembled buckling-restrained braces in a controlled environment to ensure corrosion protection and durability (courtesy of Arfen BRB).

Figure 7.19 Laboratory testing (left) and storage (right) of buckling-restrained braces before installation, verifying performance, and compliance with design specifications (courtesy of Arfen BRB).

Figure 7.20 Fabrication steps for DrBRC plates and pins, including (a) CNC oxy acetylene cutting of steel plates, (b) high-speed drilling of steel plates, (c) weld preparation of steel plates, and (d) chamfering operations (chamfered pin) (courtesy of Michalis Sofras).

Figures 7.20 and 7.21 show the main production stages of the DrBRC components described previously (dissipative and repairable design). The plates are first cut and drilled using standard CNC oxy-acetylene and high-speed drilling machines, followed by manual weld preparation and chamfering of pins. The parts are then assembled, welded, painted, and packed for shipment.

Construction Stage (A4–A5): Construction stage involves installing the specialized dissipative and replaceable components. These often use bolted connections, which can simplify and speed up assembly compared to the complex on-site welding of some ductile systems. Quality control focuses on ensuring these components are installed correctly to function as designed.

Figure 7.21 Assembly and finishing of DrBRC components: (a) manual assembly, (b) semi-automatic welding, (c) painting, and (d) packing and shipping for installation (courtesy of Michalis Sofras).

As a different example with respect to buckling-restrained frames described earlier, the case in Figure 7.22 employs another type of damage-controlled system based on minimal-repair dissipative connections. The construction stage involves the installation of these devices within the structural frame. Each element is designed to yield or slide under seismic loading, protecting the main structure from permanent deformation. The assembly process primarily relies on bolted connections, which not only ensure faster on-site installation but also allow for straightforward inspection and replacement after an event. Once installed, sensors and displacement transducers can be used to monitor the behavior of the devices, forming part of an integrated quality-control strategy. Altogether, this approach demonstrates how damage-controlled systems can align structural resilience with life-cycle performance objectives by facilitating maintenance, reuse, and deconstruction.

Use and Repair Stage (B): This is where the damage-controlled philosophy provides major sustainability benefits with respect to the conventional seismic design. After an earthquake, damage is isolated to the replaceable fuses. Instead of extensive, carbon-intensive repairs to the entire frame, maintenance involves simply unbolting and replacing the damaged devices. This dramatically reduces repair costs, downtime, and, most importantly, the associated carbon emissions. Furthermore, technologies like self-centering systems are designed to return the building to its original vertical position after shaking, eliminating the permanent tilt (residual drift) that often condemns otherwise repairable buildings.

Figure 7.22 Installation of damage-controlled devices in the C3 building: top left – completed building; top right – beam–column connection with dissipative detail; bottom left – steel frame assembly; bottom right – instrumented bolted joint for in-service monitoring (courtesy of Massimo Latour).

In systems like DrBRC (Dissipative and Replaceable Bracing Connection), developed in the DISSIPABLE RFCS project (introduced earlier as an example during A1–A3 stage explanation), this philosophy is realized through its simple pin-and-plate configuration. As shown in Figure 7.23, repair and replacement operations can be carried out directly on-site using standard tools and equipment. After a seismic event, the first step involves assessing the elongation or shortening of the diagonal bracing elements caused by the deformation of the replaceable connections. These length variations indicate which DrBRCs have yielded during the energy-dissipation process. Technicians then proceed with a sequence of localized replacement operations: removing and replacing the lower connection of the elongated diagonal, repeating the same for the opposite diagonal, and then addressing the upper connections in a controlled order.

During reassembly, shims are inserted to restore the nominal length of the diagonals, ensuring proper tension and alignment. Once all new DrBRCs are installed, locking devices are removed, and the system is ready for service again. Figure 7.23 shows the replacement sequence within a typical multi-story configuration. The process proceeds floor by floor, typically starting from the top, allowing quick and

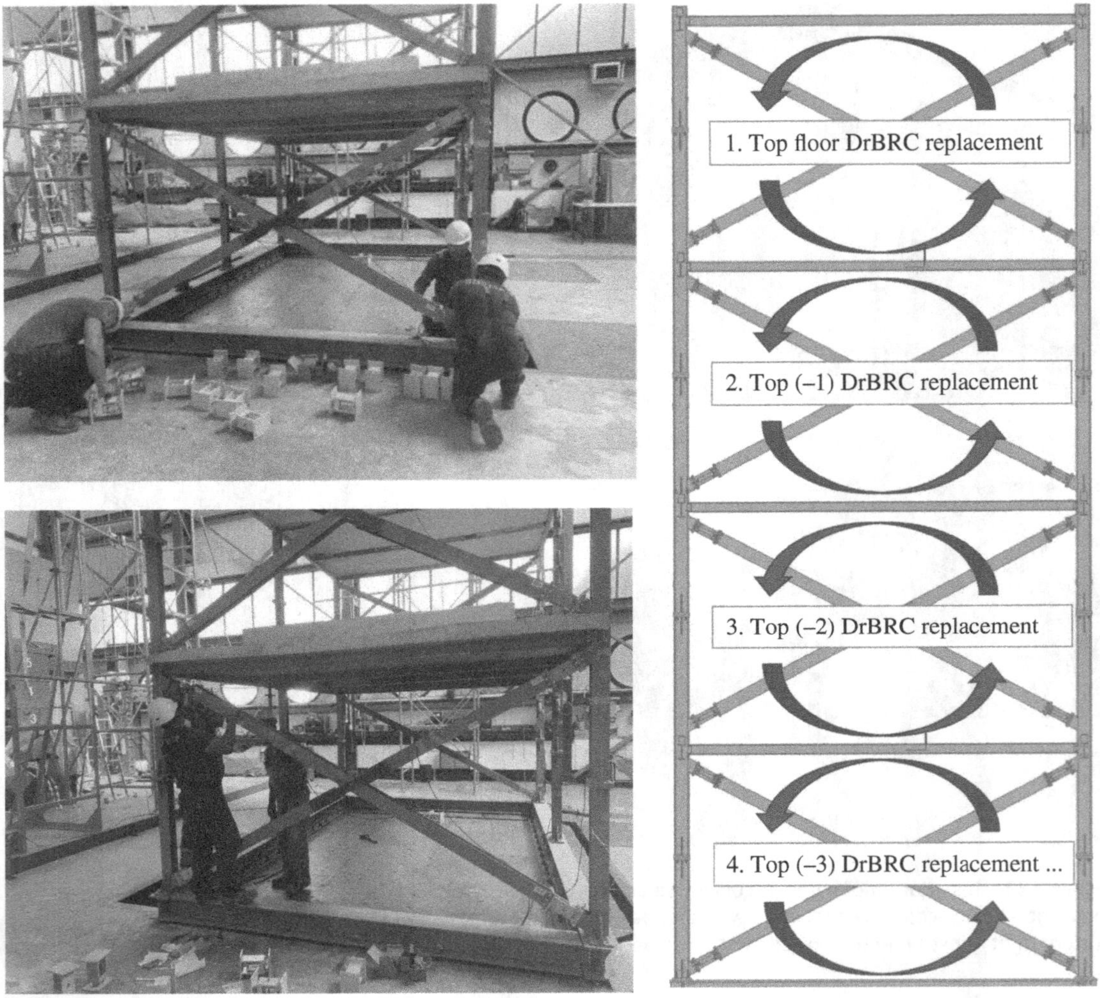

Figure 7.23 Repair and replacement operations for the DrBRC system following a simulated seismic event, showing step-by-step restoration of damaged connections (on the left). Replacement sequence in multi-story frames equipped with dissipative DrBRC components, showing floor-by-floor recovery process after an earthquake (on the right) (courtesy of Michalis Sofras).

safe restoration of structural functionality. Unlike traditional retrofitting or frame repair, no welding or large-scale demolition is needed. The entire operation can be completed within days, and the structure remains largely intact, with the main frame unaffected.

End-of-life and Reuse Potential (C and D): By preserving the primary structure, damage-controlled design ensures the building can reach its intended service life. At the end of its life, the bolted connections and modular nature of the fuses facilitate design for deconstruction. The main structural elements remain intact and can be easily disassembled for high-value reuse in other projects, while only the small fuse elements need to be recycled.

Figure 7.24 Base isolators at hospital under construction near Kahramanmaraş (courtesy of Tunc Deniz Uludag).

7.1.3 No-damage Seismic Design (Using Seismic Isolators)

No-damage design employs seismic isolation systems or advanced vibration control to physically decouple the structure from ground motion or to significantly reduce the structural demand. Seismic isolation uses special bearings to allow the ground to move underneath while the building remains relatively still. Advanced damping systems absorb and dissipate seismic energy, reducing the structural response and minimizing damage to the primary members. This principle is exemplified in Figure 7.24, which shows a hospital under construction in a highly seismic zone (in Kahramanmaras, Turkey) being built upon a grid of base isolators. These bearings, installed between the foundation piers and the main superstructure, create a flexible interface. During an earthquake, the isolators allow the ground to move beneath them while the hospital itself remains largely stationary. This strategy not only prevents structural damage but also protects vital nonstructural systems, sensitive medical equipment, and occupants, ensuring a critical facility can remain fully operational when it is needed most.

The life-cycle implications of this seismic design philosophy are:

Production Stage (A1–A3): The isolation bearings are complex (in general "proprietary") devices made of steel and/or special rubber compounds, potentially carrying their own upfront embodied carbon footprint. However, because they reduce dramatically the seismic forces transmitted to the superstructure, the building above can be designed with significantly less material. Columns, beams, and walls can be lighter, and reinforcement requirements can be reduced. Therefore, the extra carbon footprint of the isolators is commonly offset by the large material savings in the superstructure.

Figures 7.25–7.28 show the manufacturing process of friction pendulum isolators, a class of seismic isolation devices that dissipate energy through controlled sliding. These isolators consist of concave steel surfaces, polished sliders, and internal mechanisms that allow the superstructure to move horizontally during an earthquake while returning to its original position through gravity. Their fabrication involves multiple precision steps to achieve the required surface geometry, friction characteristics, and long-term durability.

The process begins with the preparation of thick steel plates (Figure 7.25), which are cut, machined, and shaped to form the concave bearing surfaces. CNC milling machines are used to achieve the high smoothness and curvature tolerances essential for consistent frictional behavior (Figure 7.26). Subsequent

Figure 7.25 Preparation and cutting of thick steel plates used for the base and top components of seismic isolation bearings (courtesy of Bahadir Sadan).

Figure 7.26 Welding and machining of bearing components to achieve the required geometry and surface quality (courtesy of Teknolojik Izolator Sistemleri).

Figure 7.27 Dimensional and surface inspection of isolator components using robotic and laser-based metrology systems (courtesy of Teknolojik Izolator Sistemleri).

Figure 7.28 Assembly and packaging area of isolation bearings, showing organized workflow, quality control, and readiness for shipment (courtesy of Teknolojik Izolator Sistemleri).

stages include welding, polishing, and assembly of the sliding components, often incorporating specialized materials such as PTFE (polytetrafluoroethylene) layers or stainless-steel coatings to control friction coefficients. Dimensional and surface inspections are performed using robotic or laser-based metrology systems to ensure manufacturing accuracy (Figure 7.27).

Although the production of these isolators requires energy-intensive machining and material processing, their ability to drastically reduce the seismic demand on the superstructure leads to major material savings. Lighter beams, columns, and reinforcement requirements result in a lower overall embodied carbon footprint at the building level. Buildings equipped with friction pendulum isolators typically experience minimal or no structural damage even after strong earthquakes, allowing immediate re-occupancy and avoiding costly repairs. This resilience substantially extends the service life of the structure and reduces long-term environmental impact. Furthermore, friction pendulum isolators can be maintained or replaced individually, ensuring continuous performance and improving sustainability over the building's entire life cycle.

Construction Stage (A4–A5): The installation of seismic isolators is a specialized task requiring high precision and heavy-duty cranes. The bearings must be placed perfectly at the foundation level, demanding expert supervision and rigorous quality control. While a critical task, it is a localized process that doesn't necessarily complicate the rest of the building's construction.

As an example of the construction stage, Figures 7.29–7.31 show the transport from the factory and the on-site installation of seismic isolation systems. Figure 7.29 shows the handling and placement of friction pendulum isolators. These isolators are delivered fully assembled and sealed to protect the precision-machined concave surfaces. On-site (Figure 7.30), cranes position each unit onto its reinforced concrete foundation with high accuracy, under the supervision of specialized engineers. Proper leveling,

Figure 7.29 Packaging, lifting, and loading of friction pendulum isolators for delivery from the factory to the construction site (courtesy of Teknolojik Izolator Sistemleri).

Figure 7.30 On-site placement of isolators on concrete plinths, showing lifting, alignment, and anchoring operations conducted under technical supervision (courtesy of Bahadir Sadan).

Figure 7.31 Installed elastomeric isolators integrated within the foundation before superstructure assembly, illustrating final positioning and protection during construction (courtesy of Bahadir Sadan).

surface cleaning, and alignment are verified before anchoring, ensuring uniform load transfer and friction behavior during seismic events. Figure 7.31 shows a different system (elastomeric bearings) installed at the foundation level. These bearings combine rubber and steel layers to accommodate horizontal displacements and vertical loads, providing an alternative low-damage isolation strategy. Their installation follows similar precision requirements, with careful positioning and anchorage to guarantee reliable long-term performance.

Use and Repair Stage (B): By preventing strong shaking from reaching the structure, seismic isolation minimizes damage to both structural and nonstructural components (such as cladding, partitions, and windows). This means virtually zero-carbon emissions from post-earthquake repairs. For critical facilities like hospitals, data centers, or emergency response centers, the ability to remain fully operational is the primary objective. This resilience also protects sensitive and valuable equipment within the building.

End-of-Life and Reuse Potential (C and D): A building that survives its service life without significant structural damage is an ideal candidate for adaptive reuse or clean deconstruction. The initial embodied carbon investment is fully preserved. The structural elements, having never been stressed to their limits, can be confidently disassembled and reused, maximizing material circularity.

In conclusion, the relationship between seismic design philosophy and embodied carbon is complex and extends far beyond initial material quantities. While traditional ductility-based design can minimize upfront carbon through reduced material use on the structural system, it creates significant sustainability challenges through planned damage and subsequent repair requirements. Low-damage design represents an emerging middle ground, accepting modest increases in initial embodied carbon to dramatically reduce maintenance and repair impacts over the building lifecycle. This approach increasingly shows promise for achieving both seismic resilience and environmental sustainability. No-damage design, while requires often a moderate initial carbon investment (the isolation system adds material, but reduced structural demands in the superstructure partially compensate for this), offers the most sustainable long-term solution by preserving the initial material investment and enabling maximum reuse potential. As the cost of specialized isolation and damping systems decreases and environmental regulations tighten, this approach may become increasingly attractive. Therefore, structural resilience and sustainability can be considered as complementary goals in seismic design. A building that survives earthquakes with minimal damage is inherently more sustainable than one that repeatedly requires resource-intensive repairs or replacement. The three philosophies represent different points on the resilience-sustainability spectrum (Table 7.2).

7.2 Role of Construction Materials on Seismic Design with Life-cycle Thinking

The choice of a primary structural material has profound implications for a building's life-cycle carbon footprint, but the most sustainable option is not always straightforward and requires moving beyond common misconceptions. A holistic analysis is needed to balance the trade-offs between a material's upfront carbon, its required use in a resilient design, and its end-of-life potential, as Table 7.3 shows for steel, reinforced concrete, and timber.

Table 7.2 Trade-off among the three seismic design philosophies.

Seismic design philosophy	Upfront carbon (A1–A5)	Maintenance carbon (B2, B5)	Reuse potential	Whole life-cycle carbon
Ductility-based	Low	High	Low	Moderate to high
Damage-controlled	Moderate	Low	Moderate	Low to moderate
No-damage	Moderate	Minimal	High	Low

Table 7.3 Seismic design characteristics of common construction materials and systems within LCA framework.

Seismic design/ life-cycle item	Steel structures	Reinforced concrete (RC) structures	Timber structures
A1–A3 (Production)	Potential reduction of member sizes in case of conventional seismic design (causing significant post-earthquake repair). Seismic requirements (overstrength, ductile connections) can increase the amount of steel needed locally. High-strength grades can optimize material use. Using dissipative devices such as buckling-restrained braces or dampers can reduce demand on the primary frame, potentially decreasing overall steel tonnage despite the material required for the devices themselves.	Potential reduction of member sizes in case of conventional seismic design (causing significant post-earthquake repair) Ductile design requires additional cement and steel rebar for confinement and strength, increasing production-stage emissions High-strength grades can optimize material use Seismic base isolation can lower forces on the superstructure, allowing for a significant reduction in member sizes and reinforcement, though this requires manufacturing specialized isolator bearings.	Commonly considered a low-carbon solution, but seismic design often requires adding steel bracing or RC shear walls, which increases the overall initial embodied carbon Modern mass timber systems may also incorporate replaceable, high-performance energy dissipators (e.g. steel fuses or friction dampers), which can reduce the need for oversized timber members and connections.
A4–A5 (Construction)	High-performance seismic designs may require specialized on-site welding or bolting and heavier cranes, increasing construction energy use.	Dense, ductile rebar detailing makes on-site assembly more complex and can extend construction time and equipment use.	Often prefabricated, leading to faster construction and less on-site equipment use. However, seismic designs require metal connectors, steel bracings, or reinforced concrete shear walls.
C (End-of-life Process)	Demolition is straightforward; steel elements can be easily cut and separated for recycling.	Demolition is energy-intensive. Concrete must be crushed to separate the rebar, generating waste, and requiring significant processing.	Components can be salvaged if undamaged. If not reusable, it can be used for biomass fuel, but this releases their stored carbon.
D (reuse and recycling potential)	Steel is highly recyclable with well-established processes. Bolted connections in "design for deconstruction" allow for direct reuse of elements.	Crushed concrete is typically downcycled into low-grade aggregate or fill. Rebars are highly recyclable if separated from the concrete.	Undamaged timber elements have reuse potential. Designing with demountable connections is key to realizing this value.

Beyond these primary materials, other systems have unique seismic and life-cycle profiles:

- Composite structures (e.g. steel-reinforced concrete, steel-timber) are highly efficient for seismic resistance, often leading to smaller members than pure RC or steel systems, but their end-of-life process is challenging as separating the bonded materials for high-value recycling is energy-intensive.
- Cold-formed steel offers a lightweight alternative where seismic performance relies on system behavior (e.g. sheathed shear walls); it has a lower material mass and excellent end-of-life potential, as screwed connections make it ideal for disassembly and reuse.
- Masonry presents a critical distinction: while its embodied carbon is primarily tied to the firing process of clay bricks or the extraction of natural stone, its exceptional durability, potential for local sourcing, and very long service life with low maintenance needs can result in a favorable whole life-cycle carbon performance. While modern reinforced masonry can provide ductile, safe performance, it is generally limited to low-to-mid-rise construction and offers less architectural flexibility for creating large open spaces compared to framed systems.

7.3 Resilience of Nonstructural Elements Under Earthquakes

While the primary structural system is essential for preventing collapse, it represents only one part of a building's seismic performance and life-cycle impact. After earthquakes, a great financial damage and a massive source of repair-related embodied carbon may come from the failure of nonstructural components (NSCs). These elements include everything other than the main load-bearing frame:

Architectural Components: Facades (cladding), interior partition walls, suspended ceilings, and windows.

MEP Systems: Mechanical (HVAC), electrical, and plumbing systems, including pipes, ducts, and critical equipment.

Figure 7.32 shows an example of nonstructural damage caused during a seismic event: in the photos, the building's concrete frame beams and columns are largely intact. However, the unreinforced masonry

Figure 7.32 Example of nonstructural component damage followed by an earthquake (courtesy of Tunc Deniz Uludag).

(URM) infill walls experienced an out-of-plane collapse. This type of nonstructural damage is often a primary source of repair costs and embodied carbon, even when the main structure performs as intended.

The primary cause of NSC damage is inter-story drift, which is the relative movement between adjacent floors as the building sways during an earthquake. As the structure deforms, these rigid and brittle components, which are attached to the frame, are stretched, compressed, or sheared until they crack, buckle, or detach completely. The seismic design philosophy chosen by the engineer has a direct impact on the performance of these components. In a conventional (ductility-based) design, the structure is designed to drift significantly to dissipate energy. Consequently, widespread NSC damage is an expected (and allowed) outcome. A building might remain standing and be deemed "safe," but be rendered completely nonfunctional by shattered windows, fallen ceilings, broken water pipes, and a damaged facade. The carbon footprint of manufacturing and installing replacements for these systems (Module B of the LCA) can be immense, often consuming the initial carbon savings of a leaner structural frame. Damage-controlled and no-damage designs control and minimize inter-story drift. A base-isolated building, for example, moves as a rigid block with very little internal deformation, protecting its nonstructural components. By preventing damage to walls, pipes, and cladding, these advanced systems can minimize or eliminate the need for most post-earthquake repairs. This not only ensures the building remains operational but also avoids a huge source of embodied carbon emissions over its life cycle.

7.4 Seismic Retrofitting of Existing Building Stock

While designing resilient new buildings is critical, the greatest seismic vulnerability in many of the world's cities lies within the vast stock of existing buildings (the original design philosophy of such buildings is typically based on a 50-year service life assumption). Many of these structures were built before the adoption of modern seismic codes and are highly vulnerable. From a life-cycle-driven perspective, addressing this challenge through seismic retrofitting can be an effective sustainability strategy available to engineers.

The core environmental benefit of retrofitting lies in the preservation of a building's existing embodied carbon. Demolishing a structure means that all the carbon emitted during its original material production and construction is effectively wasted. A replacement building then starts the carbon clock over again with a new, massive injection of embodied carbon in its "Production Stage" (Module A).

By strengthening an existing building, engineers preserve its immense carbon investment. The embodied carbon of the new retrofit materials is typically only a small fraction of what would be required for a completely new structure. Common retrofitting are explained in the next subchapters.

7.4.1 Retrofitting with Reinforced Concrete Shear Walls and Steel-braced Frames

Adding new reinforced concrete shear walls or steel-braced frames is one of the most widespread retrofitting strategies, largely because they rely on well-established construction techniques that are familiar to most contractors, require no proprietary or specialized products, and can be designed using standard code provisions available in most national seismic regulations. Their structural behavior is well understood and backed by decades of experimental and field evidence, which gives engineers and building

Figure 7.33 Examples of reinforced concrete shear wall strengthening (Antoniou, 2023).

authorities confidence in their performance. However, these methods also carry notable disadvantages: reinforced concrete shear walls require substantial foundation work (often including new footings or ground anchoring to transfer the concentrated forces to the soil), generate significant construction disruption including noise, dust, and temporary loss of usable floor area, and add considerable mass to the building. Steel-braced frames offer several practical advantages over reinforced concrete shear walls in a retrofitting context. Their significantly lower self-weight reduces the additional seismic demand imposed on the existing foundations, often avoiding the need for costly foundation upgrades. Steel elements can be prefabricated off-site and assembled rapidly on-site with bolted connections, minimizing construction time and disruption to the building's occupants. When applied externally (as exoskeletons), steel bracing systems can strengthen and stiffen the structure without requiring major intervention inside the building, preserving interior layouts and allowing continued use of the building during construction. Furthermore, steel bracing systems are inherently more adaptable: they can be designed for future disassembly and reuse, and if combined with damage-controlled connection details (such as friction-based or replaceable fuse elements, as discussed in Section 7.1.2), they can also provide a repairable seismic response. On the other hand, steel-braced retrofits still require careful attention to the interface between new steel elements and the existing structure (particularly at connection points with concrete or masonry members, where local strengthening may be necessary), and exposed external bracing may alter the building's architectural appearance, which can be a concern in historically or aesthetically sensitive contexts. As shown in Figure 7.33, new reinforced concrete walls can be inserted in selected bays of existing frames, often replacing windows or infill walls, while Figure 7.34 shows the external application of steel bracing systems to strengthen and stiffen the building without significantly disturbing internal spaces.

7.4.2 Confinement with Fiber-reinforced Polymer (FRP) Wraps

Wrapping vulnerable reinforced concrete columns shear walls and beams with advanced materials such as FRP wrap provides a lightweight yet highly effective retrofit option. The FRP encasement confines the column, preventing brittle shear or crushing failures and enabling the column to deform plastically

Figure 7.34 Seismic retrofitting example with steel bracings (courtesy of Andrea Imbrenda).

in a controlled manner during seismic loading. This confinement enhances both strength and ductility, effectively extending the service life of structural members while minimally affecting building geometry or weight (Figure 7.35).

7.4.3 Seismic Isolation Retrofitting of Existing Buildings

In the most advanced retrofits, a building can be separated from its foundation and placed on new seismic isolators. This technique requires carefully cutting the reinforced concrete columns and temporarily supporting the structure, after which isolators are inserted to decouple the superstructure from ground shaking. Figure 7.36 shows this process involving retrofitting an existing reinforced concrete building by introducing seismic isolators (left) while ensuring stability during installation through temporary supports (right). Once complete, the isolators significantly reduce seismic forces transmitted to the building, transforming a vulnerable structure into one capable of sustaining earthquakes with little or no damage. This approach not only preserves the embodied carbon of the original building but

Figure 7.35 Seismic retrofitting example with FRP (left: column strengthening, right: beam and joint strengthening) (courtesy of Infigor).

Figure 7.36 Retrofitting of an existing reinforced concrete building: insertion of seismic isolators (left) and temporary supporting of the reinforced concrete columns (right) (courtesy of Massimo Latour).

also safeguards occupants and critical functions, making it an exemplary strategy for resilience-focused retrofits. However, seismic isolation of existing buildings remains as a technically demanding retrofitting intervention: it requires highly specialized engineering expertise, detailed knowledge of the existing structural system, precise temporary load transfer during column cutting, and disruption to the building's ground floor or basement level. Its application is therefore typically reserved for critical facilities (such as hospitals, emergency response centers, or buildings of historical importance) where the long-term benefits of near-complete damage prevention justify the complexity and cost of the intervention.

In summary, seismic retrofit is a direct application of circular economy principles: it extends a building's service life, avoids landfill waste from demolition, and prevents the unnecessary emission of tons of carbon. In some cases, however, demolition and rebuilding may prove more cost- and time-effective, and under specific conditions even more favorable in terms of embodied carbon performance, particularly when the existing structure is severely damaged or weak or when new construction can achieve substantially higher (operational carbon) efficiency standards.

7.5 The Influence of Seismic Design Codes on Sustainable and Resilient Structures

In practice, structural engineers rely on building codes to guide seismic design. These codes, such as ASCE 7 in the US and Eurocode 8 in Europe, establish minimum requirements for seismic loading, detailing, and performance. Their primary goal is life safety and collapse prevention, not always minimizing damage or ensuring rapid post-earthquake recovery.

However, conventional code compliance alone is often insufficient for achieving broader resilience and sustainability goals because of the following limitations:

- They do not fully address in detail the cumulative damage from multiple earthquakes or major aftershocks.
- They often do not mandate repairability or the use of replaceable components, which are essential for reducing the embodied carbon associated with repairs and minimizing downtime.
- They rarely incorporate life-cycle performance metrics, such as expected repair costs or whole-life embodied carbon.

Table 7.4 summarizes how various regions implement their codes and are beginning to address the gap between mandated safety and the goal of sustainability. The summary draws on the relevant structural and environmental standards for each region: ACI 318-19 (2022), ANSI/AISC 341, ASCE 7, and FEMA P-58 (United States); Eurocode 8 (EN 1998-1) and EN 15978 (Europe); GB 50011 (China); NZS 1170.5 (New Zealand); the Building Standard Law and CASBEE (Japan, 2014); the Türkiye Building Earthquake Code (TBDY, 2018); IS 1893 (Bureau of Indian Standards, 2016); the ASHRAE Fundamentals handbook; and the comparative analysis of green-building certification in China by Zou (2019).

Recognizing these limitations, international standardization bodies are actively developing frameworks to better integrate seismic resilience with sustainability. For example, the development of the second-generation Eurocodes aims to enhance rules for robustness and performance-based approaches (*Second Generation of the Eurocodes|Eurocodes: Building the Future*, n.d.). In parallel, European standards EN 15643 and EN 15978 provide a formal methodology for assessing the life-cycle environmental impact of buildings, enabling embodied carbon to be treated as a quantifiable engineering

Table 7.4 Worldwide design codes and their approach to sustainability.

Region	Governing codes/ standards	Primary approach to seismic design	Integration with sustainability and carbon footprint	Key trends
Europe	Structural: Eurocode 8 (EN 1998) Environmental: EN 15978 (for LCA)	Follows prescriptive rules for ductility classes and capacity design, with a strong emphasis on structural regularity	Structural safety (EC8) and carbon reporting (national rules like RE2020) are separate requirements that designers must reconcile	Strong focus on retrofitting existing buildings (EC8 Part 3), which is inherently sustainable Growing use of hybrid systems (timber–steel) to meet both seismic and carbon targets
US	Structural: IBC, ASCE 7, ACI 318, AISC 341 Environmental: LEED, ASHRAE (voluntary)	Similar to Europe with prescriptive rules (R-factors for ductility)	Driven by voluntary initiatives and advanced methods. Initiatives like SE 2050 push for embodied carbon reduction, while tools like FEMA P-58 allow for quantifying environmental impacts of damage	Leader in Performance-based Earthquake Engineering (PBEE), linking performance directly to life-cycle impacts Common use of base isolation and dampers in high-value projects, preventing damage and reconstruction
Japan	Structural: Building Standard Law Environmental: CASBEE (voluntary)	Very high seismic resistance required by strict prescriptive codes. Widespread, common use of advanced technologies like base isolation and dampers in standard practice	Resilience is standard, but sustainability is challenged by a culture of shorter building lifespans (frequent replacement), which increases carbon footprint	Highest level of built-in resilience, but faces a "disposable building" culture Push for long-life housing to align excellent structural performance with sustainability goals
China	Structural: GB 50011, Environmental: Three Star Rating System	Prescriptive, three-level design goals ("no damage" in minor quakes to "no collapse" in major ones). PBD is used for complex, tall buildings	Government mandates for green buildings operate separately from seismic codes	Massive scale of construction drives rapid evolution of both code types; a huge testing ground for reconciling safety and carbon goals
New Zealand	Structural: NZS 1170.5, Environmental: MBIE Building for Climate Change	Focus on low-damage seismic design (LDSD), moving beyond life safety to focus on asset protection and continued use after an earthquake	Direct and mandatory integration. From 2025, carbon reporting is mandatory for building consent, with carbon caps to be enforced, forcing a unified design approach	The most progressive jurisdiction in formally codifying both seismic performance and embodied carbon, setting a global precedent

Türkiye	Structural: TBDY 2018 (TEC) Environmental: Emerging regulations referring to EPBD (2024/1275)	Modern, PBD-influenced code with mandatory performance targets (from limited damage to collapse prevention) based on building use	Life safety is priority. Sustainability is secondary but gaining traction, especially in the context of massive post-earthquake reconstruction	Immense focus on code enforcement and seismic retrofitting of a vast, vulnerable building stock. The need for resilient, rapid reconstruction is a major challenge. Post-1999 policy and TBDY 2018 provisions have led to systematic adoption of seismic isolation for healthcare facilities, particularly large public hospitals in high seismic regions
India	Structural: IS 1893, Environmental: GRIHA, IGBC (voluntary)	Prescriptive code focused on preventing collapse. The main challenge is ensuring compliance and addressing nonengineered construction	Building environmental rating systems are gaining popularity but are voluntary and not integrated with mandatory seismic safety rules yet	Key challenge is bridging the gap between code and common practice. Huge vulnerability in existing unreinforced masonry buildings
Latin America (e.g. Chile, Mexico)	(*Chilean Standard NCh433.Of96*, 2016) NTC-S (Mexico), Environmental: Various voluntary green building councils	Robust prescriptive codes born from experience. Chilean code, in particular, is known for its effectiveness in limiting damage	Primarily focused on resilience. Sustainability is an emerging topic, but robust, damage-limiting design is seen as the first step toward sustainable communities	Chile is a model for effective code enforcement and a design culture that prioritizes stiffness and damage control, leading to high resilience in practice
Performance-Based Design (PBD) at Global level	ATC-58/FEMA P-58, PEER, etc.	Goal-oriented. Sets explicit performance targets (e.g. limit damage, ensure immediate occupancy) and designs to meet them, often using advanced analysis	Directly enables optimization, which can allow designers to quantify the carbon cost of future repairs and balance it against upfront material investment to find the lowest whole-life carbon solution	The most advanced method for unifying resilience and sustainability Moves beyond "will it collapse?" to "what will be the full environmental and economic cost over its life?"

metric (EN 15643, 2021; EN 15978, 2026). Furthermore, frameworks such as FEMA P-58 offer a standardized procedure to connect seismic performance directly to repair costs, downtime, and environmental impacts, moving toward a whole-life assessment of resilience (*FEMA P-58*, n.d.).

In conclusion, codes set the baseline for seismic safety, and they inevitably influence material use (hence carbon footprint). Integrating seismic performance with environmental impact metrics is increasingly feasible worldwide. Yet in current practice, structural codes and sustainability standards still tend to operate in parallel. The integration is not complete yet: one day we might have a unified design code that includes carbon limits alongside safety factors. Until this integration is fully codified, it is the designer's responsibility to reconcile these goals. Engineers must use every available tool to deliver buildings that are safe, resilient, and have a minimal environmental impact throughout their life: modern codes, performance-based design methods, damage-controlled and no-damage methods, and LCA.

7.6 Earthquakes and Community Resilience

The seismic design philosophy chosen by an engineer has consequences that extend far beyond the building itself, directly shaping a community's ability to survive and recover. Life-cycle-driven design is therefore not just a technical or environmental consideration; it is a critical factor in social resilience.

7.6.1 Immediate Aftermath and Emergency Response

The immediate impact of an earthquake is heavily influenced by when it strikes. An event occurring at night, when most people are in residential buildings, can have a much higher casualty rate than one during the day (Jaiswal & Wald, 2010). The primary goal of any seismic code is to prevent collapse and save lives in these initial moments. However, a design focused solely on this life safety objective often accepts the complete functional loss of the building, setting the stage for subsequent community-wide disruption that can last for years. From a life-cycle perspective, a building that survives but cannot function triggers a chain of carbon-intensive consequences: emergency demolition, debris removal, temporary sheltering, and eventual reconstruction, none of which appear in a conventional upfront carbon assessment. The social cost is equally severe, as displaced families, interrupted education, lost livelihoods, and prolonged community fragmentation disproportionately affect the most vulnerable populations. A life-cycle-driven design framework recognizes that these post-earthquake costs (environmental, economic, and social) must be anticipated at the design stage, not discovered after the event.

7.6.2 Post-earthquake Safety Evaluation Challenges

In the immediate aftermath, post-earthquake safety evaluations determine the habitability of the building stock (Comfort et al., 2024). Engineers and inspectors can use a tagging system to classify structures (*Geotechnical Extreme Events Reconnaissance (GEER) Association*, n.d.): green-tagged buildings are safe to occupy, yellow-tagged buildings have restricted access pending repairs, and red-tagged buildings are deemed unsafe and must be evacuated. A city filled with red-tagged buildings (even those that successfully prevented fatalities) faces mass displacement, straining emergency shelters, and social services.

In contrast, damage-controlled and no-damage designs aim for a green tag, allowing residents to shelter in place, businesses to continue operating, and critical facilities like hospitals to remain functional when they are needed most.

7.6.3 Long-term Recovery and Reconstruction

Long-term recovery is the most challenging phase. Widespread structural damage paralyzes a region's economy, leading to business closures and job losses (*GAR2022*, 2022; Ozkula et al., 2023). The social fabric disintegrates as displaced populations are forced to relocate, sometimes permanently. This process can worsen inequality, as the financial burden of rebuilding or retrofitting falls heavily on residents, potentially leading to gentrification as those who cannot afford to rebuild are forced to sell (Worldbank, 2023). A life-cycle approach that invests in higher performance objectives (minimizing damage and ensuring rapid re-occupancy) is a direct investment in a community's future. By reducing displacement and speeding economic recovery, resilient design (Uludag, 2022) helps preserve the social and cultural identity of a place, moving beyond mere survival to ensure long-term prosperity.

7.7 Rapid, Low-cost, and Low-impact Post-earthquake Assessment

The preceding sections have established the principles of life-cycle-driven seismic design and retrofit, underlining the importance of considering environmental impacts, durability, and resilience across a structure's entire service life. Yet resilience is only as strong as our ability to verify it: when disasters hit, the critical question becomes how quickly and reliably we can assess which buildings remain safe for occupancy. Dense urban environments with complex building inventories, such as large metropolitan areas (Figure 7.37), highlight the urgent need for scalable, rapid post-disaster assessment frameworks. This section examines how Structural Health Monitoring (SHM) technologies can transform post-earthquake assessment from a resource-intensive manual process into a scalable, data-driven decision support system that aligns with life-cycle design principles. We introduce the concept of dual-purposing telecommunications infrastructure for SHM, an approach that could transform buildings from passive structures into responsive systems that continuously sense and report their own health through existing fiber-optic connections.

7.7.1 Urban-scale Post-earthquake Assessment Challenge

When an earthquake hits a city, chaos unfolds on multiple fronts simultaneously: transportation networks, airports, and roads are often damaged, entire cities may be disrupted, and collapsed buildings create severe logistical bottlenecks with qualified inspectors missing. Today, post-earthquake structural integrity assessments still rely heavily on manual visual inspections, such as those outlined in ATC-20 (Applied Technology Council, 1995). Evaluating all buildings in an earthquake-affected community with those methods is extremely time-consuming (Federal Emergency Management Agency, 2021), particularly when access to damaged areas is limited.

Following earthquakes in Türkiye (Özkula et al., 2023), Greece (Baumont et al., 2004), and Italy (Encyclopaedia Britannica, 2025; Dolce and Goretti, 2015), inspecting thousands of structures took

Figure 7.37 New York City skyline aerial view (Songquan Deng/Shutterstock).

weeks to months. This prolonged displacement affected thousands of people, even though many buildings (~60% in the 2023 Türkiye earthquake (Yilmaz et al., 2024; Alejandro et al., n.d.; Özkula et al., 2023)) were undamaged, yet their safety was not evaluated, thus the occupants were not informed. The 2016–2017 Central Italy earthquake sequence (Di Ludovico et al., 2021) highlighted significant limitations of existing emergency assessment protocols, as multiple aftershocks required repeated inspections, leading to substantial delays in recovery (Italian Civil Protection Department, n.d.).

Our cities still lack a scalable and cost-effective way to rapidly screen and prioritize buildings for post-earthquake safety evaluation. From a life-cycle perspective, this gap has serious implications. First, damage is not always visible: internal fractures in beam–column joints, yielded reinforcement, or degraded connections may leave a building appearing intact while its capacity is compromised. In addition, buildings demolished due to uncertain damage status (rather than verified structural failure) represent wasted embodied carbon and materials invested during construction. Conservative red-tagging of repairable structures contradicts the repair-over-replace philosophy central to sustainable construction.

7.7.2 Key Structural Parameters for Structural Safety Assessment

In earthquake engineering, structural safety is not described by a single measured quantity. Instead, engineers interpret structural performance through a set of complementary indicators that describe how a building vibrates, deforms, and changes stiffness during seismic excitation. Among these, three especially informative quantities are natural frequencies, stiffness-related indicators, and inter-story drift. In practice, fiber-based monitoring systems usually do not measure these quantities directly; rather, they estimate them from measured strain or vibration response using system-identification and structural-interpretation models. This distinction is important: sensors provide physical observations, whereas engineering parameters are obtained through the interpretation of those observations.

Natural frequencies represent the natural vibration rates at which a building responds to dynamic excitation. Each structure possesses characteristic frequencies determined by its mass distribution, stiffness, and geometric configuration. Structural damage typically manifests as decreased natural frequencies: a building with cracked concrete or yielded steel connections becomes more flexible, lowering its natural frequencies. Tracking modal frequency changes over time provides a global indicator of structural condition without requiring knowledge of specific damage locations. However, changes in frequency are not uniquely caused by damage. Environmental and operational variations, including temperature, humidity, and changes in excitation, can also modify the natural frequencies and may mimic damage-related effects. For that reason, frequencies should be presented as sensitive but non-specific indicators, and therefore, they need to be interpreted together with other information rather than as standalone proof of structural damage.

Stiffness quantifies a structure's resistance to deformation under applied loads and is more directly linked to structural integrity than natural frequencies alone. Reduction in stiffness indicates material degradation, connection damage, or loss of structural integrity. While related to natural frequency (frequency squared is proportional to stiffness-to-mass ratio), stiffness changes can be localized to specific structural systems or stories, providing diagnostic information about damage distribution. In practice, however, stiffness is usually not measured directly in structural monitoring. Instead, it is estimated from identified modal parameters, displacement data, or model-updating procedures. This makes stiffness especially valuable for interpretation: while frequency trends may reveal that the structure has changed, stiffness-related indicators help explain the mechanical meaning of that change. When combined with spatially distributed measurements and an appropriate structural model, they may also assist in identifying which story or subsystem is most affected.

Inter-story drift measures the relative horizontal displacement between consecutive floors during building motion, commonly normalized by the story height. This parameter correlates most directly with structural and non-structural damage in seismic codes and serves as the primary demand parameter for performance-based design (Curadelli et al., 2008; Panteliou et al., 2001; Frizzarin et al., 2010; Demarie and Sabia, 2011; Moaveni et al., 2011; Park and Ang, 1985; Cao et al., 2017). Building codes specify inter-story drift limits corresponding to different performance levels: immediate occupancy, life safety, and collapse prevention. By comparing estimated drift demands against these code-specified thresholds, monitoring systems can provide actionable information for post-earthquake decisions.

These three parameters provide complementary information for seismic safety assessment. Natural frequencies are useful for detecting global changes in dynamic behavior; stiffness-related indicators assist in interpreting the mechanical significance of such changes; and inter-story drift provides a direct link between structural response and expected damage. For this reason, they may be considered among the most informative parameters for translating measured dynamic response into engineering assessment. It is, however, important to emphasize that they do not exhaust the description of seismic performance, which may also depend on other demand and damage measures depending on the assessment framework adopted.

7.7.3 Structural Health Monitoring Approaches

Structural health monitoring (SHM) techniques, such as those introduced in the literature (Pirrò and Gentile, 2025; Sohn et al., 2001; Farrar and Worden, 2013; Malekloo et al., 2022) allow for the early detection of damage, enabling timely maintenance and repair to mitigate the risk of catastrophic failures. Traditional SHM systems (Doebling et al., 1996; Farrar and Lieven, 2007; Peeters and De Roeck, 2001) achieve reliable dimensional reconstruction at the building scale, yet their large-scale (e.g. city-wide) application remains impractical. Among optical-fiber-based approaches, fiber Bragg grating (FBG) (Chan et al., 2006) provides strain measurements at discrete or quasi-discrete sensing locations, whereas Brillouin scattering systems (Ohno et al., 2001; Motil et al., 2016; Galindez-Jamioy and López-Higuera, 2012) provide distributed strain and temperature measurements along the fiber. Distributed acoustic sensing (DAS) (Hartog, 2017; Wang et al., 2020; Zhan, 2020) extends this family of techniques by enabling distributed measurement of dynamic strain along an optical fiber, making it attractive for applications requiring continuous spatial coverage. Despite these advantages, FBG, Brillouin, and DAS systems generally require either dedicated fiber installation or access to suitable existing fiber routes, together with specialized interrogation equipment. Although they have been successfully deployed on some bridges, tunnels, and critical infrastructure (Wang et al., 2025; Gołdyn and Zdanowicz, 2025; Ohno et al., 2001; Ye et al., 2014), their widespread implementation in ordinary buildings remains constrained by installation requirements, instrumentation cost, and data-management complexity (Li et al., 2004; Soga and Luo, 2018), and are expensive to deploy at scale (Ceylan et al., 2025; Carrión et al., 2017; Jang et al., 2010) (FBG arrays cost €10 000+ per building). Table 7.5 summarizes and compares traditional SHM technologies for building applications, highlighting their measurement characteristics, costs, and limitations in urban scalability. Satellite methods (Giordano et al., 2023) are valuable for regional deformation mapping, but they cannot capture interior floor behavior and modal response immediately after an earthquake (their accuracy also depends strongly on the underlying spatial resolution, which limits their ability to identify local structural changes). The fundamental limitation shared by all traditional SHM approaches is scalability. Individual building instrumentation with custom sensors cannot economically extend to thousands of structures across an urban area. This reality has confined SHM application to critical infrastructure and landmark buildings, leaving most of the building stock without continuous monitoring capability.

Table 7.5 Comparison of structural health monitoring technologies for building applications.

Technology	Measurement type	Spatial resolution	Cost per building	Urban scalability
Accelerometers	Acceleration	Point	€5000–15 000	Limited
Fiber Bragg grating	Strain	Point	€10 000+	Limited
Brillouin scattering	Strain	Distributed	€50 000+	Limited
Distributed Acoustic Sensing (DAS)	Strain	Distributed	€100 000	Very limited
Laser Interferometry (LI)	Strain	Integrated	<€10 000	High

Understanding structural monitoring through a life-cycle thinking reveals why this scalability gap matters for the topics discussed in this book:

Embodied carbon implications: Buildings demolished (LCA stage "C") due to uncertain damage assessment, rather than verified structural failure, represent wasted embodied carbon and materials. Rapid, reliable assessment can prevent unnecessary demolition of repairable structures.

Repair optimization: Early damage detection enables targeted repair before degradation propagates, extending service life (LCA stage "B") and avoiding costly comprehensive retrofitting.

Resource allocation efficiency: Prioritizing inspection resources toward buildings with monitoring-identified damage indicators reduces total assessment time and economic disruption (LCA stage "B").

Insurance and financing: Continuous monitoring data (LCA stage "B") can support performance-based insurance models and inform post-event financial decisions with quantified uncertainty.

7.7.4 Dual-purposing Telecommunications Infrastructure Using Laser Interferometry

Distributed Fiber Optic Sensing (DFOS) offers continuous monitoring along the entire length of an optical fiber (Gue et al., 2015; Bado and Casas, 2021). Techniques such as **Distributed Acoustic Sensing (DAS)** (Liu et al., 2023) and **Laser Interferometry (LI)** (Ceylan et al., 2025; Lv et al., 2025; Marra et al., 2018; Donadello et al., 2024) on deployed fibers enable the fiber itself to act as sensors, providing rich spatially resolved strain (DAS) and integrated strain (LI) information along the fiber path. Unlike FBGs or accelerometers, which provide point measurements and therefore require careful placement at suspected vulnerable locations due to their limited spatial resolution with a high cost, DFOS can detect anomalies along the fiber path, reducing the number of discrete sensors, installation effort, and prior identification of critical points.

LI measures the variation of the optical path length experienced by laser radiation as it travels a fiber subjected to strain, capable of measuring microscopic length changes of deployed fiber-optic (FO) cables with a resolution below 1 μm **under live internet traffic** (Donadello et al., 2024; Marra et al., 2018; Marra et al., 2022). LI offers two key advantages for a potential urban-scale deployment: (i) substantially lower interrogator costs (DAS typically costs around a hundred thousand euros per unit), enabling economical scaling to thousands of fiber paths; (ii) an integrated measurement that directly captures global structural response without requiring a large number of spatial data interpretations (Donadello et al., 2024). Operationally, LI requires only an interrogator unit connected to one fiber end, enabling long-term continuous monitoring without disrupting data transmission or requiring on-site instrumentation and maintenance. A massive network of FO telecom cables already connects buildings and infrastructure. Unlike alternatives like Starlink (Starlink, n.d.), better suited for remote areas with limited infrastructure, FTTH offers greater bandwidth and scalability. If we could interpret these signals for structural integrity purposes, we could enable city-scale, continuous screening and be able to prioritize buildings for a rapid post-earthquake safety evaluation. Figure 7.38 shows conceptually how LI can be applied to measure strain in buildings that host FO for internet connectivity in the FTTH architecture.

A fundamentally different approach to urban-scale SHM emerges from recognizing that cities already possess extensive fiber-optic networks connecting buildings for telecommunications. Fiber-To-The-Home

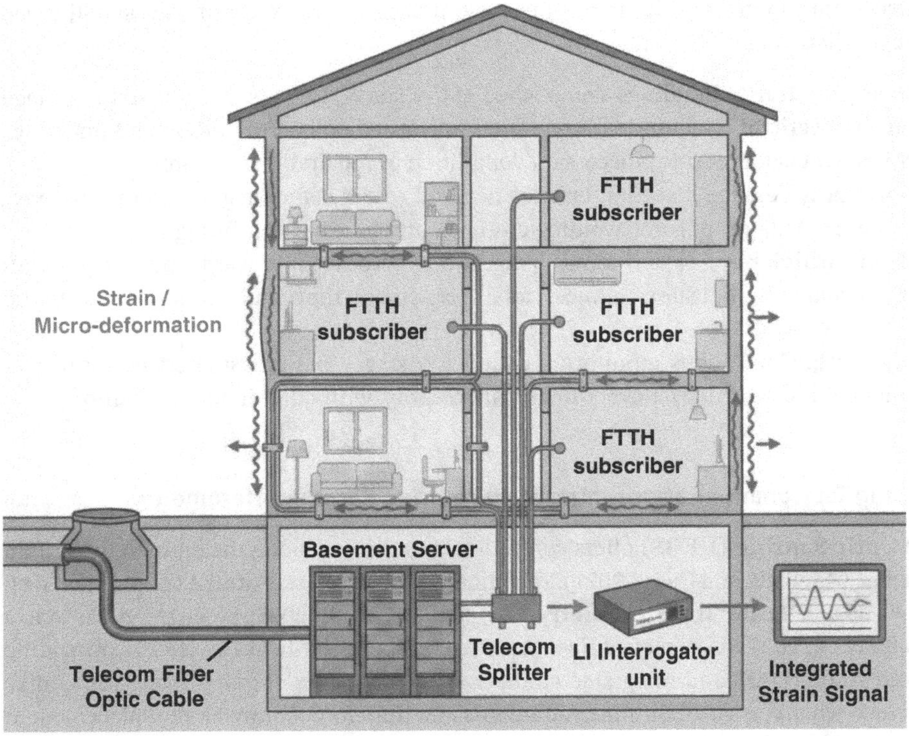

Figure 7.38 Conceptual illustration of building-scale monitoring using laser interferometry (LI) on existing FTTH fiber-optic networks. Micro-deformations induced by structural response are captured as integrated strain variations along in-building fiber paths, with signals interrogated from a remote unit without disrupting telecom services.

(FTTH) networks have expanded dramatically, reaching 74.6% coverage in the European Union by 2024 (FTTH Council Europe, n.d.) and continuing to grow globally. These optical fibers experience the same vibrations as their hosting structures (strain induced by building motion propagates through cables routed within structures).

The concept of dual-purposing this existing infrastructure can transform the economics of urban-scale monitoring (Figure 7.39). Rather than installing dedicated sensors in each building, structural information could potentially be extracted from signals already traveling through telecommunications fibers. This approach would not require new cables, on-site instrumentation, or disruption to building operations.

It is important to appreciate the magnitude of the challenge involved. Unlike purpose-built sensing fibers, telecom cables present extra complexity; signals are shaped by unknown routing, variable coupling with structures, and multiple noise sources. LI measures the cumulative optical path length change over the entire fiber span (a single, integrated strain value representing the net elongation or contraction of the full cable route). This “1D” measurement collects all deformations along potentially complex cable paths into one signal, encoding structural motion but in a form that cannot be directly decomposed without inverse modeling. Translating this 1D strain information into reliable structural performance indicators (e.g. modal properties, inter-story drift, and damage states) is an open research problem that

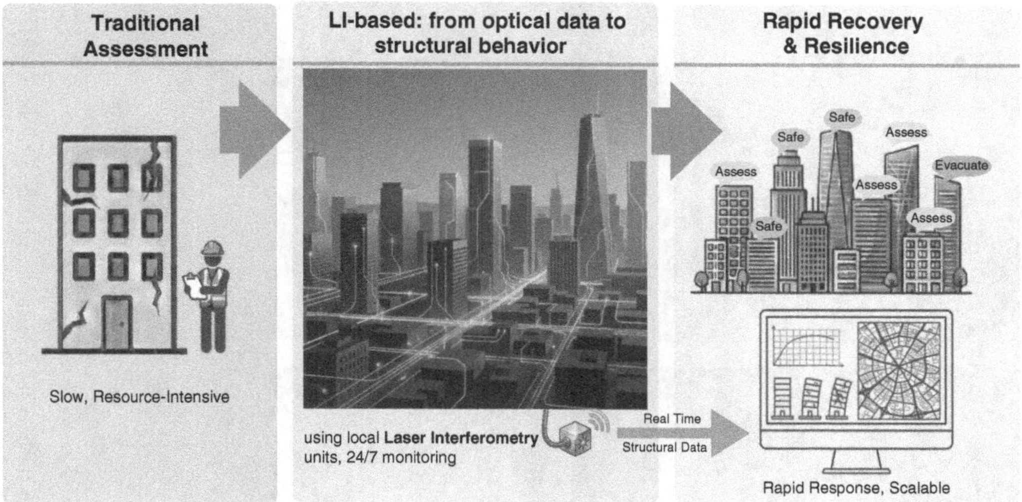

Figure 7.39 From traditional inspections to real-time, scalable structural monitoring using laser interferometry on existing fiber-optic networks.

requires advances at the intersection of structural dynamics, optical physics, and statistical inference. Key uncertainties that must be addressed include:

Fiber routing geometry: Cable paths through buildings are typically undocumented, following conduits and cable trays rather than structurally optimal routes. The relationship between structural deformation and fiber strain depends on this geometry.

Fiber-structure coupling: The mechanical connection between optical cables and building elements varies with installation method, cable type, and attachment points. Coupling may be partial, nonlinear, or position-dependent.

Signal quality under live traffic: Telecommunications fibers carry data traffic simultaneously with strain sensing. Maintaining an adequate signal-to-noise ratio for structural measurement while preserving communication function requires understanding interference sources.

Environmental effects: Temperature variations, humidity changes, and other environmental factors affect optical path length independently of structural deformation, creating measurement artifacts that must be distinguished from structural response.

Addressing these challenges requires integrating knowledge from structural dynamics, optical physics, and statistical inference (disciplines that have historically developed independently). A mechanics-based framework must link optical signal theory to building structural dynamics, enabling extraction of modal properties and damage indicators from inherently ambiguous measurements.

Recent studies have explored FO- and LI-based sensing for vibration monitoring (Wang et al., 2025; Liu et al., 2023; Nakamura et al., 2024; Zhang et al., 2024; Mahbubi et al., 2024). Existing approaches either rely on dedicated fibers and costly interrogators that are limited to vibration detection without spatial resolution or strain quantification or require centralized, power-intensive setups and controlled excitation. Recent work by the National Metrology Institute of Italy (INRIM, n.d.) demonstrated the feasibility

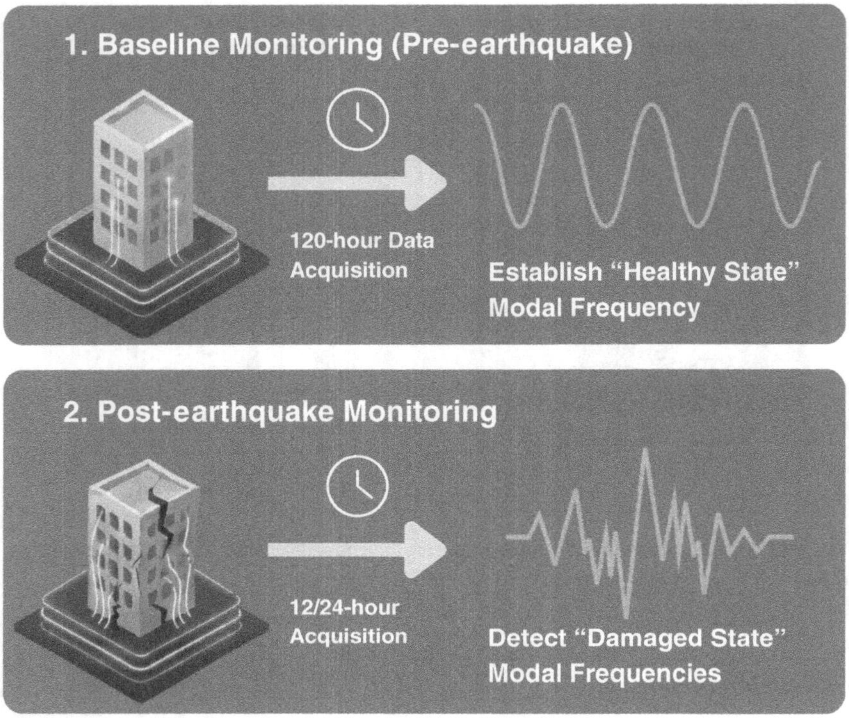

Figure 7.40 Operational concept of LI-based structural health monitoring: (1) pre-earthquake baseline acquisition to define the building's "healthy" modal properties and (2) post-earthquake monitoring to identify deviations and detect damage through changes in modal response.

of using underground telecom cables as vibration sensors for earthquake-induced ground motion via LI (Donadello et al., 2024), proving the potential of dual-purposing fiber infrastructure. However, detecting ground motion is fundamentally different from assessing structural condition. This approach can be extended to detect structural damage by comparing baseline modal properties (measured under ambient vibrations before an earthquake) with post-event recordings once the structure has returned to rest (Figure 7.40).

The authors' research group at Politecnico di Milano (S+ Lab, in the Department of Architecture, Built Environment, and Construction Engineering) is actively investigating this approach in collaboration with INRIM and telecommunications industry partners (Ceylan et al., 2026; Menghini et al., 2026). The work described below represents ongoing research and should be understood as preliminary evidence of feasibility rather than a validated operational methodology.

As part of this ongoing research at Politecnico di Milano, we performed preliminary tests on a 5-story rack system (40 × 90 × 180 cm; 20 kg) by instrumenting a single column with an FO cable and installing an accelerometer on the top floor. An additional 20 kg was subsequently added to the top floor (Figure 7.41). The rack was chosen as a simplified laboratory analogue of a multi-story frame: its bolted steel structure exhibits clearly identifiable bending modes, while its small scale allows controlled mass and stiffness modifications in a repeatable manner. It is important to note that this is a

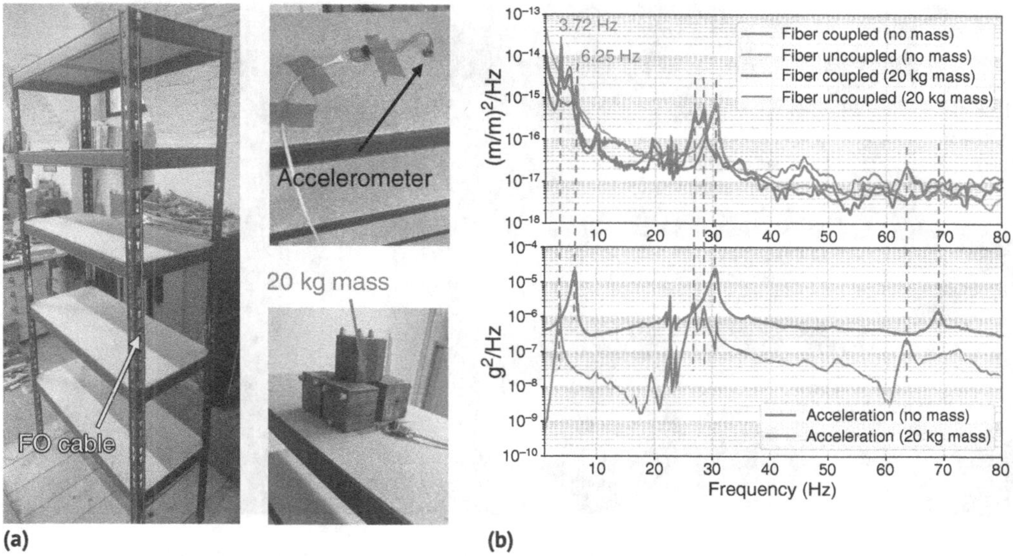

Figure 7.41 (a) The rack setup; (b) comparison of PSD spectra of the responses obtained.

highly idealized setup; a real building presents far greater complexity in terms of fiber routing, structural irregularity, noise sources, and coupling conditions. The rack was excited by hammer impacts to target down-aisle bending modes, and tests were repeated under coupled and uncoupled fiber conditions for both mass and no-mass cases, using the same excitation protocol. Figure 7.41 compares the results of these tests, revealing that FO-derived modal frequencies matched those identified from the accelerometer within 0.12–1.3% across the investigated configurations. While encouraging, these results must be validated on real buildings with realistic fiber installations before conclusions about operational performance can be drawn. Ongoing research (by the author) is currently addressing this need through experiments and deployments on real structures.

Practical implementation of urban-scale monitoring requires matching computational effort to decision criticality. Not all buildings require the same assessment depth, and available information varies with monitoring system sophistication. A tiered decision support framework, aligned with the Rytter damage assessment hierarchy (Rytter, 1993), provides appropriate responses across this spectrum (Figure 7.42):

Level 1 – Damage Detection (City-wide Screening): For rapid assessment across thousands of structures, systems rely on direct features extracted from fiber recordings to flag abnormal dynamic behavior. Statistical comparison of pre-event and post-event signal characteristics identifies buildings requiring further evaluation without building-specific numerical models. This approach enables broad screening with minimal computational overhead.

Level 2–3 – Damage Localization and Quantification (Critical structures): For buildings already flagged in Level 1 screening, physics-informed inference estimates modal properties, stiffness changes, and drift-related demands with quantified uncertainty. Observation operators and inverse models account for uncertain fiber–structure coupling, providing actionable damage indicators.

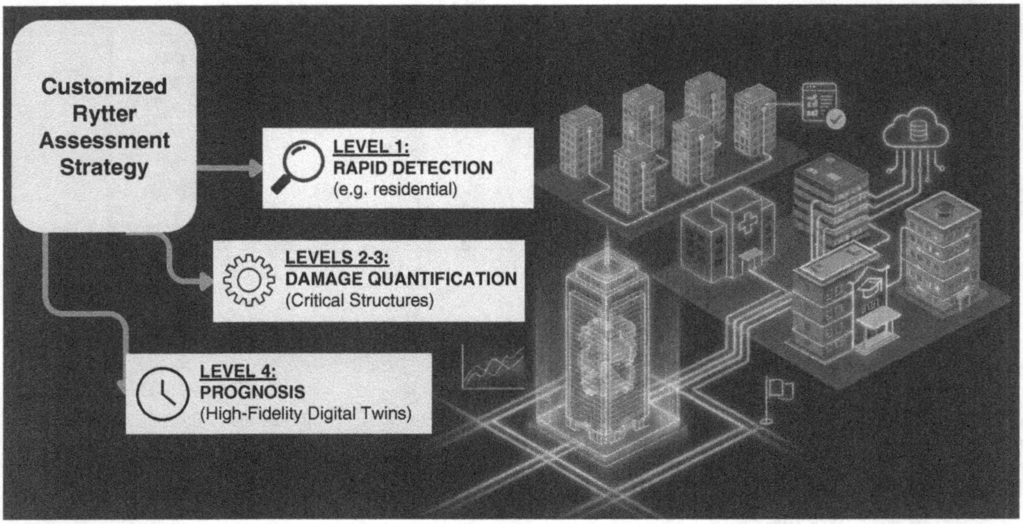

Figure 7.42 Multi-level assessment strategy for urban-scale monitoring, where computational effort is matched to decision criticality. Fiber-based measurements support rapid screening of the building stock (Level 1), followed by physics-informed damage quantification for selected structures (Levels 2–3) and digital twin-based prognosis for critical assets (Level 4).

Level 4 – Damage Prognosis (High-fidelity digital twins): For essential facilities where high-confidence decisions are required, high-fidelity digital twins are calibrated and continuously updated using fiber measurements, validated against conventional instrumentation. These models enable remaining capacity assessment and informed repair planning.

This tiered approach ensures computational resources scale with consequence severity, providing rapid screening for the building stock while enabling detailed assessment for structures requiring high-confidence decisions.

Translating this framework into operational capability will also require addressing several practical considerations. Effective damage detection depends on comparison with pre-event structural characteristics, which means monitoring systems must be operational before seismic events occur. Monitoring outputs must interface with existing post-earthquake decision protocols (such as the ATC-20 tagging system discussed in Section 7.6.2), and future systems may combine LI with emerging lower-cost DAS interrogators to provide both global response measurement and spatially resolved strain information.

7.7.5 From Buildings to Urban Infrastructure

The vision of buildings as responsive entities (continuously sensing and reporting structural health through their existing fiber connections) represents a potential transformation in how cities monitor, maintain, and protect their building infrastructure. If successfully realized, this approach would extend beyond post-earthquake assessment of buildings to life-cycle condition monitoring of bridges, tunnels, and other critical infrastructure.

Its dual-use approach supports the EU's climate neutrality, resilience, and digital transformation objectives (European Commission, 2025a; n.d.-c). Furthermore, it can enhance the growing investment in FO telecommunication systems (McKinsey & Company, n.d.), and EU smart city priorities (European Commission, 2025a, b).

The ability to monitor both sudden damages and gradual deteriorations will mark a fundamental shift from reactive to proactive urban infrastructure management. This approach further contributes to encouraging the expansion of FO networks as dual-purpose assets for communication and sensing, aligning with global trends toward smart cities.

7.7.6 Concluding Remarks

SHM represents a critical enabling technology for life-cycle-driven earthquake design. While traditional SHM approaches have proven valuable for critical infrastructure, their cost and complexity have prevented urban-scale deployment. The emerging possibility of dual-purposing existing telecommunications fiber networks offers a pathway to city-wide monitoring that aligns with life-cycle principles: minimizing additional material inputs while maximizing resilience benefits.

The scientific challenges are substantial: unknown fiber routing, uncertain coupling, and indirect measurements all complicate the inference problem, and much of the foundational research remains to be completed. Yet the potential benefits justify sustained research investment: a future where rapid post-earthquake assessment prevents unnecessary displacement, enables targeted repair, and ultimately reduces the human and economic costs of seismic events.

From a life-cycle perspective, continuous monitoring transforms buildings from passive structures into responsive systems that communicate their condition. This paradigm shift supports sustainable construction by extending service life, optimizing maintenance, and preventing premature demolition. As fiber-optic networks continue expanding globally, the infrastructure for urban-scale structural monitoring is being installed alongside internet connectivity, creating an opportunity to enhance urban resilience without proportional increases in resource consumption.

7.8 Conclusion

Sustainable seismic design is about seeing the bigger picture: not just what a building needs today, but how it will perform and endure across decades and through potentially devastating events. As we've explored throughout this chapter, every stage of a building's life cycle (from material selection and construction to repairability and end-of-life) can be shaped by seismic design decisions. And when those decisions are driven by life-cycle thinking, the result is often a structure that is both safe and environmentally responsible.

We have shown that resilience and sustainability are not opposing goals. Designs that limit damage (to both structural and nonstructural elements), not only ensure a building remains functional but also prevent the massive carbon footprint and material waste associated with post-earthquake repairs. This life-cycle logic finds its most powerful application in retrofitting, which preserves the vast embodied carbon of existing structures and stands as a cornerstone of a circular economy in the built environment. While resilience measures can increase upfront impacts and sustainability measures might raise questions about structural performance, life-cycle-driven seismic design reconciles these through strategies that yield structures that are both robust and low-carbon as explained in Table 7.6.

As codes and sustainability standards evolve, future engineers will increasingly be asked not only "Is it safe?" but also "Is it sustainable over its life?" Performance-based design, informed by life-cycle impact assessments, offers a path forward. For students and future professionals, the mindset must shift from isolated decisions to systems thinking. Can we reduce residual drift without overdesigning? Could small upfront investments in self-centering or modular systems prevent massive reconstruction later? These are the questions that will define structural engineering in a carbon-conscious world.

Table 7.6 Resilience vs. sustainability trade-offs: repairability, reuse, and material intensity.

	Strategy and trade-offs	**Problem**	**Approach**	**Example**
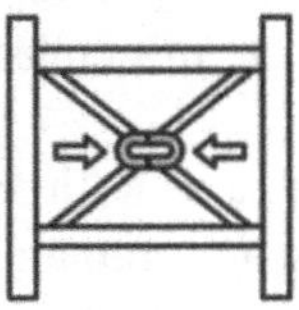	Embodied carbon vs. repair frequency	A "lean" low-carbon design may require carbon-intensive repairs after an earthquake, offsetting initial savings	Design for repairability: limit damage to specific, easily replaceable "fuses" or components	Using replaceable steel links in a frame (means only the links are replaced, not the entire structure)
	Reuse and design for deconstruction	Standard demolition after earthquake damage creates waste and loses the embodied carbon invested in the materials, a cycle that can repeat multiple times in seismically active regions	Design for disassembly: use modular components and bolted connections that can be quickly replaced, with inelastic demand concentrated in replaceable fuse elements to keep primary members undamaged and reusable	Timber or steel members with bolted connections can be reused in a retrofit or new project if the damage-controlled design has kept them within the elastic range during the seismic event
	Material intensity vs. innovative systems	The traditional assumption that achieving resilience requires using more material, which directly increases upfront carbon footprint	Use materials smarter: employ hybrid systems that combine materials to leverage their strengths and minimize overall usage	Hybrid steel-and-timber systems can provide high seismic performance with lower overall embodied carbon than a single-material design
	Economic and social trade-offs	Resilient and sustainable solutions can have higher upfront costs, creating an economic barrier for decision-makers	Use holistic analysis: employ tools like multi-criteria decision analysis to weigh long-term benefits against initial costs	Holistic design can enhance resilience and sustainability by evaluating the full life-cycle costs and benefits

Furthermore, the life-cycle perspective extends beyond design and construction to how we assess buildings after seismic events. When structural safety cannot be verified rapidly, repairable buildings risk unnecessary demolition, wasting the embodied carbon invested during their construction. Emerging monitoring approaches that leverage existing infrastructure, such as telecommunications fiber-optic networks, could help close this gap by enabling buildings to continuously communicate their structural condition.

7.9 Questions

Answer the following questions to enhance your understanding of the concepts covered in this chapter.

Discussion Questions

1 Explain why seismic regions are particularly important for global embodied carbon considerations.

2 Compare the initial embodied carbon requirements (A1–A3 stage) for the three seismic design philosophies. Explain why each approach requires different material quantities.

3 Explain the fundamental trade-off in a conventional ductile design. How does this philosophy reduce upfront embodied carbon (Module A1–A3), and what is the potential long-term environmental consequence during the building's use phase (Module B)?

4 Describe the primary goal of a damage-controlled design philosophy. How do systems like replaceable "fuses" or self-centering braces shift the embodied carbon impact away from post-earthquake repairs and toward the end-of-life/reuse stages (Modules C and D)?

5 You are the lead engineer for a new hospital in a high-seismic zone. Which of the three design philosophies (conventional, damage-controlled, or high-performance) would you advocate for? Justify your decision by considering not only structural performance but also the building's operational continuity and its total life-cycle carbon footprint.

6 Compare and contrast a seismic isolation system with a conventional ductile frame for a 10-story office building. Discuss the likely differences in embodied carbon at the life-cycle stages.

7 Compare how steel, reinforced concrete, and timber structures perform differently within each of the three seismic design philosophies, considering both resilience and sustainability factors.

8 What are the primary challenges during the construction stage (A4–A5) for a structure with high-ductility detailing (e.g. dense reinforcement) vs. one with a seismic isolation system?

9 Discuss the relationship between structural resilience and sustainability. Can a building be considered truly sustainable if it is not resilient to predictable hazards like earthquakes? Explain your reasoning.

10 You are designing a hospital in a high seismic zone with strict sustainability requirements. The client is concerned about the higher initial cost of no-damage design but wants to minimize long-term environmental impact. Develop a comprehensive argument for your recommended design philosophy, considering life-cycle carbon implication, post-earthquake functionality requirements, long-term economic and environmental costs, and social responsibility factors.

11 Explain how the failure of nonstructural components (NSCs) during an earthquake can significantly increase a building's total life-cycle carbon footprint, even if the primary structure remains safe and repairable. How does a damage-controlled design philosophy mitigate this specific environmental impact?

12 From a life-cycle and circular economy perspective, argue why seismically retrofitting an existing, vulnerable building is often a more sustainable choice than demolishing it and constructing a new, code-compliant one. What is the primary environmental benefit of this approach? What could be the counter-arguments?

13 Imagine two identical buildings after a moderate earthquake. Building A, using a conventional ductile design, has no structural damage but suffers extensive damage to its facade, partition walls, and plumbing. Building B, using a base isolation system, is entirely undamaged. Discuss the hidden environmental and economic costs for Building A and explain why Building B represents a more holistically sustainable outcome.

14 A city must decide how to manage a historic but seismically vulnerable concrete building. Option 1 is to demolish it and build a new, highly efficient timber structure in its place. Option 2 is to perform a seismic retrofit using steel-braced frames and FRP wrapping. Compare the life-cycle carbon implications of these two options.

15 Damage-controlled systems like self-centering braces, buckling-restrained braces, and friction dampers promise minimal post-earthquake repair. However, many of these are proprietary products manufactured by specialized companies, potentially creating supply chain vulnerabilities and cost premiums. Compare this to open-source solutions using standard steel plates fabricable by any qualified manufacturer. From a life-cycle and resilience perspective, what are the trade-offs between innovation (often driven by proprietary development) and accessibility?

16 No-damage design using seismic isolation can offer the lowest whole-life carbon and protects both structural and nonstructural components. Yet it remains largely confined to critical facilities like hospitals and data centers. Why hasn't base isolation become standard practice for ordinary buildings in high-seismic zones? Analyze the barriers (cost, expertise, code complexity, insurance

frameworks) and propose policy mechanisms that could democratize access to this technology. Should governments subsidize isolation systems for residential buildings, similar to energy efficiency incentives?

17 Earthquakes rarely occur in isolation (they often trigger fires from damaged gas lines, tsunamis in coastal areas, or landslides in mountainous regions). A building optimized for seismic performance may perform very differently when subjected to these cascading hazards. Discuss how life-cycle seismic design should account for multi-hazard scenarios. Does optimizing for seismic performance sometimes increase vulnerability to other hazards? For example, might a lightweight timber structure optimized for seismic and carbon performance be more vulnerable to post-earthquake fire than a concrete alternative? How should designers balance these competing risks?

18 Performance-based seismic design typically assumes a 50-year service life. However, many buildings remain in service far longer, and climate change is altering hazard patterns through induced seismicity, sea-level rise affecting coastal foundations, and changing soil conditions. Should life-cycle-driven seismic design adopt longer time horizons (100+ years) that account for evolving hazard profiles? Discuss the implications for material selection, design philosophy, and the concept of "planned obsolescence" in construction. How does uncertainty about future seismic hazards complicate life-cycle carbon calculations?

19 Seismic retrofitting preserves embodied carbon and is generally considered more sustainable than demolition and replacement. However, in some cases (particularly when existing structures are severely degraded or when new construction can achieve substantially higher operational efficiency), replacement might yield lower life-cycle carbon. Develop a decision framework that identifies the "tipping point" at which replacement becomes more sustainable than retrofit. What factors beyond embodied carbon should influence this decision (e.g. operational energy, heritage value, urban density, social displacement)? Is there a moral obligation to preserve existing buildings even when replacement is marginally more carbon-efficient?

20 Consider a city where 60% of buildings are undamaged after a major earthquake, yet their occupants remain displaced for weeks because manual inspections cannot keep pace with demand. Discuss how continuous structural monitoring using existing telecommunications fiber-optic networks could change this scenario from a life-cycle perspective, addressing both the embodied carbon implications of unnecessary demolition and the social consequences of prolonged displacement. What are the main technical and practical barriers that must be overcome before such a system could become operational at urban scale?

Scenario-based Questions

Scenario 1: A midrise reinforced concrete structure in a high-seismic region was originally designed with ductile detailing and concrete shear walls. Its production-stage embodied carbon (A1–A3) is estimated at 1200 t CO_2e. After a magnitude 6.5 earthquake, repairs to beams, columns, and nonstructural elements add an extra 300 t CO_2e.

- Identify and quantify the life-cycle embodied carbon trade-off between the original ductile design and a damage-controlled system (e.g. base isolation) that might raise initial A1–A3 carbon by 15 %.
- Recommend at least two design changes that could bring total life-cycle embodied carbon below the original 1500 t CO_2e while maintaining equivalent seismic performance.

Scenario 2: A six-story hybrid steel-timber building employs self-centering link devices in its eccentrically braced frames. Its upfront embodied carbon is 800 t CO_2e. Life-cycle analysis projects maintenance-stage repairs of 100 t CO_2e over 50 years under code-minimum detailing.

- Calculate the total embodied carbon over its service life.
- Propose an alternative detailing or system (e.g. demountable bolted connections or damage-controlled braces) that could reduce life-cycle embodied carbon by at least 10% and explain how it achieves both resilience and lower carbon.

Scenario 3: A four-story office building in a seismic zone is being evaluated for retrofit vs. replacement.

- **Option A (Retrofit):** Add carbon-fiber-reinforced polymer (CFRP) wrapping to existing columns and install supplemental braces. This retrofit incurs an additional 200 t CO_2e (all counted in modules A4–A5 and B2–B5).
- **Option B (Replace):** Demolish the building (causing 150 t CO_2e in demolition and waste processing) and construct a new mass-timber lateral system carrying an upfront embodied carbon of 1000 t CO_2e (A1–A3). Anticipated maintenance (B2–B5) adds 50 t CO_2e over 50 years.
  1. Calculate the total life-cycle embodied carbon for both Options A and B.
  2. Based on your results, which option yields the lower embodied carbon? Discuss any additional factors (e.g. end-of-life recycling potential, repairability, seismic performance) that might influence the final decision.

Scenario 4: A client in a seismically active region must choose between two proposals for a new critical facility (e.g. a hospital):

Proposal A: A standard, code-compliant steel frame building with a lower initial construction cost.
Proposal B: A building designed with specialized, repairable seismic "fuses" that concentrate damage. This has a 10% higher initial construction cost.

Question: As a sustainability consultant, which proposal would you likely recommend? Justify your choice using the principles of Life-cycle Assessment (LCA).

Scenario 5: An engineering firm has been commissioned to design a new 15-story commercial office building in a high-seismic region like Izmir, Türkiye. The client is interested in creating a "green" building but is also concerned about the initial construction budget. The firm presents two primary design options for the lateral-force-resisting system:

Design A (The "Code-minimum" Approach): A conventional reinforced concrete special moment frame designed strictly according to the ductile design philosophy. This approach is optimized to meet life-safety requirements with the minimum amount of material, resulting in a lower upfront embodied carbon footprint (Modules A1–A3).
Design B (The "High-resilience" Approach): A steel frame equipped with a combination of seismic base isolation and replaceable, self-centering bracing systems (like SC-links). This design aims for

immediate re-occupancy after a design-level earthquake. The initial embodied carbon is 20% higher than Design A due to the specialized systems and connections.

As a sustainability consultant on the project, you are asked to compare the two designs. Your analysis should:

a. Discuss the likely outcomes for both buildings across their life cycle, specifically addressing the Use and Repair Stage (B) and the End-of-life Stage (C), assuming a major earthquake occurs 30 years into their 60-year service life.
b. Explain the primary trade-off between upfront embodied carbon and whole-life carbon. Use the concept of "wasted" embodied carbon (as seen in the Christchurch example) to argue for your preferred design.
c. Conclude by recommending which design better aligns with the "Integrated resilience + sustainability" model shown in Table 7.1 of the chapter and justify your choice.

Scenario 6: A 20-story steel office building is being designed for a high seismic zone. Given the following simplified data:

Ductility-based design: 2000 tons steel, 25% repair carbon after design earthquake.
Damage-controlled design: 2200 tons steel + dissipative devices (50 tons), 5% repair carbon.
No-damage design: 1800 tons steel + isolation system (80 tons), 1% repair carbon.

Assuming steel embodied carbon of 2.5 tons CO_2e per ton steel, and 50% probability of design earthquake in building lifetime, calculate the expected total embodied carbon for each approach. Which is most sustainable? Redo the calculations assuming the carbon factor of steel as 0.6 tons CO_2e per ton, and compare the results.

References

ACI CODE-318-19(22): Building Code Requirements for Structural Concrete and Commentary (Reapproved 2022) (n.d.). Retrieved June 27, 2025, from https://www.concrete.org/store/productdetail.aspx?ItemID=318U19&Language=English&Units=US_Units

Alejandro, B., Schaefer, J., Cox, A., Dionne Daniell, K., Esper, S., Gunasekera, R., Ishizawa Escudero, O. A., Khazai, B., Macabuag, J. L. D. C., Otálora, S. G., Pomonis, A., & Romero Hernandez, R. (n.d.). Global Rapid Post-Disaster Damage Estimation (GRADE) report: February 6, 2023 Kahramanmaraş earthquakes – Türkiye report. World Bank. Retrieved November 11, 2025, from https://documents.worldbank.org/en/publication/documents-reports/documentdetail/099022723021250141

ANSI/AISC 341|American Institute of Steel Construction (n.d.). Retrieved June 27, 2025, from https://www.aisc.org/publications/steel-standards/aisc-341/

Antoniou, S. (2023). *Seismic Retrofit of Existing Reinforced Concrete Buildings*. Wiley-Blackwell.

Applied Technology Council. (1995). *ATC-20: Procedures for Post-earthquake Safety Evaluation of Buildings*. Redwood City, CA: Applied Technology Council.

ASCE 7 standard|ASCE (n.d.). Retrieved June 27, 2025, from https://www.asce.org/publications-and-news/asce-7

Atlas of the Human Planet 2017 – how exposed are we to natural hazards? – European Commission (n.d.). Retrieved June 9, 2025, from https://joint-research-centre.ec.europa.eu/jrc-news-and-updates/atlas-human-planet-2017-how-exposed-are-we-natural-hazards-2017-05-24_en

Bado, M. F., & Casas, J. R. (2021). A review of recent distributed optical fiber sensors applications for civil engineering structural health monitoring. *Sensors*, *21*(5), 5. https://doi.org/10.3390/s21051818

Baumont, D., Scotti, O., Courboulex, F., & Melis, N. (2004). Complex kinematic rupture of the Mw 5.9, 1999 Athens earthquake as revealed by the joint inversion of regional seismological and SAR data. *Geophysical Journal International*, *158*(3), 1078–1087. https://doi.org/10.1111/j.1365-246X.2004.02374.x

Bureau of Indian Standards (2016). *IS 1893-1 (2002): Criteria for Earthquake Resistant Design of Structures, Part 1: General Provisions and Buildings*. Bureau of Indian Standards.

Cao, M. S., Sha, G. G., Gao, Y. F., & Ostachowicz, W. (2017). Structural damage identification using damping: A compendium of uses and features. *Smart Materials and Structures*, *26*(4), 043001. https://doi.org/10.1088/1361-665X/aa550a

Carrión, F. J., Quintana, J. A., & Crespo, S. E. (2017). Techno-economical and practical considerations for SHM systems. *Journal of Civil Structural Health Monitoring*, *7*(2), 207–215. https://doi.org/10.1007/s13349-017-0215-x

CASBEE Certification System (n.d.). Retrieved June 27, 2025, from https://www.ibecs.or.jp/CASBEE/english/certificationE.htm

Ceylan, H., Kanyilmaz, A., Donadello, S., Clivati, C., & Gentile, C. (2026). Experimental modal identification using interferometric sensing for dual use of telecom fiber infrastructure in structural health monitoring. SSRN Scholarly Paper No. 6313502. Social Science Research Network, February 27. https://doi.org/10.2139/ssrn.6313502

Ceylan, H., Kanyilmaz, A., Donadello, S., Clivati, C., Loiacono, D., Herrero, A., Levi, F., & Calonico, D. (2025). Preliminary feasibility study on using fiber-optic infrastructure with an interferometric approach for structural health monitoring. In E. Caetano & Á. Cunha (Eds.), *Experimental Vibration Analysis for Civil Engineering Structures* (vol. 674. Lecture Notes in Civil Engineering). Springer Nature Switzerland. https://doi.org/10.1007/978-3-031-96110-6_60

Chan, T. H. T., Yu, L., Tam, H. Y., Ni, Y. Q., Liu, S. Y., Chung, W. H., & Cheng, L. K. (2006). Fiber Bragg grating sensors for structural health monitoring of Tsing Ma Bridge: Background and experimental observation. *Engineering Structures*, *28*(5), 648–659. https://doi.org/10.1016/j.engstruct.2005.09.018

Chilean Standard NCh433.Of96: Seismic Design of Buildings (2016). https://doi.org/10.1061/9780784413647.ap01

Comfort, L., Celik, S., & Basbug, B. (2024). *Communication and Coordination Networks in the 2023 Kahramanmaras Earthquakes*. https://hazards.colorado.edu/quick-response-report/communication-and-coordination-networks-in-the-2023-kahramanmaras-earthquakes

Curadelli, R. O., Riera, J. D., Ambrosini, D., & Amani, M. G. (2008). Damage detection by means of structural damping identification. *Engineering Structures*, *30*(12), 3497–3504. https://doi.org/10.1016/j.engstruct.2008.05.024

Demarie, G. V., & Sabia, D. (2011). Non-linear damping and frequency identification in a progressively damaged R.C. element. *Experimental Mechanics*, *51*(2), 229–245. https://doi.org/10.1007/s11340-010-9360-4

Di Ludovico, M., De Martino, G., Prota, A., Manfredi, G., & Dolce, M. (2021). Damage assessment in Italy, and experiences after recent earthquakes on reparability and repair costs. In S. Akkar,

A. Ilki, C. Goksu, & M. Erdik (Eds.), *Advances in Assessment and Modeling of Earthquake Loss*. Springer International Publishing. https://doi.org/10.1007/978-3-030-68813-4_4

Doebling, S. W., Farrar, C. R., Prime, M. B., & Shevitz, D. W. (1996). Damage identification and health monitoring of structural and mechanical systems from changes in their vibration characteristics: A literature review. LA--13070-MS, 249299. https://doi.org/10.2172/249299

Dolce, M., & Goretti, A. (2015). Building damage assessment after the 2009 Abruzzi earthquake. *Bulletin of Earthquake Engineering, 13*(8), 2241–2264. https://doi.org/10.1007/s10518-015-9723-4

Donadello, S., Clivati, C., Govoni, A., Marra, G., Levi, F., & Calonico, D. (2024). Seismic monitoring using the telecom fiber network. *Communications Earth & Environment*, *5*(1), 178. https://doi.org/10.1038/s43247-024-01338-2

Elettore, E., Francavilla, A., Di Benedetto, S., Latour, M., Freddi, F., & Rizzano, G. (2025). *Experimental Response of Friction-Based Dissipative Braces for Steel Seismic-Resilient Concentrically Braced Frames.* 8–10 September, Barcelona, Spain.

EN 15643 (2021). *EN 15643: 2021. Sustainability of Construction Works. Framework for Assessment of Buildings and Civil Engineering Works.*

EN 1998-1. *Eurocode 8: Design of structures for earthquake resistance – Part 1: General rules and seismic action.* European Committee for Standardization.

Encyclopaedia Britannica. (2025). L'Aquila Earthquake of 2009 | Causes, Damage, & Facts | Britannica. https://www.britannica.com/event/LAquila-earthquake-of-2009

Eurocode 8: Design of structures for earthquake resistance | Eurocodes: Building the future (n.d.). Retrieved June 27, 2025, from https://eurocodes.jrc.ec.europa.eu/EN-Eurocodes/eurocode-8-design-structures-earthquake-resistance

European Commission (n.d.-a). Design REsearch, implementation And Monitoring of Emerging technologies for a new generation of Resilient Steel buildings (DREAMERS). Retrieved October 28, 2025, from https://ec.europa.eu/info/funding-tenders/opportunities/portal/screen/opportunities/projects-details/31061225/101034015/RFCS?order=DESC&pageNumber=1&pageSize=10&sortBy=relevance&keywords=dreamers

European Commission (n.d.-b). Technologies for Immutable Multi-hazard rEversibLE Steel Structures (TIMELESS). Retrieved October 28, 2025, from https://ec.europa.eu/info/funding-tenders/opportunities/portal/screen/opportunities/projects-details/43252449/101216417/RFCS2027?order=DESC&pageNumber=1&pageSize=10&sortBy=relevance&keywords=timeless

European Commission (n.d.-c). Green Digital Sector | Shaping Europe's Digital Future'. Accessed 25 November 2025. https://digital-strategy.ec.europa.eu/en/policies/green-digital

European Commission (2025a). Climate-Neutral and Smart Cities - Research and Innovation. https://research-and-innovation.ec.europa.eu/funding/funding-opportunities/funding-programmes-and-open-calls/horizon-europe/eu-missions-horizon-europe/climate-neutral-and-smart-cities_en

European Commission (2025b). Smart Cities and Communities | Shaping Europe's Digital Future. https://digital-strategy.ec.europa.eu/en/policies/smart-cities-and-communities

Farrar, C. R., & Lieven, N. A. J. (2007). Damage prognosis: The future of structural health monitoring. *Philosophical Transactions of the Royal Society A: Mathematical, Physical and Engineering Sciences, 365*(1851), 623–632. https://doi.org/10.1098/rsta.2006.1927

Farrar, C. R. & Worden, K. (2013). Structural Health Monitoring: A Machine Learning Perspective. Online-Ausg. Wiley.

Federal Emergency Management Agency (FEMA) (2021). FEMA P-2055, Post-Disaster Building Safety Evaluation Guidance | FEMA.Gov. https://www.fema.gov/emergency-managers/risk-management/earthquake/training/fema-p-2055

FEMA P-58 (n.d.). Retrieved June 27, 2025, from https://femap58.atcouncil.org/

Frizzarin, M., Feng, M. Q., Franchetti, P., Soyoz, S., & Modena, C. (2010). Damage detection based on damping analysis of ambient vibration data. *Structural Control and Health Monitoring*, *17*(4), 368–385. https://doi.org/10.1002/stc.296

FTTH Council Europe (n.d.). European FTTH/B market panorama 2024. Accessed January 8, 2026. https://www.ftthcouncil.eu/committees/market-intelligence/2043/european-ftth-b-market-panorama-2024

FTTH Council Europe (n.d.). Why and When FTTH or Fibre Optics in the Home? | FTTH Council. Accessed 9 January 2026. https://www.ftthcouncil.eu/resources/what-is-ftth

Fundamentals Volume Index (n.d.). Retrieved June 27, 2025, from https://www.ashrae.org/advertising/handbook-advertising/fundamentals

Galindez-Jamioy, C. A., & López-Higuera, J. M. (2012). Brillouin distributed fiber sensors: An overview and applications. *Journal of Sensors*, *2012*(1), 204121. https://doi.org/10.1155/2012/204121

GAR2022: Our World at Risk (GAR) | UNDRR (2022, April 26). https://www.undrr.org/gar/gar2022-our-world-risk-gar

GB 50011-2016 English PDF (GB 50011-2010: Older version) (n.d.). Retrieved June 27, 2025, from https://www.chinesestandard.net/PDF/English.aspx/GB50011-2016

Geotechnical Extreme Events Reconnaissance (GEER) Association (n.d.). Retrieved October 7, 2025, from https://www.geerassociation.org/

Giardini, D., Grünthal, G., Shedlock, K. M., & Zhang, P. (1999). The GSHAP global seismic hazard map. *Annals of Geophysics*. https://doi.org/10.4401/ag-3784

Giordano, P. F., Miraglia, G., Lenticchia, E., Ceravolo, R., & Limongelli, M. P. (2023). Satellite interferometric data for seismic damage assessment. *Procedia Structural Integrity*, *44*, 1570–1577. https://doi.org/10.1016/j.prostr.2023.01

Gołdyn, M., & Zdanowicz, K. (2025). Distributed fiber optic sensing in bridge structural health monitoring: Insights from real-life implementations in Germany. Paper presented at the NDT-CE 2025 – International Symposium on Nondestructive Testing in Civil Engineering, September 24.

Gonzalez, R. E., Stephens, M. T., Toma, C., Elwood, K. J., & Dowdell, D. (2021). Quantifying the embodied carbon cost from demolitions following the canterbury earthquakes. *New Zealand Society for Earthquake Engineering*.

Gue, C. Y., Wilcock, M., Alhaddad, M. M., Elshafie, M. Z. E. B., Soga, K., & Mair, R. J. (2015). The monitoring of an existing cast iron tunnel with Distributed Fibre Optic Sensing (DFOS). *Journal of Civil Structural Health Monitoring*, *5*(5), 573–586. https://doi.org/10.1007/s13349-015-0109-8

Hartog, A. H. (2017). *An Introduction to Distributed Optical Fibre Sensors*. CRC Press. https://doi.org/10.1201/9781315119014

INRIM (n.d.). Homepage | INRIM. Accessed 31 December 2025. https://www.inrim.it/en

Italian Civil Protection Department (n.d.). Il piano di analisi della vulnerabilità delle zone edificate nell'area del bradisismo dei Campi Flegrei. Accessed 8 January 2026. https://rischi.protezionecivile.gov.it/it/vulcanico/vulcani-italia/campi-flegrei/il-piano-di-analisi-della-vulnerabilita-delle-zone-edificate-nellarea-del-bradisismo-dei-campi-flegrei/

Jaiswal, K., & Wald, D. (2010). An empirical model for global earthquake fatality estimation. *Earthquake Spectra*, *26*(4), 1017–1037. https://doi.org/10.1193/1.3480331

Jang, S., Jo, H., Cho, S., Kim, H., Spencer Jr., B. F., & Yun, C. B. (2010). Structural health monitoring of a cable-stayed bridge using smart sensor technology: Deployment and evaluation. *Smart Structures and Systems*, *6* (5–6), 439–459. https://doi.org/10.12989/sss.2010.6.5_6.439

Latour, M., D'Aniello, M., Zimbru, M., Rizzano, G., Piluso, V., & Landolfo, R. (2018). Removable friction dampers for low-damage steel beam-to-column joints. *Soil Dynamics and Earthquake Engineering*, *115*, 66–81. https://doi.org/10.1016/j.soildyn.2018.08.002

Li, H.-N., Li, D.-S., & Song, G.-B. (2004). Recent applications of fiber optic sensors to health monitoring in civil engineering. *Engineering Structures*, *26*(11), 1647–1657. https://doi.org/10.1016/j.engstruct.2004.05.018

Liu, J., Yuan, S., Luo, B., Biondi, B., & Noh, H. Y. (2023). Turning telecommunication fiber-optic cables into distributed acoustic sensors for vibration-based bridge health monitoring. *Structural Control and Health Monitoring*, *2023*, 1–14. https://doi.org/10.1155/2023/3902306

Lv, B., Peng, Y., Du, C., Tian, Y., & Wu, J. (2025). Review of Brillouin distributed sensing for structural monitoring in transportation infrastructure. *Infrastructures*, *10*(6), 148. https://doi.org/10.3390/infrastructures10060148

Mahbubi, K., Zaki, A., & Nugroho, G. (2024). Bibliometric and scientometric trends in structural health monitoring using fiber-optic sensors: A comprehensive review. *Journal of Civil and Hydraulic Engineering*, *2*(1), 51–64. https://doi.org/10.56578/jche020104

Malekloo, A., Ozer, E., AlHamaydeh, M., & Girolami, M. (2022). Machine learning and structural health monitoring overview with emerging technology and high-dimensional data source highlights. *Structural Health Monitoring*, *21*(4), 1906–1955. https://doi.org/10.1177/14759217211036880

Marra, G., Clivati, C., Luckett, R., Tampellini, A., Kronjäger, J., Wright, L., Mura, A., Levi, F., Robinson, S., Xuereb, A., Baptie, B., & Calonico, D. (2018). Ultrastable laser interferometry for earthquake detection with terrestrial and submarine cables. *Science*, *361*(6401), 486–490. https://doi.org/10.1126/science.aat4458

Marra, G., Fairweather, D. M., Kamalov, V., Gaynor, P., Cantono, M., Mulholland, S., Baptie, B., Castellanos, J.C., Vagenas, G., Gaudron, J.O., & Kronjäger, J. (2022). Optical interferometry–based array of seafloor environmental sensors using a transoceanic submarine cable. *Science*, *376*(6595), 874–879. https://doi.org/10.1126/science.abo1939

McKinsey & Company. (n.d.). Tech, Media, and Telecom Spur Europe's Comeback | McKinsey. Accessed 25 November 2025. https://www.mckinsey.com/industries/technology-media-and-telecommunications/our-insights/technology-media-and-telecom-in-europe-the-new-growth-engine-or-another-decade-of-missing-out

Menghini, A., Ceylan, H., Kanyilmaz, A., Donadello, S., Clivati, C., & Loiacono, D. (2026). Stress-field reconstruction from laser interferometric strain measurements. *Computers and Structures*. https://doi.org/10.1016/j.compstruc.2026.108251

Menghini, A., Kanyilmaz, A., Castiglioni, C. A., & Calado, L. (2025). A dissipative replaceable bracing component for steel-composite building frames. *Structures*, *76*, 108941. https://doi.org/10.1016/j.istruc.2025.108941

Menghini, A., Mouzakis, C. P., Kanyilmaz, A., Karapitta, L., Panoutsopoulou, L., & Castiglioni, C. (2025). Seismic response of new and repaired steel-concrete composite structures using bi-axial shaking table tests designed with similitude law. *Engineering Structures*, *336*, 120447. https://doi.org/10.1016/j.engstruct.2025.120447

Menna, C., Asprone, D., Jalayer, F., Prota, A., & Manfredi, G. (2013). Assessment of ecological sustainability of a building subjected to potential seismic events during its lifetime. *The International Journal of Life Cycle Assessment*, *18*, 504–515. https://doi.org/10.1007/s11367-012-0477-9

Moaveni, B., He, X., Conte, J. P., Restrepo, J. I., & Panagiotou, M. (2011). System identification study of a 7-story full-scale building slice tested on the UCSD-NEES shake table. *Journal of Structural Engineering*, *137*(6), 705–717. https://doi.org/10.1061/(ASCE)ST.1943-541X.0000300

Motil, A., Bergman, A., & Tur, M. (2016). State of the art of Brillouin fiber-optic distributed sensing. *Optics & Laser Technology*, *78*, 81–103. https://doi.org/10.1016/j.optlastec.2015.09.013

Nakamura, A., Manabe, T., Morikane, K., & Koshikiya, Y. (2024). Inter-modal interferometry using two-mode region of single-mode fibers for vibration sensing. *Journal of Lightwave Technology*, *42*(18), 6431–6436. https://doi.org/10.1109/JLT.2024.3387392

NZS 1170 Structural Design Actions Part 5 Earthquake Actions—Suspended ceilings: Standards New Zealand (n.d.). Retrieved June 27, 2025, from https://www.standards.govt.nz/news-and-updates/nzs-1170-structural-design-actions-part-5-earthquake-actions-suspended-ceilings

Ohno, H., Naruse, H., Kihara, M., & Shimada, A. (2001). Industrial applications of the BOTDR optical fiber strain sensor. *Optical Fiber Technology*, *7*(1), 45–64. https://doi.org/10.1006/ofte.2000.0344

Ozkula, G., Dowell, R. K., Baser, T., Lin, J.-L., Numanoglu, O. A., Ilhan, O., Olgun, C. G., Huang, C.-W., & Uludag, T. D. (2023). Field reconnaissance and observations from the February 6, 2023, Turkey earthquake sequence. *Natural Hazards*, *119*(1), 663–700. https://doi.org/10.1007/s11069-023-06143-2

Panteliou, S. D., Chondros, T. G., Argyrakis, V. C., & Dimarogonas, A. D. (2001). Damping factor as an indicator of crack severity. *Journal of Sound and Vibration*, *241*(2), 235–245. https://doi.org/10.1006/jsvi.2000.3299

Park, Y.-J., & Ang, A. H.-S. (1985). Mechanistic seismic damage model for reinforced concrete. *Journal of Structural Engineering*, *111*(4), 722–739. https://doi.org/10.1061/(ASCE)0733-9445(1985)111:4(722)

Peeters, B., & De Roeck, G. (2001). Stochastic system identification for operational modal analysis: A review. *Journal of Dynamic Systems, Measurement, and Control*, *123*(4), 659–667. https://doi.org/10.1115/1.1410370

Pesaresi, M., Ehrlich, D., Kemper, T., Siragusa, A., Florczyk, A., Freire, S., & Corban, C. (2017). *Atlas of the Human Planet 2017: Global Exposure to Natural Hazards* (EUR 28556 EN). Publications Office of the European Union. https://doi.org/10.2760/19837

Pirrò, M., & Gentile, C. (2025). Detection and localization of anomalies in bridges using accelerometer data and sparse auto-encoders. *Developments in the Built Environment*, *23*, 100715. https://doi.org/10.1016/j.dibe.2025.100715

Rytter, A. (1993). *Vibrational Based Inspection of Civil Engineering Structures*. (Aalborg), Fracture and Dynamics.

Second Generation of the Eurocodes | Eurocodes: Building the Future (n.d.). Retrieved October 7, 2025, from https://eurocodes.jrc.ec.europa.eu/second-generation-eurocodes

Soga, K. & Luo, L. (2018). Distributed fiber optic sensors for civil engineering infrastructure sensing. *Journal of Structural Integrity and Maintenance*, *3*(1), 1–21. https://doi.org/10.1080/24705314.2018.1426138

Sohn, H., Farrar, C. R., Hunter, N. F., & Worden, K. (2001). Structural health monitoring using statistical pattern recognition techniques. *Journal of Dynamic Systems, Measurement, and Control*, *123*(4), 706–711. https://doi.org/10.1115/1.1410933

Starlink. (n.d.). Starlink. Accessed 21 November 2025. https://starlink.com/

Structural: Building Standard Law Environmental: CASBEE (2014). Ministry of Land, Infrastructure, Transport and Tourism (MLIT), Japan. https://www.ibecs.or.jp/CASBEE/english/download/CASBEE-BD(NC)e_2014manual.pdf

Sustainability of Construction Works—Assessment of Environmental Impact of Buildings—Requirements and Guidance (No. EN 15978:2026). CEN.

Türkiye Bina Deprem Yönetmeliği—2018 (2018). T.C. İçişleri Bakanlığı Afet ve Acil Durum Yönetimi Başkanlığı. https://www.resmigazete.gov.tr/eskiler/2018/03/20180318M1-2-1.pdf

Uludag, T. D. (2022). *Seismic Analysis, Behavior, and Design of Diaphragm Collectors Using Ductile Replaceable Fuse Elements—ProQuest* [PhD]. https://www.proquest.com/openview/0dbb589a7f8c5225516435faa7911f41/1?pq-origsite=gscholar&cbl=18750&diss=y

Wang, G., Song, D., Pang, Z., Wang, F., Dai, H., Li, W., & Wang, B. (2025). Laser interferometry for high-speed railway health inspection using telecom fiber along the line. *Nature Communications*, *16*, 4129. https://doi.org/10.1038/s41467-025-59507-6

Wang, Z., Lu, B., Ye, Q., & Cai, H. (2020). Recent progress in distributed fiber acoustic sensing with Φ-OTDR. *Sensors*, *20*(22), 6594. https://doi.org/10.3390/s20226594

Worldbank (2023). *Türkiye Earthquakes Recovery and Reconstruction Assessment*.

Ye, X. W., Su, Y. H., & Han, J. P. (2014). Structural health monitoring of civil infrastructure using optical fiber sensing technology: A comprehensive review. *The Scientific World Journal*, *2014*, 652329. https://doi.org/10.1155/2014/652329

Yilmaz, Z., Altunişik, A. C., Taciroglu, E., Günaydin, M., Okur, F.Y., Sunca, F., Şişman, R., Aslan, B., & Sezdirmez, T. (2024). Regional building damage survey data on the 2023 Kahramanmaraş, Türkiye earthquakes. *ASCE OPEN: Multidisciplinary Journal of Civil Engineering*, *2*(1), 04024009. https://doi.org/10.1061/AOMJAH.AOENG-0041

Zhan, Z. (2020). Distributed acoustic sensing turns fiber-optic cables into sensitive seismic antennas. *Seismological Research Letters*, *91*(1), 1–15. https://doi.org/10.1785/0220190112

Zhang, C., Yin, Z., & Qin, R. (2024). Attention-enhanced co-interactive fusion network (AECIF-Net) for automated structural condition assessment in visual inspection. *arXiv preprint arXiv:2307.07643*. https://doi.org/10.48550/arXiv.2307.07643

Zou, Y. (2019). Certifying green buildings in China: LEED vs. 3-star. *Journal of Cleaner Production*, *208*, 880–888. https://doi.org/10.1016/j.jclepro.2018.10.204

8

Real-world Applications of Life-cycle-driven Structures

LIST OF CONTENTS

This chapter presents 25 projects from 15 countries (6 continents) that translate the methods and principles discussed throughout the book into practice, reflecting a wide geographical range of contexts, materials, and regulatory environments (Figure 8.1). They respond to the scale of construction growth and the embodied-carbon challenge outlined in Chapter 1, where structural choices lock in long-term impact, and the share of embodied emissions rises as operational energy improves. Each case study demonstrates how the life-cycle-driven structures framework can guide structural decisions, from concept to construction, across diverse materials, scales, and contexts.

The projects are grouped into five complementary themes, though each case study integrates elements from several of these strategies (Tables 8.1–8.5):

1. Adaptive reuse and circular construction
2. Material efficiency and high-tech solutions
3. Hybrid, composite, and modular systems
4. Regenerative design in harsh soil and seismic conditions
5. Resilient infrastructure (airport, bridge) design

Figure 8.1 Countries of the selected case studies.

Table 8.1 presents projects that prioritize retention of existing structures, circular material flows, and design for disassembly. These case studies demonstrate how extending the life of structural elements, specifying reclaimed steel, and planning for future material recovery can dramatically reduce upfront embodied carbon while preserving architectural heritage and minimizing waste.

Table 8.2 presents structures optimized through advanced computational design, high-utilization ratios, and strategic material specification. Projects in this group achieve substantial carbon reductions

Table 8.1 Adaptive reuse and circular construction.

Title	Location	Core concept	Key outcome	Provided by
Transforming a shopping mall into an ice center with adaptive reuse	Tulsa (USA)	Conversion of vacant mall into sports facility using adaptive reuse	76% ↓ global warming potential (GWP), 60% ↓ steel, 75% ↓ concrete, 18-month delivery	Martin/ Martin, Inc.
Adaptive reuse of reinforced concrete cores	Brussels (Belgium)	Circular reuse of reinforced concrete towers with new mixed-use program	85% structure retained, ≈50% ↓ upfront carbon	Bureau Greisch
Reclaimed steel in practice	London (UK)	Reuse of reclaimed steel in hybrid steel–CLT extension	25 t reused steel, 45 tCO_2e saved, cost-neutral reuse validated	Cleveland Steel & Tubes
Transforming an incomplete structure into sustainable headquarters	Pisa (Italy)	Adaptive reuse of unfinished tower through comparative LCA and fire safety simulation	51% ↓ embodied carbon (258 $kgCO_2e/m^2$), safe reuse validated	ATI Project

Table 8.2 Material efficiency and technology.

Title	Location	Core concept	Key outcome	Provided by
Centrifuged high-strength concrete with recycled aggregates	Geneva (Switzerland)	Optimized concrete mixes with recycled aggregates and centrifuged elements	≈30% ↓ carbon, local aggregate reuse, high durability	Redesco Progetti
BIM-driven industrialized construction for a next-generation hospital	Odense (Denmark)	Industrialized hospital built with prefabricated Building Information Modeling (BIM)-managed system	Faster assembly, minimal waste, future-proof layout	ATI Project
Embodied carbon reduction through cement replacement and innovation	Boston (USA)	Low-carbon RC frame using SCMs and electrified cement	≈25% ↓ GWP vs. baseline, LEED Platinum, net-zero operation	Buro Happold
Reducing embodied carbon in a mega-stadium through lightweight tensile structures on reclaimed industrial land	Dalian (China)	Cable-tension roof system with tensile membrane façade for a 63 000-seat stadium on reclaimed industrial land	Reduced steel tonnage vs. conventional roof, reduced foundation demands, brownfield regeneration into community destination	Buro Happold

Table 8.3 Hybrid, composite, and modular systems.

Title	Location	Core concept	Key outcome	Provided by
Vertical expansion with mass timber and off-site fabrication	Washington, D.C. (USA)	Vertical adaptive reuse adding mass-timber floors to existing RC building	≈40% ↓ embodied carbon, reduced time/noise, biophilic workplace	Columbia Property Trust
Architectural ambition achieved with a steel–concrete composite frame	London (UK)	Steel–concrete composite office building with long spans	>25% ↓ GWP vs. baseline, LETI 2020 compliant	Pell Frischmann
Modular mass-timber system with reinforced concrete cores	Trondheim (Norway)	Hybrid RC-core with timber modules for rapid assembly	30% faster build, minimal waste, improved safety	Ergodomus
Modular steel construction for a complex geometry	Copenhagen (Denmark)	Modular industrial steel system with reversible bolted assembly	Improved buildability, future reuse potential, reduced site risk	Gregull & Spang
Mass timber landmark assisted with concrete and steel	Toronto (Canada)	Hybrid structural system to achieve long spans, vibration control, and net-zero operational performance	TEUI 54.2 vs. 65, GHGI 2.5 vs. 5 $kgCO_2e/m^2$, validated hybrid method	Moriyama Teshima Architects
Temporary, reusable hybrid steel–CLT offices for the construction site of Sydney's metro stations	Sydney (Australia)	Hybrid steel–CLT superstructure designed for disassembly and reassembly across multiple construction project sites over a 50-year design life	11% ↓ steel tonnage vs. reference design, 2,534 tCO_2e saved over five lifecycles, superstructure erected in 25 working days	Robert Bird Group

by designing leaner sections, leveraging high-strength materials, and employing digital fabrication to minimize waste and construction tolerances.

Table 8.3 presents projects that strategically combine timber, steel, and concrete (assigning each material to its optimal structural role). These case studies show how hybrid, composite and modular approaches can reduce structural weight, accelerate construction through prefabrication, and achieve superior environmental performance compared to single-material solutions.

Table 8.4 includes projects that address challenging site and seismic conditions through integrated geotechnical and structural design strategies. These examples show how early ground investigation, foundation optimization, and resilient structural systems can minimize carbon impacts while ensuring long-term performance in contaminated soils or high-seismic regions.

Table 8.5 includes large-scale infrastructure projects where durability, adaptability, and operational continuity are dominant. These case studies demonstrate how life-cycle thinking applied to airports, bridges, and data centers can extend service life, reduce maintenance interventions, and plan for future capacity increases with minimal environmental impact.

Table 8.4 Regenerative design in harsh soil and seismic conditions.

Title	Location	Core concept	Key outcome	Provided by
Geotechnical optimization on the same soil of Pisa Tower	Pisa (Italy)	Lightweight steel frame on micropiles and ribbed mat foundation over weak soil	38% ↓ concrete, 1080 tCO_2 saved, 53% ↓ seismic forces	ATI Project
Low-carbon structural design on a contaminated site	Newcastle (UK)	Early-stage optioneering to achieve low-carbon design in difficult soil conditions	>50% ↓ GWP vs. baseline, LETI 2025 target met, optimized remediation, lean construction	Pell Frischmann
Regenerative hybrid system in a high-seismic region	Los Angeles (USA)	Hybrid CLT–steel–concrete structure combining resilience and biophilic design	≈50% lighter than RC, 3× lower GWP, 23 tCO_2 sequestered, accelerated schedule	RIOS
A modular timber–steel hybrid mosque rooted in Anatolian tradition	Bolu (Türkiye)	Timber–steel hybrid system using modular glulam cells with concealed steel frame to reinterpret the traditional Anatolian timber mosque in a high seismic zone	Maximized biogenic carbon storage, modular 5 m × 5 m cell system enabling efficient fabrication, seismic performance at 0.4 g PGA with minimal visible steel	ATTEC
How seismic isolation kept hospitals operational through the 2023 Türkiye earthquakes	Malatya (Türkiye)	Friction pendulum seismic isolation system decoupling hospital from ground motion to achieve continued functionality after earthquakes	Zero structural damage, full post-earthquake operability (served as public shelter), entire embodied carbon investment preserved	Turkish Association for Seismic Isolation

Table 8.5 Resilient infrastructure (airport, bridge, data center) design.

Title	Location	Core concept	Key outcome	Provided by
Bridge design optioneering with high-strength steel	Luxembourg (Luxemburg)	Comparative bridge design using High Strength Steel (S690) and fatigue-resistant detailing	32% ↓ steel weight, 50% ↓ weld volume, improved durability and resilience	GRID International
Sustainability ahead of its time for Bridge Viaducts of Sardinia	Sardinia (Italy)	Early composite steel–concrete viaduct design for material efficiency	92 kg steel/m^2, low maintenance, fast construction, long life	Redesco Progetti
Mastering airport design through optimization, shock transmitters, and testing	Rabat (Morocco)	Parametric optimization of complex steel roof geometry, shock transmitters, and wind tunnel testing	≈40% ↓ steel weight, no expansion joints, durable landmark	Proges Engineering

(Continued)

Table 8.5 (Continued)

Title	Location	Core concept	Key outcome	Provided by
Integrating precision and sustainability in the erection of an arch bridge	Arequipa (Peru)	Rampant assembly with strand-jack rotation and Finite Element Model (FEM)-based precision modeling for a 175 m steel arch	≈2500 t steel erected without large cranes; lower fuel use, high accuracy, long-term durability	Cimolai S.p.A.
Rebuilding an airport as a symbol of post-earthquake recovery through hybrid timber–steel construction	Kumamoto (Japan)	Hybrid steel–timber structure with locally sourced 2 × 4 lumber roof trusses replacing conventional steel/concrete elements for seismic resilience and carbon reduction	Eliminated vulnerable ceiling systems, reduced roof embodied carbon through regional timber, supported local forestry economy, post-disaster symbol of recovery	Nikken Sekkei
Data Center design for Seismic Resilience	Italy	Steel-framed data hall and gantry structures combined with precast concrete office building, designed to remain fully operational under extreme seismic events through zoned load optimization and fluid viscous dampers	Guaranteed full operational continuity at TR = 1 950 years, eliminated post-earthquake structural repair, reduced embodied carbon through material allocation and load mapping, extended service life through replaceable dissipative devices	Lombardini22 Spa, SPS S.r.l.

The following subsections examine these five groups of case studies in detail, highlighting their design strategies, key outcomes, and contributions to life-cycle-driven structures philosophy.

8.1 How to Read the Case Studies

8.1.1 Units and Scope

Metrics are presented in SI units. Imperial units are often included parenthetically for reference.

GWP is reported as $kgCO_2e$ (and its derivatives). When results are structure-only, this is stated; whole-building values are noted when available.

When GWP metric could not be reported, sustainability benefits are shown through other indicators such as reductions in structural mass, longer service life, lower maintenance and replacement needs, faster assembly with less waste, and provisions for reuse or disassembly.

8.1.2 Life-cycle Assessment (LCA) Modules

A1–A3 product stage, A4 transport to site, A5 construction and installation, B use stage (maintenance, repairs, energy), C end-of-life, D benefits beyond system boundary (reuse, recycling, energy recovery). Credits in B or D are shown only when quantified in the case data.

8.1.3 Baselines and Comparisons

Where a project compared options, the baseline is the project's declared reference (for example, a new build in RC or steel, an all-new structure vs. reuse, or a standard grade vs. high strength steel). Some project references include published targets such as LETI 2025 for upfront embodied carbon.

8.1.4 Sustainability Strategy Matrix

Each case study is concluded with a sustainability strategy matrix linking the project's main strategies to specific life-cycle stages, methods, and outcomes.

It is important to note that this chapter does not attempt to rank or compare these projects against one another; each responds to distinct contexts, climates, regulatory frameworks, and client priorities, and the objective is to show diverse pathways to life-cycle-driven design rather than to judge relative performance.

8.2 Adaptive Reuse and Circular Construction

8.2.1 Transforming a Shopping Mall into an Ice Center with Adaptive Reuse

Across the United States, a growing number of shopping malls are falling into disuse, with few viable strategies in place to repurpose them. A common response in this case would be demolition of the mall to build a new facility, an approach that demands considerable expenditure of capital and embodied carbon. The WeStreet Ice Center project (in Tulsa, USA) offers a compelling alternative. It transformed a vacant 1995 Macy's department store into a vibrant community space with twin ice rinks, now serving as the home of the Tulsa Oilers minor league hockey team (Figure 8.2). Opened in March 2024, the project demonstrates how design and engineering can unlock the potential of disused structures.

The core concept of the project is "adaptive reuse": by transforming an existing structure, this approach led to significant reductions in new material use thanks to retaining the existing exterior façade, roofing membrane, and the entire structural frame. The project achieved a 75% reduction in new construction concrete volume and a 60% reduction in steel tonnage, directly lowering both costs and embodied carbon involved. The total steel mass for the building renovations amounted to 462.66 tons (510 tons).

Beyond material savings, adaptive reuse provided a substantial schedule advantage, allowing the facility to open approximately 18 months from project kick-off, with critical truss erection beginning only 8 months into design. It also reduced labor time required for installing new utilities and infrastructure.

At the early design stage, optioneering (comparing different design scenarios) has been essential for identifying the most sustainable and cost-effective approach before committing to a single solution. Three design scenarios were analyzed: i) new-built, ii) adaptive reuse, and iii) roof replacement. A detailed life-cycle analysis (LCA) using Tally and an EC3 Revit plug-in was conducted to calculate

Figure 8.2 The Tulsa WeStreet Ice Center (courtesy of Martin/Martin, Inc.).

GWP for each of the following options. The options are compared in Table 8.6, highlighting the final choice of adaptive reuse. Figure 8.3 shows a comparison of the embodied carbon sensitivity of different options. Please note that the results are also sensitive to the selection of carbon factors and underlying database assumptions (such as EPD sources, regional energy mixes), and should therefore be interpreted as a relative comparison between the design options rather than as absolute carbon values.

Transforming the department store into an ice center posed the following structural, logistical, and regulatory challenges:

Challenge 1: Since the ice center needed much larger free space than a shopping mall, the free span of the existing roof had to be doubled. This substantial increase in roof span required:

- Converting existing 8.5 m by 17.1 m (28 ft by 56 ft) bays into 8.5 m by 34.1 m (28 ft by 112 ft) bays over each rink (Figures 8.4 and 8.5).
- Removing 2136.8 m^2 (23 000 ft^2) of second-floor framing (Figures 8.6 and 8.7) and composite slab per rink (over half the total existing floor area).
- Removing 36 steel columns within the rink footprint, which supported the second floor (18 of which also supported the roof, Figure 8.4).

Table 8.6 Comparison of different options.

Option	Description	Total GWP ($kgCO_2e$)	GWP reduction vs. new-built	Note
New-built	Baseline scenario assuming construction of a completely new structure	2 756 600	–	Full material use; highest concrete and steel quantities
Adaptive reuse (selected option)	Reuse of the existing structure with minimal new material input	651 400	76.4%	75% less new concrete, 60% less steel
Roof replacement	Full removal and replacement of the roof, retaining other existing elements	776 114	71.9%	Significant reuse benefit despite full roof replacement

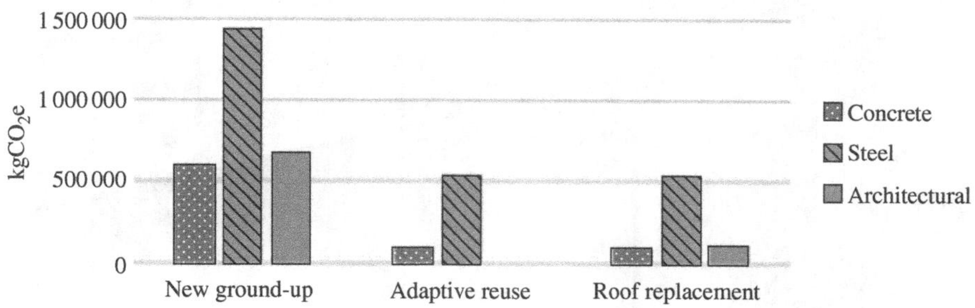

(a)

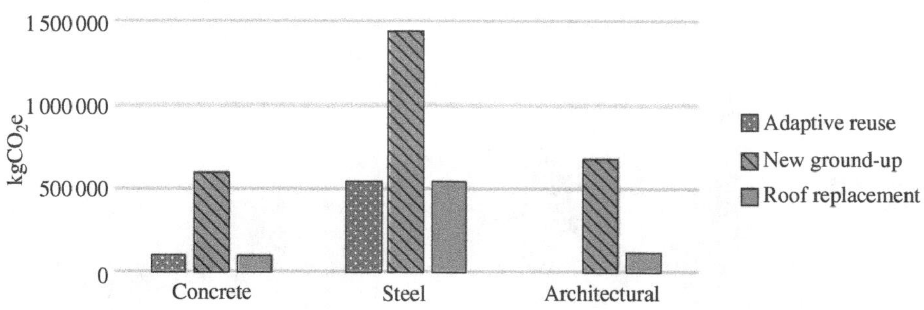

(b)

Figure 8.3 Comparison of different scenarios in terms of environmental impact: (a) total embodied carbon emissions per design option; (b) total embodied carbon emissions per material (reproduced from Martin/Martin, Inc. data).

Challenge 2: The existing braced frames of the shopping mall were inadequate for the new ice-skating facility because:

- The new building codes required application of higher lateral loads (wind and seismic, due to a transition from dated UBC codes to IBC2018 (International Code Council, 2017)/ ASCE 7-16 (American Society of Civil Engineers, 2017): the conversion from a shopping mall

Figure 8.4 Selective demolition during the initial conversion of the shopping mall. Interior columns, beams, and the second-floor slab have been removed, while the primary roof structure remains preserved (courtesy of Martin/Martin, Inc.).

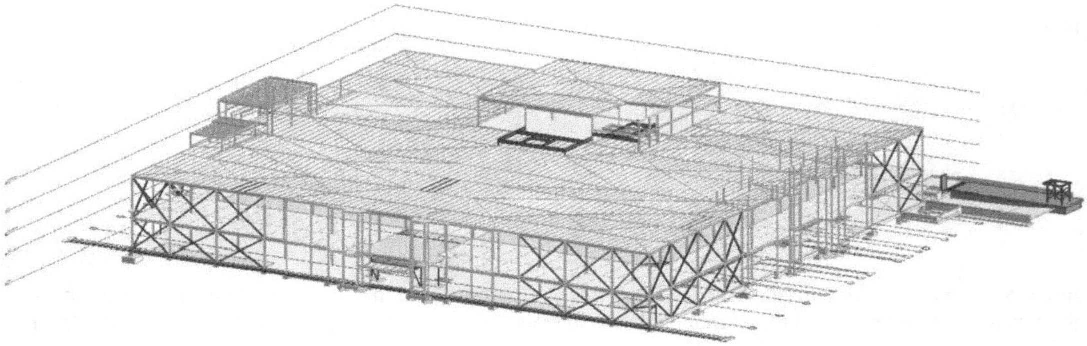

Figure 8.5 The model of The Tulsa WeStreet Ice Center project (courtesy of Martin/Martin Inc.).

(Risk Category II) to an ice skating center (Risk Category III based on occupancy numbers) subjected the design of the structure to a higher design wind speed and a 25% increase in seismic forces due to a seismic Importance Factor of 1.25 (according to Table 1.5-2 of ASCE 7-16 indicating importance factor values per Risk Category).

- The removal of two large second-floor slab openings (Figure 8.6) disrupted the original lateral force path between the diaphragm and the vertical lateral elements.
- The new lateral load path risked concentrating forces on limited foundation elements, potentially requiring expensive perimeter foundation upgrades.

Challenge 3: The existing building (shopping mall)'s interior height challenged the seating construction due to issues with moving the large construction elements (such structures would conventionally use large precast concrete elements) within the existing building space (which is the construction site of the new ice center).

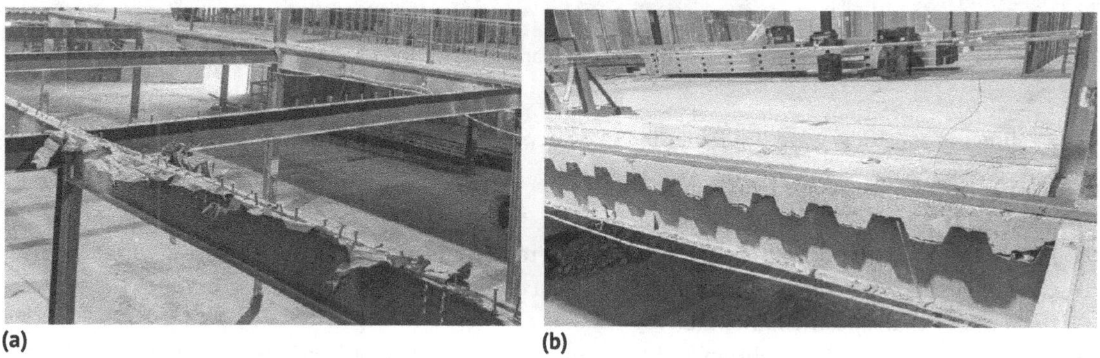

Figure 8.6 The slab removal process, showing: (a) the exposed steel beams with shear connectors and (b) a cut section of the demolished slab (courtesy of Martin/Martin, Inc.).

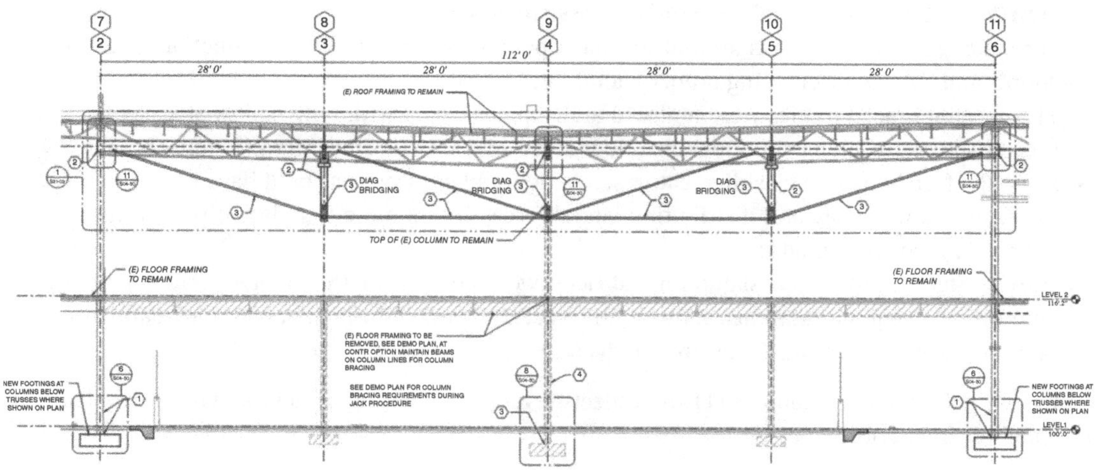

Figure 8.7 Preserved roof framing, new truss, and removed story (courtesy of Martin/Martin Inc.).

Figure 8.8 New trusses during installation (courtesy of Martin/Martin, Inc.).

The following solutions were implemented to overcome the previously mentioned challenges:

Solution 1: New roof truss girders (Figure 8.8):

- Six new 34.1 m (112 ft) steel trusses were installed per rink. These were designed to integrate existing columns as vertical web members, requiring replacement of 18 footings (Figure 8.10a).
- Customized connections for the trusses provided crucial field tolerance using built-up rod connections rather than specific hardware.
- A closely coordinated demolition and truss erection sequence involved precise jacking and shoring plans. Target jack heights ranging from 5.1 to 8.9 cm (2–3.5 inches) were provided to aid the erector in developing their jacking and shoring plan. This sequence allowed the existing roof structure to remain fully intact while erecting each truss.

Solution 2: The building's lateral force resisting system was upgraded:

- Perimeter beam connections around the slab openings were retrofitted to function as diaphragm chords and collectors, enabling proper lateral force transfer.
- The existing braced frames were replaced and supplemented with new hollow steel section (HSS) tube bracing bays, along with upgraded floor and roof framing connections (Figures 8.5 and 8.9).
- Since the façade had no clashing features, a large number of braces could have been easily added, increasing redundancy and distributing lateral forces across more foundation points, eliminating the need for costly upgrades.

Solution 3: Steel framing and slab-on-metal deck were chosen for the stepped stadia over precast concrete due to height limitations, allowing the erector to work with smaller equipment and providing greater flexibility in forming the desired stadia shape (Figure 8.10b).

The new roof structure (Figure 8.11) also reflects several engineering strategies that contributed to its sustainability and performance:

- Loads were applied directly to the trusses at panel points without altering the way existing roof joists and joist girders function to carry loads.

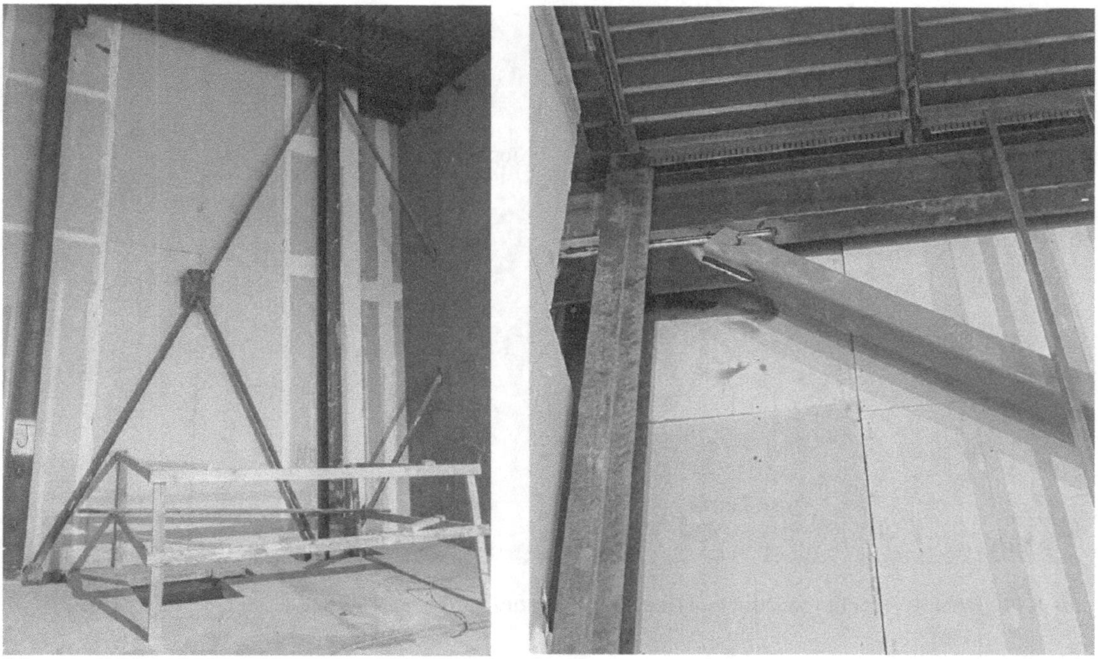

(a) (b)

Figure 8.9 During construction works: (a) old bracings removed; (b) new bracings under installation (courtesy of Martin/Martin, Inc.).

(a) (b)

Figure 8.10 During construction works: (a) foundation replacement operation; (b) steel frame for the seating units (courtesy of Martin/Martin, Inc.).

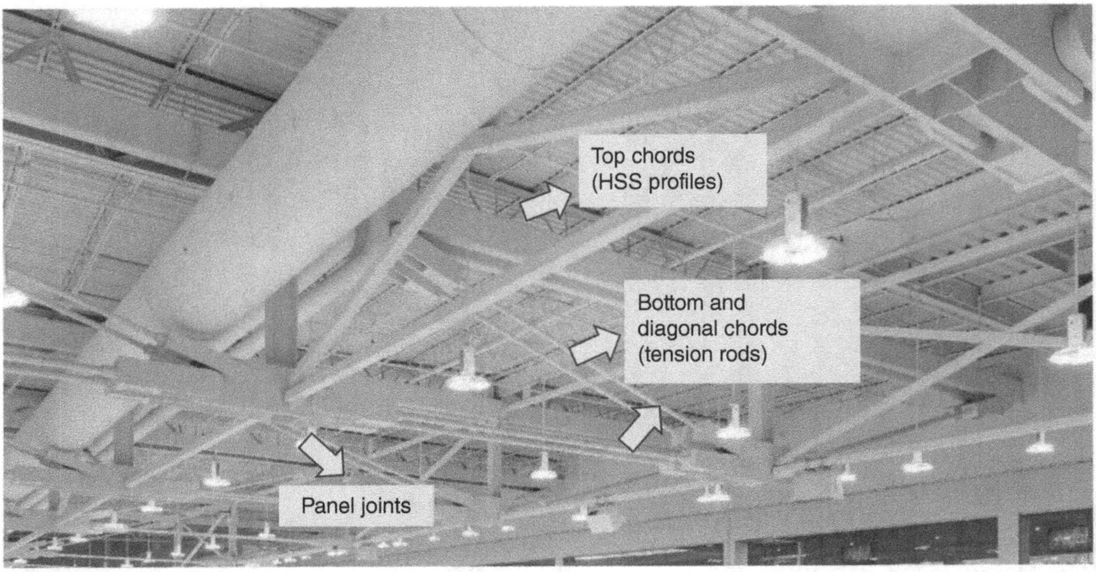

Figure 8.11 The newly built roof structure (courtesy of Martin/Martin, Inc.).

- The inverted bowstring truss shape was selected for both structural and architectural efficiency: the flat top chord sits as high as possible beneath the roof deck, while the curved bottom chord preserves optimal sightlines from the sides of each rink.
- Each rink truss was custom-designed to suit the unique constraints of the ice facility, including the geometry of chords and web members and the detailing of truss connections.
- The top chords (mostly under compression) are made of deep HSS (hollow structural section) tubes filled with grout, adding mass to resist wind uplift forces.
- The bottom chords and diagonal web members are composed of high-strength rods of 5.1–7.6 cm (2–3 inches) in diameter. Their high tensile capacity meant that design was governed by truss deflection limits rather than strength, helping to maintain consistent roof slopes for drainage and prevent overloading the existing structure.
- The roof structure uses a combination of steel grades (A992, A500, and A372) chosen to match material properties with the demands of different components. This multi-grade approach allowed for material optimization by assigning higher-strength steel to components where it brings the greatest structural and economic benefits.

In conclusion, this project demonstrated the sustainability benefits of adaptive reuse, offering also a solution for empty shopping malls beyond demolition. Adaptive reuse can be a critical tool for engineers and builders to achieve sustainability goals, effectively reducing embodied carbon, cost, and construction duration while delivering an equivalent product. Table 8.7 shows a sustainability strategy matrix for the Tulsa project that summarizes the four core lesson categories by mapping each targeted method to its specific life-cycle module and the resulting material- and carbon-saving outcomes.

Table 8.7 Sustainability strategy matrix of Tulsa project.

Strategy	Most relevant LC stages	Key method(s) used	Sustainability outcome
Preserve and reuse	A1–A3, A4, A5	Retain façade, roof membrane, structural frame	75% ↓ new concrete 60% ↓ new steel Major embodied-carbon cut
Tailored structural innovation	A5	Inverted-bowstring trusses Steel-framed seating Demolition and jacking/shoring sequence	Long spans with minimal new steel Roof kept intact Schedule shortened
Lateral system and service-life upgrade	A5, D	HSS bracing elements Retrofit collector beams	Enhanced wind/seismic performance Credit for avoided future demolition
Early optioneering and LCA	A1–A5	Comparative LCA (Tally + EC3) of "Reuse" vs. "New build" vs. "Roof only" scenarios	76% GWP reduction vs. new build Data-driven minimization across all upstream modules

Project Team

Owner and Owner's Representative: Andy Scurto, Tulsa Oilers, and International Coliseums Company (ICC)
General Contractor: Thompson Construction, Inc.
Architect: Perkins & Will
Structural Engineer: Martin/Martin, Inc.
Steel Team Fabricator and Erector: Bennett Steel, Inc.
Steel Detailer: Bennett Steel, Inc. and CADeploy

8.2.2 Adaptive Reuse of Reinforced Concrete Cores

In major cities across the globe, aging, single-purpose business districts start to become obsolete, creating sterile urban landscapes. One example is the obsolete WTC I and II towers in Brussels' North District. These towers are relics of the 1970s "Manhattan Plan," a project that displaced over 11 000 residents to build a modern economic hub. However, when the development stopped, it left a mono-functional district isolated from its surroundings. The typical response for such obsolete towers would be demolition, a solution that comes with immense costs, embodied carbon emissions, and waste. The ZIN complex in Brussels offered a compelling alternative: it tackled the challenge by transforming the former WTC I and II towers in the city's North District, not by demolishing them, but by giving a new life to their reinforced concrete core (Figure 8.12).

The new ZIN complex involved repurposing 110 000 m^2 above ground (1 184 000 ft^2) and 35 000 m^2 below ground (376 700 ft^2) into offices, housing, a hotel, shops, sports facilities, and green areas (Figure 8.13). The project retained 85% of the existing structure (cores and basements) and introduced new flexible volumes (such as the large-span, double-height "Capable Volume" connector

Figure 8.12 View of the former WTC I and II towers in Brussels' North District during the transformation into the ZIN complex. The project preserved the original reinforced concrete cores while replacing the obsolete office structures with a mixed-use development combining work, housing, and public spaces (courtesy of Bureau Greisch).

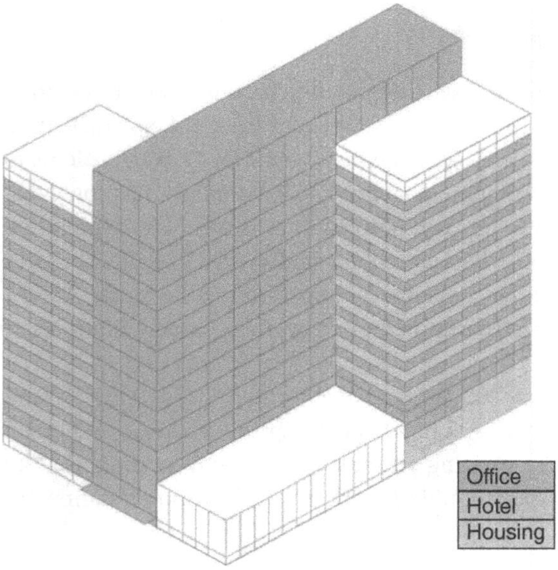

Figure 8.13 Design concept (51N4E) (courtesy of 51N4E).

between towers) to support the transformation of a mono-functional district into a sustainable, mixed-use urban hub.

The ZIN project's core concept is "hybrid use" and adaptability to minimize demolition, reduce waste, and extend service life. Circularity played a central role in this project by incorporating 94% of the building's structural materials through reuse and recycling: 62% of materials, including the existing cores and basements, were reused directly on-site, while 32% of materials, primarily concrete floor slabs and steel members, were recycled into new elements. This approach demonstrates how structural preservation and material valorization can significantly reduce the carbon footprint of a large-scale high-rise renovation (Figure 8.14).

Transforming the towers involved the following structural, logistical, and regulatory challenges:

Challenge 1: Reinforcing 35-m-deep barrettes required drilling new piles alongside existing foundations. This complex operation demanded strict waterproofing measures and constant geotechnical monitoring to ensure the building's stability (Figure 8.15).

Challenge 2: The original structural cores had to be re-evaluated to determine if they could withstand the aerodynamic forces of wind as calculated under Eurocode standards.

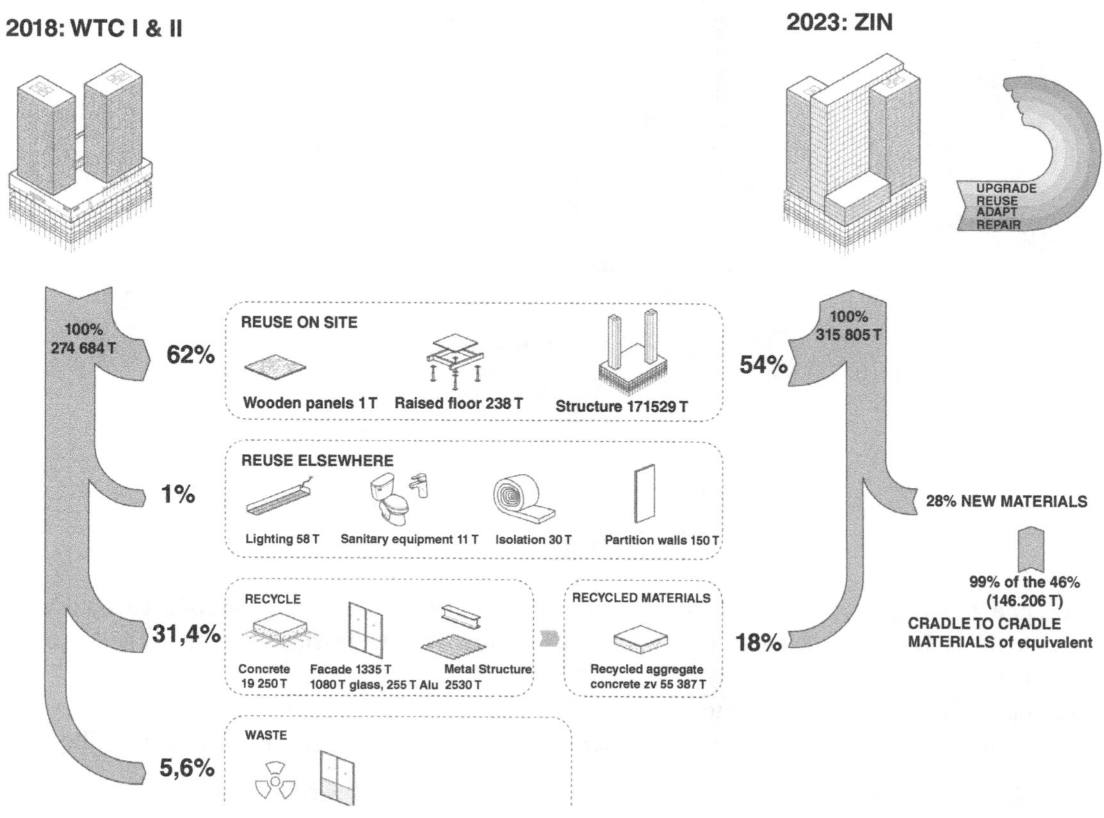

Figure 8.14 Material flow – reuse, recycle, Cradle-to-Cradle (courtesy of Bureau Greisch).

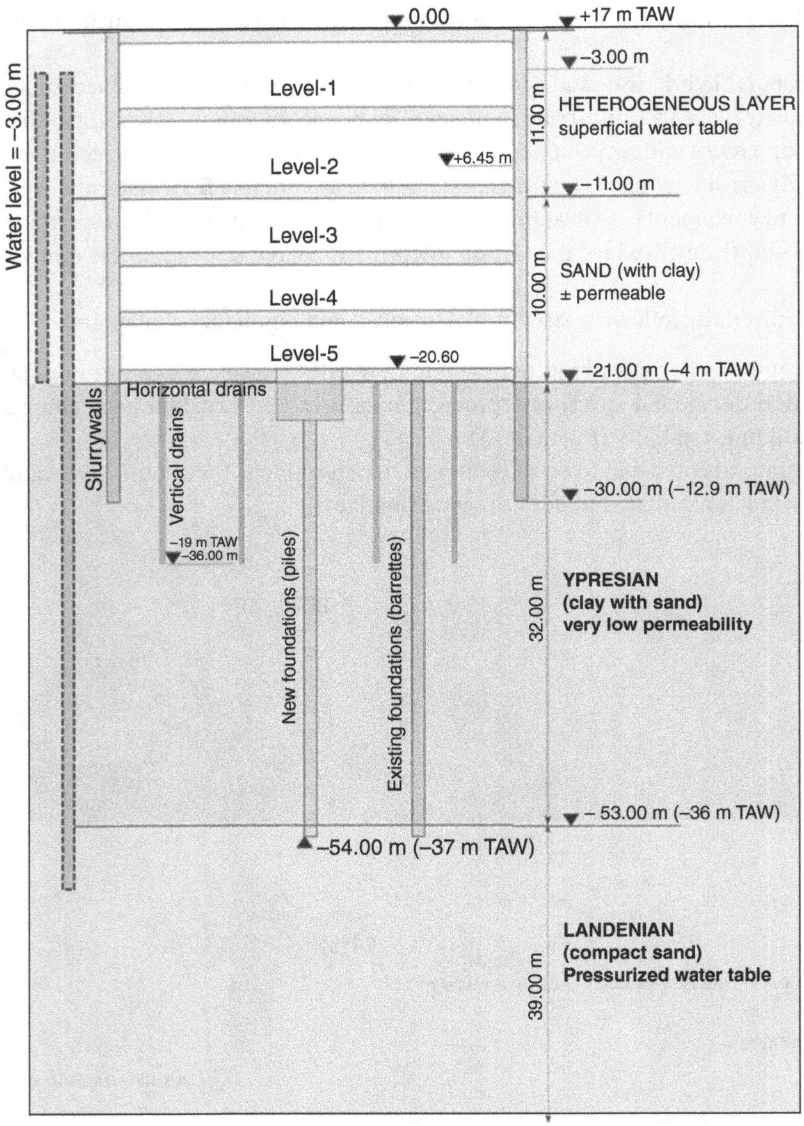

Figure 8.15 Geotechnical profile and basement plan (courtesy of Bureau Greisch).

Challenge 3: A primary difficulty was integrating a large new volume, which introduced significant additional loads that needed to be transferred safely onto the existing structure without causing overstress or failure.

Challenge 4: The new additions had to be designed for future disassembly in a way that would not damage any of the original, preserved structural elements, ensuring the building's long-term adaptability.

The following solutions were implemented for the previously mentioned challenges:

Solution 1a: Existing barrettes were reused, and new piles were drilled under geotechnical and waterproofing control from the −5 m level.

Solution 1b: Post-tensioned load-deviation structures were developed to minimize foundation interventions.

Solution 2a: Wind-tunnel testing and modeling guided the strategic coupling of the towers via the Capable Volume.

Solution 2b: Connector geometry was engineered to redistribute torsional and lateral loads across the cores.

Solution 3: Post-tensioned load-deviation structures (already noted under foundations) also allowed staged construction and avoided overloading the existing frame.

Solution 4: Full reversibility of all new structural and service connections is ensured.

In addition, a collaborative process between structural and architectural teams turned fixed constraints (such as limited floor heights) into design opportunities, as further illustrated in the reconfiguration stages of WTC I and II into the ZIN complex (Figure 8.16).

LCA showed a total embodied carbon of 572 $kgCO_2/m^2$ (Modules A: 377 $kgCO_2/m^2$, Module B: 135 $kgCO_2/m^2$, Module C: 60 $kgCO_2/m^2$) and 501 $kgCO_2/m^2$ when accounting for Module D benefits (building foundation is included in the calculation). By retaining 85% of the existing structure, the project avoided approximately 21 400 tons of upfront (module A) carbon emissions (equivalent to 184 $kgCO_2e/m^2$), resulting in a 50% reduction in embodied carbon compared to a benchmark structural system of a new construction. Whole-building embodied carbon (including façade, finishes, and mechanical, electrical, and plumbing [MEP]) was reduced by 33% compared to new construction alternative.

Figure 8.16 Reconfiguration stages of WTC I/II into ZIN: the existing reinforced concrete towers retained (left) and the new volume inserted between them with new structural framing (right) (courtesy of Bureau Greisch).

Table 8.8 Sustainability strategy matrix of ZIN project.

Strategy	Most relevant LC stages	Key method(s) used	Sustainability outcome
Preserve and reuse	A1–A3, A4, A5	Retained 85% of reinforced concrete and basements	21 400 tons CO_2 avoided; 50% reduction in upfront embodied carbon
Structural adaptation	A5	Load-deviation structures; load redistribution via Capable Volume	Reduced need for new foundations; preserved performance of existing cores under new load conditions
Design for future adaptability and circularity	A5, D	Material reuse loop; adaptable new volumes; minimal demolition	94% of materials reused or recycled; 30 000 tons of concrete reused as certified aggregates
Early optioneering and LCA	A1–D	LCA comparison of reuse vs. new-build scenarios	33% lower whole-building embodied carbon; 572 $kgCO_2/m^2$ (A–C); 501 $kgCO_2/m^2$ (with Module D benefits)

By establishing one of the first large-scale recycled aggregate supply chains in the Benelux region, 30 000 tons of demolished slabs were processed and reintroduced as new, high-quality concrete within a 15 km reuse loop, significantly reducing the project's demand for raw materials.

In conclusion, the adaptive reuse of the WTC I and II towers into the ZIN complex shows how structural retention, design for reversibility, and integrated engineering strategies can transform outdated high-rise buildings into sustainable, mixed-use buildings. Table 8.8 shows a sustainability strategy matrix for the ZIN project, summarizing four core strategy areas by mapping each targeted method to its specific life-cycle stage and the resulting material and carbon savings.

Project Team

Owner: Befimmo
General Contractor: OpenMinds (BPC/Vanlaere/VMA)
Architect: Jaspers-Eyers Architects/51N4E/AUC
Structural Engineer: Bureau Greisch

8.2.3 Reclaimed Steel in Practice

Reusing structural steel is one of the most effective strategies for reducing the embodied carbon in building projects, yet its practical application remains a significant industry challenge. The Holbein Gardens project in London provides a compelling case study on overcoming these hurdles. The project involved a three-story extension to an existing commercial property in London, commissioned by Grosvenor Estates. Contractors BHC Limited and Eurban Limited carried out the work, with Heyne Tillet Ltd. as the structural engineers and Cleveland Steel and Tubes Ltd. (CST) as the fabricator and reuse supplier. Figure 8.17 shows reused steel elements incorporated into the project.

Figure 8.17 Reused steel elements installed on-site, forming part of the primary structure at Holbein Gardens (courtesy of Roy Fishwick, Cleveland Steel).

The goal was to extend the building while avoiding the carbon impact of demolition, requiring a redesign of the existing floors and the addition of three new levels. A low-carbon approach was a priority, with the project targeting the lowest CO_2/m^2 value of any development at that time. The extension was designed as a hybrid structure combining steel and cross-laminated timber (CLT). This combination reduced the additional weight on the existing concrete structure while allowing for efficient integration. A flexible material strategy helped reduce embodied carbon while maintaining compatibility with the existing structure.

However, pioneering a circular approach with reclaimed steel presented a series of distinct logistical and technical challenges:

Challenge 1: Integrating reclaimed steel into the design process from the start, ensuring material suitability and clear standards without causing delays or creating conflicts with new fabrication requirements.

Challenge 2: Securing a consistent supply of suitable reclaimed steel sections for the project's specific needs, given the unpredictable nature and variable availability of salvaged materials.

Challenge 3: Overcoming potential fabricator reluctance to work with reclaimed materials and managing the complex certification, quality control, and logistical coordination required for reused steel.

Challenge 4: Ensuring seamless communication between different specialist contractors, particularly the CLT installer and steel supplier, to resolve evolving connection details and material requirements in real time.

Challenge 5: Aligning the procurement of specific reused sections with finalized connection designs, especially for complex joints where multiple sections converged, without risking material incompatibility or project delays.

Challenge 6: Managing the logistical complexities of small-batch steel orders, including inefficient partial-load deliveries and limited on-site storage, without disrupting the construction schedule.

The project team addressed these hurdles through the following set of solutions:

Solution 1: From the early stages, the engineers worked with CST to explore the availability of reused steel. The collaboration helped establish material suitability, develop clear reuse protocols, and define cosmetic standards. The project team agreed that existing holes and prior fabrications could remain where they had no structural impact. CST reviewed component drawings to ensure that existing features did not interfere with new connections, and cutting plans were adjusted to maximize material efficiency and prevent conflicts. Engaging with a reuse supplier early in the process helped streamline coordination, optimize material use, and avoid later design conflicts.

Solution 2: The reuse strategy was based on material availability rather than structural limitations, as all elements were deemed suitable for reclaimed steel. Steel was sourced through three channels:

- A portion came from demolition at Holbein Gardens but was used later due to scheduling changes (closed-loop circularity).
- Some material was identified from another client's site, reviewed for suitability, and incorporated into the design.
- The remainder was supplied by CST, which, as an EN1090-accredited stockist and fabricator, handled both remedial work and new fabrication.

Where reuse was unavailable, new steel was specified, with the option to substitute reused material if more became available. A design and procurement process that allowed substitutions ensured that reuse opportunities could be maximized without delaying the project.

Solution 3: Having CST act as both fabricator and supplier simplified certification, quality control, and logistical coordination. Some fabricators hesitate to work with reused steel, but this arrangement avoided potential concerns about certification and compliance. Combining the roles of supplier and fabricator reduced administrative complexity and made reuse more practical.

Solution 4: As the project progressed, the CLT installer and main contractors worked closely with CST to address material concerns. Maintaining open communication prevented misunderstandings that could have affected the reuse strategy. During contract execution, it became clear that the CLT installer was responsible for connection detailing, requiring adjustments to some section sizes.

Solution 5: Some connections involved multiple sections converging at a single point, and in some cases, reused sections were not available in the required dimensions. Delaying material procurement until connection details were finalized prevented the risk of sourcing steel that would later become unusable. Tight lead times posed challenges, but scheduling delays provided more time for sourcing and delivery.

Solution 6: The project required relatively small steel quantities across three floors, making full-load deliveries impractical. Limited on-site storage meant materials had to be held at CST, increasing the number of deliveries to over three times what would typically be expected in a full-load scenario. Planning for storage and transport logistics helped manage delivery inefficiencies caused by small batch sizes. Once on-site, the steel was installed without complications, and the reuse process had no impact on construction sequencing. Once erected, the reused elements were indistinguishable from new steel. The erectors and main contractors reported no issues with installation. Figure 8.18 presents an image of the on-site material delivery and storage area.

Figure 8.18 Reclaimed and prepared steel sections stored at the supplier's facility before being delivered to the site (courtesy of Roy Fishwick, Cleveland Steel).

Table 8.9 Sustainability matrix of Holbein Gardens project.

Strategy	Most relevant LC stages	Key method(s) used	Sustainability outcome
Early supplier collaboration	A1–A5	Joint development of reuse protocols Defining cosmetic standards Collaborative drawing review	Streamlined material integration Optimized material efficiency De-risked design and procurement
Flexible and adaptive sourcing	A1, A4	Sourcing from multiple channels Design allowing material substitutions Procurement based on availability	Maximized reuse content (25 tons) Avoided supply-related delays Increased project resilience
Integrated supplier-fabricator	A3–A5	Single entity for supply and fabrication In-house quality control and certification	Simplified logistics and administration Reduced fabricator resistance Ensured compliance and quality
Proactive logistical planning	A4, A5	Off-site storage by supplier Coordinated just-in-time deliveries Planning for small-batch transport	Avoided on-site storage issues Prevented construction disruptions Ensured smooth installation sequencing

The project successfully incorporated 25 tons of reused steel, resulting in a 45-ton reduction in embodied carbon while remaining cost-neutral. A strategy based on flexibility rather than strict targets made reuse feasible within the project's constraints. Table 8.9 summarizes the core sustainability strategies by mapping each method to its life-cycle contribution and resulting outcomes.

In conclusion, lessons learned from the project include:

- Early collaboration between engineers, suppliers, and fabricators improved material integration and prevented late-stage design conflicts.
- Allowing reuse to be incorporated throughout the project timeline provided more opportunities to source reclaimed materials.
- Having a combined supplier-fabricator simplified certification and fabrication, reducing resistance to using reclaimed steel.
- Considering logistics and storage from the outset prevented inefficiencies caused by fragmented deliveries and space limitations.
- By maintaining a flexible sourcing strategy and ensuring continuous coordination among stakeholders, the project successfully integrated reused steel while maintaining structural integrity and cost efficiency.

Paper Credits: Roy Fishwick, Cleveland Steel.

8.2.4 Transforming an Incomplete Structure into Sustainable Headquarters

Across our cities during urban regeneration, some buildings are left incomplete, their raw skeletons standing as monuments to wasted resources and unrealized potential. The typical action in these cases often involves demolition, a process that not only wastes the significant embodied carbon already invested in the structure but also requires vast new inputs of materials and energy. The North Tower, located in Pisa's "Parco delle Torri" complex (Figure 8.19), offers a prominent alternative. Developed by ATI Project, this intervention transforms an unfinished structure in a strategic, developing area into a new multifunctional hub connecting the historic city with its scientific and hospital district.

Originally designed for residential use, the tower was acquired in its raw structural state. Rather than demolishing it, ATI Project chose to upgrade the existing steel frame and transform it into their new headquarters. This decision showcases a powerful application of life-cycle-driven design, significantly reducing the project's carbon footprint.

Figure 8.19 "Parco delle Torri" complex (courtesy of ATI Project).

Figure 8.20 The incomplete twin steel towers in Pisa's *Parco delle Torri* were re-engineered rather than demolished, conserving embodied carbon (courtesy of ATI Project).

The core concept of the project is "adaptive reuse," where the decision to rehabilitate the existing structure (steel columns, beams, and steel–concrete composite slabs) is validated through a data-driven analysis covering structural and fire safety aspects while aiming to minimize environmental impact. Figure 8.20 shows the original incomplete towers whose existing steel frames formed the basis for the adaptive reuse strategy instead of demolition. Figure 8.21 highlights the composite steel–concrete floor system and the reinforced concrete cores, confirming the structure's suitability for rehabilitation and office conversion.

Figure 8.22 presents the structural grid used to evaluate the existing load paths and guide the integration of new strengthening and service systems. The rehabilitated tower now houses the ATI Project headquarters, flexible tenant offices, and community-focused amenities, demonstrating how to unlock the full life-cycle value of existing structural assets.

The project involves the renovation of the unfinished building, consisting of the elements shown in Table 8.10. All these components are preserved and reused in the new building.

Figure 8.23 illustrates the distribution of circular hollow section (CHS) profiles used in the steel structure.

Figure 8.24 shows the Gross Internal Area (GIA) calculation scheme for the North Tower, totaling 5667.34 m^2. This area includes all enclosed floor spaces and interior walls but excludes exterior walls,

Figure 8.21 The composite steel–concrete slabs and reinforced concrete cores, structurally sound for adaptive reuse and retrofit (courtesy of ATI Project).

external balcony areas, and service shafts. The ground floor is included in the GIA. The roof level is excluded, with the exception of the enclosed service shaft (39.40 m^2), which is counted as it constitutes a closed space, as seen in Figure 8.25. This GIA value is used as the functional unit denominator (per m^2 of GIA) for all embodied carbon comparisons that follow.

Transforming the incomplete tower from its original residential design into a high-occupancy business center posed the following key environmental and life-safety challenges:

Challenge 1: To prove that keeping the existing structure was the right decision in achieving the standards of circular design and LEED (Leadership in Energy and Environmental Design; the most widely used green building rating system in the world), the team had to demonstrate, in quantifiable terms, that reusing the raw steel frame would yield significant environmental benefits compared to the alternative of demolishing it and starting over with a new steel or concrete building.

Challenge 2: The (existing) building's fire safety and evacuation systems, particularly the main stairwell, were not designed for the density of a commercial hub (a high-occupancy office and business center). The team faced the critical challenge of verifying that the existing escape route could perform safely under the much greater occupant load, ensuring compliance with stringent Italian fire safety codes without requiring a costly and invasive structural redesign.

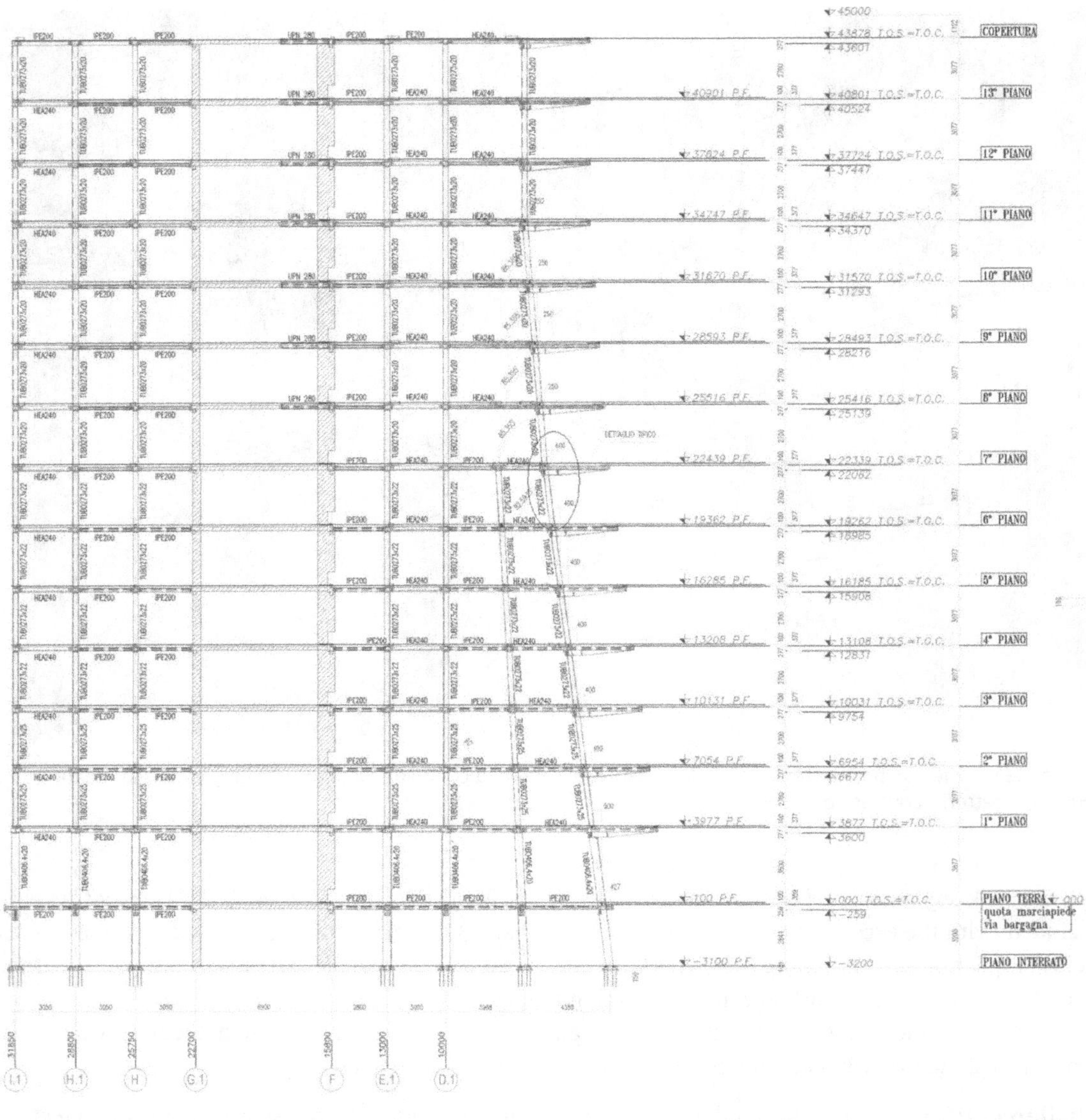

Figure 8.22 Structural layout showing beam and column grid of the existing frame (courtesy of ATI project).

Solution 1: ATI Project deployed a detailed comparative LCA. Using data extracted directly from the project's Building Information Modeling (BIM) models, the LCA study followed UNI EN15978:2011 and UNI EN15643:2021 standards and included:

Scope: load-bearing structure, envelope, and interior finishes

Time Frame: 50-year building lifespan

Life-cycle Stages: from A1 (raw material extraction) to C4 (end-of-life). Data sources: quantities extracted from BIM models, EPDs, and data included in One Click LCA database.

Environmental Indicators: GWP, ODP, AP, EU, POCP, nonrenewable resource

Table 8.10 Elements of the existing building.

Structural element	Material/type	Details
Main columns	CHS steel	Grade S355. Ranging from 406.4 × 20 mm (base) to 273 × 6.3 mm (top)
Underground columns	Reinforced concrete	Located on the underground parking floor
Beams	Steel	Grade S275: HE240A, HE260A, UPN 280, IPE 180-200-240 Grade S355: Tapered beams for balconies and HE276 connections
Floor slabs	Composite	Profiled Steel Decking: Genus 160 (0.75 mm) and A55/P600 HI-BOND (1.2 mm) Reinforcement: Welded mesh Ø6/200 Concrete: Rck 300 kg/cm^2
Central core	Reinforced concrete	Houses the stairs and elevator
Foundations	Reinforced concrete raft and piles	Raft Slab: C30/37 concrete with a 90 kg/m^3 reinforcement ratio Bored Piles: 37 piles, each 42 m long, with diameters of Ø1200–Ø1500 mm

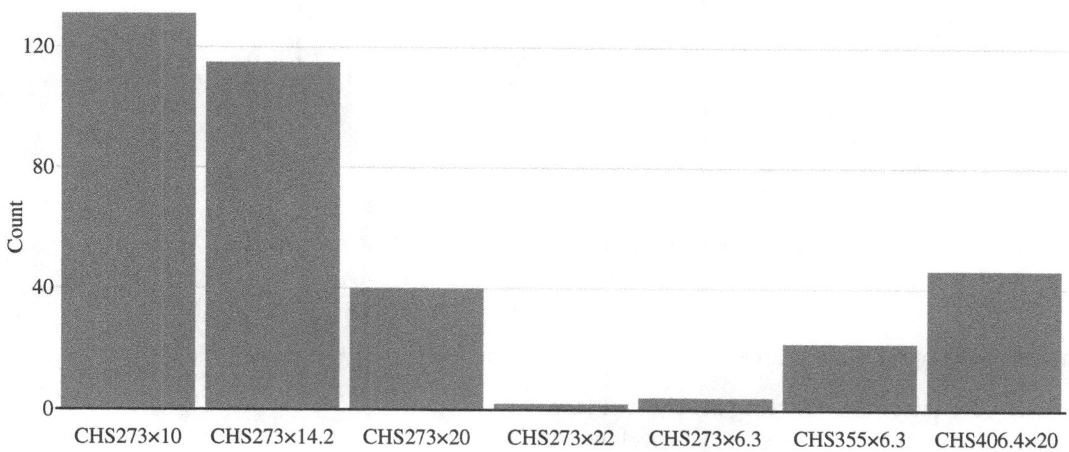

Figure 8.23 Distribution of circular hollow section (CHS) profiles used in the existing steel structure.

The following carbon factors are used for the sensitivity analysis:

Reinforcement steel 0.62 $kgCO_2e/kg$ (generic value assuming 97% recycled content)
Structural steel profiles 0.74 $kgCO_2e/kg$ (generic value assuming 90% recycled content) for steel classes S235, S275, S355
Ready-mix Concrete

- Columns (C35/45, CEM IV) 381 $kgCO_2e/m^2$
- Slabs (C35/45) 368 $kgCO_2e/m^2$
- Foundations (C30/37, CEM III/B) 148.77 $kgCO_2e/m^2$

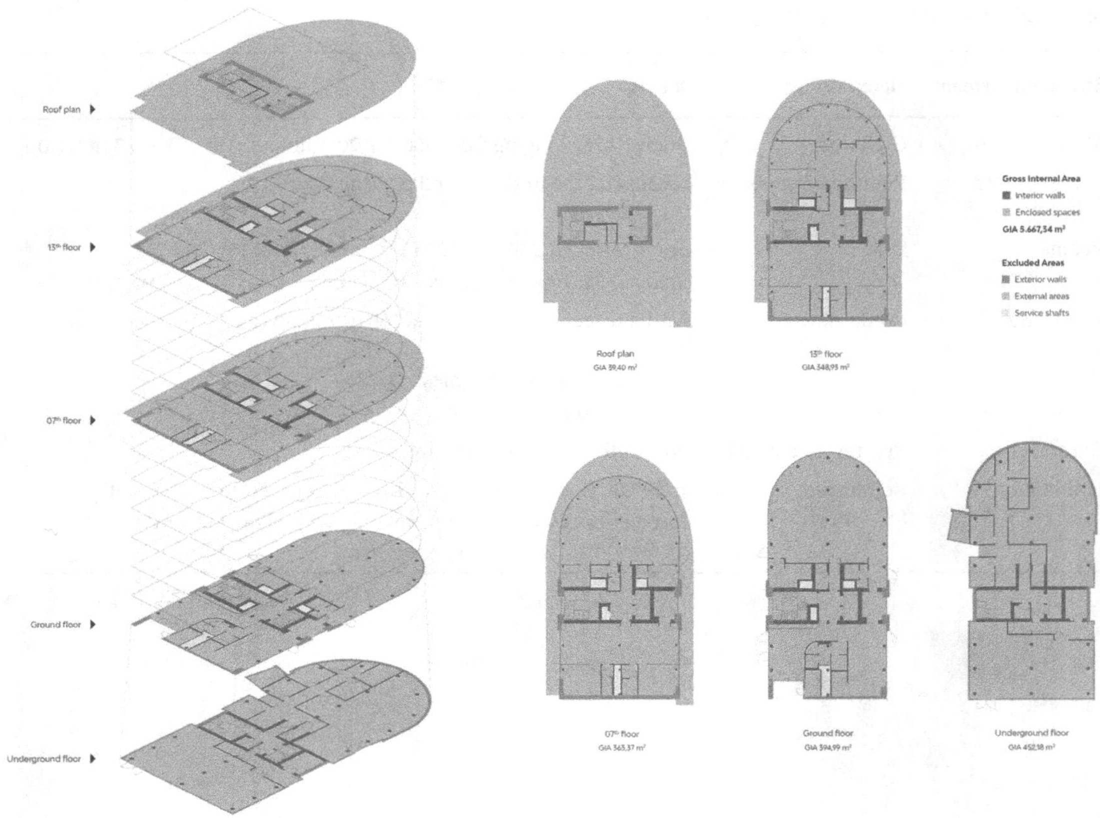

Figure 8.24 Gross Internal Area (GIA) breakdown of the North Tower by floor level (courtesy of ATI Project).

The following options are studied:

Option A – Reuse of the Existing Steel Structure

- Embodied carbon: 258 $kgCO_2e/m^2$ (envelope + finishes)
- Rating: Category A (<300 $kgCO_2e/m^2$)
- Reduction vs. new steel structure: −51%

Option A assumes that the entire existing steel structural system and foundations are kept in place. Only the new envelope and finishes are added, with no replacement of structural elements.

Option B – New Steel Structure

- Embodied carbon: 532 $kgCO_2e/m^2$ (structural system + foundation + envelope + finishes)
- Rating: Category D (460–540 $kgCO_2e/m^2$)

Option B represents the scenario where the existing building is assumed to be newly constructed in steel, with the full system (structure, foundations, envelope, and finishes) considered.

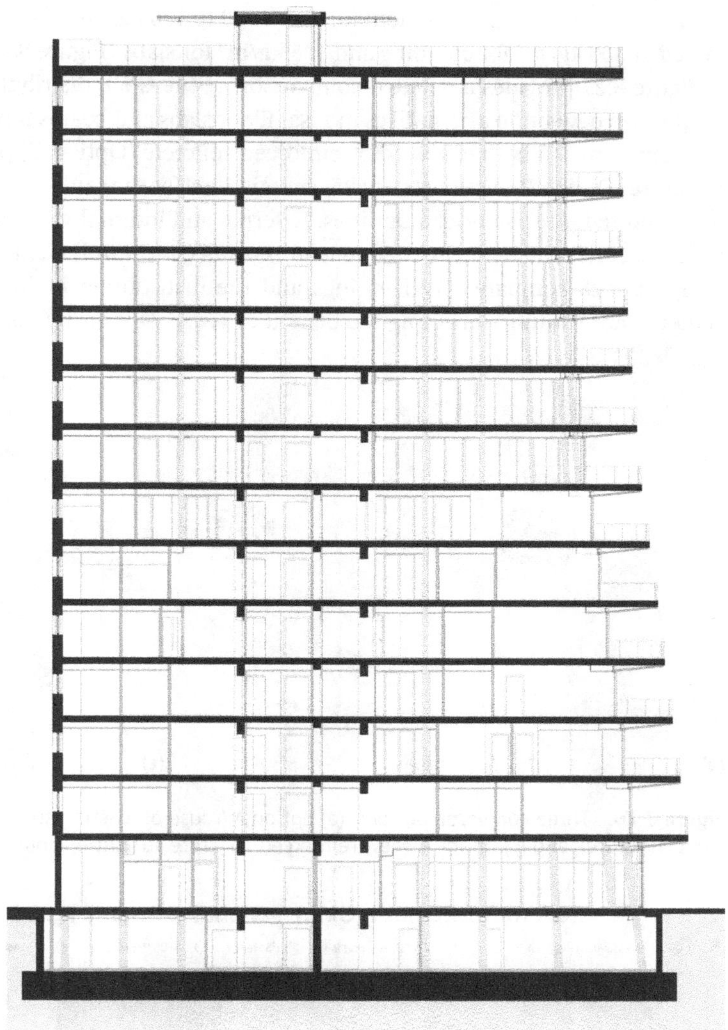

Figure 8.25 Cross-section (Section A–A) of the North Tower showing the full vertical arrangement from the underground parking level to the roof, and the enclosed service shaft at the roof level (courtesy of ATI Project).

Option C – New Reinforced Concrete Structure

- Embodied carbon: 585 $kgCO_2e/m^2$
- Rating: Category E (540–620 $kgCO_2e/m^2$)
- Increase vs. new steel: +10%
- Increase vs. reuse: +46%

In Option C, the columns, beams, and slabs are redesigned in reinforced concrete with dimensional variations to optimize load distribution at different building heights. The reinforced concrete columns (C35/45) vary in diameter from Ø500 mm at the base of the tower to Ø350 mm at the top. The slab systems are designed as 240 mm thick reinforced concrete plates, continuous in cantilever in all balcony areas. The foundations include 9 additional piles of type Ø1200 mm with 12Ø20 reinforcement, together with an increase in reinforcement ratio from 90 to 130 kg/m³. The average reinforcement content is estimated at 300 kg/m³ for columns and 125 kg/m³ for slabs (Figure 8.26).

Figure 8.27 provides a detailed comparison of the GWP distribution across major building elements for the two new-build alternatives in Pisa. Floor slabs and beams represent the dominant contributors to total emissions in both cases, with reinforced concrete (Option C) producing a higher impact. The steel structure (Option B) demonstrates lower emissions for the same components, benefiting from lighter sections and reduced material quantities. External and internal walls also play a role, though their relative contribution is smaller compared to load-bearing (structural) elements.

Figure 8.28 compares GWP ratings and the distribution of impacts across LCA stages. Option A achieves an A rating, reflecting the benefit of reuse. Option B falls in the D band, and Option C in the

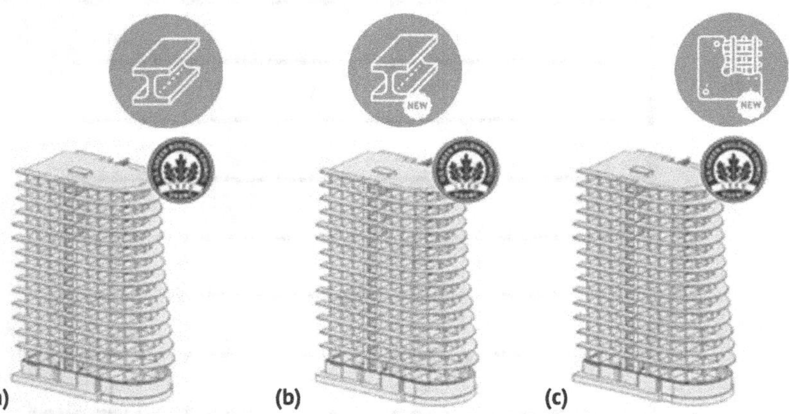

Figure 8.26 Three compared options: (a) option A (reuse of existing steel structure); (b) option B (new steel structure); and (c) option C (new reinforced concrete structure) (courtesy of ATI Project).

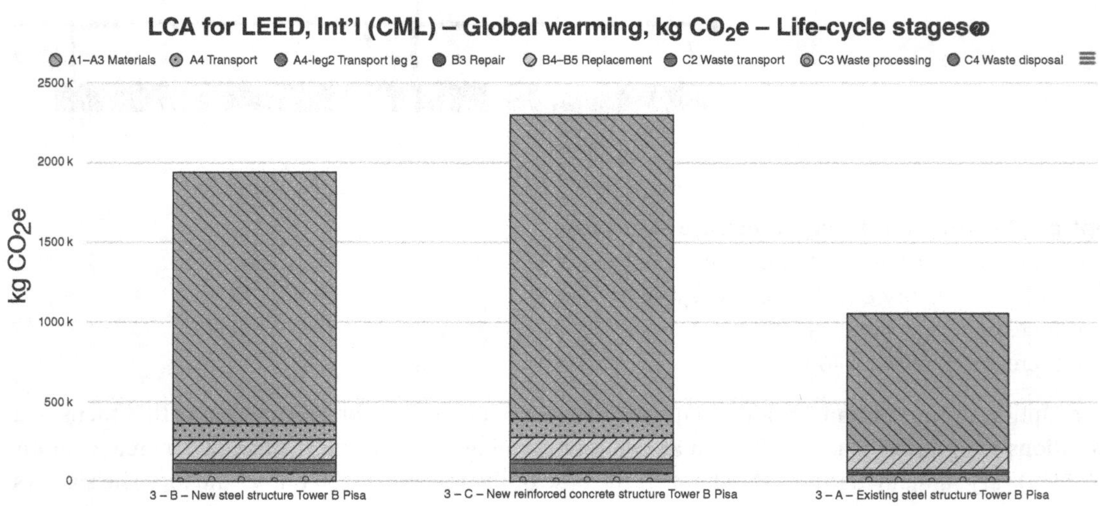

Figure 8.27 Breakdown of the total carbon impact by building element (GWP) (image produced using data provided by ATI Project).

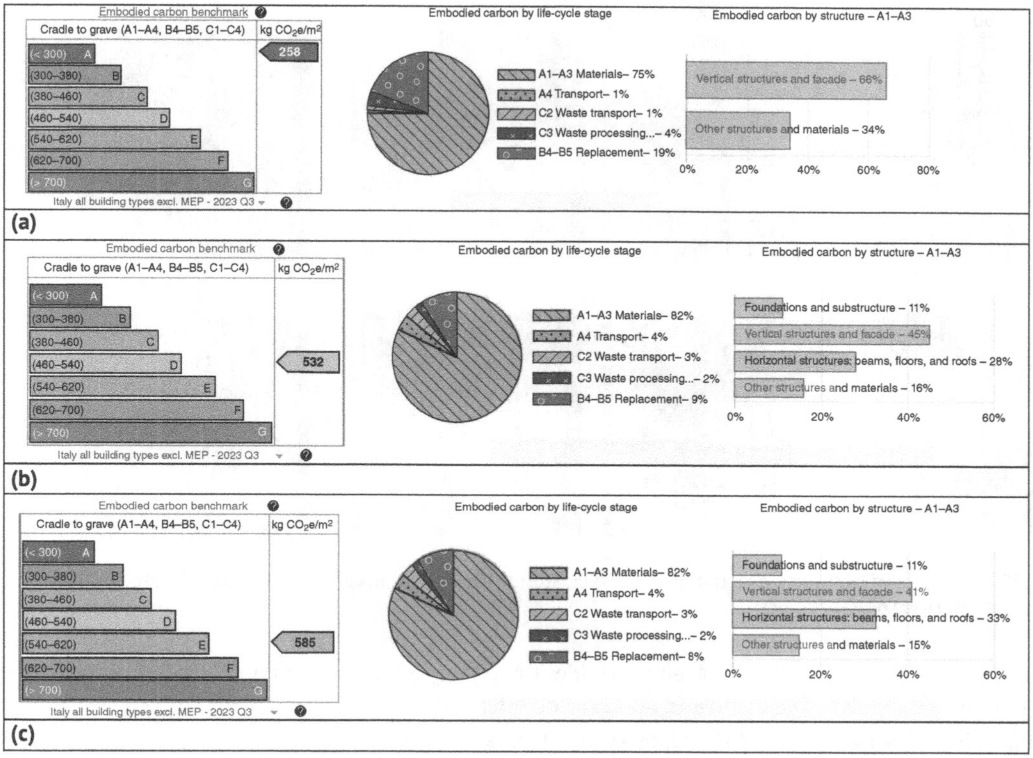

Figure 8.28 Benchmark analysis (according to the One Click LCA "The Carbon Heroes Benchmark") of all three options: (a) option A "reuse of existing steel structure," (b) option B "new steel structure," and (c) option C "new reinforced concrete structure." The benchmark accounts for embodied impacts relating to all building parts, except building technology and external areas (courtesy of ATI Project).

E band. Across all options, the product stage A1–A3 is dominant (the rating categories are defined based on One Click LCA "The Carbon Heroes Benchmark", year 2023). Option A shows a smaller A1–A3 share and a larger B4–B5 replacement share than the new-build cases, since refurbishment items over the study period become more visible once major new production is avoided. In Options B and C, the A1–A3 share grows further, while A4 transport, C2 transport to waste, and C3 waste processing remain comparatively minor in every case. Please note that the results are also sensitive to the selection of carbon factors and underlying database assumptions (such as EPD sources, regional energy mixes), and should therefore be interpreted as a relative comparison between the design options rather than as absolute carbon values.

Solution 2: ATI Project carried out a detailed study to verify compliance of the building with Italian fire safety codes. A dynamic evacuation flow simulation was conducted along the tower's main stairwell to assess its adequacy under full-occupancy conditions.

About evacuation management, a flow analysis was carried out along the vertical escape route to assess its compatibility with an adequate level of safety for the activities housed in the tower (Figure 8.29). The study analyzed the expected occupancy conditions within the building and observed the evacuation

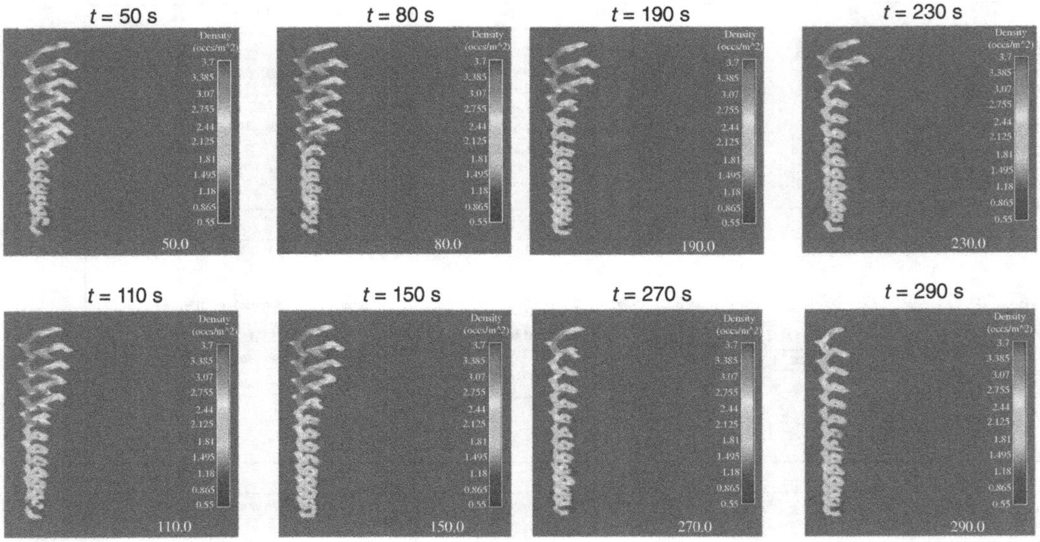

Figure 8.29 Analysis of occupant evacuation dynamics along the vertical escape route to verify safety performance (courtesy of ATI Project).

dynamics of the occupants, with the objective of identifying any additional or complementary measures needed to reduce risk levels and ensure an appropriate safety standard.

The analysis performed on the stairwell of Tower B aims to represent the occupancy density along the escape route, focusing on occupant density during movement and evaluating waiting times in case of localized overcrowding. The stairwell is designed as a smoke-proof enclosure, and the study has been conducted under the assumption that it constitutes a safe zone – comparable to protected floor lobbies. Consequently, the flow analysis focuses on assessing whether queueing times and occupant density within the stairwell remain within thresholds considered acceptable for maintaining comfort conditions during evacuation.

The underlying assumptions of the study are as follows:

- the maximum occupancy levels of both activities are considered;
- all occupants use the escape route simultaneously;
- a single occupant behavior profile is defined, with a normal distribution of walking speeds;
- occupant movement is locally halted when a density of 3.7 persons/m^2 is reached.

For the purposes of the analysis, assuming a normal distribution of walking speeds allows for the introduction of discontinuities in the flow regularity, thereby increasing the likelihood of the formation of localized zones where peak crowd densities may occur (Figure 8.30). The occupant flow analysis shows that, under the modeling assumptions, localized overcrowding occurs, leading to the interruption of evacuation flow due to the critical crowd density threshold of 3.7 persons/m^2 being exceeded in specific areas. As occupants from different floors converge toward the stairwell, localized congestion points develop at the junctions of flows from adjacent floors. This results in a slowdown of movement on the stairs, up to a complete halt, for the time required to resolve the blockage. Over time, occupants from the lower floors continue moving downward, gradually relieving the local congestion caused by

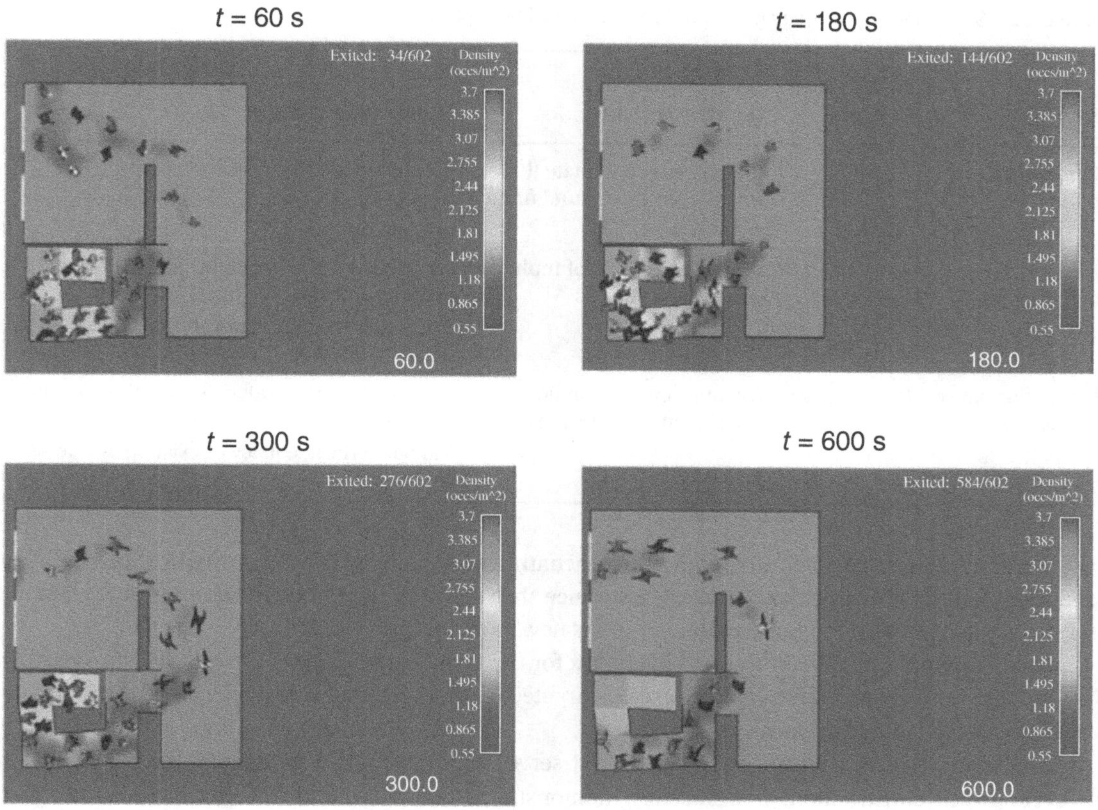

Figure 8.30 Effect of normally distributed walking speeds on pedestrian flow regularity and the emergence of localized peak crowd densities (courtesy of ATI Project).

the inflow of evacuees from the immediately higher floor. This process reduces crowd density below the critical threshold, and the effect propagates upstream along the route until the entire escape path reaches a steady-state condition with crowd density consistently below the stop-flow threshold.

It is important to highlight that the number and width of the final exits located in the ground floor lobby are sufficient to ensure the evacuation of occupants to the outside without the occurrence of localized overcrowding.

The assumptions adopted for the flow model along the vertical escape route also apply to this part of the analysis, which represents the final phase of the flow along the staircase. Therefore, it is again assumed that the building is at maximum occupancy, regardless of simultaneity considerations, and that all occupants begin the evacuation process at the same time.

Therefore, over time, there is no significant increase in crowding within the lobby, despite the continuous flow of occupants arriving from the stairwell. This indicates that the occupants encounter no obstruction in reaching the exterior through the two designated exits.

The simulation revealed that while minor, temporary points of congestion would occur at landings as occupants merged, the occupant density never reached hazardous levels. The flow, while momentarily slowed, remained safe and orderly. Furthermore, the analysis confirmed that the ground-floor lobby and

Table 8.11 Sustainability strategy matrix of Pisa Business Tower project.

Strategy	Most relevant LC stages	Key method(s) used	Sustainability outcome
Preserve and reuse	A1–A5	Retain and rehabilitate the existing steel structural frame	51% reduction in embodied carbon Avoided demolition waste and new material production
Data-driven optioneering	A1–C4	Comparative LCA of multiple structural scenarios	Quantified and verified the environmental benefits of reuse Enabled informed, data-backed decision-making
Fire performance verification	Building use	Dynamic evacuation flow simulation to assess life safety	Ensured occupant safety for a higher-density use case Validated the adequacy of existing escape routes

exits were more than capable of handling the evacuating occupants without creating bottlenecks. This advanced analysis provided the necessary assurance that the repurposed design was not only efficient and sustainable but also fundamentally safe for its new, more intensive use.

Table 8.11 shows a sustainability strategy matrix for the Pisa Business Tower project that summarizes the four core lesson categories by mapping each targeted method to its specific life-cycle module and the resulting material- and carbon-saving outcomes.

In conclusion, the Pisa Business Tower project serves as a powerful testament to the principles of life-cycle-driven construction. It masterfully demonstrates that a structure's value includes not just its physical materials, but also the embodied energy and carbon it contains. Through a meticulous, data-driven approach combining LCA and advanced safety simulations, ATI Project not only created a sustainable and functional headquarters but also forged a clear, replicable model for urban regeneration. This case study proves that the most innovative and sustainable new building is often the one that is already standing. The rehabilitated tower is now a multifunctional business center. The upper floors (7–13) house the headquarters of ATI Project, while the lower floors (1–6) offer flexible office space for tenants. The ground floor fosters a sense of community with a shared lobby, lounge, bar, and restaurant, with parking located in the basement. This adaptive reuse project demonstrates how to unlock the potential of incomplete or abandoned structures, turning a potential urban blight into a valuable asset.

Project Team

Owner and Owner's Representative: ATI Invest
Designer: ATI Project
Paper Credits: Eng. Anna Continanza, Eng. Michele Fascilla, Eng. Alessio Bacci, Eng. Marianna Sbarganti

8.3 Material Efficiency and High-tech Solutions

8.3.1 Centrifuged High-strength Concrete with Recycled Aggregates

Building a massive, multi-building complex, including a 90 m tower, is a challenge in itself. Beyond this challenge, building it on poor soil that requires deep, extensive foundations would create a huge carbon footprint before the first floor is even poured. This was the complex reality facing the Campus Pictet de Rochemont project (Figure 8.31), a large-scale development in Geneva consisting of four buildings, including a 90 m tower, 54 000 m^2 of offices, and 10 000 m^2 of residential space (100 apartments), all supported by a reinforced concrete structure with five underground levels and deep foundations that represented a significant environmental liability. Figure 8.32 shows the piles installed to stabilize the poor soil conditions and transfer the structural loads to deeper, more competent ground layers.

The core concept of this project is holistic carbon reduction through material science and supply chain optimization. Rather than accepting the high embodied carbon of a conventional concrete structure, the strategy included optimizing concrete mix designs, off-site fabrication, and supply chain management to achieve a 30% reduction in the project's overall carbon footprint compared to a standard project.

From the earliest stages, parametric simulations were used to identify optimization opportunities. This data-driven approach continued through the executive design phase, where a technical dialogue with

Figure 8.31 The Campus Pictet de Rochemont project (courtesy of dl-a, designlab-architecture sa/Nightnurse).

Figure 8.32 Deep foundations (courtesy of Redesco Progetti).

suppliers was initiated to ensure the selection of materials and components with the lowest possible environmental impact.

In particular, the Campus Pictet de Rochemont project presented a series of interconnected sustainability challenges:

Challenge 1: As a predominantly reinforced concrete structure with deep foundations, the project's expected upfront embodied carbon was bound to be very high.

Challenge 2: Accommodating the spatial requirements of the architectural design while minimizing the environmental impact of the structural system presented a significant optimization challenge. The team needed to reduce material quantities and optimize structural efficiency throughout the building without compromising architectural intent or structural performance.

Challenge 3: The presence of a large auditorium on the first basement level required innovative structural solutions to span significant distances and create column-free space. This meant designing transfer structures capable of redistributing loads from the tower above while managing the structural complexity and additional material requirements.

The project team implemented the following solutions that targeted every stage of the design and construction process to systematically reduce the building's carbon footprint:

Solution 1: The first challenge has been addressed by combining several approaches: optimizing concrete mix design through increased GGBS (Ground Granulated Blast-furnace Slag) content, the use of recycled aggregates, using high-strength concrete classes to reduce material quantities and structural dimensions while improving durability, selecting concrete and steel suppliers with GWP values lower than the national reference baseline, and employing materials from controlled demolition processes located in proximity to the construction site. In particular:

i. High-strength concrete and steel classes have been used where possible, especially for columns. While the high strength concrete mix has a higher GWP per cubic meter, it allowed for smaller columns, reducing the total volume of concrete required and leading to a lower overall carbon

Table 8.12 Material types and quantities used in the project.

Material	Class	Percentage	Quantity
Foundation			
Lean concrete	C25/30	40%	2250 m^3
Diaphragm walls	C35/45	25%	23 000 m^3
Slabs and retaining walls	C25/30	25%	32 000 m^3
Superstructure			
Cast-in-place r.c. structures	C35/45	25%	49 000 m^3
Precast r.c. columns	C75/85	25%	850 m^3
Precast r.c. stairs	C35/45	40%	800 m^3
Reinforcement steel	B450C		5000 ton
Structural steel	S355J0		300 ton
Structural steel	S460		200 ton

footprint when considering the entire structure. Table 8.12 shows the concrete and steel types and their quantities in the project.

ii. Achieving a significant impact reduction required looking beyond just the structural design. The team had to address the entire life cycle, from the sourcing of raw materials and the selection of suppliers to the efficiency of on-site construction processes and energy use. The team engaged directly with suppliers to select concrete and steel producers with verified low-GWP products (such as GGBS, a cement substitute that reduces concrete's carbon footprint by replacing a portion of traditional Portland cement with industrial byproducts). Priority was given to local suppliers.

iii. The team specified concrete mixes that maximized the use of recycled aggregates, using the highest percentage allowed by the SN EN 206-1 standard for the concrete to be classified as "compliant" (up to 25% in structural concrete and 40% in nonstructural elements). This choice allowed for the use of concrete with lower GWP compared to traditional concrete. Furthermore, it aligns with a circular system approach: considerable volumes of material, if suitable, are diverted from landfills and reused as construction material. It should also be acknowledged that the environmental benefits of recycled aggregates are dependent on the proximity of the source to the construction site, as transportation over long distances can offset the carbon savings achieved through material reuse, making local sourcing a critical factor in the overall sustainability assessment. In this project, the aggregates are obtained from a site within a distance of circa 50 km (from controlled demolition sites near the project). While aggregates contribute minimally to concrete's overall carbon emissions, their recycling demonstrates a commitment to resource conservation at every level of the project.

Solution 2: To optimize the construction process and reduce waste, the project used standardized, prefabricated components (Figure 8.33) made of:

i. High-strength centrifuged concrete columns reaching up to class C75/85. Centrifugation maximizes the mechanical performance of the materials by compacting the concrete and eliminating

Figure 8.33 Prefabricated beams (on the left) and centrifuged high-strength columns (on the right) (courtesy of Redesco Progetti).

internal voids. This process allows for a reduced structural cross-section, improving usable space and increasing load-bearing capacity compared to traditional solutions. The result is an extremely uniform surface finish, with no visible porosity and excellent long-term durability.

ii. Hybrid steel–concrete floor system manufactured off-site. These composite floors combine steel with reinforced concrete, reducing material usage and construction time while maintaining structural performance and enabling faster assembly.

Solution 3: Design included the presence of large transfer structures (beams with a 20 m span) that transfer the load transmitted from the core of the building to the perimeter vertical structures of the auditorium hall (Figure 8.34). These deep transfer beams redistribute vertical loads horizontally, allowing open column-free spaces below while supporting the weight of the tower structure above the auditorium.

Table 8.13 summarizes the core sustainability strategies by mapping each method to its life-cycle contribution and resulting outcomes.

The Campus Pictet de Rochemont project demonstrated that it is possible to build large-scale, concrete-intensive structures with a significantly reduced environmental impact. The key to its success was a holistic approach that went far beyond conventional design. By integrating advanced material science, strategic prefabrication, and supply chain management, the project achieved a significant reduction in its total embodied carbon footprint.

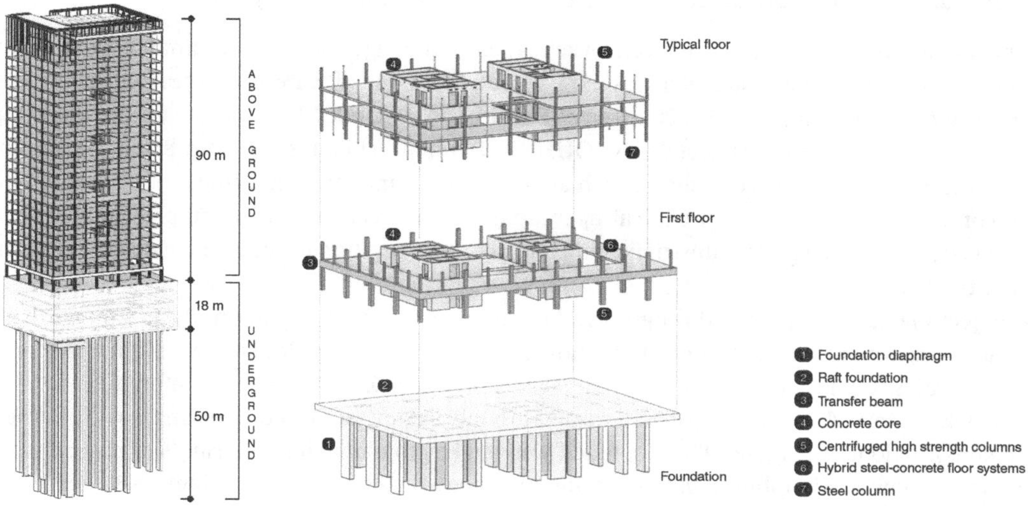

Figure 8.34 Concept of building showing the transfer structure above the foundation level (courtesy of Redesco Progetti).

Table 8.13 Sustainability strategy matrix of Campus Pictet de Rochemont project.

Strategy	Most relevant LC stages	Key method(s) used	Sustainability outcome
Optimized concrete technology	A1–A3 (Materials)	Use of recycled aggregates and GGBS	Lowered GWP of concrete mixes
		High-strength concrete	Reduced total concrete volume
Design for manufacturing and assembly (DfMA)	A5 (Construction)	Prefabricated centrifuged columns	Reduced on-site construction time and waste
		Hybrid steel–concrete floor systems	Improved quality and efficiency
Proactive carbon management and procurement	A1–A4 (Materials and Transport)	Selection of low-GWP suppliers	Minimized transport emissions
		Prioritization of local sourcing	Ensured lowest-impact material choices
			Achieved 30% overall carbon reduction compared to standard projects

8.3.2 BIM-driven Industrialized Construction for a Next-generation Hospital

While most large construction projects overrun schedules and budgets due to coordination complexity, supply chain issues, and late design changes, how do you build one of Europe's largest and most technologically advanced hospitals (a 250 000 m^2 city of health) on schedule? This was the critical challenge of the New Odense University Hospital (Nyt OUH) in Denmark (Figures 8.35 and 8.36). The facility also houses vibration-sensitive equipment, such as MRI devices and labs, requiring strict vibration and acoustic control with uninterrupted clinical operations. ATI Project turned this unprecedented scale and complexity into a strategy of industrialization and digital integration to succeed where conventional methods would fail.

This project masters complexity through prefabrication and digital integration. Rather than relying on traditional cast-in-situ construction, the team made a crucial early decision to adopt a prefabricated structural system. This choice was managed and executed through a comprehensive BIM workflow, which served as the project's digital backbone from design and coordination to fabrication and construction sequencing. This approach was the key to managing thousands of components, coordinating dozens of disciplines, and delivering this mega-project efficiently. Figure 8.37 shows the federated BIM model.

The sheer scale of the project required a system-based approach. The design was broken down into several interconnected subprojects, including a central "Knowledge Axis" and multiple treatment and inpatient clusters (Table 8.14). The decision to use prefabrication was made after a value-engineering phase that compared it directly with traditional construction, with speed being the deciding factor.

The design and construction of the Nyt OUH presented a convergence of three challenges that pushed the boundaries of project management.

Figure 8.35 The new Odense University Hospital (Nyt OUH), a building complex designed as an integrated center for treatment, research, and education (courtesy of ATI Project).

Figure 8.36 Construction site pictures (courtesy of ATI Project).

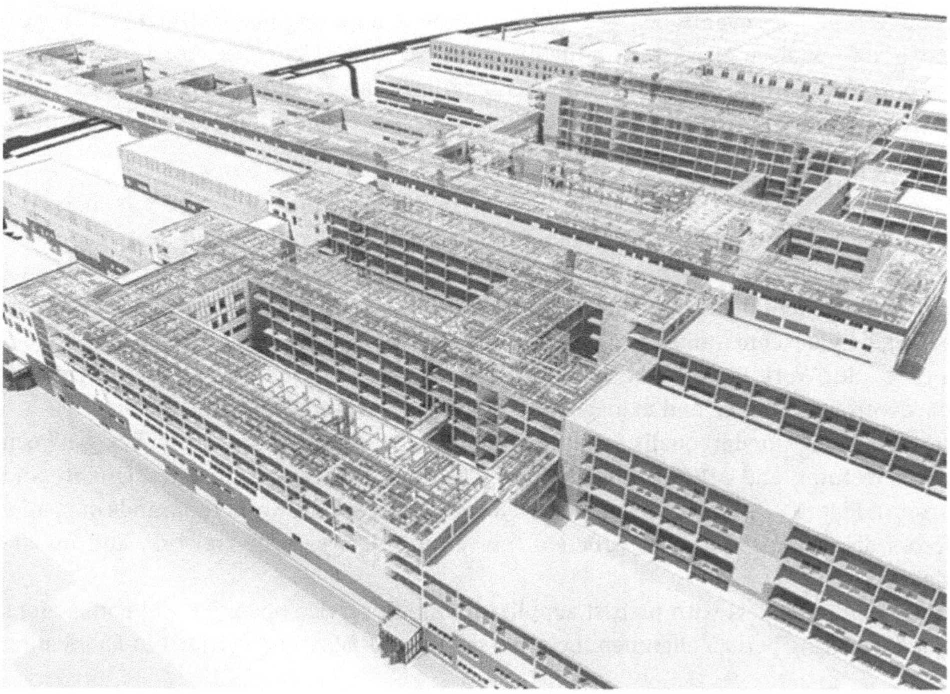

Figure 8.37 The federated building information modeling (BIM) model, which served as the single source of truth for coordinating all design disciplines and managing the project's immense complexity (courtesy of ATI Project).

Table 8.14 The Nyt OUH project by the numbers.

Component	Quantity
Total area	250 000 m^2
Inpatient rooms	590
Operating theaters	71
Intensive care units	121
Prefabricated steel–concrete columns	29 608
Prefabricated steel–concrete beams	54 964
Prefabricated facades	60 000 m^2

Challenge 1: With a footprint equivalent to a small town and a structure comprising nearly 30 000 columns and 55 000 beams, managing the design, fabrication, and assembly was a monumental logistical task. The project involved over 725 final Revit models, and coordinating this volume of information across dozens of disciplines was impossible with a traditional approach.

Challenge 2: The project timeline demanded a rapid construction speed that traditional methods could not achieve. Simultaneously, the design had to be "future-proof," allowing the hospital to adapt its departments and technologies over its 50+ year lifespan without requiring major structural changes.

Challenge 3: Modern hospitals are packed with highly specialized technology. The structure had to be designed to support the immense weight of MRI machines, linear accelerators, and cyclotrons, while also meeting stringent vibration criteria to ensure their proper function. In nuclear medicine areas, massive concrete walls up to 2 m thick were required for radiation shielding.

Solution 1: BIM workflow was the central "nervous system" of the entire project. Figure 8.38 shows the progressive detailing of the structural BIM model, from conceptual to construction-ready geometry, integrating supplier data directly into the coordinated model. Using a centralized database for spatial programming and equipment data, shared in real-time between the client and contractor, the team managed a federated model containing over 700 sub-models. This allowed for:

- Integrating schedule (Work Breakdown Structure) and cost (Bill of Quantities) data to simulate construction, control the budget, and manage logistics in real-time (4D/5D Simulation).
- Continuously checking model quality and identifying thousands of interferences between structural, architectural, and MEP systems before construction began (Automated Quality and Clash Detection). Figure 8.39 shows how automated clash detection resolved thousands of spatial conflicts across disciplines before construction, preventing delays, costly rework, and on-site coordination errors.
- Sharing precise digital models with precast suppliers, with all service openings and connections fully integrated, ensuring perfect alignment between design and fabrication (Direct-to-Fabrication Data).

Solution 2: The backbone of the project was a prefabricated structure primarily using steel–concrete composite beams, precast columns, and hollow-core slabs. Manufacturing these components off-site in a controlled factory environment ensured higher quality, reduced on-site labor, and dramatically

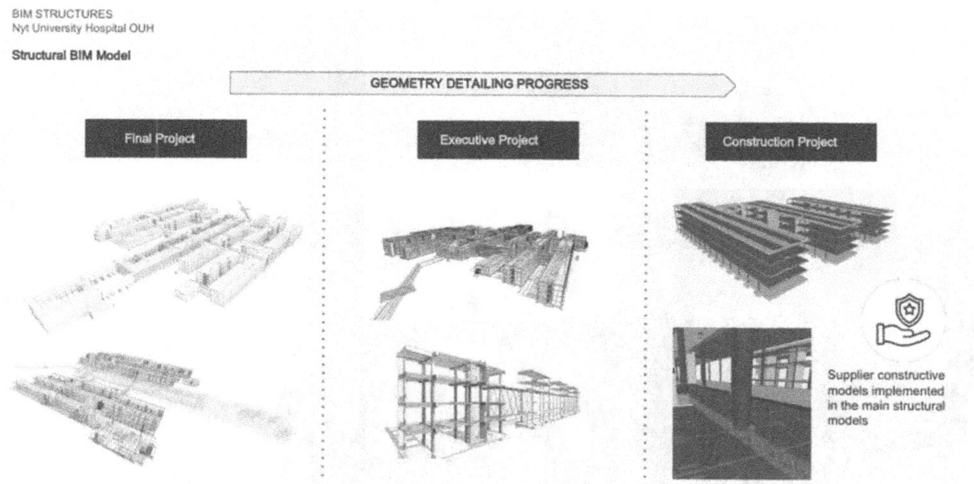

Figure 8.38 Structural BIM models (courtesy of ATI Project).

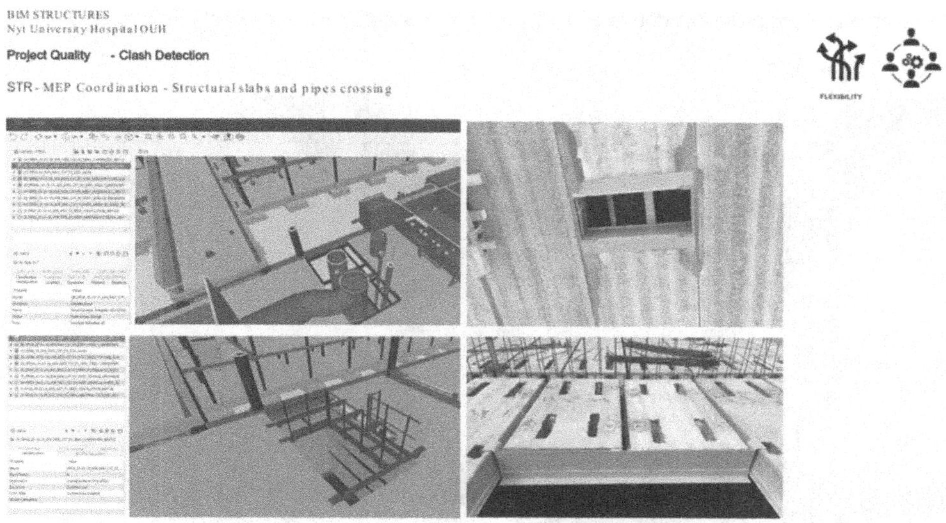

Figure 8.39 Clash detection in the BIM model allowed the team to resolve thousands of potential conflicts digitally before they could cause delays and costly rework on site (courtesy of ATI Project).

accelerated the construction schedule. The entire supply was split between two producers to mitigate risk and maintain pace. Figure 8.40 shows prefabricated steel–concrete composite beams stored and labeled on site, ready for just-in-time installation directly from off-site factories. Figure 8.41 shows the early assembly phase, where precast columns and beams are rapidly connected to form the building frame with minimal temporary supports. Figure 8.42 shows the progressing superstructure, demonstrating the repetitive assembly of composite floors and columns, which enabled fast, high-quality, and low-waste construction.

Figure 8.40 Prefabricated composite beams delivered and stored for installation (courtesy of ATI Project).

Figure 8.41 Assembly of precast columns and composite beams on site (courtesy of ATI project).

Figure 8.42 Assembly of composite floors and columns (courtesy of ATI Project).

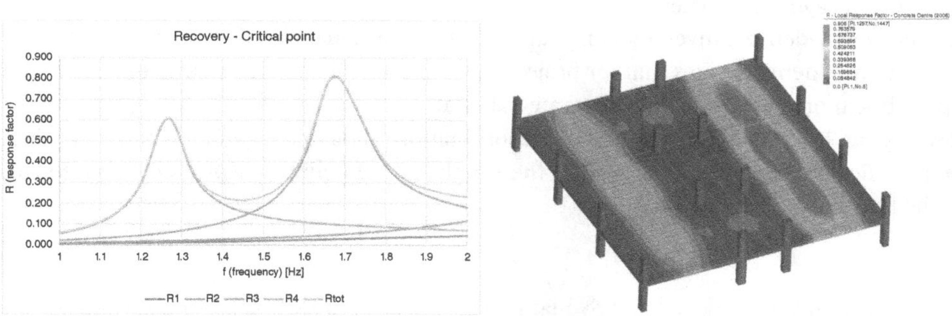

Figure 8.43 Verification results demonstrating that all structural checks satisfy $R < 1$, indicating compliant and safe performance (courtesy of ATI Project).

Solution 3: To meet the hospital's technical requirements, the team conducted detailed vibrational analyses to guarantee the stiffness of the floor systems. Figure 8.43 shows the verification results of the structural elements, where all checks meet the required safety conditions with utilization ratios below one, confirming full compliance of the design. For heavy equipment, detailed load maps were created to ensure the structure was reinforced precisely where needed. To ensure future flexibility, a custom ceiling hanger system was designed to be installed and removed entirely from below, allowing for easy changes without disrupting the floor above.

Table 8.15 Sustainability strategy matrix of Nyt OUH project.

Strategy	Most relevant LC stages	Key method(s) used	Sustainability and efficiency outcome
Design for speed and quality	A5 (Construction)	Prefabricated structural system (steel-composite and precast concrete)	Dramatically reduced on-site construction time Higher quality control from factory production
Digital integration and coordination	A1–A5	BIM workflow (4D/5D, clash detection)	Minimized on-site conflicts and material waste Speed up information flow between all stakeholders
Design for performance and flexibility	B (Use), D (End of Life)	Vibration analysis for sensitive equipment Adaptable hanger systems for future changes	Ensured high performance for critical medical functions Enhanced long-term usability and adaptability of the facility

Table 8.15 shows a sustainability strategy matrix for the Nyt OUH project that summarizes the three core lesson categories by mapping each targeted method to its specific life-cycle module and the resulting material- and carbon-saving outcomes.

In conclusion, the New Odense University Hospital project is a landmark achievement in managing large-scale construction. It demonstrates that for projects of immense complexity, a strategy combining industrialized prefabrication with a deeply integrated digital workflow is not just an option, but a necessity. By leveraging BIM as a central hub for coordination, data management, and planning, the team was able to de-risk the project, accelerate the timeline, and deliver a world-class healthcare facility built for the future.

Project Team

Owner: Region of Southern Denmark (Region Syddanmark)
General Contractor: OHPT (A Joint Venture of CMB and ITINERA)
Architect and Structural Engineer: ATI Project
Paper Credits: Eng. Michele Fascilla, Eng. Anna Continanza, Eng. Alessio Bacci

8.3.3 Embodied Carbon Reduction Through Cement Replacement and Innovation

Today, the focus of the building industry is expanding from operational efficiency to the critical issue of embodied carbon. For large-scale commercial high-rise projects, this shift presents a major hurdle since those structures require immense amounts of carbon-intensive steel and concrete. The One Boston Wharf Road project (Figure 8.44) demonstrates how both goals can be tackled together. This 17-story, 707 000-square-feet (65 684 m^2) tower is not only Boston's largest net-zero operational carbon building but also stands out for its efforts to cut embodied carbon related to the production stage (A1–A3) of concrete.

Figure 8.44 One Boston Wharf Road project in Boston (courtesy of WS Development).

The core concept of the project is an integrated, holistic, net-zero carbon design. From its inception, every decision was driven by the goal of minimizing energy demand and meeting the remaining needs with renewable sources, eliminating all on-site and off-site use of fossil fuels. This approach has resulted in a landmark building that achieved LEED Platinum certification and has set a new standard for performance, comfort, and environmental responsibility, reducing its operational carbon footprint by 100% and eliminating 5.1 million pounds of CO_2 emissions annually.

While achieving operational net-zero was a major milestone, the project also took steps to reduce its embodied carbon: the design team concentrated on the structural system (the primary contributor to embodied carbon in high-rise buildings) and addressed the following challenge:

Challenge: For a building of this scale, a conventional concrete and steel frame (Figure 8.45) carries a massive embodied carbon footprint. The team was tasked with finding solution that could reduce this upfront carbon load without compromising performance or budget.

Solution: An as-built evaluation of the embodied carbon in the concrete within the One Boston Wharf project confirms a 25% reduction in GWP when compared to the NRMCA 2022 Baseline. This was achieved by increasing the available Supplementary Cementitious Materials (SCMs) in the concrete mix. It should be noted that while SCMs (such as ground granulated blast furnace slag, fly ash) offer carbon savings by replacing clinker, their global availability is limited and unevenly distributed.

Figure 8.45 One Boston Wharf Road project in Boston, structural system view (courtesy of WS Development).

Nevertheless, using SCMs wherever feasible remains an effective strategy for lowering concrete's embodied carbon. Additionally, the project features the first commercial application of Sublime Systems' zero-fossil-fuel cement. "Sublime Cement" is a replacement for ordinary portland cement (OPC) that offers inherent advantages in durability, color, and consistency. This ASTM-compliant cement is created through a fully electrified manufacturing, which electrochemically extracts reactive calcium and silicates from abundant feedstocks, avoiding some part of the emissions, waste, and inherent impurities from limestone and combustion, and producing the cement.

This project is an example of a carbon reduction strategy focused on materials. Such an approach is valuable when there isn't significant room for further improvement or optimization of the structural system itself. While a holistic design approach is always ideal, the Sublime installation still provides a powerful example of how material technology is moving forward. It effectively demonstrates the industry's progress in trying to reduce the carbon impact of cement, one of the world's biggest pollutants, paving the way for a more sustainable future in construction.

Table 8.16 shows a sustainability strategy matrix for the project that summarizes the three core lesson categories by mapping each targeted method to its specific life-cycle module and the resulting material- and carbon-saving outcomes.

Table 8.16 Sustainability strategy matrix of One Boston Wharf Road project.

Strategy	Most relevant LC stages	Key method(s) used	Sustainability outcome
Embodied carbon reduction	A1–A3	Supplementary cementitious materials (SCM) A new type of low-carbon cement	Reduced upfront carbon footprint
Operational decarbonization	B6, B7	All-electric systems Energy recovery	Net zero operational carbon Elimination of on-site fossil fuels Enhanced energy security
Holistic systems integration	A1–A5, B6	Integrated design process focused on total carbon reduction	Synergy between systems Maximized efficiency and performance, achieved ambitious LEED and Net-Zero targets

Project Team

Owner/Developer: WS Development
Design Architect: Henning Larsen
Architect of Record: Gensler
MEP and Sustainability Engineer: Buro Happold
Structural Engineer: McNamara Salvia
Lighting: OneLux Studio

8.3.4 Reducing Embodied Carbon in a Mega-stadium Through Lightweight Tensile Structures on Reclaimed Industrial Land

Large-span sports venues are among the most material-intensive building types, with roofs that can span over 250 m and enclose tens of thousands of spectators. Conventional roof structure materials such as steel can result in significant quantities of embodied carbon as measured per square meter or per seat of the building. The Dalian Suoyuwan Football Stadium project (in Dalian, China) adopted a cable-tension roof system combined with a tensile fabric façade. Through this and other measures, the design team achieved a cost and carbon-efficient long span roof while providing the necessary performance requirements of cover and shade for this 63 000-seat venue on reclaimed industrial land at the city's waterfront (Figure 8.46). Completed in 2024, the stadium is now home to Chinese League One side Dalian Youngboy F.C. and serves as the centerpiece of the City Marina Park urban regeneration program.

The core concept of the project is "touching the ground lightly": by replacing a conventional heavy steel roof with a lightweight cable-tension structural system (Figures 8.47 and 8.48), and wrapping the entire building in tensile membrane materials rather than conventional cladding, the design minimized both the embodied and operational carbon of the superstructure and foundations. The near-circular geometry of the stadium (with plan axes of approximately 338 m by 294 m) naturally lent itself to a spoke-type wheel cable structure, which efficiently transfers loads through tension rather than bending and compression. The result is a roof that spans 253 m using a cable net only 19.5 m thick, covered by

Figure 8.46 The Dalian Suoyuwan Football Stadium on its waterfront peninsula site (*Source:* https://commons.wikimedia.org/wiki/File:Dalian_Suoyuwan_Football_Stadium.jpg).

approximately 50 000 m^2 of PTFE membrane formed from 252 individual membrane units, while the outer curtain wall comprises approximately 37 000 m^2 of single-layer ETFE film in over 20 shades of blue. The lightweight tensile approach delivered significant carbon savings compared to a traditional steel cantilever structure, while simultaneously reducing construction costs and enabling a swifter construction sequence on site.

The project also addressed a broader urban sustainability objective: the stadium is built on reclaimed industrial land on a peninsula west of the port, transforming a disused brownfield site into a community destination. The surrounding masterplan includes outdoor training pitches supporting youth academies, active travel routes, and public realm areas, making the entire development a model for brownfield regeneration through sport and public life.

Designing and constructing a 63 000-seat stadium with a lightweight cable-tension roof on a reclaimed coastal peninsula posed the following structural, logistical, and environmental challenges:

Challenge 1: The stadium required a roof covering the entire seating bowl (approximately 50 000 m^2 of canopy area) with a span of over 250 m, yet with minimal structural depth to preserve spectator sightlines and the architectural profile. Achieving this demanded:

- A structural system capable of spanning such distances without relying on intermediate supports that would obstruct views or clutter the bowl interior.

Figure 8.47 Interior view of the completed stadium, showing the 63 000-seat bowl and the lightweight cable-tension roof canopy with its central circular opening over the pitch (*Source:* https://commons.wikimedia.org/wiki/File:Dalian_Yingbo_vs_Beijing_Guoan_2025-05-05.jpg).

Figure 8.48 Close-up view of the cable net roof structure from below, showing the crossed upper-layer cables, radial lower-layer cables, and PTFE membrane panels spanning between them (*Source:* https://commons.wikimedia.org/wiki/File:Dalian_Yingbo_tifo_2025-05-05.jpg).

- Keeping the self-weight of the roof as low as possible, since the stadium sits on reclaimed industrial land where foundation capacity is limited and additional dead load would require costly ground improvement.

Challenge 2: The site on a coastal peninsula in northern China imposed severe environmental and geotechnical constraints:

- The construction program placed the most precision-sensitive roof erection activities during winter months, when Dalian experiences sub-zero temperatures and high coastal winds that affect material behavior, crane operations, and dimensional tolerances.
- The reclaimed land foundations had limited bearing capacity, requiring the entire superstructure (roof, façade, and seating) to be as lightweight as possible to avoid expensive deep foundation solutions.

Challenge 3: The architectural concept required a complex double-curved façade composed of three interlocking shell forms that appear to float and twist within each other, integrated with a 1.6 km double-helix walking ramp spiraling from the podium to the roof:

- The façade had to achieve a wave-like aesthetic while remaining structurally integrated with both the roof and the spiral walkway, creating a highly irregular geometry with thousands of uniquely shaped panels.
- The envelope needed to provide passive environmental performance (wind protection, winter warming, summer shading) without adding significant weight or requiring heavy secondary steelwork.

The following solutions were implemented to overcome the above-mentioned challenges:

Solution 1: A cable and tension structural system was adopted for the roof, exploiting the near-circular geometry to achieve long spans at minimal weight.

- The lightweight cable roof system carries loads primarily through tension rather than bending and compression, dramatically reducing the steel tonnage compared to a conventional roof structure. The complexity and repeatability of the structural forms allowed them to be delivered at low cost, low weight, and with low embodied carbon.
- The roof was covered with approximately 50 000 m^2 of PTFE membranes, formed as a succession of geometric waves across the entire roof surface. These translucent technical fabrics allow natural light to enter the stadium while ensuring optimal fire resistance. Their high mechanical resistance facilitated longer spans, thereby minimizing the need for secondary structures and further reducing the weight of the overall building.

Solution 2: The lightweight tensile approach reduced foundation loads to a level suitable for the reclaimed land, eliminating the need for costly deep foundation upgrades and allowing the construction program to proceed efficiently within the compressed schedule.

Solution 3: The façade was realized as an integrated tensile membrane envelope:

- The outer envelope adopted tensile fabric technology that could follow the complex double-curved geometry of the three interlocking shell forms while minimizing secondary structure.
- The wrapping effect of the tensile membrane façade and roof delivers the required passive environmental benefits: keeping the wind off spectators, delivering passive warming in winter, and providing shade in the summer months, all without heavy secondary steelwork.

Table 8.17 Sustainability strategy matrix of the Dalian Suoyuwan Football Stadium project.

Strategy	Most relevant LC stages	Key method(s) used	Sustainability outcome
Lightweight tensile roof structure	A1–A3, A4, A5	Spoke-type wheel cable net system with PTFE membrane covering (253 m span, 19.5 m depth)	Significant reduction in structural steel tonnage vs. conventional construction; reduced foundation demands on reclaimed land
Tensile membrane façade	A1–A3, A4, A5, B	Single-layer ETFE film on cable net (37 000 m^2, ~95% lighter than glass per unit area)	Major weight and embodied carbon reduction; passive environmental performance (wind shelter, thermal comfort)
Brownfield regeneration	A1, B	Stadium and community park built on reclaimed industrial peninsula	Avoided greenfield development; restored disused land to productive community use
Structural optioneering	A1–A5	Comparative analysis at early design stage for best economy and mechanical performance	Optimized material use
Material durability and low maintenance	B2, B3	PTFE membranes (25–35+ year life); self-cleaning ETFE surfaces	Extended service life; reduced maintenance and replacement embodied carbon

In conclusion, the Dalian Suoyuwan Stadium demonstrates that lightweight tensile structural systems can transform the sustainability profile of large-span sports venues, achieving major reductions in embodied carbon, structural weight, and construction cost compared to conventional steel roof construction. The project is also notable for its contribution to urban regeneration, converting disused industrial land into a major public destination. Table 8.17 shows a sustainability strategy matrix for the Dalian project that summarizes the core lesson categories by mapping each targeted method to its specific life-cycle module and the resulting sustainability outcomes.

Project Team

Owner: Dalian Municipal Government/Wanda Group
Architecture: BDP
Structural, Façade, MEP, and Sustainability Engineer: Buro Happold
Local Design Institute (Delivery): The Dalian Architectural Research and Design Institute
Collaborating Design Institute: The Harbin Institute of Technology, Architectural Design and Research Institute
General Contractor: China Construction Eighth Engineering Division Corp.
Membrane Supplier: Serge Ferrari Group (PTFE membranes)

8.4 Hybrid, Composite, and Modular Systems

8.4.1 Vertical Expansion with Mass Timber and Off-site Fabrication

As cities grow denser, the challenge of sustainably expanding existing building stock becomes critical. In this context, the common approach of demolishing the old and rebuilding new building carries an immense cost and environmental impact. As an alternative approach, the 80 M Street project in Washington, D.C. (Figure 8.49) transformed an existing seven-story reinforced concrete office building by adding a three-story vertical extension made entirely of mass timber. Completed in summer 2022, the project demonstrates how to unlock new potential in existing structures.

The core concept of the project is "vertical adaptive reuse" through mass timber. The project added 9700 m^2 of new office space and nearly 370 m^2 of outdoor terrace space without overloading the original structure. The project team carried out early-stage comparisons and confirmed that the selected option would reduce embodied carbon of the structure by approximately 40% compared to alternative solutions, thanks to the use of wood (obtained from sustainably managed forests) and reduced on-site construction schedule.

The project not only delivered a high-performance office space but also gained a significant market advantage, attracting tenants with its unique aesthetic and wellness-focused environment, leading to higher lease rates and a faster lease-up. Figure 8.50 highlights the timber structure as a design driver. The exposed mass-timber column reads clearly as a primary vertical element and support biophilic comfort.

Figure 8.49 80 M Street project in Washington, D.C. (photo by RON BLUNT).

Figure 8.50 View from inside: 10th floor penthouse amenity (photo by RON BLUNT).

Adding three new floors to an existing and already occupied building in a dense urban environment presented a unique set of challenges:

Challenge 1: How to add three floors of office space onto a building not originally designed to carry such significant additional vertical load? A conventional concrete or steel addition would have required extensive (and likely unfeasible) reinforcement of the existing columns and foundations.

Challenge 2: In the competitive Washington, D.C. Trophy office market, there is increasing tenant demand for sustainable, wellness-focused, and unique workspaces. A standard renovation would not be enough to differentiate the property and command premium rents.

Challenge 3: The project site and occupied building (all mass timber vertical expansion construction had to take place during normal business hours while the building was occupied by existing tenants) located in a busy neighborhood, demanded a construction methodology that minimized on-site disruption, noise, traffic, and overall construction time.

The following solutions were implemented to overcome these challenges:

Solution 1: The system used Glue-Laminated (Glulam) timber for the columns and beams and CLT panels for the floors and roof, creating a robust and lightweight superstructure (Figure 8.51):

- The full timber assembly resulted about half of what it would have been out of reinforced concrete frame.

Figure 8.51 View from construction site showing installed GLT columns and beams, and CLT floors (courtesy of Columbia Property Trust).

- The timber frame could be added without overloading the existing building's foundation and requiring invasive reinforcing.

Solution 2: The use of exposed wood CLT throughout the new floors (Figure 8.52) not only reduced embodied carbon by 40% compared to alternative solutions but also introduced a biophilic design element, bringing the qualities of natural materials into the workplace and distinguishing it from typical office interiors.

The new mass timber structure also enabled superior architectural features that enhance the tenant experience (Figure 8.53):

- The strength and relatively light weight of the CLT panels allowed for large spans, creating more open and flexible floors: the glulam beams span 30 ft (9 m) between columns; the CLT panels span 10 ft (3 m) between beams.
- This structural efficiency made it possible to achieve 15 ft (4.6 m) ceiling heights, greater than the D.C. standard.

Solution 3: The Glulam and CLT components were precision-manufactured off-site to exact specifications. This process of prefabrication drastically reduced the time, labor, and waste generated on-site. The components were delivered and erected in a sequence that was faster, quieter, and less disruptive than a conventional build. Figure 8.54 shows prefabricated GLT columns on site with factory-installed plate steel connectors ready for rapid bolting to beams and bases. Figure 8.55 shows the same columns erected and temporarily braced, where the concealed steel parts provide the required stiffness and strength at the frame joints.

The steel-braced frame shown in Figure 8.55 was used in back-of-house areas where the structure would not be exposed and major service chases were located. These frames also served as the vertical lateral resisting system.

Figure 8.52 Installation of CLT floors (courtesy of Columbia Property Trust).

Figure 8.53 View from inside: 9th floor vacant space with long spans and 4.6 m high ceilings (photo: @RON BLUNT).

Figure 8.54 Prefabricated GLT columns ready for installation (courtesy of Columbia Property Trust).

Figure 8.55 Prefabricated GLT columns installed (courtesy of Columbia Property Trust).

Figure 8.56 shows factory-made steel connectors for GLT beam–column joints, including plates and brackets with predrilled holes sized to transfer shear and bending moment while remaining largely concealed after installation (these joints enable quick dry assembly and clear load paths to the CLT diaphragm). Figure 8.57 shows the installation sequence for the prefabricated GLT beams: a crane lifts each wrapped beam into position, crews seat it on the column's plate assembly, align the parts, then bolt and screw the joint to achieve quick, accurate erection.

Besides embodied carbon benefits, the project also showed a commitment to sustainable operations and a superior occupant experience. The tall ceilings allowed for expansive floor-to-ceiling windows,

Figure 8.56 Steel connectors for GLT beam–column joints (courtesy of Columbia Property Trust).

Figure 8.57 Prefabricated GLT beams during installation on site (courtesy of Columbia Property Trust).

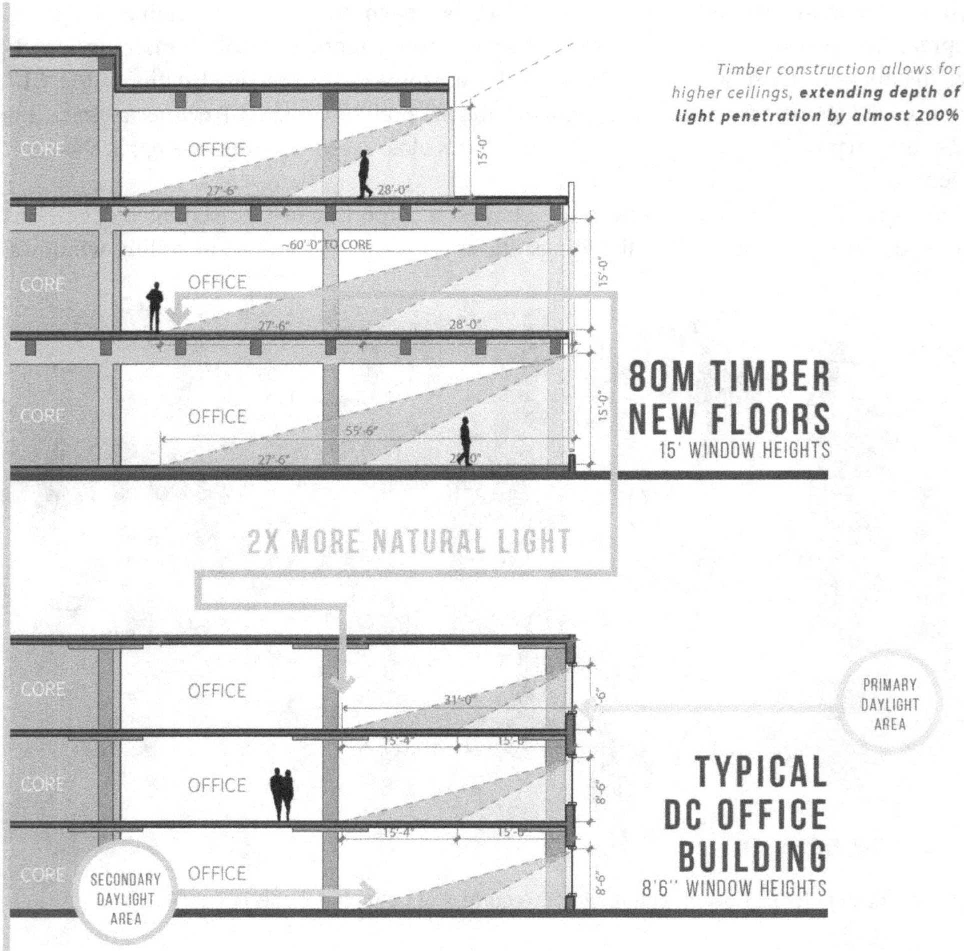

Figure 8.58 Comparison of daylight penetration between typical DC office floors and 80 M timber new floors (courtesy of Hickok Cole Architects and Columbia Property Trust).

which dramatically increase the penetration of natural light deep into the office space, reducing the need for artificial lighting and enhancing employee well-being. Figure 8.58 shows how the selected construction technique nearly doubled the daylight penetration compared to a typical office building.

In conclusion, the 80 M Street project showed the transformative potential of mass timber in urban development. While mass timber can add complexity through moisture control, fire protection, and insurance requirements, which could normally raise upfront costs, this project successfully overcame those obstacles and delivered a beautiful, sustainable, and highly valuable commercial asset, with pricing nearly equal to a steel alternative.

Table 8.18 shows a sustainability strategy matrix for the project that summarizes the four core lesson categories by mapping each targeted method to its specific life-cycle module and the resulting material- and carbon-saving outcomes.

Table 8.18 Sustainability strategy matrix of the 80 M Street project.

Strategy	Most relevant LC stages	Key method(s) used	Sustainability outcome
Preserve and expand	A1–A5	Retain existing 7-story structure and its foundation	Avoided emissions from demolition and new construction
Mass timber from sustainably managed forests	A1–A3, A5, C	Mass timber (Glulam and CLT) superstructure Wood obtained from sustainably managed forest	40% ↓ embodied carbon vs. alternative solutions
Design for real estate market needs	B1–B6	Exposed natural wood Enhanced natural daylighting	Improved tenant well-being and productivity Reduced need for artificial lighting Created unique market identity Increased asset value and lease rates, longer leases
Off-site manufacturing	A4, A5	Prefabrication of timber components	Reduced on-site construction schedule, Minimized neighborhood disruption, Less construction waste

Project Team

Owner/Developer: Columbia Property Trust
Architect: Hickok Cole Architects
Structural Engineer: Arup

8.4.2 Architectural Ambition Achieved with a Steel–Concrete Composite Frame

As the traditional office model becomes outdated, society demands a new generation of workplaces that are both architecturally and functionally compelling, besides meeting stringent environmental requirements through resource efficiency. Delivering such buildings requires a life-cycle-driven design approach because environmental performance, structural efficiency, and long-term adaptability must be considered from the earliest design stages to minimize embodied carbon, extend service life, and reduce future retrofit needs. The 80 Fenchurch Street project (Figure 8.59) presents a successful example. Opened in 2020, it is a 15-story, 250 000 ft^2 (23 225 m^2) office building in London, offering a BREEAM-excellent rated, contemporary workspace.

The team's primary goal was to enable the building's striking architectural features (Figure 8.60) using the leanest possible steel and concrete tonnage. This was accomplished through an early-stage, research-led approach to structural design and a deep understanding of the design team for low-carbon steel-framed buildings. Both client and architect were also committed to a structural solution that offered optimal environmental performance and best value, without compromising the unique architectural vision.

Figure 8.59 The 80 Fenchurch Street project (courtesy of Pell Frischmann).

(a)

(b)

Figure 8.60 Project's architectural features: (a) centrally-located atrium which spirals up the building's floors; (b) double-height glass box, surrounded by six landscaped terraces which are elevated at different heights (courtesy of Pell Frischmann).

Delivering a landmark building with market-leading sustainability credentials presented the following challenges:

Challenge 1: The project's ambitious architectural vision (featuring a central spiraling atrium, a double-height glass box, and six staggered terraces) required structural solutions that were robust yet able to create large, flexible, and open-plan office spaces.

Challenge 2: The building's structural system was to be designed to cope with the high environmental ambitions of the scheme. The structural design team have taken the LETI 2025 upfront embodied carbon target (LETI, n.d.), as a benchmark to support decision making, which required meeting strict embodied carbon limits, alongside delivering the project ahead of schedule and under budget, despite the highly volatile steel market at the time.

The following solutions were implemented to overcome the above mentioned challenges:

Solution 1: A steel–concrete composite steel frame (Figure 8.61) has been designed with a minimal number of internal columns. This solution successfully supported the complex atrium and terrace

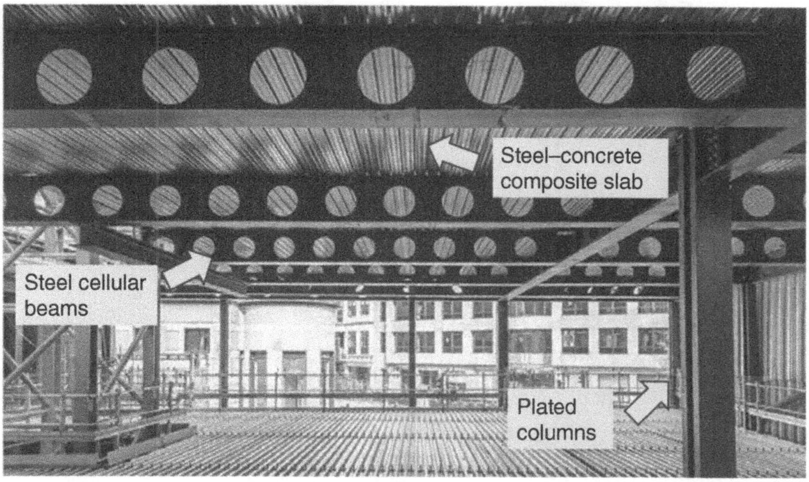

Figure 8.61 From the construction site of the project (courtesy of Pell Frischmann).

while delivering a wide 15 m × 12 m structural grid, creating the flexible and large spaces of the architectural brief.

Solution 2: The sustainability and commercial objectives were achieved through a rigorously lean structural design composed of the following outcomes:

- **i.** Substantial savings in material quantities were achieved by designing structural elements for utilization ratios of over 95%.
- **ii.** The lean superstructure design directly reduced the loads on the foundations, which in turn reduced the size of the substructure and its associated embodied carbon.
- **iii.** The efficient design enabled the superstructure to be erected ahead of schedule and significantly under the cost plan. Pell Frischmann completed 70% of the steelwork design ahead of schedule, enabling the client to procure it in advance at a favorable cost, given the steel market fluctuations at the time. This initiative also enabled the superstructure to begin construction before the basement was fully completed.
- **iv.** Plated structural elements were favored to enable a lighter steel frame, a critical decision to minimize commercial risk given the highly volatile steel market at the time.
- **v.** Cellular plated steel beams helped create a lighter structure with large web openings that also allow for easy integration of building services within the floor depth, thus maximizing ceiling heights without adding to the overall height of the building.
- **vi.** Plated S460 (a higher-strength grade than those typically used in standard steel construction) columns provided high strength with a slender profile, reducing the sizes at lower levels and enabling an expansive open plan.

The superstructure's low-carbon design was driven by a strategic material palette shown in Figure 8.62. Steel formed the largest component, primarily in the form of plated sections, representing around 65% of the superstructure upfront embodied carbon (plated sections are generally considered to be more carbon-intensive than standard hot-rolled members because their production involves multiple flat plates; typically manufactured from Basic Oxygen Furnace routes; that are cut, welded, and

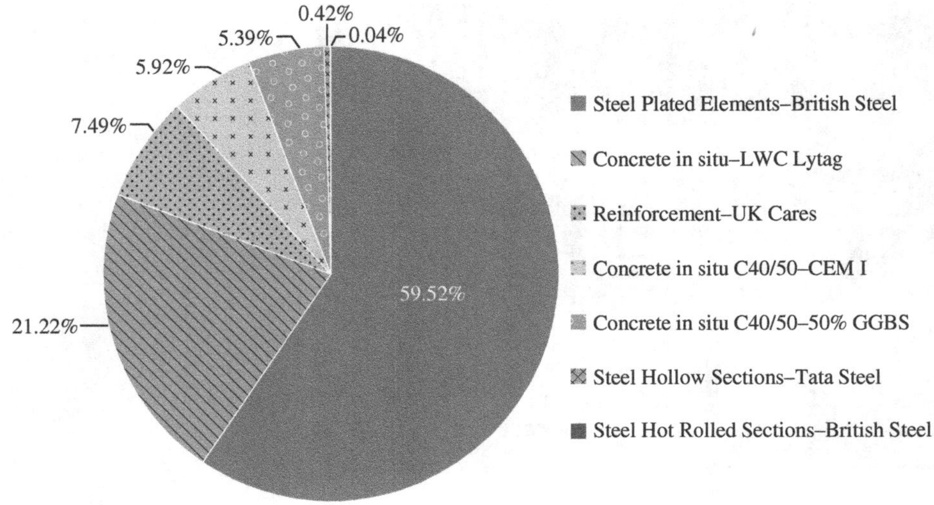

Figure 8.62 Upfront embodied carbon breakdown of the superstructure (reproduced with permission of Pell Frischmann).

assembled into built-up profiles, whereas most hot-rolled sections can be produced directly through Electric Arc Furnace processes with higher recycled content and lower energy demand). As discussed in Chapters 2 and 3, the manufacturing route and processing intensity significantly influence a steel product's embodied carbon. However, project-specific and supply-chain constraints dictated that plated sections would better serve the scheme by offering significant steelwork tonnage savings. Lightweight aggregate concrete formed the second largest component (21.22%), reducing the self-weight of the floors and minimizing foundation and frame loads. Furthermore, the concrete specification for other elements favored a 50% GGBS cement-replacement mix over pure cement to cut emissions. While the long-term sustainability of these materials is a subject of industry debate (the carbon benefit of recycled steel is linked to grid decarbonization, and global GGBS supply is limited), the decision to maximize their use remains the right choice, delivering immediate and substantial carbon reductions at project level while encouraging the development of scalable alternatives to GGBS by the supply chain due to increased demand. The superstructure's upfront embodied carbon intensity was assessed at 373 $kgCO_2e/m_2$, which is higher than the initial LETI 2025 benchmark target but responded well to the project's challenges and ambition. The LETI 2020 benchmark target for the superstructure was achieved, delivering a 25% reduction compared to the baseline.

Table 8.19 summarizes the core sustainability strategies by mapping each method to its life-cycle contribution and resulting outcomes. In conclusion, the 80 Fenchurch Street project demonstrates that ambitious architectural expression and exceptional low-carbon performance can be achieved together. Through an early-stage, research-led design process and a commitment to material efficiency, the team delivered a structurally lean and commercially successful building that met the client's ambitious demands without compromising its unique design. This case study also proves that today's workplaces can be simultaneously beautiful, sustainable, and economically viable.

Table 8.19 Sustainability strategy matrix of the project.

Strategy	Most relevant LC stages	Key method(s) used	Sustainability outcome
Research-driven and early-stage optioneering	A1–A5	Research into low-carbon steel solutions Close collaboration between client, architect, and engineer from project outset LETI 2025 embodied-carbon target as a benchmark to support decision-making	Best value combining cost, embodied carbon, and speed of construction LETI 2020 embodied carbon target for the superstructure achieved Superstructure upfront embodied carbon intensity of 373 $kgCO_2e/m^2$ ≈ 25% GWP reduction vs. baseline
Structural frame with high material utilization	A1–A3, A4–A5	Composite steel–concrete frame construction Design of elements for >95% utilization ratio	Superstructure erected ahead of schedule and under cost plan Minimized steel and concrete tonnage and embodied carbon Plated section to achieve a lightweight frame to cope with supply chain constraints
Long-span floors	A5, B1	15 m × 12 m structural grid Composite steel frame with fewer internal columns	Flexible, open-plan workspace Enabled complex architectural features (atrium, terraces)
Substructure and foundation optimization	A4–A5	Lean superstructure design reduced overall building weight	Reduced size of foundations Lowered substructure embodied carbon

Project Team

Client: YardNine
Partners: TP Bennett, Skanska
Civil Engineering|Digital Engineering and BIM|Geotechnical Engineering|Structural Engineering: Pell Frischmann

8.4.3 Modular Mass-timber System with Reinforced Concrete Cores

Traditional multi-story construction is often a slow process, causing on-site waste, unexpected site delays, and complex logistics. The Heimdalsporten project in Trondheim, Norway (completed in 2020), instead of building from the ground up in the conventional way, employed a hybrid modular system (Figure 8.63). Commissioned by Heimdals Porten, this project consists of four towers, with two reaching seven stories and the other two eight stories. The advancement of volumetric modular construction for multi-story buildings increasingly relies on hybrid systems that optimize both structural efficiency and the speed of on-site assembly. This project serves as a significant case study in this methodology.

The design utilizes reinforced concrete cores to provide lateral stability for the four towers, which reach up to eight stories (Figure 8.64). Prefabricated timber-based modular units were then assembled around these cores. The modular system enabled rapid construction, with the first 7-story tower completed in 10 days and the second assembled in 7 days.

Figure 8.63 The Heimdalsporten project in Trondheim, Norway (courtesy of ergodomus).

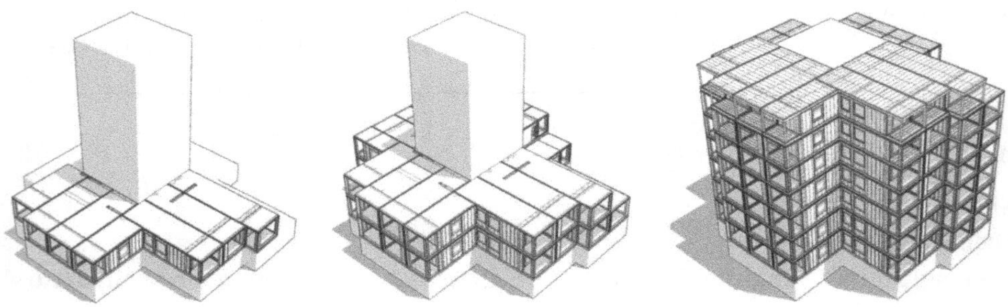

Figure 8.64 Sequential models showing the hybrid modular construction of the Heimdalsporten project: the reinforced concrete core (left), partial assembly of prefabricated timber modules around the core (center), and the completed structure (right) (courtesy of ergodomus).

The project applied DfMA principles through tailor-made connections to optimize on-site assembly and reduce errors. The project is notable not only as one of the tallest modular developments in Norway but for its extensive level of preassembly, where modules arrived on-site with nearly all internal finishes and systems installed. This method drastically reduces on-site construction time and showcases a highly integrated approach to modular design and execution.

The Heimdalsporten project encountered several challenges across structural, fabrication, logistics, and construction aspects, for which innovative solutions were implemented:

Challenge 1: Predicting and preventing structural failures caused by timber's natural shrinkage and expansion due to moisture changes, which could create unforeseen internal tensions and compromise connections.

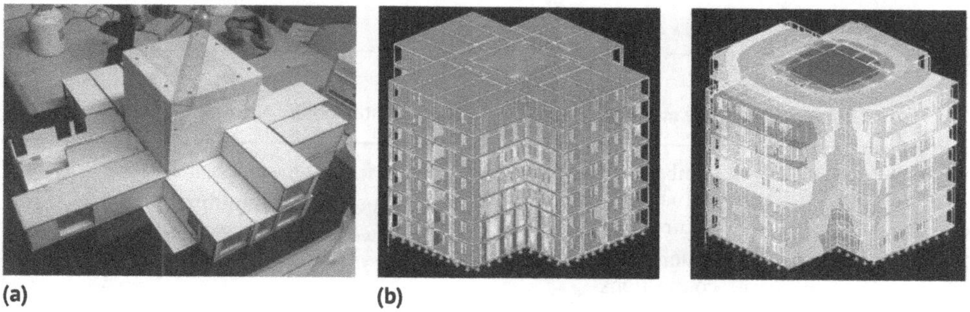

Figure 8.65 (a) Mock-up and (b) finite element models (FEMs) of the building (courtesy of ergodomus).

Challenge 2: Ensuring the stability of tall modular towers against wind loads in a way that accounts for the building's unique geometry and modular construction system.

Challenge 3: Perfecting the complex assembly sequence and specialized connections before arriving on-site to minimize errors and construction delays in cold Norwegian conditions.

Challenge 4: Coordinating the transportation and assembly sequence of 110–120 modules by ship from Poland to Trondheim to ensure efficient unloading and construction flow.

The project team implemented the following solutions to systematically address these challenges:

Solution 1: A Finite Element Model (FEM) was also used to predict real-world behavior and changes, enabling the calculation of correct dimensions and optimal design of elements and connections, preventing breakage from unforeseen internal tension (due to timber shrinkage and expansion due to moisture changes) (Figure 8.65b).

Solution 2: A virtual wind tunnel simulation using Computational Fluid Dynamics (CFD) analysis was performed to analyze wind behavior and guarantee optimal structural stability.

Solution 3: To improve on-site work, a 1 : 1 scale model of the special connections was studied in Poland. Furthermore, a timber scale model of the shaft and modules was created to analyze the exact assembly/fixing sequence in the office, allowing for improvements to the on-site process (Figure 8.65a).

Solution 4: The loading process was carefully planned so that the first module loaded onto the ship was the last to be assembled, streamlining the unloading and assembly sequence upon arrival and optimizing the construction timeline.

In summary, by employing FEM and CFD analyses, full-scale connection prototypes, timber scale models, and sequenced logistics, the Heimdalsporten project overcame its structural and climatic challenges. These advanced simulations and transport and assembly planning validated timber modular construction's overall efficiency, stability, and adaptability for rapid, high-rise residential development in harsh environments.

Table 8.20 shows a sustainability strategy matrix for this project that summarizes the four core lesson categories by mapping each targeted method to its specific life-cycle module and the resulting material- and carbon-saving outcomes.

Table 8.20 Sustainability strategy matrix.

Strategy	Most relevant LC stages	Key method(s) used	Sustainability outcome
Off-site prefabrication and DfMA	A1–A3, A5	Timber modules delivered virtually complete (MEP, fixtures, finishes) Tailor-made, error-proof connections	30% faster tower assembly On-site waste minimized Lower embodied carbon
Precision simulation and prototyping	A5, D	FEM for shrinkage/expansion and connection design CFD wind-tunnel checks Full-scale and scale mock-ups	Rework nearly eliminated Enhanced structural safety Service-life credit
Sequenced logistics and transport	A4, A5	Ship-load planning so last-on first-off modules align with site sequence	Handling time cut sharply Transport emissions reduced Streamlined site flow
Rapid assembly in harsh climate	A5	First 7-story tower in 10 days; second 30% faster Limited on-site work in winter conditions	Schedule shortened dramatically Quality control under factory conditions Improved worker safety

8.4.4 Modular Steel Construction for a Complex Geometry

Amager Bakke (literally "Hill of Amager") (Figure 8.66), completed in 2017 in Copenhagen, is a waste incineration plant processing up to 400 000 tons of waste annually. It provides district heating to approximately 160 000 households and electricity to 60 000 homes. A common way to build such industrial structures would be to locate the plant as a self-contained industrial facility (fenced off and hidden from view) an approach that misses the chance to enrich the urban fabric. In this case, instead of hiding its industrial purpose, the design team placed an artificial ski slope atop its sloped roof and mounted an 85-m climbing wall on its façade (Amager Bakke, 2025; From waste to energy, n.d.).

The structural system is characterized by a highly irregular geometry: both the roof and floor plan curve continuously with virtually no right angles. This generated a high volume of geometrically unique components with spatial complexity and tight fabrication tolerances. Planning and production were challenged by constant dimensional variation across all axes, increasing risk in both manufacturing and site assembly (Figure 8.67).

Key structural and logistical challenges included:

Challenge 1: Achieving continuous three-dimensional curvature across both roof and floor planes without an orthogonal grid required exceptional precision in design coordination, as each steel component had a unique geometry and position within the global model.

Challenge 2: The combination of complex geometry and the large number of bespoke elements made the control of fabrication and assembly tolerances extremely demanding. Even small deviations could

Figure 8.66 Amager Bakke with ski slope and climbing wall (courtesy of Gregull).

Figure 8.67 Preassembled steel components (courtesy of Gregull + Spang).

propagate through the structure, risking misalignments during installation and requiring rigorous dimensional management throughout production.

Challenge 3: The erection of curved and heavy structural elements at considerable height posed major logistical and safety challenges. The limited space and high elevation reduced the possibility of on-site corrections, demanding careful sequencing, accurate surveying, and precise lifting operations to ensure geometric continuity and stability during assembly. The following solutions were implemented:

Solution 1: To realize the complex geometry, a significant and varied inventory of steel components has been prepared, totaling approximately 6400 tons. The primary steel grade used was S355 for plates and larger profiles, with S235 employed for smaller sections. The fabrication process for these elements spanned 15 months, with the on-site installation also requiring a 15-month period. The specific breakdown of steel components, which highlights the material diversity needed for the structure, is detailed in Table 8.21.

Solution 2: The largest possible structural subsegments were preassembled off-site and lifted into position as complete units. As shown in Figure 8.68, the steel members were erected using prealigned modules, allowing accurate fit-up and quick stabilization once lifted into place. Figure 8.69 shows one of these volumetric modules assembled with integrated envelope panels and internal framing before

Table 8.21 Breakdown of steel elements used in the project.

Component type	Specification/size range	Total tonnage (approx.)
Plates	Thickness: 10–90 mm	4000 t
Hot-rolled profiles	IPE 200 to HEM 700	1400 t
Circular hollow sections (CHS)	CHS 33.7 × 3.2 mm to CHS 610 × 20 mm	700 t
Rectangular hollow sections (RHS)	RHS 120 × 60 × 5 mm to RHS 500 × 300 × 16 mm	300 t

Figure 8.68 Erection of the steel structure (courtesy of Gregull + Spang).

Figure 8.69 Preassembled building segment (courtesy of Gregull + Spang).

transportation to site. This modular strategy optimized logistics, ensured consistent quality through controlled factory conditions, and significantly reduced installation time on site while maintaining the geometric accuracy required by the project's complex architecture.

Solution 3: All on-site connections were executed with bolted joints, allowing for field tolerance and enabling full future disassembly without damage to components.

While formal embodied carbon assessment data has not been published, the project incorporates circular design principles by enabling full disassembly of the primary structure (e.g. the use of bolted joints allows for future reuse of steel components) (Figure 8.70).

In conclusion, the Amager Bakke project demonstrates how off-site prefabrication and reversible connection strategies can be used to realize highly complex geometries at industrial scale. The structure's bolted assembly method enables future adaptability and disassembly, aligning with circular construction principles. Despite the structural complexity, buildability was enhanced through targeted preassembly and geometric coordination, offering a replicable model for integrating industrial infrastructure with long-term material retention. Table 8.22 shows a sustainability strategy matrix for the Amager Bakke project that summarizes the three core lesson categories by mapping each targeted method to its specific life-cycle module and the resulting material- and carbon-saving outcomes.

Figure 8.70 Preassembled modular units on-site (courtesy of Gregull + Spang).

Table 8.22 Sustainability strategy matrix of Amager Bakke project.

Strategy	Most relevant LC stages	Key method(s) used	Sustainability outcome
Modular Off-site prefabrication	A5	Preassembly of largest possible steel segments Lifting complete units into position	Improved buildability at height Reduced on-site risk Minimized cumulative geometric error
Reversible bolted assembly	A5, C, D	All connections designed with bolted joints No welding at site	Structure can be disassembled without damage Components retain reuse or recycling value
Circularity	D	Future reuse potential of structural steel via disassembly	Long-term material retention Alignment with circular economy strategies despite no formal LCA published

Project Team

Owner: ARC Amager Resource Center, Vindmøllevej 6, 2300 København S
General Contractor: Züblin Stahlbau GmbH, Bahnhofstraße 13, 01996 Hosena
Architect: BIG – Bjarke Ingels Group

Structural Engineer

Basic Engineering: MOE – Rambøll
Detail Engineering: Construction Stages and Assembly Processes + Connection Design: Gregull and Spang GmbH
Steel Team Fabricator and Erector: Züblin Stahlbau GmbH, Bahnhofstraße 13, 01996 Hosena
Steel Detailer: Gregull and Spang GmbH

8.4.5 Mass-timber Landmark Assisted with Concrete and Steel

To have a meaningful impact on the urgent climate emergency, building design must not only innovate but also aspire to change the culture and expectations of all those connected to it. Limberlost Place, a 10-story addition to George Brown Polytechnic's Toronto campus (Figure 8.71), embodies this philosophy. Designed by Moriyama Teshima Architects in joint venture with Acton Ostry Architects, the 225 000 ft^2 (20 903 m^2) building represents an ecological innovation. Completed in August 2025, it sets a new precedent

Figure 8.71 Limberlost Place, a 10-story addition to George Brown Polytechnic's Toronto campus (photo by Double Space Photography).

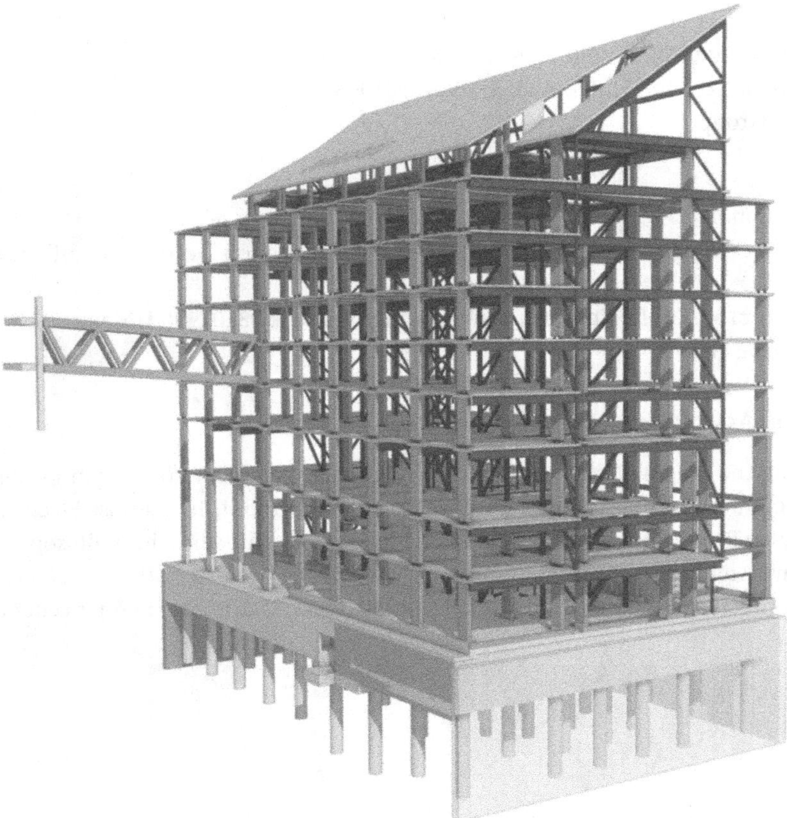

Figure 8.72 A view from inside the building showing GLT beams, steel beams, and columns, and CLT slabs (courtesy of Moriyama Teshima Architects).

for mass timber construction in North America by achieving Toronto's stringent 2030 emissions targets years in advance, demonstrating a replicable model for a new generation of high-performance buildings.

The core concept of Limberlost Place is the integration of a "Lean Design Process" with a hybrid structural system (Figure 8.72) made of mass timber columns and beams, steel bracings, and composite CLT-reinforced concrete slabs to eliminate waste and maximize performance across the building's entire life cycle (the reinforced concrete in the project is limited to the caissons, raft slab, and foundation walls, along with a 100 mm concrete topping acting in composite action with the slab bands). The building's form and "3-bar" plan places dark programs and services in a central core to maximize daylight for perimeter classrooms, which is a direct response to this (Figure 8.73). By involving specialists from different disciplines early and committing to prefabrication and just-in-time delivery, the project optimizes quality, accelerates construction, and drastically reduces on-site waste, providing the framework needed to successfully execute the building's ambitious ecological goals.

Achieving this vision required overcoming significant structural, regulatory, and logistical challenges:

Challenge 1: The design required 9-m column-free spans for flexible classroom design but was constrained by a strict 38-m street wall height limit, making conventional deep-beam mass timber systems unfeasible.

Figure 8.73 Cross-section of the building (courtesy of Moriyama Teshima Architects).

Challenge 2: Enclosing a 10-story timber structure quickly, while meeting ambitious Passive House energy targets, posed a significant logistical and performance challenge.

Challenge 3: The goal of net-zero operational carbon required a design that could function passively for much of the year, reducing reliance on mechanical systems.

Challenge 4: The 10-story structure required a robust lateral stability system. While all-timber or reinforced concrete core options were explored, they presented challenges with spatial efficiency, structural performance, and integration with the extensive mechanical services required in the central core.

Challenge 5: Long-span mass timber floors can be susceptible to unacceptable vibrations due to their lower density compared to reinforced concrete slabs that could impact occupant comfort.

Challenge 6: As the first 10-story mass timber building in Ontario, Limberlost Place faced a complex approval process with no established precedent.

Solution 1: The structural team engineered an innovative, shallow floor system comprised of GLT columns and timber–concrete composite (TCC) "slab bands" spanning 9.2 m (Figure 8.74) with standard CLT panels infilling between them. This design maximized headroom for mechanical systems while meeting the span requirements. The system's viability was confirmed through an extensive testing program. For further details, see Shahnewaz et al. (2022), which optimized the connection details between the concrete and timber (Figure 8.75).

Solution 2: The team developed a vertically oriented prefabricated paneling system (Figure 8.76), with panels spanning two stories. This approach allowed for just-in-time delivery and rapid installation by a small crew, minimizing neighborhood disruption and enhancing site safety. Manufacturing the panels in a controlled factory setting ensured superior quality control and insulation (R-30), creating a highly sealed envelope critical for the building's energy performance (Figure 8.77).

Figure 8.74 A view showing long-span floors (courtesy of Salina Kassam).

Figure 8.75 Lab tests to measure structural performance of the CLT-reinforced concrete floor system (courtesy of Moriyama Teshima Architects).

Figure 8.76 Building facade being installed (courtesy of Salina Kassam).

Figure 8.77 CLT slabs being installed on the steel frame (courtesy of Salina Kassam).

Figure 8.78 Hybrid steel and mass timber roof system (courtesy of Salina Kassam).

Solution 3: The angled roof slopes (Figure 8.78) to provide an optimal framework for solar PV arrays and stretches north for better daylighting. The east and west façades incorporate large solar chimneys that drive a natural ventilation system, drawing fresh air from operable windows, through classrooms and corridors, and exhausting it out the roof, all without fans. This system is complemented by "breathing rooms" (multi-story social spaces at the perimeter that animate the building and enhance the occupant experience).

Solution 4: The team selected a central core of concentric steel-braced frames (Figure 8.79). This solution was optimal for several reasons: the slender steel braces provided greater flexibility for service openings than bulky timber; the system offered superior ductility for a more efficient seismic design; and it could be erected simultaneously with the mass timber components, accelerating the overall construction schedule. Around the perimeter, a continuous steel angle contributes to the diaphragm action of the floor plate (Figure 8.72), while lateral forces are transferred back to the central core through concealed steel drag straps. This arrangement eliminates the need for lateral braces within occupiable spaces, maintaining open floor areas and maximizing views and daylight. The steel structure of the solar chimneys further complements the system, integrating architectural and structural functions efficiently.

Solution 5: The project team engaged authorities having jurisdiction early and proactively. Using detailed 3D models, virtual fly-throughs, and frequent collaborative meetings, they clearly communicated the design's integrity and safety features. This high level of engagement not only secured approvals but also contributed to new research, helping to advance the collective understanding of mass timber

Figure 8.79 Concentrically braced frames for lateral resistance (courtesy of Salina Kassam).

systems for future applications, fostering a shared knowledge base that strengthens confidence in large-scale mass timber construction.

Solution 6: The team extended their testing from the lab to the field. Ambient and experimental vibration tests conducted on the fully constructed composite floor system within Limberlost Place confirmed its performance was acceptable. This holistic, in situ validation proved the system's real-world viability and provided crucial data for future long-span TCC designs.

Limberlost Place stands as a benchmark for measurable sustainability, significantly outperforming municipal requirements and setting new standards for low-carbon construction. Table 8.23 compares the energy and carbon performance of Limberlost Place with the highest level of the Toronto Green Standard (Tier 4). The building surpasses all three benchmarks: its total energy use intensity (TEUI) is lower than the limit (54.2 vs. 65 $kWh/m^2/year$), its thermal energy demand (TEDI) is reduced (12.8 vs. 15 $kWh/m^2/year$), and its greenhouse gas intensity (GHGI) is cut in half (2.5 vs. 5 $kgCO_2e/m^2/year$). In short, the project not only meets but exceeds Tier 4 targets, achieving higher efficiency and much lower operational carbon emissions.

In conclusion, Limberlost Place demonstrates how sustainable construction can move beyond simply meeting performance targets to actively reshaping building culture. It transforms the educational building into a living laboratory that makes sustainability intuitive for its occupants. The project proves that a holistic, lean design process can deliver a net-zero carbon building that is not only years ahead of regulatory standards but also pioneers new, economical structural systems. The integration of a steel-braced frame for lateral stability further illustrates how hybrid designs can solve complex spatial and seismic challenges, increasing the replicability of tall timber construction. Table 8.24 shows a sustainability strategy matrix for the Limberlost project that summarizes the four core lesson categories by mapping each targeted method to its specific life-cycle module and the resulting material- and carbon-saving outcomes.

Table 8.23 Comparison of project's performance against Toronto Green Standard Tier 4 targets.

Performance metric	Toronto Green Standard (Tier 4)	Limberlost place performance
TEUI (Total Energy Use Intensity)	65 kWh/m^2/year	54.2 kWh/m^2/year
TEDI (Thermal Energy Demand)	15 kWh/m^2/year	12.8 kWh/m^2/year
GHGI (Greenhouse Gas Intensity)	5 kgCO_2e/m^2/year	2.5 kgCO_2e/m^2/year

Table 8.24 Sustainability strategy matrix of Limberlost project.

Strategy	Most relevant LC stages	Key method(s) used	Sustainability outcome
Hybrid use of construction materials	A1–A3, A4, A5	Reinforced concrete for foundations and a 100-mm raft slab acting in composite action CLT for floor systems GLT for beams and columns Steel-braced frame for lateral stability	Structural efficiency Reduced material waste High construction speed Lower embodied carbon (with sustainable timber sourcing)
Lean design and prefabrication	A4, A5	Design-assist partners Prefabricated TCC and Envelope systems Just-in-time delivery	Elimination of on-site waste Enhanced quality control Accelerated schedule
Long-span floors	A1–A3, A5	Mass timber TCC slab-band system CLT infill panels and GLT columns	9.2 m column-free spans Maximized MEP headroom
High-performance passive systems	B, D	Form shaped for solar optimization Solar chimneys and natural ventilation Prefabricated envelope	Net-zero operational carbon Exceeds 2030 energy targets Enhanced occupant well-being
Experimental testing	A1–A5	Full-scale structural testing of TCC Proactive regulatory engagement with BIM Structural and vibration tests	Validated and de-risked floor system Pioneered approvals pathway Advanced industry knowledge

Project Team

Client: George Brown Polytechnic
Architects: Moriyama Teshima Architects in Joint Venture with Acton Ostry Architects
Structural Engineers: Fast + Epp
Mechanical and Electrical Engineers: Introba
Environmental Specialist: Transsolar Inc.
Building Envelope: Morrison Hershfield
Constructor: PCL Construction
Code Consultants: GHL Consultants Ltd
Fire Engineers: CHM Fire Consultants Ltd

8.4.6 Temporary, Reusable Hybrid Steel–CLT Offices for the Construction Site of Sydney's Metro Stations

For major transport infrastructure projects, construction site offices are typically delivered as clusters of stacked shipping containers that are procured, assembled, and removed for each new work site. This conventional approach presents several shortcomings: environmentally, containers are often treated as single-use assets, generating construction waste and landfill burden with each project cycle; in terms of workplace quality, they offer poor thermal performance, limited natural light, and cramped spaces unsuited to sustained office work; and economically, repeated procurement and fit-out costs accumulate across a project portfolio. The St Marys Integrated Project Office (IPO) is a temporary 200-workstation design office created to support Sydney Metro during the construction of six new metro stations connecting St Marys railway station to the future Western Sydney Aerotropolis (NSW Government, 2026). Different from the conventional temporary site-offices, it is a fully integrated, commercial-quality building designed from the outset for disassembly and reassembly across future Sydney Metro projects (Figure 8.80).

Serving as a hub for multiple project teams, the building provides four commercial levels with a glazed curtain wall façade, internal access stairs, a lift, and an outdoor terrace, delivering the amenities and spatial quality of a permanent office rather than a conventional site compound. By treating the structure as a long-term reusable asset rather than disposable infrastructure, the IPO reduces embodied carbon and construction waste while delivering the spatial quality, thermal comfort, and natural daylight of a

Figure 8.80 St Marys Integrated Project Office, Sydney (courtesy of Raw Life Studios).

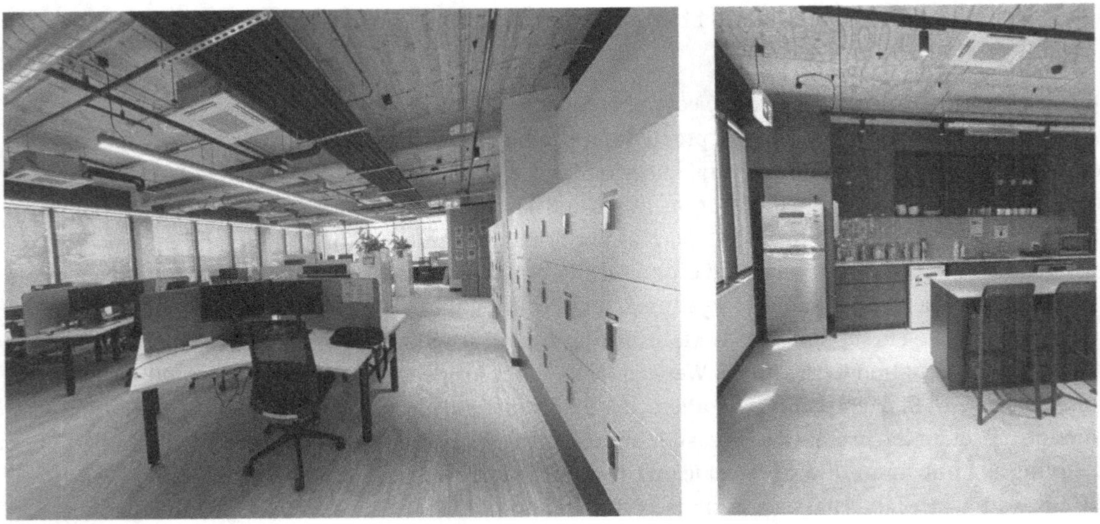

Figure 8.81 Open-plan office space with exposed CLT ceiling panels, natural lighting, and commercial-grade fit-out (left), the communal kitchen area (right) (courtesy of Robert Bird Group).

permanent office (Figure 8.81). Robert Bird Group (RBG) was engaged to deliver the structural design of the building, including foundations, a superstructure, and construction engineering services for build sequencing and temporary works.

The structural system is a hybrid steel–CLT superstructure (Figure 8.82) configured for a 50-year design life, with reassembly anticipated every 10 years as the building migrates between metro project sites.

At the early design stage, optioneering was essential for identifying a structural solution that balanced speed, sustainability, and reuse. RBG first studied the project tender design scheme and then reconfigured the structural solution, identifying an elevated structural steel frame with CLT floor panels as the most economical and appropriate construction approach that satisfied all components of the brief (Figure 8.83). This hybrid system aligned with the temporary, reusable nature of the building while still delivering a commercial-standard workspace. Once the primary materials were confirmed, the design team further refined the scheme by aligning the building grid to standard structural steel stock lengths, reducing cutting waste and fabrication time. As a direct result, approximately 10 tons of structural steel were saved compared with the reference design, corresponding to an 11% reduction in total steel tonnage. This optimization also reduced cradle-to-gate embodied carbon by 33.5 tCO_2e, a 6% decrease in the superstructure's carbon footprint relative to the reference scheme.

From an environmental performance standpoint, the superstructure's cradle-to-gate carbon (Modules A1–A3) was measured at 563 tCO_2e. Because the IPO was specifically designed to be dismantled and reconstructed for subsequent projects, the team assessed performance over five lifecycles. On that basis, the reusable design was calculated to save 2534 tCO_2e in cradle-to-gate emissions compared with a series of standard single-use temporary like-for-like structures using prime materials.

Transforming the brief into a functioning, relocatable office building presented several engineering obstacles that had to be resolved without compromising program or sustainability objectives.

Figure 8.82 Hybrid structural steel and CLT superstructure, designed for disassembly and reuse (courtesy of Robert Bird Group).

Challenge 1: Achieving an accelerated, efficient build while enabling future dismantling and relocation:

- The structure needed to be erected quickly to meet Sydney Metro's mobilization requirements, with minimal temporary works.
- The building had to be designed from the outset for dismantlement and reconstruction on future metro sites, requiring simple, consistent connection details.
- Construction sequencing had to maintain stability both during initial erection and eventual disassembly.

Challenge 2: Providing adequate fire resistance for a hybrid steel–CLT structure:

- The proposed primary materials, structural steel and CLT, do not possess the same inherent fire resistance as reinforced concrete or masonry.
- The building's structural elements needed to achieve the required fire-resisting period without undermining the simplicity and reusability of connections.

Challenge 3: Controlling floor vibration in a lightweight office structure:

- The combination of a lightweight steel frame and CLT floors raised concerns that the floor system might be overly flexible.

Figure 8.83 Typical level showing CLT floor (courtesy of Robert Bird Group).

- Excessive floor vibration could result in an uncomfortable and disruptive environment for office occupants.

The engineering team addressed these challenges through specific solutions that integrated structural design, construction methodology, and material optimization:

Solution 1: Standardized framing, elevated structure, and integrated erection methodology:

- A structural steel framing system was conceived with simple bolted connections throughout all levels, significantly accelerating erection and simplifying future dismantling and reassembly.
- Prefabricated CLT floor panels were crane-lifted and placed directly onto the steel frame on site (Figure 8.84), enabling rapid dry assembly without formwork or curing time. The standardized panel dimensions matched the repetitive steel bay geometry, allowing sequential floor-by-floor installation with minimal on-site cutting or adjustment.
- Floor beams, column sizes, connection details, and CLT floor panels were intentionally kept identical throughout the building, allowing components to be efficiently dismantled, transported, and reassembled on future sites.

Figure 8.84 On-site assembly of CLT floor panels onto the steel frame: (a) crane-lifting of a prefabricated CLT panel into position and (b) installed CLT panels (courtesy of Robert Bird Group).

Figure 8.85 Prefabricated steel lift shaft in the factory (left) and steel frame during on-site erection (right), demonstrating the standardized structural system with simple bolted connections designed for rapid construction and future disassembly (courtesy of Robert Bird Group).

- The building was conceived as an elevated ground-floor structure, eliminating excavation and allowing services to be run in the sub-floor zone, which reduced construction time.
- A full-height steel lift shaft was fabricated and erected as a single-piece assembly at the start of construction (Figure 8.85). Each floor was then framed and braced off this shaft, eliminating the need for temporary bracing and ensuring that the shaft would be the final element removed during disassembly, maintaining stability throughout both phases.

Solution 2: Fire design using CLT charring and selective steel protection:

- RBG collaborated closely with the project fire engineer and CLT supplier to determine charring rates for the CLT floor panels, enabling the design to rely on a reduced residual cross-section during fire to achieve the required fire-resisting period without additional passive protection to the timber.
- Structural steel beams were analyzed under ultimate fire load cases, and only those beams directly supporting the floors were specified to receive passive fire protection (Figure 8.86), avoiding unnecessary treatment of secondary members.
- Secondary beams that restrained columns at each level were intentionally left unprotected in the fire analysis, with the design justified by the robustness of the CLT diaphragm (which, even with reduced section from charring, was shown to provide adequate restraint to prevent column buckling).

Solution 3: Early-stage floor vibration assessment: RBG performed finite element analysis of the CLT floor system early in the design period, including natural frequency assessment, which confirmed that the structure possessed sufficient mass relative to span lengths to avoid vibration-sensitive behavior, preserving the lightweight, efficient nature of the hybrid scheme.

Figure 8.86 Interior view of the completed office showing the selective fire protection strategy: primary steel beams supporting the CLT floor panels above are clad with passive fire protection (visible as white-finished columns and protected beam sections), while secondary framing members and ceiling-level steelwork remain exposed, reflecting the rationalized approach to fire engineering that avoided unnecessary treatment of non-load-path-critical elements (courtesy of the project team).

Figure 8.87 Façade view of the building (courtesy of Robert Bird Group).

These measures enabled the contractor to complete the structural steel frame and CLT floor panels (the total superstructure) within 25 working days, with the full-height lift core erected within only 4 hours. Compared with typical temporary site offices, the IPO delivers a more comfortable work environment, with designated meeting spaces, improved site amenities, a glazed curtain wall façade (Figure 8.87) admitting natural light, and enhanced insulation, windows, doors, and roofing that help stabilize internal temperatures and reduce energy consumption.

In conclusion, this project demonstrates how a hybrid steel–CLT structure, configured from the outset for disassembly and reuse, can replace conventional demountable project offices with a more sustainable and high-quality alternative. Over multiple lifecycles, the approach significantly reduces embodied carbon compared with repeatedly constructing standard temporary offices, offering a replicable model for infrastructure delivery organizations. The project team believes this building to be the first of its kind, serving as a delivery model for designing and constructing commercial buildings with the capacity to be dismantled and relocated in emerging precincts such as Bradfield City Centre. Table 8.25 shows a sustainability strategy matrix for the St Marys IPO project that summarizes the four core lesson categories by mapping each targeted method to its specific life-cycle module and the sustainability outcomes.

Table 8.25 Sustainability strategy matrix of the St Marys IPO project.

Strategy	Most relevant LC stages	Key method(s) used	Sustainability outcome
Preserve and reuse	A1–A3, C, D	Design of hybrid steel–CLT superstructure for dismantling, reassembly every 10 years over 50 years	Enables reuse of the IPO on future metro projects instead of constructing new facilities; reduces construction waste
Tailored structural innovation	A1–A3, A5	Elevated steel frame with CLT floors, full-height single-piece lift shaft, standardized connections	Reduces material use, temporary works, and construction time while maintaining structural performance
Fire and vibration performance	A1-A3, B	CLT charring analysis, selective steel protection, early vibration finite element analysis	Achieves required safety and comfort without unnecessary material additions
Early optioneering and material efficiency	A1–A3	Alignment of grids to stock lengths, early redesign from tender scheme	11% reduction in steel tonnage, 33.5 tCO_2e savings, and 2,534 tCO_2e savings over five lifecycles

Project Team

Owner/Owner's Representative: Sydney Metro
Architect: Place Studio
Structural Engineer: Robert Bird Group
Steel Fabricator/Erector: Proline Steel
CLT Supplier: XLAM
CLT Installer: Savcon

8.5 Regenerative Design in Harsh Soil and Seismic Conditions

8.5.1 Geotechnical Optimization for a Sports Center on the Same Soil of Pisa Tower

How do you build a state-of-the-art professional sports facility on land famous for its instability (the same soil that caused the world-famous Pisa Tower to lean)? This is the fundamental challenge facing ATI Project in the design of the new Pisa Training Center for the Pisa Sporting Club (Figure 8.88). The project required a design that was not only lightweight and architecturally ambitious but also resilient enough to overcome geotechnical, seismic, and hydraulic risks.

The core concept of this project is overcoming the soil challenges through integrated geotechnical and structural design. An extensive geotechnical investigation confirmed the poor soil conditions and high seismic and hydraulic risks. The design team at ATI Project used this data to engineer a series of solutions that not only solved the technical problems but also optimized material use and construction time. Faced with 20 m of loose clays, sands, and peat, the team worked on a foundation system of micropiles and ribbed mat foundations to ensure stability, and a lightweight steel moment-resisting frame to minimize the load on the poor soil. This integrated approach, managed through a BIM workflow, has been essential to safely delivering a high-performance facility on a uniquely challenging site.

The site's geotechnical and environmental conditions presented three challenges for the design team:

Challenge 1: Due to soil bearing capacity conditions, foundations needed to be designed to reach deep layers, because, as can be noted in Figure 8.89, the first 20 m are composed of a mix of loose clays and sands. In some areas, even peat layers have been found.

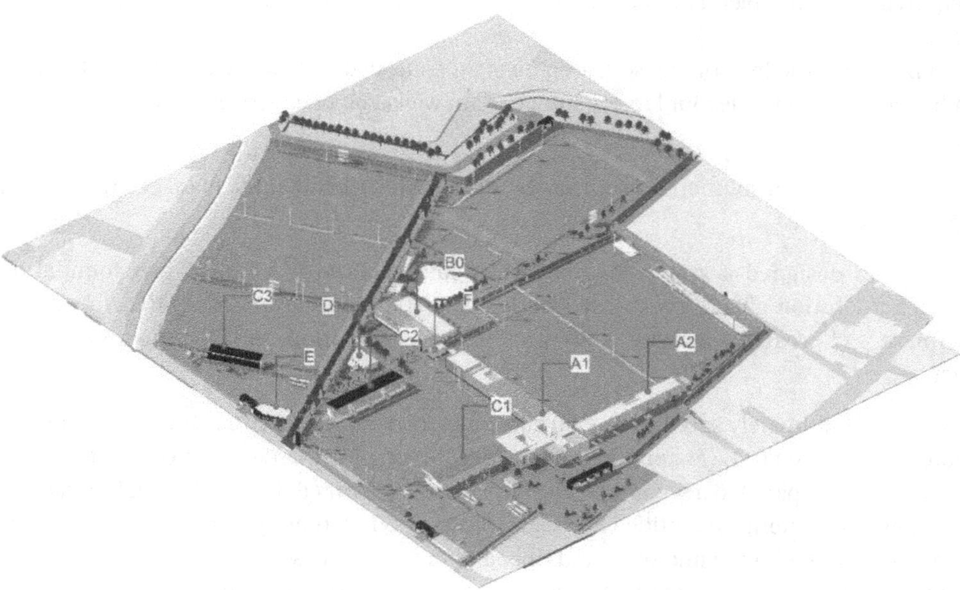

Figure 8.88 A 3D view of the Pisa Training Center, a multi-building complex designed to serve the club's first team, youth teams, and fans (courtesy of ATI Project). Letters show different buildings in the complex.

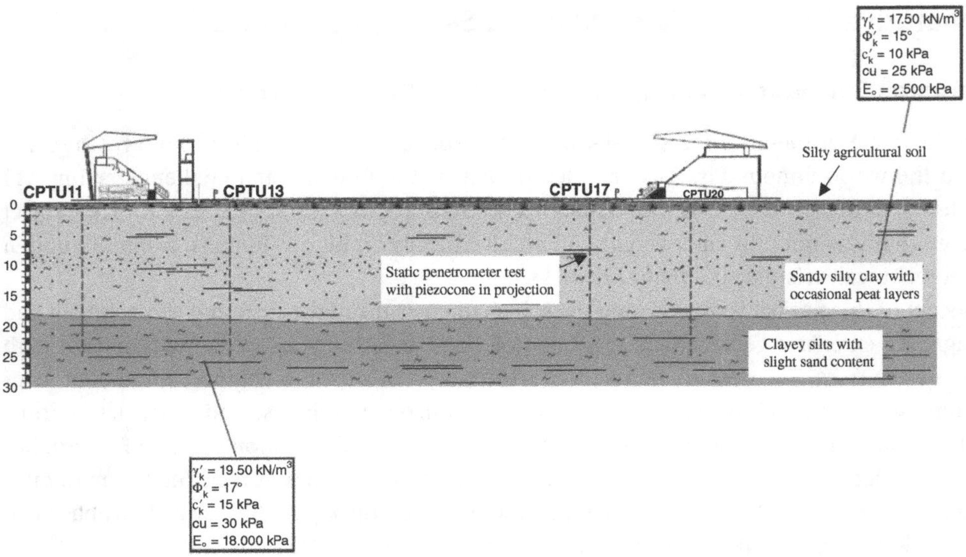

Figure 8.89 Geological stratigraphy of the project site (courtesy of ATI Project).

Challenge 2: The site was classified with a moderate-to-high seismic risk, and the soft soil (Soil Category D) was expected to amplify ground shaking significantly (a factor of 1.8 according to the simplified approach provided by Italian building standard NTC2018). Furthermore, a very high flood risk required the entire facility to be raised by 1.2 m, adding complexity to the foundation and ground-floor design.

Challenge 3: To minimize the load on the soil, a lightweight superstructure was essential. At the same time, the architectural vision called for large, glazed curtain walls and open spaces, precluding the use of visible bracing elements.

ATI Project developed a set of solutions to address the site's complex challenges, managed through an end-to-end BIM process:

Solution 1: The team designed a system of micropiles combined with a ribbed mat foundation (Figure 8.90). This solution effectively transferred loads to stable soil layers while the ribbed slab provided stiffness and created space for underground water retention tanks, compensating the hydraulic risk.

This optimized design ultimately saved over 4500 cubic meters of concrete, corresponding to approximately 1080 tons of embodied CO_2 (assuming an average factor of 240 $kgCO_2e/m^3$ of concrete), representing a 38% reduction compared to a scenario using large-diameter drilled piles and a mat foundation. Table 8.26 compares the material quantities between the two foundation systems, highlighting the significant savings achieved with the micropile and ribbed foundation slab solution.

The entire process, from design to detailing, was managed in a BIM environment using Revit and Tekla, with custom scripts developed to automate the modeling of reinforcement and complex steel connections.

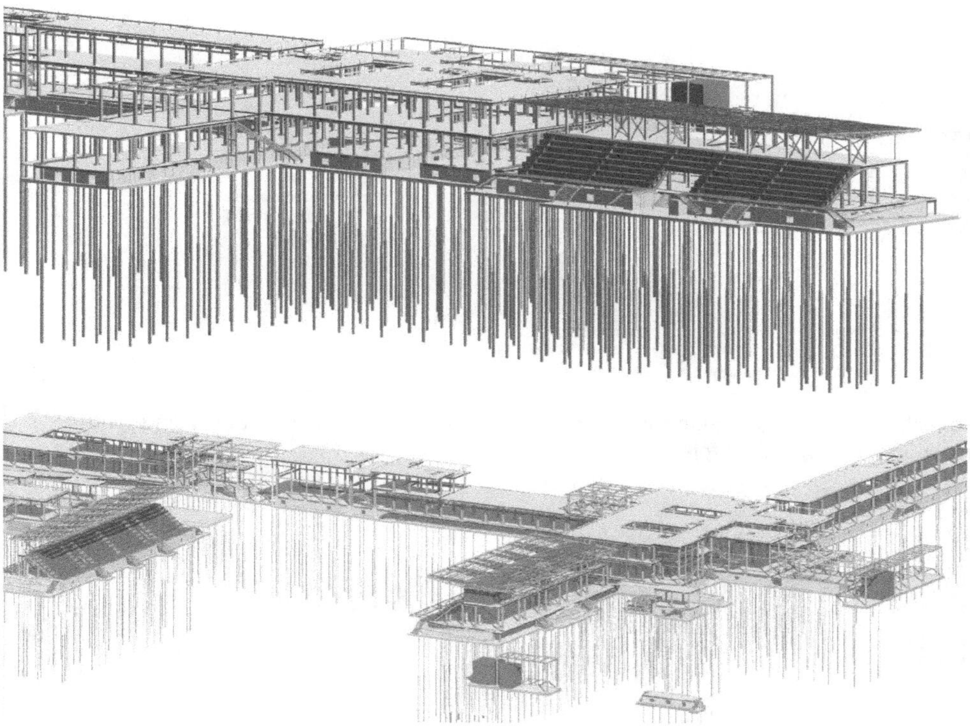

Figure 8.90 The structural BIM models, illustrating the lightweight steel frame and the deep foundation system required to ensure stability on the poor soil (courtesy of ATI Project).

Table 8.26 Foundation system material optimization.

Element	Scenario 1: Drilled piles + mat foundation	Scenario 2: micropiles + ribbed foundation slab
	Concrete volume (m^3)	Concrete volume (m^3)
Screed	1054.65	1144.92
Plate foundation	4359.41	4969.88
Pile quantity	6440.89	1215.24
Total	**11 854.95**	**7330.04**
Savings		**−4524.91 m^3 (−38.17%)**

Solution 2: Rather than using the conservative and simplified seismic design approach, which would have resulted in over-designed structures, the team performed a detailed Local Seismic Response (RSL) analysis. The spectrum of the RSL presents a maximum acceleration at the plateau of 0.43 g (Figure 8.91). In the comparison, the RSL spectrum was obtained by applying a behavior factor $q = 1.5$

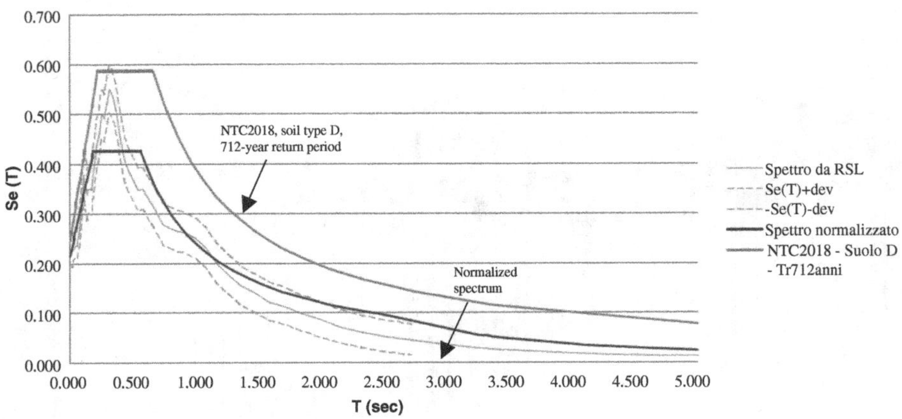

Figure 8.91 Normalized response spectrum comparison; piano campagna SLV (life safety limit state) with 712-year return period (courtesy of ATI Project).

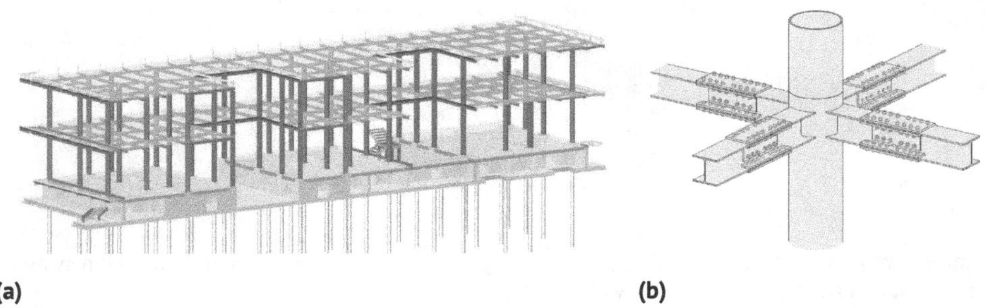

Figure 8.92 Moment-resisting frame a typical connection detail: (a) moment-resisting frame scheme; (b) I-beam to CHS connections (courtesy of ATI Project).

for non-dissipative structures. This site-specific study therefore provided a more accurate prediction of ground motion, showing that the design seismic acceleration at the plateau could be reduced by 53% compared to the elastic spectrum (NTC2018 – soil type D, 712-year period). This allowed a more efficient and well-balanced structural design, avoiding unnecessary oversizing.

Solution 3: To meet the architectural requirement for open, brace-free spaces, a moment-resisting frame (MRF) was designed (Figure 8.92). To maximize quality and speed, beam-to-column connections were welded in the factory and assembled on-site with bolted connections.

In conclusion, the Pisa Training Center project overcame site challenges through an integrated engineering. By combining advanced geotechnical analysis with an optimized, lightweight structural system, the team turned a high-risk site into a high-performance facility. A digital workflow enabled automation, precision, and seamless coordination among the team, proving that even on the most unstable ground, a data-driven approach can deliver a foundation for success. Table 8.27 shows a sustainability strategy matrix for the project that summarizes the three core lesson categories by mapping each method to its specific life-cycle module and the resulting material- and carbon-saving outcomes.

Table 8.27 Sustainability strategy matrix of Pisa Training Center project.

Strategy	Most relevant LC stages	Key method(s) used	Sustainability and efficiency outcome
Foundation optimization	A1–A3, A4, A5	Ribbed mat foundation on micropiles	Helped considering realistic seismic forces, reducing concrete use by 38%
Advanced seismic design	A1–A5	Local seismic response (RSL) analysis	Reduced design seismic forces by 53% Avoided over-design, leading to a more efficient structure
Lightweight, prefabricated superstructure	A1–A5, C, D	Steel moment-resisting frame BIM-to-fabrication workflow	Minimized building self-weight on poor soil Met architectural demands for open, brace-free spaces Accelerated construction schedule

Project Team

Owner: Pisa Sporting Club
Main Designer: ATI Project
Paper Credits: Eng. Alessio Bacci, Eng. Anna Continanza, Eng. Michele Fascilla

8.5.2 Low-carbon Structural Design on a Contaminated Site

Across Europe, many postindustrial cities are faced with the challenge of large, abandoned brownfield sites. Often, these complex areas are left undeveloped or are regenerated in a disconnected fashion that fails to create significant impact on today's needs. The Lumen project (Figure 8.93), located in Newcastle (UK), opened in 2020, presents an ambitious path forward. This eight-story, 106 000 ft^2 (9848 m^2) office building provides a state-of-the-art workspace accommodating the requirements of the current science and technology sectors. Its structure, with an upfront embodied carbon of 307 $kgCO_2e/m^2$ (about 50% GWP reduction vs. baseline) achieved the LETI 2025 (LETI, n.d.) carbon target. This is a significant achievement, as LETI targets are voluntary UK-based benchmarks that represent a best-practice, ambitious goal at the time for low-carbon buildings, requiring a design philosophy that goes far beyond standard building regulations. Moreover, the project's foundation design addressed the risks of building over disused coal mines, employing a geotechnical investigation to avoid the high cost and carbon footprint of full-site remediation.

The core strength of this project was its data-driven approach to structural optimization. The design team's primary goal was to achieve a long-span (15.5 m × 9 m) column grid using the smallest possible steel and concrete tonnage. This was accomplished through early-stage "optioneering" (a process of creating and assessing multiple structural solutions against key metrics like embodied carbon, cost, and the construction program). This data-driven method allowed the team to prove which design offered the best performance and value before committing to a final direction. This also allowed the team to discard heavy or over-designed sections before detailed design began, cutting both weight and carbon without compromising performance.

Figure 8.93 Newcastle Lumen (courtesy of Pell Frischmann).

Transforming a former industrial site into a high-tech office hub presented the following structural, geotechnical, and sustainability challenges:

Challenge 1: The project's location over disused coal mines posed a significant risk to its foundation and structural stability: the site contained shallow mine seams, two unfilled mine shafts, and (despite prior remediation) remaining contamination hotspots and hazardous ground gas.

Challenge 2: The structural design team voluntarily considered the LETI (London Energy Transformation Initiative) 2025 upfront embodied carbon target, a goal that pushes beyond standard industry practice. Meeting this target was not possible with a business-as-usual approach, since it requires carbon performance to be treated as a critical driver of design decisions from day one.

The following solutions were implemented to overcome the previously mentioned challenges:

Solution 1: Pell Frischmann's specialist geo-contamination team conducted extensive site investigations. This data was used to create a detailed site zoning plan, which allowed for the suitable placement of the building's piled foundations, mitigating the development risk. This approach avoided the high cost and carbon footprint of remediating the entire site, representing a "lean" strategy applied to geotechnical engineering.

Solution 2: Embodied carbon objectives have been achieved thanks to a high-efficiency structural system (Figure 8.94) composed of the following outcomes:

i. Through extensive research and optioneering, a composite steel was designed with a material utilization rate of 95% (the material utilization rate measures how efficiently a structural element is being used): steel beams and a reinforced concrete slab are designed to act together as a single, efficient structural unit, Figure 8.95. This solution delivered a column-free floorplate with a wide 15.5 m × 9 m structural grid.

Figure 8.94 The model of the project (courtesy of Pell Frischmann).

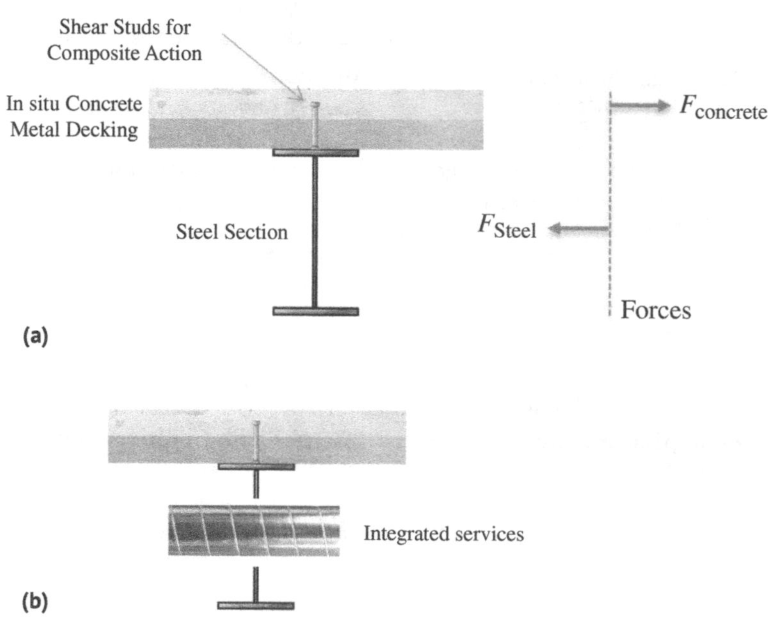

Figure 8.95 (a) Steel–concrete composite slab and (b) integration of services in cellular beams.

ii. Cellular steel beams are used, whose large web openings create a lighter, more efficient structure while allowing building services to be integrated within the floor depth, thereby maximizing ceiling heights.

Table 8.28 summarizes the core sustainability strategies by mapping each method to its life-cycle contribution and resulting outcomes. In conclusion, by embedding a “Lean and Low-Carbon” philosophy from the earliest design stages, the Pell Frischmann team demonstrated that ambitious environmental targets are achievable. The use of data-driven optioneering to inform structural design proves that it is possible to deliver high-performance, long-span commercial spaces that are both economically viable and exceptionally low in embodied carbon.

Table 8.28 Sustainability strategy matrix of Lumen project.

Strategy	Most relevant LC stages	Key method(s) used	Sustainability outcome
Geotechnical site zoning and foundation optimization	A5	Extensive site investigations by geo-contamination team Detailed zoning plan for piled foundations	Mitigated risk from shallow mine seams, unfilled shafts, and gas hotspots Optimized foundation placement, reducing over-remediation
Data-driven early-stage optioneering	A1–A5, C	Early-stage assessment of structural schemes against embodied carbon, cost, and construction program	Met LETI 2025 upfront embodied-carbon target Guided lean design decisions across all options Structure upfront embodied carbon intensity of 307 $kgCO_2e/m^2$ ≈50% GWP reduction vs baseline
Structural frame with high material utilization	A1–A3, A4–A5	High-efficiency composite steel–concrete frame	Minimized steel tonnage Material utilization rate of 95%
Long-span floors	A5, B1	15.5 m × 9 m structural grid Steel–concrete composite frame – Column-free floorplate	Flexible, open-plan workspace

Project Team

Client: Legal & General
Partners: Newcastle University, Newcastle City Council
Civil Engineering|Fire Engineering|Geotechnical Engineering|Structural Engineering: Pell Frischmann

8.5.3 Regenerative Hybrid System in a High-seismic Region

The modern commercial office is at a crossroads, often defined by carbon-intensive concrete and steel structures that are disconnected from the natural environment. The 42XX campus in Marina del Rey, California, USA (Figure 8.96) presents a new direction. Developed by The Bradmore Group and designed by RIOS, this 151 000 ft^2 (14 028 m^2) campus is a large-scale hybrid structure integrating timber, steel, and concrete, making it the first of its kind in Los Angeles with extensive landscaping (47% of the 3-acre site is dedicated to open green space). It transforms the conventional office model into a dynamic ecosystem that actively supports occupant well-being, ecological health, and community connectivity. Completed in 2024, 42XX showcases how hybrid structural systems can significantly reduce carbon emissions while creating a dynamic work environment that supports occupant well-being, ecological corridors, and community connectivity. Located in one of the highest seismic regions of the United States, the building also required a structural system capable of ensuring both safety and material efficiency under seismic demands.

The core concept of the project reimagines the traditional office structural grid as a regenerative hybrid system (Figure 8.97), by integrating CLT, glulam, steel, and reinforced concrete. The design optimizes the unique strengths of each material, creating a whole that is far more sustainable and human-centered than the sum of its parts. This approach moves beyond simple material substitution to create a truly regenerative environment.

Figure 8.96 42XX campus in Marina del Rey, California (photography by Pavel Bendov/ArchExplorer, courtesy of RIOS).

Figure 8.97 A view from inside building showing GLT beams, steel beams, columns, bracings, and CLT slabs (photography by Pavel Bendov/ArchExplorer, courtesy of RIOS).

However, pioneering such a large-scale hybrid timber system in a dense urban environment in a seismic region (Los Angeles lies within one of the most active seismic zones in the United States, influenced by the nearby San Andreas fault, where design accelerations are significantly higher than in most US cities) presented a unique set of structural, logistical, and regulatory challenges:

Challenge 1: The primary technical hurdle was integrating the different structural materials with precision. Off-site manufactured CLT and glulam components have different tolerances and natural variations (e.g. warping) compared to steel. In addition, the project's location in one of the most seismically active regions of the United States required all hybrid connections to ensure both stiffness and ductility under lateral loads. Ensuring a structurally sound connection between these disparate elements on-site was a significant logistical and engineering puzzle that risked delays and costly rework.

Challenge 2: The three-acre site was tightly constrained by physical obstacles, including nearby residences, overhead power lines, and narrow access roads. These conditions made the delivery and maneuvering of large structural components exceptionally difficult and created a high risk of neighborhood disruption and construction delays.

Challenge 3: As one of the first large-scale mass timber projects in the city, the project faced intense examination from the Los Angeles Department of Building & Safety, with no established precedent for the approval process.

Figure 8.98 Hybrid steel and mass timber roof system of 42XX campus in Marina del Rey project (photography by Pavel Bendov/ArchExplorer, courtesy of RIOS).

Challenge 4: A key goal was to create a warm, biophilic environment by leaving the timber structure exposed. This created the challenge of how to integrate the extensive mechanical, electrical, and plumbing (MEP) systems without cluttering the clean aesthetic of the wood surfaces with visible pipes and conduits.

Solution 1: The team addressed the structural integration challenge of the hybrid system (Figure 8.98) by developing a responsive fabrication process: they took precise, as-built measurements on-site, and fed this data back to the workshop. This allowed for the off-site prefabrication of custom steel brackets that were specifically designed to connect the timber and steel components seamlessly, accommodating any geometrical variations and ensuring a good fit during on-site assembly. These joints were also optimized to provide continuity under seismic loading, ensuring both safety and efficiency in a high-seismic environment.

Solution 2: To navigate the physical constraints, the team implemented a strategy of methodical on-site planning. This involved carefully orchestrated scheduling, just-in-time deliveries to avoid storage issues on the site, and the use of compact lifting equipment to operate within the limited space. Figure 8.99 shows the construction site during hybrid assembly. The steel gravity framing is visible

Figure 8.99 Aerial view of the 42XX campus during construction, showing the steel gravity framing (beams and columns) with diagonal Buckling Restrained Braces (BRBs) serving as the seismic-resisting system for the steel levels (courtesy of RIOS).

across the lower levels, with diagonal Buckling Restrained Braces (BRBs) (also seen in Figure 8.100) acting as the seismic-resisting system and distributing lateral forces down to the reinforced concrete shear walls. CLT panels are being installed over the steel frame in the upper portion, showing the hybrid construction sequence.

Solution 3: The team engaged in close and proactive collaboration with Los Angeles Department of Building & Safety officials from the early stages of the project, helping to establish a new approval framework that can now also benefit future projects.

Solution 4: The design team left intentional gaps between CLT panels to conceal mechanical systems, reducing visual clutter while keeping the focus on exposed wood surfaces.

The innovative approach of the 42XX campus resulted in significant and quantifiable sustainability benefits, validating its regenerative design philosophy.

- The hybrid structure reduced overall building weight by 50% with respect to a fully reinforced concrete option, significantly lowering foundation demands and material use.
- Reduced embodied carbon emissions offset emissions equivalent to removing 262 cars from the road annually.
- Exposed timber elements contribute to a tactile, human-centered aesthetic and support occupant well-being through biophilic design.

Figure 8.100 Interior view of the completed hybrid framing system, with a focus on the steel Buckling Restrained Braces integrated into the open floor plan while maintaining the biophilic aesthetic of exposed timber surfaces (courtesy of RIOS).

- Off-site prefabrication of mass timber components speeds up construction and cuts down on on-site waste and energy use (through three-month mass timber installation vs. traditional lengthy concrete curing periods).
- The landscape strategy (featuring 34 mature trees) turned the site into an active carbon sink: they sequester 23 tons of CO_2, representing one-third of the building's embodied emissions (Figure 8.101).

A comparative analysis of structural systems performed during the early design stages highlights the GWP advantages of the selected hybrid structural system (Figure 8.102). When benchmarked against a traditional steel frame with concrete over metal deck and a full concrete frame and slab, the CLT/Steel Hybrid system demonstrates a dramatically lower embodied carbon at the product stage (~45 000 $kgCO_2$ eq). The steel system shows a 1.6× higher carbon impact, while the reinforced concrete system's impact is three times greater during the LCA product stages (A1–A3), while construction emissions (A4–A5) are

Figure 8.101 The 34 mature trees sequester 23 tons of CO_2 during the service life of the project (photography by Pavel Bendov/ArchExplorer, courtesy of RIOS).

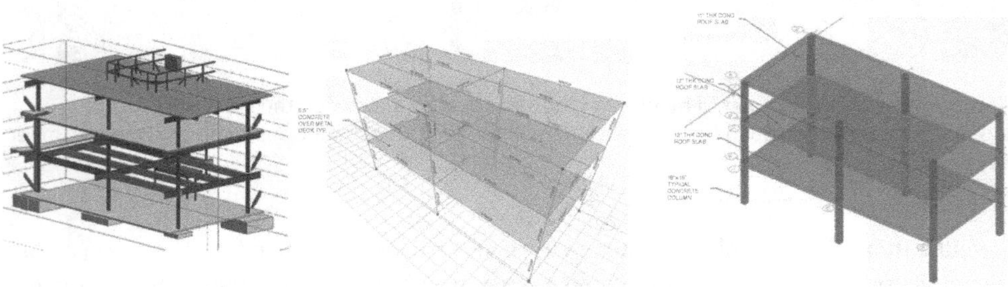

(a) (b) (c)

Figure 8.102 Comparison of the three structural system alternatives studied during 42XX at early-design stage: (a) hybrid steel and mass timber system, (b) steel frame with reinforced concrete over metal deck, and (c) reinforced concrete frame and slab (photography by Pavel Bendov/ArchExplorer, courtesy of RIOS).

similar across systems (Figure 8.103). The significant weight reduction also directly translates to superior seismic performance, as a lighter building generates a lower design seismic base shear, reducing structural demands (Figure 8.104).

Altogether, the project shows how sustainable construction can move beyond material substitution to create regenerative environments that support both people and the urban ecosystem. This transforms the building from a static structure into a dynamic ecosystem actively contributing to urban environmental

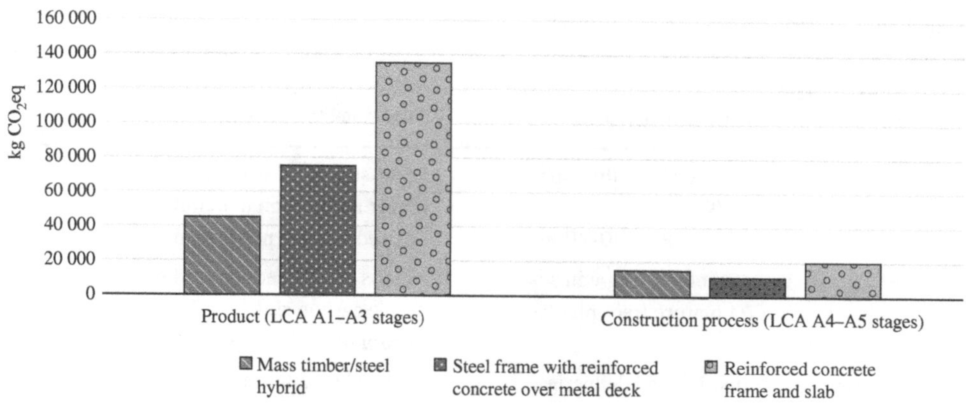

Figure 8.103 Comparison of GWP at product (A1–A3) and construction process (A4–A5) stages for the three early-stage alternatives: mass timber/steel hybrid, steel frame with reinforced concrete over metal deck, and reinforced concrete frame and slab.

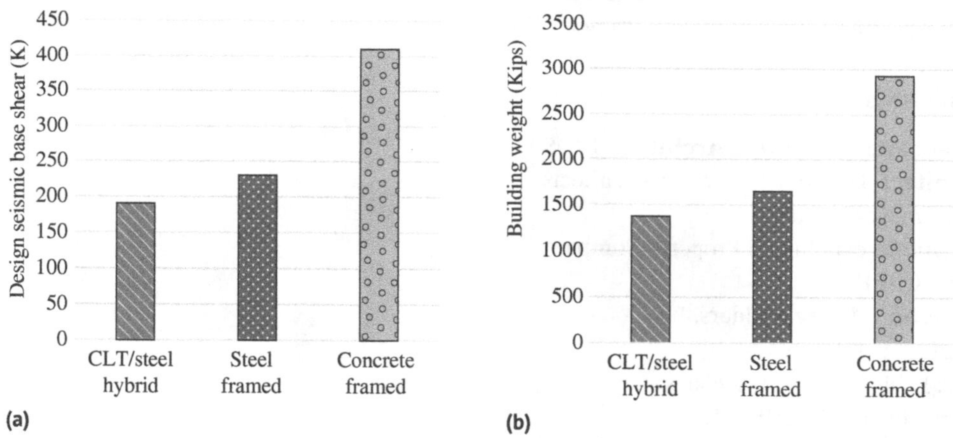

Figure 8.104 Comparison of the three structural systems studied during 42XX at early-design stage: mass timber/steel hybrid, steel frame with reinforced concrete over metal deck, and reinforced concrete frame and slab: (a) design seismic base shear (K); (b) building weight (Kips) (photography by Pavel Bendov/ArchExplorer, courtesy of RIOS).

health. The project demonstrates that hybrid mass timber and steel construction can achieve cost parity with conventional methods when considering reduced foundation requirements, accelerated timelines, and operational efficiency gains.

Table 8.29 shows a sustainability strategy matrix for the 42XX campus project that summarizes the four core lesson categories by mapping each targeted method to its specific life-cycle module and the resulting material- and carbon-saving outcomes.

Table 8.29 Sustainability strategy matrix of 42XX project.

Strategy	Most relevant LC stages	Key method(s) used	Sustainability outcome
Regenerative hybrid system	A1–A3, A5	Hybrid CLT/Glulam/Steel system Off-site prefabrication	50% ↓ structural weight Lower foundation demand Desired seismic performance
Integrated regenerative landscape	B	47% of site as green space 34 mature trees planted	23 tons of CO_2 sequestered over service life Active carbon sink Occupant well-being
Design for assembly and logistics	A4, A5	Just-in-time deliveries Custom prefabricated connection brackets	Reduced on-site waste Accelerated construction Seamless on-site assembly
Early optioneering and LCA	A1–A3	Comparative LCA of Hybrid vs. Steel vs. Concrete scenarios	Up to 3× ↓ GWP vs. concrete Hybrid approach validated

Source: Photography by Pavel Bendov/ArchExplorer, courtesy of RIOS.

Project Collaborators

Design Architect and Landscape Architect: RIOS
Executive Architect: House & Robertson Architects
Owner: The Bradmore Group
Development Manager: Lincoln Property Company
Leasing Broker: CBRE
General Contractor: Morley Builders, Inc.
Structural Engineer: Holmes US
Civil Engineer: Labib Funk + Associates
MEP Engineer: ARC Engineering
Photography: Pavel Bendov/ArchExplorer
Paper Credits: Golnaz Ighany, Sustainability Director at RIOS

8.5.4 A Modular Timber–Steel Hybrid Mosque Rooted in Anatolian Tradition

Contemporary religious architecture often faces a conflict between adhering to traditional aesthetic forms and meeting modern structural requirements. Today, this challenge carries an additional dimension: the growing imperative to reduce embodied carbon and consider the environmental impact of material choices across a building's life cycle. Timber, with its capacity for biogenic carbon storage and relatively low processing energy, offers a path toward both cultural continuity and environmental responsibility (provided it can be engineered to meet contemporary performance demands). In the case of the Mosque of Hayreddin Tokadi Külliye (currently under construction, Figure 8.105), the task was to reinterpret the traditional Anatolian timber mosque (characterized by repeated, rhythmic timber columns defining a grid of spatial cells) using modern engineering to ensure every visual component served a structural purpose. Rather than concealing the load-bearing system behind decorative cladding, the design team

Figure 8.105 Mosque of Hayreddin Tokadi Külliye project render (courtesy of Abdurrahman Çekim).

pursued a "timber-priority hybrid" approach. This 980 m^2 mosque, part of a larger social complex including a library, soup kitchen, and guesthouse, uses a modular "cell" system to achieve a vault-like interior. By integrating slender glulam elements with a hidden steel backbone, the project demonstrates how hybrid systems can honor building traditions, maximize architectural expression, and optimize material efficiency within a life-cycle-driven design framework.

The mosque plan is organized as a field of identical 5 × 5 m "cells," each defined by four 360 × 360 mm block-glulam columns rising 9 m from ground floor to roof (Figure 8.106). This modular approach allowed for the optimization of a "timber-priority" design: timber is used for the primary visual and compressive elements, while minimized steel components handle high-tension zones and connections. The cells are spanned in one direction by a steel truss with a central spine-arch, and in the perpendicular direction by a dense array of slender 60 × 160 mm glulam braces following parabolic curves. The result is a contemporary interpretation of the traditional Anatolian timber vault (achieved not through massive, curved members but through the aggregation of many small, straight elements). Unlike conventional framing, where structure and decoration are separate, here the architectural desire was for all visual elements to carry the loads. This ambition had to be realized in one of Türkiye's highest seismic zones, with a design peak ground acceleration of 0.4 g (a level associated with frequent devastating earthquakes). The lightweight nature of timber works in favor of seismic performance, but the slender braces and minimal steel connections still had to be engineered to resist significant lateral forces while preserving the visual clarity of the timber-priority concept.

At the early design stage, the primary question was not comparing standard code-minimum options, but rather determining how to make the architectural ambition physically viable, while simultaneously addressing the seismic demands of a 0.4 g zone and the environmental imperatives in achieving low

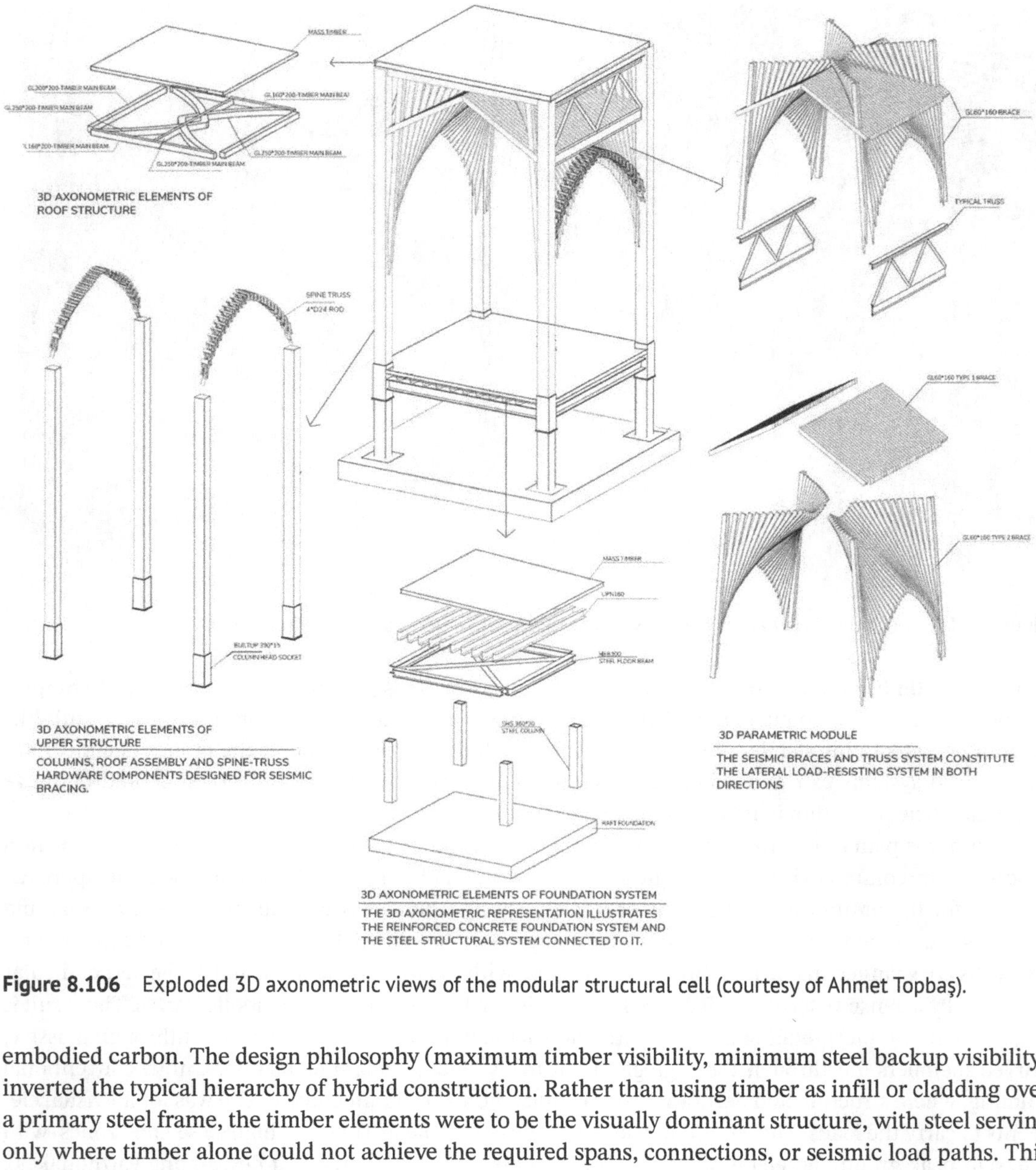

Figure 8.106 Exploded 3D axonometric views of the modular structural cell (courtesy of Ahmet Topbaş).

embodied carbon. The design philosophy (maximum timber visibility, minimum steel backup visibility) inverted the typical hierarchy of hybrid construction. Rather than using timber as infill or cladding over a primary steel frame, the timber elements were to be the visually dominant structure, with steel serving only where timber alone could not achieve the required spans, connections, or seismic load paths. This led to a distinctive set of challenges:

Challenge 1: Achieving a 9 m clear height with slender timber columns and minimal visible steel:

- The steel structure supporting the columns must be concealed below floor level, with only timber visible in the mosque interior.
- Column base fixity must be achieved without bulky visible connections.

Challenge 2: Providing lateral and seismic resistance:

- Brace connections must transfer significant tension and compression loads while remaining visually unobtrusive.
- The system must behave predictably despite the geometric complexity of hundreds of braces at varying angles.

Challenge 3: Forming a rigid roof diaphragm while maintaining the visual rhythm of the brace system:

- The roof must act as a stiff diaphragm to distribute lateral (seismic) loads to the vertical bracing systems.
- The diaphragm structure must integrate with the varying brace connection points without disrupting the parabolic geometry.

Challenge 4: Ensuring constructability of hundreds of geometrically unique connections:

- The parabolic brace geometry means no two connections are identical, creating a coordination challenge for fabrication and erection.
- Screw placements must be resolved in advance to avoid clashes and ensure field assembly is feasible.
- The architectural intent to make "decorative" knee braces structurally active requires each connection to reliably transfer load, not merely provide visual rhythm.

The engineering team addressed these challenges through a layered hybrid strategy, concealing steel where necessary for capacity, exposing timber where possible for expression, and coordinating hundreds of unique connections through detailed 3D modeling:

Solution 1: The 9 m glulam columns are supported by a steel moment-frame basement structure concealed below the ground floor slab (Figure 8.107). SHS 360 × 20 tubular steel columns with rigidly

Figure 8.107 Steel basement and timber superstructure (courtesy of Ahmet Topbaş).

Figure 8.108 Steel basement moment resisting frame connections (courtesy of Ahmet Topbaş).

connected HEB 300 beams form the basement frame, founded on a reinforced concrete raft. The block-glulam columns are socketed 900 mm into the tops of the steel tubes, creating rotational fixity at the column base without any visible steel above floor level (Figure 8.108).

Solution 2: The lateral and seismic resistance is provided by a hybrid "spine-arch" system (Figure 8.106). In the direction vertical to the steel truss, the slender glulam braces are connected continuously to a central spine. The roof acts as a rigid diaphragm, using a 250 mm × 300 mm glulam brace system below the solid 80 mm glulam roof plate. This diaphragm distributes lateral forces evenly across the cells. Each of the hundreds of slender glulam braces connects to the spine-arch via screws, with connection capacity designed for up to 15 kN in tension and compression under seismic load combinations. Given the varying angles of the parabolic braces, no two connections are geometrically identical. The engineering team developed a full 3D coordination model to resolve constructability, ensuring that screw patterns did not clash and that each connection could be installed in sequence. The slender 60 × 160 mm brace sections, combined with the screw connections, allow the braces to act as continuous tension-compression elements transferring lateral loads to the spine and truss system (Figure 8.109).

Figure 8.109 Mosque interior during construction, showing the timber-priority hybrid system: 360 × 360 mm glulam columns, parabolic 60 × 160 mm braces converging on the steel spine-arch, and the glulam roof diaphragm above (courtesy of Ahmet Topbaş).

Solution 3: The roof structure consists of an 80 mm solid glulam plate laid sideways as a continuous diaphragm, supplemented by a 250 × 300 mm glulam brace grid below. This secondary grid serves two purposes: it stiffens the roof diaphragm for lateral load distribution, and it provides a geometric "guideline" for locating the varying connection points of the parabolic wall braces. The roof brace grid thus mediates between the regular column grid and the irregular brace geometry, ensuring that loads flow cleanly into the vertical structure (Figure 8.110).

Figure 8.110 Steel spine-arch with trapezoidal gusset plates receiving the parabolic glulam braces; the 80 mm solid glulam roof diaphragm and 250 × 300 mm brace grid are visible above (courtesy of Ahmet Topbaş).

Figure 8.111 Structural cell before (left) and after (right) installation of the 80 mm glulam floor plate (courtesy of Ahmet Topbaş).

Solution 4: All connections were accurately coordinated in 3D to ensure screw placement was feasible (Figure 8.111). These connections are designed to transfer up to 15 kN of tension/compression forces, turning the decorative "knee brace logic" into an active load-bearing system.

Given the extreme number of slender braces acting at varying angles, a full-building finite element model would have been computationally intensive and difficult to interpret. Instead, the engineering team adopted a "unit cell" approach: each 5 × 5 m cell was modeled and analyzed independently, with boundary conditions representing the continuity provided by adjacent cells (Figure 8.112). This simplified model captured the essential structural behavior (column buckling, brace axial forces, spine-arch bending, diaphragm stiffness) while remaining tractable for iterative design refinement.

As of early 2026, the mosque superstructure is substantially complete, with the timber-priority hybrid system fully erected and the vault-like interior clearly visible (Figure 8.113).

In conclusion, the Mosque of Hayreddin Tokadi Külliye demonstrates that contemporary engineering can honor traditional building cultures while addressing today's seismic and environmental challenges. By treating timber as the primary visual and structural material (and designing steel elements for minimum visibility), the project achieves a spatial quality rooted in Anatolian precedent while meeting modern seismic, environmental, and serviceability requirements. The aggregation of hundreds of slender glulam braces into a vault-like system, the concealed steel basement frame, and the meticulous 3D coordination of screw connections all reflect a design philosophy where architecture, structure, and life-cycle performance are inseparable. Table 8.30 shows the sustainability strategy matrix of this project.

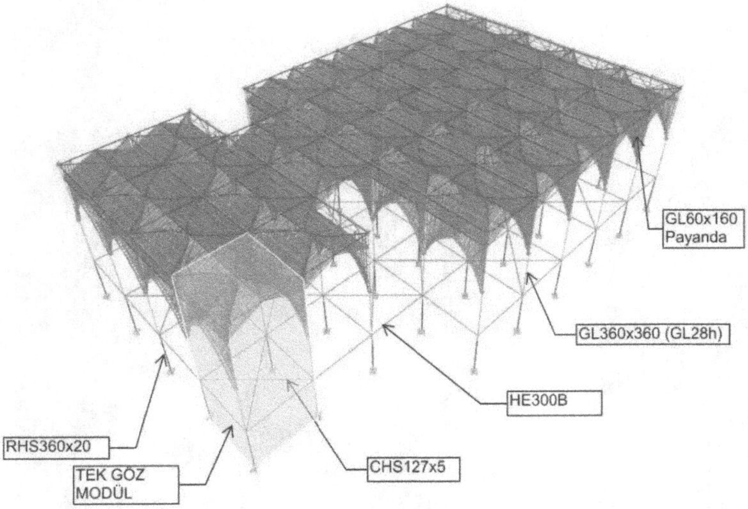

Figure 8.112 Finite element model of the structure (courtesy of Ahmet Topbaş).

Figure 8.113 Current construction status of the mosque, showing the completed timber superstructure with parabolic glulam braces forming the vault-like interior. Exterior cladding and finishing works are underway (courtesy of Hacer Bozkurt).

Table 8.30 Sustainability strategy matrix of Mosque of Hayreddin Tokadi Külliye project.

Strategy	Most relevant LC stages	Key method(s) used	Sustainability outcome
Material efficiency	A1–A3	Hybrid steel/timber composition	Maximizes biogenic carbon storage; reduces visual and material reliance on steel
Modular design	A4, A5	Repetition of 5 × 5 m structural cells; standardized column, brace, and connection details	Simplifies fabrication and erection; enables unit-cell analysis approach
Integrated roof diaphragm	A1–A3	80 mm solid glulam plate, and 250 × 300 mm brace grid	Provides lateral load distribution while guiding parabolic brace geometry
Design for disassembly	C, D	Mechanical screw connections-Socketed bases	Potential for component separation and reuse

Project Team

Design: Abdurrahman Çekim, Ahmet Topbaş, Enver Çakar & Behlül Kıryolcu
Assistant Architects: Hale Nur Çakar, Zehra Karakoç, Nefise Betül Tuncel Akdi & Berkay Öztürk
Offices: Baraka, Attec & CAS
Timber Specialist Fabricator: Attwood
Mechanical Engineering: Tanröver Mühendislik
Electrical Engineering: Zenith Proje Mühendislik
Facade Consultant: Alüma Cephe Danışmanlık
Lighting Consultant: Kitoko

8.5.5 How Seismic Isolation Kept Hospitals Operational Through the 2023 Türkiye Earthquakes

In earthquake-prone regions worldwide, hospitals face a critical vulnerability: conventional construction methods target immediate occupancy performance levels (a design objective requiring that the structure remain safe to occupy immediately after an earthquake, with only minor damage to the load-bearing system), yet often leave facilities nonoperational precisely when they are needed most. This is because conventional seismic design, even when aiming for immediate occupancy, focuses primarily on the structural system while largely overlooking nonstructural elements (such as ceilings, partitions, mechanical equipment, and piping) that can sustain extensive damage and render a hospital unusable after a major seismic event. The typical response to this outcome has been accepting that hospitals may require costly repairs and downtime after major earthquakes and rebuilding or extensively retrofitting them afterward (an approach that demands considerable time, capital, and embodied carbon while leaving communities without critical healthcare during recovery periods). The environmental cost of this cycle is substantial: reconstructing a single mid-size hospital can generate upward of tens of thousands of tons of CO_2e in embodied carbon from new materials alone, in addition to the operational carbon penalty of running temporary medical facilities during the multi-year rebuilding period. Recognizing this gap, the Turkish Ministry of Health released a regulation in 2013 requiring that all new hospitals with a bed capacity greater than 100 be constructed using seismic isolation (Erdik et al., 2018) making Türkiye one of the leading countries in mandating this technology for critical healthcare infrastructure. The Malatya Battalgazi State Hospital project (300 beds, 10 stories, reinforced concrete frame, and infill walls) offers a compelling demonstration of the effectiveness of this policy. On February 6, 2023, two devastating earthquakes (Mw 7.8 and Mw 7.5) struck southeastern Türkiye within nine hours of each other, subjecting the region's infrastructure to extreme seismic demands. While conventionally built hospitals across the region sustained damage to both structural and (more critically) nonstructural components, forcing them to cease operations, the Battalgazi Hospital (equipped with a seismic isolation system positioned between its foundation and superstructure) remained fully operational throughout the sequence and its aftermath, continuing to treat patients and serving as a public shelter for displaced residents (Figure 8.114). The project demonstrates how seismic isolation technology can

Figure 8.114 Malatya Battalgazi State Hospital after the earthquake sequence, showing the seismically isolated facility remained fully operational with minimal damage despite experiencing approximately 17 cm maximum displacement (courtesy of Bahadir Sadan).

transform the resilience paradigm from "immediate occupancy on paper" to "continued functionality in practice," by protecting not only the structural system but also the nonstructural elements that are essential for hospital operations. Figure 8.115 shows the severe damage to conventional buildings caused by the same earthquakes.

The core concept of the project is seismic isolation: by decoupling the building from ground motion using friction pendulum isolators positioned between the foundation and superstructure, this approach dramatically reduces the seismic forces transmitted to the structure above. For the Battalgazi Hospital, 222 Double Concave Friction Slider (DCFS) isolators were designed with an effective radius of 6900 mm and a displacement capacity of 450 mm. The isolation system was designed to achieve a long effective period (~4.0 seconds), shifting the building's dynamic response away from the dominant frequencies of earthquake ground motion and thereby reducing structural demands to levels compatible with continued operation. At the early design stage, the engineering team established performance objectives that exceeded conventional code requirements. The isolation system design parameters were selected to ensure both structural protection and continued post-earthquake functionality. Two design approaches were evaluated to understand the benefits of seismic isolation over conventional construction methods. Table 8.31 compares these approaches, highlighting the advantages of the selected solution.

Figure 8.115 Damage to residential buildings in the earthquake-affected region, in clear contrast to the continued functionality of the seismically isolated hospital shown in Figure 8.114.

Table 8.31 Comparison of different design approaches.

Option	Description	Structural damage	Nonstructural damage	Post-EQ operational status
Conventional fixed-base	Baseline scenario assuming code-compliant conventional construction	Minor to moderate structural damage at this seismic intensity	Severe nonstructural damage expected	Nonoperational; requires repairs before use
Seismic isolation (selected option)	222 DCFS isolators, $R_{eff} = 6900$ mm, $\mu_d = 0.055$, 450 mm capacity	Zero damage observed	Near-zero damage observed (minor partition wall cracks)	Fully operational; served as public shelter

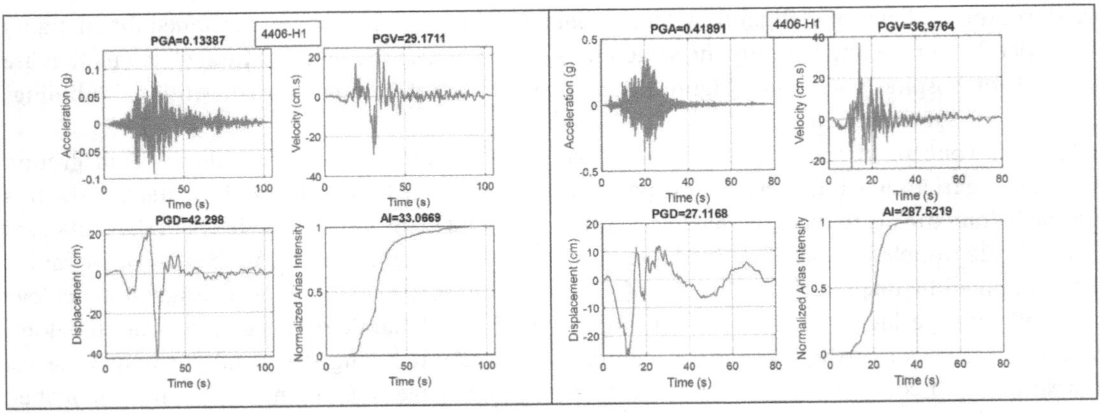

Figure 8.116 Recorded ground motion time histories at station 4406 H1 (31.5 km away from the hospital) for the Mw 7.8 Pazarcık event (left), the Mw 7.5 Elbistan event (right) (courtesy of Bahadir Sadan).

Implementing seismic isolation at Malatya Battalgazi State Hospital to achieve continued functionality posed the following structural, operational, and construction challenges:

Challenge 1: The hospital faced extreme seismic demands from two major earthquakes occurring only nine hours apart (Mw 7.8 at 04:17 and Mw 7.5 at 13:24 on February 6, 2023), with epicenters at approximately 160 km and 100 km respectively (Figure 8.116):

- The fully operational hospital had patients, medical staff, and critical life-support equipment in place during both events, meaning any significant structural damage or system failure would force immediate evacuation and disrupt critical medical care for the surrounding region.
- Sequential loading raised concerns about residual displacement accumulation from the first earthquake compounding during the second event, with hundreds of subsequent aftershocks creating additional cumulative demand (no precedent existed for how friction pendulum isolators would perform through such a closely spaced major earthquake sequence).

Challenge 2: Maintaining full hospital functionality after large isolator displacements (up to 450 mm design capacity) required rigorous detailing of all systems crossing the isolation plane:

- Seismic gaps at building perimeters needed to accommodate the full 450 mm displacement capacity without contact, while construction tolerances, debris accumulation, or improper installation could reduce effective clearances. Experience from other isolated hospitals in the region demonstrated that improperly detailed gaps could restrict isolator movement and compromise system performance.
- All utility systems crossing the isolation plane (water, medical gas, wastewater, electrical, HVAC, fire suppression, and data) required flexible connections designed to accommodate the full displacement capacity. Since these systems were detailed by different MEP disciplines with limited structural coordination, the risk existed that rigid connections or insufficient flexibility could rupture during a seismic event, forcing hospital evacuation even if the structure itself remained intact.

The following solutions were implemented to overcome the above-mentioned challenges:

Solution 1: A robust friction pendulum isolation system with optimized geometric properties and substantial displacement margin (Figures 8.117 and 8.118):

Figure 8.117 General view of isolation basement after the earthquakes (courtesy of Bahadir Sadan).

- The 222 DCFS isolators (with $R_{eff} = 6900$ mm) provided a long isolation period (~4.0 seconds) that shifted the building response away from peak ground motion energy, while the 450 mm displacement capacity created substantial margin (the observed 17 cm maximum displacement represented only 38% utilization of the system's capacity).
- The design friction coefficient ($\mu_d = 0.055$), calibrated from prototype testing, provided consistent energy dissipation across both earthquake events. The curved sliding geometry created gravity-induced restoring forces that limited residual displacement to only 1.5 cm (less than 1% of capacity), preventing accumulation during the 9-hour earthquake sequence and over hundreds of subsequent aftershocks (Figures 8.117 and 8.118).
- Nonlinear time history analysis of a representative model using recorded ground motions confirmed limited isolator displacements, very low inter-story drifts, and significant reductions in structural accelerations compared to a fixed-base system.

Solution 2: Attention to the nonstructural system design with rigorous construction quality control and inspection protocols:

- Seismic gap detailing maintained 450 mm clearances with flexible cover plates and drainage channels (Figure 8.119). An inspection protocol (including photographic documentation during construction and final debris removal before occupancy) was implemented. Post-earthquake inspection confirmed zero gap closure or contact occurred during either event.

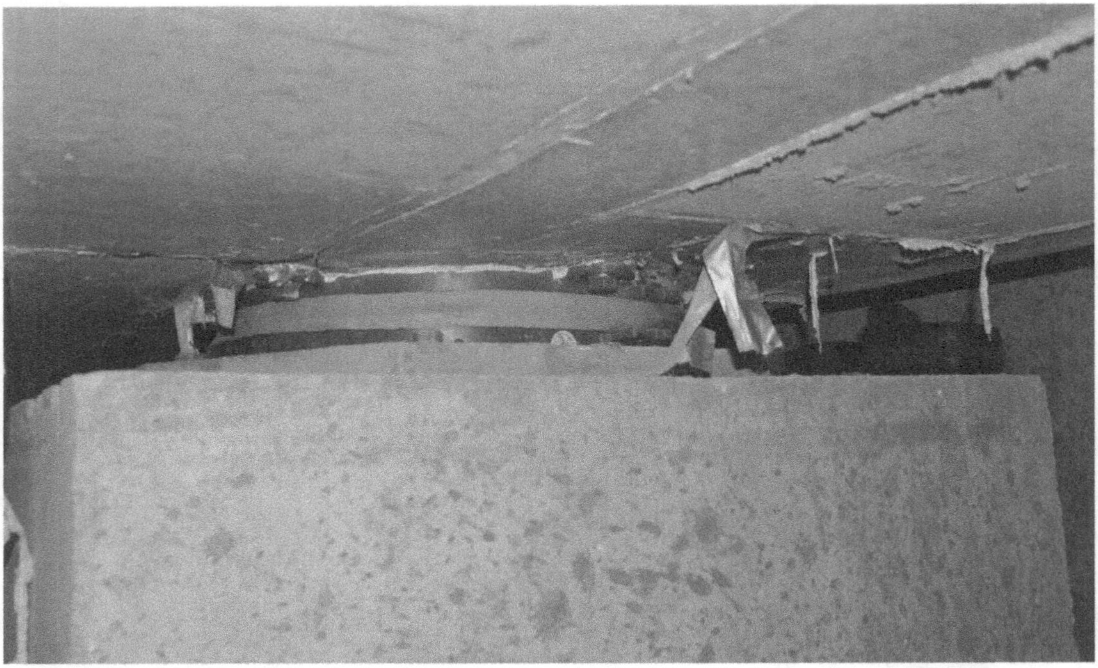

Figure 8.118 Close-up view of a Double Concave Friction Slider (DCFS) isolator positioned between the foundation footing and the superstructure slab, photographed after the 2023 earthquake sequence. The isolation devices show no visible damage, confirming the integrity of the isolation system following both seismic events (courtesy of Bahadir Sadan).

- Flexible utility connections were designed, detailed, and factory-tested for 450 mm displacement capacity across all systems crossing the isolation plane: flexible hoses for water and medical gas, slack cables for electrical systems, and fabric bellows for HVAC ducts (Figure 8.120). MEP engineers coordinated with the structural team through shop drawing reviews to ensure compatibility.
- Post-earthquake performance confirmed all utility systems remained fully functional with no leaks or service interruptions, enabling the hospital to immediately serve as a public shelter for displaced residents.

Beyond structural performance, the Battalgazi Hospital's survival carries significant sustainability implications. Across the earthquake-affected region, other conventionally built hospitals without seismic isolation suffered partial or severe damage, with several requiring complete demolition and reconstruction. Each reconstruction repeats the full embodied carbon expenditure of modules A1 through A5 (raw material extraction, manufacturing, transport, and construction), while also generating substantial demolition waste. By contrast, the seismically isolated hospital preserved its entire embodied carbon investment (estimated at several tens of thousands of tons of CO_2e for a facility of this scale) and avoided any material expenditure for post-earthquake repair. When considered across all 11 isolated hospitals that survived the earthquake sequence, the cumulative avoided reconstruction represents a significant reduction in life-cycle carbon emissions for Türkiye's healthcare infrastructure.

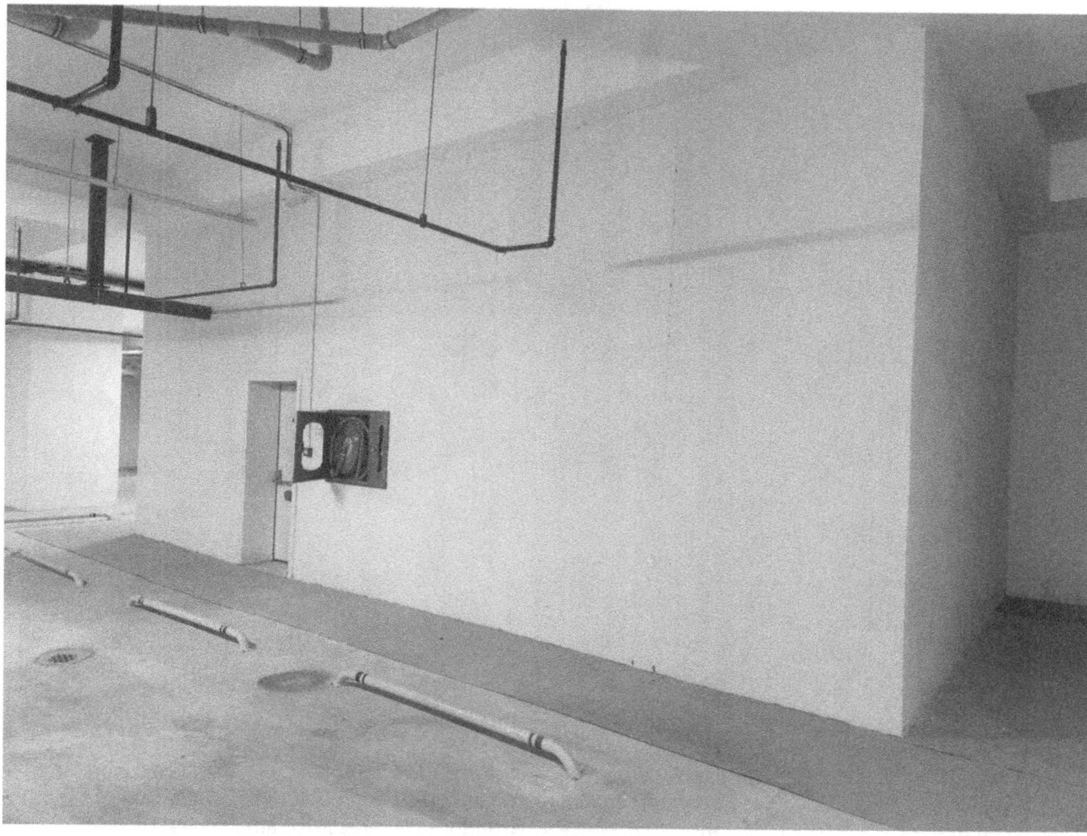

Figure 8.119 Seismic gap detailing at the building perimeter, showing the 450 mm clearance maintained between the isolated superstructure and the surrounding retaining wall, with flexible cover plates at floor level. Post-earthquake inspection confirmed zero gap closure or contact during either event (courtesy of Bahadir Sadan).

In conclusion, this project demonstrated that seismic isolation can successfully protect critical healthcare facilities under extreme real-world conditions, shifting the design paradigm from immediate occupancy to continued functionality. Seismic isolation can be a critical tool for engineers and designers to achieve both resilience and sustainability goals for essential facilities, effectively eliminating structural damage, maintaining operational capacity, preserving the full embodied carbon investment of the original structure, and enabling immediate post-disaster service when the surrounding community is in greatest need. The broader validation from the 2023 Kahramanmaraş earthquake sequence (where all eleven seismically isolated hospitals in the affected region maintained their structural integrity) provides additional confirmation that this technology delivers reliable performance at scale (while collectively avoiding the reconstruction carbon footprint that would have resulted from conventional design). Table 8.32 shows a resilience and sustainability strategy matrix for the Malatya Battalgazi project that summarizes the core approaches by mapping each method to its engineering principle and the resulting resilience outcome.

Figure 8.120 Utility systems crossing the isolation plane in the isolation basement, including fire suppression piping and cable tray supports, all designed and detailed to accommodate the full 450 mm displacement capacity, all suspended to the isolated floor. No damage or service interruptions were recorded after the earthquake sequence (courtesy of Bahadir Sadan).

Table 8.32 Resilience and sustainability strategy matrix.

Strategy	Most relevant LC stages	Key method(s) used	Sustainability outcome
Seismic isolation technology	A1–A3, B	Double concave friction sliders with large displacement capacities	Prevention of structural damage in all eleven hospitals; extended asset service life
Nonstructural design	A1–A3, B	Robust detailing of partition walls, ceilings, utilities, and seismic gaps	Reduction of functional disruption; ensuring continued hospital operability

Project Team
Structural Engineer (Structural and Seismic Isolation Design): Bensum Muhendislik
Seismic Isolator Supplier: TIS Teknolojik İzolatör Sistemleri (TIS)
Isolator Testing Facility: ESQUAKE isolator testing facility

8.6 Resilient Infrastructure Design

8.6.1 Bridge Design Optioneering with High-strength Steel

For decades, the design of steel–concrete composite highway bridges has relied on standard steel grades. While such materials offer proven reliability, they often lead to heavy superstructures, thick plates, extensive weld volumes, and designs predominantly governed by strength and rigidity criteria. A recent research project OPTIBRI, funded by Research Fund for Coal and Steel (European Commission. Directorate General for Research and Innovation., 2019) challenged this paradigm by investigating the holistic implications of employing High-Strength Steel (HSS) (class S690) in a major 360-m-long highway bridge. The objective was to quantify the potential benefits and identify the new engineering challenges that arise when extending the boundaries of material utilization.

Figure 8.121 presents the longitudinal and cross-sectional views of the case-study bridge – a five-span continuous composite structure with a total deck width of 21.5 m, accommodating multiple traffic lanes. The geometry reflects both the functional requirements of a modern highway crossing and the structural opportunities for material optimization enabled by the adoption of HSS.

The core principle of the project was a comparative design study to assess the potential advantages of HSS in bridge engineering. Three parallel designs were developed in accordance with Eurocode standards (Figure 8.122):

- **Design A:** Standard Steel S355
- **Design B:** High-Strength Steel S690
- **Design C:** High-Strength Steel S690 with enhanced fatigue detailing and weld treatments

This methodology enabled a comprehensive evaluation of how HSS redefines the governing design criteria (shifting the focus from standard strength-based limitations to issues primarily associated with fatigue and plate buckling) thereby establishing a clear framework for its efficient and safe application in future bridge design.

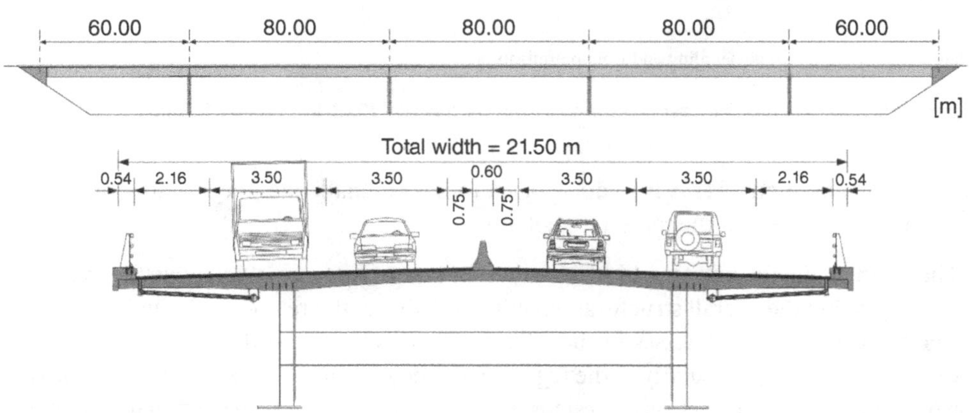

Figure 8.121 Longitudinal view and deck cross-section with highway platform data (Pedro et al., 2018/ John Wiley & Sons).

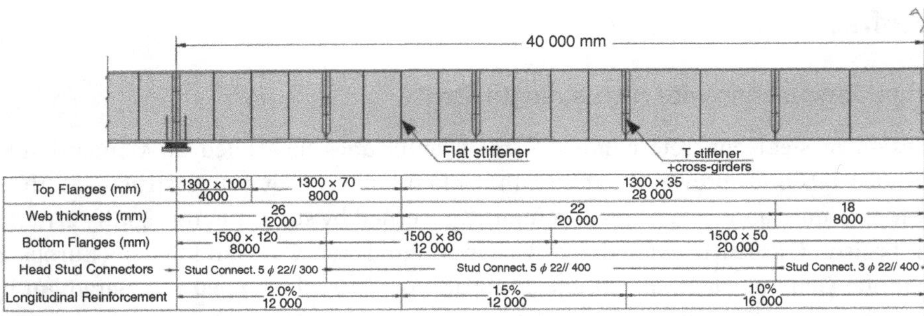

Design A: Standard Steel S355

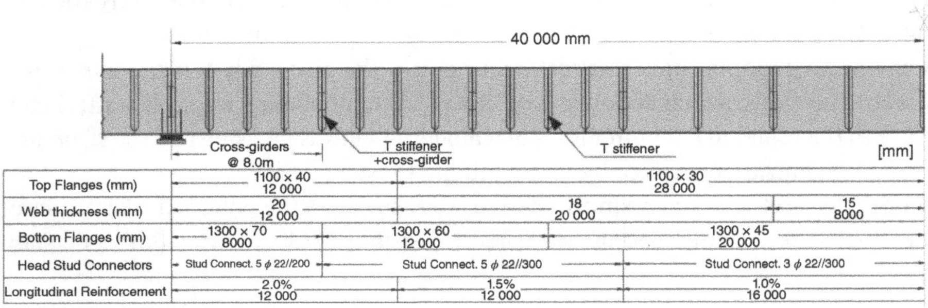

Design B: HSS S690

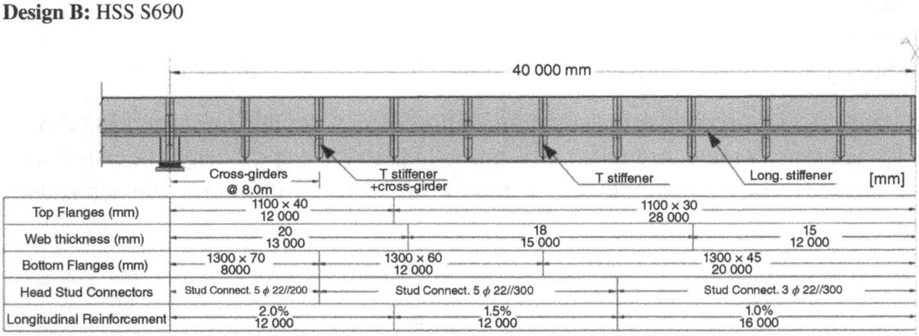

Design C: HSS S690 with advanced fatigue detailing and weld optimization

Figure 8.122 Comparison of designs for a typical inner span (Pedro et al., 2018/John Wiley & Sons).

At the same time, the adoption of HSS introduced a set of design challenges, demanding innovative engineering solutions to fully exploit the material's potential:

Challenge 1: The main challenge was to achieve a lighter and more resource-efficient structure, while maintaining or improving the overall structural reliability relative to the reference design.

Challenge 2: The reduced plate thicknesses in the HSS alternatives increased the susceptibility of the girder webs to shear buckling, particularly in the regions near the supports. The conventional approach of employing only transverse stiffeners, as in Designs A and B, proved inadequate to mitigate this effect.

Challenge 3: The higher strength of HSS results in greater stress ranges under equivalent loading conditions, making fatigue even more critical (and often governing) design consideration. The fatigue behavior of welded details, particularly the connection between transverse stiffeners and the tension flange, emerged as the primary limiting factor in the design.

Challenge 4: The standard S355 design required very thick plates, with thicknesses reaching up to 120 mm. Such plate dimensions presented a significant risk of brittle fracture at the project's low reference temperature of −30 °C, and their justification within the Eurocode framework was challenging. Moreover, welding operations for plates of this thickness are considerably more demanding, time-consuming, and costly, further compromising fabrication efficiency and overall project economy.

To address these challenges, several design alternatives were examined, leading to the following solutions:

Solution 1: By adopting S690 steel, the design team achieved a substantial reduction in the required thickness of the main girder flanges and webs. The final optimized configuration (Design C) reduced the total structural steel weight from 219 kg/m^2 in the S355 design to 149 kg/m^2, representing a 32% reduction. This optimization also resulted in an approximately 50% decrease in the volume of full penetration welds, providing notable fabrication and cost benefits (Table 8.33 and Figures 8.123 and 8.124).

Solution 2: Whereas Design B adopted more closely spaced transverse stiffeners, Design C introduced a more efficient alternative: a single, robust trapezoidal longitudinal stiffener positioned externally on each web. This configuration proved more effective in enhancing the shear buckling resistance of the slender high-strength steel (HSS) webs, allowing for an efficient use of material without compromising structural safety (Figure 8.125).

Table 8.33 Comparison of the three design steel quantities.

Steel weight (average)	Design A (S355)	Design B (S690)	Design C (S690 optimized)
Total steel ratio	219 kg/m^2	165 kg/m^2	149 kg/m^2
Main girder sections	186 kg/m^2	123 kg/m^2	114 kg/m^2

Source: Adapted from Pedro et al. (2018).

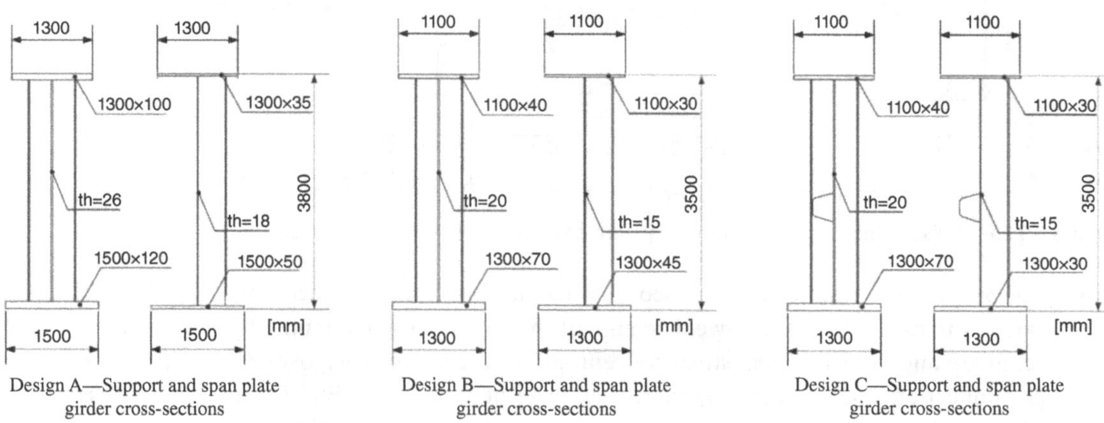

Figure 8.123 A visual showing of "Solution 1" (Pedro et al., 2018/John Wiley & Sons).

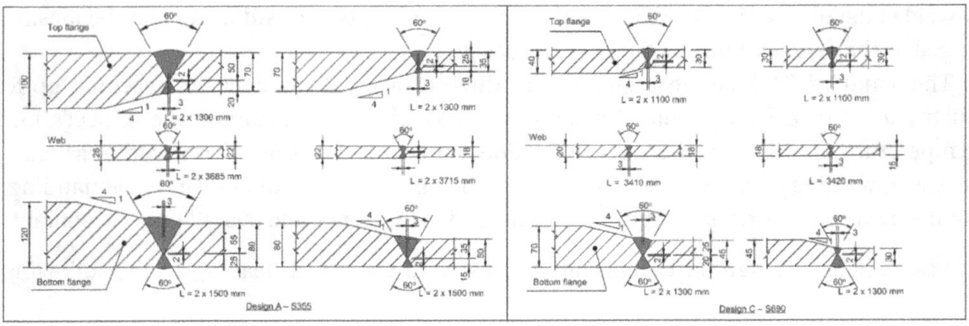

Figure 8.124 A visual showing the point of "Challenge 4" (Pedro et al., 2018/John Wiley & Sons).

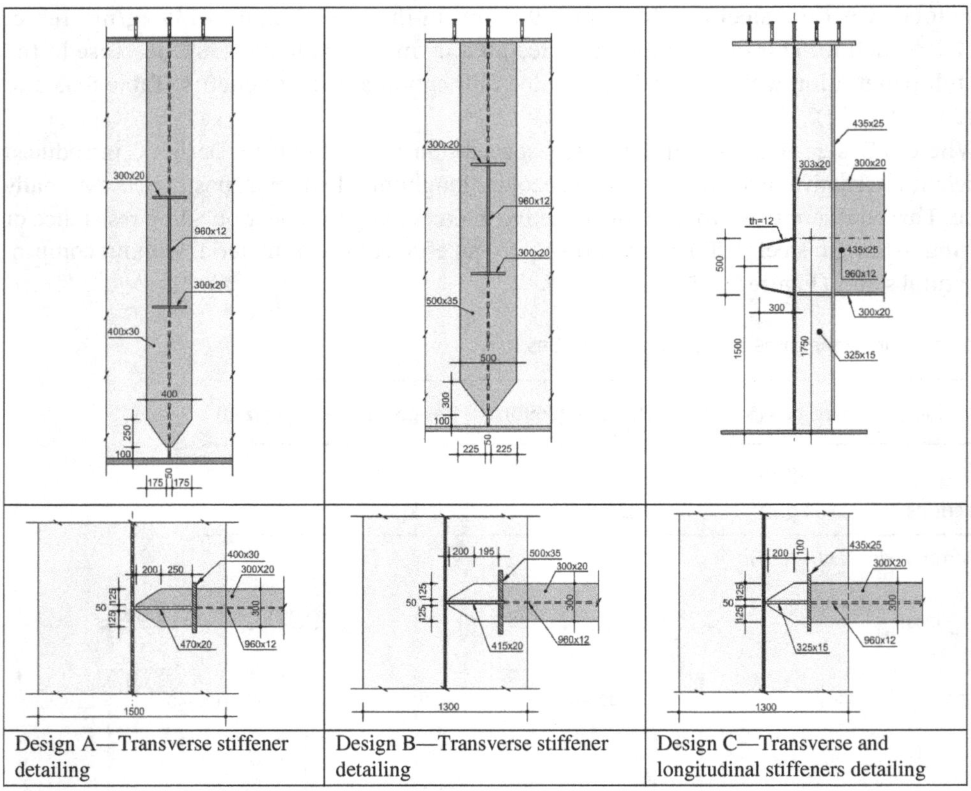

Design A—Transverse stiffener detailing	Design B—Transverse stiffener detailing	Design C—Transverse and longitudinal stiffeners detailing

Figure 8.125 A visual showing of "Solution 2" (Pedro et al., 2018/John Wiley & Sons).

Solution 3: Design C directly addressed the fatigue concerns by incorporating high-frequency mechanical impact (HFMI) post-weld treatment in the most critical joints. By improving the weld toe geometry and reducing local stress concentrations, the fatigue category (FAT) of the stiffener-to-flange connection was substantially increased. (The new proposed FAT for the joint between bottom flange and transverse stiffeners according to research project OPTIBRI is shown in Figure 8.126

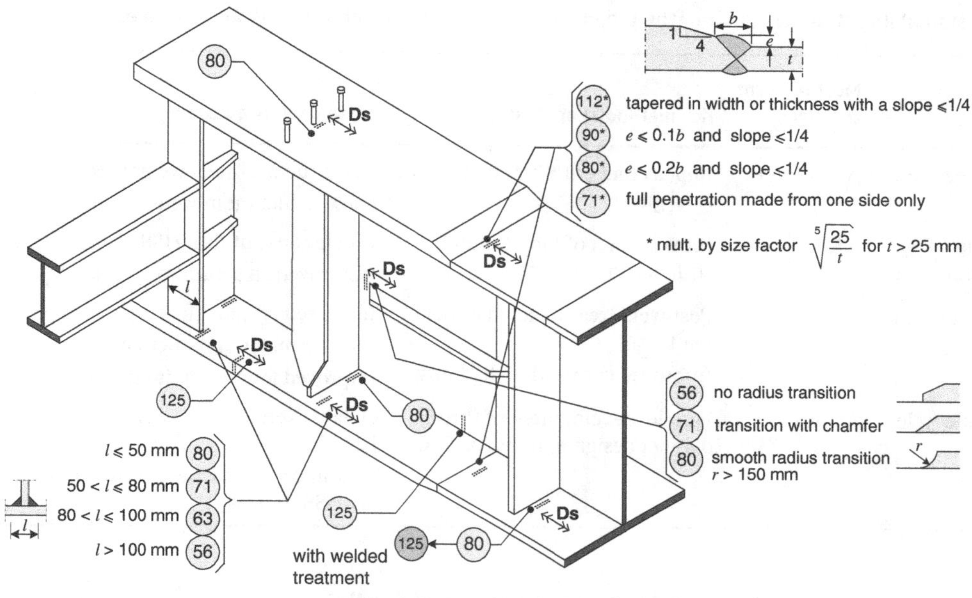

Figure 8.126 A visual showing of "Solution 3" – typical FAT detail categories and possible improvement by high-frequency mechanical impact (HFMI) post-weld treatment (reproduced based on Pedro et al., 2018/ John Wiley & Sons).

[Pedro et al., 2018].) This targeted enhancement enabled a further reduction in the bottom flange thickness, achieving an optimized configuration beyond what was attainable using HSS alone.

Solution 4: The adoption of HSS also inherently resolved the issue associated with excessively thick plates (Figure 8.124). The higher material strength allowed for significantly thinner plate sections (up to 70 mm maximum), which exhibit improved toughness and fall within the Eurocode limits for brittle fracture resistance, even when using the high-performance QL1 grade.

Table 8.34 presents the sustainability strategy matrix developed for the project, summarizing the four core categories of lessons learned. Each targeted strategy is mapped to its corresponding life-cycle (LC) stages, along with the applied methods and the resulting material and carbon savings (Lemma et al., 2020; Pedro et al., 2018).

This comparative study demonstrates that HSS is not merely a stronger substitute for conventional structural steel but the foundation of a new design philosophy in bridge engineering. While HSS offers substantial reductions in material usage and structural weight, it also shifts the main design focus from conventional strength limitations to new challenges, particularly those related to buckling and fatigue, which were already important in all bridge designs. The research project OPTIBRI confirmed that, through the application of sound engineering solutions (such as longitudinal stiffeners and post-weld treatments) such challenges can be effectively mitigated. The resulting outcomes of this research project establish a validated and replicable framework, enabling engineers to confidently specify HSS for the development of more efficient, economical, and durable bridge infrastructure.

Further details about this project can be found in Lemma et al. (2020) and Pedro et al. (2018).

Table 8.34 Sustainability strategy matrix of the project according to Lemma et al. (2020) and Pedro et al. (2018).

Strategy	Most relevant LC stages	Key method(s) applied	Sustainability outcome
Material optimization with HSS	A1–A3, A4	Replacement of S355 with S690 steel grade	32% structural steel weight reduction ~50% reduction in weld volumes
Advanced plate buckling control	A5, B	Introduction of longitudinal stiffeners on slender webs	Efficient use of thin HSS plates Optimized structural behavior
Enhanced fatigue and durability	B, D	Post-weld treatments in critical joints Improved connection detailing	Increased fatigue life Further material reduction Improved long-term durability
Comparative lifecycle design	A1–A5	Parallel development of three distinct design options (A, B, C)	Data-driven validation of HSS benefits Definition of new design guidelines for HSS bridges

8.6.2 Sustainability Ahead of Its Time for Bridge Viaducts of Sardinia

At a time when sustainability was not yet a dominant theme, bridge engineering pioneers were already pursuing optimization goals that society today recognizes as fundamental to reducing environmental impact. The viaducts built in 2003 for the SS125 Orientale Sarda in Sardinia, Italy (Figure 8.127) are an example of this philosophy. Anticipating the debate on sustainability by decades, this project demonstrates how innovation and the pursuit of material efficiency (in the tradition of great masters like Pier Luigi Nervi) constitute the essence of an intrinsically sustainable approach.

The core concept of the project is a patented, three-dimensional continuous composite steel–concrete girder, applied over approximately 2 km of viaducts in Sardinia, Italy, including the Rio Masone Murtas (977 m), Gutturu Frascu (221 m), Pispisa (137 m), and others. Using only 92 kg/m^2 of structural steel for 40-m spans, this solution represents a benchmark for material efficiency in infrastructure projects.

The project faced complex challenges, typical of large infrastructure works in sensitive contexts:

Challenge 1: The need to build nearly 2 km of viaducts required a solution that minimized the use of raw materials (and consequently, embodied carbon), as well as optimizing the structure shape for better landscape integration.

Challenge 2: Road infrastructure is exposed to aggressive atmospheric agents. Ensuring a long service life with reduced maintenance costs was a fundamental requirement to ensure the economic and functional sustainability of the work over time.

Challenge 3: Bridge construction required rapid assembly and launching methods to reduce construction time and minimize interference with the territory.

Figure 8.127 SS125 Orientale Sarda in Sardinia (courtesy of Redesco Progetti).

The following engineering solutions were implemented to solve previously mentioned challenges:

Solution 1: A composite steel–concrete girder was designed combining a central built-up steel beam with a tubular truss structure. This system uses material optimally: the steel works mainly in tension and shear, while the reinforced concrete is subjected to compression. The result is a lightweight and highly efficient structure, which drastically reduces the load on the piers and foundations.

In particular, the structural efficiency of this bridge is achieved through the following elements (Figures 8.128 and 8.129):

- The Y-shaped reinforced concrete pier provides a stable base and an optimized path to transfer loads to the foundation.
- The three-dimensional (weathering, i.e. corten) steel truss uses the inherent load transfer efficiency of a truss and buckling-resistant tubular members to create a robust, stiff, and lightweight superstructure.
- The main girders provide the primary bending resistance in an optimized I-beam shape.
- The shear studs enable highly efficient composite action between the reinforced concrete deck and the steel frame.
- The lateral bracing provides global stability against lateral and torsional forces.

Figure 8.128 Elements of the bridge providing its structural and material efficiency (courtesy of Redesco Progetti).

Solution 2: The use of weathering ("corten") steel eliminates the need for anti-corrosion treatments, reducing life-cycle maintenance costs.

Solution 3: The construction details were designed for easy on-site assembly (Figure 8.130). The open geometry of the truss structure, protected from direct solar radiation by the deck, reduces the effects of thermal changes, improving the overall stability and durability of the structure. Furthermore, the lower truss elements, made from press-bent plates with an inverted "OMEGA" section, serve as both a structural element and permanent formwork for casting the reinforced concrete chords, simplifying construction.

Table 8.35 summarizes the core sustainability strategies by mapping each method to its life-cycle contribution and resulting outcomes. In conclusion, the viaducts of the SS125 demonstrate how the search for structural efficiency inherently leads to sustainable results, reducing the use of resources and environmental impact. This approach, rooted in the Italian engineering tradition, produced a state-of-the-art solution in 2003 that remains a benchmark today.

1 Bracings

2 Shear studs

3 Main steel girder

Figure 8.129 Elements of the steel part of the bridge deck providing its structural and material efficiency (courtesy of Redesco Progetti).

(a) **(b)**

Figure 8.130 Construction site: (a) cast-in-place reinforced concrete piers; (b) installation of the steel deck (courtesy of Redesco Progetti).

Table 8.35 Sustainability strategy matrix of SS125 Orientale Sarda in Sardinia project.

Strategy	Most relevant LC stages	Key method(s) used	Sustainability outcome
Structural optimization	A1–A3	A patented composite steel and reinforced concrete girder Analysis to minimize self-weight Tubular steel profiles	92 kg/m^2 of steel (much lower than alternatives) Reduced loads on piers and foundations
Off-site fabrication	A4–A5, C1–C4	Lightweight prefabricated elements Design for rapid assembly and disassembly	Reduced launching and construction times Less impact from site activities
Innovation and R&D	A1–A5	Development of a patented system Integration of functions (structure and formwork)	Construction efficiency Replicable solution for multiple viaducts Driving innovation in the sector
Durability	B2, B3	Use of weathering (Corten) steel Design for thermal stability	Elimination of protective coatings Reduced maintenance costs Longer service life

8.6.3 Mastering Airport Design Through Optimization, Shock Transmitters, and Testing

Airport terminals are gateways that shape a nation's identity. For the New Passenger Terminal at Rabat-Salé Airport in Morocco (Figure 8.131), the design had to be both highly functional and symbolically powerful. The architectural vision was ambitious, featuring a vast, sinuous roof, soaring 39-m-tall sculptural columns, and sweeping, low-arched entrances. The engineering challenge was immense: how to translate this fluid, nonstandard geometry into a safe, efficient, and buildable structure?

The core concept of this project is enabling architectural freedom through parametric design and advanced steel engineering. Using parametric design tools as the central nervous system, the team was able to model, analyze, and optimize the complex geometry, manage huge thermal forces with innovative joints, and deliver a structure that remained completely true to the ambitious architectural intent. The structural system is organized into several distinct levels and features, driven by the architectural choices as shown in Table 8.36.

The terminal combines reinforced concrete and steel in a clear, purpose-driven hierarchy: reinforced concrete forms the underground level with its pillars and transfer beams and continues into the ground-floor slab, while the two intermediate decks at +4.00 m and +8.50 m are built as composite concrete-on-steel floors. Primary load-bearing elements – the interior and façade columns – are steel, and they in turn support the elaborate three-dimensional truss roof entirely in steel, marrying lightweight spans with the rigidity needed for long cantilevers.

The terminal's unique architectural vision created a series of formidable engineering challenges that required principles thinking and innovative solutions. The main challenge was to reconcile architectural ambition with sustainable performance. This was addressed through early-stage optimization using a decision matrix leading to "good enough" trade-offs between cost, feasibility, and sustainability

Figure 8.131 Render of the project (courtesy of Andrea Imbrenda, Proges Engineering).

Table 8.36 Main characteristics of the building.

Component/level	Area/dimension
Building footprint	250 m × 120 m
Roof footprint	300 m × 170 m (50 000 m^2)
Total usable area	95 000 m^2
Basement (−6.00 m)	15 430 m^2
Ground floor (±0.00 m)	30 600 m^2
Mezzanines (+4.00 m and +8.50 m)	Boarding and retail zones
Maximum column height	39 m

(especially for the columns and the roof). The roof design combined a wide variety of profiles with wind-tunnel data to enhance both efficiency and safety. Other specific challenges were:

Challenge 1: The design featured highly complex elements that could not be engineered with traditional methods. The roof's elegant double-curved surface, cantilevering up to 36 m beyond its supports and spanning a 24 m × 24 m column grid, required advanced parametric modeling and iterative optimization to balance structural efficiency with architectural expression.

Challenge 2: The original three-branched, X-shaped façade column design required advanced press-forming technology for 20–25 mm thick steel sheets, which was unavailable locally in Morocco and became unfeasible due to COVID-19 import restrictions.

Challenge 3: The 50 000 m^2 steel roof structure would naturally expand and contract with temperature changes. Introducing traditional expansion joints would have compromised the seamless architectural aesthetic and created potential points of failure. The team needed a way to allow the roof to "breathe" without breaking it into smaller sections.

Challenge 4: The terminal's vast, undulating roof has a unique aerodynamic profile that standard building codes cannot adequately model. Relying on simplified code-based pressures would lead to either an unsafe or a grossly over-designed structure, making physical testing and advanced analysis essential.

The project team implemented the following solutions for the abovementioned challenges:

Solution 1: The entire project was managed through a parametric BIM. This digital core (Figure 8.132) served as the single source of truth and enabled a seamless workflow:

- Connected directly with various structural analysis software packages for rapid design iteration.
- Included custom routines to process complex wind tunnel data to accurately model wind loads on the unique roof shape.
- Facilitated structural verification by allowing for easy changes to element types, and generated a detailed production model for the fabricators, ensuring perfect alignment between design and construction.

The roof structure was progressively optimized by expanding the range of section sizes and integrating wind tunnel test results. This led to substantial weight savings resulting in a notable reduction in GHG emissions. Figure 8.133 shows how expanding the profile catalogue from 4 to 22 section types delivers a weight saving of approximately 40%. The vertical line marks the point where wind-tunnel data is applied, yielding an additional 10% reduction beyond the 22-section baseline.

Solution 2: Along the perimeter, "X"-shaped façade columns spaced at 12 m provide additional bracing against wind and seismic forces (Figure 8.134). The columns were originally conceived as three-branched, X-shaped configuration formed from 20 to 25 mm press-formed sheets (an optimal choice for minimal material use, reduced labor hours, and high structural efficiency). Although this forming technology is common in Europe, no Moroccan fabricator could handle sheets of that thickness, and COVID-19 travel and trade restrictions made importing the semi-finished profiles impractical. To keep the project on schedule, the design was revised to use H-shaped sections cut from flat

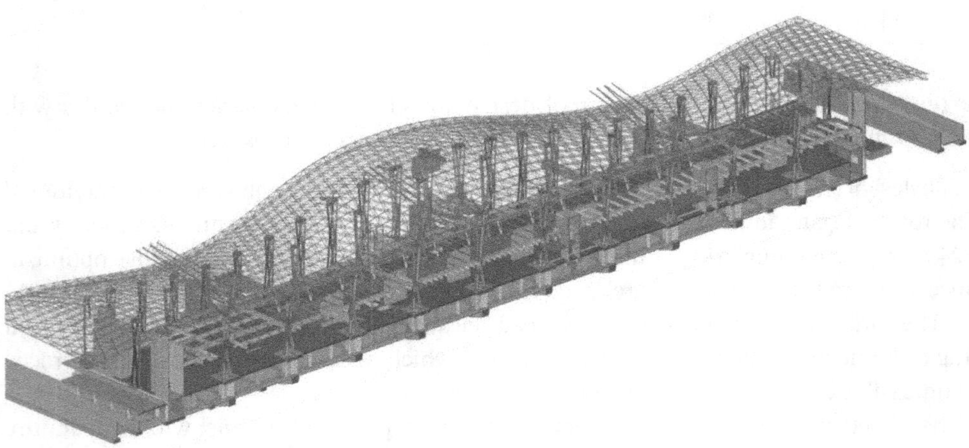

Figure 8.132 BIM model (courtesy of Andrea Imbrenda, Proges Engineering).

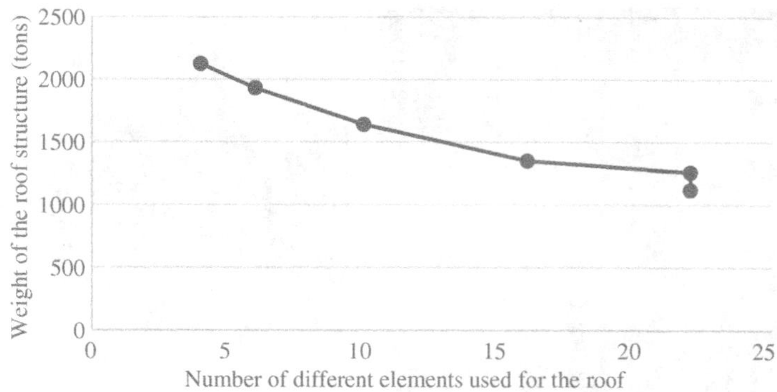

Figure 8.133 Weight trend as the number of sections used increases.

Figure 8.134 X-shaped columns (courtesy of Andrea Imbrenda, Proges Engineering).

plates. While this "good" solution demands roughly 15% more steel and added manual fabrication time, it guarantees local producibility and uninterrupted progress. Although these solutions look very steel-heavy with respect to standard buildings, large-span airport terminals require substantially more steel due to their need for column-free spaces, long cantilevers, and complex geometries that preclude

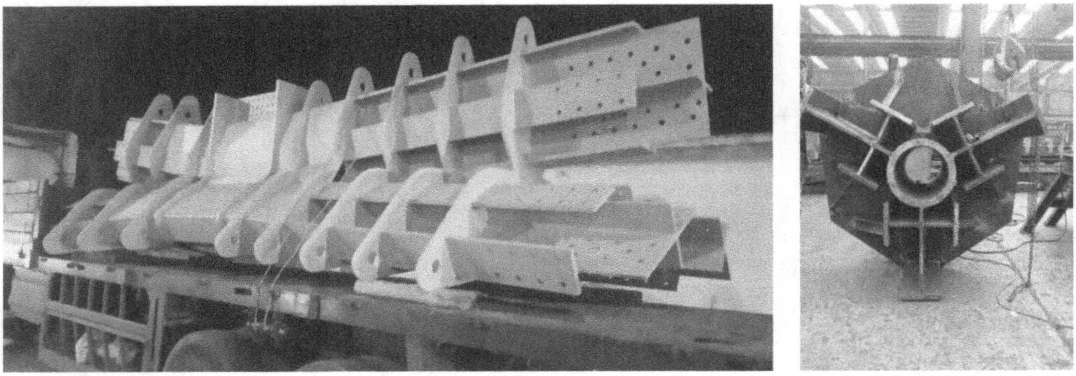

Figure 8.135 Columns during off-site fabrication (courtesy of Andrea Imbrenda, Proges Engineering).

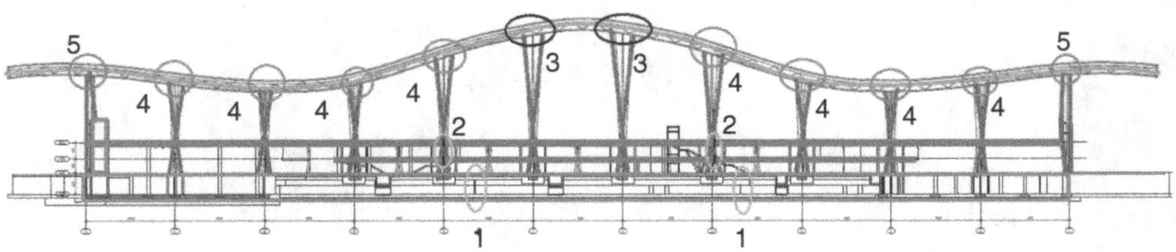

Figure 8.136 Thermal joints (1 – expansion joints for underground structures, 2 – expansion joints for steel structures, 3 – fixed supports, 4 – vertical supports and shock transmitters, 5 – bilateral support and shock transmitters) (courtesy of Andrea Imbrenda, Proges Engineering).

typical concrete solutions. Steel's high strength-to-weight ratio and ability to span large distances make it essential for achieving the architectural vision while managing dead loads efficiently. These columns were completely manufactured off-site (Figure 8.135).

Solution 3: To accommodate temperature changes without compromising strength, the roof "floats" on Shock Transmitter Units (STUs) (stiff under wind or seismic forces but flexible under slow thermal shifts), while four central connections remain fixed to maintain balanced movement and rigid behavior when needed. Four "fixed points" were established in the center of the structure to provide lateral stability (Figure 8.136). All other column supports were designed with bearings to allow for slow thermal movement, but were equipped with shock-transmitters (Figure 8.137). These devices remain flexible under slow temperature-induced movement but lock up instantly during rapid, dynamic events like an earthquake, ensuring the structure is both flexible and resilient.

Although technically complex, these shock-transmitter solutions eliminate the need for traditional expansion joints throughout the structure, avoiding additional structural materials, aesthetic compromises, and ongoing maintenance requirements. This approach reduces long-term operational costs while maintaining structural integrity and the seamless architectural design.

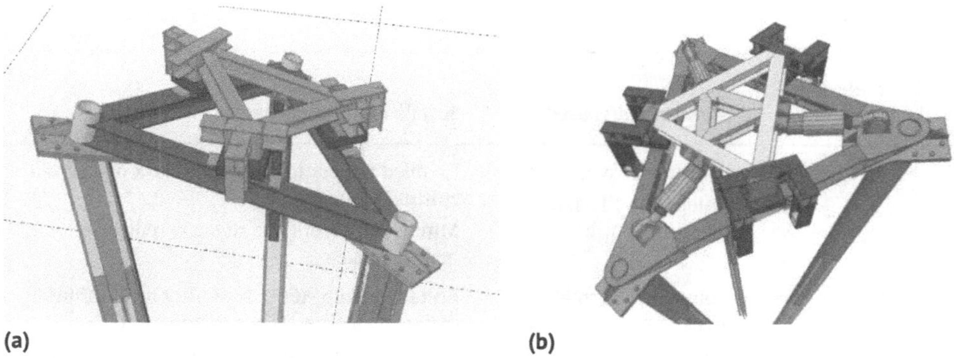

Figure 8.137 Different types of column joints (from BIM models): (a) fixed joint; (b) joint with shock transmitters (courtesy of Andrea Imbrenda, Proges Engineering).

Figure 8.138 Wind tunnel tests (courtesy of Andrea Imbrenda, Proges Engineering).

Solution 4: The workflow included custom routines to process complex data from physical wind tunnel tests, allowing the team to apply highly accurate, real-world wind loads directly to the digital model for precise analysis (Figure 8.138). These experimental wind tunnel tests provide site-specific pressure distributions that are far more accurate than code-prescribed values, enabling engineers to optimize member sizes precisely where needed. This results in significant material savings by avoiding the conservative over-design required when using generic code values for nonstandard geometries.

Table 8.37 shows a sustainability strategy matrix for the project that summarizes the three core strategies used to achieve the desired outcomes.

In summary, this project demonstrates how bold architectural intent and environmental responsibility can advance together. Early-stage optioneering ensured balanced choices between cost, buildability and sustainability; supply-chain adaptation kept fabrication local and on time; and rigorous roof optimization cut structural weight by half and trimmed GHG emissions.

Table 8.37 Sustainability strategy matrix of the project.

Strategy	Most relevant LC stages	Key method(s) used	Sustainability outcome
Design for architectural longevity	A5, B (Use)	Parametric workflow Integrated FEM/BIM analysis	Enabled realization of a landmark design with cultural value Minimized errors and rework, reducing material waste
Resilience and durability	B (Use)	Joint-free thermal expansion system with shock-transmitters	Enhanced long-term durability by eliminating maintenance-heavy expansion joints Ensured resilience against both thermal and seismic forces
Sustainable construction methods	A5 (Construction)	Phased concrete pouring with closure strips	Minimized material stress (cracking) and potential for future repairs Optimized material use and ensured structural longevity

8.6.4 Integrating Precision and Sustainability in the Erection of an Arch Bridge

Located near the city of Arequipa in southern Peru, the La Joya Bridge is an extraordinary example of technical ingenuity in steel construction. Designed by Cons. Ing. Arequipa – Cesma Ingenieros, with steelwork by Cimolai S.p.A., this arch bridge crosses a steep and inaccessible gorge over 120 m deep (Figure 8.139). The project required specific solutions for both assembly and erection, given the complex topography and preexisting concrete foundations that could not be modified. Beyond its structural and logistical complexity, the project integrates several sustainability-oriented strategies that demonstrate how precision engineering and construction planning can reduce resource use and environmental impact across the bridge's life cycle.

The bridge is a steel through arch with a span of 175 m between two sides (Figure 8.140), formed by tubular members of 2250 mm diameter and varying thicknesses, weighing 1350 tons. The upper viaduct, consisting of a composite steel–concrete deck, extends for 371 m and adds another 1150 tons of steel, bringing the total steel weight to approximately 2500 tons. The structure is distinctive for its negative vertical alignment, visible to the naked eye, and for the paired arrangement of the arch ribs supporting twin superstructures.

Transforming the design into a finished structure over the deep gorge posed several structural, logistical, and design challenges:

Challenge 1: The gorge beneath the site had extremely limited access, with only short segments reachable from the top through steep, narrow service roads. The contractor also found the foundations already built, composed of a combination of short piles, long piles, and direct rock footings, which could not be altered. These fixed conditions imposed significant constraints on the assembly sequence and equipment selection.

Figure 8.139 View of the La Joya Arch Bridge and the deep gorge near Arequipa, Peru (courtesy of CIMOLAI S.p.A).

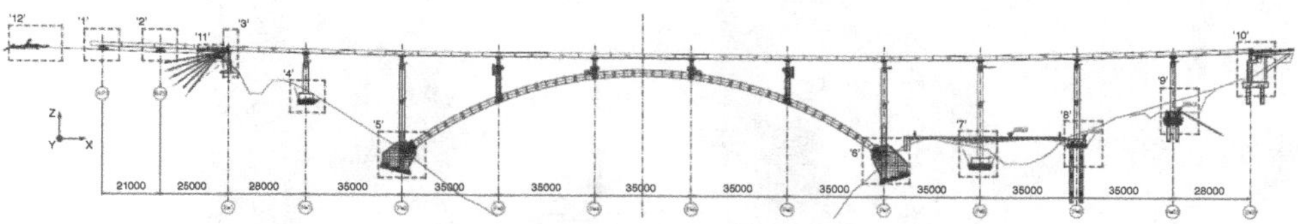

Figure 8.140 Longitudinal elevation and main geometric dimensions of the bridge (courtesy of CIMOLAI S.p.A).

Challenge 2: Erecting long, slender tubular sections in a high-wind environment required careful analysis. The site's dominant wind axis, approximately perpendicular to the Pacific Ocean, imposed potential dynamic excitation during construction (Figure 8.141).

Challenge 3: Once the two half-arches were connected in the tilted (rampant) position, they needed to be rotated into their final vertical configuration (a highly delicate operation requiring millimetric control) (Figure 8.142). Following the completion of the arch rotation and joint welding according to EN 1090, the two upper decks had to be installed simultaneously. Launching each deck independently would have produced unbalanced reactions on the arch and required excessive crane reach.

The following solutions were implemented to overcome these challenges:

Solution 1: Instead of using cranes or cableways (which were impractical due to weight and reach) the engineering team developed a rampant assembly method based on temporary hinges at the arch bases (Figure 8.143). Four half-arches were prefabricated and flange-connected in a tilted position, allowing the structure to be gradually closed by rotation around the hinges. The hinges were later filled with

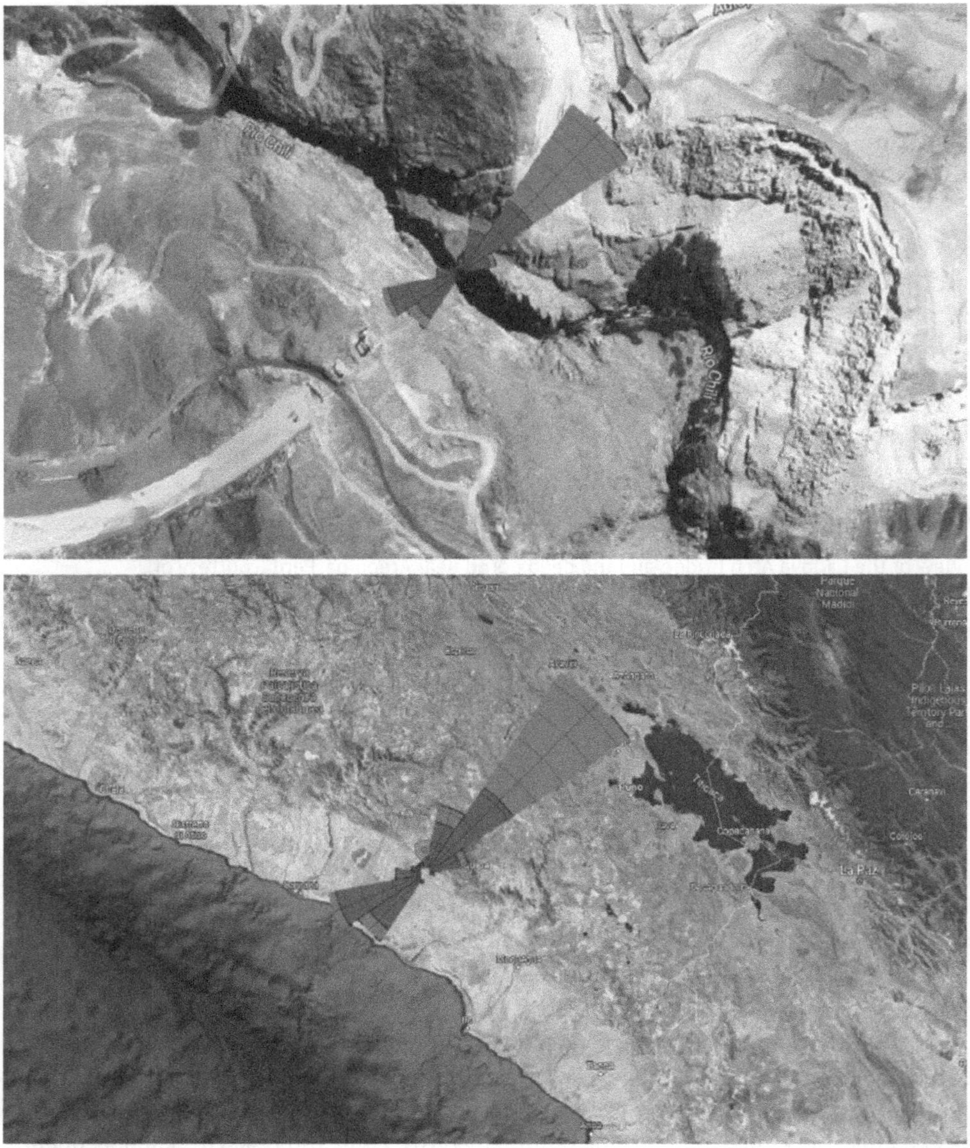

Figure 8.141 Wind exposure at meso and macro scale (courtesy of CIMOLAI S.p.A).

concrete to create the final fixed connection. The system was equipped with fine-tuning adjustment devices, ensuring positional accuracy during both temporary and final stages.

Solution 2: To ensure stability, engineers performed a wind and vortex-shedding analysis, considering each construction stage's geometry and modal frequencies (Figure 8.144). A steel portal frame anchored to existing concrete supports provided lateral restraint. On the southern side, additional tiebacks grouted into rock counteracted uplift forces. The entire assembly was thus secured against overturning and dynamic instabilities throughout the erection phase.

Figure 8.142 Rampant assembly of arch halves on temporary hinges during erection, detail view (courtesy of CIMOLAI S.p.A).

Figure 8.143 Rampant assembly of arch halves on temporary hinges during erection, overview (courtesy of CIMOLAI S.p.A).

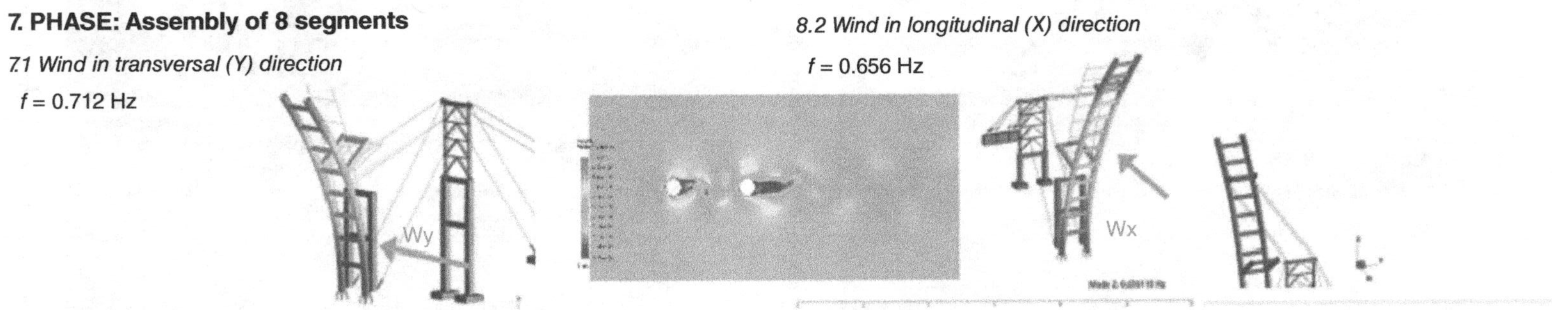

Figure 8.144 Wind exposure and vortex-shedding assessment for construction stages (courtesy of CIMOLAI S.p.A).

Figure 8.145 Controlled pronation of the arch ribs from ramped to final position (courtesy of CIMOLAI S.p.A).

Solution 3: The rotation was executed using four 185-ton strand jacks per side, anchored to the abutments (Figure 8.145). The operation was modeled and monitored in real time, with predicted deformations and portal reactions validated against field measurements. The three-dimensional counter-camber and control of degrees of freedom were designed so precisely that the two arches closed perfectly without any shim adjustment at the 60-mm-thick central flanges. The precision of fabrication and geometry control ensured long-term durability by minimizing stress concentrations and corrosion risks. A synchronized twin launching system was developed, reusing the same strand jacks employed for arch rotation (Figure 8.145). Each deck was advanced on roller bearings using a twin nose-and-tail arrangement to control deflection. This paired launch avoided unbalanced loading, reduced equipment needs, and allowed the completion of the composite deck with the concrete topping poured in situ.

A nonlinear staged FEM of the full structure was developed using Straus7 (Figure 8.146). The model simulated the entire erection sequence with beam elements, allowing the team to assess stress evolution, deformation compatibility, and wind sensitivity at every stage. Local finite element sub-models were used to verify the thin-walled box sections, stiffener layouts, and flange connections. These analyses confirmed full alignment between computational predictions and field results, supporting the safety and precision of the erection process.

La Joya exemplifies resource-efficient construction through fabrication precision, reuse of existing foundations, and minimized temporary works. The erection method avoided large cranes and aerial cable systems typically used for deep gorges, reducing both fuel consumption and installation time. The integrated

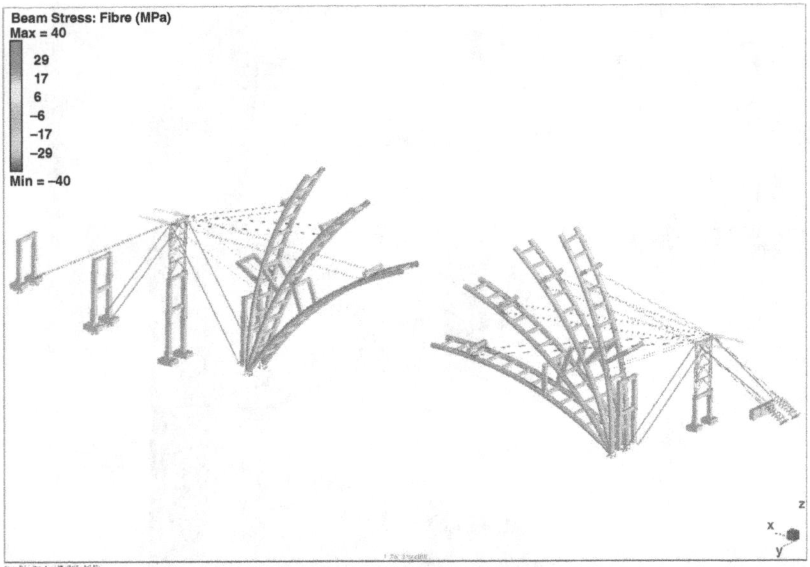

Figure 8.146 FEM of staged construction and deformation control (courtesy of CIMOLAI S.p.A).

Table 8.38 Sustainability strategy matrix of La Joya project.

Strategy	Most relevant LC stages	Key method(s) used	Sustainability outcome
Reuse of existing foundations	A1–A3, A5	Utilization of prebuilt concrete abutments and portals	Avoided new concrete works and excavation
Innovative erection sequence	A5	Rampant assembly on hinges and strand-jack rotation	Lower equipment energy use; reduced on-site risk
Precision fabrication and modeling	A1–A5	FEM-based counter-camber calibration and full-scale validation	No shimming; minimized rework and waste
Twin deck launching	A5	Simultaneous push with reused strand jacks	Reduced time, fuel, and auxiliary steel for temporary works
Integrated durability design	B, C	Internal access routes for inspection and maintenance	Extended service life; reduced maintenance interventions

approach (combining design, analysis, and construction know-how) also contributes to lower maintenance needs throughout the bridge's life cycle. Table 8.38 summarizes the sustainability strategies applied throughout design, fabrication, and erection.

Project Team

Client: Regional Government of Arequipa, Peru
General Contractor: Consorcio La Joya II – Pizzarotti S.p.A. & Eralma

Structural Executive Design (Steel): Cons. Ing. Arequipa – Cesma Ingenieros
Main Steel Contractor and Erection Design: Cimolai S.p.A.
Erection and Launching Equipment Design: Cimolai S.p.A.
Erection Supervision and Control: JNR Consultores S.A., Peru
Erection Class (Cimolai internal protocol): ERC 3b
Text Written in Collaboration with: Alessandro Catanzano, former R&D Technical Director in Cimolai S.p.A. and now CTO in Santo Engineering S.p.A.

8.6.5 Rebuilding an Airport as a Symbol of Post-earthquake Recovery Through Hybrid Timber–Steel Construction

In April 2016, the Kumamoto region of Kyushu, Japan, was struck by a devastating earthquake sequence: a magnitude 6.5 foreshock on April 14, followed just 28 hours later by a magnitude 7.3 mainshock on April 16. Kumamoto Airport, located just 4 km from the foreshock epicenter, was immediately closed to all but emergency flights. The existing terminal building (which had received the prestigious BELCA Award following a 2012 renovation by Nikken Sekkei) withstood the main shocks with only minimal structural damage. However, repeated aftershocks caused cumulative damage to nonstructural elements, particularly the cavity ceiling, raising the possibility of collapse and making it extremely difficult to guarantee occupant safety. Due to severe damage to the boundary sections of the addition and the lack of reserve seismic capacity in the original design, the Ministry of Land, Infrastructure, Transport, and Tourism (MLIT) recommended a rebuild.

Rather than simply replacing the damaged terminal with a conventional steel-and-concrete structure, the project team (led by Nikken Sekkei as architect and Taisei Corporation as contractor) reimagined the airport as a symbol of Kumamoto's life-cycle-driven post-earthquake recovery. Opened in Phase 1 on March 23, 2023 (with Phase 2 completed in September 2024), the new 39,507 m^2 terminal features a hybrid steel–timber structure with a roof system built with timber using locally sourced Kumamoto wood. This project demonstrates how structural design can simultaneously address seismic resilience, embodied carbon reduction, and regional identity (Figure 8.147).

The core concept of the Kumamoto Airport terminal is the integration of a hybrid timber–steel structural system that serves multiple objectives simultaneously (Figure 8.148). The primary steel frame provides the seismic resistance required for an essential facility in one of Japan's most seismically active regions, while the timber roof structure (composed of locally sourced 2 × 4 lumber assembled into 10-meter-span trusses) addresses embodied carbon reduction, thermal performance, and cultural symbolism. The decision to use residential-grade 2 × 4 lumber (rather than engineered mass timber products such as CLT or glulam) demonstrates that locally available, low-tech wood products can be assembled into high-performance structural systems for large public buildings, supporting the regional forestry economy while reducing both transportation emissions and material cost.

The roof sheathing consists of 100 mm-thick insulated wooden sandwich panels, which combine structural diaphragm action, thermal insulation, and an interior finish in a single element. The cavities within the truss system were utilized as duct routes and equipment pathways, integrating structure and building services into a unified assembly that eliminated the need for a suspended ceiling system entirely. This decision directly addressed lessons from the 2016 earthquake sequence, when repeated

Figure 8.147 View of the rebuilt Aso Kumamoto Airport terminal, completed in phases between 2023 and 2024, featuring a 39,507 m^2 hybrid steel–timber structure with an extended timber canopy constructed from locally sourced Kumamoto timber (courtesy of SS Co., Ltd.)

Figure 8.148 Interior view of the main departure hall showing the hybrid structural system: a primary steel frame supporting 10-meter span timber trusses fabricated from locally sourced 2 × 4 lumber, with exposed plywood roof panels forming the finished ceiling surface above the central commercial zone (courtesy of SS Co., Ltd.)

Figure 8.149 Exterior canopy and exposed timber roof structure showing 100 mm insulated wooden sandwich panels spanning between steel members (courtesy of SS Co., Ltd.)

aftershocks caused progressive damage to ceiling systems across Kumamoto. As visible in Figure 8.149, a cantilevered canopy combines a steel primary frame with lightweight timber decking, achieving long column-free overhangs while reducing self-weight and expressing structural clarity through exposed hybrid construction.

Rebuilding the terminal as a hybrid timber–steel structure posed the following challenges:

Challenge 1: The terminal had to be reconstructed in a region of intense seismic activity, requiring a structural system capable of withstanding major earthquakes and repeated aftershocks.

Challenge 2: The previous terminal revealed that suspended ceilings and hanging services were highly vulnerable during earthquakes, creating life-safety risks and operational disruption that had to be fully eliminated in the new design. Therefore, the new design needed to eliminate the risk of ceiling collapse entirely while still achieving the acoustic, thermal, and aesthetic performance expected of a modern passenger terminal.

Challenge 3: The project needed to achieve long-term economic feasibility and environmental responsibility while rebuilding critical airport infrastructure under tight budgetary and operational constraints.

Challenge 4: Although timber is often perceived as low carbon, its actual environmental impact can increase significantly through processing, transportation, and engineered products, requiring careful material sourcing and structural strategy.

Figure 8.150 Construction view of the hybrid timber–steel roof system showing 10-meter span timber trusses integrated with the primary steel frame (courtesy of SS Co., Ltd.).

Solution 1: The structural design philosophy prioritized clarity, regularity, and reduced mass. The primary system consists of a steel frame with no basement level. As visible during construction (Figure 8.150), the depth of the timber trusses accommodates both structural action and service integration within a single coordinated assembly. By eliminating heavy suspended ceilings and minimizing irregularities in plan and section, the building achieves seismic resilience through simplification

Solution 2: The cavity ceiling was entirely replaced by using the wooden sandwich panels (100-mm-thick insulated panels) as both the roof sheathing and the interior finish surface (Figures 8.151 and 8.152). This removed the main vulnerable nonstructural element identified during the 2016 earthquake sequence, ensuring that repeated aftershocks cannot cause progressive ceiling damage or sudden collapse. The truss cavities were designed as integrated pathways for mechanical ducts and equipment routes, combining structure and building services into a single system. This eliminated the conventional approach of hanging services below the structure (which itself requires a ceiling to conceal), achieving both seismic safety and cost efficiency.

Figure 8.151 Departure lounge showing exposed timber trusses and plywood roof panels integrated with the steel frame (courtesy of SS Co., Ltd.)

Solution 3: The cross section was organized as a two-floor configuration with an intermediate mezzanine dedicated to mechanical equipment rooms (Figure 8.152). This "sandwich" arrangement efficiently serves both upper and lower floors while eliminating the need for underground mechanical spaces. Major utility facilities were relocated outdoors, allowing simplified construction, easier inspection, and future adaptability. Structural spans were optimized for economy. The resulting cross section, combined with dynamic zoning, maintained clear passenger flow despite the reduced visibility typically associated with shorter spans.

Solution 4: Locally sourced Kumamoto 2 × 4 lumber was assembled into 10-meter-span trusses to support the roof sheathing (Figure 8.153), which supported the regional forestry economy, minimized transportation emissions, reduced material cost, and conveyed the post-earthquake recovery message directly and symbolically to every passenger. Lateral buckling of the timber trusses was prevented by structurally engineered plywood with decorative wood grain sourced from Oguni Town (a renowned timber-producing area in Kumamoto Prefecture). This plywood served the dual function of structural bracing and architectural finish, maintaining the exposed timber aesthetic while providing the lateral restraint required for the long-span truss configuration.

Figure 8.152 Axonometric diagram of hybrid steel–timber canopy at Kumamoto Airport, showing the two-floor "sandwich" cross-section with intermediate mechanical mezzanine and the integrated wooden sandwich-panel roof (courtesy of SS Co., Ltd.).

In conclusion, the Kumamoto Airport Passenger Terminal demonstrates how post-disaster reconstruction can exceed mere replacement to become an opportunity for innovation in structural design, sustainability, and regional identity. By integrating locally sourced timber into a hybrid steel–timber structural system, the project achieved simultaneous improvements in seismic resilience (through elimination of

Figure 8.153 Detail of 10-meter span timber trusses (courtesy of SS Co., Ltd.).

Table 8.39 Sustainability strategy matrix of the Kumamoto Airport Terminal project.

Strategy	Most relevant LC stages	Key method(s) used	Sustainability outcome
Hybrid timber–steel structure	A1–A3, A4	Locally sourced 2 × 4 lumber trusses replacing steel/concrete roof elements.	Reduced embodied carbon in roof system; regional forestry economy support; carbon sequestration in biogenic material
Ceiling elimination and structural integration	A5, B4, D	Timber trusses as combined structure, insulation, finish, and services pathway; no suspended ceiling system	Eliminated most vulnerable nonstructural element under repeated seismic loading; reduced construction material and labor; avoided future ceiling maintenance and replacement cycles
Rational spatial planning	A1–A5	Regular structural grid, economical spans, no basement.	Minimized structural footprint and material quantities; enabled phased expansion without structural modification; reduced construction duration
Seismic resilience through simplicity	B4, D	Increase earthquake resistance exceeding code requirements	Building designed to withstand repeated major earthquakes; credit for avoided future demolition and rebuild; operational continuity as critical infrastructure

vulnerable ceiling systems), embodied carbon reduction (through substitution of steel and concrete with renewable timber in the roof structure), thermal performance (through integrated wooden sandwich panel insulation), and cultural meaning (through the visible and tactile presence of Kumamoto timber conveying the message of creative recovery to every visitor). Table 8.39 presents a sustainability strategy matrix mapping each core design strategy to relevant life-cycle stages, methods, and outcomes.

Client: Kyushu Kumamoto International Airport Co., Ltd.
Lead Architect: Nikken Sekkei Ltd
Supervision (Phase 1): Collaborative supervision: AZUSA SEKKEI CO., LTD.
Construction Contractor: Taisei Corporation
Construction Management: Nikken Sekkei Construction Management, Inc.
Supervision: Collaborative supervision: AZUSA SEKKEI CO., LTD. (Phase 1 only)
Design Collaborators:
Wayfinding: i Design Inc.
Interior Design: D-Brain Co., Ltd.
Baggage Handling System (BHS): AGP CORPORATION
Construction Plan: Nikken Sekkei Construction Management, Inc.
Landscaping: Miraifukeisha LTD.
Photo Credits: SS Co., Ltd.

8.6.6 Data Center Design for Seismic Resilience

Data centers are becoming the backbone of the digital economy, housing the servers and network equipment that sustain cloud computing, financial transactions, and communication networks worldwide (Figure 8.154). Unlike conventional commercial buildings, a data center must always remain fully operational: any interruption can cascade into massive economic losses and service disruptions affecting millions of users. This places extraordinary demands on structural design, particularly in seismically

Figure 8.154 A data center example from metropolitan area of Milan, Italy (courtesy of Lombardini22 S.P.A).

active regions where earthquake-induced damage can compromise sensitive equipment, break cable routes, and trigger costly downtime.

At its core, a data center contains three categories of IT equipment: servers (powerful computers that run applications and process data), storage systems (devices such as hard drives that hold large volumes of information), and network equipment (routers, switches, and firewalls that connect systems and manage data flow). These components are the reason the building exists, and every architectural and structural decision must ultimately serve their protection and uninterrupted operation.

Beyond the IT hardware itself, a data center relies on five major supporting infrastructures with direct implications for structural design:

i. **Uninterruptible power supply (UPS) units and backup generators:** Ensure continuous electricity during grid outages. Their substantial mass, often several tons per unit, introduces significant concentrated loads on floor slabs and supporting beams.
ii. **Cooling systems:** Including computer room air handling units, chillers, cooling towers, and associated piping. These systems dissipate the large heat output of densely packed servers and, when located on rooftops or in plant rooms, impose considerable additional dead loads on roof and upper-floor structures.
iii. **Power distribution units (PDUs) and energy storage systems:** Add further localized loads and contribute to complex load patterns across the floor plate.
iv. **Cable tray networks:** Dense overhead and underfloor routing systems increase distributed loads and require careful coordination with the structural frame.
v. **Security systems:** Physical and digital security infrastructure, such as access control, monitoring equipment, and perimeter barriers, influence the building envelope and structural layout.

The combination of these systems means that a data center imposes floor loads, vibration sensitivity requirements, and environmental control demands that far exceed those of conventional commercial or office buildings. Several factors distinguish data center structural design from conventional building design:

a. **Heavy and variable floor loads:** Server racks, UPS systems, and battery arrays impose floor live loads in the range of 5–13 kN/m^2 (and sometimes higher), compared to typical office live loads of 2.0–3.0 kN/m^2. These loads are not uniformly distributed: they vary zone by zone depending on equipment density, creating complex loading patterns that the structural engineer must consider in detail.
b. **Vibration sensitivity:** IT equipment (particularly hard drives and high-precision optical components) is sensitive to floor vibrations. The structural floor system must meet stringent vibration criteria (typically very low acceleration limits) that go beyond the serviceability requirements of standard commercial buildings. This often drives the selection of stiffer floor framing systems or shorter spans than would otherwise be structurally necessary.
c. **Absolute operational continuity:** Unlike most buildings, where temporary closure after a natural disaster is inconvenient but manageable, a data center outage can cascade into economic losses measured in millions of euros per hour, disruption to critical services (healthcare, finance, communications), and potential data loss. This zero-downtime requirement elevates the seismic, wind, and flood performance targets well beyond code minimums.

d. **Dense and complex MEP integration:** The structural frame must accommodate extensive mechanical, electrical, and plumbing (MEP) systems, including overhead cable trays, underfloor cooling plenums, raised-floor support systems, rooftop HVAC plant, and heavy transformer rooms (Figure 8.155). Structural penetrations (sleeves for pipes and cables through beams and slabs) must be coordinated meticulously to avoid clashes that compromise structural integrity, and generous floor-to-floor heights are needed to house the layered services above and below the server floor.
e. **Adaptability for future expansion:** Data center operators frequently upgrade or densify their IT equipment over the building's service life. The structure must be designed with sufficient reserve capacity (in both gravity and lateral load paths) to accommodate heavier future equipment without requiring expensive structural modifications. This forward-looking approach has important implications for initial material quantities and embodied carbon.

To appreciate the engineering decisions in this case study, it is helpful to understand the structured design process through which a data center moves from initial concept to operational building. The

Figure 8.155 Interior of the data hall showing installed server cabinets, overhead cable trays, and integrated mechanical and electrical services infrastructure (courtesy of Lombardini22 S.P.A).

RIBA Plan of Work (The Royal Institute of British Architects 2020) provides a widely used framework of eight stages (0 through 7) that organizes this progression (Table 8.40). For a data center, each stage carries specific structural considerations that differ from those of a conventional building.

For a data center, the early stages are especially consequential: decisions about the column grid, floor loading strategy, and seismic performance philosophy made at Stages 0–2 fundamentally shape the

Table 8.40 RIBA Plan of Work stages applied to data center structural design (The Royal Institute of British Architects 2020).

Stage	Name	Design output	Data center-specific considerations
0	Strategic definition	High-level concept statement on structural needs.	• Determine the required scale (number of racks, total area). • Select the general structural system (e.g. steel frame for a multi-story facility, pre-engineered for a single-y module). • Identify the maximum floor load for server racks and equipment.
1	Preparation and briefing	Preliminary structural scheme and concept sketches.	• Define the column grid based on the required rack layout. • Establish initial floor-to-floor heights to accommodate overhead cable trays, cooling ducts, and raised-floor systems. • Select the structural material (steel, RC, or hybrid). • Initial assessment of seismic and wind load zones.
2	Concept design	Structural concept drawings and outline specification.	• Initial plans showing column, beam, and slab locations. Indicate preliminary cores (stairs, elevators, pipe risers). • Document design codes (e.g. Eurocodes, IBC), target concrete strength, and steel grades. • Confirm vibration criteria for IT equipment sensitivity.
3	Spatial coordination	Coordinated structural model (BIM) and design freeze.	• Finalize structural penetrations (sleeves for pipes/cables) to avoid clashes with beams and columns. • Ensure clearances for large HVAC units. Detailed load analysis including permanent loads (structure, MEP), imposed loads (racks, people), and dynamic loads (earthquake, wind).
4	Technical design	Fully detailed structural drawings and calculations.	• Detailed foundation design (piles, raft, pad footings), rebar detailing for concrete elements, connection details for steelwork. • Every dimension, section, and note required for construction tender. • Final structural plans submitted for building control/permit approval.
5	Manufacturing and construction	Site inspections, quality checks, and site queries.	• Structural engineer conducts site visits to inspect formwork, rebar placement, concrete pouring, and steel erection. • Review material test reports (concrete cube tests, steel mill certificates) to ensure compliance. • Address unforeseen conditions via site instructions.

(Continued)

Table 8.40 (Continued)

Stage	Name	Design output	Data center-specific considerations
6	Handover	As-built drawings and structural safety file.	• Update drawings to reflect changes made during construction. • Compile all material certificates, inspection reports, calculations, and maintenance advice for the building's structural safety file (critical component of the Building Control Final Certificate).
7	Use	Post-occupancy evaluation (POE) and asset management support.	• Assess structural performance in operation, including vibration monitoring and thermal performance checks. • Provide advice for future modifications (adding heavier equipment, installing additional plant), confirming the structure's capacity to handle proposed changes.

building's structural system, material quantities, embodied carbon, and adaptability for the remainder of its service life.

A new data center complex currently under construction (in the metropolitan area of Milan, Italy, under the project and construction management of Lombardini22 Spa) shows how advanced structural engineering can address the above-mentioned challenges. The complex comprises interconnected buildings:

- a large data hall;
- two parallel gantry structures that house mechanical and electrical services;
- and an office building for administrative functions.

The data hall and gantries are built with steel frames, while the office building uses precast concrete, reflecting the different functional and structural requirements of each component (Figure 8.156).

The defining feature of this project is its seismic performance philosophy. Conventional buildings are designed to survive major earthquakes by allowing controlled structural damage (through plastic deformation of beams, columns, or braces) while preventing collapse. This approach, codified in standards such as Eurocode 8 and the Italian NTC, accepts that a building may need significant repair after a strong earthquake. For a data center, however, post-earthquake repair is not acceptable: the facility must remain operational without interruption.

Designing a data center complex to remain elastic under extreme seismic loads posed the following structural, operational, and analytical challenges:

Challenge 1: Data centers impose significantly heavier floor loads than typical commercial or office buildings, owing to the weight of server racks, uninterruptible power supply (UPS) units, cooling infrastructure, and dense cabling. In this project, superimposed dead loads (SID) on the data hall floors reach 7.50 kN/m^2 (accounting for hung services and raised-floor systems), while live loads (LL) vary across different zones from 5.00 to 12.80 kN/m^2 depending on the equipment density. These

Figure 8.156 Overview of a data center complex showing the data hall area (courtesy of Lombardini22 S.P.A).

Table 8.41 Seismic performance targets under the Italian code compared with the data center requirements.

Performance level	Standard commercial building	Hospital/essential facility	This data center (client requirement)
Operational	$T_R = 30$ years	$T_R = 60$ years	$T_R = 1{,}950$ years
No collapse	$T_R = 475$ years	$T_R = 1{,}950$ years	$T_R = 2{,}475$ years
Design implication	Controlled damage accepted; post-earthquake repair expected.	Limited damage; essential function preserved.	Entire structure remains elastic; full operation guaranteed without repair.

loads are substantially higher than those of standard office occupancy (typically 2.0–3.0 kN/m^2 for live loads) and place heavy demands on the floor framing, columns, and foundations.

Challenge 2: The client specified seismic performance requirements that significantly exceed those mandated even for essential facilities such as hospitals. Under Italian seismic code, hospital-class structures (essential facilities) must satisfy an operational limit state (SLO) for earthquakes with a return period of TR = 60 years and a no-collapse limit state (SLC) for TR = 1950 years. The data center client required full operational status at TR = 1950 years and no-collapse performance at TR = 2475 years. In practical terms, this means the structure must remain essentially elastic (with no damage to structural members) under ground shaking that would be considered the collapse-prevention scenario for a hospital. Table 8.41 compares these performance targets.

To meet these demands, two complementary strategies were adopted:

Solution 1: Material and construction system optimization

The choice of steel framing for the data hall and gantries, where long spans, flexibility in service routing, and rapid construction are critical (Figures 8.157 and 8.158) vs. precast concrete for the

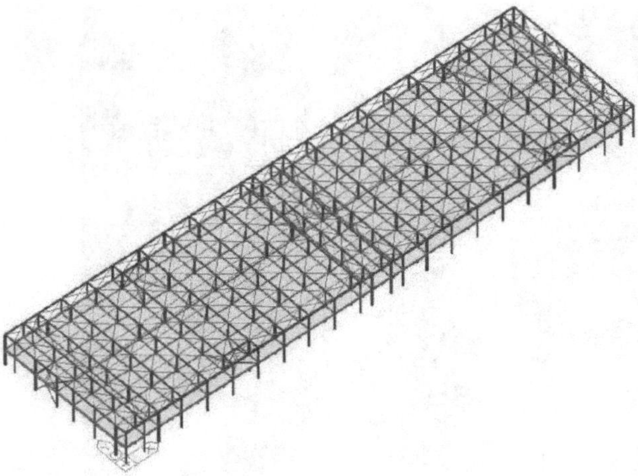

Figure 8.157 Data hall (FEM model) (courtesy of Lombardini22 S.P.A, Studio Progetti Strutturali SPS S.r.l.).

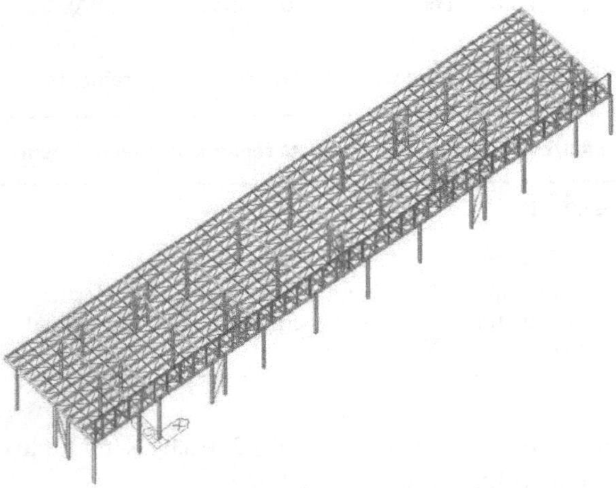

Figure 8.158 Gantry (FEM model) (courtesy of Lombardini22 S.P.A, Studio Progetti Strutturali SPS S.r.l.).

office building (Figures 8.159 and 8.160), where shorter spans and conventional occupancy loads prevail, allowed each component to be optimized for its specific functional requirements. This material-allocation strategy avoided the inefficiency of applying a single structural typology to the entire complex.

Rather than applying a uniform design load across the entire floor plate, the engineering team mapped superimposed dead loads and live loads zone by zone, reflecting the actual equipment layout. SID values ranged from 1.00 (lightweight zones) to 7.50 kN/m^2 (fully loaded data halls), while live loads ranged from 1.20 to 12.80 kN/m^2. This zoned approach prevented the unnecessary over-specification of structural members in lighter-loaded areas, reducing both steel tonnage and embodied carbon.

Figure 8.159 Construction phase of DC building showing precast concrete structural elements and on-site reinforcement works during assembly (courtesy of Lombardini22 S.P.A).

Figure 8.160 Office building (FEM Model) (courtesy of Lombardini22 S.P.A, Studio Progetti Strutturali SPS S.r.L).

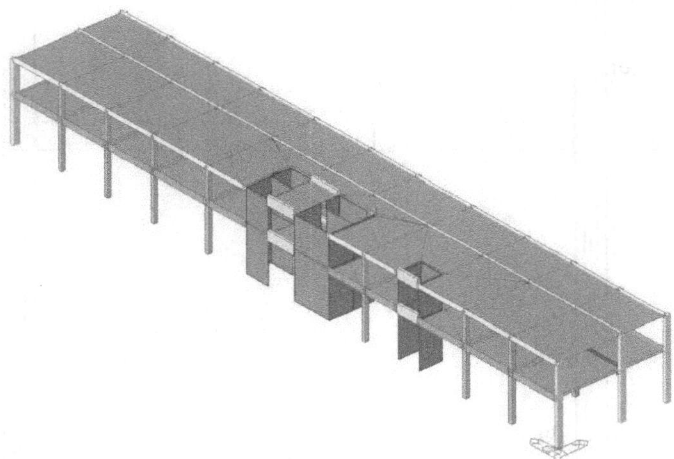

Solution 2: Dissipative steel bracing with fluid viscous dampers

Under conventional seismic design practice, achieving very high-performance targets while keeping the structure essentially elastic would typically require a material-intensive approach. In practical terms, this means substantially increasing the size, stiffness, and strength of structural members so that they can resist extreme seismic forces without yielding. Columns and beams would need larger cross-sections and higher reinforcement ratios, steel members would require thicker plates and stronger connections, and foundations would have to be enlarged to transfer the increased forces safely to the ground. Although this strategy can reduce damage, it results in significantly higher quantities of steel and concrete, increased construction costs, greater embodied carbon, and potential constraints on architectural flexibility. In contrast, the engineering strategy adopted in this project concentrates seismic energy dissipation in specially designed anti-seismic devices, namely fluid viscous dampers, while ensuring that all primary structural members, including columns, beams, braces, and foundations, remain within their elastic range. This philosophy separates structural survival from structural damage. Rather than dissipating earthquake energy through yielding steel or cracking concrete, the building channels seismic forces into replaceable dampers. After a major event, these devices can be inspected and, if necessary, replaced without affecting the integrity or functionality of the main structure, allowing the building to remain operational with minimal intervention (Figure 8.161).

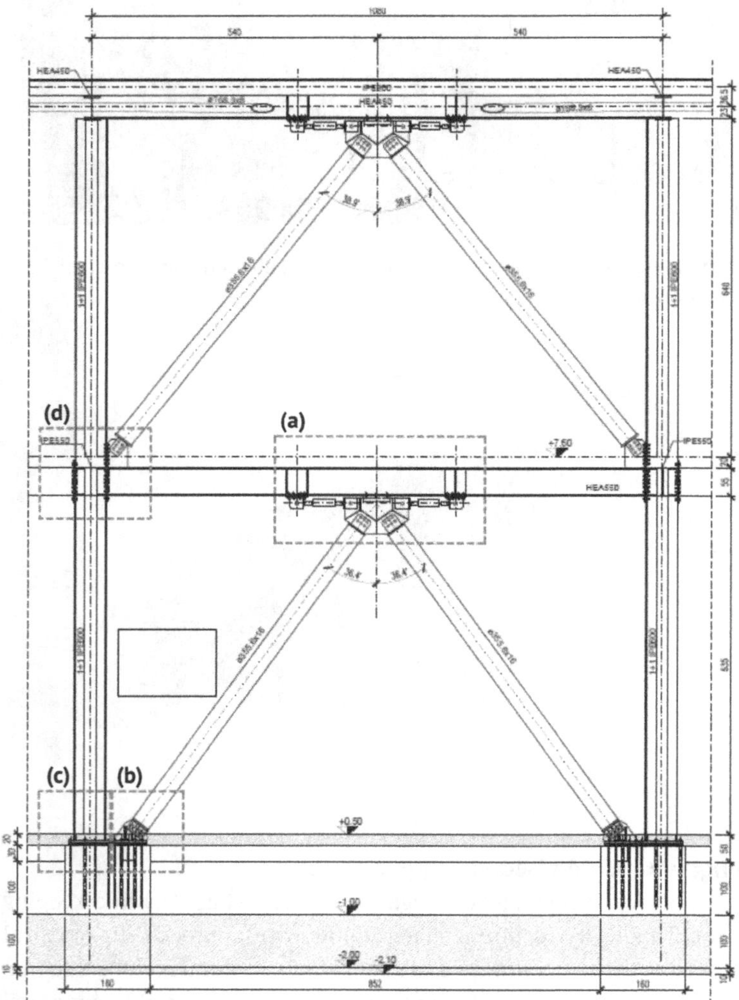

Figure 8.161 Detail of the dissipative bracing system with fluid viscous dampers installed in the gantry structure. (a) Central brace-to-beam connection with fluid viscous damper assembly. (b) Lower brace-to-foundation connection. (c) Column base connection to foundation. (d) Beam-to-column moment connection (courtesy of Lombardini22 S.P.A and Studio Progetti Strutturali SPS S.r.l.).

Table 8.42 Sustainability strategy matrix of the data-center project.

Strategy	Most relevant LC stages	Key method(s) used	Sustainability outcome
Material-allocation optimization	A1–A3 (Production Stage)	Steel for data hall/gantry (long spans, flexibility); precast concrete for office. Zone-specific load mapping to avoid over-specification.	• Reduced steel tonnage and embodied carbon in lighter-loaded zones • Each building optimized for its function.
Replaceable dissipative devices	B4 (Replacement), D (beyond life cycle)	Fluid viscous dampers concentrate dissipation; all structural members remain elastic under extreme earthquakes. Dampers are discrete, inspectable, and replaceable without structural modification.	• Avoided post-earthquake structural repair or demolition • Preservation of embodied carbon investment • Extended service life of the primary structure • Reduced whole-life material consumption • Maintained operational continuity after seismic events

This project demonstrates how data center facilities require a structural approach that goes beyond conventional code compliance. Exceptional equipment loads, stringent vibration criteria, and dense service integration were addressed through specific material allocation and detailed load zoning. The seismic strategy shifts from controlled damage to performance preservation: energy dissipation is concentrated in fluid viscous dampers, allowing the primary structural system to remain within the elastic range even under extreme ground motions. The result is a robust, low-carbon structural solution capable of safeguarding digital infrastructure while minimizing repair needs and lifecycle resource consumption. Table 8.42 shows the sustainability strategy matrix of this project.

Project team

Design, Project Management, and Construction Management lead: Lombardini22 S.P.A (https://www.lombardini22.com/)
Structural design: Studio Progetti Strutturali SPS S.r.l. (www.spssrl-mi.it)
[For confidentiality reasons, the project name, exact location, and client cannot be disclosed]

8.7 Conclusion

The case studies of this chapter highlight an evolution in structural practice. Taken together, they show the four pillars from Chapter 1 in action: being open-minded and transparent about options, building with the right resources, designing to last, and using technology to cut waste and error.

The 25 projects (from 15 countries, 6 continents) gathered in this chapter point in the same direction. Teams that treat carbon, cost, risk, and buildability as a linked problem from day one deliver higher performance with fewer materials, faster programs, and better buildings. What changes outcomes is not a single product or patent but a way of working that makes decisions early, measures impact consistently, and keeps options open until the right trade-offs are visible. Across these case studies, the ten reduction strategies introduced in Chapter 3 reappear as following specific practical design moves:

i. Design for and with circularity by prioritizing adaptive reuse where cores, basements, and frames retain residual capacity, treating reclaimed steel as a procurement workflow (with QA, substitutions, and small-batch logistics), and configuring new work for reversible disassembly.

ii. Use high-strength materials where effective by selecting higher steel grades, high-performance concretes, or other advanced materials in members where strength governs the design, after verifying that the resulting volume and weight savings outweigh the production penalty per ton and that secondary effects (fatigue, buckling, deflection) are properly addressed.

iii. Explore conceptual options early by co-developing form, layout, and structural system rather than handing them off sequentially, comparing structural schemes (including form-active and lightweight typologies alongside conventional bending systems) and foundation strategies on the actual soil profile, and testing each against LCA, cost, program, seismic, and vibration targets before geometry is frozen.

iv. Adopt hybrid and composite systems by assigning timber, steel, and concrete to the roles they serve best, and locking interfaces early, including MEP penetrations and tolerance stacks.

v. Employ prefabrication and modular delivery by letting transport limits, site access, and crane reach inform structural choices from concept stage, and using a BIM framework, mock-ups, just-in-time logistics, the largest safe preassemblies, and dry bolted site joints for speed, safety, and future disassembly.

vi. Source low-carbon materials by procuring verified lower-GWP suppliers, shortening haul distances, aligning pours and deliveries with the schedule, confirming at-scale availability and curing times for SCMs and recycled aggregates, and using reclaimed products wherever the section catalogue allows.

vii. Engage stakeholders early in design by bringing clients, contractors, fabricators, and fire engineers into the room from concept stage, especially for precedent-setting projects, using 3D models, simulations, and physical mock-ups to evidence safety and accelerate approvals.

viii. Apply digital, AI, and automation tools across the full project life, driving geometry through parametric BIM, simulating structural behavior, fire and evacuation, wind, vibration, and the construction sequence itself through FEM, CFD, and 3D coordination, so that on-site assembly proceeds without rework.

ix. Design for durability and adaptability by specifying weathering (or stainless) steels or protective systems where exposure demands it, routing services through accessible cavities, leaving reserve structural capacity for future loads, and ensuring maintenance access while quantifying avoided repairs and end-of-life credits.

x. Design for robustness by modeling local seismic, wind, and vibration conditions, using replaceable fuses, viscous dampers, or base isolation to keep primary members elastic under extreme events, and providing thermal-movement strategies so that critical assets remain operational and repairable.

While the case studies cover different countries, materials, and project types, they are not intended for direct comparison: each emerged from unique site conditions, regulatory environments, budgets, and performance requirements.

It should also be acknowledged that data quantity and quality, regional material availability, and the limited reporting of end-of-life benefits may introduce some uncertainty when comparing results across projects. Yet, even with these variations, the overall direction of progress remains clear.

In summary, these cases demonstrate the life-cycle-driven framework in practice. The results came when teams set targets early, kept options alive long enough to see the trade-offs, and aligned procurement, detailing, construction sequencing, and logistics with their embodied carbon goals. In most cases, the payoff is more than lower GWP: shorter programs, safer and more efficient construction stages, leaner foundations clearer space planning, faster construction, structures that remain durable and serviceable with planned maintenance access, resilience to earthquakes, wind, and extreme events that prevents the carbon penalty of premature reconstruction, and assets that can be adapted, dismantled, or reused rather than demolished. The result is a practice that is analytical and regenerative at once, where performance is measured across the full life cycle and decisions avoid narrow focus. Life-cycle thinking turns design, construction, adaptation, maintenance, and reuse into a single iterative process, and is the most credible route the built environment has toward real sustainability.

Looking ahead, these findings point toward how future standards, education, and procurement frameworks could evolve: rewarding teams that integrate carbon, cost, and constructability metrics from the earliest design stages and sustain this alignment throughout the building's life.

8.8 Questions

Answer the following questions to enhance your understanding of the concepts covered in this chapter.

Scenario 1: A city owns a 25-year-old office block on a brownfield site. The client wants a low-carbon, adaptable workspace with a tight budget and a 20-month program. Compare two routes: retrofit with partial structural reuse vs. new build with a high-efficiency composite frame. Outline the key checks you would run in concept design (LCA scope, structural utilization targets, geotechnical constraints, services integration, fire strategy, acoustic/vibration), the early metrics you would track, and what would make you recommend one route over the other.

Scenario 2: A university needs a mid-rise teaching building with 9 m clear spans, high daylight, and strict comfort criteria. Propose two viable hybrids that could meet performance goals: a mass-timber solution with selective steel or concrete assist, and a steel–concrete composite option with modular elements. For each, state how you would control vibration, route services, manage approvals, plan prefabrication, and evidence whole-life carbon benefits.

Scenario 3: A hospital extension must be built next to live facilities, on soft soils with a high water table. Schedule risk and site disruption are critical. Set out an approach that combines early geotechnical optimization, prefabricated structure, and a resilience plan for seismic and wind. Explain your decision path for foundations, structural system, and logistics, and how you would balance A1–A3 impacts with A5 construction efficiency and B stage durability.

Scenario 4: A regional transport hub needs a free-form roof with 36 m cantilevers and a continuous architectural surface. Compare two strategies: segmented roof with expansion joints vs. continuous roof on bearings with shock transmitters. Define the parametric workflow, wind assessment plan, thermal movement checks, joint or bearing detailing, local fabrication limits, and the criteria you would use to select the final approach.

Scenario 5: A public agency targets 500 apartments with volumetric modular construction through a winter season in a Nordic city. Set out module design rules, moisture control, prototype testing, tolerance strategy, logistics sequencing from factory to site, crane planning, and how you would combine RC cores with timber or steel modules to meet program and carbon targets.

Scenario 6: A highway authority must upgrade a river crossing. Two routes are on the table: strengthen the existing composite bridge or build a new span in S690. For both, list the governing checks and decisions: plate buckling near supports, fatigue of welded details, brittle fracture at low temperature, weld volume and access, inspection planning, and LCA impacts from A1 to D. State what evidence would lead you to recommend one route.

Scenario 7: You are the structural lead in a client meeting. In 90 seconds, justify choosing a hybrid system over an all-timber or all-steel alternative using three measurable points.

Scenario 8: Build a 10-item risk register for a modular build in winter conditions. For each risk, add a trigger, mitigation, and an owner.

Scenario 9: Your preferred option misses the embodied carbon target by 8%. Show one change in spans, one in mix design, and one in connection strategy that could close the gap.

Scenario 10: Name one decision you would make earlier on your last project to cut A5 impacts, and how you would prove it to the contractor.

Scenario 11: A municipal authority owns a pair of 1980s reinforced concrete-framed office towers in a central business district. The towers are structurally sound but functionally obsolete (inefficient floor plates, outdated MEP systems, single-use zoning). A developer proposes full demolition and replacement with a new mixed-use complex. As the life-cycle consultant, you are asked to evaluate an adaptive reuse alternative that retains the reinforced concrete cores and basements while replacing the floor slabs and façades. Describe the structural investigations you would need to perform on the existing cores before committing to retention, explain how current wind loading codes (which may be more demanding than the original design codes) could affect the feasibility of the reuse strategy, and discuss the embodied carbon implications of each approach across modules A1 through D.

Scenario 12: You are the structural engineer for a new sports training facility on a site with 20 m of soft clay and peat overlying competent bearing strata. The project architect has proposed a conventional reinforced concrete frame on large-diameter bored piles with a thick mat foundation. Your preliminary analysis suggests this approach is structurally conservative but material-intensive. Propose an alternative integrated geotechnical and structural strategy that reduces both foundation concrete volumes and superstructure weight. Explain how performing a site-specific seismic response analysis (rather than relying on the simplified code approach for soft soils) could further reduce material quantities, and identify which life-cycle modules benefit most from each optimization.

Scenario 13: A national highway authority is planning a 400 m-long steel-concrete composite bridge. The reference design uses conventional S355 steel, resulting in main girder flange thicknesses of up to 110 mm. You are asked to evaluate an alternative using high-strength steel (S690). Identify at least three design challenges that will shift from strength-governed to other failure modes when adopting HSS, explain the engineering solutions available for each, and discuss the life-cycle implications (including fabrication, maintenance, and fatigue durability) of the lighter but more slender HSS design.

Scenario 14: An engineering firm has been commissioned to build an arch bridge spanning a 150 m deep gorge in a remote mountainous region with limited access roads and high wind exposure. Conventional crane erection from the gorge floor is not feasible. Describe an erection methodology that avoids heavy lifting equipment at the gorge base, explain how you would address wind-induced dynamic instabilities during the construction sequence (before the arch is fully closed), and discuss how precision fabrication and geometry control during erection contribute to long-term durability and reduced maintenance (and therefore lower life-cycle carbon).

Scenario 15: A developer has acquired an incomplete 12-story steel-framed tower (originally designed for residential use) and wishes to convert it into a high-occupancy office headquarters. The existing structure was never completed or occupied. As the design lead, describe the two principal investigations you must conduct before confirming that adaptive reuse is viable: one related to structural capacity under the new loading and code requirements, and one related to life-safety compliance under the higher occupancy density. Explain how a comparative LCA of the reuse option vs. new-build alternatives can provide the quantitative justification for the decision.

Scenario 16: A city government plans to build a 40 000-seat athletics stadium on a former shipyard site where the soil has low bearing capacity due to decades of landfill. The initial design proposes a conventional steel cantilever roof spanning 200 m. As the life-cycle design consultant, you are asked to propose an alternative roof strategy that reduces both embodied carbon and foundation demands. Describe the structural principle you would recommend, explain why it is better suited to the site's geotechnical constraints, and identify which life-cycle modules (A1–A5, B, D) benefit most from the change.

Scenario 17: A regional health authority must build a new 250-bed hospital in a seismic zone with a design peak ground acceleration of 0.35 g. The project budget allows either (a) a conventionally designed fixed-base reinforced concrete structure with an immediate occupancy performance objective, or (b) a seismically isolated structure with a continued functionality objective at a moderately higher initial cost. The authority asks you to prepare a life-cycle argument comparing the two options. In your answer, explain the fundamental difference between immediate occupancy and continued functionality, describe how nonstructural elements factor into each scenario, and discuss the embodied carbon implications if a major earthquake strikes within the building's service life.

Scenario 18: A national railway operator needs temporary office accommodation for 150 staff at each of five successive construction sites over a 15-year program. The default approach is to procure new modular shipping-container offices for each site and dispose of them at the end of each phase. You are asked to propose a design-for-disassembly alternative. Outline the key structural design decisions (connection types, material choices, dimensional coordination) that would enable repeated relocation, and estimate qualitatively how the cumulative embodied carbon of your reusable solution compares with the single-use approach over the full program.

Scenario 19: An architect proposes a large community building in a high-seismicity region (0.4 g design PGA) using a timber-priority hybrid system, where glulam columns and braces are visually dominant, and steel is concealed. The structural engineer raises concerns about lateral resistance and connection complexity. Describe at least two strategies by which a concealed steel sub-system can provide the required seismic capacity while preserving the architectural intent of maximum timber visibility. Explain how such an approach also relates to embodied carbon objectives.

Scenario 20: Following a devastating earthquake, the government decides to rebuild a damaged airport terminal. One faction advocates for a rapid like-for-like replacement using conventional steel and concrete. Another proposes a hybrid timber–steel structure using locally sourced wood. Discuss the trade-offs between these two approaches in terms of seismic resilience, embodied carbon, construction speed, and regional socioeconomic impact. Under what conditions might the hybrid approach actually accelerate (rather than delay) construction?

Scenario 21: An existing 1990s industrial warehouse is being considered for conversion into an edge data center. The building consists of precast concrete columns and roof beams with limited lateral resistance. Describe the structural investigations required to determine whether the existing frame can support increased floor loads, concentrated rack loads, and enhanced vibration criteria. Propose strengthening strategies that minimize embodied carbon, and explain how an adaptive reuse approach compares to demolition and new build when assessed across Modules A1–D.

Scenario 22: A client is comparing two structural options for a new data center:

a. a conventional reinforced concrete frame with high thermal mass, and

b. a steel-braced frame with modular construction.

Describe how each option interacts differently with cooling strategies, vibration performance, construction speed, and end-of-life recyclability. In your answer, compare their likely performance across Modules A, B, C, and D, and explain how operational carbon dominance in data centers complicates purely embodied carbon optimization.

True/False Questions

1 Early-stage optioneering compares schemes on embodied carbon, cost, and program.

2 Long-span grids always increase embodied carbon because they need deeper beams in every case.

3 Composite steel–concrete floors can reduce structural depth while keeping services within the zone.

4 Using higher-strength steel can shift checks from strength to fatigue and buckling.

5 Material impacts mostly occur in use phase, so product stage emissions are minor in most projects.

6 Prefabrication and DfMA usually cut on-site waste and improve quality control.

7 Bolted site connections support future disassembly more than site welding.

8 Adaptive reuse often shortens schedules and reduces upfront carbon vs. complete replacement.

9 Plated built-up sections always have lower embodied carbon than hot-rolled EAF sections.

10 Increasing GGBS or other SCMs in concrete can lower A1–A3 impacts, with trade-offs like setting time and supply limits.

11 Lighter superstructures can reduce foundation sizes and their embodied carbon.

12 For buildings with unusual roof or facade geometry, code wind pressures are generally precise enough, so wind tunnel or CFD studies are not needed. Local seismic response studies can refine spectra and help avoid oversizing.

13 Hybrid timber steel concrete systems are heavier than single-material frames.

14 Prefabrication invariably raises transport emissions, so the total embodied carbon is higher than site-cast work.

15 Reporting LCA by modules A1–A3, A4, A5, B, C, and D helps locate where reductions occur.

16 Bolted connections reduce structural reliability compared with welded connections in all situations.

17 Digital workflows linking BIM, analysis, quantities, and clash detection reduce rework and support lower emissions.

18 Adaptive reuse rarely helps program because surveys take longer than new build in most cases.

19 Cellular beams provide no benefit for MEP routing and typically increase structural weight.

20 In a hybrid steel–timber structure designed for disassembly and reuse, using standardized bolted connections throughout the steel frame (rather than welded connections) is beneficial because bolted connections can be reversed without damaging the members, enabling efficient dismantling and reassembly.

21 A cable-tension roof system spanning over 200 m is typically heavier than a conventional steel cantilever roof of equivalent span because the cables require massive anchorage elements at the perimeter.

22 When a building is designed for multiple life cycles of disassembly and reassembly, the embodied carbon savings are calculated only for the structural materials (modules A1–A3) and do not extend to construction-phase emissions (modules A4–A5).

23 In a seismically isolated hospital, ensuring continued post-earthquake functionality depends solely on the performance of the isolation devices; the detailing of nonstructural elements and utility crossings at the isolation plane has a negligible effect on operational continuity.

24 Using locally sourced, minimally processed timber (such as standard dimensional lumber) in a hybrid structural system always results in lower embodied carbon than using engineered mass timber products (such as CLT or glulam), regardless of the structural application.

25 In life-cycle assessment, the avoided carbon emissions from not having to demolish and rebuild a seismically resilient building after an earthquake are typically captured in Module B (use stage) of the life-cycle framework.

26 When an existing reinforced concrete core is retained during an adaptive reuse project, the embodied carbon associated with that core's original construction (modules A1–A3) is counted as zero in the LCA of the new project, because no new material was produced for those elements.

27 Performing a site-specific local seismic response analysis on soft soil always results in higher design accelerations than the simplified code approach, because it captures amplification effects that simplified spectra underestimate.

28 When high-strength steel (S690) is used instead of conventional steel (S355) in a bridge design, the primary design challenge shifts from strength-governed to fatigue- and buckling-governed behavior, because the thinner plates required to exploit the higher strength are more susceptible to instability and cyclic stress.

29 Recycling demolished concrete from an existing building into aggregate for new concrete on the same site eliminates all embodied carbon associated with the concrete production for the new elements.

30 In adaptive reuse of an existing building for a higher-occupancy use, verifying structural adequacy under the new loads is sufficient to confirm the building is safe for its new function; fire safety and evacuation capacity are not affected by the change of use.

31 Using a BIM-integrated design workflow for foundation and superstructure modeling primarily benefits the architectural design process and has a limited impact on structural material quantities or construction efficiency.

32 Introducing base isolation in a data center reduces seismic forces transmitted to the structure but may increase displacement demands at the isolation level, requiring careful detailing of service connections and cable trays.

33 In a seismically isolated hospital or data center, achieving continued functionality requires coordinated design of structural systems, nonstructural components, and backup utilities, not just enhanced performance of the primary load-resisting system.

Discussion Questions

1 How should early-stage optioneering be structured so that LCA, cost, program, and risk are compared on the same footing, and which two or three metrics would you present to decision makers first?

2 What threshold questions should be answered before choosing adaptive reuse over new construction, and how would you treat uncertainty in existing material properties and hidden defects in early estimates?

3 When is a hybrid system justified over a single-material solution? List the primary triggers such as span, vibration, fire, MEP routing, architectural depth limits, and site logistics.

4 What are practical ways to raise material utilization in steel and composite frames without compromising robustness, and how do you communicate the trade-offs to the client and contractor?

5 Which prefabrication choices most influence carbon and quality at A5? Discuss module size limits, transport and cranage, tolerance strategy, reversible joints, and site sequencing.

6 How do you integrate geotechnical findings into structural carbon decisions? Give examples of how soil conditions, local spectra, and flood risk can change foundation type, superstructure weight, and repair scenarios.

7 What durability strategies meaningfully reduce whole-life impacts? Compare options such as weathering steel, protection systems, exposure-driven detailing, and planned inspection intervals.

8 If reclaimed structural steel or other reclaimed components are considered, what verification, traceability, and certification steps are needed, and how should design teams plan for variability in section availability?

9 How do you design now for future change and end-of-life? Discuss reversible connections, accessible load paths, clear material passports, and how to claim credible benefits at stages C and D.

10 What processes help de-risk approvals for nonstandard systems? Outline the role of testing, prototypes, digital models shared with authorities, and staged performance evidence for fire, vibration, and acoustics.

11 In several of the case studies presented in this chapter, the vulnerability of nonstructural elements (ceilings, partitions, utilities, façade systems) proved to be at least as consequential as structural damage. Argue why life-cycle-driven design should require explicit performance objectives for nonstructural systems in essential facilities.

12 Timber is often described as a “low-carbon” material due to biogenic carbon storage. However, its actual environmental footprint depends on processing intensity, transportation distance, and product type. Discuss the conditions under which a timber structural element truly delivers a carbon advantage over steel or concrete.

13 Design for disassembly (DfD) promises significant life-cycle carbon savings by enabling structures to be reused across multiple service periods. However, DfD imposes constraints on connection design, material selection, and structural form. Discuss the engineering trade-offs involved in designing a multi-story building for repeated disassembly, and consider whether DfD is better suited to certain building typologies or occupancy durations than others.

14 Several projects in this chapter demonstrate that lightweight structural systems (tensile cable nets, timber trusses, hybrid steel–timber frames) can simultaneously reduce embodied carbon and improve seismic performance by lowering the building’s mass. Discuss the general structural principle linking reduced mass to improved seismic behavior, and identify at least two situations where pursuing minimum weight might create unintended design challenges (such as vibration, uplift, or dynamic instability).

15 Brownfield regeneration and adaptive reuse are presented in this chapter as sustainability strategies that avoid the embodied carbon of greenfield development or full demolition. Discuss the limitations of these approaches: under what structural, environmental, or programmatic conditions might adaptive reuse or brownfield development actually increase life-cycle carbon compared with a purpose-built new facility on a greenfield site?

16 Discuss the types of structural deficiencies most commonly revealed when re-evaluating existing buildings under modern codes, and argue whether regulatory frameworks should provide a more flexible pathway for adaptive reuse projects that can demonstrate acceptable life-cycle performance even if they do not fully comply with new-build code provisions.

17 High-strength steel (HSS) offers significant weight reductions compared to conventional grades, yet its adoption in bridge engineering remains limited. Discuss the engineering and institutional barriers to wider HSS adoption, considering factors such as the shift from strength-governed to fatigue- and buckling-governed design, the availability of post-weld treatment technologies, and code coverage. Under what project conditions does HSS offer the most compelling life-cycle advantage?

18 Discuss whether life-cycle-driven design should extend to specifying or constraining erection methods, and consider the contractual, practical, and safety implications of doing so.

19 Discuss the limitations of LCA as a design tool, including sensitivity to assumed carbon factors, system boundary choices, service life assumptions, and the treatment of biogenic carbon. How should engineers communicate LCA uncertainty to clients and decision-makers without undermining confidence in the results?

20 Data centers are often described as operationally carbon-intensive, leading some stakeholders to prioritize energy efficiency over embodied carbon reduction. Discuss whether this framing risks underestimating the importance of structural optimization, especially as grids decarbonize. At what point might embodied carbon become a dominant consideration in mission-critical facilities?

21 In high-seismic regions, essential facilities such as hospitals and data centers are increasingly designed for continued functionality rather than life safety alone. Discuss how shifting the performance objective from collapse prevention to operational continuity alters structural system selection, nonstructural detailing, and foundation design. How does this shift affect whole-life carbon risk?

22 In mission-critical facilities, structural vibration performance may govern member sizing more than ultimate strength. Discuss how serviceability-driven design changes the carbon optimization problem compared to strength-driven design. Are there structural forms that better reconcile vibration control with material efficiency?

References

American Society of Civil Engineers. (2017). Minimum design loads and associated criteria for buildings and other structures (ASCE/SEI 7-16). https://doi.org/10.1061/9780784414248

Erdik, M., Ülker, Ö., Şadan, B. & Tüzün, C. (2018). Seismic isolation code developments and significant applications in Turkey. *Soil Dynamics and Earthquake Engineering, 115* (December), 413–437. https://doi.org/10.1016/j.soildyn.2018.09.009

European Commission. Directorate General for Research and Innovation. (2019). *Optimal use of high strength steel grades within bridge (OPTIBRI): Final report*. Publications Office. https://data.europa.eu/doi/10.2777/93807

International Code Council. (2017). *2018 International Building Code (IBC)*. International Code Council.

Lemma, M. S., Gervásio, H., Pedro, J. O., Rigueiro, C., & Da Silva, L. S. (2020). Enhancement of the life-cycle performance of bridges using high-strength steel. *Structure and Infrastructure Engineering, 16*(4), 772–786. https://doi.org/10.1080/15732479.2019.1662067

LETI. (2020). Climate Emergency Design Guide. https://www.leti.uk/cedg

NSW Government. (2026, April 15). *Western Sydney Aerotropolis Investment Takes Off*. https://www.nsw.gov.au/ministerial-releases/western-sydney-aerotropolis-investment-takes-off

Pedro, J. J. O., Reis, A. J., & Baptista, C. (2018). High Strength Steel S690 in highway bridges: General guidelines for design. *Stahlbau, 87*(6), 555–564. https://doi.org/10.1002/stab.201810615

Royal Institute of British Architects. (2020). *RIBA plan of work 2020 overview*. https://www.riba.org/media/syneeeto/2020ribaplanofworkoverviewpdf.pdf

Robert Bird Group. (n.d.). *St Marys IPO*. Retrieved May 11, 2026, from https://www.robertbird.com/rbg-projects/st-marys-ipo/

Shahnewaz, M., Dickof, C., Zhou, J., & Tannert, T. (2022). Vibration and flexural performance of cross-laminated timber–glulam composite floors. *Composite Structures, 292*, 115682. https://doi.org/10.1016/j.compstruct.2022.115682

U.S. Green Building Council. (n.d.). LEED certification. Retrieved from https://support.usgbc.org/hc/en-us/p/leed-certification

9

Concluding Remarks

LIST OF CONTENTS

Every structural design decision is a carbon decision. When an engineer defines the overall structural system, sets the primary grid, or commits to a long-span transfer structure with the architect, the embodied carbon of that choice will not change for the life of the structure. This book has presented the methods to make those decisions visible, quantifiable, and improvable.

9.1 What We Have Established

We began by establishing the scale of the challenge. Construction consumes a vast share of the world's raw materials and generates a similar proportion of global greenhouse gas emissions. As buildings become more energy efficient, embodied carbon, the emissions embedded in materials, fabrication, transport, construction, maintenance, and end-of-life processing, increasingly dominates the life-cycle equation and can represent a major portion of a high-performance building's total carbon footprint. Structures and foundations alone can account for more than half of that embodied share. With the wave of new construction expected in the coming years (much of it concentrated in Asia and Africa where awareness of construction-related carbon emissions remains less developed), the cumulative carbon consequences of these design decisions will only intensify, highlighting the urgent need for the global construction community to act.

The technical vocabulary and the analytical framework have been developed. Life-cycle assessment (LCA), with its specific stages from A0 through D, provides a common language for tracking carbon across the full life of a structure. The detailed treatment of steel and concrete production routes, and timber's relationship with biogenic carbon storage has shown that no material is inherently "green" or "dirty." Context, sourcing, and specification are what determine environmental performance. So too do the secondary systems that each material demands: steel typically requires fire protection (intumescent coatings or fireboards) whose carbon cost is often overlooked, clinker replacements such as GGBS face global supply constraints that limit their universal application, and structural timber is rarely recycled in practice, with the vast majority currently ending in landfill or incineration. The weight of the chosen system also propagates into foundations, meaning that a comparison confined to the superstructure alone can be misleading. Steel, concrete, and timber systems each have configurations in which they may outperform the others, depending on span, height, seismic demand, sourcing, and end-of-life route. The most sustainable structural solution is not a material; it is a result of rigorous, data-driven comparison.

The practical methodology for measuring embodied carbon (in structural systems) has been introduced. It presented the fundamental calculation equation linking material quantities from the Bill of Quantities with carbon factors from Environmental Product Declarations (EPDs) and explained the role of generic versus project-specific data. The variability within a single material category (captured by the range of over 1000 published EPDs analyzed in this book) is often as large as the difference between material categories. This means that statements about which material is "best" for the environment are unreliable without project-specific, verified data. The transition from generic database values to product-specific EPDs, and eventually to supplier-verified data is a prerequisite for credible decision making. As EPD availability continues to grow, engineers have both the opportunity and the responsibility to demand better data. The most sophisticated optimization algorithm produces misleading results if fed with unverified carbon factors. At the same time, assessments must be complete: items routinely underestimated or omitted (steel connections, fire protection, reinforcement detailing) can considerably alter the result. Region-specific EPDs and appropriate contingency factors are essential to ensure that early-stage estimates remain credible as design progresses.

To make comparisons systematic, this book introduced life-cycle parameter analysis (LCPA), which extends the sensitivity analysis required by ISO 14044 into a method that actively drives structural design decisions. At the building level, the LCPA across different design configurations demonstrated that early-stage decisions (e.g. span length, composite action, material grade, and column section type) and building geometry itself (structural grid dimensions, basement inclusion, floor-to-floor height) can significantly reshape the embodied carbon profile of a project, often outweighing later optimizations at the element or detailing level. A representative portion approach can be conducted rapidly at the conceptual stage, with results that closely approximate full-building assessments. Single-value estimates of embodied carbon can be misleading. Uncertainty propagates through every calculation, and responsible practice demands that results be presented with appropriate ranges and confidence intervals. By the time detailed design begins, the conceptual choices that determine most of a project's environmental outcome have already been made. Life-cycle thinking must therefore enter the design process at its earliest stages, when the ability to influence outcomes is highest and the cost of change is lowest. Annex 1 provides a quick-start checklist that translates these principles into a practical, stage-by-stage workflow for design teams.

Life-cycle optimization (LCO) method showed that the design space for a building contains hundreds of feasible solutions, where cost and embodied carbon do not always move

in the same direction. Pareto front identified regions where trade-offs must be navigated with stakeholder input while guarding against carbon tunnel vision so that reductions in embodied carbon are not achieved at the expense of robustness, service life, resilience, or whole life cost. The parametric studies on steel recycling, GGBS content, span length, building height, and building shape provided quantitative evidence for how sensitive these trade-offs are to evolving material technologies and economic conditions. At its core, the most effective carbon strategy is not replacement but retention: extending the service life of existing structures, preserving structural capacity already embodied in materials, and treating the existing building stock as a resource.

The treatment of earthquake-resistant design within a life-cycle framework addressed what is yet a blind spot in conventional practice. The three seismic design philosophies (ductility-based, damage-controlled, and no-damage) represent fundamentally different long-term carbon strategies. A conventionally designed ductile structure minimizes upfront material but creates a building whose post-earthquake repair carbon can add much more to its initial environmental impact. Damage-controlled systems concentrate energy dissipation in replaceable fuses, dramatically reducing repair emissions. Seismic isolation, while carrying higher initial investment, preserves the entire material investment by preventing damage altogether. The implication is that resilience and sustainability are complementary, not competing. A building that survives an earthquake without damage preserves its embodied carbon investment. A building that can be quickly repaired through the replacement of modular fuses avoids the massive carbon cost of reconstruction. A seismically retrofitted building extends the useful life of materials already produced. In every case, designing for resilience under exceptional loads is also designing for environmental responsibility. Yet resilience is only as strong as our ability to verify it after an event. The emerging possibility of dual-purposing existing telecommunications fiber-optic networks for city-scale post-earthquake assessment using laser interferometry, without disrupting data traffic, extends this logic from design into operation. Because buildings demolished due to uncertain damage status (rather than verified structural failure) represent wasted embodied carbon, transforming post-earthquake evaluation from a slow manual process into a scalable, data-driven one is itself a life-cycle imperative.

The 25 case studies from 15 countries and 6 continents (offices, hospitals, stadiums, cultural buildings, airports, bridges, viaducts, shopping malls, mixed-use towers, residential buildings, data centers, and temporary construction facilities) demonstrated that the life-cycle-driven structural framework is not a theoretical ambition. Projects achieving reductions in global warming potential through adaptive reuse, retained concrete cores, reclaimed steel, hybrid timber–steel systems, seismic isolation, modular construction, and reduced steel weight through high-strength materials confirm that life-cycle-driven design is already delivering measurable results across diverse materials, scales, typologies, and regulatory environments.

The case studies also made clear that circularity is a design decision, not an afterthought. Design for deconstruction, bolted connections, modular components, reclaimed steel, seismic isolation, and adaptive reuse must be integrated from the conceptual stage with an awareness that circularity extends beyond the project boundary. Over-reliance on a supply-limited low-carbon resource on one site may simply displace carbon to another. Circularity should be designed in from the outset, not considered only when a building reaches the end of its service life. Beyond carbon, the evidence further indicates that teams adopting life-cycle-driven approaches consistently achieve faster construction programs, reduced material quantities, and buildings that are easier to adapt to future uses.

Annex 2 maps the current global landscape of embodied carbon regulations, benchmarks, and reporting requirements across these rules.

9.2 What Remains to Be Done

This book has aimed to provide a comprehensive foundation, but the field of life-cycle-driven structural design is evolving rapidly. Several frontiers deserve attention from researchers, practitioners, and standardization bodies:

9.2.1 Integration of Operational and Embodied Carbon in Decision Making

This book has focused primarily on embodied carbon of structures because they account for the largest share of a building's carbon footprint (one that, unlike operational emissions, cannot be diminished through energy retrofits or cleaner grids after construction). However, the interaction between structural and operational carbon (e.g. the thermal mass benefits of concrete, the air tightness implications of different framing systems, or the energy penalties of deeper floor zones) deserves an integrated treatment. Structural engineers optimize for strength and material efficiency, while energy consultants model operational performance separately, often assuming a fixed building geometry that was never tested for its embodied consequences. Bridging that gap will require shared modeling environments where structural and energy variables can be evaluated simultaneously, with objective functions that account for both carbon streams under the same time horizon. Whole-life carbon optimization that covers both domains remains an open challenge, and it is the logical next step from the methods presented in this book.

9.2.2 Scaling LCA Standardization from Leading Building Codes to Global Practice

While standards such as EN 15978 and ISO 21930 provide methodological frameworks, and regulators in Denmark, the Netherlands, and New Zealand are pioneering mandatory carbon limits, most structural design codes worldwide still operate without reference to environmental performance. The Paris Agreement established the first legally binding international commitment to limit global warming, yet translating that commitment into structural design practice remains uneven. Increasingly, it is not only national governments but also large organizations and private clients who are setting embodied carbon targets, driven by funding requirements, reporting obligations, or institutional values. Rating frameworks such as SCORS already provide a practical scale for comparing embodied carbon intensity across projects; embedding similar benchmarks within design codes would give engineers a clear target rather than an open-ended aspiration. The growing adoption of mandatory Scope 3 reporting means that embodied carbon in structural systems is no longer invisible to building owners and investors, creating a market-driven incentive that reinforces regulatory action.

9.2.3 Dynamic and Probabilistic LCA

The carbon factors used throughout this book represent snapshots of current production technologies. As electricity grids decarbonize, manufacturing processes evolve, and recycling infrastructure expands, these factors will change. Dynamic LCA methods account for the temporal evolution of carbon intensities over a building's service life. The decarbonization of steel and cement production (through hydrogen-based direct reduction, carbon capture, new supplementary cementitious materials, and the progressive renovation of electricity grids with renewable resources) means that the carbon factors used in today's and future assessments may look very different within a single building's service life. Probabilistic approaches that systematically quantify uncertainty will become increasingly important for long-horizon decisions.

9.2.4 Artificial Intelligence and Data-driven Life-cycle-driven Structural Design

The genetic algorithm tool presented in Chapter 6 represents one approach to navigating complex, multi-objective design spaces. The rapid development of neural networks, reinforcement learning, and generative design tools offers the potential to dramatically accelerate early-stage carbon optimization, integrate lateral stability and seismic performance assessment with reduced computational cost, and make life-cycle-driven design accessible through interactive, web-based platforms integrated with industry-standard design software. These same technologies can support data-driven design in a broader sense: ensuring that real performance data (from concrete cube tests, pile load trials, and in-service monitoring) is captured, stored, and made available to future designers so that buildings can be adapted with reduced uncertainty and lower factors of safety. As these tools mature, however, the profession will need to establish clear standards for transparency, validation, and accountability.

9.2.5 Scaling Life-cycle Thinking to Infrastructure and Urban Systems

While this book has focused primarily on buildings, with attention to the infrastructure (e.g. bridge, airport, and data-center case studies), the principles of life-cycle-driven design apply equally to transportation networks, energy infrastructure, water systems, and urban development at the district and city scale. Extending these methods to infrastructure portfolios and urban master plans is a natural and necessary next step. Such an expansion must also broaden the environmental lens beyond carbon to include resource depletion, biodiversity loss, and the feedback loops that connect construction activity to ecosystem degradation. At the urban scale, the interactions between individual structures become as important as the structures themselves: shared foundations, district energy systems, and coordinated material banks for reuse all create opportunities for carbon reduction that are invisible at the single-building level.

9.2.6 Digitalization of Fabrication, Erection, and Maintenance

Current structural design frameworks consider material quantities but remain largely disconnected from the processes that transform those materials into built structures. Cutting, drilling, and welding (which together can represent roughly a third of total structural cost, with carbon following a similar proportion) are neither modeled nor tracked in most structural design workflows. The link between structural analysis, connection detailing, and enterprise resource planning systems that measure both cost and carbon is largely absent, meaning that a single poorly detailed connection can double the required weld volume and alter the carbon profile of an entire project. Until design platforms close this loop, a significant share of embodied carbon will remain invisible to the engineer whose decisions determine it.

Erection strategy is also consequential. Maximizing workshop fabrication and minimizing site operations reduces waste, energy consumption, and environmental disruption. Large-scale prefabrication projects have shown that assembling major components offsite and installing them in a single operation can deliver substantial savings across material, fabrication, transport, and erection. Realizing these benefits, however, requires designers trained to conceive the construction sequence as an integral part of the structural concept.

Maintenance belongs to the same challenge. A structure left without maintenance degrades faster, shortening its service life and triggering repair or replacement that consumes additional materials and energy. Proactive maintenance is therefore a carbon strategy that protects the embodied investment already made. Ensuring that data generated during design and construction (material test results, as-built connection details, geotechnical records) is preserved and accessible for future interventions is essential to realizing the service lives on which LCA calculations depend.

9.2.7 Education for High-performance Structural Design

Some of the carbon reduction strategies presented in this book (high-strength steels, optimized connections, slender composite systems, advanced erection methods) demand engineering skills that go beyond standard curricula. Adopting S460 or S690 steel to reduce material weight is an effective carbon strategy, but it produces slender members governed by buckling, connection zones with concentrated forces, and lightweight assemblies that are sensitive to erection loads and tolerances. Designing, detailing, and safely erecting such structures remains, at present, a competence confined to a small number of highly experienced engineers. If life-cycle-driven design is to scale beyond landmark projects, university programs and continuing professional development must equip a much broader base of engineers with fluency in stability design, advanced connection behavior (including fatigue and fracture), and construction-sequence planning. Without that preparation, the profession risks specifying high-performance materials without the design competence to use them safely and efficiently.

9.3 A Final Word

The transition to a better built environment requires an update in how we educate engineers, how we structure design teams, how we write contracts, and how we define value. It requires, above all, that structural engineers and architects accept a broader definition of their professional responsibility, one that extends beyond the safety of a building's occupants to include the environmental consequences of the structural strategy established at the earliest design stage. Vitruvius defined architecture by firmness, utility, and beauty; our generation is adding a fourth requirement: environmental accountability across the full life cycle. It is a requirement that does not compete with the original three but redefines what each of them means in an era of climate constraint. This book has attempted to provide a framework for meeting that requirement.

For university instructors who wish to structure a course around this material, Annex 3 offers adaptable syllabi that align with the progressive logic of the chapters. Annex 4 lists recommendations for using this book together with generative artificial intelligence.

The tools are here. The data is improving. The regulatory momentum is building. The knowledge base is no longer a barrier. What remains is the will to act, and I have every confidence that the next generation of AEC professionals will bring it.

Annex 1

Life-cycle-driven Structural Design

Quick-start Checklist for Practitioners

A step-by-step workflow for integrating life-cycle thinking into structural design, from project brief to end-of-life. Each item maps to the relevant life-cycle module and chapter of *Life Cycle Driven Structures* (Kanyılmaz, Wiley 2026).

1. Project brief and target setting	**Pre-design**
• Set an embodied carbon target (e.g. $kgCO_2e/m^2$, $kgCO_2e/m^2$/year) using published benchmarks (or standards) and a rating system (e.g. SCORS)	*Chs. 2, 3*
• Define performance objectives beyond code minimum: continued functionality, adaptability, robustness, and resilience	*Chs. 1, 7*
• Define the minimum acceptable loading and vibration performance criteria, and agree these with the client	*Chs. 1, 8*
• Identify how adaptive reuse, partial retention, or new build could serve the design brief. Quantify the carbon cost of demolition before ruling out existing structures	*Chs. 1, 3, 8*
• Establish embodied carbon as a project KPI alongside cost, program, and safety	*Ch. 1*

2. Early design and optioneering	**Modules A1–A5**
• Compare at least three structural schemes (e.g. reinforced concrete flat slab, steel-concrete composite deck, timber beams, hybrid systems) across at least two structural grids against LCA, cost, program, and resilience criteria before freezing geometry. Use automation tools where available to expand the option space	*Chs. 3, 4, 5, 8*
• Account for the weight of each system and its effect on foundations: a lighter superstructure may reduce foundation carbon enough to offset a higher carbon intensity at floor level	*Chs. 5, 6, 8*
• Assess future flexibility: evaluate whether each scheme can accommodate changes in building services, partition layouts, and load redistribution without major structural intervention	*Chs. 5, 6, 8*

2. Early design and optioneering	Modules A1–A5
• Run component-level parameter analysis: test how column and beam type, span, and supply-chain scenario shift embodied carbon	*Ch. 4*
• Run building-level parameter analysis (e.g. evaluate steel sourcing [virgin, scrap, and reclaimed], transport distance, connection type, and hybrid options)	*Ch. 5*
• Include allowances for items not yet detailed: add a percentage of contribution to steel beam and column weights to account for connections, and include fire protection, reinforcement anchorages, and other secondary items that are often omitted at this stage but can considerably alter the carbon total	*Chs. 4, 5, 8*
• For seismic zones, compare conventional, damage-controlled, and isolation strategies through a life-cycle lens	*Ch. 7*
• For soft or challenging soils, perform site-specific geotechnical and seismic response analysis rather than relying on simplified code approaches	*Ch. 8*
• Prioritize hybrid systems: assign timber, steel, and concrete to the roles they serve best	*Chs. 5, 8*
• Apply contingency factors to early-stage carbon estimates (e.g. 10–15% at concept design, reducing to around 5% at detailed design) to account for quantities that may increase as the design develops	*Chs. 3, 4, 5, 6*

3. Material specification and procurement	Modules A1–A3
• Collect and compare Environmental Product Declarations (EPDs) for all primary structural materials. Use product-specific data where available. At the early design stage (when suppliers have not yet been selected), generic carbon factors are acceptable, but use ranges rather than single values to capture the uncertainty in supply-chain assumptions	*Ch. 3*
• Use region-specific data wherever possible (EPDs where available, region-appropriate generic factors otherwise), as production methods, energy mixes, and transport distances vary significantly between regions	*Ch. 3*
• Establish what is locally available first: map regional suppliers, transport distances, and material sourcing options before comparing products	*Chs. 2, 3*
• Select verified lower-GWP suppliers: EAF or DRI-EAF steel routes, SCM-blended cements, and locally sourced timber with chain-of-custody certification	*Chs. 2, 3*
• Confirm at-scale availability and curing times for SCMs and recycled aggregates before specifying them	*Chs. 2, 8*
• Explore reclaimed steel or recycled aggregate as procurement options, treating reuse as a workflow	*Chs. 3, 8*
• Evaluate the use of higher strength materials where structurally justified, ensuring that reductions in material quantity are not offset by fabrication constraints and availability limits	*Chs. 3, 8*

4. Detailed design	Modules A1–A5
• Design connections for disassembly where the brief supports future reuse: prefer bolted over welded joints, and standardize member sizes and bay geometry	*Chs. 3, 8*
• Verify that connection detailing is consistent with the fabrication and erection strategy: weld volumes, bolt counts, and access for inspection all carry carbon and cost consequences that compound across a project. Invest in experienced review of connection designs, as a poorly detailed node can multiply fabrication effort and carbon	*Chs. 4, 8*
• Where high-strength steels (S460, S690) are specified, ensure the design team has the expertise to address the stability, buckling, and erection challenges of slender members that these grades produce	*Ch. 8*
• Align the structural elements to standard stock lengths to reduce cutting waste and fabrication time	*Ch. 8*
• Detail nonstructural elements (ceilings, partitions, and utilities) to a comparable performance standard as the structure, especially in seismic zones	*Chs. 7, 8*
• Coordinate all systems crossing movement joints or isolation planes (seismic gaps, flexible utility connections) through integrated shop drawing reviews	*Chs. 7, 8*
• Size foundations to match the final design superstructure loads (not conservative estimates from earlier stages). Reassess after every weight reduction	*Ch. 8*
• Carry out site-specific geotechnical investigations or testing at this stage to reduce uncertainty in ground conditions: better data often allows smaller, less carbon-intensive foundations	*Ch. 8*

5. Construction and industrialized delivery	Modules A4 and A5
• Maximize prefabrication and preassembly: use the largest subassemblies, dry-bolted site joints, and crane-placed elements to reduce site time and waste	*Ch. 8*
• Design the erection sequence to maximize workshop fabrication and minimize site welding, temporary works, and environmental disruption. Material, fabrication, transport, and erection each carry significant shares of total cost and carbon; an intelligently planned construction method can reduce all four simultaneously	*Ch. 8*
• Develop a BIM-integrated construction model: automate reinforcement detailing, clash detection, and erection sequencing	*Ch. 8*
• Plan erection sequences that maintain structural stability at every intermediate stage and minimize temporary works	*Ch. 8*
• Align concrete pours, steel deliveries, and timber shipments with the construction schedule to reduce site storage and double handling	*Chs. 3, 8*
• Implement quality control and inspection protocols (especially for fire protection, seismic gap clearances, and isolation system details)	*Chs. 7, 8*

6. Use phase, maintenance, and adaptation	**Modules B1–B5**
• Specify durable, low-maintenance materials	*Chs. 2, 8*
• Establish a proactive maintenance plan from the outset: a structure left without maintenance degrades faster, shortening its service life and triggering repair or replacement carbon that erodes the life-cycle case	*Chs. 2, 7, 8*
• Preserve and hand over all design and construction data (material test results, connection details, geotechnical records) to support future interventions	*Ch. 8*
• Provide accessible replaceable elements (fuses, bearings, and expansion joints) and document replacement intervals	*Ch. 7*
• Design floor plates, structural grids, and service zones for future change of use	*Chs. 1, 8*
• Quantify avoided repair and replacement carbon (Modules B4 and B5) as part of the whole-life carbon case	*Chs. 2, 7*

7. End-of-life and beyond	**Modules C1–C4 and D**
• Document the structural system, connection details, and material grades in a building passport to enable future disassembly or material recovery	*Ch. 3*
• Quantify Module D credits (reuse potential, recycling recovery rates) and include them in the embodied carbon assessment	*Chs. 2, 3*
• Design for selective demolition: ensure that structural elements can be separated from finishes and services without contamination	*Chs. 3, 8*
• Consider the carbon cost of not designing for resilience	*Ch. 7*

This checklist accompanies Life Cycle Driven Structures by Alper Kanyılmaz (Wiley 2026). Chapter references indicate where to find the detailed methods, worked examples, and case studies behind each action item. The checklist is intended as a starting framework; teams should adapt it to the specific regulatory, regional, and programmatic context of each project.

Annex 2

Global Regulatory Mapping of Embodied Carbon in Buildings

Region	Country	Framework	Scope	Requirement
EU-LEVEL FRAMEWORKS				
EU	All MS	EPBD (2024/1275)	Buildings	Life-cycle GWP calculation
EU	All MS	Level(s) Indicator 1.2	Buildings	Common EU WLC calculation methodology ($kgCO_2e/m^2$, 50-yr RSP); reporting framework for life-cycle GWP; no binding limits; referenced by EPBD and EU Taxonomy as the baseline assessment method
EU	All MS	CPR revision (2024/3110)	Construction Products	GWP and environmental indicators declared in DoPC; EPD data integrated into product documentation; Digital Product Passports for priority product categories; phased environmental reporting (GWP from 2026, additional indicators by 2030)
EU	All MS	EU Taxonomy Climate Delegated Act (2021/2139)	Buildings >5000 m^2	WLC calculation and disclosure to investors/clients on demand; no absolute GWP performance threshold

Threshold/limit	Ref. Standard	Date published	References	Notes
>1000 m^2 from 2028; all new buildings from 2030; numerical limits set by each MS via roadmaps	EN 15978	2024	Directive (EU) 2024/1275[1]	A mandatory EU law requiring all new buildings to report their total carbon footprint. The EPBD mandates the introduction of whole-life carbon limit values, to be defined by Member States through national roadmaps
No limit values (benchmarking and disclosure only)	EN 15978, EN 15804+A2	2020	EC Level(s) page[2]	Level(s): The method showing how to calculate carbon footprint EPBD 2024/1275: The mandate telling when and for which buildings the method must be used (The Level(s) framework for sustainability performance assessments has inspired and supported the revision of the EPBD to mandate the assessment of whole-life carbon for new buildings)
No building-level limits (product-level declarations only)	EN 15804+A2	2025	ECOS analysis[3]	CPR 2024/3110 is relevant to the supply side. It ensures that product-level environmental data (GWP, EPDs) is declared by manufacturers, so that building-level WLC calculations required by the EPBD can be performed using reliable, standardized inputs
No GWP limit values (disclosure only)	EN 15978, Level(s)	2022	ECOS analysis[3]	It defines what counts as a "sustainable" building investment by requiring WLC disclosure for large new buildings, but without setting carbon performance limits. While Level(s) provides the calculation method, EPBD mandates its use for all new buildings, and CPR ensures product data availability, the Taxonomy links these to sustainable finance (green bonds, ESG reporting, CSRD). It creates market demand for WLC data by making it a prerequisite for green investment labeling

(Continued)

Region	Country	Framework	Scope	Requirement
EUROPE				
EU	France	RE2020	Buildings	Binding thresholds tightening in 4 phases (2022/2025/2028/2031); dynamic LCA with temporal weighting of emissions; 50-yr RSP LCA stages A1–A5, B1–B4, C1–C4, D via RE2020 indicators
EU	Denmark	BR18	Buildings	Typology-specific annual WLC limits; LCA stages A1–A5, B4, B6, C3–C4 (D reported but excluded from compliance); separate A4+A5 construction-phase limit of 1.5 $kgCO_2e/m^2/yr$; tightening in 4 phases (2023/2025/2027/2029)
EU	Netherlands	MPG	Buildings	Shadow cost metric (€/m^2 GFA/yr). LCA stages A1–A5, B1–B4, C1–C4, D
EU	Sweden	Klimatdeklaration (Act 2021:787)	Buildings	Climate declaration of GHG emissions from construction stage (A1–A5) using GWP-GHG (excl. biogenic carbon).

Threshold/limit	Ref. Standard	Date published	References	Notes
Houses: 640 $kgCO_2e/m^2$ (2022), 530 $kgCO_2e/m^2$ (2025), 475 $kgCO_2e/m^2$ (2028), 415 $kgCO_2e/m^2$ (2031); Apartments: 740 $kgCO_2e/m^2$ (2022), 650 $kgCO_2e/m^2$ (2025), 580 $kgCO_2e/m^2$ (2028), 490 $kgCO_2e/m^2$ (2031); Offices: 810 $kgCO_2e/m^2$ (2025), 600 $kgCO_2e/m^2$ (2031); Schools: 770 $kgCO_2e/m^2$ (2025), 590 $kgCO_2e/m^2$ (2031).	EN 15804+A2, INIES/FDES, EN 15978	2022	Réglementation environnementale RE2020[4]	France is the first EU country to enforce binding whole-life carbon limits on new buildings, going beyond disclosure to set progressively tightening performance thresholds. It also introduces two distinctive features: dynamic LCA (which weights emissions by their timing, favoring materials that delay carbon release) and I_c énergie (a separate cap on operational energy carbon), effectively decoupling the two sources of building emissions
12.0 $kgCO_2e/m^2/yr$ (2023, all types); 7.5 $kgCO_2e/m^2/yr$ apts/offices, 7.1 $kgCO_2e/m^2/yr$ avg (2025); further reductions in 2027 and 2029 (target range 3.2–6.4 $kgCO_2e/m^2/yr$ by 2029 depending on typology)	EN 15804+A2, EN 15978	2023	Buro Happold summary[5]	Denmark was among the first countries to introduce mandatory embodied carbon limits into building regulations (2023, one year after France's RE2020). Unlike France's RE2020 (which uses $kgCO_2e/m^2$ over a 50-year RSP), Denmark expresses limits as annualized values ($kgCO_2e/m^2/yr$), Stage D (reuse/recycling potential) must be reported separately
Housing: 0.8 €/m²/yr (tightening to 0.5 postponed pending cost review); Offices: 1.0 €/m²/yr (tightening to 0.85 proceeding)	EN 15978, EN 15804+A2, Nationale Milieudatabase	2025	RVO MPG page[6]	Unlike France and Denmark (which set limits in $kgCO_2e/m^2$), the Netherlands uses a monetized environmental cost metric (€/m² GFA/yr) that aggregates multiple impact categories (including but not limited to GWP) into a single shadow cost score. This makes it a broader environmental performance indicator rather than a pure carbon metric.
No limits yet; Boverket developing reference values from national building data.	EN 15804, Boverket generic DB	2022	Boverket climate declaration[7]	Sweden chose to mandate disclosure before limits (unlike France and Denmark which launched with binding thresholds). The logic is to build a national database of real building data before setting evidence-based limits. The current scope is limited to A1–A5. Sweden also uses GWP-GHG (which excludes biogenic carbon uptake and release) rather than GWP-total, a methodological choice that avoids the contentious accounting of carbon stored in bio-based materials

(Continued)

Region	Country	Framework	Scope	Requirement
EU	Finland	Rakentamislaki 751/2023; Decree 1027/2024	Buildings	Climate declaration covering carbon footprint & handprint; building product list in machine-readable format at permit application; national limit values by building type under preparation. LCA stages A1–A5, B4, C1–C4.
EU	Germany	QNG/DGNB	Buildings	QNG quality seal requires min. DGNB Silver. LCA stages A1–A5, B4, C1–C4.
EU	Norway	TEK17 §17-1	Buildings	LCA stages A1–A5 (A5 limited to waste), B2, B4; no limit values; biogenic carbon excluded from calculations
EU	Italy	CAM (DM 256/2022; updated DM Nov 2025)	Buildings	EPD-based product environmental assessment for all public tenders; LCA as rewarding criterion (criterio premiante) in tender evaluation

Threshold/limit	Ref. Standard	Date published	References	Notes
No national limits yet. Helsinki municipal requirement: 14.0 $kgCO_2e/m^2/yr$ for apartments	EN 15978, CO2data.fi	2025	Nordic LCA overview[8]	Like Sweden, Finland chose a declaration-first approach before introducing limits. A distinctive feature is the "carbon handprint" concept (alongside the carbon footprint), which accounts for beneficial climate effects beyond the building's system boundary (e.g. exported renewable energy, carbon storage in bio-based materials). Finland also requires a machine-readable building product list at the permit stage, enabling digital integration with BIM workflows
Residential QNG-PLUS: ≤24 $kgCO_2e/m^2/yr$, ≤96 $kWh/m^2/yr$; Residential QNG-PREMIUM: ≤20 $kgCO_2e/m^2/yr$, ≤64 $kWh/m^2/yr$; Non-residential: project-specific	EN 15804, ÖKOBAUDAT	2022	Federal Ministry for Housing, Urban Development and Building[9]	The Subsidy-Driven Approach. Unlike France and Denmark (which impose binding limits through building regulations), Germany uses financial incentives to drive WLC assessment. The QNG seal (Quality Seal for Sustainable Buildings) is voluntary for now. This creates a market pull rather than a regulatory push. While Germany has not yet introduced binding embodied carbon limits in building codes, the QNG framework is widely seen as a pre-regulatory step, positioning the industry for the mandatory WLC assessments expected under the EPBD from 2028/2030
No limit values; national benchmark values under development	NS 3720:2018	2022	One Click LCA TEK17[10]	The standard is currently under revision to improve comparability (introducing a proposal to implement limit values measured in $kgCO_2e/m^2$
No building-level $kgCO_2e/m^2$ cap; product-level EPD and recycled content thresholds	EN 15804, EN 15978, national EPDs	2022	Baker McKenzie EU guide[11]	Italy addresses it through mandatory Green Public Procurement criteria (CAM). Since all public construction must comply with CAM, and public works represent a significant share of Italian construction (especially with PNRR/Recovery Plan investments), the impact is substantial. However, LCA is currently a rewarding criterion rather than a mandatory threshold, meaning there is no building-level carbon cap.

(Continued)

Region	Country	Framework	Scope	Requirement
EU	Austria	klimaaktiv/OI3	Buildings	Ökoindex OI3 composite score includes a weighted score of GWP, primary energy depletion, and acidification potential with a fixed scale of performance.
EU	Belgium	TOTEM tool	Buildings	TOTEM calculates aggregated environmental score at material and building level across full life cycle; B-EPD database feeds product-specific data into TOTEM.
TR	Türkiye	EPBD (2024/1275)	Buildings	Building life-cycle analysis mandatory for designated projects from 2027. Bep-TR Regulations and rating of energy performance in buildings will be used.

Threshold/limit	Ref. Standard	Date published	References	Notes
No uniform national $kgCO_2e/m^2$ cap	EN 15978, OIB Guideline 6	2020	SE 2050 klimaaktiv[12]	Austria's approach is unique in that building regulation is a provincial (Länder) competence, not federal. The OI3 index and klimaaktiv certification provide a common national methodology, but implementation varies across the nine provinces through their respective housing subsidy schemes. The OI3 is distinctive in combining three environmental indicators (GWP, acidification, primary energy depletion)
No binding building-level EC limits; TOTEM provides benchmarking, not regulatory thresholds	EN 15978, EN 15804, Belgian EPDs	2018	Baker McKenzie EU guide[11]	Belgium's federal structure means building regulation is a regional competence (Flanders, Wallonia, Brussels), yet TOTEM is a rare example of cross-regional collaboration on a common methodology. Since the EU's EPBD (2024/1275) is now law, the Belgian regions are currently (in 2026) drafting the legislation that will move TOTEM from a voluntary optimization tool to a mandatory reporting tool for the Energy Performance Certificate (EPC) by 2028
No binding EC limits	National methodology under development	2026	Ministry of Environment, Urbanization and Climate Change. (2025, March 11). Planlı Alanlar İmar Yönetmeliğinde Değişiklik Yapılmasına Dair Yönetmelik [Regulation on the Amendment of the Planned Areas Zoning Regulation]. *T.C. Resmî Gazete*[13]	Although Türkiye is not an EU Member State and has no formal transposition obligation, the March 2025 amendment to the "Planlı Alanlar İmar Yönetmeliği" introduces building life-cycle analysis requirements that are expected to align with the EPBD framework (WLC calculation methodology, EN 15978, Level(s)). A national EPD system and construction materials database are under development to provide the product-level data infrastructure needed for building-level LCA.

(Continued)

Region	Country	Framework	Scope	Requirement
UNITED KINGDOM				
UK	UK-wide	Net Zero Carbon Building Standard (NZCBS)	Buildings	Upfront embodied carbon limits (A1–A5) by building type and year of commencement.
UK	London	GLA Policy SI 2(F)	Buildings	Submission of a WLC assessment (LCA stages A1–A5, B1–B4, C1–C4, D [reported separately]) using the GLA Excel template at three stages: pre-application, planning application submission (RIBA Stage 2/3), and post-construction.

Threshold/limit	Ref. Standard	Date published	References	Notes
New homes: 500 $kgCO_2e/m^2$ (2025), tightening to ~350 $kgCO_2e/m^2$ (2031); New offices: ~600 $kgCO_2e/m^2$ (2025); Retrofit: ~50% of new-build limits. All tightening progressively to 2050	RICS WLC Professional Standard (2nd ed), EN 15978, EPDs	2024	UK Net Zero Carbon Buildings Standard: guidance on embodied carbon[14]	The UK introduced a voluntary whole-life carbon framework (the Net Zero Carbon Buildings Standard, NZCBS) for major developments, reinforcing its position as one of the leading regulatory environments for embodied carbon. The NZCBS sets upfront embodied carbon limits (A1–A5) differentiated by building type and year of commencement (e.g. 500 $kgCO_2e/m^2$ for new homes and ~600 $kgCO_2e/m^2$ for new offices in 2025, tightening to ~350 $kgCO_2e/m^2$ for homes by 2031, with retrofit limits at ~50% of new-build values, and progressive tightening to 2050), alongside whole-life carbon reporting (modules A–D, without binding limits). The Greater London Authority already requires whole-life carbon assessments for referable planning applications, and national-level policy is progressing toward stricter limits aligned with the country's legally binding net-zero 2050 target (LETI).
Residential A1–A5: <850 $kgCO_2e/m^2$ (benchmark), <500 $kgCO_2e/m^2$ (aspirational); Offices A1–A5: <950 $kgCO_2e/m^2$ (benchmark); <600 $kgCO_2e/m^2$ (aspirational), Schools, Universities A1–A5: <750 $kgCO_2e/m^2$ (benchmark), <500 $kgCO_2e/m^2$ (aspirational); Retail A1–A5: <850 $kgCO_2e/m^2$ (benchmark), <550 $kgCO_2e/m^2$ (aspirational). In these values biogenic carbon is excluded.	BS EN 15978, RICS WLC Professional Standard (1st ed), 60-yr RSP	2022	GLA WLC guidance (PDF)[15]	London's Policy SI 2(F) is the most advanced local WLC policy in the UK, requiring full WLC assessments for Mayor-referable planning applications since 2022. A distinctive feature is the mandatory post-construction WLC assessment (not just design-stage), enabling the GLA to build a database of actual building performance

(Continued)

Region	Country	Framework	Scope	Requirement
UK	UK-wide	Part Z (campaign)	Buildings	Proposed Building Regulation amendment: Z1 requires mandatory WLC assessment and disclosure for all major new buildings and major renovations; Z2 introduces upfront embodied carbon limits (A1–A5) by building type, phased in after initial data-gathering period
UK	UK-wide	RICS WLC Standard, 2nd Ed.	Buildings + Infrastructure	Full life-cycle WLC assessment methodology (modules A0–D); expanded to infrastructure via EN 17472; module A0 optional for buildings, required for infrastructure where significant, 60-yr RSP buildings, 120-yr infrastructure
UK	UK-wide	PAS 2080:2023	Buildings + Infrastructure	Carbon management framework across entire value chain and whole life cycle; carbon hierarchy (avoid, switch, improve); governance, leadership and collaboration requirements; no quantification methodology itself (references BS EN 15978, EN 17472, EN 15804); third-party certification via accredited bodies

Threshold/limit	Ref. Standard	Date published	References	Notes
Draft limits by building type (not yet enacted); intended to align with NZCBS benchmarks and LETI/RIBA targets	BS EN 15978, RICS WLC Professional Standard	2024	ASBP Part Z[16]	Part Z is an industry-led campaign proposing a new section in UK Building Regulations to make WLC assessment mandatory. It comprises two parts: Z1 (mandatory WLC assessment and disclosure for all major new buildings) and Z2 (upfront embodied carbon limits by building type, phased in after a data-gathering period). Part Z is designed as the regulatory complement to the voluntary NZCBS: while the NZCBS demonstrates that limits are feasible and defines best practice, Part Z would make measurement and reporting a legal requirement for all projects (not only those voluntarily opting in)
No binding limits (methodology standard, not regulation); provides benchmarks by building type	BS EN 15978, EN 17472, EN 15643:2021, EN 15804+A1/A2	2024	Circular Ecology RICS[17]	The RICS WLC Standard (2nd edition, effective July 2024) is the de facto methodology behind almost all UK WLC practice (the GLA policy, NZCBS, and Part Z all reference it). While it does not set binding limits (it is a methodology, not a regulation), its benchmarks by building type inform the limits set by other UK frameworks
No quantitative limits; process-based (requires carbon reduction plans, hotspot identification, year-on-year improvement)	BS EN 15978, EN 17472	2023	ICE PAS 2080 guidance[18]	Unlike the other UK frameworks (which focus on calculating and limiting carbon at building level), PAS 2080 is an organizational carbon management standard for the entire value chain. Originally published in 2016 for infrastructure only, the 2023 revision expanded to buildings and aligned with the 1.5 °C scenario. It does not set quantitative carbon limits or provide a calculation methodology (it references EN 15978 and EN 17472 for that). Instead, it requires organizations to embed carbon governance, set reduction targets, identify hotspots, and demonstrate year-on-year improvement

(Continued)

Region	Country	Framework	Scope	Requirement
NORTH AMERICA				
N. America	US (Federal)	Buy Clean (rescinded)	Federal projects	IRA low-carbon materials funds rescinded by One Big Beautiful Bill Act in July 2025. In this annex, selected states are presented that offer unique features. A full list of states is available in the source.[19]
N. America	US (California)	CALGreen §5.409	Non-res Buildings	Three compliance routes: (1) Building Reuse (≥45% of existing structure/ enclosure retained), (2) Performance WBLCA (60-yr cradle-to-grave, 10% GWP reduction vs baseline), or (3) Prescriptive EPDs (product-specific EPDs for steel, glass, mineral wool, concrete at ≤175% of industry average GWP). Voluntary tiers: Tier 1 (≤150% industry avg) and Tier 2 (20% WBLCA reduction)
N. America	US (California)	Buy Clean CA (AB 262)	State public works	GWP limits on structural steel, rebar, flat glass, insulation (A1–A3)

Threshold/limit	Ref. Standard	Date published	References	Notes
EPD requirements + GWP limits were developed by GSA for specific material categories (cement, concrete, steel, flat glass, asphalt); limits were in effect for obligated projects but program is no longer funding new work	Various (EPA's Interim Determination on Low Carbon Materials, specific PCRs, ISO 14025, EN 15804)	2021–2025	Columbia Law analysis[20]	The IRA (2022) represented the most significant US federal investment in embodied carbon. However, the One Big Beautiful Bill Act (signed July 4, 2025) rescinded unobligated IRA funds (including for EPA's EPD Assistance grant program and the development of a low carbon label) and accelerated phase-outs of clean energy tax credits, effectively ending the federal Buy Clean program for new projects. Construction projects with already-obligated funding continue under existing rules. This federal rollback shifts the center of gravity for US embodied carbon policy to state and city level (California Buy Clean Act, New York City LL97, Colorado, Oregon, and others).
Performance: 10% GWP reduction vs. reference baseline (mandatory), 20% (Tier 2); Prescriptive: ≤175% of industry average GWP per material (mandatory), ≤150% (Tier 1), ≤100% (Tier 2)	ISO 14044, EN 15978, ASTM E2921-22	2024	CALGreen[21]	California became the first US state to mandate embodied carbon reduction in its building code. Unlike European approaches that set absolute $kgCO_2e/m^2$ limits, CALGreen uses a relative 10% improvement pathway (against a project-specific baseline), combined with a prescriptive EPD route and a building reuse pathway. CALGreen currently applies only to non-residential occupancies; a petition to extend to residential (under HCD jurisdiction) is under consideration for the 2025 code cycle
Material-specific GWP limits ($kgCO_2e$ per unit); e.g. flat glass: 1430 $kgCO_2e/t$; limits set at industry average of facility-specific EPDs; concrete added via subsequent legislation. Limits can only tighten, never increase	ISO 14025, EPDs	2017	Buy Clean California Act, A.B. 262, 2017–2018 Leg., Reg. Sess. (Cal. 2017)[22]	It operates at the material level (not building level), requiring facility-specific EPDs and setting GWP limits at the industry average for four material categories. Unlike CALGreen §5.409 (which applies to all non-residential buildings above a size threshold), Buy Clean CA applies only to state-funded public works. It also served as the template for similar legislation in Colorado, Oregon, New York, and the (now rescinded) federal Buy Clean initiative. The addition of concrete via subsequent legislation expanded the program's material coverage to the single largest contributor to construction embodied carbon

(Continued)

Region	Country	Framework	Scope	Requirement
N. America	US (California)	AB 2446	All new Buildings	California Air Resources Board (CARB) to develop framework for measuring and reducing average carbon intensity of all building materials (not limited to specific categories)
N. America	US (Colorado)	Buy Clean Colorado Act	State public building projects	LCA data for eligible material products (asphalt, cement/concrete, glass, post-tension steel, reinforcing steel, structural steel, and wood), covering LCA stages A1–A3, must demonstrate that products meet the maximum acceptable GWP values set for their category.
N. America	Canada	Toronto Green Standard v4	All new development in Toronto (low-rise residential, mid/high-rise residential, non-residential, city-owned facilities)	Tier-based system: Tier 1 (mandatory) requires whole-building energy modeling and GHG intensity limits; Tiers 2–3 (voluntary, incentivized) require upfront embodied emissions assessment (A1–A5) for structure and envelope per CaGBC ZCB Standard methodology, with absolute $kgCO_2e/m^2$ caps.
N. America	Canada (federal)	Greening Government Strategy / Standard on Embodied Carbon in Construction	Federal projects	Two-track approach: (1) WBLCA to estimate and reduce embodied carbon footprint of major construction projects; (2) Material-specific disclosure and reduction for structural materials: ready-mix concrete (since Dec 2022) and construction steel (from Sep 2025) with EPD-based GWP limits

Threshold/limit	Ref. Standard	Date published	References	Notes
20% net GWP reduction by 2030 (interim, adjustable by CARB); 40% net GWP reduction by 2035 (mandatory); baseline to be established from 2026 industry-average EPD data; targets begin no sooner than Jan 2027	TBD by CARB (ISO 14040 series, EPDs)	2022	CA DGS Buy Clean[23]	While Buy Clean CA (AB 262) sets material-specific GWP limits for state procurement and CALGreen §5.409 mandates building-level assessment for large non-residential projects, AB 2446 goes further by directing CARB to develop a comprehensive framework for reducing embodied carbon across all building materials in all new construction (including residential). The legislation applies to all building materials (concrete, wood, metal, windows, asphalt, insulation), not just the four categories covered by Buy Clean CA. Together, AB 262, CALGreen §5.409, and AB 2446 form a three-layered California framework: procurement limits (AB 262), building-code compliance (CALGreen), and economy-wide reduction targets (AB 2446)
Material-specific maximum acceptable GWP values (e.g. Fabricated hot-rolled steel: 1,220 kg CO_2eq/1 metric ton)	ISO 14025 Type III EPDs	2024	Buy Clean Colorado Act Policy[24]	This is a Buy Clean procurement requirement, not a whole-building LCA requirement. Compliance is assessed at the product or material level using LCA data for stages A1–A3. The intent is to ensure that eligible materials procured for the project meet the applicable maximum GWP limits for their product category.
Tier 2: ≤350 $kgCO_2e/m^2$ residential/commercial, ≤400 $kgCO_2e/m^2$ other types (A1–A5); Tier 3: ≤250 $kgCO_2e/m^2$ residential/commercial, ≤275 $kgCO_2e/m^2$ other types; Low-rise residential Tier 2: ≤250 $kgCO_2e/m^2$ (A1–A3); city-owned: mandatory ≤350 $kgCO_2e/m^2$	CaGBC Zero Carbon Building Standard v2/v3, NRC National WB-LCA Practitioner's Guide	2022	Toronto Green Standard Version 4[25]	Toronto is the first jurisdiction in North America to set absolute whole-building embodied carbon caps ($kgCO_2e/m^2$) for new city-owned buildings, a building-level approach closer to European models (France, Denmark) than to the material-specific US Buy Clean programs. The Toronto Green Standard's tier system is a performance framework that divides sustainable building requirements into a mandatory baseline (Tier 1) and voluntary upper levels (Tiers 2, 3, 4)
30% reduction in total embodied carbon for the WBLCA Track; material-specific GWP limits for concrete and steel by product subcategory	CSA Z782, ISO 14025, ISO 14044, ISO 21930	2025	TBS Standard (archived)[26]	Canada's approach is a hybrid between the US and European models: like the US Buy Clean programs (California, Colorado), it operates through federal procurement with material-specific GWP limits, but it also requires whole-building LCA for major projects, which is closer to the European building-level approach. Provincial initiatives (British Columbia, Québec) are developing complementary frameworks

(Continued)

Region	Country	Framework	Scope	Requirement
N. America	Canada (Vancouver)	Climate Emergency Action Plan/ Vancouver Building By-law (VBBL)	Buildings (larger multi-story buildings)	Embodied carbon reporting mandatory since 2023; reduction requirements from 2025; two compliance routes absolute ($kgCO_2e/m^2$) or relative (% reduction from project-specific baseline); cradle-to-grave scope (A1–A5, B1–B5, C1–C4; module D excluded from compliance but reported separately)
ASIA–PACIFIC				
Asia–Pacific	New Zealand	Building for Climate Change (BfCC) programme (Building Code amendments under MBIE)	Buildings	WLC reporting from 2025; mandatory caps from 2026, tightened through 2030. All modules A1–A5, B1–B5 (excluding B6 operational energy and B7 operational water), C1–C4, and D. Full whole-of-life scope
Asia–Pacific	Singapore	Green Mark 2021	Buildings	WLC assessment section (A1–A5, B, C; D reported separately); EC benchmarks by typology

Threshold/limit	Ref. Standard	Date published	References	Notes
Building-type-specific limits	EN 15978, City of Vancouver Embodied Carbon Guidelines, CaGBC ZCB methodology	2025	Vancouver EC guidelines (PDF)[27]	Vancouver is one of the first cities in North America to embed mandatory embodied carbon requirements directly in its building by-law (not just procurement or voluntary certification). Vancouver's requirements apply to all Part 3 buildings (public and private). The two-route compliance system (absolute $kgCO_2e/m^2$ or relative % reduction) offers flexibility similar to California's CALGreen, while the cradle-to-grave scope (A1–C4) is broader than most North American requirements and closer to European practice
No specific numerical limit values published yet. Caps to be derived from 2025 reporting data (the intent is to set levels that approximately two-thirds of new construction can already meet). Progressively tightened on a transparent schedule through 2030 in line with the National Emissions Reduction Plan	EN 15978, EN 15804(+A2), ISO 21930, ISO 21931-1; NZGBC Embodied Carbon Methodology	2025	HERA. (n.d.). Reports. HERA. Retrieved April 15, 2026, from https://hera.org.nz/technical-support/resources/reports/[28]	The assessment scope is notably comprehensive, covering all life-cycle modules A through D (excluding only operational energy B6 and water B7), and applies to all new buildings regardless of typology or size. The MBIE technical methodology aligns with European standards (EN 15978, EN 15804+A2) but relies on a hybrid data ecosystem combining NZGBC's ~1300-product database, BRANZ generic data, and product-specific EPDs. A "reference building" compliance mechanism (similar in concept to a notional building approach) will compare each design against a baseline with predefined carbon intensities. One important caveat: while the timeline was set under the previous Labour government's first Emissions Reduction Plan (2022–2025), the current National-led coalition has shown a less aggressive posture on climate policy, and the second ERP (December 2024, covering 2026–2030) is notably less specific on building sector actions. Whether the caps will be introduced on schedule in 2026, and at what stringency, remains to be confirmed
Benchmarks, no absolute caps	BS EN 15978, RICS	2021	One Click LCA Green Mark[29]	Includes a specific Whole Life Carbon (WLC) section. It requires projects to calculate upfront carbon (A1-A5) alongside operational emissions. Instead of fixed regulatory caps, the Building and Construction Authority (BCA) uses typology-specific reference baselines. Projects earn certification points by demonstrating percentage reductions against baselines

(Continued)

Region	Country	Framework	Scope	Requirement
Asia–Pacific	Australia	NABERS Embodied Carbon/NCC 2025/ASIC	Buildings and Corporate	Upfront carbon (A1–A5) measurement; mandatory corporate climate disclosures from Jan 2025

Current to early 2026. Regulations are evolving rapidly; readers should verify the current status via the reference links provided.
Abbreviations: EC = Embodied Carbon; WLC = Whole-Life Carbon; RSP = Reference Study Period; EPD = Environmental Product Declaration; GWP = Global Warming Potential; MS = Member States.
From: Life Cycle Driven Structures (Kanyılmaz, Wiley 2026).

1 Directive (EU) 2024/1275 of the European Parliament and of the Council of 24 April 2024 on the energy performance of buildings (recast). Official Journal of the European Union, L 2024/1275. http://data.europa.eu/eli/dir/2024/1275/oj
2 EC Level(s) page: https://green-forum.ec.europa.eu/green-business/levels/quick-introduction-levels_en
3 ECOS briefing: https://ecostandard.org/wp-content/uploads/2024/02/2024-02-15_ECOS_EU-taxonomy-construction_FINAL.pdf
4 Réglementation environnementale RE2020, https://www.ecologie.gouv.fr/politiques-publiques/reglementation-environnementale-re2020
5 Buro Happold summary: https://www.burohappold.com/news/denmark-sets-stricter-co2-emissions-rules-for-construction-what-you-need-to-know/
6 RVO MPG page: https://www.rvo.nl/onderwerpen/wetten-en-regels-gebouwen/milieuprestatie-gebouwen-mpg
7 Boverket climate declaration: https://www.boverket.se/en/start/laws-and-regulations/climate-declaration/
8 Nordic LCA overview: https://anavitor.ai/en/ecobuild-lca-perspectives-for-the-built-environment/life-cycle-assessment-requirements-in-nordic-building-regulations
9 Federal Ministry for Housing, Urban Development and Building (BMWSB). (n.d.). QNG Requirements. QNG - Quality Seal for Sustainable Buildings. Retrieved April 9, 2026, from https://www.qng.info/en/qng-certification/qng-requirements/
10 One Click LCA TEK17: https://help.oneclicklca.com/en/articles/275709-norway-greenhouse-gas-accounting-tek17

Threshold/limit	Ref. Standard	Date published	References	Notes
Benchmarks and ratings; no absolute caps	EN 15978	2025	Australian Securities and Investments Commission. (2024, September 18). 24-205MR ASIC urges businesses to prepare for mandatory climate reporting[30]	Australia uses a hybrid approach combining corporate mandates with building-sector rating systems. Mandatory climate-related financial disclosures for large businesses, enforced by Australian Securities & Investments Commission (ASIC), commence in January 2025. In the building sector, the National Construction Code (NCC) 2025 establishes a voluntary pathway for commercial buildings to measure and report upfront embodied carbon (life-cycle stages A1–A5). The primary assessment methodology is the NABERS Embodied Carbon rating tool, which requires a bill of quantities and verified product carbon claims to calculate a certified carbon intensity measure. The framework currently relies on benchmarking rather than absolute regulatory limits. These sector mandates are driven by state and territory GHG targets, which universally aim for net-zero by 2050. Interim 2030 reduction targets vary by jurisdiction (e.g. 50% in Victoria and NSW; 30% in Queensland; 80% below 2020 levels in WA; maintaining net-zero in Tasmania)

11 Baker McKenzie EU guide: https://resourcehub.bakermckenzie.com/en/resources/global-sustainable-buildings/europe-middle-east-and-africa/italy/topics/green-certification
12 SE 2050 klimaaktiv: https://se2050.org/klimaaktiv-austria-version-2020/
13 Ministry of Environment, Urbanization and Climate Change. (2025, March 11). Planlı Alanlar İmar Yönetmeliğinde Değişiklik Yapılmasına Dair Yönetmelik [Regulation on the Amendment of the Planned Areas Zoning Regulation]. *T.C. Resmî Gazete* https://www.resmigazete.gov.tr/eskiler/2025/03/20250311-1.htm
14 UK Net Zero Carbon Buildings Standard: guidance on embodied carbon
15 GLA WLC guidance (PDF): https://www.london.gov.uk/sites/default/files/lpg_-_wlca_guidance.pdf
16 ASBP Part Z: https://asbp.org.uk/asbp-news/part-z
17 Circular Ecology RICS: https://circularecology.com/news/rics-whole-life-carbon-assessment-2nd-edition-now-in-effect
18 ICE PAS 2080 guidance: https://www.ice.org.uk/areas-of-interest/decarbonisation/guidance-document-pas2080
19 https://www.builtcold.com/
20 Columbia Law analysis: https://blogs.law.columbia.edu/climatechange/2025/03/26/cancelling-the-buy-clean-program/
21 NBI CAL Green: https://newbuildings.org/news/california-building-standards-commission-approves-landmark-embodied-carbon-requirement/
22 Buy Clean California Act, A.B. 262, 2017–2018 Leg., Reg. Sess. (Cal. 2017). https://leginfo.legislature.ca.gov/faces/billCompareClient.xhtml?bill_id=201720180AB262
23 CA DGS Buy Clean: https://www.dgs.ca.gov/pd/resources/page-content/procurement-division-resources-list-folder/buy-clean-california-act
24 https://osa.colorado.gov/energy-environment/buy-clean-colorado-act

25 City of Toronto. (n.d.). Toronto Green Standard version 4. https://www.toronto.ca/city-government/planning-development/official-plan-guidelines/toronto-green-standard/toronto-green-standard-version-4/
26 TBS Standard (archived): https://www.tbs-sct.canada.ca/pol/doc-eng.aspx?id=32742
27 Vancouver EC guidelines (PDF): https://vancouver.ca/files/cov/embodied-carbon-guidelines.pdf
28 HERA. (n.d.). Reports. HERA. Retrieved April 15, 2026, from https://hera.org.nz/technical-support/resources/reports/
29 One Click LCA Green Mark: https://help.oneclicklca.com/en/articles/275726-singapore-green-mark-building-rating-system
30 Australian Securities and Investments Commission. (2024, September 18). 24-205MR ASIC Urges Businesses to Prepare for Mandatory Climate Reporting. https://asic.gov.au/about-asic/news-centre/find-a-media-release/2024-releases/24-205mr-asic-urges-businesses-to-prepare-for-mandatory-climate-reporting/

Annex 3

Life-cycle-driven Structures

Suggested Course Syllabi for Instructors

Two model syllabi for teaching from *Life Cycle Driven Structures* (Kanyılmaz, Wiley 2026). Option A is a single-semester course (14 weeks) suitable for a dedicated module on sustainable structural design or as a companion to a traditional structural design course. Option B is a two-semester sequence (28 weeks) that allows deeper coverage of parametric analysis, optimization, seismic life-cycle-driven design, and a capstone project. Both options include suggested assignments, assessment suggestions, and guest lecture slots. Instructors are encouraged to modify and adapt the content, pacing, and emphasis to their program context.

Option A: One-semester Course (14 WEEKS)

Course title (suggested): Life-cycle-driven Structural Design
Level: Upper undergraduate/MSc
Prerequisites: Knowledge of conceptual design of steel, concrete, or timber structures[1]
Credits: Adaptable to 5–6 ECTS (EU)
Class size assumption: Up to 50 students[2]

1 If students lack this background, the instructor should incorporate conceptual structural design material into the first weeks of the course.

2 This is the number at which group project work can be supervised effectively; for larger classes, the group project scope and deliverables should be adapted to maintain pedagogical quality.

Weekly Schedule[3]

Week	Topic	Chapter(s)
1	The climate context for structures. Four pillars of life-cycle-driven design. Student background test[4]	1
2	Life-cycle assessment (LCA): Modules A0 through D. System boundaries and functional units	2
3	Global Warming Potential profiles of steel, concrete, and timber	2
4	Measuring embodied carbon	3
5	Strategies for carbon reduction. Design for circularity, robustness, and material efficiency	3
6	Component-level life-cycle parameter analysis	4
7	Building-level life-cycle parameter analysis	5
8	Mid-semester test, capstone project[5] kick-off	—
9	Life-cycle optimization (LCO)	6
10	Seismic design through a life-cycle lens: three philosophies and their carbon profiles	7
11	Resilience of nonstructural elements, retrofitting as circular economy, and community recovery	7
12	Case studies: adaptive reuse, hybrid systems, and material efficiency (selected from Chapter 8)	8
13	Guest lecture[6] from industry	8
14	Synthesis: from analysis to practice. The quick-start checklist	1–8

3 If teaching hours are split across multiple sessions per week, the weekly content outlined below can be divided accordingly across those sessions.
4 A short background test in the first session helps the instructor identify knowledge gaps in structural analysis and design, allowing early content to be calibrated so that all students can engage with the course.
5 A team-based assignment in which students develop the conceptual design of a real or hypothetical building and perform an LCA comparing at least three structural alternatives. The work includes embodied carbon calculations and interpretation of life-cycle parameters. The design and assessment are progressively refined during the weekly exercise hours.
6 Instructors are encouraged to invite contributors (practitioners with experience) to present their work and engage directly with students. The 25 case studies in Chapter 8 were contributed by practicing engineers from international firms and can serve as a framework for instructors when shaping the content and scope of these invited lectures.

Suggested Assessment Breakdown

Component	Weight (%)	Description
Individual assignment	20	Life-cycle-driven conceptual design of structures, embodied carbon calculations, and parameter analysis interpretation
Capstone project	40	LCA of a real or hypothetical building comparing 3+ structural options (report + 15-min presentation)
Final written (individual) exam	40	Written exam to evaluate individual performance

Capstone Project Brief

Teams of 3–4 students are assigned a building typology (e.g. office, school, hospital, sports facility, data center, or bridge) and a site context (climate zone, seismic zone, and/or soil conditions). Using the methods from Chapters 3 to 6, they develop and compare at least three structural design options through LCA, covering Modules A1–A5 at minimum and extending to Modules B and D where data permits. The final deliverable is a written report and a 15-minute presentation. Grading criteria: rigor of LCA methodology, quality of optioneering, critical interpretation of results (including sensitivity to carbon factor assumptions), and clarity of recommendations.

Suggested project ideas: (a) Compare a new build vs. adaptive reuse scenario for an existing structure. (b) Design a hybrid timber–steel school and compare it against all-steel and all-reinforced concrete alternatives. (c) Evaluate three foundation strategies for a building on soft soil, quantifying carbon savings from geotechnical optimization. (d) Compare conventional vs. seismically isolated design options over a 50-year service life including earthquake repair scenarios.

Intended Learning Outcomes (ILO)

By the end of the course, students will be able to:

- Describe the climate context for structural engineering and articulate how the four pillars of life-cycle-driven design connect material efficiency, longevity, circularity, and resilience.
- Explain the LCA framework (Modules A0 through D) and apply it to quantify the embodied carbon of structural systems using EPD data and carbon factor databases.
- Conduct component-level and building-level life-cycle parameter analyses to compare structural design alternatives across materials and supply-chain scenarios.
- Interpret Pareto fronts from multi-objective optimization and use cost-carbon trade-off surfaces to inform early-stage design decisions.
- Critically evaluate structural systems through a life-cycle lens, recognizing how material choice and construction method influence environmental outcomes.
- Integrate resilience and circularity considerations into structural design, understanding how seismic performance, adaptive reuse, and design for disassembly affect whole-life carbon.

Option B: Two-semester Sequence (28 Weeks)

Course title (suggested): Life-cycle-driven Structural Design
Level: Upper undergraduate/MSc
Prerequisites: Knowledge of conceptual design of steel, concrete, or timber structures[7]
Credits: Adaptable to 9–10 ECTS (EU)
Class size assumption: Up to 50 students[8]

Semester 1: Foundations of LCA for Structures

Week	Topic	Chapter(s)
1	The climate context for structures. Four pillars of life-cycle-driven design. Student background test[9]	1
2	LCA: Modules A0 through D. System boundaries and functional units	2
3	Steel Global Warming Potential profiles: BF-BOF, EAF, and DRI-EAF routes. Scrap rates and regional variation	2
4	Concrete Global Warming Potential profiles: cement types, SCMs, and water-to-cement ratio effects	2
5	Timber Global Warming Potential profiles: biogenic carbon accounting, adhesive contributions, and end-of-life	2
6	Measuring embodied carbon. Reading EPDs. Regional databases and generic data sources worldwide	2, 3
7	Normalization of embodied carbon. Embodied carbon intensity ratings	3
8	Guest lecture: EPD provider or material supplier on data quality and availability	—
9	Embodied carbon reduction strategies (Part 1): circularity, high-strength materials, and hybrid systems	3
10	Embodied carbon reduction strategies (Part 2): digital methods, robustness, and procurement	3
11	Component-level life-cycle parameter analysis: column types across heights and supply-chain scenarios	4
12	Component-level life-cycle parameter analysis: beam types across spans. Sensitivity to carbon factor inputs	4
13	Building-level life-cycle parameter analysis	5
14	Semester 1 synthesis and review	1–5

7 If students lack this background, the instructor should incorporate conceptual structural design material into the first weeks of the course.
8 This is the number at which group project work can be supervised effectively; for larger classes, the group project scope and deliverables should be adapted to maintain pedagogical quality.
9 A short background test in the first session helps the instructor identify knowledge gaps in structural analysis and design, allowing early content to be calibrated so that all students can engage with the course.

Semester 2: Advanced Life-cycle-driven Design: Optimization, Resilience, and Practice

Week	Topic	Chapter(s)
1	Life-cycle optimization	5, 6
2	Genetic algorithms for structural design	6
3	Pareto front analysis	6
4	Parametric studies: building height, shape, recycling rates, and GGBS content	6
5	Seismic design and LCA (Part 1): three philosophies (conventional, damage-controlled, and no-damage)	7
6	Seismic design and LCA (Part 2): material trade-offs, repair carbon, and isolation system carbon	7
7	Resilience of nonstructural elements, retrofitting as circular economy, and community recovery	7
8	Guest lecture: seismic engineer or isolation specialist on real-world applications	7
9	Probabilistic life-cycle carbon under seismic hazard: expected annual losses and expected annual carbon	7
10	Case studies (Part 1): adaptive reuse, circular construction, and material efficiency	8
11	Case studies (Part 2): hybrid and modular systems, design for disassembly	8
12	Case studies (Part 3): seismic resilience, geotechnical optimization, and sustainable infrastructure	8
13	Guest lecture[10]	8
14	Synthesis: the future of life-cycle-driven practice. Capstone presentations	1–8

Assessment (Both Semesters)

Component	Weight (%)	Description
Individual assignment	20	Embodied carbon calculations, parameter analysis interpretation, and Pareto front evaluation
Capstone project	40	LCA of a real or hypothetical building comparing 3+ structural options (report + 20-min presentation)
Final written (individual) exam	40	Written exam to evaluate individual performance

Capstone Project Brief (Semester 2)

Teams of 3–4 students undertake a comprehensive life-cycle design study integrating methods from the full book. The project must include: (a) a clearly defined building or infrastructure typology

10 Instructors are encouraged to invite contributors (practitioners with experience) to present their work and engage directly with students. The 25 case studies in Chapter 8 were contributed by practicing engineers from international firms and can serve as a framework for instructors when shaping the content and scope of these invited lectures.

and site context; (b) early-stage optioneering comparing at least three structural systems using LCA; (c) a parametric or optimization study using the tools from Chapters 4–6; (d) consideration of at least one resilience dimension (seismic, adaptive reuse, or design for disassembly); and (e) a sustainability strategy matrix mapping design decisions to life-cycle modules and outcomes (following the format used in the Chapter 8 case studies). The final deliverable is a written report and a 20-minute presentation.

Suggested capstone topics: (a) Design a reusable, relocatable construction-site office and compare its five-cycle carbon footprint against single-use alternatives. (b) Redesign a bridge using high-strength steel (e.g. S460, S690) and quantify the life-cycle trade-offs against the original design in S355 steel. (c) Propose a post-earthquake reconstruction strategy for a hospital, comparing new-build, retrofit, and seismically isolated options over a 75-year service life. (d) Transform an obsolete commercial building through adaptive reuse, retaining the structural core and quantifying avoided carbon against demolition and new construction. (e) Design a large-span sports venue on a brownfield site, comparing conventional steel, cable-tension, and hybrid timber–steel roof systems.

Intended Learning Outcomes

By the end of Semester 1, students will be able to:

- Describe the climate context for structural engineering and articulate how the four pillars of life-cycle-driven design connect material efficiency, longevity, circularity, and resilience.
- Explain the LCA framework (Modules A0 through D) and apply it to quantify the embodied carbon of structural systems using EPD data and carbon factor databases.
- Conduct component-level and building-level life-cycle parameter analyses to compare structural design alternatives across materials and supply-chain scenarios.
- Critically evaluate structural systems through a life-cycle lens, recognizing how material choice and construction method influence environmental outcomes.

By the end of Semester 2, students will be able to:

- Formulate structural design as a multi-objective optimization problem and interpret Pareto fronts linking cost, carbon, and structural performance.
- Evaluate seismic design philosophies (conventional, damage-controlled, and no-damage) through a life-cycle lens, quantifying repair carbon and the carbon cost of resilience strategies including seismic isolation.
- Assess the role of nonstructural components, retrofitting, and adaptive reuse in whole-life carbon performance, applying circular economy principles to existing building stock.
- Critically analyze real-world case studies, identifying how engineering decisions map to specific life-cycle modules and sustainability outcomes using a strategy matrix.
- Develop and present a comprehensive life-cycle design study that integrates optioneering, parametric analysis, resilience considerations, and a sustainability strategy matrix for a chosen building or infrastructure typology.

Notes for Instructors

Guest lectures

Instructors are encouraged to invite contributors from industry to present their work and engage directly with students. The 25 case studies in Chapter 8 were contributed by practicing engineers from international firms and can serve as a framework for instructors when shaping the content and scope of these invited lectures.

Flexibility

Both syllabi are designed as frameworks, not prescriptions. Instructors teaching in programs with strong seismic design content may expand the Chapter 7 coverage. The two-semester option can also be taught as two independent courses (Semester 1 as a standalone introductory module, Semester 2 as an advanced elective) if program structures require it.

LCA completeness

A recurring challenge in practice is that early-stage LCA omits items that can considerably alter results. Instructors are encouraged to dedicate time to what should be measured in a complete structural LCA: for steel buildings this includes connections (typically 10–15% of beam and column weight), intumescent coatings or fireboards, secondary steelwork, and shear studs; for concrete buildings it includes reinforcement, formwork, and curing-related emissions. Building this awareness early helps students avoid systematic underestimation and prepares them for credible professional practice.

Companion materials

The book includes worked examples, review questions, and discussion prompts in every chapter. The practitioner checklist (available as a separate PDF, annex 1) can be used as a teaching aid in the final weeks of either course option. Each chapter also cites a wide range of references spanning research papers, technical standards, and industry reports, which instructors can use to build supplementary reading lists. Instructors seeking additional resources are welcome to contact the author.

Adaptations for different programs

For architecture programs, instructors may wish to give greater weight to the case studies and design optioneering sections (Chapters 1, 3, and 8) while reducing the depth of parametric and optimization analysis. For construction management programs, the procurement, logistics, and industrialized delivery aspects of Chapters 3 and 8 can be emphasized. For earthquake engineering courses, Chapter 7 can be expanded into a multi-week module with additional seismic-LCA exercises.

These syllabi accompany Life Cycle Driven Structures by Alper Kanyılmaz (Wiley 2026). They are offered as a starting point for course design and may be freely adapted by instructors.

Annex 4

Using This Book with Generative AI

As discussed in the Preface, this book and generative AI serve complementary roles. The book builds structured understanding of life-cycle-driven design (guiding you through concepts in a deliberate sequence, exposing you to items you would not yet know to ask about), while AI can act as a powerful amplifier once that foundation is in place. This appendix offers practical guidance on pairing the two, whether you are reading a printed copy or a digital PDF.

Option A: Reading the Print Edition

With a physical copy, your interaction with AI happens in a separate window (a laptop, tablet, or phone beside the book). The strength of print lies in peripheral learning: flipping pages, scanning figures, and absorbing context that surrounds the section you are reading. Use AI as a study companion that responds to what you encounter on the page. The recommended workflow is:

1. read the chapter or section first without interruption, letting the narrative sequence do its work;
2. open a conversation with an AI assistant and use the prompts (proposed in the next page) to deepen your understanding, test alternative scenarios, or clarify specific passages;
3. return to the book to verify AI outputs against the methods, tables, and case studies in the text.

This loop (read, query, and verify) builds critical judgment.

Option B: Reading the Digital PDF Edition

In this case, you can upload specific pages, figures, or tables directly into the conversation and ask the AI to interpret, extend, or critique them. The recommended workflow is:

1. upload the relevant chapter (or a specific page range) to your AI assistant at the start of a study session;
2. ask the AI to work directly with the content (re-derive a calculation using different assumptions, reinterpret a figure for a new material, or extend a case study to a different location);

3. use side-by-side viewing (PDF on one half of the screen, AI chat on the other) to maintain the contextual browsing that the Preface described as peripheral learning.

Prompt Ideas

The table below provides example prompts organized around the book's themes. Each prompt is designed to be used after you have read the relevant section, so that the AI amplifies the understanding you have already built. Adapt the specifics (material type, location, building height, and seismic zone) to match your own project or coursework.

When you want to...	Try this prompt
Understand climate context (Ch. 1)	"Using the four-pillar framework from Chapter 1, evaluate a [building type] in [your city]. Which of the twelve strategies would yield the greatest carbon reduction given the local climate and energy grid?"
Explore LCA stages (Ch. 2)	"I have read the A1–A3 production-stage profiles for steel (BF-BOF vs. EAF). Recalculate the embodied carbon of a 500-ton steel frame if the EAF plant sources 80% renewable electricity. Show each step."
Work with EPDs (Ch. 3)	"Here is an EPD for [product]. Walk me through how to extract the A1–A3 GWP value and compare it against the generic data in the ICE Database. Flag any assumptions I should question."
Run component-level what-ifs (Ch. 4)	"For a HEB300 column at 4 m height, compare the embodied carbon using three EPD sources: a European EAF mill, a North American BF-BOF mill, and a generic value."
Scale to building-level LCPA (Ch. 5)	"Replicate the Chapter 5 parametric study for a 10-story building but change the location to [your city]. Adjust transportation distances and local steel carbon factors. How do the rankings shift?"
Interpret optimization results (Ch. 6)	"Explain the Pareto front from Chapter 6. If my client prioritizes open-plan office space over minimum cost, which region should I focus on, and what is the embodied carbon penalty?"
Integrate seismic + life-cycle thinking (Ch. 7)	"Compare the life-cycle carbon of a conventional ductile frame versus a damage-controlled frame with replaceable fuses for a hospital in a high-seismicity zone. Include B5 repair impacts over 50 years."
Learn from case studies (Ch. 8)	"I have just read the [case study name] from Chapter 8. Adapt its sustainability strategy matrix to a similar project in [your country], adjusting for local materials, codes, and carbon factors."
Conceptual clarification	"Explain the embodied carbon equation in Chapter 3 using a simple numerical example for a 6-story steel frame." "Compare BF-BOF and EAF steel routes in terms of system boundaries and sensitivity to electricity carbon factors." "Why does the conclusion of an LCPA study depend strongly on the assumed carbon factor? Provide a scenario illustrating this." "Translate the four-pillar framework from Chapter 1 into a decision checklist for early schematic design."

When you want to...	Try this prompt
Testing alternative assumptions	"If transportation distance doubles in the Chapter 5 10-story case study, how might that shift the ranking of configurations?" "How would increasing GGBS content from 10 percent to 30 percent affect trade-offs in Life-cycle Optimization?" "What happens to seismic-LCA trade-offs if repairable fuse systems reduce post-earthquake replacement by 70 percent?"
Challenging your reasoning	"Here is my reasoning about hybrid timber–steel systems in mid-rise buildings. Identify weaknesses or missing considerations." "Review my interpretation of Module D benefits. Where might I be overestimating credits?" "I concluded that material substitution dominates sourcing effects. Argue the opposite position."
Dialog with extracted content	"Rewrite this explanation at two levels: one for first-year students and one for experienced engineers." "Convert this description of LCPA into a step-by-step workflow." "Turn this case study into a decision tree diagram (described textually)."
Data expansion	"Using the embodied carbon equation structure from Chapter 3, construct a hypothetical dataset comparing three beam spans at 6 m, 9 m, and 12 m." "Generate a simplified Pareto front example balancing cost and carbon for five candidate designs." "Create a sensitivity table showing how steel carbon factor variations between 0.5 and 2.5 $kgCO_2e/kg$ change total building impact."
Cross-chapter synthesis	"Connect the four-pillar framework in Chapter 1 with the seismic philosophies in Chapter 7." "How do the case studies in Chapter 8 demonstrate the reduction strategies listed in Chapter 3?" "Map Life-cycle Optimization outcomes to the embodied carbon rating systems introduced earlier."

Tips for Effective AI-assisted Study

- **Always read first, prompt second.** The book's sequenced progression builds the judgment you need to evaluate AI-generated answers critically.
- **Provide context in every prompt.** Reference the specific chapter, table, or figure. The more precisely you frame a question, the more reliable the response.
- **Verify numerical outputs.** AI can make computational errors. Cross-check carbon calculations against the methods and data sources in Chapters 2–6.
- **Use AI to explore, not to conclude.** Treat AI outputs as draft explorations. Professional judgment for final design decisions comes from the structured understanding this book provides.

- **Ask AI to challenge your reasoning.** After working through an exercise, try: "What assumptions in my analysis are most likely to change the conclusion?"
- **Extend case studies to your own context.** Use AI to adapt a Chapter 8 sustainability strategy matrix to your local codes, materials, and supply chains.
- **Enable web search**, and ask the AI to provide source links for any substantive claim or numerical value, so you can trace it back to the original reference.
- **Consider setting up a dedicated project in your AI assistant** (most current platforms offer a "Projects" feature). You can configure it once with the book title, a short description, and your reading goals, so every new chat inherits the same context and avoids the overload that long single conversations tend to suffer.

Let this book teach you how to think about the problem, and let AI help you think further.

Before Asking AI

1. Read the section fully.
2. Summarize it in your own words on paper.
3. Solve at least one example or thought experiment independently.
4. Identify what you still find unclear.

Only then begin a structured AI dialog.

Common Pitfalls

Try to avoid the following:

Asking for shortcuts before building foundations
Without the conceptual sequence developed across the book, AI outputs can appear convincing yet lack internal coherence.

Accepting fabricated data
AI may generate plausible but incorrect emission factors or code references. Always verify numerical inputs against EPDs, databases, or cited sources. Requiring source links from the assistant (with web search enabled) can make this verification routine.

Over-automation of judgment
Optimization tools generate options; they do not define values. Decisions about resilience, circularity, and community impact require human evaluation.

Ignoring regional context
Current AI models tend to default to the most common data in their training (often US or Northern European). A prompt for technical specifications may return standards, materials, or parameter values that do not reflect local supply chains or practices. Always anchor AI outputs to the regional EPD databases and regulatory frameworks discussed in Chapter 3.

Index

Note: Page numbers in *italics* and **bold** refers to figures and tables respectively.

d

f

g

t